Manual of Techniques in Insect Pathology

BIOLOGICAL TECHNIQUES

A Series of Practical Guides to New Methods in Modern Biology

Series Editor
DAVID B SATTELLE

Computer Analysis of Electrophysiological Signals
J Dempster (published December 1992)
Fluorescent and Luminescent Probes for Biological Activity
WT Mason (Editor) (published May 1993)
Planar Lipid Bilayers
W Hanke and W-R Schlue (published October 1993)
***In Situ* Hybridization Protocols for the Brain**
W Wisden and BJ Morris (Editors) (published May 1994)

CLASSIC TITLES IN THE SERIES

Microelectrode Methods for Intracellular Recording and Ionophoresis
RD Purves
Immunochemical Methods in Cell and Molecular Biology
RJ Mayer and JH Walker

BIOLOGICAL TECHNIQUES

Manual of Techniques in Insect Pathology

Edited by

LAWRENCE A. LACEY

Yakima Agricultural Research Laboratory,
USDA-ARS,
5230 Konnowac Pass Road,
Wapato,
WA 98951,
USA.

ACADEMIC PRESS
An imprint of Elsevier Science
Amsterdam · Boston · London · New York · Oxford
Paris · San Diego · San Francisco · Singapore · Sydney · Tokyo

This book is printed on acid-free paper.

Academic Press, Inc.
An imprint of Elsevier Science
525 B Street, Suite 1900, San Diego, California 92101-4495, USA
http://www.apnet.com

Academic Press Limited
An imprint of Elsevier Science
84 Theobald's Road, London W1CX 8RR
http://www.hbuk.co.uk/ap/

ISBN 0-12-432555-6

A catalogue record for this book is available from the British Library

Typeset by Phoenix Photosetting, Chatham, Kent
Printed in Great Britain by The Bath Press, Bath
02 EB 9 8 7 6 5 4 3 2

Series preface

The rate at which a particular aspect of modern biology is advancing can be gauged, to a large extent, by the range of techniques that can be applied successfully to its central questions. When a novel technique first emerges, it is only accessible to those involved in its development. As the new method starts to become more widely appreciated, and therefore adopted by scientists with a diversity of backgrounds, there is a demand for a clear, concise, authoritative volume to disseminate the essential practical details.

Biological Techniques is a series of volumes aimed at introducing to a wide audience the latest advances in methodology. The pitfalls and problems of new techniques are given due consideration, as are those small but vital details that are not always explicit in the methods sections of journal papers. The books will be of value to advanced researchers and graduate students seeking to learn and apply new techniques, and will be useful to teachers of advanced undergraduate courses, especially those involving practical and/or project work.

When the series first began under the editorship of Dr John E Treherne and Dr Philip H Rubery, many of the titles were in fields such as physiological monitoring, immunology, biochemistry and ecology. In recent years, most biological laboratories have been invaded by computers and a wealth of new DNA technology. This is reflected in the titles that will appear as the series is relaunched, with volumes covering topics such as computer analysis of electrophysiological signals, planar lipid bilayers, optical probes in cell and molecular biology, gene expression, and *in situ* hybridization. Titles will nevertheless continue to appear in more established fields as technical developments are made.

As leading authorities in their chosen field, authors are often surprised on being approached to write about topics that to them are second nature. It is fortunate for the rest of us that they have been persuaded to do so. I am pleased to have this opportunity to thank all authors in the series for their contributions and their excellent co-operation.

David B Sattelle ScD

Dedication

This manual is dedicated to Professor Yoshinori 'Joe' Tanada. His scientific discoveries, books and other publications, and personal interest in the development of students and other colleagues, have been an inspiration to all of those who know him and/or who have benefited from his work. During his scientific career of over 50 years he has worked on virtually every major pathogen group, and helped to further develop and promote the field of Invertebrate Pathology worldwide.

Contributors

J.J. Becnel *CMAVE, USDA, ARS, PO Box 14565, Gainesville, FL 32604, USA*

W.M. Brooks *Department of Entomology, North Carolina State University, Raleigh, NC 27695-7613, USA*

T.M. Butt *Rothamsted Experimental Station, AFRC, Harpenden, Herts AL5 2JQ, UK*

H.F. Evans *Forest Research Station, Alice Holt Lodge, Wrecclesham, Farnham, Surrey GU10 4LH, UK*

E. Frachon *Unité des Bactéries Entomopathogénes, 25 et 28, Rue du Dr Roux, Institut Pasteur, 75724 Paris CEDEX 15, France*

L.J. Galan-Wong *Departamento de Microbiologia, Facultad de Ciencias Biologicas, Universidad Autonoma de Nuevo Leon, S. Nicolas de los Garza, Nuevo Leon AP 2790 64450, Mexico*

M. Goettel *Agriculture Canada Research Centre, PO Box Main, Lethbridge, Alberta T1J 4B1, Canada*

A.E. Hajek *Department of Entomology, Cornell University, Ithaca, NY 14853, USA*

R.A. Humber *Plant, Soil and Nutrition Laboratory, USDA, ARS, Tower Road, Ithaca, NY 14853, USA*

D. Inglis *Agriculture Canada Research Centre, PO Box Main, Lethbridge, Alberta T1J 4B1, Canada*

L. Joshi *Boyce Thompson Institute, Tower Road, Cornell University, Ithaca, NY 14853, USA*

H.K. Kaya *Department of Nematology, University of California, Davis, CA 95616, USA*

J.L. Kerwin *Department of Botany KB-15, University of Washington, Seattle, WA 98195, USA*

M.G. Klein *USDA-Agricultural Research Service, Application Technology Research Unit, OSU-OARDC, 1680 Madison Ave., Wooster, OH44691, USA*

L.A. Lacey *Yakima Agricultural Research Laboratory, USDA-ARS, 5230 Konnowac Pass Road, WA 98951, USA*

M.R. McGuire *USDA-ARS, National Center for Agricultural Utilization Research, 1815 N. University St, Peoria, IL 61604, USA*

B. Papierok *Entomopathogenes, Lutte Biologique, 25 et 28 Rue du Dr Roux, Institut Pasteur, 75724 Paris Cedex 15, France*

E.E. Petersen *Department of Botany KB-15, University of Washington, Seattle, WA 98195, USA*

M. Shapiro *USDA-ARS, Insect Biocontrol Lab, Bldg 011A, BARC-West, Beltsville, MD 20705, USA*

J.P. Siegel *Department of Economic Entomology, 607 E. Peabody Drive, 172 Natural Resources Bldg, Champaign, IL 61820, USA*

R.J. St Leger *Boyce Thompson Institute, Tower Road, Cornell University, Ithaca, NY 14853, USA*

S.P. Stock *Facultad de Ciencias Naturales y Museo, Centro de Estudios Parasitologicos y Vectores, Universidad Nacional de La Plata, La Plata 1900, Buenos Aires, Argentina*

P. Tamez-Guerra *Departamento de Microbiologia, Facultad de Ciencias Biologicas, Universidad Autonoma de Nuevo Leon, S. Nicolas de los Garza, Nuevo Leon, AP 2790 64450, Mexico*

I. Thiery *Unité des Bacteries Entomopathogénes, 25 et 28, Rue du Dr Roux, Institut Pasteur, 75724 Paris Cedex 15, France*

A.H. Undeen *Medical and Veterinary Entomology Research Laboratory, USDA/ARS, PO Box 14565, Gainsville, FL 32604, USA*

J. Vavra *Faculty of Sciences, Department of Parasitology and Hydrobiology, Charles University, Vinicna 7, Prague, Czech Republic*

Preface

The beginnings of practical insect pathology can be traced into antiquity to work with beneficial insects. As a discipline, however, it is a fairly young branch of science. There are several accounts of published research in insect pathology in the early literature dating from Bassi's incrimination of *Beauveria bassiana* as a pathogen of the silkworm in 1835, but the field came into its own in the 1940s and 1950s with the development of formal course work in insect pathology and the publication of *Principles of Insect Pathology* (Steinhaus, 1949). Over the past two decades, interest in the use of alternative insecticides due to environmental and human health concerns has stimulated increased efforts in the development of microbial control agents as components of integrated pest management systems.

The literature in book form is a veritable cornucopia of analytical information covering both basic and applied aspects of invertebrate pathology. Atlases and manuals are available that enable the identification of a wide range of insect pathogens. Some of these have also included a limited amount of techniques to be used predominantly for preparing entomopathogens for isolation and identification. A large number of techniques for the isolation, identification, production and evaluation of insect pathogens are scattered throughout the literature. However, a single comprehensive manual of techniques in insect pathology has heretofore not been available. When confronted with the need to work on a new pathogen group, those of us trained in a particular area of invertebrate pathology must scour the literature in search of instructions for working with the new organisms.

In this *Manual* an international group of experts have brought together a broad array of techniques for the identification, isolation, propagation/cultivation, bioassay and storage of the major groups of entomopathogens. This *Manual* was designed to provide general and specific background to experienced insect pathologists, biologists and entomologists who are beginning work with pathogen groups that are new to them. It will also be useful as a laboratory manual for courses in insect pathology and biological control and related areas of study. It is our hope that this *Manual* will also provide practical information to other researchers, students, biotechnology personnel, entomologists working in integrated pest management, and government regulators concerned with the more technical side of regulatory issues. Chapters on safety testing of entomopathogens in mammals and complementary techniques for the preparation of entomopathogens and diseased specimens for more detailed study using microscopy and molecular techniques, broaden the subject matter of the *Manual* beyond classical insect pathology. To provide in-depth background to the user, this *Manual* will be an ideal complement to the book, *Insect Pathology* recently published by Tanada and Kaya (1993).

The style of presentation differs somewhat between pathogen groups due to the variety of backgrounds of the authors and diversity of subject matter and also inherent differences between the individual pathogen groups. We have concentrated on the 'how to' aspects of the techniques, but have also tried to provide the reader with an appreciation for why they are used as well as to provide a spectrum of supplemental literature and recipes for media, fixatives and stains. Owing to the extensive possibilities resulting from the diversity of pathogens and their hosts and a finite page limit for the *Manual*, we have had to be somewhat selective in the number of techniques that were covered and the amount of literature that could be referenced.

Lawrence A. Lacey
May 1996
Montpellier, France

Steinhaus, E. (1949) *Principles of Insect Pathology*, McGraw Hill, New York, 757 pp.
Tanada, Y. & H. K. Kaya (1993) *Insect Pathology*, Academic Press, New York, 666 pp.

Acknowledgements

I thank the contributing authors for their efforts and attention to detail. A number of colleagues reviewed the *Manual* during its preparation. We are indebted to the following individuals for their review of one or more chapters: Ray Akhurst, Wayne Brooks, Tariq Butt, Douglas Dingman, Jacques Fargues, Lindsey Flexner, Mark Goettel, Ann Hajek, Richard Humber, Harry Kaya, Lloyd Knutson, Tad Poprawski, Ole Skovmand, Grover Smart, Natalie Smits, Donald Stahly, Anthony Sweeney, Isabel Thiery, Albert Undeen, John Vandenberg and Allen Yousten. I thank David Sattelle, editor of the biological techniques series and Elizabeth Davidson, past president of the Society for Invertebrate Pathology for encouraging me to proceed with production of the *Manual*. I am especially grateful to Cindy Lacey for her overall help with the *Manual* and for her constant encouragement and support.

Contents

CHAPTER I

Initial handling and diagnosis of diseased insects

LAWRENCE A. LACEY* & WAYNE M. BROOKS†

Yakima Agricultural Research Laboratory, USDA-ARS, 5230 Konnowac Pass Road, Wapato, WA98951, USA.
† Department of Entomology, North Carolina State University, Raleigh, NC 27695-7613, USA

1 INTRODUCTION

The increased potential of microbial control of insect pests over the past 50 years has been largely the result of the discovery and development of new species and strains of entomopathogens. Some of these discoveries have been serendipitous, but most have been due to systematic and exhaustive surveys. Despite the successes of the past, there is a continuing need to discover and develop new entomopathogens if we are to meet the future needs of increased food and fibre production with concomitant reduction in the use of chemical pesticides. Sustainable agriculture in the 21st century will rely increasingly on microbial control and other alternative interventions for pest management that are environmentally friendly (Lacey & Goettel, 1995). This chapter will provide general guidelines for the recognition, handling and initial diagnosis of diseased insects and the identification of major entomopathogen groups. Some of the key terms in insect *pathology* that are used in this chapter are italicized in the text and defined in a glossary at the end of this chapter. For an introduction to the principles and thorough coverage of the field of insect pathology, the reader is referred to *Insect Pathology* (Tanada & Kaya, 1993).

Insects are associated with a broad diversity of microorganisms in a variety of *symbiotic* relationships including: *commensalism*, *mutualism*, and *parasitism*. Internal mutualistic organisms are critical to the survival of the *host*, such as *symbiotes* which are found in *mycetocytes* and *mycetomes* within many insect species. Although mutualistic organisms may be abundant in the insect, such as the protozoa associated with termites, they are not pathogenic to the host insect. *Pathogens* on the other hand, result in a variety of conditions in host insects that are distinctly to subtly unfavourable to the host. *Entomopathogens* cause *disease* in insects through the effects of infection, parasitism and/or *toxaemia*.

There is an astronomical number of entomopathogens which cause *disease* and as great a number of insect hosts in which to find them. Distinguishing one disease from another based on

MANUAL OF TECHNIQUES IN INSECT PATHOLOGY
ISBN 0–12–432555–6

the *signs* and *symptoms* of the disease, its *aetiology*, *pathogenesis* and other characteristics is the process of *diagnosis*. Steinhaus (1963a) described the diagnosis of insect diseases as one of the most important and complex branches of pathology. The elements and background of diagnosis of insect disease are presented in detail by Steinhaus & Marsh (1962) and Steinhaus (1963a). Essentially, diagnosis is divided into two main categories: the gathering of facts concerning insect disease and their analysis.

Disease and death in insects is not always an indication of infection with entomopathogens. Information on non-infectious diseases in insects due to non-microbial causes (poisoning, mechanical and physical injuries, and diseases of nutrition and metabolism) is presented by Steinhaus (1949), several authors in Steinhaus (1963b) and Tanada and Kaya (1993). This chapter focuses on infectious diseases of insects that are caused by entomopathogens and presents general guidelines for their recognition and initial diagnosis. Greater diagnostic detail using a variety of microscopic techniques as well as microbiological, biochemical and other procedures for the identification of specific pathogens will be provided in subsequent chapters.

2 COLLECTION OF DISEASED INSECTS AND ENTOMOPATHOGENS

Both living and dead insects that are patently infected with entomopathogens can be found in virtually every setting inhabited by insects including natural terrestrial and aquatic ecosystems, agroecosystems and in laboratory and commercial insect colonies. It is often the researcher or technician who is most familiar with healthy insects who is the first to notice that something is not right with a diseased insect. Visual search of habitats of interest for individual insects which stand out from normally appearing members of the same population is one of the more common means of collecting diseased insects. Although many pathogens are usually present at low levels in insect populations (as *enzootic diseases*) they are most easily discovered during *epizootics* when there is an unusual abundance of diseased insects. Hand picking of specimens allows selectivity and conservation of space, if that is an important consideration. Collection of large numbers of living insects using standard insect collecting techniques with subsequent screening in the field or lab is another strategy. Basically the same methods of collection used for the survey of healthy insect populations are utilized (traps, sweeping, hand picking, aquatic nets and dippers).

The recognition of diseased insects in the field or subsequently in the lab will initially rely on *gross pathology* and *patent infections*.

A Recognition of diseased insects – gross pathology

Insects that are patently infected with entomopathogens often manifest characteristic *symptoms* and *signs* (*syndrome*) of disease, e.g. striking colour changes, luxuriant growth of the pathogen on the outside of the cadaver, signs of the *pathogen* or *aetiological agent* inside the host (visible through the cuticle), *dysentery*, peculiar behaviour including lack of feeding or unusual position on host plants, tremors, *mummification*, fragility or hardening of the integument, noticeable difference in size, and other signs and symptoms. In some cases, symptoms of disease may be very subtle or not initially apparent (sublethal effects such as *parasitic castration*, reduced longevity, etc.). It is even possible for some pathogens to be *occult* (see *occult virus*).

Some of the most common aspects of gross pathology are described below.

1. Colour changes

Often one of the first symptoms of diseased insects to be noticed in a population is coloration that sets them apart from healthy members of their cohort. This phenomenon is observed in both living and dead diseased insects. Colour changes due to entomopathogens in living insects are usually associated with those insects with transparent to semi-transparent integuments, such as in white grubs (Plate 1) and larval forms of weevils, hymenopterous larvae, and many groups of aquatic insects, most notably the Diptera (Plate 2). The shift in colour from that of the normal variation observed in the insect may be subtle to drastically different. Blue iridescence, for example in beetle grubs, mosquito larvae (Plate 2), and certain lepidopterous larvae is not associated with the colour of healthy larvae and indicates an iridescent virus infection. In such infections, iridescence rapidly disappears with the death of the host.

In dead insects a broad array of colour changes are seen ranging from white (Plates 3 and 4), grey, red (Plates 5 and 6), orange, yellow, blue, green (Plate 7) to brown and black. *Melanized* areas in the cuticle of insects (usually in the form of black spots) are often due to immune responses of the host as a result of invading nematodes or fungi.

2. Physical signs of the entomopathogen

Often the causal agent of the disease can be observed directly in association with the infected insect or cadaver. Fungi, for example, frequently produce luxuriant and colourful growth over the surface of the insect (Plates 3, 7, 8 and 9). Infective and reproductive forms of nematodes can be observed in the haemocoel in living (Plate 10) and dead hosts (Plate 6). Virus infections and several species of fungi may be seen through the cuticle of living insects (Plates 11, 12 and 13). In insects with transparent cuticles, certain protozoan infections may result in *hypertrophied* tissues that are abnormally opaque and/or coloured (Plates 14 and 15). The location of infected tissues may also be characteristic for certain entomopathogens and may be visible through the cuticle. For example, the cytoplasmic polyhedrosis virus infection shown in the black fly larva in Plate 11 is characteristically found in the gastric caeca and the posterior portion of the midgut. On the other hand, nuclear polyhedrosis virus infections of mosquito larvae are found throughout the entire midgut.

3. Aberrant behaviour

This includes a lack of feeding in normally voracious insects, irritability, and unusual dispersal or aggregation. For example, insects infected with certain fungi or viruses often climb to high points on host plants and become attached to the plant just prior to death (Plate 4).

4. Changes in form and texture

A variety of physical changes occur in diseased insects, most often after death. The cadavers may become mummified, soft or liquified, firm, 'cheese-like' internally, or leathery and dry. Anatomical abnormalities such as prolapsed rectum, a characteristic sign of certain viral infections and other morphological aberrations are observed in living, *moribund* and dead insects.

5. Odour

In certain cases, cadavers may become odiferous. This is usually associated with insects whose body tissues are liquified. For example, European foulbrood of the honey bee is associated with a sour, rotten-meat odour of infected larvae.

B Initial handling of specimens

When an insect is suspected of being infected with an entomopathogen, it should be examined as soon as possible after collection. The invasion of cadavers by fast-growing saprophytic organisms may complicate the determination of the true *aetiological agent.* In the field, diseased insects should be removed from the substrate upon which they are found with fine forceps and placed individually, if possible, in clean dry containers in the case of terrestrial insects. To avoid damaging cadavers that are tightly attached to the substrate, the portion of the host plant upon which they are fixed should be collected with the insect attached. Minute insects, such as scales and whiteflies, can also be collected in this manner. The collecting container should be capped in a manner that allows gas exchange and prevents condensation or the retention of excess moisture. The addition of a drying agent, such as silica gel, to the container used for temporary storage will slow or prevent germination of entomopathogenic fungi and help to eliminate the growth of saprophytic fungi on specimens (Figure 1). In the case of specimens containing nematodes and certain protozoan *parasites*, the specimens should not be allowed to dry. Weiser & Briggs (1971) suggest that fragile cadavers, especially those from aquatic habitats, could be placed on filter paper or a microscope slide and allowed to dry or placed in a drop of 4% formalin in a vial. When possible, diseased insects should be held at low temperatures (e.g. in an ice chest or refrigerator) until they can be examined later in the laboratory.

Patently infected living insects should be kept in the medium in which they are found (i.e. on foliage, in soil or water) and transported to the laboratory as soon as possible. Aquatic insects may also be

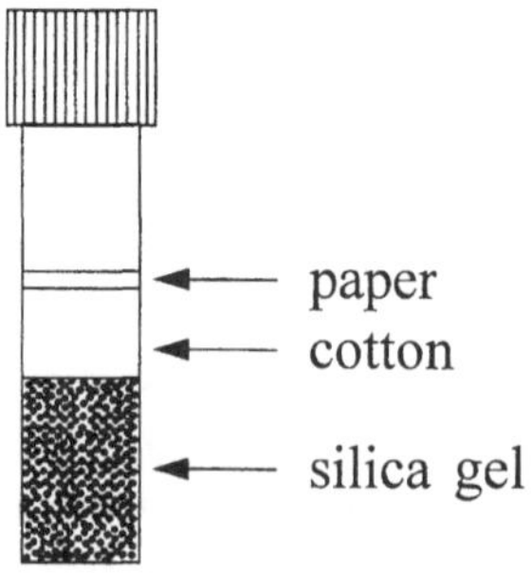

Figure 1 Collecting tube for dry preservation of specimens.

transported on damp aquatic plants (see Chapter III-2). Cool temperatures will help reduce *stress* on the organisms and retard unwanted microbial growth until they can be examined. Healthy insects should also be collected for comparative purposes and held separately under conditions that enable good survival.

Where diseased individuals are rare or infections are inapparent, it may be useful to collect large numbers of apparently healthy individuals for rearing and observation in the laboratory. The stress of laboratory rearing may accelerate the *incubation* period and the development of pathogens that are present at low levels, occult or in an *eclipse period* at the time of collection. Stressing insects by crowding, starving, or other conditions may result in an overt infection through the *induction* of an occult virus or the appearance of other diseases that might not be apparent in the field. However, Weiser & Briggs (1971) caution that crowding and/or inclusion of excess food may cause asphyxia or stimulate the development of otherwise saprophytic bacteria. In addition, some species of insects become aggressive or even cannibalistic when crowded.

Large numbers of apparently healthy larvae may also be mass processed by trituration followed by differential centrifugation (Chapter IV) to detect pathogens that may be present at low densities.

C Collecting of entomopathogens from insect habitats

It is possible to collect entomopathogens without ever having to find a diseased natural host. The use of baiting with *surrogate* host insects has been used commonly for the isolation of entomopathogenic fungi and nematodes. Details on techniques using the wax moth, *Galleria mellonella*, as bait are presented in Chapters V and VI. Selective media and procedures for the isolation of bacteria and fungi from insect habitats are presented in Chapters III and V. Methods for centrifugation of water from mosquito habitats for the isolation of microsporidia and other pathogens are presented by Avery & Undeen (1987).

D Collection of field data

At the time of collection, as detailed information as possible should be recorded regarding:

1. the insect (species, stage, gross pathology, aberrant behaviour, location in the environment);
2. host plant or habitat;
3. *prevalence* of disease (was there an epizootic or were diseased insects less numerous?);
4. elevation and climatic conditions;
5. suggested additional information pertaining to specific pathogen groups is presented in subsequent chapters.

3 EXAMINATION IN THE LABORATORY

Diagnosis based on gross pathology alone can be quite misleading if not followed up with microscopic examination. Several classes of signs and symptoms are common to different groups of entomopathogens (colour, odour, behaviour). Details of general laboratory conditions (cleanliness, instruments, etc.) that are suitable for examination of diseased insects are presented by Wittig (1963), Weiser & Briggs (1971) and Thomas (1974).

A Logging diseased specimens

A system for complete record keeping was devised by Steinhaus & Marsh (1962; also published in Steinhaus 1963a, Thomas 1974) to cover the spectrum of information related to the accession of diseased insects, their examination and eventually for coming to and the recording of diagnostic conclusions. Together with the field data, and other collection information (date, collector, number of specimens, etc.), an accession number is assigned to the specimens. The accession number is the means by which the specimens are tracked through the various examinations and disease diagnosis.

B Preparation of specimens for initial laboratory examination and/or isolation of suspected pathogens

To enable detailed examination or isolation of the causal agent, the specimen or specimens will usually require further preparation. The following are procedures for entomopathogens in general.

1. Surface sterilization

When a non-contaminated sample of diseased tissues will be used for inoculating media or injecting into healthy insects, surface sterilization of diseased insects will usually be necessary. Figure 2 shows a typical set-up for surface sterilization of insects. The sequence used in our lab is as follows:

1. place insect in 70% alcohol for a few seconds to facilitate wetting of the specimen;
2. rinse briefly in distilled water;
3. place in dilute sodium hypochlorite (NaClO) for 1 min or longer (commercially available bleach, such as clorox which is approximately 5% NaClO, can be diluted to the appropriate concentration, which depending on the size and state of the insect is 0.5–1% NaClO);
4. rinse briefly in 2–3 changes of sterile water;
5. blot dry with sterile filter paper.

Figure 2 Typical set-up for surface sterilization of insects.

Variations of the above method include more or less time in alcohol, greater or lower concentration of NaClO or fewer rinses in water. Due to the hydrophobic nature of insect cuticle, the addition of a surfactant such as Tween 80 to the NaClO may increase its effectiveness. Alternative substances such as Hyamine and Zephiran chloride are also used for surface sterilization. Procedures for and advantages of their use are presented by Martignoni & Milstead (1960). Other germicides are presented by Wittig (1963).

Surface sterilization of very small insects will kill the entomopathogen inside of the host (e.g. whiteflies infected with entomopathogenic fungi). In this case, the removal of spores from the tips of *conidiophores* (see glossary Chapter V-1) is accomplished by touching the most distal spores to a minute amount of sterile media on the tip of a sterile minuten pin (mounted on a match stick) and then inoculating an appropriate medium in a Petri plate.

2. Dissection

To examine individual organs and tissues, careful dissection of diseased specimens will be necessary. Dissecting instruments such as fine-tipped forceps, microscalpels, iris scissors, stainless-steel minuten pins mounted on match sticks and the like are useful for this type of precision dissection. It may be desirable to dissect the insect in a drop of fluid. Quarter strength Ringer's solution provides a medium that is more osmotically compatible with tissues and pathogens than water. Ringer's solution and other dissecting fluids are presented in Chapter VIII-1.

3. Preparation of slides

a. Unstained

Whole small insects, organs, blood and other tissues can be mounted on slides in quarter strength Ringer's solution, water or other medium for observation using phase contrast microscopy. Drops of regurgitate or diarrhoea should also be placed on a slide and covered with a coverslip for subsequent observation. When examining wet-mount preparations of various tissues, organs, or other body components, care should be taken to make the preparation as thin as practical to assist visualization under phase microscopy. To prevent the rapid drying out of these preparations, the edges of the coverslip can be sealed with melted paraffin or mineral oil using a small camel-hair brush.

b. Stained slides

The same materials mentioned above can also be stained with a variety of materials that enable coloration of insect cells and entomopathogens.

In addition to wet-mounts, cellular debris and other materials can be spread thinly on the slide using a pair of fine-tipped forceps, allowed to air dry, followed by fixation if required and staining with a differential stain. Stains and procedures for their use for specific pathogen groups are presented in Chapters II, III-1, IV, V, VI, VIII-1 and VIII-2.

4. Preparation of tissues for histological sections and subsequent molecular studies

Often more detailed examination will be required regarding the *histopathology* and pathogenesis of disease in order to make an accurate diagnosis. The non-occluded viruses in particular will require examination using electron microscopy before identification can be made. Fixatives and procedures for preparing tissues for light and electron microscopy are presented in Chapter VIII-1. Procedures used in molecular studies are presented in Chapter VIII-3.

5. Preparation of inoculum for transmission studies

Satisfying Koch's postulates is the final step in the diagnosis process (see Section 3 E). The preparation of inoculum for such tests may require isolation and culture of the suspected pathogen, or where this is not possible (pathogens that can only be produced *in vivo*), purification of inoculum from field-collected or lab-infected hosts. Some obligate parasites will require production in an *intermediate host.* Procedures for isolation, cultivation/propagation, and determining the *pathogenicity* and *virulence* of specific micro-organisms or nematodes is covered in Chapters II-VI. General procedures for sterile technique are covered by Thomas (1974), and in a variety of microbiology manuals such as *Bergey's Manual* (Holt 1977).

C Microscopic examination

1. External and internal examination using the dissecting microscope

Preliminary examination of diseased and healthy specimens with the dissecting microscope (mag. 6–50) can be conducted in conjunction with dissection. Before making an incision in the insect, observe and record any abnormal behaviour, signs of regurgitation, dysentery, external lesions or growths, abnormal morphology or coloration and presence of structures seen through the cuticle that are not present in healthy hosts. Upon opening the host, note any changes in colour, size (*atrophy* or *hypertrophy*) or structure (*hypoplasia* or *hyperplasia*) of organs.

2. Comparison of healthy and diseased tissues using light microscopy

A prerequisite for recognizing diseased organs and tissues is to become familiar with corresponding tissues in healthy insects. Diseased tissues may be differently coloured, *atrophied* or even missing (see *aplasia*), hypertrophied or undergo an increase in cells (see *hyperplasia*) from that seen in healthy insects. The intestine (especially the midgut), fat body, malphigian tubules and blood of diseased insects often demonstrate patent signs of disease and are good starting points for comparison with healthy insects. By using the slides that were made with diseased and healthy tissues, the general condition of tissues from diseased insects should be observed using light microscopy (especially phase contrast microscopy) and compared and contrasted with healthy tissues. Note colour, external and internal evidence of aetiological agents or abnormalities in organs.

The phase contrast microscope not only provides an excellent means of observing non-stained specimens, but also produces characteristic *refringence* in certain pathogens such as microsporidian spores. When observed through polarizing filters, uric acid crystals commonly present in the malphigian tubules of insects are *birefringent* whereas superficially similar viral *polyhedra* are not birefringent. Fat cells may sometimes be confused with similarly shaped polyhedra. After staining with Sudan III fat cells become red whereas polyhedra do not take up the stain (see Chapter VIII-1).

3. Localization of infection

The specific organs or tissues that are infected may be characteristic of a particular disease. For example the cytoplasmic polyhedrosis viruses are usually restricted to certain portions of the midgut of host insects (see Plate 11). The location of pathogens within cells can also be used to distinguish the type of entomopathogen.

4. Size and shape of the causal agent

The size ranges and shapes of the various pathogen groups are presented in Section 3 D. *2* and in greater detail in the following chapters. In addition to measuring and recording the size and shape of suspected pathogen stages, record the presence of inclusions within cells and associated crystals and other structures.

D Identification of major entomopathogen groups

1. Key to the major entomopathogen groups

1a. Distinct external growth on insect 2

1b. No external growth on insect 3

2a. Mass of wormlike (non-segmented) organisms over surface of insect which may be red or cream to light brown in colour, many with a distinct second body covering (Plate 16) NEMATODES

2b. Growth or organisms on surface of insect, often powdery (Plates 3 and 7) (white, green or red) and sometimes limited to intersegmental areas, or growth clublike (Plates 8 and 9) FUNGI

3a. Insect usually normal in appearance but may be stunted or slightly malformed; upon dissection body may contain one or more multicellular organisms with many segments and a distinct to indistinct head region with mandibles, possess a tracheal system marked by various types and arrangements of spiracles; in some cases body of organism may protrude through host's integument or even may be feeding externally on host tissues. PARASITOIDS

3b. Insect may be normal or abnormal in appearance but upon dissection does not contain metazoan parasites, may contain wormlike, non-segmented organisms or small, unicellular particles represented by motile or non-motile rod-shaped to spherical organisms or various life-cycle stages of other micro-organisms including spores, cysts, inclusion bodies or hyphal-like structures. 4

4a. Aquatic or other insect with essentially transparent cuticle (integument) 5

4b. Terrestrial insects or those with basically non-transparent cuticle 12

5a. Insect iridescent or specific tissues (especially fat body) iridescent 6

5b. Insect non-iridescent 7

6a. Insect (scarabaeid grubs) white-bluish to blue-greyish in coloration, containing crystals and minute bacterial-like forms, often pleomorphic in shape that are just visible by light microscopy RICKETTSIAE

6b. Insect orange to green to blue in coloration, infectious agent not visible with light microscopy VIRUSES

7a. Hemolymph (blood) as seen through cuticle milky in coloration, rod-shaped, motile cells often with refringent spore giving footprint appearance under phase microcopy BACTERIA

7b. Hemolymph clear, essentially normal in appearance 8

8a. Wormlike or rapidly motile, ciliated organisms visible through cuticle at low microscope magnification 9

8b. Nematodes or ciliated organisms absent 10

9a. Organisms wormlike and elongate, non-segented (Plates 6, 10 and 16) NEMATODES

9b. Organisms pyriform and motile by cilia PROTOZOA

10a. Intestine (gut) abnormally opaque white (Plate 11), particles polyhedral in shape, visible with phase microscopy in cytoplasm of gut cells VIRUSES

10b. Intestine essentially normal in appearance 11

11a. Abnormal whitish masses in haemocoel

associated with various tissues (e.g. fat body) (Plates 14 and 15) or in haemolymph itself, masses composed of ovoid to pyriform spores refringent under phase microscopy PROTOZOA

11b. Haemocoel (especially of mosquito larvae) filled with hyaline hyphal-like bodies or rust-coloured oval spores with sculptured walls (Plate 12) FUNGI

12a. Insect body usually hardened, mummified and cheesy in consistency, filled with hyaline hyphae, hyphal-like bodies or spherical resting spores FUNGI

12b. Insect not hardened or cheesy, may be stunted or with malformed body parts 13

13a. Insect flaccid and usually discoloured, integument may be fragile 15

13b. Insect may be stunted or malformed, integument usually normal in appearance 14

14a. Various body tissues and cells containing refringent, non-motile spores or cysts (oval, pyriform, navicular or spherical in shape) best visualized by phase contrast microscopy; in some cases intestine may contain motile, flagellated organisms or relatively large, slow moving, septate organisms often occurring in pairs or chains PROTOZOA

14b. Various body tissues and cells containing minute, non-motile, bacterial- to pleomorphic-like cells just visible by light microscopy, cells usually exhibit Brownian movement and are highly refringent, often occurring in pairs or chain-like structures, crystals may or may not be present RICKETTSIAE

15a. Wormlike, non-segmented organisms in body tissues which may be liquified and creamy, greyish to reddish in coloration (Plate 6) NEMATODES

15b. Nematodes not present in body tissues 16

16a. Insect often with putrid odour, usually brown, black or reddish in coloration, body tissues may be liquified with rod-shaped, motile organisms that may contain refringent spores evident under phase microscopy BACTERIA

16b. Insect flaccid and discoloured, integument may be very fragile, liquified body tissues filled with refringent spherical to polyhedral-shaped inclusions which may also occur in the cytoplasm or nuclei of intact cells, inclusions usually dissolved by a weak solution of NaOH, cadaver (caterpillar) may be attached to host plant in inverted v-shaped manner hanging by abdominal prolegs (Plate 4) VIRUSES

A number of other keys are available for the identification of the major pathogen groups (Weiser & Briggs, 1971; Poinar & Thomas, 1984) including keys in Portuguese (Alves, 1986) and Italian (Deseö-Kovács & Rovesti, 1992).

2. *General characteristics of insect disease caused by the major groups of entomopathogens*

Brief descriptions of signs and symptoms of insect disease caused by the major groups of entomopathogens and additional information are provided below to aid in initial diagnosis. Detailed descriptions of each group are provided in subsequent chapters and by Tanada & Kaya (1993).

a. *Viruses*

Viruses are reported from virtually every insect order and are the smallest of the entomopathogens. Virions of the non-occluded forms range in size from 0.01 to 0.3 μm whereas the polyhedra and other inclusion bodies (IBs) which occlude the *virions* of the *occluded* viruses range from 1.0 to 15 μm in size. Some of the more virulent viruses produce widespread epizootics resulting in dramatic collapses in host populations. Most of the non-occluded (virions not occluded in a protein matrix) or non-aggregated viruses are not visible under light microscopy. The occluded viruses: the nuclear polyhedrosis viruses (NPV), granulosis viruses (GV), entomopoxviruses (EPV) and cytoplasmic polyhedrosis viruses (CPV) are the most commonly observed due to the incorporation of the virus particles into a protein matrix which is large enough to be visible under light microscopy. The protein matrix of the IBs of NPV, GV, CPV dissolve in basic solutions such as 1N KOH and NaOH, enabling their separation from other crystalline structures such as uric acid crystals. While the infected insect is alive, IBs are sometimes shed in drops of diarrhoea or regurgitate.

Colour changes due to virus infections are observed in both dead and living insects. The blue coloration associated with iridescent viruses may be pale to a deep blue–purple with a distinct iridescence. Less commonly, orange iridescent virus may also be observed in mosquito larvae (Plate 2). Chalky white zones in the midgut and fat body are observed with some viruses. In aquatic Diptera, NPV in mosquitoes and CPV in mosquitoes, black flies and others are distinctly observed in the midgut region as chalky white areas (Plate 11). Colour changes in Lepidoptera due to patent infections of NPV and GV may result in a change in colour to white, grey or light brown. The normal colour of the integument may fade somewhat and the normally translucent areas of the integument (e.g. the prolegs, ventrum and cervix) become milky due to the presence of IBs in the haemolymph. Insects infected with these viruses often die attached to substrates by the prolegs (Plate 4). At this point the insect is flaccid, the integument is easily ruptured and the insects appear to disintegrate.

b. Rickettsiae

Less commonly observed with the naked eye than many of the viral infections in insects, rickettsial infections may occasionally stand out in certain insect populations. Rickettsiae are found in a broad insect host range. They are small (0.2–0.6 μm) rod shaped, Gram-negative organisms that look like bacteria and behave like viruses (i.e. are obligate intracellular pathogens). Species in the genus *Wolbachia* produce inapparent infections in insects and are seldom harmful to their hosts. Species in the genus *Rickettsiella* are pathogenic for insects and are reported from Coleoptera, Diptera, Lepidoptera and Orthoptera. *Rickettsiella* that infect larvae of several scarab species produce a bluish cast to the infected fat body. Krieg (1963) describes the colour as a bluish-green iridescence similar to blue iridescent virus, but not as intensive. Also known as blue disease, these rickettsioses are somewhat chronic (Krieg, 1963). Large birefringent crystals are an accompanying characteristic of some of the *Rickettsiella* infections.

Some important human pathogens in the genus *Rickettsia* (causal agents for typhus, Rocky mountain spotted fever and others) are transmitted by arthropod intermediate hosts and are pathogenic for the arthropod vectors as well as humans. *Rickettsiella melolonthae* has also been reported as being pathogenic for mammals (Krieg, 1963, 1971). Due to some doubt regarding their specificity and potential danger for humans and the need for *in vivo* production, *Rickettsiella* pathogens of insects have not been developed as microbial control agents and will not be treated in greater detail in this volume. For more information on techniques for their isolation, identification, cultivation, bioassay and storage refer to Krieg (1963, 1971) and Poinar & Thomas (1984).

c. Bacteria

Bacteria found in insects include both spore-forming and non-spore forming varieties. Entomopathogenic bacteria and related organisms come in a range of shapes (rods, cocci, spiral and pleomorphic) and sizes (0.5–50 μm), occur singly or in chains, are Gram-negative or Gram-positive and are aerobic or anaerobic. Unfortunately, dead insects make excellent media for a broad diversity of saprophytic species. Even insects that have been killed as a result of some of the other entomopathogens may be invaded by non-pathogenic bacteria. These invasions, however, usually result in a population of mixed species. When only one or a predominant species is found, it is an indication that the insect has possibly been killed by bacteria. Insects that have been recently invaded by nematodes with bacterial symbiotes may also be filled with a single bacterial species.

Some of the bacterial entomopathogens are not initially lethal to their insect hosts and signs and symptoms may be observed in living insects (e.g. *Bacillus popilliae*, one of the bacteria that causes milky disease in scarabs; Plate 1; and *Serratia entomophila*, the species that causes honey disease in scarabs). Both of these are covered in detail in Chapter III-4.

Changes in colour due to some of the bacterioses of insects are quite distinct. Infected larvae can be white (Plate 1), red (Plate 5), amber, black or brown. Recently killed insects may be odiferous, flaccid and fragile. Cadavers that have aged somewhat usually shrivel and dry into a hard scale.

The most commonly used microbial control agent, *Bacillus thuringiensis*, may produce a range of symptoms in insects depending on the variety of the bacterium and the target insects. Because its predominant mode of action is as a stomach toxin, insects may be killed due to toxaemia with or without subsequent reproduction of the bacterium in the haemocoel (see *Bacteraemia* and *Septicaemia*). Prior to death many species stop feeding and may wander from their original feeding site or even from the host plant.

d. Protozoa

Unlike most of the other types of entomopathogens, protozoa usually produce chronic infections manifested by such signs and symptoms as irregular growth, sluggishness, loss of appetite, malformed pupae or adults, or adults with reduced vigour, fecundity and longevity. Such characteristics are seldom *pathognomonic* in nature, although black, pepper-like spots on the integument of silk worm larvae are essentially diagnostic for the microsporidian disease known as Pebrine. In insects with transparent cuticles (mosquitos and other aquatic insects), whitish masses of microsporidian spores may be visible scattered throughout the haemocoel (Plate 15). In the majority of hosts infected with protozoa, however, one must dissect and examine various tissues for the presence of vegetative forms and cysts or spores, the latter ranging in size from 2 to 20 µm. Although the reproductive forms (spores or cysts) are readily recognized when examined in wet-mount preparations by phase microscopy, it is usually necessary to stain wet-mounted, impression smears of various tissues to examine the vegetative stages of development. The life cycles of microsporidia and neogregarines are often extremely complex, sometimes involving an intermediate host, and specific identifications can only be made with the assistance of a specialist.

Protozoa, especially the microsporidia, are relatively host specific and can usually be found within specific host species. Under laboratory conditions, however, many species can be cross transmitted to a wide range of hosts that may be helpful in carrying out infectivity tests involved in Koch's postulates. Almost all the entomophilic protozoa are obligate parasites and cannot be grown on artificial media. Detailed investigations of most species will require examination of infected hosts by transmission electron microscopy as presented in Chapter VIII-1.

e. Fungi

Some of the most spectacular infections in insects are produced by fungal entomopathogens and many result in colourful (Plate 7) and/or striking outgrowths of the fungus (Plates 8 and 9). Some fungal species that forcibly discharge spores from the host or grow onto the substrate from the host, may produce a distinct halo around the infected insect. The entomopathogenic fungi are a broad and diverse group taxonomically and biologically and infect virtually every insect Order. Due to the mode of entry through the host cuticle by most species of entomopathogenic fungi, they are the only entomopathogens found in sucking insects (Homoptera and Hemiptera). Infectious propagules come in a broad array of shapes and sizes (5 µm to several centimetres) and may be motile, projected from host cadavers, wind-borne, dispersed by water or by the insect hosts themselves.

Most of the entomopathogenic fungi kill their hosts relatively soon after infection. Following death, infectious spores are usually produced on the surface of the insect. Larger, thick walled resting spores of many fungal species can also be found in the host. Insect cadavers are often mummified due to mycoses and may persist in the environment for several weeks, enabling isolation of the pathogen long after death of the host.

Developmental and reproductive stages of entomopathogenic fungi can also be found in living hosts, most commonly in larvae of aquatic Diptera (Plates 12 and 13).

f. Nematodes

Except in the early stages of infection, signs of nematode infections are readily apparent to the observer; one or several nematodes may be seen through the cuticle. With many of the more commonly observed nematode species, the host may be alive up to the moment the nematode emerges (Plate 10) or is killed shortly after infective forms invade the host (Plate 6). The coloration of host insects that are attacked by heterorhabditids or steinernematids changes to red (Plate 6), orange, to honey or cream-coloured due to the presence of bacterial symbiotes. Infective forms of these nematodes have a distinct second cuticle (Plate 16). In addition to causing distinctive coloration, many species of the symbiotes of heterorhabditid nematodes, *Photorhabdus* spp., are luminescent. The size of nematodes found in insects ranges from less than 1 mm to several centimetres.

Identification of most species of entomopathogenic nematodes requires the adult stage. Stages leaving the host are usually not the adult stage and must be held under appropriate conditions until the pre-adult stages mature (see Chapter VI).

Several atlases of insect diseases provide colour and black and white photographs of diseased insects that can aid in the recognition of diseased insects in the field. These include: Weiser (1969, 1977), Poinar & Thomas (1978, 1984), Samson *et al.* (1988), Adams & Bonami (1991).

E Conclusive Diagnosis – Satisfying Koch's postulates

The positive identification of a suspected pathogen from a diseased insect does not always incriminate the organism as the causal agent of the disease. Careful analysis of the facts gathered in the field, from laboratory examinations, study of the progress of the disease and other aetiological information outlined by Steinhaus (1963a) will be necessary when the diagnosis is critical or other information on hand is not conclusive. Satisfying Koch's postulates is the most definitive way to make a conclusive diagnosis. The following steps are taken to confirm that the isolated micro-organism is the causal agent of the disease (modified from Steinhaus, 1963a and Agrios, 1988):

1. The pathogen must be isolated from all of the diseased insects examined, and the signs and/or symptoms of the disease recorded.
2. The pathogen must be grown in *axenic culture* on a nutrient medium (for non-obligate pathogens) or in a susceptible insect (obligate pathogens), and it must be identified and/or characterized.
3. The pathogen must be inoculated on/in healthy insects of the same or a similar species to the original, and signs and symptoms of disease must be the same.
4. The pathogen must be isolated in *axenic culture* again and its characteristics must be exactly like those observed in Step 2.

When it is possible to culture a suspected pathogen on artificial media and produce infectious propagules, satisfying Koch's postulates is a relatively straightforward process. However, many organisms are obligate pathogens (all viruses and Rickettsiae, most protozoa, many fungi and nematodes and some bacteria). In these cases, the infectious agent is produced *in vivo* and purified using methods presented in Chapters II, III-4, IV, V-2, and VI. It should be noted that some organisms that were previously regarded as impossible to produce on artificial media have since been successfully cultured on complex media that satisfy specific nutritional requirements that enable production of infectious propagules (see *Lagenidium giganteum*, Chapter V-4). Obligate parasites that require intermediate hosts (some protozoa and fungi) may be even more problematic (see Chapters IV and V-4) especially if the requirement is suspected and the intermediate host is not yet known.

With pathogens that are obligate parasites or are submicroscopic in size, one must use a variety of techniques in carrying out Koch's postulates. The use of the electron microscope is essential in detecting and characterizing such intracellular entomopathogens as viruses, Rickettsiae, and protozoa. In addition, identification may involve the use of various serological or molecular techniques such as enzyme-linked immunosorbent assays, sodium dodecyl sulphate (SDS)–polyacrylamide gel electrophoresis, restriction endonuclease analyses of DNA, randomly amplified polymorphic DNA technology or biochemical analyses (see Chapter VIII-3). It is also more difficult to obtain pure cultures of such obligate entomopathogens and various techniques such as sucrose-gradient or rate-zonal centrifugation must be utilized (see Chapters II and IV). Tissue cultures can also be used to obtain pure cultures of viruses or protozoa but care must be taken to avoid contamination.

4 SAFETY CONSIDERATIONS

Although most pathogens found in insects are selective for insects, care should be taken when handling these organisms until identifications are made and their safety determined. With the exception of the Rickettsiae, the safety to non-target organisms of each of the entomopathogen groups covered in this *Manual* are discussed by several authors in Laird *et al.* (1990).

ACKNOWLEDGEMENTS

We thank Mark Goettel, Tad Poprawski, John Vandenberg, Fernando Vega and Sam Yang for reviewing the manuscript. We are also grateful to the several colleagues who furnished photographs for the colour plates and to Guy Mercadier and Claire Vidal for preparation of the other figures. We thank James Harper for providing the computer file of the glossary. Cynthia Lacey and Michelle Kellogg helped with preparation of the manuscript.

REFERENCES

Adams, J. R. & Bonami, J. R. (1991) *Atlas of invertebrate viruses*. CRC Press, Boca Raton.

Agrios, G. N. (1988) *Plant pathology*, 3rd. edn. Academic Press, San Diego.

Alves, S. B. (ed.) (1986) *Controle microbiano de insetos*. Editora Manole Ltda., São Paulo.

Avery, S. W. & Undeen, A. H. (1987) The isolation of microsporidia and other pathogens from concentrated ditch water. *J. Am. Mosq. Control Assoc.* **3**, 54–58.

Deseö-Kovács, K. V. & Rovesti, L. (1992) *Lotta microbiologica control i fitofagi teoria e pratica*. Edagricole-Edizioni Agricole, Bologna.

Holt, J. G. (1977) *The shorter Bergey's manual of determinative bacteriology*, 8th edn. Williams and Wilkins, Baltimore.

Krieg, A. (1963) Rickettsiae and rickettsioses. In *Insect pathology, an advanced treatise*, Vol. 1. (ed. E. A. Steinhaus) pp. 577– 617. Academic Press, New York.

Krieg, A. (1971) Possible use of Rickettsiae for microbial control of insects. In *Microbial control of insects and mites* (eds. H. Burges & N. W. Hussey), pp. 173–179. Academic Press, New York.

Lacey, L. A. & Goettel, M. (1995) Current developments in microbial control of insect pests and prospects for the early 21st century. *Entomophaga*, **40**, 3–28.

Laird, M., Lacey, L. A. & Davidson, E. W. (eds.) (1990) *Safety of microbial insecticides*. CRC Press, Boca Raton.

Martignoni, M. E. & Milstead, (1960) Quaternary ammonium compounds for the surface sterilization of insects. *J Insect Pathol* **2**, 124–133.

Martignoni, M. E., Krieg, A., Rossmore, H. W. & Vago, C. (1984) *Terms used in invertebrate pathology in five languages: English, French, German, Italian, Spanish*. Publ. PNW-169, US Dept. Agric., Forest Serv., Portland.

Poinar, G. O., Jr & Thomas, G. M. (1978) *Diagnostic manual for the identification of insect pathogens*. Plenum Press, New York.

Poinar, G. O., Jr & Thomas, G. M. (1984) *Laboratory guide to insect pathogens and parasites*. Plenum Press, New York.

Samson, R. A., Evans, H. C. & Latgé, J.-P. (1988) *Atlas of entomopathogenic fungi*. Springer-Verlag, Berlin.

Steinhaus, E. (1949) *Principles of insect pathology*. McGraw Hill, New York.

Steinhaus, E. A. (1963a) Background for the diagnosis of insect diseases. In *Insect pathology, an advanced treatise*, Vol. 2 (ed. E. A. Steinhaus) pp. 549–589. Academic Press, New York.

Steinhaus, E. A. (ed.) (1963b) *Insect pathology, an advanced treatise*, Vol. 1. Academic Press, New York.

Steinhaus, E. A. & Marsh, G. A. (1962) Report of diagnoses of diseased insects 1951–1961. *Hilgardia* **33**, 349–490.

Steinhaus, E. A. & Martignoni, M. E. (1970) *An abridged glossary of terms used in invertebrate pathology*, 2nd edn, USDA Forest Service, Pacific NW Forest and Range Experiment Station.

Tanada, Y. & Kaya, H. K. (1993) *Insect pathology*. Academic Press, San Diego.

Thomas, G. M. (1974) Diagnostic techniques. In *Insect Diseases*, Vol. 1. (ed. G. E. Cantwell), pp. 1–48. Marcel Dekker, New York.

Weiser, J. (1969) *An Atlas of Insect Diseases*. Academia, Prague.

Weiser, J. (1977) *An Atlas of Insect Diseases*. Academia, Prague.

Weiser, J. & Briggs, J. D. (1971) Identification of pathogens. In *Microbial control of insects and mites*. (eds. H. Burges & N. W. Hussey), pp. 13–66. Academic Press, New York.

Wittig, G. (1963) Techniques in insect pathology. In *Insect pathology, an advanced treatise*, Vol. 2 (ed. E. A. Steinhaus), pp. 591–636. Academic Press, New York.

GLOSSARY

Most of the following terms have been selected from the glossary prepared by Steinhaus & Martignoni (1970). Additional words used in this chapter have also been added. Various terms in invertebrate pathology are also defined in the multilinguistic glossary by Martignoni *et al.* (1984). Additional glossaries for specific terms appearing in this Manual are provided in Chapters II, V-1, VIII-2 and VIII-3.

Aetiological agent. The pathogen responsible, also referred to as the causal agent.

Aetiology. The study of the causes of disease.

Aplasia. The entire failure of organs or tissues to develop. The congenital absence of an organ or tissue.

Atrophy. (1) Decrease in size of a tissue, organ, or part after full development has been obtained. A wasting of tissues, organs, or entire body from disuse, old age, injury, or disease. A condition in which the affected cells undergo degenerative and autolytic changes, become smaller, and have a lessened functional capacity. (2) If there is destruction of some of the cells in a tissue we speak of 'quantitative atrophy'. (See Hypoplasia (2)).

Axenic culture. The rearing of one or more individuals of a single species in or on a non-living medium.

Bacteraemia. The presence of bacteria in the haemolymph or blood of invertebrates and other animals, without production of harmful toxins or other deleterious effects.

Birefringent. Refracting twice, splitting a ray of light in two. (See Refringent).

Commensalism. A symbiotic relationship in which one of the two partner species benefits, without apparent effects on the other species. (See also Symbiosis).

Diagnosis. To distinguish one disease from another. The determination of a disease from its signs, symptoms, aetiology, pathogenesis, physiopathology, morphopathology, etc. Also, the decision reached.

Disease. (See also Syndrome). 'Lack of ease.' Departure from the state of health or normality. Condition or process (not a thing) that represents the response of an animal's body to injury or insult. A disturbance of function or structure of a tissue or organ of the body, or of the body in general. (A healthy animal is one so well-adjusted in its internal milieu and to its external environment that it is capable of carrying on all the functions ultimately necessary for its maintenance, growth and multiplication with the least expenditure of energy.) There are several additional definitions of the term disease presented by Steinhaus & Martignoni (1970).

Dysentery. A term given to a number of disorders marked by lesions of the alimentary canal and often attended by abnormal frequency and liquidity of faecal discharges. In sericultural practice the term flacherie has been used for certain forms of dysentery of the silkworm larvae.

Eclipse period. In the developmental cycle of viruses, a phase or period, occurring immediately after infection (i.e., immediately after a virus enters the host cell), in which infective particles cannot be detected. The phase during which the infected host cell contains no material capable of infecting another cell or host.

Entomopathogen. A micro-organism or nematode that causes disease in insects. (See Pathogen).

Enzootic disease. A disease (usually in low prevalence) which is constantly present in a population.

Epizootic. An outbreak of disease in which there is an unusually large number of cases. A disease or a phase of a disease of high morbidity and one that is only irregularly present in recognizable form. (See also Enzootic and Panzootic).

Gross pathology. The study of macroscopic structural lesions. Abnormalities of gross structure.

Histopathology. The study of abnormal microscopic changes in the tissue structure of plants and animals.

Host. A host in which the pathogenic micro-organism (or parasite) is commonly found and in which the pathogen can complete its development. The term 'natural host' implies that the host is the usual one and is synonymous with 'typical host.'

Hyperplasia. An increase in the number of functional units of an organ (organelles, cells, tissues), excluding tumour formation, whereby the bulk of the organ is increased in response to increased functional demands. (See also Hypertrophy).

Hypertrophy. An increase in size (weight) and functional capacity or an organ or tissue, without an increase in the number of structural units upon which their functions depend. Hypertrophy is usually stimulated by increased functional demands. (See also Hyperplasia).

Hypoplasia. (1) A defective or incomplete development of an organ or tissue. (2) Sometimes used to indicate an atrophy caused by the destruction of some of the elements (e.g. cells) rather than a general reduction in size.

Incidence (of a disease). The number of new cases of a particular disease within a given period of time, in a population being studied. Compare with Prevalence (of a disease).

Incubation period. The period of time elapsing between entrance or introduction of micro-organisms in the animal body and the development of symptoms and signs of an infectious disease.

Induction. The activation of an occult pathogen, leading to progressive (overt or patent) infection and disease. In particular, the provoked transformation of a provirus into a virulent (cytocidal) virus.

Infection. The introduction or entry of a pathogenic micro-organism into a susceptible host, resulting in the presence of the micro-organism within the body of the host, whether or not this causes detectable pathological effects (or overt disease). The term infection has also been used by some authors to indicate the invasion of tissues by living pathogenic micro-organisms in such a way that their proliferation, growth, and/or toxin production injure the tissues or cells involved.

Acute infection. Of short duration. Characterized by sharpness or severity. As 'acute disease'. (See patent infection).

Attenuated infection. An infection which is not immediately followed by overt disease, but may follow a phase of overt disease. Usually three types of attenuated infections are recognized: Microbial persistence, Latent infection and the Carrier state.

Chronic infection. Of long duration. Not acute. As 'chronic disease'.

Inapparent infection. An infection which gives no overt sign of its presence.

In vitro. In the 'test tube', or other artificial environment. Outside a living organism.

In vivo. In the living organism.

Melanization. Deposition of the black pigment, melanin, and associated materials on the surfaces of foreign objects, both biotic and abiotic. Often accomplished by haemocytes as a response to injury or to the presence of a parasite. Common in arthropods.

Moribund. Dying. Near death.

Mutualism. A symbiotic relationship between two different species in which both jointly benefit. Usually obligatory.

Mycetocyte. A cell containing intracellular mutualistic and commensalistic microsymbiotes. One of many cells making up the mycetome.

Mycetome. In various invertebrate animals, the structure or organ which houses symbiotes. The cells making up the mycetome and containing the symbiotes are known as mycetocytes.

Occluded. Said of those viruses in which the virions are occluded in a dense protein crystal, large enough to be visible in the light microscope (e.g. polyhedrosis viruses, granulosis viruses).

Occult virus. A special phase of some viruses, characteristic of latent infections, in which the pathogenic agent is presumed to differ from the infective phase, and in which virions cannot be detected. Synonymous with but preferable to 'hidden virus' and 'masked virus' (see Latent infection). The occult phase of a virus should not be confused with the eclipse, which is a normal phenomenon during viral replication.

Panzootic. Denoting a disease affecting all, or a large proportion of the animals of a region. Extensively epizootic.

Parasite. An organism that lives at its host's expense, obtaining nutriment from the living substance of the latter, depriving it of useful substance, or exerting other harmful influence upon it. Some authors distinguish 'parasite' from 'parasitoid', the latter having among others the following two characteristics: (a) the development of an individual destroys its host; (b) it is parasitic as a larva only, the adult being free-living e.g. the entomophagous Hymenoptera are parasitoids.

Parasitic castration. Any process that interferes with or inhibits the production of mature ova or spermatozoa in the gonads of an organism. (The term is not limited to meaning the sudden and complete extirpation of the gonads.)

Parasitism. A symbiotic relationship between two different species in which one (the parasite) benefits at the expense of the other (the host). (See Parasite).

Patent infection. An overt infection with distinct signs and symptoms of disease.

Pathogen. A specific cause of disease. A micro-organism capable of producing disease under normal conditions of host resistance and rarely living in close association with the host without producing disease. Any micro-organism, virus, substance, or factor causing disease.

Pathogenesis. The origination and development of a disease or morbid process.

Pathogenicity. The quality or state of being pathogenic. The potential ability to produce disease. Applied to groups or species of micro-organisms, whereas virulence is used in the sense of degree of pathogenicity within the group or species. Some authors regard pathogenicity as the genetically determined ability to produce disease, and virulence as disease-producing ability that is not genetically determined. (See also Virulence).

Pathognomonic. A pathognomonic (diagnostic) symptom or sign is one that points with certainty to a particular disease or malfunction. Such a special symptom or sign indicates an aberration or disturbance of a particular nature by which a disease may be definitely recognized.

Pathology. The science that deals with all aspects of disease. The study of the cause, nature, processes, and effects of disease. Any branch of science, or any technique or method or body of facts that contributes to our knowledge of the nature and constitution of disease belongs in the broad realm of pathology. In a more limited sense, pathology refers to the structural

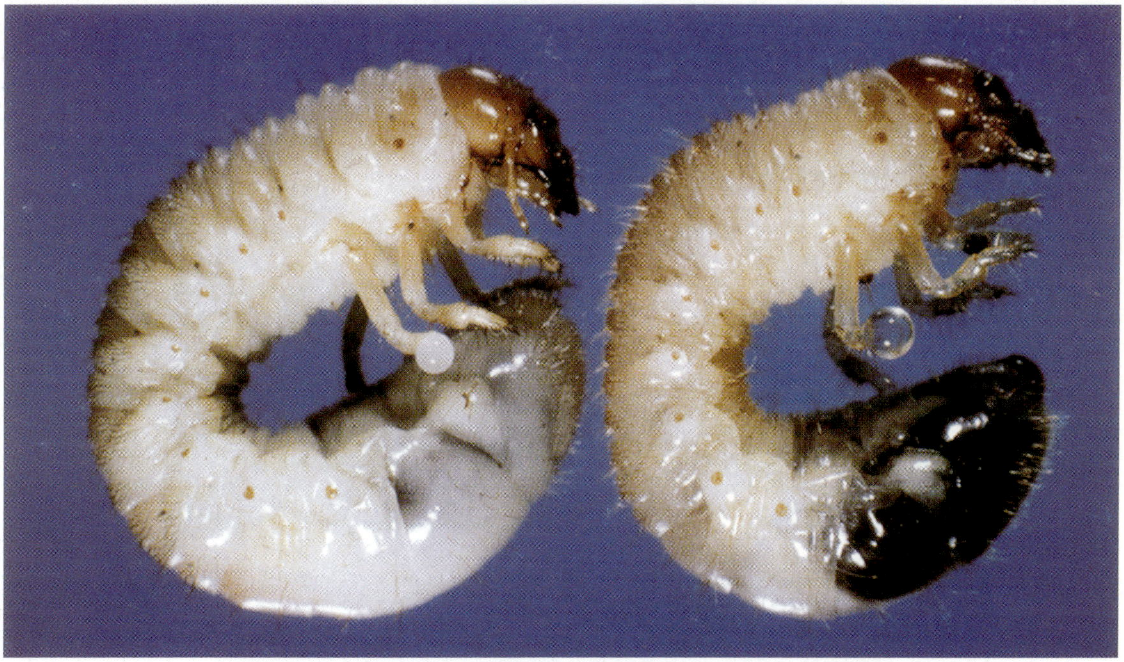

Plate 1. Living larvae of *Popollia japonica* with (left) and without (right) milky disease caused by the bacterium, *Bacillus popilliae*. The third right prolegs have been cut to demonstrate the milkiness of the haemolymph in the infected larva compared to that of the healthy larva. (Courtesy of Michael Klein.)

Plate 2. Living larvae of *Aedes taeniorhynchus* infected with blue (also referred to as T-MIV) and orange (also referred to as R-MIV) iridescent virus. (Courtesy of Tokuo Fukuda.)

Plate 3. Cicada adult infected with the fungus, *Entomophthora* sp. (Courtesy of Harry Evans.)

Plate 4. Larva of *Sabulodes aegrotacta* infected with a granulosis virus. Note the attachment to the host plant by the prolegs. (Courtesy of Brian Federici.)

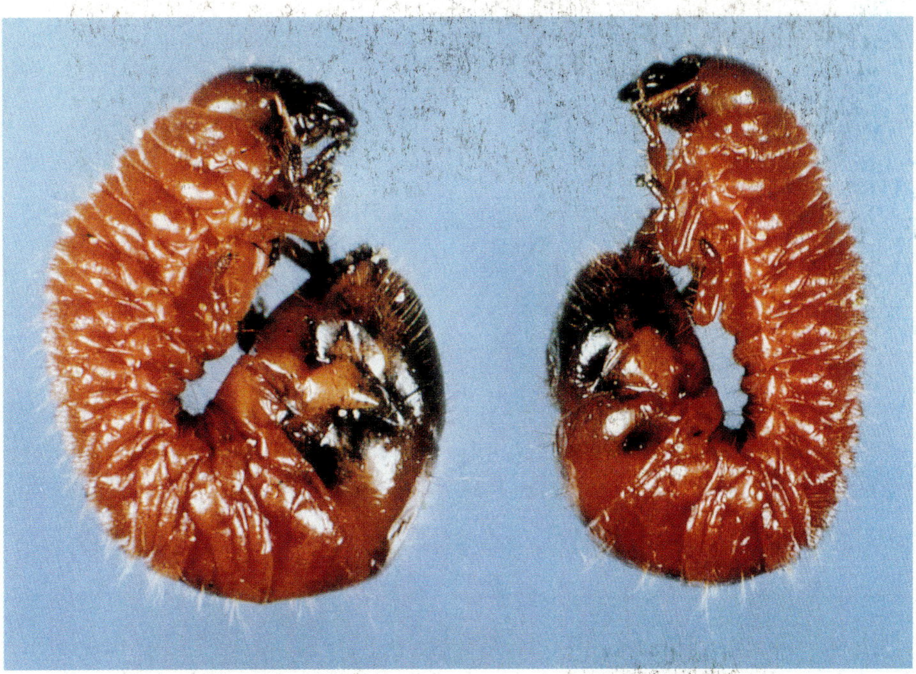

Plate 5. Larvae of *Popillia japonica* infected with the bacterium, *Serratia marcesans*. (Courtesy of Michael Klein.)

Plate 6. Larva of the black vine weevil, infected with the nematode, *Heterorhabditis bacteriophora*. (Courtesy of Robin Bedding.)

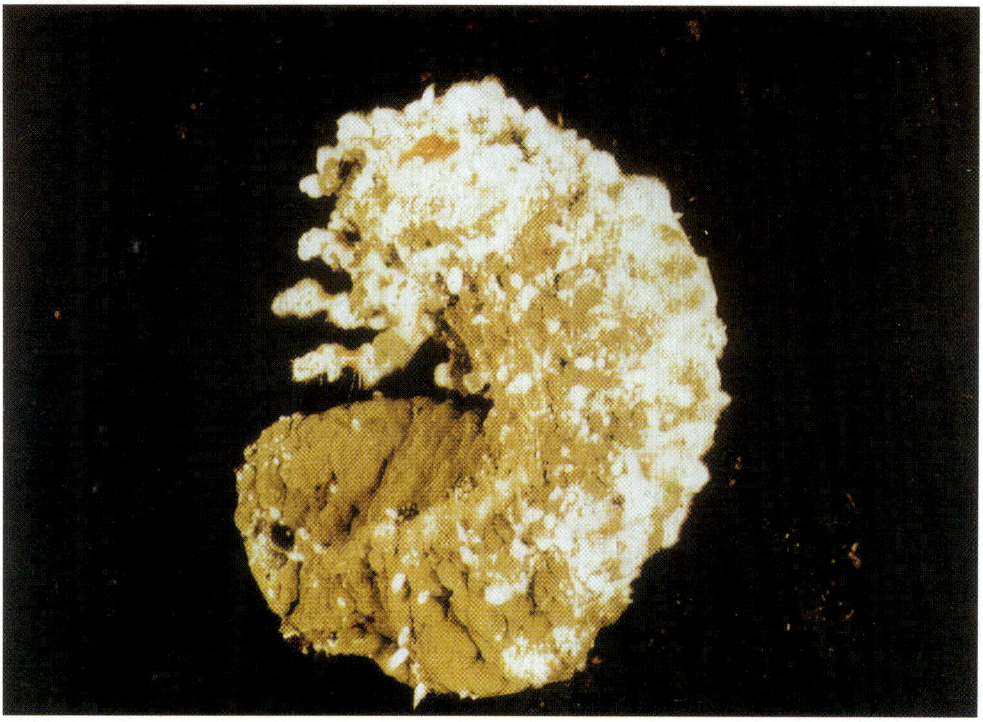

Plate 7. Larva of *Popillia japonica* infected with *Metarhizium anisopliae*. (Courtesy of Michael Klein.)

Plate 8. Adult forest locust, *Schistocerca* sp., infected with the fungus, *Cordyceps* sp. (Courtesy of Harry Evans.)

Plate 9. Adult tabanid fly infected with the fungus, *Cordyceps dipterigena.* (Courtesy of Harry Evans.)

Plate 10. Pre-adults of the mermithid nematode, *Romanomermis culicivorax,* as seen emerging from and through the cuticle of *Culex quinquefasciatus* larvae. (Courtesy of Tokuo Fukada.)

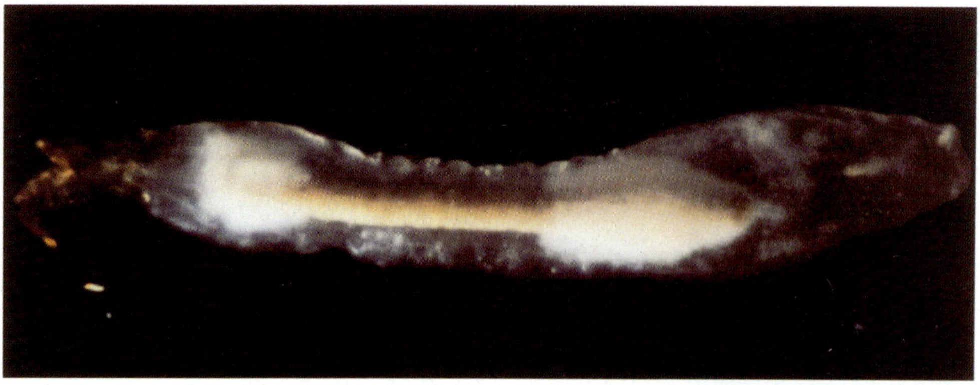

Plate 11. *Simulium vittatum* larva infected with a cytoplasmic polyhedrosis virus. The larva has been cleared to facilitate viewing the characteristic chalkiness in the gastric caecae and in the posterior portion of the midgut. (Courtesy of Dan Molloy.)

Plate 12. Sporangia of the fungus, *Coelomomyces*, as seen through the integument of a larval mosquito. (Courtesy of Brian Federici.)

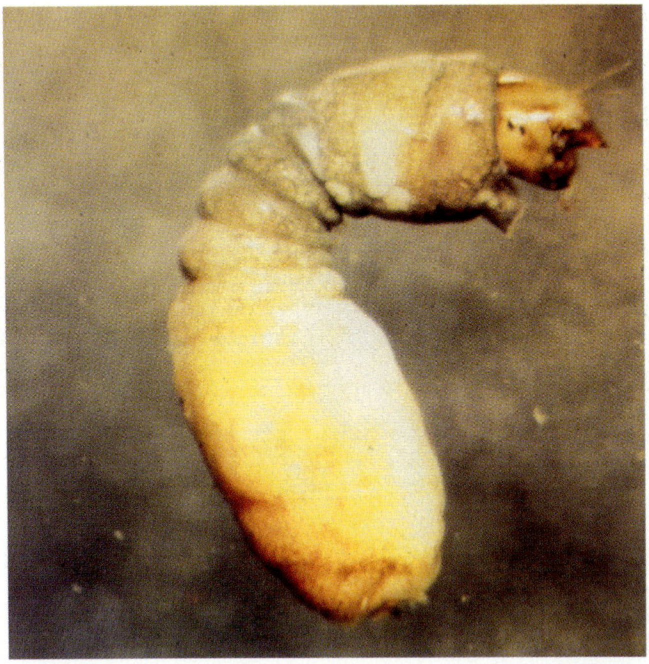

Plate 13. Spherical cysts of the fungus, *Coelomycidium simulii*, visible through the integument of a *Simulium* larva. (Courtesy of Dan Molloy.)

Plate 14. Larva of *Popillia japonica* infected with the protozoan *Pseudomonocystis* sp. The large spherical cysts containing spores are clearly visible in the posterior of the larva above the rectal sac. Contrast with the rectal sac of the healthy larva in Plate 1. (Courtesy of Michael Klein.)

Plate 15. Larva of the mosquito *Culiseta melanura* infected with the microsporidium *Hyalinocysta chapmani*. (Courtesy of Ted Andreadis.)

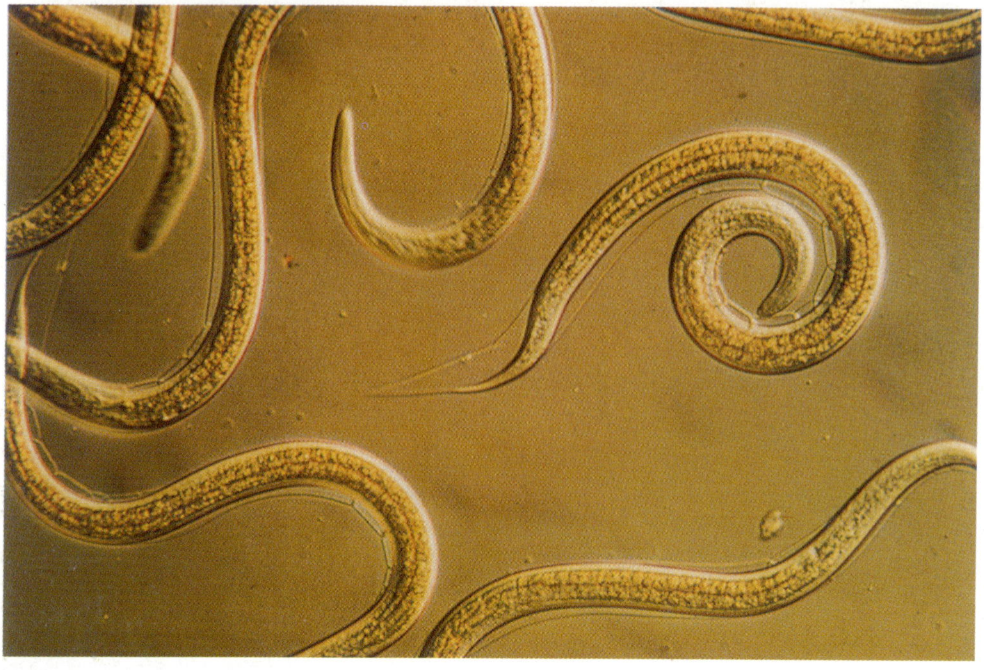

Plate 16. Infective juveniles of the nematode, *Steinernema carpocapsae*. (Courtesy of Robin Bedding.)

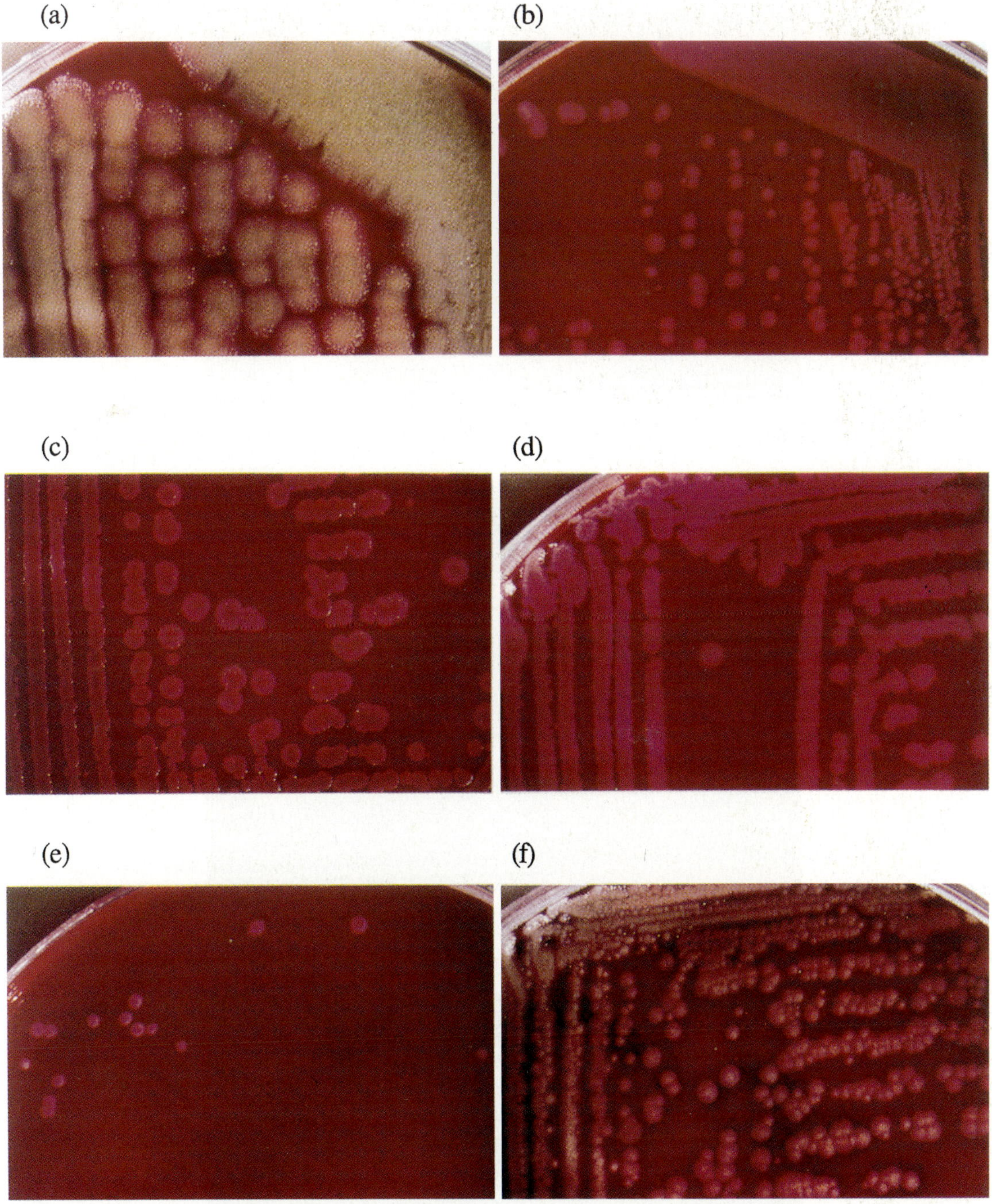

Plate 17. Photographs of *Bacillus* sp. colonies grown on trypace soy horse blood agar (24 h, at 30°C): (a) *B. thuringiensis* var. *israelensis*, (b) *B. sphaericus*, (c) *B. subtilis*, (d) *B. licheniformis*, (e) *B. circulans*, (f) *B. panthotenticus*.

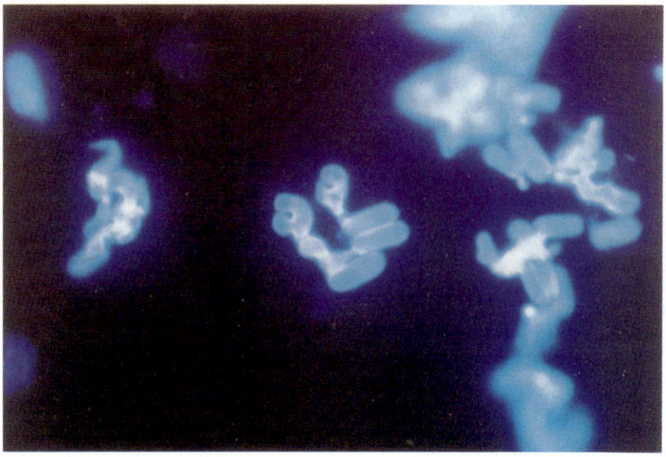

Plate 18. Germlings of *Metarhizium anisopliae* stained with Calcofluor. × 823.

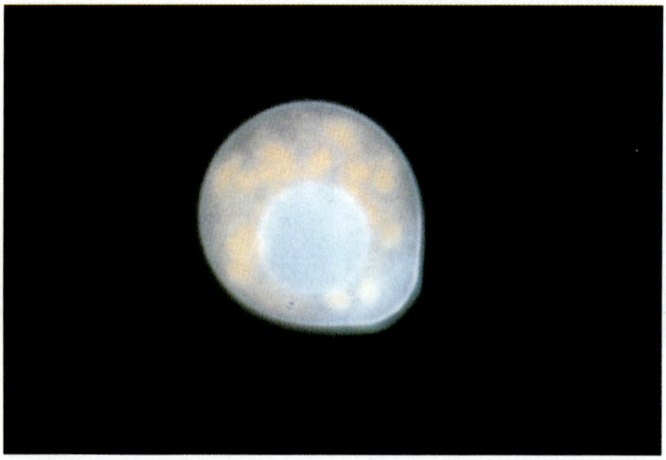

Plate 19. Conidium of *Entomophaga maimaiga* stained with primulin and hydroethidine which stain the cell wall and nuclei, respectively. Hydroethidine also stains the lipid blue. × 335.

Plate 20. DAPI-stained nuclei of *Neozygites* sp. The nuclei are at various stages of mitosis. The nuclei in the central hyphal body are in metaphase whereas those in the adjacent cell are in interphase. × 823.

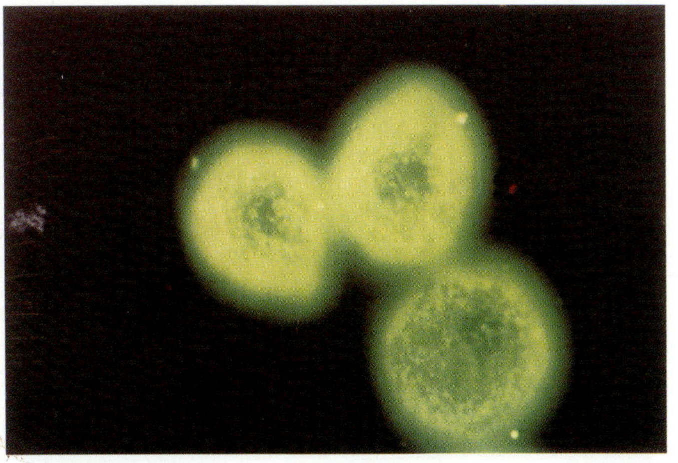

Plate 21. Conidia of *Entomophaga maimaiga* stained with $DiOC_6$ showing the endoplasmic reticulum and mitochondria. × 335.

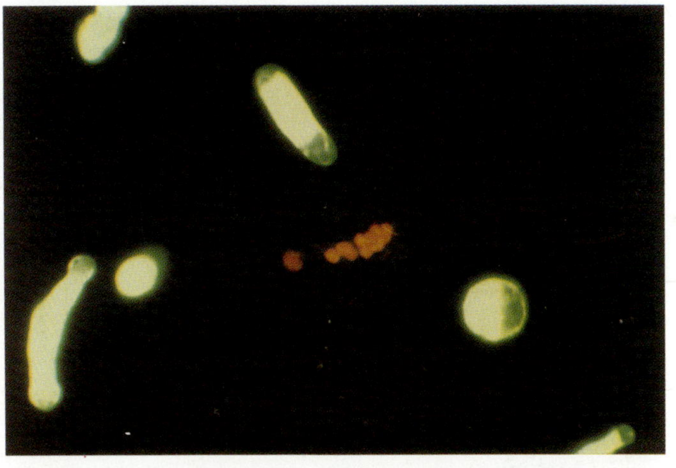

Plate 22. Hyphal bodies of *Massospora cicadina* stained with propidium iodide and fluorescein diacetate. The nuclei in the dead cells stain orange–red. Healthy cells fluoresce yellow–green. × 149.

Plate 23. Actin localization in hyphal bodies of *Neozygites* with Rh-phalloidin, showing perinuclear actin and actin plaques. × 823.

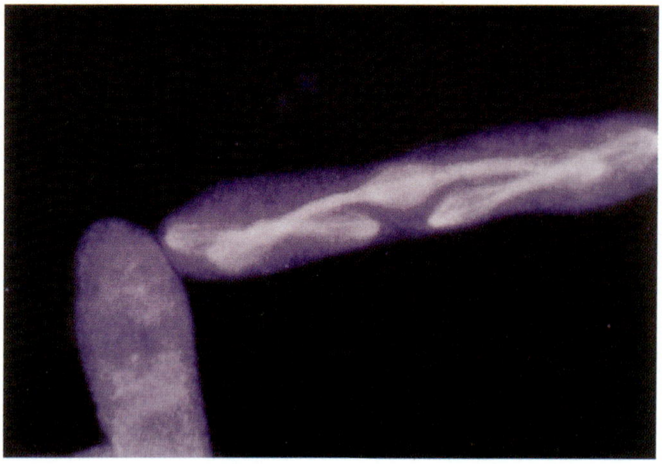

Plate 24. Hyphal bodies of *Neozygites* showing immunofluorescent localization of tubulin in anaphase spindle. × 882.

Plate 25. Appearance of *Bombyx mori* larvae infected with transformed and untransformed *M. anisopliae* strains. At 72 h after infection, the larvae were injected under the cuticle with 50 μg X-gluc. (in 25 μl DMSO) and incubated at 37°C for 30 min. (A) uninfected *B. mori*; (B) stable Ben^R transformant; (C) stable GUS expressing Ben^R transformant. Expression of the GUS gene is visible through the cuticle as blue staining regions.

and functional changes from the normal. 'Invertebrate pathology' refers to all aspects of disease (including abnormalities) which occur in invertebrate animals. Similarly, 'insect pathology' is that branch of entomology or invertebrate pathology that embraces the general principles of pathology as they may be applied to insects.

Polyhedra. Plural of polyhedron.

Polyhedron. Crystal-like inclusion body that occludes virions produced in the cells of tissues affected by certain insect viruses. Synonymous with polyhedral inclusion body.

Prevalence (of a disease). The total number of cases of a particular disease at a given moment of time, in a given population. (Compare with Incidence of a disease).

Refringent. Deflection of a ray of light when it passes from one medium into another of different optical density.

Septicaemia. The invasion of host haemocoel or tissues by bacteria or other micro-organisms with subsequent multiplication, production of toxins and death of the host. A morbid condition caused by multiplication of micro-organisms in the blood. (See also Bacteraemia).

Sign. Any objective aberration or manifestation of disease indicated by a change in structure. Also, physical presence of pathogen.

Stress. A state manifested by a syndrome or bodily changes, caused by some force, condition, or circumstance (i.e. by a stressor) in or on an organism or on one of its physiological or anatomical systems.

Surrogate host. An insect that is substituted for the natural host.

Symbiosis. The living together of individuals of two different species. Especially the living together of dissimilar organisms in a more or less intimate association (as in Mutualism, Commensalism and Parasitism).

Symbiote. An organism living in symbiosis. Usually the smaller member of a symbiotic pair of dissimilar size (also called Microsymbiote). Frequently, those micro-organisms associated in a regular mutualistic manner with insects and other invertebrates. Usually preferred to 'symbiont'.

Symptom. Any objective aberration in function (including behaviour), indicating disease. (See also Sign).

Syndrome. A group of signs and symptoms characteristic of a particular disease. A running together or concurrence of symptoms and signs associated with any morbid process. There is a trend toward considering as a 'disease entity' any morbid process that has a specific cause, while a 'syndrome' reflects not so much specific disease factors as a chain of disrupted physiological processes. Thus, the same syndrome may arise from many different causes.

Toxaemia. A condition produced by the dissemination of toxins in the blood. Certain entomopathogens, such as *Bacillus thuringiensis*, are not invasive *per se* and kill the host through the destructive effects of toxins on the midgut epithelium.

Virion. A morphologically complete virus particle. It can be either a naked or enveloped nucleocapsid.

Virulence. The quality or property of being virulent; the quality of being poisonous; the disease-producing power of a micro-organism.

CHAPTER II

Viruses

HUGH EVANS* & MARTIN SHAPIRO†

* Forest Research Station, Alice Holt Lodge, Wrecclesham, Farnham, Surrey GU10 4LH, UK
† USDA-ARS, Insect Biocontrol Lab, Bldg 011A, BARC-West, Beltsville, MD 20705, USA

1 INTRODUCTION

Insect viruses have been studied for many years, due to an intrinsic interest in the general study of diseases of invertebrates and, more particularly, because of their potential as environmentally benign pest management agents (Evans, 1986). During the early history of their identification and use in pest management, their pathology was based only on symptoms, giving rise to various descriptive names of disease aetiology (Benz, 1986). For example, the *grasserie* of silkworm was a good French descriptor of nuclear polyhedrosis virus (NPV) (Baculoviridae) infection which resulted in liquefaction and disintegration of the affected insects. The NPV of nun moth (*Lymantria monacha*) causes changes in infected larvae that gives rise to aberrant behaviour involving larvae climbing upwards to die in the topmost branches of trees. This was described in German as *wipfelkrankheit* or *tree top disease* (Hofman, 1891). The wilt disease of gypsy moth, *Lymantria dispar*, was also described during early studies on the ecology of this insect imported into the USA (Jones, 1910).

Recognition of occluded viruses as the causative agents of these evocative diseases came at around the turn of the century but it was only during the late 1940s and 1950s, led by the pioneering work of Steinhaus (1956), that the study of the ecology of viruses and their potential use for pest management began.

Early techniques of study, based on the recognition of symptoms, were augmented by more sophisticated light and, later, electron microscopy (EM). In this way it was possible to recognize various virus groups, particularly in those families where virions were occluded within proteinaceous inclusion bodies (IBs). With the use of EM, it was also realized that many non-occluded viruses could be present and these too were described according to combinations of symptoms and morphological characteristics at the organ and cellular levels. The development of sophisticated biochemical and molecular techniques from the 1960s and the continuing refinement of those techniques have opened the way for a more

MANUAL OF TECHNIQUES IN INSECT PATHOLOGY
ISBN 0-12-432555-6

detailed taxonomy and, at least for those laboratories equipped to handle the techniques, the prospect of relatively rapid and extremely accurate methods of identification.

Recognizing that not all laboratories will be able to utilize highly refined techniques, and also that the majority of researchers using pathogenic viruses for pest management will be more interested in contamination than the fine detail of identification, this chapter deals with those techniques that, with minimal outlay, should be usable in virtually all laboratories. Production of field-scale quantities of virus, however, requires a large investment.

2 A BRIEF INTRODUCTION TO THE INSECT VIRUS GROUPS

Although details of how to identify the principal virus groups are given in Section 3, we provide here an introduction to the key characteristics of the majority of viruses, some of which will not be covered in the later sections. Each group will be described under a 'family' heading, although it should be noted that the taxonomy of insect viruses is still evolving, particularly on the basis of biochemical characteristics, and, therefore, the groupings should not be regarded as definitive.

A Baculoviridae

The baculoviruses have been studied intensively, initially reflecting their potential as pest control agents and, more recently, their prominent roles as expression vectors for a wide range of biologically active genes (Summers, 1991). The basic characteristic of the family is the presence of a double-stranded DNA (dsDNA) genome that is covalently closed. Bacilliform virions are composed of nucleocapsids that may be singly or multiply enveloped. The Nuclear Polyhedrosis Viruses (NPV) and Granulosis Viruses (GV) have virions occluded within IBs of crystalline protein called polyhedrin and granulin, respectively. IBs range in size from 0.3 μm to 15 μm in diameter in the NPVs and 0.3 μm to 0.5 μm in the GVs. Virion dimensions are in the size range (40–140 nm) × (250–400 nm) for NPVs and (30–60 nm) × (260– 360 nm) for GVs.

The most typical symptoms are noted in the larval stages where either whitening or yellowing of the gut and/or the remainder of the body organs is associated with infection and replication. After death, rapid melanization takes place, leading to blackening of the body and, linked to infection of the hypodermis, weakening of the outer skin which ruptures easily, releasing the liquefied body contents.

The baculoviruses are the most extensively studied of all the virus groups, arising from their potential as microbial control agents and, more recently, their role as expression vectors for a wide range of genes.

On the basis of its dsDNA genome and specificity to a single family of Coleoptera, the non-occluded virus of coconut palm rhinoceros beetle, *Oryctes rhinoceros*, was previously included in the baculoviruses. It has now been transferred to its own family.

For a detailed treatise on the baculoviruses, see the two-volume work edited by Granados & Federici (1986a,b) and the paper by Adams & McLintock (1991).

B Reoviridae

The Cytoplasmic Polyhedrosis Viruses (CPV) bear a close morphological resemblance to the NPVs. However, reflecting their similarity to the family Reoviridae, they are double-stranded RNA (dsRNA) viruses having ten segments on the genome. The virions (68–69 nm) have a characteristic 12 spikes on the icosahedral particles within the IBs (0.2–10 μm). At least 12 types of CPV have been recognized, based on the segmentation of the genome.

CPV infection is restricted to the gut which may become white or yellow. Other symptoms include extended development and reduced feeding, leading to lowered longevity and breeding performance in infected adults. Transmission by adults is common in this virus, infections often being noted in laboratory-reared colonies of insects.

A useful reference is Hukuhara & Bonami (1991).

C Entomopoxviridae

The third major family of occluded viruses, the entomopoxviruses, are included in the family

Poxviridae but have not been shown to exhibit cross-infection to any vertebrate hosts. Virions, composed of dsDNA, are characteristically brick shaped and are occluded within the paracrystalline protein (spheroidin) of the IB. Other, spindle-shaped occlusions may also be present, especially in the Lepidoptera and Coleoptera. These secondary occlusions tend to be absent in the Diptera and Orthoptera. IB size ranges from 1.0 μm to 24 μm whereas the virions range from 350 × 250 nm to 450 × 250 nm.

Symptoms may be manifested in specific organs, particularly the fat body, or be found in most organs of the body. Colour changes associated with infection may include white or light blue body but the most striking characteristic is the extremely extended longevity of infected individuals.

Key references are Arif (1984) and Goodwin *et al.* (1991).

D Iridoviridae

The major characteristic of this family is the presence of iridescent blue, green or purple coloration in heavily infected individuals, although the mosquito iridescent virus of *Aedes taeniorhynchus* produces orange to brown iridescence. The virions, composed of a dsDNA core, range in size from 130 nm to 180 nm and, when arranged in paracrystalline arrays, produce the characteristic iridescence of this family. At least 32 types of IV have been recognized, based on size and serological relationships. Key references are Hall (1985) and Anthony & Comps (1991).

E Ascoviridae

This unusual group of non-occluded dsDNA viruses has been named on the basis of virion-filled vesicles that are formed when the nucleus of the infected cell ruptures. Virions are large (130 × 400 nm) with a complex structure. They have been isolated only from the Noctuidae (Lepidoptera). Pathology is not strongly developed and may be manifested in lightening of the larval body colour or difficulty in completing a larval moult. The most noticeable effect is, therefore, extremely extended development. Key reference is Federici *et al.* (1991).

F Birnaviridae

This family has a single representative, called *Drosophila* X virus, in the Diptera, having been isolated from laboratory-reared *Drosophila melanogaster*. The virions measure 72 nm × 62 nm and are icosahedral in shape. It is a dsRNA non-enveloped virus. Key reference is Bonami & Adams (1991).

G Caliciviridae

The Caliciviridae were first isolated from the navel orangeworm, *Amyelois transitella*, and were subsequently shown to be composed of ssRNA. The virions have characteristic cuplike morphology that are typical of caliciviruses which, on proteolytic cleavage, form smooth particles about 28 nm in diameter. Little is known about their biology. See Evans & Entwistle (1987) for description and key references.

H Nodaviridae

The nodaviruses are single-stranded RNA viruses with virions around 29 nm in diameter and having two segments of RNA. Most information has been gained from *Nodamura* virus isolated originally from mosquitoes in Japan, and black beetle virus from the beetle *Heteronychus arator* (Coleoptera: Scarabaeidae) in New Zealand. Another notable isolation is flock house virus from the scarab beetle, *Costelytra zealandica*. Symptoms may be absent or be manifested only in slightly extended development or reduced egg survival. However, *H. arator* larvae may become flaccid and lose pigmentation of the hypodermis. Nodaviruses are morphologically identical to picornaviruses of insects and are, thus, difficult to distinguish by electron microscopy. Garzon & Charpentier (1991) provide a good description of the characteristics of the family.

I Parvoviridae

This family consists of single-stranded DNA viruses packaged in virions ranging in size from 19 to 24 nm. The type genus is *Densovirus*, giving rise to the common name of Densonucleosis Virus (DNV). On the basis of virion morphology and symptoms of infection, the family has been divided into Type 1 (acute infection and rapid death with all tissues except the gut infected) and Type 2 (chronic infection and relatively slow death, affecting the gut only). Key reference is Tijssen & Arella (1991).

J Picornaviridae

This group of ssRNA viruses has spherical virus particles, 22–30 nm in diameter. The three best described members of this family are cricket paralysis virus, *Drosophila* C virus and *Gonometa* virus. There is insufficient detail in electron microscope examinations to determine the family characteristics. Main distinguishing features are the presence of three major and two minor polypeptides combined with resistance to acid. Symptoms range from flaccidity of the body to paralysis depending on the particular isolates of the virus. Adults may show shortened longevity and reduced fecundity and fertility. Moore & Eley (1991) provide a full description of the family.

K Polydnaviridae

This family of viruses replicates exclusively in the calyx fluid of parasitic Hymenoptera. As the name implies, a major characteristic of the family is the presence of polydispersed superhelical dsDNA. Polydnaviruses from Ichneumonidae are distinguishable from those isolated from Braconidae. The ovoid virus particles range in size up to 150 × 350 nm in the Ichneumonidae and are smaller in the Braconidae. There are no obvious pathological effects on the parasitoid hosts in which the viruses replicate. Instead, it is thought that the viruses influence the survival of the parasitized host larvae thus contributing to the efficiency of the parasitoid in a mutualistic way. For further reading see Krell (1991).

L Rhabdoviridae

The best studied virus from this family of ssRNA viruses is sigma virus of *Drosophila*. This virus has bacilliform virus particles measuring 75 nm × 200 nm, with surface spikes 8 nm in length. The only known symptom of this virus is the lethal sensitivity of adult flies to CO_2. Moore (1985) and Brun (1991) provide fuller description.

M Tetraviridae

The Tetraviridae are a family of ssRNA viruses having icosahedral virions with diameters of 35–38 nm. The best studied is *Nudaurelia* β virus, isolated from *Nudaurelia cytherea capensis* (Lepidoptera Saturniidae). Young infected larvae become chronically infected giving rise to underweight larvae and pupae. Infections initiated in later stage larvae appear to have no effect but can persist as inapparent infections. See Reinganum (1991) for further reading.

N Other non-occluded viruses

Other families of ssRNA viruses include the Togaviridae (virus particles 60–65 nm diameter), Flaviviridae (35–45 nm diameter) and Bunyaviridae (90–100 nm diameter), all having spherical morphology with surface peplomers. These are all arboviruses, linked to arthropod transmission between hosts and are, therefore, beyond the scope of this chapter.

3 IDENTIFICATION

A Preparation for identification

In the majority of cases the basis for identification of viral pathogens is the availability of invertebrates showing symptoms of infection. Symptoms arise from many different causes and there may well be complications in visual identification of a particular pathogen group, depending particularly on whether the specimen in question has already died, which may lead to indeterminate symptomology. However, regardless of which virus group is concerned, or, to a very great extent, of which pathogen group, the basic methodology for specimen preparation is similar, being refined only after

Table 1 Checklist for preliminary diagnosis for the presence of insect pathogenic viruses.

Criterion	*Record*	*Main points to note*
Life stage of specimen	Life stage, ideally to instar for larvae	Most infections tend to occur in the larval stage but pupal, adult and, rarely, the egg stage should also be assessed.
Size	Body length, width, head capsule width	This may indicate abnormalities relative to the equivalent healthy life stage.
Duration of life stages	Note time in each known life stage relative to normal development	Some virus groups, especially CPVs and EPVs, induce extended development.
Behaviour	General movement, feeding activity	A useful characteristic if there is good knowledge of normal, healthy behaviour. Increased activity or paralysis represent the extremes of this characteristic.
Appearance	Note body colour and any visible internal organs, especially gut, fat body, muscle and hypodermis.	Massive development of viruses in hosts can result in major colour changes in internal organs before the skin eventually changes colour. In some cases the gut changes colour.

the initial diagnostic tests have been carried out. The following is, therefore, a step-wise procedure that should enable basic identification to be carried out, the final prognosis depending on the results of intermediate diagnosis.

1 External symptoms

Examination of invertebrates for the presence of diseases is aided if there is a good knowledge of the appearance of healthy life stages so that any unusual symptoms can be compared with the normal appearance of that life stage. This is easiest for laboratory-reared insects where the appearance, rate of development and general behaviour should be well known. Invertebrates, whether live or dead, collected from the field will be more difficult to diagnose and it may not be possible to provide cross-reference to healthy individuals. Examination should be carried out on each specimen available and should concentrate on the characteristics in Table 1, which provides a checklist of items to look for in intact live or recently dead specimens.

External examination and careful recording of symptoms can aid diagnosis and point to particular follow-up regimes at the organ and cellular levels.

a. Key to identification based on external symptoms
Information derived from external examination (see Table 1) can differentiate between many of the virus groups. This is only possible if sufficient specimens are available over an extended period to enable the full spectrum of symptoms to be assessed. Figure 1 provides a simple key to the major virus groups that can be distinguished on the basis of external symptoms. This can provide valuable guidance but further confirmation will be required for definitive diagnosis. Further guidance can be obtained through knowledge of the known host ranges by order of the invertebrate viruses. Table 2 provides a checklist of this information.

2 Light microscopy

The initial assessment of specimens using light microscopy is an important tool in differentiating between virus groups and, in many cases, may be sufficient to confirm the presence or absence of a given group. Procedures for specimen preparation are similar, but staining regimes will differ, if different virus groups are being diagnosed on the same microscope slide. Adams & Bonami (1991a) provide more detailed overviews of diagnostic techniques for insect viruses. Becnel (Chapter VIII-1) provides a useful guide to preparation of specimens for general microscopic examination of diseased insects.

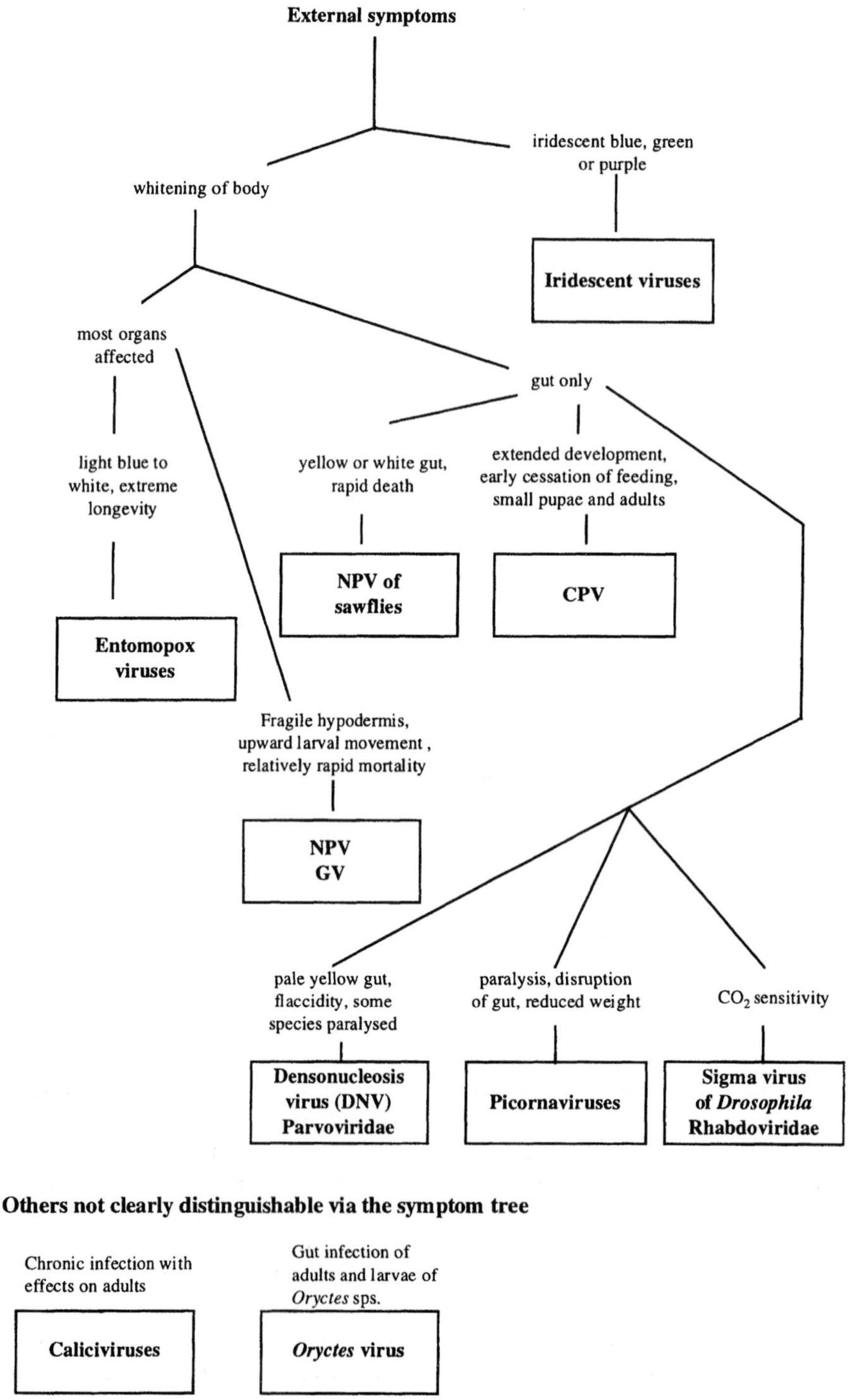

Figure 1 Flow chart for identification of the principal virus groups based on external symptoms.

Table 2 Recorded host ranges by insect order of the principal virus groups.

Virus family	*Recorded host orders*	*Usual host stage*
Baculoviridae: NPV and GV	Coleoptera, Diptera, Hymenoptera, Lepidoptera, Neuroptera, Siphonaptera, Thysanura, Trichoptera	Larvae, sometimes pupae or adult
Reoviridae: CPV	Diptera, Hymenoptera, Lepidoptera	Larvae, pupae, adults
Entomopoxviridae: EPV	Coleoptera, Diptera, Hymenoptera, Lepidoptera, Orthoptera	
Iridoviridae: IV	Range of insect and other invertebrate families	Larvae
Ascoviridae	Lepidoptera (Noctuidae only)	Larvae
Birnaviridae	Diptera (recorded in genus *Drosophila* only)	Adults
Caliciviridae	Lepidoptera (Noctuidae only)	Larvae
Nodaviridae	Diptera, Coleoptera, Lepidoptera	Larvae, adults
Parvoviridae: DNV	Diptera, Blattoideae, Lepidoptera, Odonata, Orthoptera	Larvae, pupae, adults
Picornaviridae	Diptera, Lepidoptera, Orthoptera and wide range of insect families	Larvae, adults
Polydnaviridae	Parasitic Hymenoptera	Adults
Rhabdoviridae	Diptera	Adults
Tetraviridae	Lepidoptera	Larvae
Oryctes virus	Coleoptera	Larvae, adults

a. Preparation of microscope slides

During the early stages of diagnosis the normal procedure is to prepare a smear from the whole insect body, ensuring that both gut and internal organs are included. Depending on the stain procedure, the smear should be localized or spread across the width of the slide. It is essential that the smear is not too thick, the ideal being a monolayer of cells enabling nuclear and cytoplasmic detail to be viewed. The choice of equipment for preparation of smears is wide but, essentially, consists of forceps to handle and tease apart the specimens and mounted needles or disposable wooden slivers to aid spreading of body contents. In most cases, unless dissection of specific organs is being carried out, it is not necessary to use dissecting fluids to prevent desiccation. However, if the specimen is already somewhat desiccated or is very small and likely to dry out quickly on the slide, it may be necessary to use a saline solution (see Chapter VIII-1). Slides should be labelled with a permanent marking system such as diamond marker or, for frosted slides, with a marker that is insensitive to any of the reagents employed in the staining procedure.

Prevention of contamination between smears is important and the equipment should be decontaminated by rinsing in alcohol and wiping or flaming off. Preparations should be air dried before staining. Use of a fixative is dependent on the stain to be used, but is not always necessary.

b. Staining methods for light microscopy

Among the many possible stains for light microscopy, two principal methods are commonly employed to aid diagnosis of viruses, particularly those producing IBs. Buffalo Black 12B and Giemsa's stain offer simplicity in use and rapid diagnosis without the necessity for complex fixation and mounting procedures. Both are virtually permanent stains, enabling slides to be stored in light-tight containers for many years without significant loss of detail.

Buffalo Black 12B stains protein blue–black and thus is a positive stain that allows crystalline protein to be distinguished from a range of different backgrounds. Giemsa's stain is, for the majority of occluded viruses, a negative stain in that the IBs can remain unstained while the background stains in blues and reds. It is a particularly useful 'all-round' stain that can aid diagnosis of some bacteria, fungal spores and, particularly, Microsporidia as well as occluded insect viruses.

(i) *Buffalo Black 12B*. Buffalo Black 12B is also known as Naphthalene Black 12B or Amido Schwartz or Acid Black 1.

1. Air dry the preparation to be stained.
2. Heat the Buffalo Black solution to 40–45 °C in a staining rack on a hotplate.
3. Immerse the slide in the Buffalo Black solution for 5 min.
4. Wash the slide under running tap water for 10 s.
5. Dry the slide and examine under oil immersion for the presence of inclusion bodies.

(ii) *Giemsa's stain*. Giemsa's stain is a differential stain that clearly distinguishes nuclear and cytoplasmic cellular details and, thence, aids in the diagnosis of site of replication of various virus groups. Procedures for Giemsa staining are as follows:

1. Immerse slides with air dried smears for 2 min in Giemsa's fixative.
2. Rinse slides under running tap water for 10 s.
3. Stain for 45 min in 10% Giemsa stain in 0.02M phosphate buffer, pH 6.9. Gurrs Improved R66 Giemsa has been extensively tested and is known to work well.
4. Rinse under running tap water for 10 s. If the slide appears to be very red (overstained) immerse in 0.02M buffer until the red colour disappears. Rinse again in running tap water.
5. Air dry the slide and examine under oil immersion.

c. Differential Giemsa staining to distinguish NPV from CPV IBs

Using different prefixatives, it is possible to change the characteristics of CPV to take up stain (they remain colourless under normal Giemsa's stain) while leaving NPV IBs colourless and EPV IBs stained their normal blue colour. This is very useful if it is suspected that laboratory cultures may be contaminated with CPV – a very common occurrence. The differential staining regime below, based on the method of Wigley (1980b), although complicated, provides a means of distinguishing occluded viruses in pure, semi-pure or crude preparations.

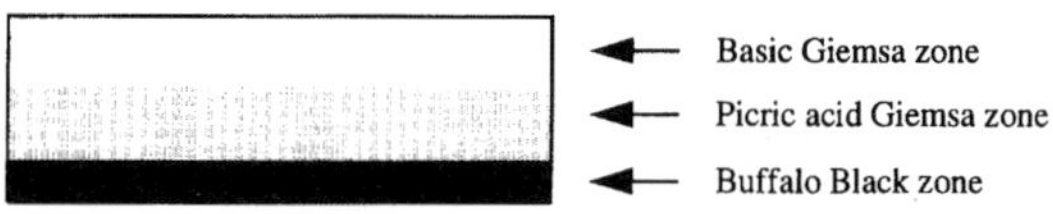

Figure 2 Schematic representation of the three zones for differential staining of specimens for occluded insect viruses on standard glass microscope slides.

The slide is stained in three zones, making it essential that a slide rack and staining dish system be used to control the height up the slide width to which the various solutions reach. The three zones are illustrated in Figure 2.

A number of staining dishes containing the different reagent solutions should, therefore, be employed, and a rack staining system is essential so that the slides enter the stain horizontally.

1. Make thin smears entirely across the width of the slides. It is important to use the full width because the staining depends on using solutions that cover some or all of the slide longitudinally (see Figure 2).
2. Heat the slides in the rack to 75 °C. This can be achieved on a hotplate, covering the rack with aluminium foil.
3. Completely immerse the slides in a fixative solution (90% absolute alcohol, 10% formalin solution) for 3 min.
4. Rinse in absolute alcohol for 30 s and dry.
5. Place saturated picric acid solution in a staining dish so that, when the slide rack is placed in the staining dish, the liquid reaches to two-thirds of the width of the slide. Heat the picric acid to 40 °C.
6. Immerse the slide rack in the picric acid solution for 2 min.
7. Rinse for 20 s in a staining dish under running tap water.
8. Immerse the wet slide rack in Giemsa's fixative for 60 s.
9. Rinse for 60 s in a staining dish under running tap water.
10. Immerse the slide rack in 0.02M phosphate buffer for 2 min.
11. Drain the slide rack for 2 min on absorbent paper.
12. Stain the entire slide rack to its full width for

45 min in 10% Giemsa's stain in 0.02M phosphate buffer.

13. Rinse for 30 s in a staining dish under running tap water.
14. Immerse for 60 s in 0.02M phosphate buffer.
13. Dry the slides and heat on a hot plate (covering the rack with aluminium foil to retain heat) at 65 °C for 5 min.
16. Prepare a staining dish with Buffalo Black solution to reach to ¼ of the width of the slides (see Figure 2).
17. Immerse the slide rack for 1 min in the Buffalo Black solution heated to 45 °C.
18. Rinse the slides under running tap water until the water runs clear (approximately 60 s).
19. Drain the slides on absorbent paper then remove and dry individually.
20. Examine under oil immersion at a magnification of at least 900×.

Other stains such as modified Azan and the Sudan II staining techniques (see Chapter VIII-1), can be employed but, for all stains, it is important to test using known preparations of viruses.

d. Diagnostic features under light microscopy

Detail of infected cells can be distinguished using a differential stain such as Giemsa's stain (see Section 3 A *2b*). If the specimen is prepared from fresh material, then nuclei and cytoplasm can be distinguished clearly and compared with the normal appearance for the tissues being examined. Loss of detail, particularly the breaking down of the nucleus and appearance of dense material in either nucleus or cytoplasm may indicate infection and development of a virogenic stroma. However, the occluded viruses are the only groups that can be determined with reasonable certainty, ideally using the differential staining schedule described in Section 3 A *2c*.

Using Giemsa's stain, nuclei appear red, cytoplasm blue while IBs of NPV or CPV remain colourless but with a distinct edge. EPV IBs stain a light blue. When the IBs are still in the nucleus it is easy to be certain that what is seen are NPVs (GVs are too small to be distinguished with certainty). When the nuclei are broken up, it is not so easy to confirm that the non-staining crystalline bodies are NPV IBs. However, by examining a range of slides, including those where presence of IBs in nuclei confirms the presence of NPV, it should be possible to recognize NPV in all situations. Similarly the presence of CPV and EPV IBs in the cytoplasm is aided by examination of intact cells. However, it is not easy to be certain that CPV, in particular, is present. This can be aided by differential staining using the method developed by Wigley (1980b).

Diagnosis using the differential staining method depends on the appearance of virus inclusion bodies in the different staining zones. The main diagnostic features are summarized in Table 3.

Although these diagnostic features are clearly distinguishable under bright field illumination, further confirmation can be gained from use of phase contrast and dark field illumination. Phase contrast, whether from a Heiné or a Zernike condenser, gives distinctive birefringence of the crystalline protein of IBs. For example, under Heiné phase the IBs of NPVs and CPVs show as light to dark purple with distinct bright areas on the surface, indicating the presence of virions that are not fully enclosed by the polyhedrin. Under Zernike phase the IBs are orange. IBs shine brightly under dark field and it is possible to distinguish CPVs, EPVs, NPVs and GVs, although the latter are difficult to identify with certainty.

Table 3 Diagnostic features of inclusion body viruses using the differential staining technique of Wigley (1980b).

	Appearance of inclusion bodies		
Virus group	*Full Giemsa zone*	*Picric acid Giemsa zone*	*Buffalo black zone*
NPV polyhedra	Colourless	Colourless or yellow	Black
CPV polyhedra	Colourless	Range of colours from red-blue, blue-grey to deep purple	Black
GV granules	Colourless	Colourless	Black
EPV inclusions	Blue	Blue	Black

The essential features of the differential staining method are the differential take-up of stain by picric acid-treated CPV IBs which contrast to NPV IBs that remain unstained or slightly yellow in that zone and EPV IBs that stain blue regardless of Giemsa zone. Buffalo Black staining provides confirmation that the inclusions being assessed are actually proteinaceous. All proteinaceous bodies stain black. This includes all the occluded viruses, including NPV, GV, CPV and Pox viruses. However, in a whole body smear, it is usually possible to see the IBs of NPVs still inside the nuclear membrane of the cells, whereas CPVs would be found in the cytoplasm of the cell. EPV IBs (spheroids) are also found in the cytoplasm of host cells.

3 Electron microscopy

The major drawback of light microscopy is the inability to distinguish non-occluded viruses and small IBs (GV) with certainty. In such cases, useful further diagnosis can be provided by scanning or, particularly, transmission electron microscopy (EM). Use of thin sections or direct layering of virions on grids and examination under the EM enables virion structure to be examined which can provide considerable insight into the groups of viruses that might be present.

a. Preparation of specimens

Detailed protocols for preparation of specimens for electron microscopy are provided in Chapter VIII-1. Additional specific procedures for handling insect viruses are summarized here.

In brief summary, the normal sequence is tissue preparation (fixation, dehydration, embedding in resin, staining), sectioning and, finally, placement on a grid for examination. Further staining to improve contrast may be carried out just prior to the examination stage.

(i) *Direct examination of viruses.* It is not always necessary to section material; purified or semi-purified pellets of virus (see Section 4) can be placed directly onto formvar or other grids and layered with carbon. These can be examined without further treatment or stained with either positive (e.g. 1% (w/v) aqueous ammonium molybdate) or negative (a heavier stain of aqueous ammonium molybdate) staining. Adams & Bonami (1991b) provide a useful summary of these procedures and a full set of more detailed references. Complete removal of sucrose after sucrose gradient centrifugation to purify virus has been solved by continuous washing of grids on filter paper either individually (Webb, 1973) or in groups (Adams & Bonami, 1991b).

(ii) *Embedding and sectioning of tissues or purified virus suspensions.* Methods suitable for general electron microscopy are dealt with in Chapter VIII-1 and are reviewed by Adams & Bonami (1991b) with further information in Appendix 1 of Adams & Bonami (1991a). We do not propose, therefore, to provide a further account of the precise procedure for fixation and embedding of specimens, particularly bearing in mind that laboratories equipped with electron microscopes will already have procedures in place that will have been refined for local use.

Sectioning normally aims to cut to thicknesses of 90–150 nm, thus enabling fine details of cellular organization to be distinguished.

b. Diagnostic features using electron microscopy

Electron microscopy provides an effective screening tool to distinguish the major features of the virus groups. Thin sections of body tissues or purified viral preparations can reveal most diagnostic features necessary for identification. Alternatively, purified virions can be layered directly on to EM films and examined with or without further staining.

Description is inadequate to convey the parameters that are used for diagnosis in electron microscopy. However, the key morphological features, based on Adams (1991) are indicated in Table 4. Figures 3 and 4 (kindly supplied by Dr Jean Adams, USDA) are representative examples of each virus group. The comprehensive treatment of this subject in Adams & Bonami (1991a) is recommended for further information.

4. Biochemical and molecular techniques for identification

Although light and electron microscopy can provide a great deal of information on the morphology and, therefore, basic characteristics of virus groups, precise identification relies increasingly on biochemical and molecular techniques. These are specialized procedures that may not be available in all laboratories working on insect pathogens. Indeed, it can also be assumed that any laborat-

Table 4 Principal morphological features of insect virus families. Virus particle dimensions can be determined by electron microscope examination.

Virus family	*Virus particle morphology*	*Particle dimensions (nm)*	*Inclusion body/dimension (μm)*	*Enveloped virion*
Baculoviridae	Bacilliform	NPV 40–60 × 200–400	+/0.3–15.0	–
		GV 30–60 × 260–360	+/0.3–0.5	–
Reoviridae (CPV)	Icosahedral with 12 projections	55–69 (diameter)	+/0.2–10.0	–
Entomopoxviridae (EPV)	Brick-shaped or ovoid	165–300 × 150–470	+/1.0–24.0	+
Iridoviridae (IV)	Icosahedral	125–300	–	–
Ascoviridae	Allantoid to bacilliform	130 × 400	–	+ (2 or 3)
Birnaviridae	Icosahedral	60 (diameter)	–	–
Caliciviridae	Cup-shaped	38 (diameter)	–	–
Nodaviridae	Icosahedral	29 (diameter)	–	–
Parvoviridae	Isometric	18–26	–	–
Picornaviridae	Spherical	22–30 (diameter)	–	–
Polydnaviridae	Ovoid	150 × 350	–	+ (2)
Rhabdoviridae	Bullet-shaped or bacilliform	50–95 × 130–380	–	+
Tetraviridae	Icosahedral	35–39 (diameter)	–	–

ories working on the molecular identification of insect pathogens may already be well versed in the details of the methods employed. We intend, therefore, to provide an overview only of these techniques which are the subject of a number of books on the subject, reflecting the rapid developments in this field. Useful general overviews can be found in Ausubel *et al.* (1991), Maramorosh (1987), Padhi (1985), and St. Leger & Joshi (Chapter VIII-3).

The principal procedures employed under this category of diagnostic techniques are summarized in Table 5.

4 ISOLATION

Isolation of viruses from the host in which they have been grown is an essential step for further diagnosis or for specific purposes requiring a high degree of purity. In essence, therefore, methods of virus purification and concentration are necessary for detailed studies of all insect virus groups. This has been well reviewed by Tompkins (1991). A summary of the principal steps in extraction and purification is provided in Table 6. Here we deal with the major steps required to extract and purify virus from infected individuals of a given host species. Precise methodology for each virus group will depend on common practice in a given laboratory, especially regarding the use of sucrose gradient and caesium chloride gradient protocols that can differ in detail of the precise centrifugation times and buffering or solvent solutions. Tompkins (1991) provides a useful and comprehensive set of references to these various methodologies.

5 QUANTIFICATION OF VIRUSES

Further development beyond identification usually requires estimation of the concentration of virus. This can be achieved in a number of ways, depending on the virus groups and on the purpose of the quantification. Of particular concern is the need to have accurate counts of infectious units for use in virus propagation and bioassay and in any field programme of pest management.

There are a number of methods for counting pure or semi-pure suspensions of occluded viruses. A permanent record of a count can be obtained using the dry counting method developed by Wigley (1980a) and is the preferred method when there is likely to be some contamination of the preparation.

A Dry counting method for occluded viruses

The principle of this method, based on Wigley (1980a), is to prepare a smear of the virus preparation on a known area of a microscope slide and to count the IBs using a standard subsampling regime. The counts are

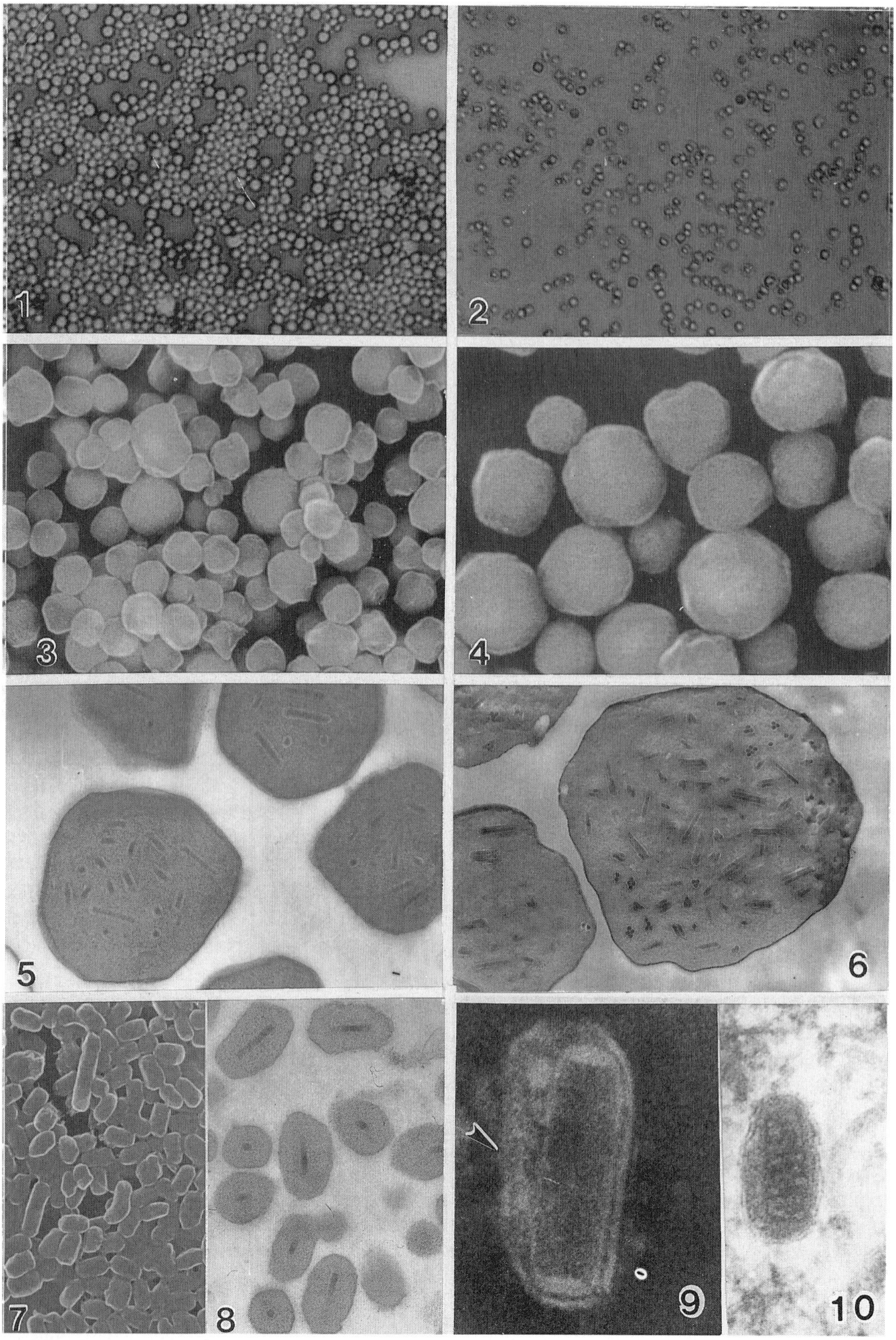
1
2
3
4
5
6
7
8
9
10

Table 5 Techniques for the biochemical/molecular identification of invertebrate viruses.

Technique/key references	*Range of techniques*	*Degree of sensitivity/specificity*
Serology	Precipitation	Relatively low
(Volkman, 1985)	Neutralization	High using monoclonal antibodies
	Radioimmunoassay (RIA) and Enzyme Linked Immunosorbent Assay (ELISA)	Highly sensitive and, using monoclonal antibodies, high specificity.
	Immunofluorescence	Useful for detection of viruses within cells. High sensitivity.
	Immunoaffinity chromatography	Highly sensitive using monoclonal antibodies.
Characterisation of viral proteins (Ausubel *et al*, 1991)	Electrophoresis (SDS-PAGE), isolectric focusing, two-dimensional electrophoresis	Useful to produce protein profiles but not definitive.
Genome mapping (Ausubel *et al*, 1991), Chapter VIII 3, this volume).	Restriction endonuclease analysis of viral DNA. Linked to complementary techniques such as Polymerase Chain Reaction (PCR) and Restriction Fragment Length Polymorphism (RFLP) to amplify DNA.	Detailed mapping of DNA sequences allowing specific comparison and construction of genome maps. Gels are used to separate bands. REN digested DNA fragments analysed further using Southern blotting.

then multiplied by a series of factors to achieve a concentration per ml of original suspension.

The procedure is outlined below.

1. Prepare a stock of diluted albumen. The easiest way is to use 0.5 g dried ovalbumin (purchased from any scientific supplier) dissolved in 5 ml of sterile water. The solution is then mixed with an equal volume of glycerol and further diluted to a 10% working solution. These aliquots can be divided into small containers (0.5 or 1.0 ml volume) and frozen until needed.
2. Take 50 µl of virus suspension and 50 µl of albumen and mix thoroughly. Use larger volumes if equipment is not accurate enough to dispense these small volumes.
3. Using an accurate pipette, dispense 5 µl of the mixture onto a microscope slide placed over a template defining a 15 mm diameter circle. For maximum accuracy this must be done under a binocular microscope. The suspension should be spread evenly over this area using a bent needle and a series of concentric circular movements to take the liquid precisely to the edge of the circle. It is important that as little suspension as possible is removed when the bent needle is lifted off. This is best achieved by turning the needle onto its tip and lifting vertically. Four circles should be made on each slide, as shown in Figure 5.
4. Air dry the slide and then heat fix for 2 min on a hotplate at approximately 80 °C.
5. Stain for 5 min in Buffalo Black solution heated to 45 °C. Wash in tap water and dry.
6. Counting requires a 10 × 10 eyepiece grid on a compound microscope with oil immersion objectives. The area covered by the grid should

Figure 3 (1) Light micrograph of *Lymantria dispar* MNPV (Abby strain) nigrosin stain × 2940. (2) Light micrograph of *Autographa californica* MNPV (no stain) × 2940. (3) Scanning electron micrograph of *Helicoverpa zea* SNPV × 5000. (4) Scanning electron micrograph of *L. dispar* MNPV × 5000. (5) Electron micrograph of sections of *H. zea* SNPV × 27 150. (6) Electron micrograph of section of *L. dispar* MNPV × 19 820. (7) Scanning electron micrograph of *Plutella xylostella* GV × 12 500. (8) Electron micrograph of sections of *Cnaphalocrocis medinalis* GV × 29 675. (9) *Oryctes rhinoceros* virus particle negatively stained with phosphotungstic acid. The nucleocapsid has clearly thickened or 'capped' ends. The tail-like protusion (arrowhead) is also visible × 200 000. (Courtesy of A. M. Huger, Darmstadt, Germany.) (10) Section of *Oryctes rhinoceros* virus particle showing the double membrane envelope, necleocapsid with helicoid structure of the nucleoprotein core, and the tail-like appendage in a unilateral dilation of the envelope × 135 000. (Courtesy of A. M. Huger, Darmstadt, Germany.)

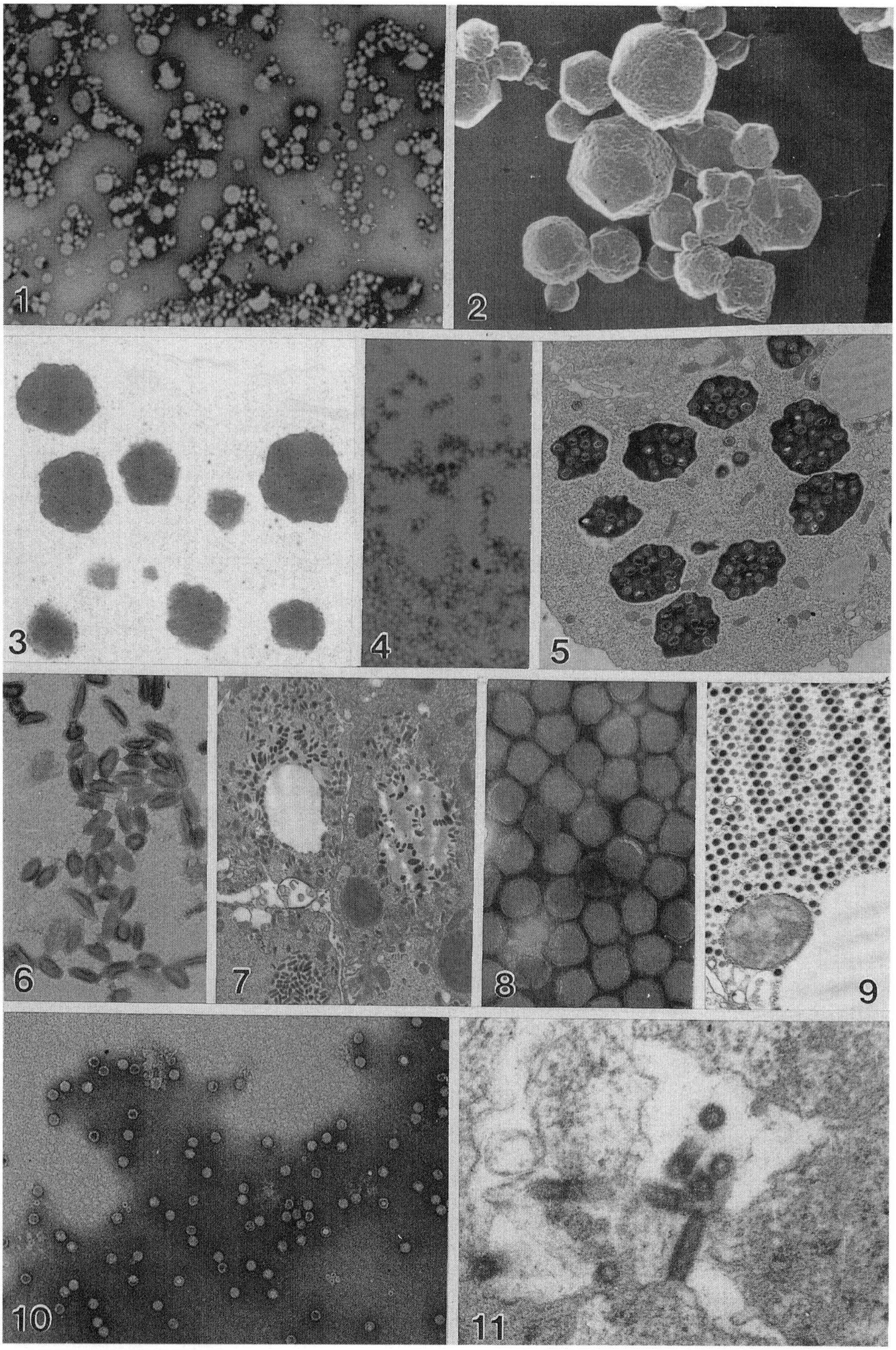
1
2
3
4
5
6
7
8
9
10
11

have been precalibrated against a stage micrometer slide.

7. Each of the four circles should be counted along a different direction as indicated by the cross lines in Figure 5.
8. Counting commences by locating the edge of the smear and lining up the edge of the eyepiece grid with it. Using the stage micrometer of the microscope move the slide 0.25 mm towards the centre of the circle. Count all IBs within the grid and also those that touch the left or top edges of the grid (ignore any touching the bottom or right edges). Record the results on the form (see Table 7).
9. Move the slide 0.5 mm each for the next eight counts and 1.0 mm for the final two, making a total of 11 counts in all. Record the number of IBs for each sector.
10. Repeat for each circle direction indicated in Figure 5.
11. Analysis of the results and conversion to total numbers of IBs is based on subsampling a given area of the circle. The outer counts are a smaller proportion of the available area than the inner counts and, therefore, a weighting factor has to be calculated for each sector. The formula for the weighting factor is:

$$\textit{Sector weight factor} = \frac{\textit{outer radius} - \textit{inner radius}}{\textit{circle radius}^2}$$

where the areas are calculated from radii defined by the outer and inner borders of each sector. These are outlined in Table 8.

The total number of IBs is estimated by multiplying the individual sector counts by the sector weighting factor and then extrapolating by the relative areas of the count grid and the volume applied to each circle.

Let w_n = sector weight factor

$\bar{x}_n$ = sector mean

$$\text{Mean count per grid} = \sum_{1-11} w_n \, \bar{x}_n$$

$$\text{Standard error of the mean} = \sum_{1-11} w_n^2 \times \frac{\bar{x}_n}{n}$$

The total number of IBs in the suspension is, therefore, calculated from:

$$\text{Total IBs per ml} = \frac{C}{ga} \times (pg \times df)$$

where C = circle area,
ga = grid area;
pg = mean number of IBs per grid;
df = dilution factor.

The most efficient dilution is to reach around 5×10^8 IBs per ml which is a suspension that is just milky to look at. This gives accurate counts with low standard error but does not involve excessive time in examining the smears. This method can be used for pure or impure preparations because the IBs, by staining black, stand out from thc background. It can even be used for assessing numbers of IBs in soil (Evans *et al.*, 1980).

It is a simple matter to set up a computer program in Basic or another computer language or to use a spreadsheet to calculate the mean counts per circle. This will then need to be multiplied by the dilution factors used to make the suspension. It is important to remember the 50 : 50 dilution with albumen as well as any dilution made from the original stock suspension.

Table 7 provides a template for completion of the results of the dry counts.

Each count should be entered in the appropriate position in the table, after which the sector mean should be calculated and multiplied by the weight factor below it to give the weighted mean value. These

Figure 4 (1) Light micrograph of *L. dispar* CPV (nigrosin stain) × 2940. (2) Scanning electron micrograph of *L. dispar* CPV × 7810. (3) Electron micrograph of sections of L. dispar CPV × 15 600. (4) Light micrograph of entomopox virus *Amsacta moorei* passed through *L. dispar* larvae (nigrosin stain) × 2940. (5) Electron micrograph of section of *Euxoa auxiliaris* larva infected with *E. auxiliaris* entomopox virus × 5315. (6) Electron micrograph of Ascovirus isolated from *H. zea* (stained lightly with 1% ammonium molybdate) × 22 100. (7) Electron micrograph of *H. zea* larval tissues infected with ascovirus × 6015. (8) Electron micrograph of iridescent virus isolated from infected *H. zea* larvae × 40 800. (9) Electron micrograph of fat body of *H. zea* larva infected with iridescent virus × 9250. (10) Densovirus isolated from *Galleria mellonella* larvae (negatively stained with 1% ammonium molybdate) × 69 425. (11) Electron micrograph of sigma virus budding from the plasma membrane of testicle tissue of *Drosophila melanogaster* × 74 000. (Courtesy of D. Teninges, CNRS, Lab de Génétique des Virus, Gif-sur-Yvette, France.)

Table 6 Methods for isolation and purification of invertebrate viruses.

Virus group	*Homogenization and filtration*	*Centrifugation*
All	Grind tissue in homogenizer or use purpose-built equipment such as a stomacher. Use de-ionized water or 0.01M Tris buffer at pH 7.3–8.0. Crude filtration through muslin or equivalent to remove cellular debris.	Occluded viruses: centrifuge at 10 000 g for 10 min, re-suspend in appropriate carrier fluid, re-pellet and suspend in de-ionized water twice. Further purification by sucrose gradient centrifugation.
NPV, GV	As above or add 0.1% w/v SDS to improve extraction efficiency.	As above. Equilibrium sucrose gradients (25% to 60% (w.w)) at 65 000 g to 96 000 g for 1 to 3 h. Remove virus band and wash out sucrose with repeated pelleting and suspension in de-ionized water. Final pellet in de-ionized water and store frozen.
CPV	Extract gut only, macerate as for baculoviruses.	As above. Sucrose gradient centrifugation at 30 000 g for 1 h.
EPV	As above.	Initial centrifugation at 12 000 g for 10 min at 4°C, re-suspend in 0.01M Tris-HCl at pH 7.5 with 3% SDS. Sucrose gradient centrifugation 40–65% step gradient at 64 000 g for 1.5 h. IBs at 58% sucrose. Remainder as above.
IV	As above or in 0.05M phosphate buffer (pH 7.0–7.4) or 0.01M sodium borate buffer (pH 7.5).	Centrifuge at 1000 g and 17 500 g to remove cellular debris. Pellet from latter re-suspended in water or buffer and centrifuged on 5% to 50% w.w sucrose gradient at 15 000 g to 29 000 g for 30 min. Virus band re-suspended and washed out of sucrose at 30 000 g for 30 min.
Non-occluded DNA viruses	Macerate larvae in phosphate buffered saline at pH 7.5. Three fluorocarbon treatments and combine with Gene solv-D and shake for 30 s.	Centrifuge at 8000 g for 15 min and collect supernatant containing virus particles. Concentrate at 80 000 g for 20 min then treat with chloroform/butanol to remove insect debris, repeat. Add ammonium sulphate for 24 h at 4°C and centrifuge at 72 000 g for 30 min to pellet.
Cricket paralysis virus	Macerate in 10 mM ammonium acetate (pH 7.0) and CCl_4 and separate by centrifugation. Homogenize supernatant with ether and then CCl_4 to remove ether.	Supernatant centrifuged at 12 000 g for 45 min and re-suspend in 10 mM ammonium acetate. Sucrose gradient (10% to 40% w/v) in 10 mM ammonium acetate at 80 000 g for 2.5 h. Dialyse against ammonium acetate buffer and centrifuge in caesium chloride density gradient at 95 000 g for 16 h. Dialyse final band against 20 mM ammonium acetate, pH 7.2.
Small RNA viruses	Homogenize larvae in 0.05M Tris buffer (pH 7.4).	Centrifuge at 2000 g for 10 min and collect supernatant. Centrifuge this at 80 000 g for 1 h, re-suspend in buffer and place on 10% to 50% sucrose gradient in buffer at 65 000 g for 90 min. Repeat procedure, then final 32% caesium chloride gradient centrifugation. Other options are use of CCl_4 for initial extraction and refinements to gradient centrifugation.

are totalled to give the total weighted number of IBs per sector. The count per ml is found as shown above.

B Counting by haemocytometer

It is only possible to use a haemocytometer on pure or semi-pure preparations where there is no danger of mistaking other particles for IBs. A good design is a 0.1 mm deep counting chamber with improved Neubauer ruling.

1. Place a cover slip over the depression in the counting chamber and press down firmly (not too hard otherwise the cover slip will break). Ideally the surface should be humidified by breathing on it to aid adhesion of the cover slip.
2. Place a drop of virus suspension at the edge of the

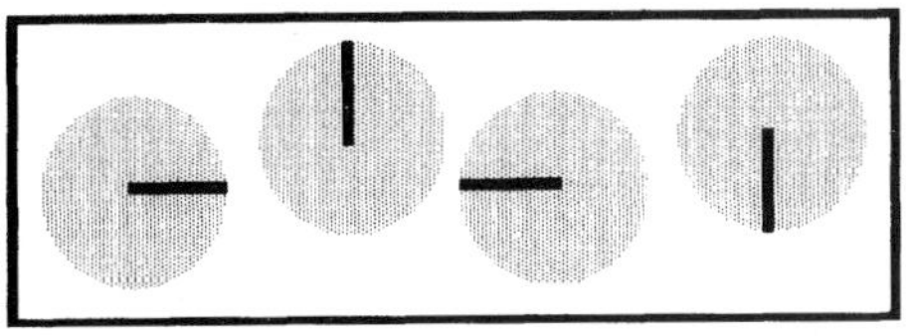

Figure 5 Positions of virus suspension circles on a microscope slide (circle diameter = 15 mm).

cover slip so that the liquid is taken up and fills the chamber under the cover slip.

3. Let the slide stand for about 20 min to reduce the amount of Brownian motion of the IBs.
4. Examine under a compound microscope and count IBs in five large squares, one at each corner of the chamber graticule and one in the centre. Count only those touching the top and right-hand sides of each square.
5. The chamber will have known area and volume so that it is possible to extrapolate from the number of IBs per square to the total concentration per ml of suspension. The typical Neubauer ruling would be:

$$\text{Area of small square} = \frac{1}{400}\ \text{mm}^2$$

Each large square has 16 small squares, thus,

$$\text{Area of small square} = \frac{16 \times 25}{400}\ 1\ \text{mm}^2$$

Depth of counting chamber = 0.1 mm

Volume of counting chamber = 0.1 mm^3, giving a multiplication factor of 10^4 for 1 ml.

Total number of IBs per ml = IBs per large square × no. of large squares × 10^4.

C Impression film technique for counting the number of IBs on plant surfaces

Without the aid of a scanning electron microscope it is very difficult to determine the numbers and distribution of IBs on plant surfaces. However, an estimate, with reasonable accuracy, can be obtained by use of double-sided adhesive tape (impression film) to remove the IBs and, thence, to stain and count them (Elleman *et al.*, 1980). It is important that the impression film is not too 'sticky' otherwise it will tend to remove the plant epidermis and obscure the IBs.

1. Method

1. Clean the microscope slide on which the impression film will be placed with alcohol.
2. Cut a piece of double-sided adhesive tape to an appropriate length and remove paper from one side only.
3. Press the adhesive surface of the tape firmly onto the slide, ensuring that no air is trapped underneath which may obscure the view when examined under the microscope. Slides prepared in this way can be stored ready for later use.
4. Remove the paper from the adhesive tape and then press the plant surface firmly onto the tape. Trial and error may be necessary to determine the correct pressure to apply for removal of IBs, making sure that the pressure is not so great that sap is exuded from the leaf or that the epidermis is stripped off.
5. Stain in Buffalo Black, making sure that the slide is not agitated too strongly, which may result in the adhesive tape coming loose. Allow to dry at air temperature.
6. Examine the adhesive tape under oil immersion. It is possible to make counts of the preparation by a series of standard counts using the eyepiece graticule. However, it is important that the distribution of IBs is taken into account in determining the positions of the counts on the tape. Examination of the tape will reveal whether there is significant clumping of the IBs, particularly along veins and between waxy and non-waxy areas, which should then be used to design the counting system.

D Electron microscope estimation

Estimation of small IBs, such as GVs, and of non-occluded viruses is best carried out under the transmission EM. This can be achieved by mixing the unknown virus preparation with a suspension of latex beads or other standardized commercial preparation so that the concentration of the unknown is derived by proportional count. Droplets of the preparation can be applied directly to formvar grids or

Table 7 Blank form for completion of dry counting procedure to calculate the number of polyhedra per ml of a given suspension of virus

Radius\Sector	1	2	3	4	5	6	7	8	9	10	11
1											
2											
3											
4											
Sector mean											
Wt. factor	0.129	0.120	0.111	0.102	0.093	0.084	0.076	0.067	0.058	0.106	0.054
Sector mean × wt. factor											
	Total of (Sector mean × wt. factor)										

Table 8 Dry counting procedure. Positions of sector counts and weighting factors for each sector.

Sector	*Distance between counts (mm)*	*Position from edge of circle)*	*Outer and inner radii of sector*	*Factor*
1	0.5	0.25	7.5–7.0	0.129
2	0.5	0.75	7.0–6.5	0.12
3	0.5	1.25	6.5–6.0	0.111
4	0.5	1.75	6.0–5.5	0.102
5	0.5	2.25	5.5–5.0	0.093
6	0.5	2.75	5.0–4.5	0.084
7	0.5	3.25	4.5–4.0	0.076
8	0.5	3.75	4.0–3.5	0.067
9	1.0	4.75	3.5–3.0	0.058
10	1.0	5.75	3.0–1.75	0.106
11		6.75	1.75–0	0.054

can be sprayed on. In either case the concentration of virus is derived by counting the numbers of both the virus and of the bead standard in a number of EM fields of view. The higher the concentration the lower the number of views that are required. DeBlois *et al.* (1978) used this technique to compare counts by different methods for various NPVs and IVs.

6 PROPAGATION

During the past several years there has been a renewed interest in the use of insect pathogenic viruses to control pest populations. This interest has been due to several factors: (1) renewed interest in the Environment and the use of 'environmentally-benign' alternatives to classic chemical pesticides; (2) advances in baculovirus genetic engineering (Miller *et al.*, 1983; O'Reilly & Miller, 1989; Hammock *et al.*, 1993; Miller, 1995) which has led to a proliferation of research and development by scientists in universities, research institutes, government and industry to develop products for insect control, as well as for use in medicine and pharmaceuticals. Moreover, the use of genetically engineered baculovirus expression vector systems for production of different insecticidal (Miller, 1995) or pharmaceutical proteins (Baker *et al*., 1993) has stimulated large companies to invest capital and manpower to develop the products.

Historically, several entomopathogenic viruses have been produced in susceptible host insects, because: (1) the insect host is an efficient virus producer (Ignoffo & Couch, 1981); (2) automation of *in vivo* rearing and *in vivo* production systems is feasible (Powell & Robertson, 1993; Bell & Hardee, 1995); (3) the research has been carried out primarily by entomologists, who have been assigned the problem of production because of their familiarity with the host insect and the insect pathogenic virus(es).

For the past 60 years, insect tissue culture (or cell culture) has been used for the study of insect viruses (Trager, 1935), as well as for the production of insect viruses (Goodwin *et al*., 1970). This approach has several inherent advantages over in host virus production: (i) absence of contaminating micro-organisms in the product; (ii) absence of insect parts, which could act as allergens (Weiss *et al*., 1994); (iii) companies with expertise and experience in cell culture technology and/or fermentation technology are attracted to *in vitro* technology to produce insect viruses as insecticides or proteins for use in medicine or pharmaceuticals; (iv) the process can be controlled, resulting in a more uniform product than can be obtained with in host production.

Within the past 20 years, much research has been expended to improve both cell production and virus production (Reid *et al*., 1994; Weiss *et al*., 1994), but no product is yet available in sufficient quantities for large-scale field trials. We are confident, however, that *in vitro* produced baculoviruses will be available for use as microbial control agents within the next five years. For the purposes of this chapter, however, we will deal entirely with in host or *in vivo* virus production for several reasons: (i) in host produced viruses have been used successfully to control insect pests (Ignoffo & Couch, 1981; Bell, 1991), (ii) research is continuing in this area to produce more efficient systems (Shapiro, 1986; Bell & Hardee, 1995), which makes this approach an economically viable one; (iii) in many areas of the world in host virus production is the only approach feasible (Katagiri, 1981; Moscardi *et al*., 1981).

In vivo virus production systems have changed little over the past 30 years. The development of semi-synthetic artificial diets by Vanderzant *et al*. (1962) resulted in rearing and virus production systems for the cotton bollworm (*Heliothis zea*), the tobacco budworm (*Heliothis virescens*) and the cabbage looper (*Trichoplusia ni*) by Ignoffo (1965). The initial rearing system was made more efficient by the introduction of disposable, multicelled plastic trays (Ignoffo & Boening, 1970), automation in rearing and automation in virus inoculation and harvesting. The goals of virus production, whether *in vivo* or *in vitro*, are to obtain the greatest quantity of virus with the highest quality (= ability to kill insects) at the lowest cost (Shapiro, 1986). While the goals should be obtainable, they are dependent upon careful research and adherence to protocols. Although virus can be produced without prior research, optimal production depends on determining those factors influencing both the quantity and quality of the virus product and making production decisions based on fact(s). Optimal virus production is the result of interrelationships of host–pathogen–environment and each factor in this triad must be assessed for influence on quantity and quality of product. Research in these areas has been summarized (Shapiro *et al*., 1981a, 1986; Shapiro & Bell, 1982) and it is not our intent to review this literature. Instead, we will highlight some of these critical areas and show where they can make a significant impact.

A The host

Three factors should be considered regarding the host: (1) wild vs. colonized; (2) host biology; and (3) age-stage. Because of the developments of semi-synthetic diets, containerization and automation, laboratory-reared insects have been the hosts of choice (when feasible). The advantages of these insects are several: (1) laboratory-reared insects tend to be larger than feral insects, because of selection and adaptation to the laboratory environment (i.e. diet, temperature, humidity, photoperiod); (2) they are normally disease-free, which should result in a virus product that is free from other pathogens; (3) the growth and development of laboratory-reared insects tends to be faster than feral insects, because of selection; (4) virus yield among laboratory-reared insects tends to be greater than among feral insects, since virus yield is dependent on host biomass (Hedlund & Yendol, 1974; Shapiro *et al*., 1981a).

Although laboratory-colonized insects provide several advantages over feral insects as virus producers, feral insects have also been used successfully to produce NPVs from the potato moth (*Phthorimaea operculella*) in Australia (Matthiessen *et al.*, 1978), the velvetbean caterpillar (*Anticarsia gemmatalis*) in Brasil (Moscardi *et al.*, 1981), the European pine sawfly (*Neodiprion sertifer*) in the United States (Rollinson *et al.*, 1970) and a CPV from the pine caterpillar (*Dendrolimus spectabilis*) in Japan (Katagiri, 1981) on natural foliage.

B Maintaining disease-free cultures

An essential requirement for efficient virus production is the maintenance of disease-free cultures. This ensures that only the virus of interest is propagated during mass production and that contaminants, particularly CPV, are avoided. Although there is some evidence that chronic infections transmitted transovarially can occur, by far the most common problem arises from surface contamination of diet or of neonate larvae directly from the egg surfaces. There are many references to methods of avoiding contamination but all rely on use of sodium hypochlorite and/or formalin to inactivate pathogens directly on the egg surfaces before eclosion. The basic methods are outlined below.

1. Sodium hypochlorite treatment

Eggs, either on the substrate on which they have been laid or loose, should be immersed and agitated in a 1% sodium hypochlorite solution plus 0.025% Tween 80® for 10–15 min. Stewart (1984) has shown that agitation and decanting of floating debris is essential to remove floating eggs and, particularly, insect scales. Indeed, Stewart showed clearly that aerial movement of CPV-contaminated insect scales was the principal source of infection in rearing of pink bollworm, *Pectinophora gossypiella*. Eggs should be washed with sterile distilled water and air dried before being used for further rearing. In most cases this treatment should be sufficient to prevent contamination.

2. Formalin treatment

A stand-alone or, in severe cases, additional treatment to sodium hypochlorite, is the use of formalin solution or vapour to disinfect eggs. This should be done only after the chorion has hardened. A solution of 10% formaldehyde with the addition of 0.025% Tween 80® should be employed. If used as an addition to sodium hypochlorite, immerse for 10 min; otherwise immerse for 40 min (but test for adverse effects on the eggs first). Alternatively, place the eggs on a grid over formaldehyde solution for 6 h so that disinfection is achieved by exposure to formalin vapour. This has the advantage over immersion in that all parts of the eggs are exposed, avoiding the risk of surface tension effects preventing the solution from making direct contact with the egg surface.

C Host biology

The biology of the host is of vital importance in selecting the proper container for rearing and virus production. For example, if the insect is aggressive or cannibalistic (or solitary), it must be separated from its cohorts (Ignoffo & Anderson, 1979). If the insect is gregarious, it can be reared with others in a single container (Vail *et al.*, 1973). In some cases, insects can become aggressive or cannibalistic if the larval density is too great (Chauthani & Claussen, 1968). The determination of the most efficient containerization system must '. . . meet the physiological and ecological needs of the insect whether it be environment, space or other' (Burton & Perkins, 1984).

D Age and stage

The objective of virus production is to utilize host tissue as efficiently as possible and to obtain the greatest amount of biologically active virus possible (Ignoffo, 1966; Shapiro *et al.*, 1981a). In general, late stage larvae have been utilized for virus production, since more virus (= viral inclusion bodies) is produced in these insects than in young insects (Shapiro *et al.*, 1981a). Although the use of late stage larvae has been predominant among virus producers, these larvae have several disadvantages, which must be considered: (1) more time is required due to waiting for larvae to grow to late instars than would be the case by using younger larvae; (2) the time period from virus challenge to virus harvest is longer when late stage larvae are used than when younger larvae are used; (1) + (2) means that fewer mature larvae can be utilized within a given time period; (3) the bacterial population increases during the larval developmental period (Podgwaite & Cosenza, 1966) and during the virus production period (Shapiro *et al.*, 1981a); and (4) the most efficient virus producer may not be the most mature larva. For example, virus production (as measured by IBs per milligram of larval tissue (Shapiro *et al.*, 1986) or by IBs per insect (Teakle & Byrne, 1988) was lower for the most mature larvae.

Moreover, viral activity (as measured by insect bioassay) was lower when virus was obtained from mature larvae than when younger insects were challenged (Shapiro *et al.*, 1986; Teakle & Byrne, 1988). In the case of the gypsy moth, virus yield from 4th stage larvae was 20% lower than was virus yield from 5th stage larvae, but NPV from the younger larvae was more than four-fold more active than NPV from the older larvae. In addition, although virus yield from 3rd stage larvae was 50% lower than virus yield from 5th stage larvae, NPV from the younger larvae (L3s) was more than twice as active as NPV from the older larvae (L5s) (Shapiro *et al.*, 1986). Although this phenomenon has not been widely reported or even investigated, it should be borne in mind. Other advantages in using younger larvae would be: (i) faster time for rearing to production and faster time for harvest post virus challenge (= more larvae utilized per given period of time); and (ii) lower bacterial populations (= a cleaner product).

E The virus

Virus inoculum may be obtained from either *in vitro* cell culture or *in vivo* host production (Shapiro *et al.*, 1981a). Although virus produced in cell culture is bacteria-free and non-contaminated (in contrast to virus produced *in vivo*), this material has not been used as primary inoculum for in host virus production. In general, a large batch of virus is produced, and this virus is the primary inoculum for the entire production. In some cases, primary, secondary and tertiary inoculum is produced and used for virus production (Martignoni, 1978).

Although it is well known that significant differences exist in the biological activities of strains or isolates within a given virus species (Hamm & Styer, 1985) or within a given strain or isolate (Shapiro & Robertson, 1991), little is known of the relationship between biological activity and virus yield. By means of mutagenesis, Wood *et al.* (1981) obtained a strain of *Autographa californica* NPV that was more potent than the parent, and also produced more virus (= HOB isolate). Lynn *et al.* (1993) selected 17 clones from the Abington, MA, USA strain of the gypsy moth NPV and found differences in both production of virus (= viral inclusion bodies) as well as biological activity against *Lymantria dispar* larvae. Thus, it should be possible to examine different strains, isolates or clones and select for those with the greatest biological activity and the greatest virus production.

F Virus production scheme

1. Inoculum feed

Whether the production insect must be reared separately (= solitary) or may be reared together (= gregarious), the production scheme for both scenarios is basically the same. Insects are reared in a separate room or facility to the desired age or stage. At this time, they are then transported to the virus production area or facility, which is separated from the 'clean' rearing and colony area (Shapiro *et al.*, 1981a). The virus is then inoculated on the diet (Martignoni, 1978) or incorporated in the diet. Surface treatment is an efficient system that is easily automated and requires much less virus than does diet incorporation (Shapiro *et al.*, 1981a). Moreover, the same diet (and containers) can be used for surface treatment (without larval transfer), whereas larvae would have to be transferred to new diet which contains virus (= increase in costs). In general, the concentration of virus is adjusted to produce 90–95% host mortality. This concentration assures good larval growth during the infection cycle and maximal utilization of insect tissues for viral multiplication and production.

2. Incubation

After larvae have been challenged with virus, they are placed in a temperature controlled area (box, incubator, room) for a predetermined number of days at a predetermined temperature (usually 25–30°C). The holding time is of critical importance, as this time period can determine the quantity and quality of virus produced, and the bacterial 'load' (Shapiro & Bell, 1981). In other words, while an LC_{90-95} is given to the insects, the time of harvest (LT_0–LT_{100}) will vary according to the criteria established by the producer. If a virus product is desired with a 'minimum' load of adventitious bacteria, harvest may take place before any insects die (LT_0) or when less than 25% of the larvae have died from virus. If maximal virus activity is required, insects are harvested at a later time (LT_{75-100}). Although this virus product may be more virulent than one harvested early in the incubation cycle, this product will contain a higher number of bacteria, which may render the product unacceptable from a quality control standpoint (Podgwaite *et al.*, 1983).

3. *Harvest*

Harvest is the most time-consuming and expensive procedure in virus production (Shapiro, 1986), and can be influenced by such parameters as container, and time of harvest. If insects are reared singly, it is more efficient for these insects to be contained within multicellular cubicles (Bell & Hardee, 1995) than within individual containers (Ignoffo, 1966). Even for insects reared gregariously, the choice of container has a great impact on the efficiency of virus production. For the gypsy moth, ten larvae were reared within a 180-ml ice cream container. Larval biomass and the yield of virus per container was optimal at this larval density (Shapiro *et al.*, 1981a). The manual operation of removing lids (from 1700 cups/day) and then removing virus-killed larvae (>75%), living-infected larvae (>20%) and male pupae (>5%) required more time than any other procedure (Shapiro, 1986). If larger containers could have been used, each with greater numbers of larvae, harvest would have been more efficient. Moreover, manual collection (= handpicking) has invariably been very time consuming, regardless of the container system used. The efficiency of harvest should be increased by the collection of virus-killed larvae by vacuum (Bell & Hardee, 1995).

The time of harvest is also a critical factor in the production scheme. Although optimal viral activity (= virulence) can be obtained by waiting until all insects die (Shapiro & Bell, 1982), these insects may be difficult to harvest without significant loss of harvestable virus (i.e. insects wilt and virus soaks into the diet). Moreover, bacterial build-up may be so great as to present a quality control problem and potential safety hazard. Several approaches have been utilized to minimize or circumvent these problems. Living infected larvae were hand-collected >24 h prior to death, and were placed in containers and refrigerated (4 °C) (Lewis, 1971) or were freeze dried (Cunningham *et al.*, 1972). In the former case, insects were allowed to die, and the cadavers were collected and macerated. Virus was obtained following differential centrifugation. In the latter case, larvae were freeze dried and then ground to a powder. Whether virus was recovered following blending and centrifugation (Bell & Hardee, 1995) or freeze drying and milling (Shapiro *et al.*, 1981b), the goal was the same: to maximize the amount of virus recovered.

4. *Quality control*

Quality control of the technical material is required and involves determination of the microbial population present (numbers and types), safety to a vertebrate test animal (mouse), as well as biological activity (Shapiro *et al.*, 1981a). The final step in the production scheme is often very critical (i.e. storage of virus as a concentrated technical powder or as part of a virus formulation). In the past, long-term storage of virus at room temperature or, worse, at elevated temperatures, has proven to be a serious problem.

When virus is produced just prior to usage, short-term storage of unformulated virus under refrigerated conditions is quite feasible. In this case, virus is mixed with adjuvants on site and virus potency is maintained. When virus is produced the year before usage and then stored, storage conditions become critical.

G Case studies

It has not been our intention to provide a detailed review of virus production (see Shapiro, 1982, 1986), but to provide an overview and to highlight selected aspects of production. During the past several years, progress has been made in insect rearing and in virus production. It is our intent to highlight advances for a solitary insect (= cotton bollworm) and for a gregarious insect (= gypsy moth), which have led to successful *in vivo* virus productions.

1. *Heliothis baculovirus*

The development of the *Heliothis* baculovirus technology has been highlighted previously (Ignoffo &

Couch, 1981) and was developed in the 1960s and 1970s. Improvements in containerization and automation and research on virus production led to the capability of producing large numbers of larvae for virus production.

Although this research was initiated by USDA scientists led by Ignoffo, subsequent development was achieved by Ignoffo and colleagues in Industry, USDA-ARS continued to improve the rearing of *Heliothis* and the ARS facility at Mississippi State is capable of producing >130 000 multicelled trays (= 32 cells/tray at 1 larva/cell) per day (Bell & Hardee, 1995). Virus production was initiated in November 1993 and was terminated in March 1994. During this time, virus was obtained from more than 8 million larvae (= enough virus to treat >200 000 acres at 2.4 $\times 10^{11}$ IBs per acre). Virus was produced at a total cost of $185 000 ($150 000 for materials, $35 000 for labour), which is equivalent to $1.09 per acre cost of virus (Bell & Hardee, 1995). This production certainly indicates the feasibility of large-scale *in vivo* virus production and is the culmination of more than 30 years of research and effort.

2. Gypsy moth baculovirus

In the case of the gypsy moth, *in vivo* virus production is the result of research in the 1970s at both Universities and the US Department of Agriculture. While this system was efficient and virus was produced from 1 700 000 larvae in 100 days (= avg yield was >2.8 $\times 10^{9}$ IBs per larva) at a cost of $1.00–1.50 per acre (Shapiro *et al.*, 1981b). This system has been further mechanized by APHIS scientists at Otis ANGB, MA (Bernon *et al.*, 1994). The largest constraint to this system is the 180-ml container. Even at a pilot scale production of 17 000 larvae per day for 100 days, the task of opening and closing 1700 lids per day to inoculate and later to harvest was both time consuming and inefficient.

In order to summarize the in host virus production systems for *Heliothis* and *Lymantria*, Table 9 highlights the production schemes and may be useful for other insects and systems.

In many ways, the procedures are similar for both production systems, with the exception of the rearing container. For each insect, the container must be optimized and adapted to the biology of that insect. The use of multicelled trays for solitary, aggressive insects (*Heliothis*) and large or high density rearing containers for gregarious insects (*Lymantria*) is a very reasonable and cost-effective approach and should be encouraged.

H Recent developments

Although some improvements can be made to reduce costs (i.e. automation, use of agar substitutes), containerization becomes an obstacle to large-scale

Table 9 In host virus production schemes for *Heliothis* and *Lymantria*.

Insect	*Heliothis*	*Lymantria*
Inoculum	Secondary	Secondary
Virus concentration	10^5 PIB/0.1 ml	10^6 PIB/ml
Treatment	Surface of diet	Surface of diet
Insect	Third-fourth stage	Fifth stage
Temperature	26°C	29°C
Incubation	7 days	14 days
Container	Multicellular tray (32 cells)	180 ml
Larvae/cell(or container)	1	10
Harvest	Once at day 7	Once at day 14
Process	Freeze larvae	Freeze larvae
	Blend	Freeze dry
	Screen	Dehair larvae
	Dilute	Mill
	Refreeze	Freeze
		(US Forest Service follows a blend, screen, centrifugation, freeze dry, freeze regime.)

virus production. Recent advances by Hughes (1994) should optimize virus production from gregarious insects and is summarized here. The rearing system 'HeRD' (for 'High efficiency Rearing Device') is based on the behaviour of the host insect(s) and its specific physical (= space) and dietary requirements. In general terms, the rearing unit is enclosed and consists of an upper area for diet, a middle area for larvae, and a bottom area, where frass is collected. Rows of tabs are placed in the container, which provide increased feeding surfaces. Using this system, Hughes reared more than 9000 cabbage looper larvae per cubic foot and higher densities for beet armyworm larvae. This system has been utilized for other gregarious insects such as the European corn borer and gypsy moth, and can certainly be used for *in vivo* virus production (Hughes, 1994). This system appears to be promising and builds upon prior systems for large-scale rearing of lepidopterous insects (Baumhover *et al.*, 1977). An additional advance that Patana (1969) and Baumhover *et al.*, (1977) pioneered was the use of a cheap agar substitute (Gelcarin HWG), which was later used for small-scale virus production (Shapiro & Bell, 1981). Advances such as these should make in host virus production both logistically and economically feasible.

7 BIOASSAY

Assessment of the infectivity of a given virus preparation is an essential procedure for determining both the infectivity *per se* and as an aid to comparison of different isolates or batches of the same or different viruses. Assays are normally carried out under laboratory conditions in order to maintain the maximum control over variability that might affect the result of the tests. This section, therefore, deals with the main assay methods but it must be emphasized that much will depend on the insects and on the type of virus being tested.

Consideration must also be given to the form of analysis that will be employed to assess the results of the work. This has been reviewed by Hughes & Wood (1987) who evaluated the precise nature of the infection process. Although probit analysis (Finney, 1971) has been employed to analyse dosage–mortality data, it is more realistic biologically to assume that each virus particle has an equal probability of inducing infection and that they act independently of each other. Such assumptions, therefore, point to the use of other models for analysis of dosage–mortality data. For example, Ridout *et al.* (1993) describe a generalized one-hit model specifically for bioassays of occluded insect viruses. This allows both for intrinsic variability in larval susceptibility to virus, for variation in the amount of virus suspension ingested and for the distribution of virus within that amount. Although providing a better biological basis for observed data, the one-hit model does not necessarily give a significantly better fit than the usual probit methods. Thus, allowing for the fact that most laboratories have established procedures for probit analysis, we assume that this procedure will be the method of choice, provided that allowance is made for control mortality using Abbott's formula (Abbott, 1925). See Chapter III-2 for use of the formula.

The remainder of this section deals with methods for administering virus accurately for oral ingestion during *in vivo* bioassays. This is the normal approach for all occluded viruses and can also be employed for non-occluded viruses. It is also possible to administer dosages by injection but this can only be carried out after considerable trial and error and is more difficult to interpret. We therefore do not deal with this form of bioassay.

As a general rule, at least 30 larvae per dose and >2 replicates per assay, should be employed in all bioassays. Repetition is the key to consistency and serves to reduce variability within and between assays.

A Assays using semi-synthetic diets

The wide availability of semi-synthetic diets for many species of insect offers the possibility of development of simple, reproducible methods for incorporation of virus into the feeding medium. Two principal approaches can be adopted, namely surface contamination or diet incorporation methods.

1. Surface contamination assays

The principle of this method is shown in Figure 6. Depending on the amount of diet employed, the method can be used for determination of lethal dosages of virus (LD_{50-90}), where the dose is administered to a precise amount of diet that is consumed entirely, or determination of lethal concentrations (LC_{50-90}), where a known concentration of virus is applied to the surface of diet but it is not consumed entirely and, thus, the precise dosage ingested is not known. Particularly for LD_{50} (the lethal dose required to kill 50% of a given population of test organisms) the precise rate of feeding, so that the entire diet-virus aliquot is consumed in a known short time, must be ascertained by trial and error. It may not be possible to use the diet plug method for very small larvae or for larvae that generally feed gregariously.

The amount of time that larvae are allowed to feed on the diet determines the total acquisition time for virus uptake. The longer the feeding period, the greater the variability in dosage consumed, reflecting differences in feeding rate and in intrinsic

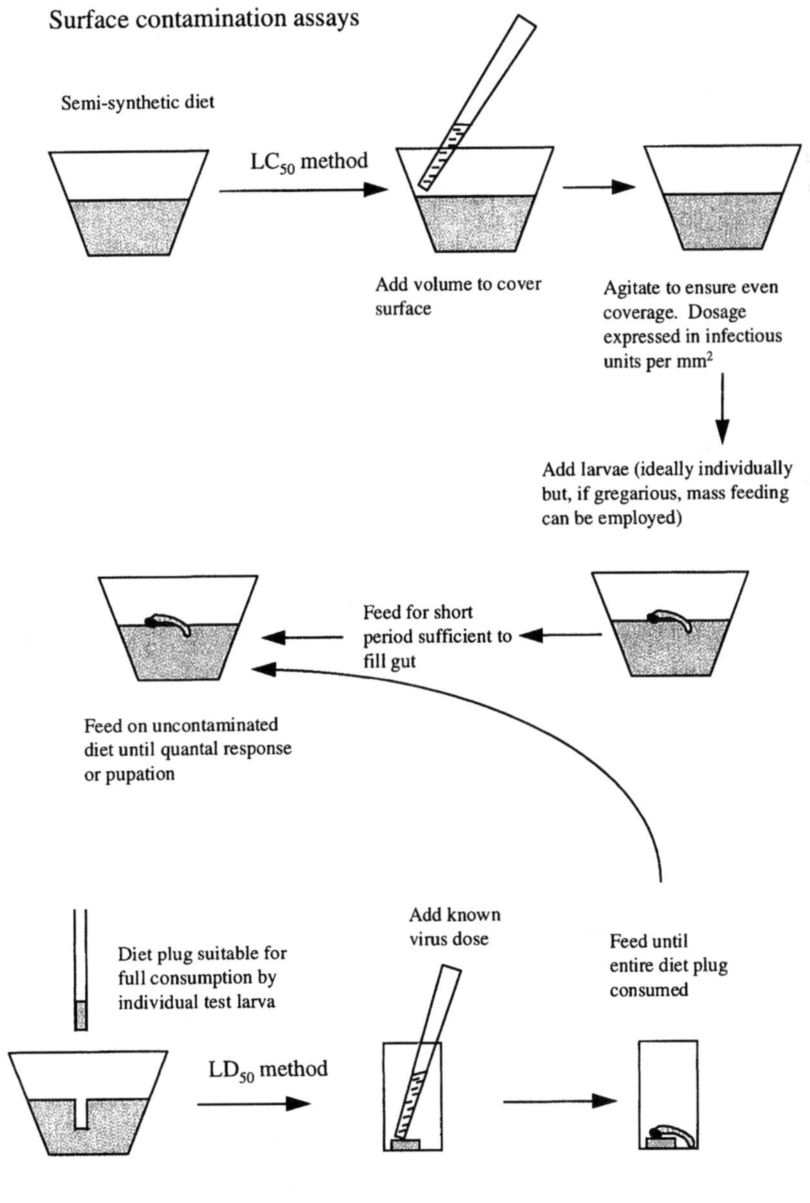

Figure 6 Bioassay procedures using surface contamination of semi-synthetic diet with suspensions of virus of known concentration.

susceptibility with increasing age of test larvae. Indeed, the dramatic change in LD_{50} observed with increased larval age, linked to their increasing weight, has been observed by many authors (for review see Evans, 1986). It is, therefore, preferable to remove larvae from the virus treated diet in the shortest time compatible with their ingesting a potentially lethal dosage. This can be determined in ranging assays (see Section F).

Dosage is normally expressed as quantity of virus/mm of diet surface. Hughes & Wood (1987) provided a useful summary of a spruce budworm, *Choristoneura occidentalis* assay using NPV. The assay employed 25 μl of diet per container each having a single larva. Two replicates of 30 larvae per dose across four dose levels were used. Observations were carried out daily and numbers of dead larvae recorded as well as the number still alive on day 14 (at 30 °C).

2. Diet incorporation assays

In general, incorporation of virus directly into the diet increases accuracy because the dosage is more evenly distributed through the medium. The methods are outlined in Figure 7, where the important steps are to ensure that the diet has cooled sufficiently to avoid the risk of thermal inactivation of the virus but still be liquid enough to be poured into appropriate containers. The method is most conveniently employed for LC_{50} estimations, particularly for larvae with gregarious feeding habits. However, removal of diet plugs of known volume which are then consumed entirely by individual test larvae would allow the method to be used for LD_{50} studies.

A good example of a diet incorporation assay is that provided by Martignoni & Ignoffo (1980) for the NPV of *H. zea*. They tested neonate, unfed larvae and incorporated virus into diet poured into a tray with individual compartments. A total of 100 larvae per dose and three dose levels were used per assay. Analysis, based on mortality after 6 days, was by probit analysis.

B. Leaf assays

Some insect species, particularly sawflies (Hymenoptera) cannot be reared on semi-synthetic diet. In these cases a leaf assay can be used, using the normal food plant of the test species. The principles are the same as a surface contamination assay but there are a number of ways of presenting the virus-treated leaf to the test insects. These are illustrated in Figure 8. It is more difficult to carry out LD_{50} studies but methods, such as those described by Evans (1981), can be used, the precise procedures being determined by the nature of the insect feeding. Evans (1981) tested instars I to VI of *M. brassicae* using templates to expose defined leaf areas with discrete dosages of NPV to individual larvae. At least 30 larvae per dose per replicate were employed. Larvae that had consumed the entire exposed leaf area were transferred to semi-synthetic diet and fed until death or pupation. Results were analysed by probit analysis.

C. Droplet feeding assays

This method, by Hughes & Wood (1981) and developed further by Hughes *et al.* (1986), relies on the reactions of many larval stages of insects to drink liquids on surfaces, especially after a short period of starvation. The principles are illustrated in Figure 9. A great advantage of the method is the ability to administer a precise dosage to an individual larva in a very short time, thus reducing variability of ingestion rate and ensuring that each larva that is used for the test has received the same dosage of virus. The method is particularly good for carrying out LD_{50} tests of neonate larvae that would otherwise not ingest a known dosage.

Dosage is based on knowledge of the rate of consumption of liquid by a given life stage. It is, therefore, necessary to determine the volumes of suspension ingested by the larvae of the species being tested. This can be done using radio-labelled suspensions (e.g. ^{32}P) or weighing batches of larvae before and after feeding. Hughes & Wood (1981) quoted data for a number of species where the mean volumes of virus suspension ingested ranged from 0.006 μl of fluid for *Trichoplusia ni* to 0.049 μl of fluid for *Estigmene acrea* neonates. Where the larvae are likely to become trapped in droplets through surface tension, the suspension can be covered with perforated parafilm through which the larvae can drink. This was used successfully for *E. acrea* neonates.

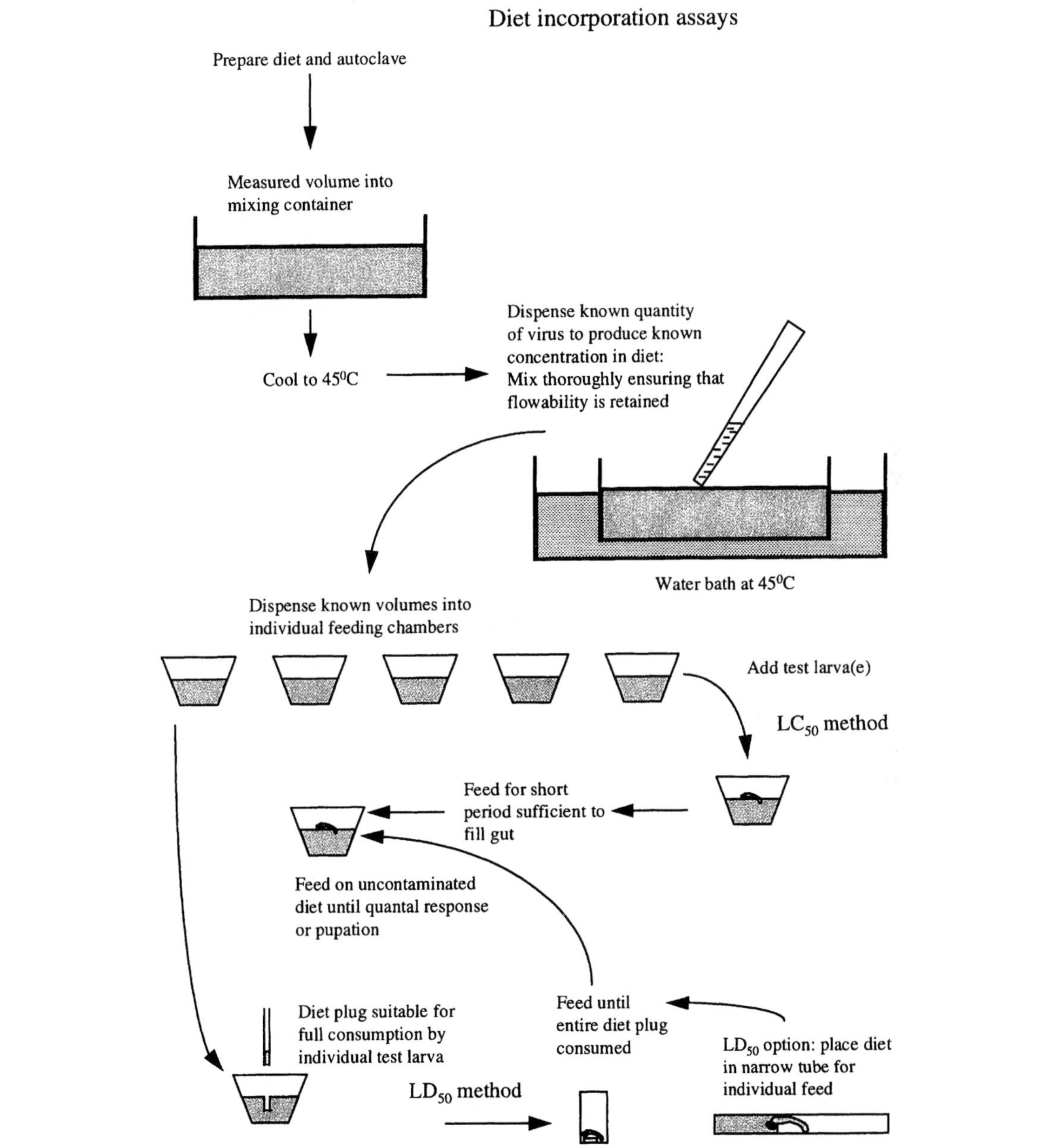

Figure 7 Bioassay procedures using incorporation of suspensions of virus of known concentration in semi-synthetic diet.

D. Dipping of eggs in virus suspension

A habit of neonate larvae is the consumption of the chorion of the egg soon after eclosion. This can be exploited, for bioassay of neonates, by dipping the eggs in a suspension of virus that is then consumed during and soon after larval eclosion. It is not possible to assess accurately the actual dosage ingested and thus, the method is suitable only for LC_{50} determination. Virus may be delivered in water alone or a small amount of wetting agent such as Tween 80® can be employed. The normal method is total immersion of eggs on an appropriate substrate, followed by air drying. Larvae acquire an unknown dosage as they emerge from the egg.

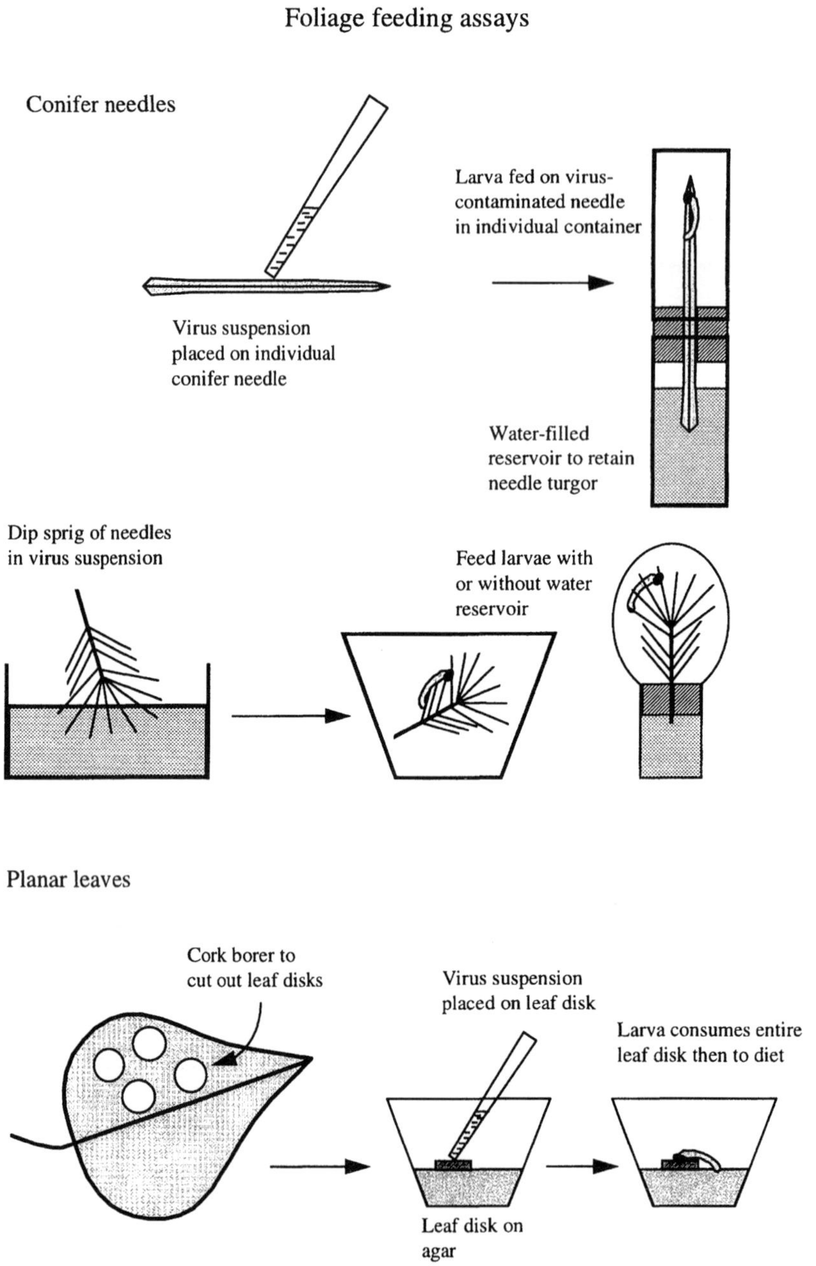

Figure 8 Bioassay procedures for dispensing virus suspensions on foliage.

E. Post-treatment handling of test larvae

Regardless of the initial method of virus administration to test larvae, it will be necessary to feed them until the end of the assay. Unless it is absolutely necessary, because of gregarious behaviour, larvae should be reared individually to avoid contamination between them. In all cases, in order to avoid contamination, commence with handling of non-treated control larvae and then transfer larvae in order of ascending virus concentration, sterilizing the handling equipment between each dosage batch. There

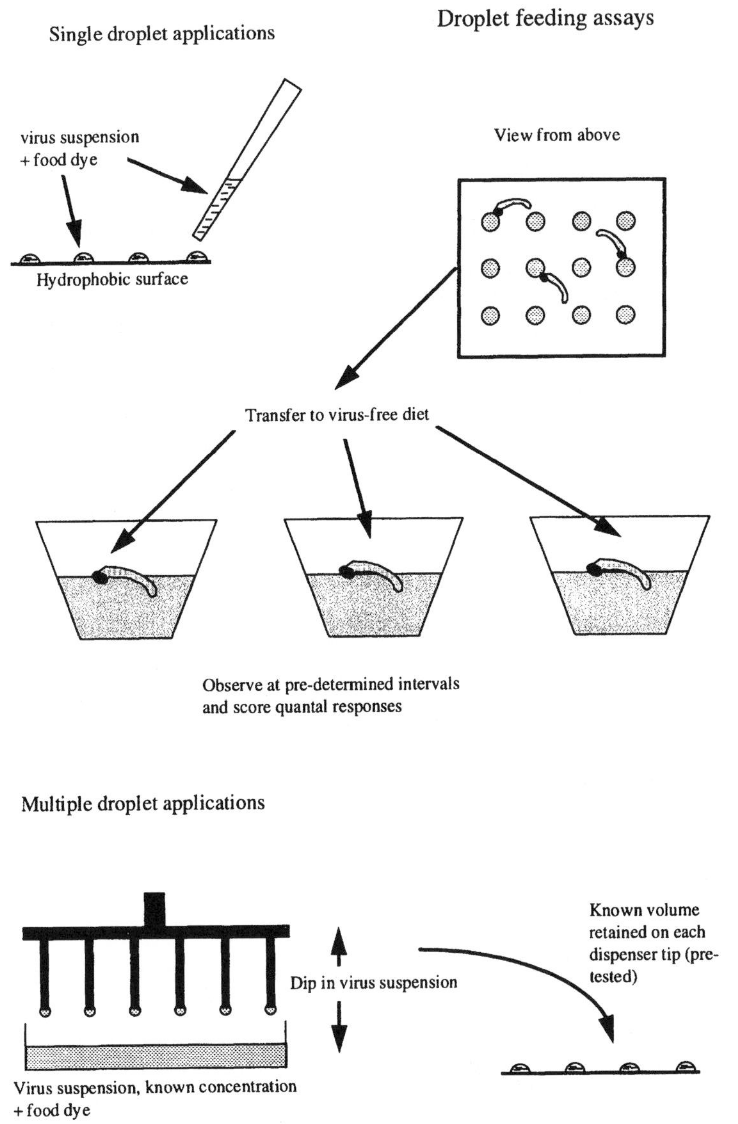

Figure 9 Bioassay procedures using the droplet feeding methods of Hughes & Wood (1981) and Hughes *et al.* (1986).

should be no possibility of contamination between the closed containers in which the treated larvae are feeding.

F. Selection of dosage range and numbers of test larvae

Unless there is already good information to indicate likely lethal dosage, it will be necessary to carry out ranging assays before commencing detailed bioassays. This can be done using widely spaced dilutions ($\log_{10}$ scale) of the virus suspension and equal numbers of test larvae per dose (a symmetric design). If there is no information to help narrow down the mortality response range, full $\log_{10}$ dilutions and at least five dilution steps should be employed. A low number of larvae per step can be used (10–20). The data can then be subjected to preliminary probit analysis and the likely LD_{50} and slope ascertained, even though the precision of the assay will be low and some dosages may give nil or 100% kill. Detailed assays can then be based on these parameters and, using methods described by Finney (1971) and elaborated by Hughes & Wood (1987), the optimal dosage steps and number of test larvae per step for a required level of statistical confidence can be ascertained. In general, the aim is to use five dosage steps centred on the LD_{50} and equally spaced on either side to the 10% and 90% mortality levels.

In practice, if comparative assays with full description of the LD_{50} and slope are required, it is normal to use at least 30 subjects per dose and to repeat the assays, ideally up to five times. However, even this high level of replication is subject to errors in reproducibility (see Section 7 H 4).

G. Recording of data

The initial design and overall purpose of the assay will determine the type and frequency of recording during the duration of the assay. For example, comparison of the population responses of different groups of insects to the same virus preparation may only require determination of the absolute level of mortality and, thus, an end-point when no more larvae die, may be sufficient. However, in the majority of cases the interest lies in both the absolute mortality exhibited by the test population and the rate at which the test organisms die. In such cases LD_{50} or LC_{50} and LT_{50} (time at which 50% mortality of the population occurred) information can be obtained, allowing greater inferences to be gained on the pathogenicity of the test virus. In some cases the quantal response may not be mortality but the appearance of particular symptoms. Provided that these are classified in advance, the assay should be accurate, although an element of subjectivity is likely in interpretation of symptoms. In such cases, the response is expressed as an *effective dose* (ED_{50}) or an *infective* or *infectious dose* (ID_{50}) producing 50% response. Unless specified otherwise, the use of the term LD_{50} in the rest of this section includes all variants of quantal response to dosage (LC_{50}, ED_{50}, EC_{50}, ID_{50}, IC_{50}, etc.).

In general, therefore, data should be recorded on a daily or more frequent basis, depending on the rate of quantal response and on whether accurate time-based information is required. Both treated and non-treated (control) test organisms should be observed and the numbers of both live and dead (or predetermined quantal response) specimens recorded against dosage received.

H. Analysis of dosage–mortality data

A number of methods are available for analysis of dosage–mortality data, most of which are aided by the use of computers or programmable calculators. Huber & Hughes (1984) and Hughes & Wood (1987) provide useful comparisons of models for analysis of quantal responses. Other methods, not requiring access to sophisticated computer equipment can also be used for basic analysis of the data and are described below.

1. Spearman–Karber analysis

This method relies on data spanning the full range of responses from 0% to 100%, bearing in mind the provisos on the statistical value of these extreme response data expressed earlier. Calculation of the LD_{50} (or other quantal response) is based on an end-point approach so that the value is determined by assessing the response at each dose, expressed as a $\log_{10}$ dilution factor. The LD_{50} is then calculated from the formula

$$\log_{10}LD_{50} = X_{p-1} + \frac{(1d)}{2} - d\,\Sigma\,p$$

where X_{p-1} = highest $\log_{10}$ dilution giving 100% quantal response,

d = $\log_{10}$ dilution factor,

p = proportion of positive responses at a given dosage,

Σp = sum of p for X_{p-1} and all higher dilutions.

Standard error of the LD_{50} can be calculated from the formula

$$SE = \sqrt{d^2}\,\frac{\Sigma\,p(1-p)}{n-1}$$

The method gives a good approximation of the LD_{50} but does not allow the slopes of the relationship between dosage and quantal response to be calculated. Although tedious, it is possible to calculate the values of LD_{50} and its standard error with a hand calculator but it is generally more convenient using a programmable calculator or computer spreadsheet.

2. Probit analysis

Although, as discussed by Hughes & Wood (1987) and Ridout *et al* (1993), probit analysis may not be based on the most appropriate biological assumptions of independent action of virus particles, the technique is so widely established for dosage–mortality analysis that it is the only method that we have included in this chapter. The majority of laboratories now have access to computers for analysis and may have software designed specifically for analysis of dosage–mortality data. For example the GENSTAT (Lawes Agricultural Trust, Rothamsted, UK) suite of statistical software includes probit analysis. Specific programs such as POLO (Russell *et al.*, 1977) are also available. In all cases, probit analysis is the log-normal transformation of the data to enable the sigmoid dosage–response curves to be linearized and compared for LD_{50} and slope value. Analysis normally uses a maximum likelihood procedure to estimate the LD_{50} iteratively using the basic probit transformation initially and then a set of calculated probits from the transformed curve. Iterations continue until the values and their standard errors stabilize. Fiducial limits (normally at the 95% level) and the degree of heterogeneity (based on chi-square estimation) are then determined. If high heterogeneity is demonstrated, the dose–response curve should be examined for any evidence of systematic deviation from the expected. If deviation is random, a heterogeneity factor can be calculated and applied to the data to allow for the observed random variability. In the absence of significant heterogeneity, different assays can be compared using chi-square, provided the slopes are parallel. In such cases, a potency ratio can be established by simple division of the LD_{50} values. If heterogeneity is high the responses should be compared using the variance ratio.

The typical response curve of untransformed mortality data illustrates an important point concerning dosage–mortality relationships. The sigmoid nature of the response curve indicates that the extremes of mortality near 0% and 100% provide little information on how the population as a whole is responding. Indeed, 100% mortality may indicate that just sufficient virus has been ingested to kill the population or that there has been an excessive dosage ingested. Although time–mortality data may provide further inference on this parameter, there is little statistical value to the 100% point within probit analysis. Comparison between assays tends, therefore, to be made at the LD_{50} point.

3. Time–mortality analysis

Relationships between dosage and quantal response are measured in terms of both absolute response and by reference to the time taken to reach a given response. The latter, usually concerned with the LT_{50}, is a useful measure of the rate of expression of response and can be used to compare assays where the LD_{50} may be similar. Analysis of the results as a plot of probit value against time can be used to visually estimate the LT_{50}. A more accurate method using transformations that allow slopes and standard errors to be compared is provided by Bliss (1937) and described in more detail, including a listing of a computer program for logit analysis, by Hughes & Wood (1987).

4. Reproducibility

Variability between bioassays can be, and usually is, very great. This may be a reflection of the methods being employed so that there is intrinsic variability in the procedures for dosage administration or in selection of test organisms. This is certainly the case in surface contamination assays where the test organisms may not consume the entire dose. There is also a great deal of variability in consistency in carrying out the assay itself so that replicate assays, even carried out by the same person, may vary widely in results even though parameters are kept as constant as possible. Fenlon (personal communication) has analysed the way in which operator variability can influence assay results. He analysed a series of 20 assays using the GV of diamond back moth, *Plutella xylostella,* carried out over a period of two years. The first five assays were carried out at infrequent intervals over the first year giving rise to wide variation in the logarithmic LD_{50} values (from 4.07 to 7.5). A further series of 11 assays was performed over a much shorter time span, accompanied by a rapid stabilization of the logarithmic LD_{50} values (mean 6.1) and a significant reduction in the sample standard deviation and probit standard error. It was concluded that assay variability was reduced as a result of increased experience, reflecting a learning process on the part of the operators. Of equal importance was the finding that the standard error of the probit model was improved, thus demonstrating that replication over time is not as accurate as replication during the same experiment. There were also important lessons in determining the range of dosages to use within a given assay. Experience over several assays helped to reduce the span of dosages required but, because it will usually not be possible to carry out such a long series of repeat assays, it is generally necessary to use a dosage range that is wider than strictly necessary. This will allow for the high variability that can be expected in carrying out a single or low number of assays.

8 PRESERVATION/ LONG-TERM STORAGE

The problems of long-term storage of viruses has been dealt with partially in Section 6. Questions of both maintenance of stock virus cultures and, if the viruses are intended for practical pest management in the field, shelf-life of the formulated product must both be addressed in ensuring storage with no loss of infectivity.

The simplest, and probably most effective, way of storing virus, whether still within the host insect or purified, is to deep freeze the preparation. This can be done in liquid nitrogen or, as a more practical long-term measure, by storage in a deep freeze at –70°C (ideally) or –20°C. Such methods, particularly liquid nitrogen or freezing at –70°C can preserve the activity and physical integrity of both occluded and non-occluded viruses.

Shelf storage at room temperatures requires more careful preservation techniques. These have been discussed in some detail for baculoviruses by Young & Yearian (1986). The primary factor in loss of activity is the effect of ultraviolet (UV) light and, to a lesser extent, high temperatures. Non-occluded viruses are particularly prone to one or both of these attrition factors and are not usually stored at room temperature.

The most effective methods for shelf storage involve flowable or, more successfully, dry powder preparations. In all cases the formulation must compromise between maintenance of biological activity of the virus and the danger of fungal or bacterial contamination that, through fermentation, could result in problems of storage. An acid pH helps to prevent bacterial growth and also reduces the risk of dissolution of occluded IBs. The choice of formulant will depend on the initial method of virus extraction from the host insect. If the preparation consists of a macerate that, at most, will have been partially purified, then refrigeration or freezing is preferable for longer-term storage. Shelf life of flowable preparations can be improved by acid stabilization and inclusion of various UV protectants. Other formulation additives that may help persistence in the field are not strictly necessary for laboratory storage. A fuller discussion on field formulation can be found in Young & Yearian (1986).

Various methods for preparation of powders for use as wettable powder field formulations are available. These include lyophilization of filtered insect macerates (Shapiro, 1982) or acetone–lactose co-precipitation (Dulmage *et al.*, 1970). A more reliable method is to use spray drying of a mixture of the NPV with various clays and other diluents (Bull, 1978). However, problems of stability can be encountered with such methods, particularly for GVs. They are not suitable for non-occluded viruses. Microencapsulation has also been attempted but with relatively limited success (Ignoffo & Batzer, 1971).

In conclusion, therefore, it would appear that storage prior to any formulation requirement is most successfully achieved by liquid nitrogen or other deep-freezing techniques. This is essential for non-occluded viruses that rapidly lose activity at room or even chilled (2–5°C) temperatures.

ACKNOWLEDGEMENTS

We would like to thank Dr Jean Adams, USDA, Otis, for her generosity in supplying the electron microscope plates illustrating the virus groups.

REFERENCES

Abbott, W. S. (1925) A method of computing the effectiveness of an insecticide. *J. Econ. Entomol.* **18**, 265–267.

Adams, J. R. (1991). Introduction and classification of viruses of invertebrates. In *Atlas of invertebrate viruses* (eds J. R. Adams & J. R. Bonami), pp. 1–8. CRC Press, Boca Raton.

Adams, J. R. & Bonami, J. R. (1991a) *Atlas of invertebrate viruses*. CRC Press, Boca Raton.

Adams, J. R. & Bonami, J. R. (1991b) Preparation of invertebrate viruses and tissues for examination. In *Atlas of invertebrate viruses* (eds J. R. Adams & J. R. Bonami) pp. 9–30. CRC Press, Boca Raton.

Adams, J. R. & McClintock, J. T. (1991) Baculoviridae. Nuclear Polyhedrosis Viruses. Part 1. Nuclear Polyhedrosis Viruses of Insects. In *Atlas of invertebrate viruses* (eds J. R. Adams & J. R. Bonami), pp. 87–204. CRC Press, Boca Raton.

Anthony, D. W. & Comps, M. (1991) Iridoviridae. In *Atlas of invertebrate viruses* (eds J. R. Adams & J. R. Bonami) pp. 55–86. CRC Press, Boca Raton.

Arif, B. M. (1984) The Entomopoxviruses. *Adv. Vir. Res.* **29**, 195.

Ausubel, F. M., Brent, R., Kingston, R. E., Moore, D. D., Seidman, J. G., Smith, J. A. & Struhl, K. (1991) *Current Protocols in Molecular Biology*. Wiley Interscience. New York.

Baker, K., Zheng, Y., Reid, S. & Greenfield, P. F. (1993) Production of multisubunit particles for use as vaccines using the baculovirus expression system (BEVS). In *Animal cell technology: basic and applied aspects* (eds S. Kaminogawa, A. Ametani & S. Hachimura). Kluwer Academic Publishers. Dordrecht, The Netherlands.

Baumhover, A. H., Cantelo, W. W., Hobgood, J. M. J., Knott, C. M. & Lam, J. J. J. (1977) An improved method for mass rearing the tobacco hornworm. *USDA Res. Serv.* **ARS-S-167**, 1–13.

Bell, M. R. (1991) Effectiveness of microbial control of *Heliothis* spp. on early season wild geraniums: field and field cage tests. *J. Econ. Entomol.* **84**, 851–854.

Bell, M. R. & Hardee, D. H. (1995) Tobacco budworm and cotton bollworm: Methodology for virus production and application in large-area management trials. Conference Proceeding. 1995 Beltwide Cotton Production Conference. San Antonio, Texas. pp. 857–858.

Benz, G. A. (1986) Introduction: Historical Perspectives. In *The biology of baculoviruses. Biological properties and molecular biology* (eds R. R. Granados & B. A. Federici) pp. 1–35. CRC Press, Boca Raton.

Bernon, G., Antell, H., Banner, D., Barton, S., Cruse, C., Davis, R., Lane, S., Levine, J., Martin, F., Tanner, J. & Terrien, H. (1994) *Mass-rearing the gypsy moth, Lymantria dispar* L. US Department of Agriculture, APHIS. Washington, DC. pp.1–122.

Bliss, C. I. (1937). The calculation of the time-mortality curve. *Ann. Appl. Biol* **24**, 815.

Bonami, J. R. & Adams, J. R. (1991) Birnaviridae. In *Atlas of invertebrate viruses* (eds J. R. Adams & J. R. Bonami) pp. 435– 442. CRC Press, Boca Raton.

Brun, G. (1991) Rhabdoviridae. In *Atlas of invertebrate viruses* (eds J. R. Adams & J. R. Bonami) pp. 443–460. CRC Press, Boca Raton.

Bull, D. L. (1978) Formulations of microbial insecticides: microencapsulation and adjuvants. *Misc. Publ. Entomol. Soc. Am.* **10**, 11.

Burton, R. L. & Perkins, W. D. (1984) Containerization for rearing insects. In *Advances and challenges in insect rearing* (eds E. G. King & N. C. Leppla) pp. 51–56. US Department of Agriculture, Agriculture Research Service. Washington, DC.

Chauthani, A. R. & Claussen, D. (1968) Rearing Douglas fir tussock moth larvae on synthetic media for the production of nuclear polyhedrosis virus. *J. Econ. Entomol.* **61**, 101–103.

Cunningham, J. C., Bird, F. T., McPhee, J. R. & Grisdale, D. (1972) The mass propagation of two viruses of the Spruce budworm, *Choristoneura fumiferana* (Clem.) (Lepidoptera: *Tortricidae*). *Information Report, Insect Pathology Research Institute, Canada* No. IP-X-1, 19 pp.

DeBlois, R. W., Uzgiris, E. E., Cluxton, D. H. & Mazzone, H. M. (1978) Comparative measurements of size and polydispersity of several insect viruses. *Annal. Biochem.* **90**, 273.

Dulmage, H. T., Martinez, A. J. & Correa, J. A. (1970) Recovery of the nuclear polyhedrosis virus of the cabbage looper, *Trichoplusia ni*, by coprecipitation with lactose. *J. Invertebr. Pathol* **16**, 80.

Elleman, C. J., Entwistle, P. F. & Hoyle, S. R. (1980) Application of the impression film technique to counting inclusion bodies of nuclear polyhedrosis viruses on plant surfaces. *J. Invertebr. Pathol.* **36**, 129–132.

Evans, H. F. (1981) Quantitative assessment of the relationships between dosage and response of the nuclear polyhedrosis virus of *Mamestra brassicae*. *J. Invertebr. Pathol.* **37**, 101–109.

Evans, H. F. (1986) Ecology and epizootiology of baculoviruses. In *Biology of baculoviruses* Vol.2. *Practical application for insect pest control* (eds R. R. Granados & B.A. Federici), pp. 89–132. CRC Press, Boca Raton.

Evans, H. F. & Entwistle, P. F. (1987) Viral diseases. In *Epizootiology of insect diseases* (eds J. R. Fuxa & Y. Tanada), pp. 257–322. Wiley, New York.

Evans, H. F., Bishop, J. M. & Page, E. A. (1980) Methods for the quantitative assessment of nuclear polyhedrosis virus in soil. *J. Invertebr. Pathol.* **35**, 1–8.

Federici, B. A., Hamm, J. J. & Styer, E. L. (1991) Ascoviridae. In *Atlas of invertebrate viruses* (eds J. R. Adams & J. R. Bonami), pp. 339–349. CRC Press, Boca Raton.

Finney, D. J. (1971) *Probit Analysis*. Cambridge University Press. London.

Garzon, S. & Charpentier, G. (1991) Nodaviridae. In *Atlas of invertebrate viruses* (eds J. R. Adams & J. R. Bonami), pp. 351– 370. CRC Press, Boca Raton.

Goodwin, R. H., Vaughn, J. L., Adams, J. R. & Louloudes, S. J. (1970) Replication of a nuclear polyhedrosis virus in an established insect cell line. *J. Invertebr. Pathol.* **16**, 284–288.

Goodwin, R. H., Milner, R. J. & Beaton, C. D. (1991) Entomopoxvirinae. In *Atlas of invertebrate viruses* (eds J. R. Adams & J. R. Bonami), pp. 259–285. CRC Press, Boca Raton.

Granados, R. R. & Federici, B. A. (1986a) *Biological properties and molecular biology. The Biology of Baculoviruses*. 1. CRC Press, Boca Raton.

Granados, R. R. & Federici, B. A. (1986b) *Practical Application for Insect Control. The Biology of Baculoviruses*. 2. CRC Press, Boca Raton.

Hall, D. W. (1985) Pathobiology of Invertebrate Icosahedral Cytoplasmic Deoxyriboviruses (Iridovirdae). In *Viral insecticides for biological control* (eds K. Maramorosch & K. E. Sherman), pp. 163–196. Academic Press. New York.

Hamm, J. J. & Styer, E. L. (1985) Comparative pathology of isolates of *Spodoptera frugiperda* nuclear polyhedrosis virus in *S. frugiperda* and *S. exigua*. *J. Gen. Virol.* **66**, 1249–1261.

Hammock, B. D., McCutchen, B. F., Beetham, J., Choudary, P., Fowler, E., Ichinose, R., Ward, V. K., Vickers, J., Bonning, B. C., Harshman, L. G., Grant, D., Uematsu, T. & Maeda, S. (1993) Development of recombinant insecticides by expression of an insect specific toxin and insect specific enzyme in nuclear polyhedrosis viruses. *Arch. Biochem. Biophys.* **22**, 315–344.

Hedlund, R. C. & Yendol, W. G. (1974) Gypsy moth nuclear-polyhedrosis virus production as related to inoculating time, dosage, and larval weight. *J. Econ. Entomol.* **67**, 61–63.

Hofman, O. (1891) Die Schlaffsucht (Flacherie) der Nonne (*Liparis monacha*) nebst einem Anhang. In *Insektentotende Pilze mit besonderer Berucksichtigung der Nonne*, Anonymous pp. 31 P. Weber, Frankfurt.

Huber, J. & Hughes, P. R. (1984) Quantitative bioassay in insect pathology. *Bull. Entomol. Soc. Am.* **30**, 31–34.

Hughes, P. R. (1994). High density rearing system for larvae. US Patent number 5,351,643.

Hughes, P. R. & Wood, H. A. (1981) A synchronous peroral technique for the bioassay of insect viruses. *J. Invertebr. Pathol.* **37**, 154–159.

Hughes, P. R. & Wood, H. A. (1987) *In vivo* and *in vitro* bioassay methods for baculoviruses. In *The biology of Baculoviruses*: Vol. II. *Practical application for insect control* (eds R. R. Granados & B. A. Federici), pp. 1–30. CRC Press, Boca Raton.

Hughes, P. R., Beek, N. A. M., Wood, H. A. & Van-Beek, N. A. M. (1986) A modified droplet feeding method for rapid assay of *Bacillus thuringiensis* and baculoviruses in noctuid larvae. *J. Invertebr. Pathol.* **48**, 187–192.

Hukuhara, T. & Bonami, J. R. (1991) Reoviridae. In *Atlas of invertebrate viruses* (eds J. R. Adams & J. R. Bonami), pp. 393–434. CRC Press, Boca Raton.

Ignoffo, C. M. (1965) The nuclear polyhedrosis virus of *Heliothis zea* (Boddie) and *Heliothis virescens* (Fabricius). I. Virus propagation and its virulence. *J. Invertebr. Pathol.* **7**, 209–216.

Ignoffo, C. M. (1966) Insect viruses. In *Insect Colonization and Mass Production* (ed. C. N. Smith), pp. 501–530. Academic Press, New York.

Ignoffo, C. M. & Anderson, R. F. (1979) Bioinsecticides. In *Microbial technology*, Anonymous, pp. 1–28. Academic Press, New York.

Ignoffo, C. M. & Batzer, O. F. (1971) Microencapsulation and ultraviolet protectants to increase sunlight stability of an insect virus. *J. Econ. Entomol.* **64**, 850–853.

Ignoffo, C. M. & Boening, O. P. (1970) Compartmented disposable trays for rearing insects. *J. Econ. Entomol.* **63**, 1696–1697.

Ignoffo, C. M. & Couch, T. L. (1981) The nucleopolyhedrosis virus of *Heliothis* species as a microbial insecticide. In *Microbial control of pests and plant diseases 1970–1980* (ed. H. D. Burges), pp. 330–361. Academic Press, New York.

Jones, H. N. (1910) Further studies on the nature of the wilt disease of the gypsy moth larvae. *Annual Report of the State Forester, Massachusetts* **43**, 101.

Katagiri, K. (1981) Pest control by cytoplasmic polyhedrosis viruses. In *Microbial control of pests and plant diseases 1970–1980* (ed. H. D. Burges), pp. 433–440. Academic Press, New York.

Krell, P. J. (1991) Polydnaviridae. In *Atlas of invertebrate viruses* (eds J. R. Adams & J. R. Bonami), pp. 321–338. CRC Press, Boca Raton.

Lewis, F. B. (1971) Conference Proceeding. IV, International Colloquium on Insect Pathology. College Park, MD, pp. 320–326.

Lynn, D. E., Shapiro, M. and Dougherty, E. M. (1993) Selection and Screening of Clonal Isolates of the Abington Strain of Gypsy Moth Nuclear Polyhedrosis Virus. *J. Invertebr. Pathol.* **62**, 191–195.

Maramorosch, K. E. (1987) *Biotechnology in invertebrate pathology and cell culture*. Academic Press, New York.

Martignoni, M. E. (1978) Production, activity and safety. In *The Douglas fir tussock moth: a synthesis* (eds M. H. Brooks, R. W. Stark & R. W. Campbell), pp. 140–147. US Department of Agriculture Technical Bulletin 1585. Washington, DC.

Martignoni, M. E. & Ignoffo, C. M. (1980) Biological activity of Baculovirus preparations: *in vivo* assay. In *Characterization, production and utilization of entomopathogenic viruses* (eds C. M. Ignoffo, M. E. Martignoni & J. L. Vaughn), pp. 138 American Society for Microbiology and National Science Foundation. Washington, DC.

Matthiessen, J. N., Christian, R. L., Grace, T. D. C. & Filshie, B. K. (1978) Large-scale field propagation and the purification of the granulosis virus of the potato moth, Phthorimaea operculella (Zeller) (Lepidoptera: Gelechiidae). *Bull. Entomol. Res.* **68**, 385–391.

Miller, L. K. (1995) Genetically engineered insect virus pesticides: present and future. *J. Invertebr. Pathol.* **65**, 211–216.

Miller, L. K., Lingg, A. J. & Bulla, L. A., Jr (1983) Bacterial, viral, and fungal insecticides. *Science* **219**, 715–721.

Moore, N. F. (1985) Pathology Associated with Small RNA Viruses of Insects. In *Viral insecticides for biological control* (eds K. Maramorosch & K. E. Sherman), pp. 233–245. Academic Press, New York.

Moore, N. F. & Eley, S. M. (1991) Picornaviridae: Picornaviruses of Invertebrates. In *Atlas of invertebrate viruses* (eds J. R. Adams & J. R. Bonami), pp. 371–386. CRC Press, Boca Raton.

Moscardi, F., Allen, G. E. & Greene, G. L. (1981) Control of the velvetbean caterpillar by nuclear polyhedrosis virus and insecticides and impact of treatments on the natural incidence of the entomopathogenic fungus Nomuraea rileyi. *J. Econ. Entomol.* **74**, 480–485.

O'Reilly, D. R. & Miller, L. K. (1989) A baculovirus blocks insect molting by producing ecdysteroid UDP-glycosyl transferase. *Science* **245**, 1110–1112.

Padhi, S. B. (1985) Viral proteins for the identification of insect viruses. In *Viral insecticides for biological control* (eds K. Maramorosch & K. E. Sherman), pp. 55–78. Academic Press, New York.

Patana, R. (1969) Rearing cotton insects in the laboratory. *USDA Res. Rept.* **108**, 1–6.

Podgwaite, J. D. & Cosenza, B. J. (1966) Bacteria of living and dead larvae of *Porthetria dispar* (L.). *US For. Ser. Res. Note* **NE50**, 1–7.

Podgwaite, J. D., Bruen, R. B. & Shapiro, M. (1983) Microorganisms associated with production lots of the nucleopolyhedrosis virus of the gypsy moth, Lymantria dispar (Lep.: Lymantriidae). *Entomophaga* **28**, 9–15.

Powell, J. E. & Robertson, J. L. (1993) Status of rearing technology for cotton insects. In *Cotton insects and mites: characterization and management* (eds E. G. King & J. M. Brown), Cotton Foundation. Memphis, TN.

Reid, S., Greenfield, P. F., Power, J., Radford, K. M., Neilson, L. K., Wong, T. K. K., Peter, C. & Chakraborty, S. (1994) An improved process for the large scale *in vitro* production of baculoviruses. In *Proceedings 1st Brisbane symposium on biopesticides: ppportunities for Australian industry* (eds C. J. Monsour, S. Reid & R. E. Teakle), pp. 64–70. CSIRO. Canberra.

Reinganum, C. (1991) Tetraviridae. In *Atlas of invertebrate viruses* (eds J. R. Adams & J. R. Bonami), pp. 387–392. CRC Press, Boca Raton.

Ridout, M. S., Fenlon, J. S. & Hughes, P. R. (1993) A generalized one-hit model for bioassays of insect viruses. *Biometrics*, **49**, 1136- 1141.

Rollinson, W. D., Hubbard, H. B. & Lewis, F. B. (1970) Mass rearing of the European pine sawfly for production of the nuclear polyhedrosis virus. *J. Econ. Entomol.* **63**, 343–344.

Russell, R. M., Robertson, J. L. & Savin, N. E. (1977) POLO: a new computer program for probit analysis. *Bull. Entomol. Soc. Am.* **23**, 209.

Shapiro, M. (1982) *In vivo* mass production of insect viruses for use as pesticides. In *Microbial and viral pesticides* (ed. E. Kurstak), pp. 463–492. Marcel Dekker, New York.

Shapiro, M. (1986) *In vivo* production of baculoviruses. In *The biology of Baculoviruses*. Vol. II. *Practical application for insect control* (eds R. R. Granados & B. A. Federici), pp. 31–61. CRC Press, Boca Raton.

Shapiro, M. & Bell, R. A. (1981) Biological activity of Lymantria dispar nucleopolyhedrosis virus from living and virus-killed larvae. *Ann. Entomol. Soc. Am.* **74**, 27–28.

Shapiro, M. & Bell, R. A. (1982) Production of gypsy moth, *Lymantria dispar* (L.), nucleopolyhedrosis virus, using carrageenans as dietary gelling agents. *Ann. Entomol. Soc. Am.* **75**, 43–45.

Shapiro, M. & Robertson, J. L. (1991) Natural variability of three geographical isolates of gypsy moth (Lepidoptera: Lymantiidae) nuclear polyhedrosis virus. *J. Econ. Entomol.* **84**, 71–75.

Shapiro, M., Bell, R. A. & Owens, C. D. (1981a) *In vivo* mass production of gypsy moth nucleopolyhedrosis virus. In *The gypsy moth: research toward integrated pest management* (eds C. C. Doane & M. L. McManus), pp. 633–655. USDA Technical Bulletin. Washington, DC.

Shapiro, M., Owens, C. D., Bell, R. A. & Wood, H. A. (1981b) Simplified, efficient system for *in vivo* mass production of gypsy moth nucleopolyhedrosis virus. *J. Econ. Entomol.* **74**, 341–343.

Shapiro, M., Robertson, J. L. & Bell, R. A. (1986) Quantitative and qualitative differences in gypsy moth (Lepidoptera: Lymantriidae) nucleopolyhedrosis virus produced in different-aged larvae. *J. Econ. Entomol.* **79**, 1174–1177.

Steinhaus, E. A. (1956) Microbial control. The emergence of an idea. A brief history of insect pathology through the nineteenth century. *Hilgardia* **26**, 107–160.

Stewart, F. D. (1984) Mass rearing of the PBW, *Pectinophora gossypiella*. In *Advances and challenges in insect rearing* (eds E. G. King & N. C. Leppla), pp. 176–187. USDA, ARS. New Orleans.

Summers, M. D. (1991) Baculovirus-directed foreign gene expression. *ACS Symp. Ser.* **453**, 237–251.

Teakle, R. E. & Byrne, Y. S. (1988) Nuclear polyhedrosis virus production in *Heliothis armigera* infected at different stages. *J. Invertebr. Pathol* **53**, 21–24.

Tijssen, P. & Arella, M. (1991) Parvovirdae. Structure and Reproduction of Densonucleosis Viruses. In *Atlas of invertebrate viruses* (eds J. R. Adams & J. R. Bonami), pp. 41–53. CRC Press, Boca Raton.

Tompkins, G. J. (1991) Purification of invertebrate viruses. In *Atlas of invertebrate viruses* (eds J. R. Adams & J. R. Bonami), pp. 31–40. CRC Press, Boca Raton.

Trager, W. (1935) Cultivation of the virus of grasserie in silkworm tissue culture. *J. Exp. Med.* **61**, 501–505.

Vail, P. V., Anderson, S. J. & Jay, D. L. (1973) New procedures for rearing cabbage loopers and other Lepidopterous larvae for propagation of nuclear polyhedrosis viruses. *Environ. Entomol.* **2**, 339–344.

Vanderzant, E. S., Richardson, C. D. & Fort, S. W. J. (1962) Rearing of the bollworm on artificial diet. *J. Econ. Entomol.* **55**, 140.

Volkman, L. E. (1985) Classification, identification, and

detection of insect viruses by serologic techniques. In *Viral insecticides for biological control* (eds K. Maramorosch & K. E. Sherman), pp. 27–53. Academic Press. New York.

Webb, M. J. W. (1973) A method for the rapid removal of sugars and salts from virus preparations on electron microscope grids. *J. Microsc.* **98**, 109.

Weiss, S. A., Thomas, D. W., Dunlop, B. F., Georgis, R., Vail, P. V. & Hoffmann, D. F. (1994) *In vitro* production of viral pesticides: key elements. In *Proceedings 1st Brisbane symposium on biopesticides: opportunities for Australian industry* (eds C. J. Monsour, S. Reid & R. E. Teakle), pp. 57–63. CSIRO. Canberra.

Wigley, P. J. (1980a) Practical: Counting micro-organisms. In *Microbial control of insect pests* (eds J. Kalmakoff & J. F. Longworth), pp. 29 New Zealand Department of Science and Industrial Research Bulletin 228. Wellington, New Zealand.

Wigley, P. J. (1980b) Practical: Diagnosis of virus infections – staining of insect inclusion body viruses. In *Microbial control of insect pests* (eds J. Kalmakoff & J. F. Longworth), pp. 35 New Zealand Department of Science and Industrial Research, Bulletin 228. Wellington, New Zealand.

Wood, H. A., Hughes, P. R., Johnston, L. B. & Langridge, W. H. R. (1981) Increased virulence of Autographa californica nuclear polyhedrosis virus by metagenesis. *J. Invertebr. Pathol.* **38**, 236–241.

Young, S. Y. & Yearian, W. C. (1986) Formulation and application of baculoviruses. In *The biology of Baculoviruses*: Vol. II. *Practical application for insect control* (eds J. R. Fuxa & Y. Tanada), pp. 157–179. CRC Press, Boca Raton.

GLOSSARY

(Courtesy of Clinton Kawanishi and Wayne Brooks)

Capsid. The protein coat or shell of a virus particle. The capsid is a 'surface crystal', built of structure units. The structure units are the smallest functionally equivalent building units of the capsid. The structure unit could be a single polypeptide chain or an aggregate of identical or different polypeptide chains.

Capsule. The inclusion body formed by members of the genus Granulosis Virus. Synonymous with *granule*.

Core. Protein structure containing the viral genome which is enclosed by the viral capsid.

Envelope. An outer lipoprotein bilayer membrane bounding the virion.

Episome. Quiescent viral genome that persists within the cell as a naked nucleic acid.

Gene. A segment of DNA which encodes the sequence of a protein or an RNA molecule. In its simplest form it is composed of a regulatory region that controls the activity of the gene and the coding region that specifies the amino acid or RNA base sequences of the gene product. Viral genes may be activated sequentially in groups such that the product of one group turns on the next in cascade fashion. Upon virus entry into a cell, cellular components activate the immediate early genes (IE). IE gene products then activate the early genes group whose gene products in turn activates the late gene group.

Genome. The genetic material of an organism.

Granule. Synonymous with capsule.

Granulin. The virus coded phosphoprotein that forms the crystalline protein matrix within which the granulosis virion is occluded to form the granule.

Icosahedron. A geometric term applied to a polyhedron with cubic symmetry which has 20 equilateral triangular faces, 12 vertices and 30 sides. This is the most common design of the capsids of 'spherical' or isometric viruses.

Inclusion body. In insect virology this generally refers to a large, virus-coded, crystalline, proteinaceous body within which are occluded the virions. Nuclear polyhedra, capsule, cytoplasmic polyhedra, entomopox spheroids.

Lateral body. A structural component of the entomopox virion located between the core and envelope.

Multipartite genome. Viral genomes divided between two or more nucleic acid molecules. These may be encapsidated in the same particle or be in separate particles. Synonymous with Segmented genome or polydispersed genome.

Nucleocapsid. The structure composed of the capsid with the enclosed viral nucleic acid.

Polyhedron. Crystalline inclusion body that occludes virions of nuclear and cytoplasmic polyhedrosis viruses.

Polyhedrin. The virus-coded phosphoprotein that forms the crystalline protein matrix of nuclear polyhedra within which the virions of the genus Nuclear Polyhedrosis Viruses are occluded.

Provirus. Virus in the form of naked nucleic acid that is integrated into cellular DNA.

Ring zone. Clear region of a nuclear polyhedrosis virus-infected nucleus observable by light microscopy that surrounds the virogenic stroma and within which nucleocapsid envelopment and virion occlusion occur.

Spheroid. Crystalline inclusion body that occludes virions of entomopox viruses.

Spheroidin. The virus-coded protein that forms the crystalline protein matrix of spheroids within which the virions of the entomopox viruses are occluded.

Spindle. Fusiform body that forms in the cytoplasm of cells along with spheroids in certain hosts infected with entomopox viruses.

Strandedness. Whether a nucleic acid molecule exists as a single strand (ss) or base paired with its colinear complementary strand to form a double strand (ds).

Uncoating. Process by which the viral genome is released from the virion within the cell.

Virion. Morphologically complete virus particle. It can be either a naked or enveloped nucleocapsid.

Virogenic stroma. A microscopically differentiable region of viroplasm that develops in virus-infected cells from which virions assemble. In the nucleoplasm of cells infected by nuclear polyhedrosis or granulosis viruses, it is a dark staining network at the edges of which nucleocapsids form. With cytoplasmic polyhedrosis virus, it is a dense granular region of the cytoplasm that develops after infection and which gives rise to the icosahedral particles.

Viroplasm. A modified region within the infected cell in which virus replication occurs, or is thought to occur.

Virus. Non-cellular entities whose genome is an element of nucleic acid, either RNA or DNA, which replicates inside living cells, and uses intracellular pools of precursor materials and cellular synthetic machinery to direct the synthesis of specialized particles, the virions, which contain the viral genome and transfers it to other cells. Replication and assembly occurs within the cellular cytoplasm or nucleoplasm and are not separated from the host cell contents by a lipoprotein bilayer membrane as with cellular pathogens.

APPENDIX: RECIPES FOR STAINS

Buffalo Black 12B (Napththalene Black 12B or Amido Schwarz or Acid Black 1) working solution

Mix the solution using the following ingredients and weights/volumes (to produce 100 ml of working solution).

Buffalo Black	1.5 g
Glacial acetic acid	40 ml
Distilled water	60 ml

Preparation of working solutions for Giemsa's stain

1. Make up 0.02M phosphate buffer solution:
   Solution A: 28.39 g of Na_2HPO_4 dissolved in 1 l of distilled water
   Solution B: 31.21 g of $Na_2HPO_4.2H_2O$ dissolved in 1 l of distilled water
   Mix 55 ml of solution A with 45 ml of solution B and make up to 1 l with distilled water, making a working solution of phosphate buffer with pH between 6.9 and 7.0.

2. Make up Giemsa's fixative:
   94% Absolute alcohol
   5% formalin solution
   1% acetic acid

CHAPTER III-1

Identification, isolation, culture and preservation of entomopathogenic bacteria

I. THIERY & E. FRACHON

Unité des Bactéries Entomopathogénes, Institut Pasteur, 25 rue du Docteur Roux, 75724 Paris cedex 15, France

1 INTRODUCTION

Entomopathogenic bacteria are found among the Gracilicutes (bacteria with a thin peptidoglycan layer), and Firmicutes (bacteria with a thick peptidoglycan layer) divisions within the kingdom Procaryotae. The most well-known bacteria pathogenic for insects are listed here in a simple overview key (Figure 1). These bacteria are either facultative or obligate entomopathogens, and are either Gram-negative, for example, *Serratia marcescens* and *Pseudomonas aeruginosa* or Gram-positive such as *Bacillus* sp. and *Clostridium* sp. The latter genera are similar in that both produce endospores. For more details on general bacterial classification, on each genus and on the role of each species in insect infections refer to the following: *Bergey's Manual of Systematic Bacteriology* (Sneath, 1986), *Microbiology* (Wistreich & Lechtman, 1988) and *The Prokaryotes, A Handbook on the Biology of Bacteria* (Stahly *et al.*, 1991). This chapter will emphasize techniques for working with entomopathogenic bacteria in the genus *Bacillus*. Descriptive information on bacteria found in soil inhabiting insects is also presented in Chapter III-4.

2 IDENTIFICATION

A Determination of the genus *Bacillus*

Although several bacterial genera are able to produce endospores, the genus *Bacillus* is recognized by being rod-shaped, usually Gram-stain positive, producing catalase and being aerobic or facultatively anaerobic. *Bacillus* cells produce an endospore on completion of growth. Gordon *et al.* (1973) arranged the species into three morphological groups based on spore shape and swelling of the sporangium.

Group I contains *Bacillus* species producing terminal oval endospores that do not cause the rod-shaped bacterial cell to swell. This group can be divided into two classes: bacterial cells with rod width greater than 0.9 μm (class 1); and those under

MANUAL OF TECHNIQUES IN INSECT PATHOLOGY
ISBN 0–12–4325556

0.9 μm width (class 2). Group II strains have oval endospores that swell the sporangium, and Group III strains contain round spores inducing a swollen sporangium (Figure 2).

Although there are some *Bacillus* isolates with a spore morphology not so easily classified within these groups, this classification method is still the most useful as it corresponds to most *Bacillus* species.

For optimal microscopic observation of bacterial morphology, the quality of the optical instruments and use of standardized conditions are very important. A good quality phase-contrast microscope is absolutely essential for spore examination. This allows observation of differentiation of spore refringence from the other components within the bacterial cell or medium.

Three steps are necessary for differentiating a *Bacillus* isolate under the microscope :

1. Early observation of the morphology of the culture during the vegetative stage;
2. After 24–72 h incubation, observation of spores and search for parasporal bodies in *Bacillus thuringiensis* and *Bacillus sphaericus* strains. Usually, in optimum conditions, spore refringence appears after 24–48 h incubation at 30°C. But appearance of spores can be slower so the culture may be incubated longer;
3. After 48–72 h observation of sporangium lysis, spore liberation into the medium, and confirmation of presence of the proteinaceous parasporal inclusion bodies, or 'crystals' (*Bacillus thuringiensis* and *Bacillus sphaericus*).

The shape of the spores and the bacterial cells may be modified if the bacteria are grown under less than optimum growth conditions. The quality of nutrients in the culture medium is important and changes

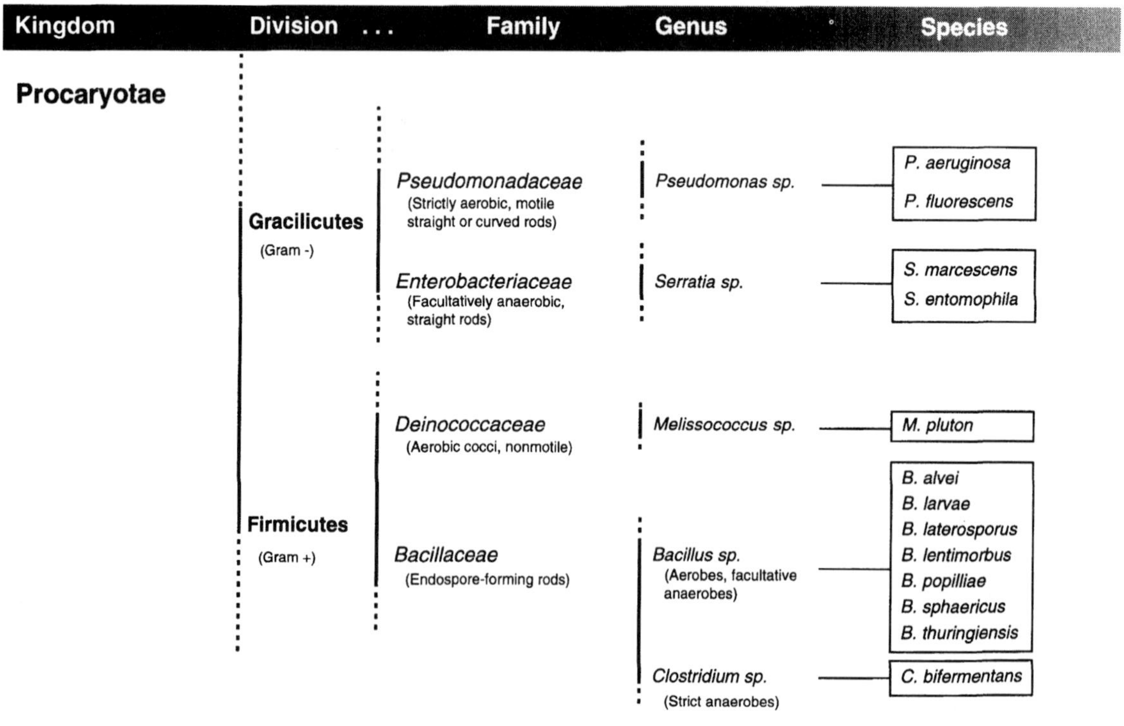

Figure 1 Classification of the most well-known entomopathogenic bacteria. After Krieg (1981), Sneath *et al.* (1986) and Wistreich & Lechtman (1988).

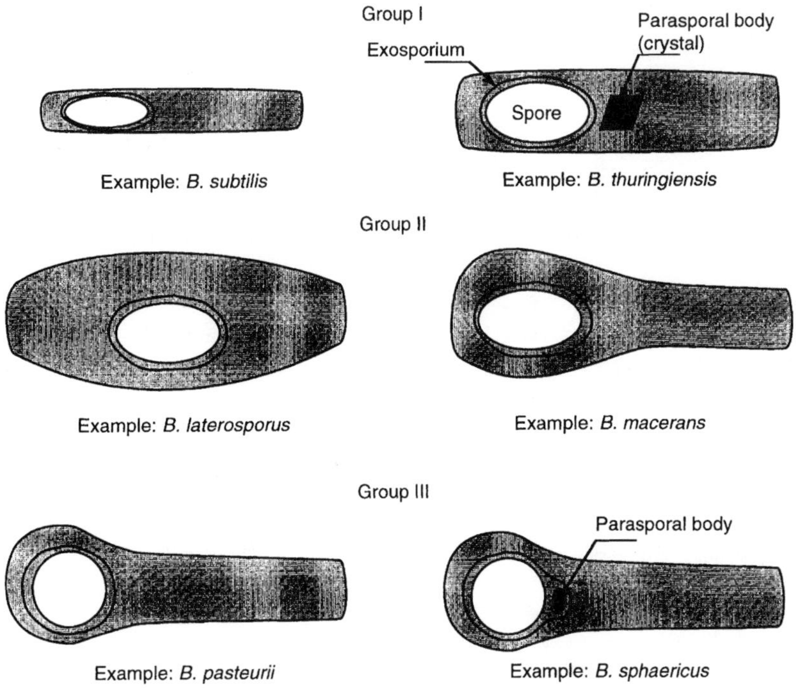

Figure 2 Morphological aspects of *Bacillus* species.

should be made if poor level of sporulated cells or slow growth are observed. Generally, depending on the oxygen requirements of *Bacillus* sp., cultures are grown at ca 30°C on a rotary shaker in UG medium (see Appendix medium no. 14). Sample origin or the need for isolation of particular bacterial strains or species also influence the choice of culture conditions. For example, for isolation of *B. thermophilus,* the culture will be grown at 45°C, for *B. coagulans* at 37°C and for *B. macquariensis* at 4°C.

Selection of Bacillus *sp.*

There are no selective media for *Bacillus* species. Heat treatment of environmental samples and aerobic incubation will allow selection of *Bacillus* from global bacterial flora. The spores (but not the vegetative cells) are heat-resistant. After heat treatment (80°C, 10 min), optimal conditions must be provided in order to induce spore germination and growth.

Note: Reasonable microbiological caution should be exercised when working with environmental samples. One should always keep in mind that the sample might contain human pathogens!

B Keys for identification of major groups of *Bacillus*

The role of identification keys is to facilitate identification of strains using a minimum of phenotypic characteristics. For simplification, 22 of the *Bacillus* species most frequently found in nature, which are well-identified and recognized worldwide are presented in Figure 3. There are, however, more than 70 *Bacillus* species according to the IJSB (International Journal of Systematic Bacteriology) validated bacterial name list. Traditional methods for keying *Bacillus* to the major species are described below (Figure 3). One can also refer to Norris *et al.* (1981) or to the key in Gordon *et al.* (1973). An excellent reference for all aspects of the phenotypic testing of bacteria is Smibert & Krieg (1994).

Note: To ensure proper identification take care to follow recipes for culture media precisely, and pay

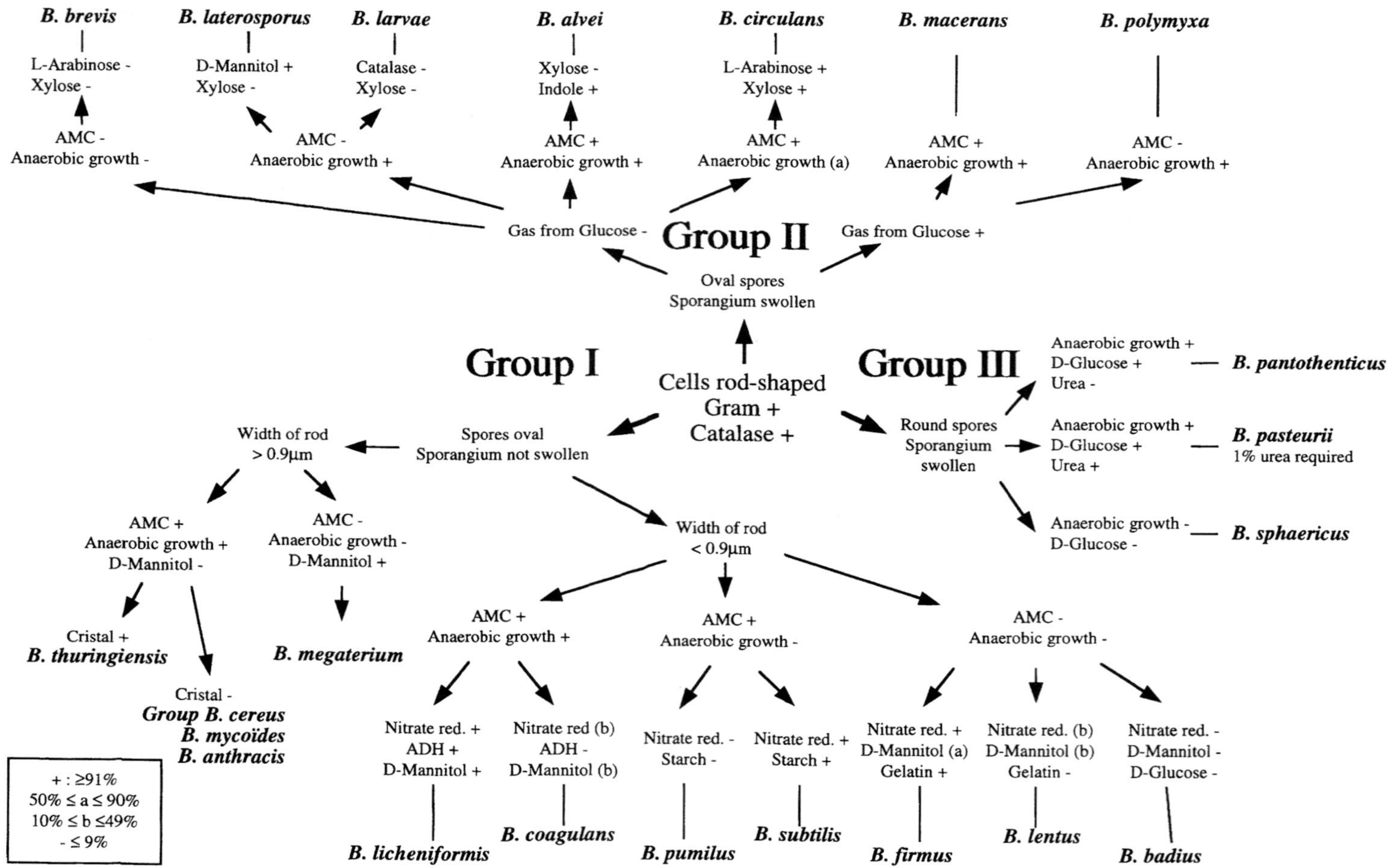

Figure 3 Key for identification of major groups of *Bacillus* species.

particular attention to substrates and incubation conditions.

1. Microscopic observation of bacterial cells

1. After isolation of various colonies from the soil sample (see Section 3), transfer a small part of one colony well-isolated on the agar plate, using a wire needle heated until red then cooled, into a sterile tube containing sporulating medium (10 ml liquid medium in a pyrex tube (diameter 22 mm × height 200 mm), see Appendix medium no. 14 or 17).
2. Grow the whole culture (WC) for 16–24 h (preferably agitated at 250 rpm).
3. Put a drop of culture between a slide and cover-slip. To avoid drying of the suspension and to protect the worker from aerosol, seal the cover-slip to the slide with hot paraffin wax as shown in Figure 4.
4. Observe using direct light microscope (× 1000) under oil immersion.
   Bacillus is a rod-shaped cell with rounded extremities. Record the dimensions and motility of the cell (motility is defined as the cell movements in distinct directions). This movement can be enhanced around air bubbles stuck between slide and cover slip.
5. Plate a drop of WC onto a Petri dish (P) containing nutrient agar as described in Figure 5 in order to isolate and ensure pure colonies. Incubate at the same temperature (30°C) until gram staining step.

Note: For gram staining, the culture must be less than 18 h old, because some strains may lose their gram properties and give false results!

2. Observation of the spores and determination of the Bacillus *group*

Re-examine the whole culture (WC) under a phase-contrast microscope after 48–72 h to study the production of spores. This step is important for determining which additional biochemical tests should be used. Spore morphology is sometimes difficult to classify within one of the three groups. Observation of swollen sporangium is not always obvious, therefore it is necessary to compare with cells still in the vegetative stage. Parasporal bodies

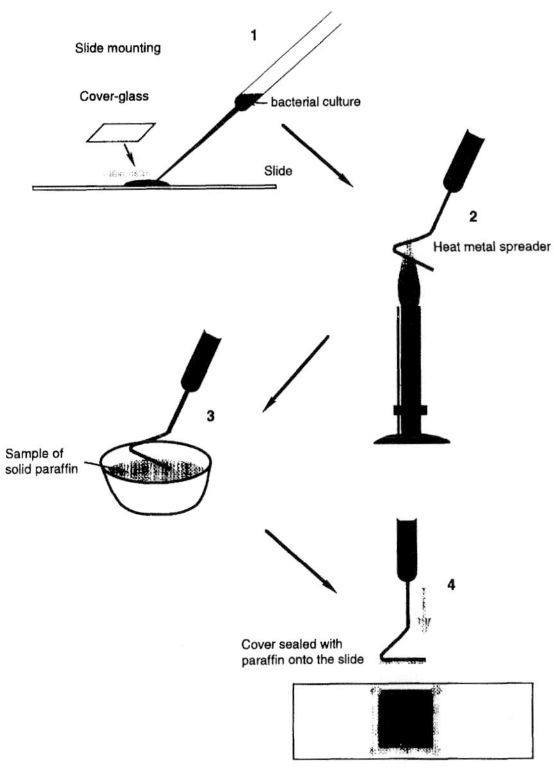

Figure 4 Method for microscopic observation of bacterial suspension.

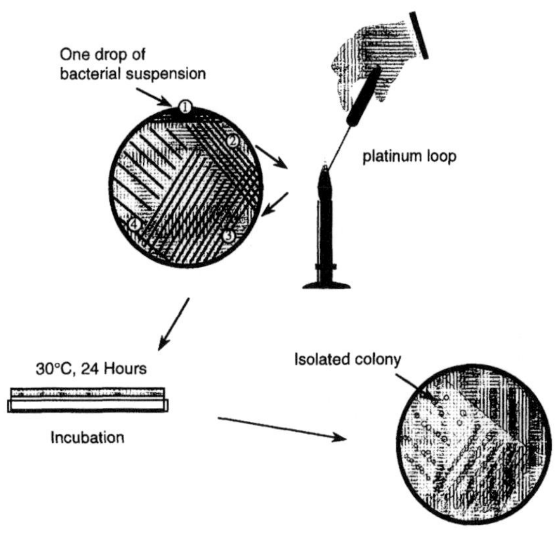

Figure 5 Example of streaking of agar plate for isolation of individual bacterial strains.

of *B. thuringiensis* and *B. sphaericus* are discussed in detail below.

3. Gram staining

1. After 18 h, while still in the exponential growth phase, sample a colony from the nutrient agar plate (P) (see Appendix medium no. 3) under sterile conditions with a wire needle, and mix it with a drop of sterile distilled water on a glass slide for Gram-staining (Figure 6).
2. Follow the diagram on Figure 6 or use the instructions contained in a commercial Gram-staining kit.
3. Observe under direct light microscope (1000 × with oil immersion). The Gram-positive bacteria appear dark violet whereas the Gram-negative bacteria are coloured pink. Most *Bacillus* species are Gram + but some irregular staining might be observed.

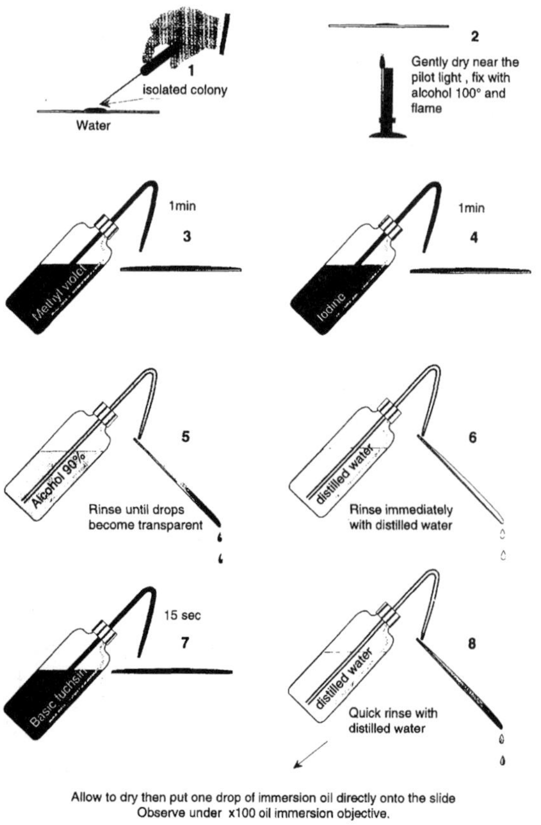

Figure 6 Procedure for Gram staining.

Note: Always use a control for gram staining with two reference strains, one Gram + and one Gram –.

4. Search for presence of catalase

1. Sample another colony from (P) and stir it gently into a Kahn tube (5 ml glass tube) containing 0.5 ml distilled water with 1% Tween 80.
2. Add 0.5 ml hydrogen peroxide (commercial solution 35%). The appearance of oxygen bubbles shows the presence of catalase (Inside the bacteria this enzyme catalyses the reaction: $2H_2O_2 \rightarrow 2H_2O + O_2$ which destroys the H_2O_2 molecules produced during aerobic respiration.

Note: Do not use colonies grown on a medium containing catalase, e.g. blood agar.

According to the results obtained and to Figure 3 other characters can be defined.

5. Culture in anaerobic conditions

It is possible to classify the micro-organisms according to their redox potential necessary to start the culture. The three principal types, easily observable, are:

1. strict anaerobic bacteria which do not grow in presence of oxygen;
2. aero-anaerobic facultative bacteria which can grow with or without oxygen;
3. strict aerobic bacteria which require oxygen to grow.

In this last case, some species are able to use electron acceptors such as nitrates, therefore the medium nutrients should not include mineral acceptors.

1. Melt a semi-solid medium (gelatin agar without nitrate) (see Appendix medium no. 9) in a boiling-water bath for 30 min to eliminate all dissolved oxygen (Figure 7).
2. Maintain the melted medium at 50–55 °C and inoculate from the liquid vegetative WC with a melted, closed Pasteur pipette, swirling from the bottom of the tube to the surface. Cool very quickly by putting the tube into cold water and let the agar solidify (Figure 7).
3. Incubate at 30 °C, for several days if necessary, until colony formation.

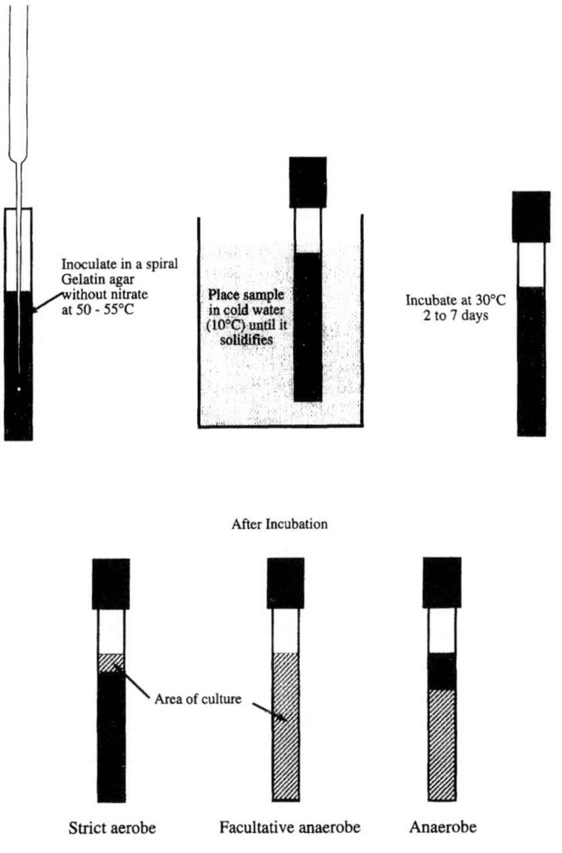

Figure 7 Determination of respiratory type.

6. Reaction of Voges-Proskauer (Barrit's method)

During the intermediate steps of glucose metabolism, acetylmethylcarbinol (AMC) is produced by certain strains of bacteria (from pyruvic acid or during the course of butylen-glycolic fermentation). Detection of this substance is a useful phenotypic test.

1. Inoculate a tube containing MRVP medium (see Appendix medium no. 18 or commercially available) with a single colony and incubate at 30°C for 48 h.
2. In a Kahn tube, mix 1 ml culture, 0.6 ml VP reagent A and add 0.2 ml VP reagent B (Appendix reagent no. 21).
3. Place the open tube on a slant to increase the contact with the air. When the surface changes to a pink or red colour within 10–30 min, the test for AMC production is positive.

7. Study of nitrate reduction (Griess' reaction)

Some bacteria can use nitrates as electron acceptors during anaerobic respiration. Nitrites produced by reduction of nitrates may be further reduced and produce NH_4^+. This test will detect the presence of nitrite.

1. Inoculate a nutrient medium containing 1/1000 of nitrate KNO_3 (see Appendix medium no. 2) with a 24 h-old bacterial suspension and incubate at 30°C until growth is abundant (24–48 h).
2. Add a few drops of reagent NO_3 A (Appendix reagent no. 20).
3. Add a few drops of reagent NO_3 B. Observe the colour change.
   The test is positive for nitrites when the mixture turns from yellow to red.
4. If there is no colour change, add a pinch of Zinc powder (Zo-Bell test). If there are still some nitrates that are not reduced and left in the medium, they will then be transformed into nitrites and the red coloration will appear, implying a negative response to the nitrate reductase test. The yellow colour of the medium proves the complete disappearance of nitrates and reduction beyond the nitrite state, thus the test is positive.

8. Study of carbohydrate metabolism

The production of acid by fermentation of sugars may be masked by liberation of ammonia from proteinaceous material in the medium. For this reason we use a semi-synthetic medium (salts, sugars and yeast) without any peptone (Appendix medium no. 11).

The tested sugars are sterilized by filtration through a 0.22 µm filter, stored at 4°C as a concentrated aqueous solution and are added aseptically to the medium after autoclaving.

Incubation can continue for 15 days. A positive test for sugar fermentation is indicated by a change from violet to yellow. Make sure the bacteria actually grew in the tube so as not to confuse absence of acidity with lack of growth.

The biochemical differentiation of certain species is based on the difference of metabolism of one or two sugars. In general, 20–50 carbohydrates can be tested.

9. Study of proteolytic metabolism (proteolysis of gelatin)

Proteolysis involves the excretion of proteases by the bacterial cells during growth. The most classic test is the activity on gelatin.

1. Inoculate a UG broth medium (Appendix medium no. 15) containing 4% (v/v) gelatin nutrient (Appendix medium no. 8) in tubes. Incubate for at least 48 h at 30°C.
2. To test for proteolysis, place tube in an ice bath. If protease is produced, the medium will not solidify.

10. Study of metabolism of aminoacids (arginine dihydrolase)

This test will determine whether arginine dihydrolase (ADH) is present. It can be produced by certain strains and activated in anaerobic, slightly acidic conditions.

After inoculation of a medium containing arginine (Appendix medium no. 10), cover with ca 1 cm of sterile vaseline oil. As a control use a tube containing the medium without arginine.

In the first step, glucose fermentation decreases the pH of the medium (purple indicator changes to yellow) in the two tubes. In the presence of ADH, arginine is broken up into alkaline products that raise the pH, changing the indicator back to purple.

The ADH test is positive at 48 h if the control tube has changed to yellow and the test tube has reverted to purple.

11. Search for presence of urease and indole

Urea can be degraded by bacteria in alkaline ammonium carbonate which changes the indicator colour. Several techniques and different media can be used to highlight the urease. These techniques differ by their sensitivity. For *Bacillus* strains, the Christensen medium (Appendix medium no. 12) or the urea–indole medium are commonly used. With the latter liquid medium, the urease reaction is quick and it also allows testing for the presence of indole produced from tryptophan. The Christensen medium contains agar, is less buffered and thus more susceptible than the urea–indole liquid medium to the presence of indole.

a. Christensen medium

1. Melt the Christensen medium in a boiling-water bath in two tubes per strain to be tested.
2. Cool to 50–55°C and add, under sterile conditions, 1 ml urea solution to one tube. Label the tube containing urea.
3. Allow to solidify on a slope.
4. Inoculate the two tubes from one colony and incubate at 30°C for 6 days.

In a positive reaction the urea medium must be redder than the control tube. In fact, in certain cases, the alkalinization is only due to the peptone. If the two tubes are a similar colour, the reaction is negative.

b. Urea–indole medium

It is also possible to use a liquid medium, such as Ferguson, which allows for testing for indole as well.

1. Inoculate 1 ml Ferguson medium with several colonies in order to obtain a very dense suspension.
2. Incubate at 30°C for 48 h. Positive reactions (red–violet colour) might be very quick – just a few minutes for some strains of *B. sphaericus.*
3. Indole production is indicated by adding drops of Kovac's reagent after 48 h incubation (Appendix medium no. 19). A red ring on the surface shows the presence of indole.

For organization and time-saving purposes, an abstract of the biochemical tests is recorded in Table 1.

C Determination of *B. thuringiensis* and *B. sphaericus* strains

B. thuringiensis, *B. sphaericus* and *B. popilliae* are the best-studied entomopathogenic bacteria. In the literature, some strains of *B. laterosporus*, *B. alvei*, *B. circulans*, *B. larvae* and *B. brevis* have shown noticeable, although low activity towards invertebrates (Singer, 1973, 1974; Favret & Yousten, 1985). These toxicities were associated with the cell mass or the culture supernatant; no production of toxic inclusion bodies has been recorded.

Within Group I, *B thuringiensis* strains are distinguished from *B. cereus*, *B. mycoïdes* and *B. anthracis* by the ability to produce parasporal crystalline inclusions (also called crystals) during sporu-

Table 1 Summary of biochemical tests.

Tests	*Incubation*	*Reagents (delay)*	*Positive reaction*	*Negative reaction*
Catalase		H_2O_2 (immediate)	Oxygen bubbles	
Anaerobiosis	3–7 days		Growth in entire tube	Growth only on top of the tube
VP	48 h	VP A + VP B (5–30 min)	Pink	Colourless
Nitrate		NO3 A + NO3 B (immediate)	Red	Yellow
		Yellow + Zn (5–15 min)	Yellow	Red
Sugars	2–15 days		Yellow	Violet
ADH	48 h		Violet (control yellow)	Yellow (test and control tube)
Urea	2–6 days		Red	Orange
Indole	48 h	Kovacs (2 min)	Red ring	Yellow ring

lation. These inclusions are responsible for entomopathogenic activity. Formation of the crystal is the criterion for distinguishing between *B. cereus* and *B. thuringiensis*, otherwise they could be considered as the same species. Within Group III, *B. sphaericus* is a strictly aerobic bacterium. Some strains also produce both spores and parasporal bodies inside the exosporium. These strains are pathogenic specifically to mosquito larvae.

For 40 years, serological assays have been carried out using antisera directed against flagella, crystal proteins and whole cells. One widely used classification system for both *B. thuringiensis* and *B. sphaericus* strains is based on the determination of the H-flagellar antigen technique described in de Barjac & Bonnefoi (1962).

1. Serological classification

This technique needs very motile bacterial cultures to prepare flagellar suspensions. These suspensions are titrated against antisera directed against *B. thuringiensis* or *B. sphaericus* reference-type strains of each serotype. Presently *B. thuringiensis* strains are classified within 50 serotypes and subdivided into 63 serovars (Table 2). More than 50 *B. sphaericus* serotypes have been determined but less than a dozen of these contain entomopathogenic strains (de Barjac *et al.*, 1985; Thiery & de Barjac, 1989; Charles *et al.*, 1996).

1. Inoculate a Craigie tube (glass cylinder inside a Pyrex tube) containing semi-solid nutrient agar with a vegetative-stage culture of a new isolate at 30°C for ca 18–24 h (Appendix medium no. 6). The most motile cells will move from the small inside cylinder to the surface of the medium outside the cylinder where they are collected.
2. Inoculate the selected motile bacilli into a 1 litre Erlenmeyer flask containing 100 ml nutrient broth (pH 7.4) (Appendix medium no. 1).
3. Incubate at 30°C with 150 rpm agitation.
4. After 4–6 h incubation ($OD_{650\ nm}$: 0.8–1), add 0.5 ml formaldehyde.

These flagellar suspensions can be kept for several months at 4°C and are used for flagellar agglutination and (when a new serotype) for eventual immunization of rabbits for production of new H-flagellar antisera.

a. Production of flagellar antisera

The H-antigen serum is produced by injection of flagellar suspension ($OD_{650\ nm}$: 1) into rabbits, first subcutaneously (0.5 ml), then intravenously in increasing dosages (1 ml, 2 ml, 4 ml).

1. Twice a week an ear vein is inoculated at progressive dosages for 3 weeks.
2. At eight days after the last injection, production of antibodies is checked by sampling 2 ml of blood from the ear vein. If the titre of antibodies against the flagellar suspension is high enough (1/25 600) and specific, blood is harvested by intracardiac puncture or by sampling from the carotid artery after the rabbit is anaesthetized.

Table 2 Classification of *Bacillus thuringiensis* strains according to H serotype.

H Antigen	*Serovar*	*Code*	*First mention and/or first valid description*
1	*thuringiensis*	THU	Berliner, 1915; Heimpel & Angus, 1958
2	*finitimus*	FIN	Heimpel & Angus, 1958
3a, 3c	*alesti*	ALE	Toumanoff & Vago, 1951; Heimpel & Angus, 1958
3a, 3b, 3c	*kurstaki*	KUR	de Barjac & Lemille, 1970
3a, 3d	*sumiyoshiensis*	SUM	Ohba & Aizawa, 1989
3a, 3d, 3e	*fukuokaensis*	FUK	Ohba & Aizawa, 1989
4a, 4b	*sotto*	SOT	Ishiwata, 1905; Heimpel & Angus, 1958
4a, 4c	*kenyae*	KEN	Bonnefoi & de Barjac, 1963
5a, 5b	*galleriae*	GAL	Shvetsova, 1989; de Barjac & Bonnefoi, 1962
5a, 5c	*canadensis*	CAN	de Barjac & Bonnefoi, 1972
6	*entomocidus*	ENT	Heimpel & Angus, 1958
7	*aizawai*	AIZ	Bonnefoi & de Barjac, 1963
8a, 8b	*morrisoni*	MOR	Bonnefoi & de Barjac, 1963
8a, 8c	*ostriniae*	OST	Gaixin *et al.*, 1975
8b, 8d	*nigeriensis*	NIG	Weiser and Prasertphon, 1984
9	*tolworthi*	TOL	Norris, 1964; de Barjac & Bonnefoi, 1968
10a, 10b	*darmstadiensis*	DAR	Krieg *et al.*, 1968
10a, 10c	*londrina*	LON	Arantes *et al.* (unpublished)
11a, 11b	*toumanoffi*	TOU	Krieg, 1969
11a, 11c	*kyushuensis*	KYU	Ohba & Aizawa, 1979
12	*thompsoni*	THO	de Barjac & Thompson, 1970
13	*pakistani*	PAK	de Barjac *et al.*, 1977
14	*israelensis*	ISR	de Barjac *et al.*, 1977
15	*dakota*	DAK	De Lucca *et al.*, 1979
16	*indiana*	IND	De Lucca *et al.*, 1979
17	*tohokuensis*	TOH	Ohba *et al.*, 1981
18a, 18b	*kumamotoensis*	KUM	Ohba *et al.*, 1981
18a, 18c	*yosoo*	YOS	Lee, H. H. (unpublished)
19	*tochigiensis*	TOC	Ohba *et al.*, 1981
20a, 20b	*yunnanensis*	YUN	Wan-Yu *et al.*, 1981
20a, 20c	*pondicheriensis*	PON	Rajagopalan *et al.* (unpublished)
21	*colmeri*	COL	De Lucca *et al.*, 1984
22	*shandongiensis*	SHA	Ying *et al.*, 1986
23	*japonensis*	JAP	Ohba & Aizawa, 1986
24a, 24b	*neoleonensis*	NEO	Rodriguez-Padilla *et al.*, 1988
24a, 24c	*novosibirsk*	NOV	Burtseva, Kalmikova *et al.* (unpublished)
25	*coreanensis*	COR	Lee *et al.*, 1994
26	*silo*	SIL	de Barjac & Lecadet (unpublished)
27	*mexicanensis*	MEX	Rodriguez-Padilla & Galan-Wong, 1988
28a, 28b	*monterrey*	MON	Rodriquez-Padilla (unpublished)
28a, 28c	*jegathesan*	JEG	Lee L. H. (unpublished)
29	*amagiensis*	AMA	Ohba (unpublished)
30	*medellin*	MED	Orduz *et al.*, 1992
31	*toguchini*	TOG	Hodirev (unpublished)
32	*cameroun*	CAM	Jacquemard, 1990; Juarez-Perez et al.
33	*leesis*	LEE	Lee *et al.*, 1994
34	*konkukian*	KON	Lee *et al.*, 1994
35	*seoulensis*	SEO	Shim (unpublished)
36	*malaysiensis*	MAL	Ho (unpublished)
37	*andaluciensis*	AND	Santiago-Alvarez *et al.* (unpublished)
38	*oswaldocruzi*	OSW	Rabinovitch *et al.* (unpublished)
39	*brasiliensis*	BRA	Rabinovitch *et al.* (unpublished)
40	*huazhongensis*	HUA	Yu Ziniu *et al.*, 1995
41	*sooncheon*	SOO	Lee (unpublished)
42	*jinghongiensis*	JIN	Rong Sen Li (unpublished)
43	*guiyangiensis*	GUI	Rong Sen Li (unpublished)
44	*higo*	HIG	Ohba (unpublished)
45	*roskildiensis*	ROS	Hinrinschen and Hansen (unpublished)
46	*chanpaisis*	CHA	Chanpaisang, 1994
47	*wratislaviensis*	WRA	Lonc, 1995
48	*balearica*	BAL	Iriarte Garcia, 1995
49	*muju*	MUJ	Park, 1995
50	*navarrensis*	NAV	Iriarte Garcia, 1995

After IEBC catalog, Unité des Bactéries Entomopathogènes, Institut Pasteur (1996)

3. Blood is left at room temperature until clot formation.
4. Blood is centrifuged for 5 min at 2000 rpm.
5. Serum is harvested, its quality is checked by immunodetection and it is stored in sterile vials at 4°C or –20°C.

b. Serological technique

Each bacterial suspension is assayed against each serum directed against all type strains of each serotype (Figure 8). This means that for *B. thuringiensis,* 50 tubes multiplied by the appropriate number of dilutions per bacterial suspension (Table 2).

1. To each 5 ml plastic tube add 1.8 ml of 0.15 M NaCl and mix with 200 µl flagellar antiserum followed by two (1/1) dilutions (Figure 8). From each tube, transfer 100 µl into a haemolysis tube containing 900 µl bacterial suspension. Three dilutions per serum are thus obtained 1 : 100; 1 : 200; 1 : 400. A control tube contains 900 µl bacterial suspension and 100 µl 0.15 M NaCl.
2. After 2 h incubation at 37°C, agglutination is observed. In that case, the supernatant is clear and a white pellet (due to sedimentation of antigen–antibody complex) is produced. If a reaction is noticed within the three tubes, titration of the bacterial suspension against a series of serum dilutions is performed as shown in Figure 8 until dilution 1 : 25 600.
3. Record whether agglutination is noticed after 2 h at 37°C.

Note: It is important not to shake or disturb the tube-holder, as this would compromise the evaluation of agglutination.

The agglutination rate is expressed when total agglutination is observed, in other words the supernatant is extremely clear.

A bacterial suspension may agglutinate with two antisera. In that case, consider the serotype to be that antiserum giving agglutination at the highest dilution. When a serum agglutinates with several bacterial strains, these possess a common antigenic factor. But they may differ by one or more other antigenic factors present in certain strains. The saturation technique of sera can then be used to differentiate these strains (de Barjac & Bonnefoi, 1962).

2. Classification of B. thuringiensis *according to protein composition*

Toxic strains from the various serotypes of *B. thuringiensis* previous to 1977 were all pathogenic towards lepidopteran larvae. In 1977 the discovery of *B. thuringiensis* serovar *israelensis* (*B.t.i.*) toxic to mosquito larvae (Goldberg & Margalit, 1977) enhanced the entomopathogenic potential of *B.t.* strains. At that time, serological classification still played an important role in pathogenic classification as only *B.t.i.* strains were toxic to Culicidae. After 1983, when a strain from *B.t. morrisoni* was found pathogenic to Coleoptera larvae (Krieg *et al.*, 1983), serological classification, although still in use as a basic method to classify *B.t.* strains, could no longer be related to pathogenicity. Therefore, it was necessary to find new classification methods related to pathotypes.

Since then, studies have shown that, within a serotype, different activity spectra can be found in diverse strains. For example, in serotype *morrisoni,* pathogenic activity against Diptera, Coleoptera or Lepidoptera occurs in different strains. These strains produced parasporal bodies which contain different (though related) proteins. When bacterial culture conditions are optimal, the crystal morphology before and after cell lysis can be observed microscopically, giving an idea of the eventual pathogenicity against a certain order of insects (Table 3).

Until recently, classification based on crystal components included five classes of Cry proteins as shown in Table 3.

With the knowledge and identification of the toxin genes of *B. thuringiensis*, one can refer to a more detailed classification system based on *cry* toxin genes (Höfte & Whiteley, 1989; Lereclus *et al.*, 1989) and their subclasses. The complexity of gene sequences has led to a new nomenclature based on the sequence identity of various *cry* genes (Crickmore *et al*, 1995).

D Overview of other techniques

1. API Gallery for Bacillus *sp identification (API 20E + 50 CHB)*

Several techniques, derived from the biochemical identification gallery, are generally used in order to avoid more time-consuming techniques.

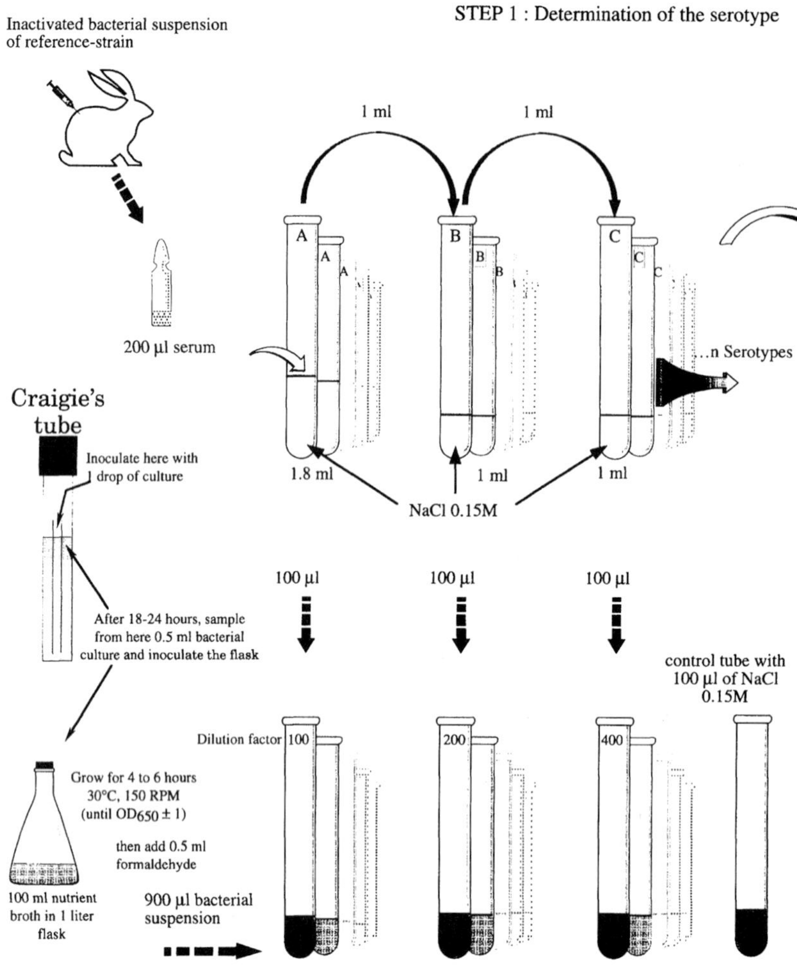

Figure 8 The H-flagellar serotyping technique used for *B. thuringiensis* and *B. sphaericus* strains.

These micro-methods become more and more mechanical which is sometimes prejudicial to the quality of results (number of identified species, false negatives, etc.) (Stager & Davis, 1992). As far as *Bacillus* species are concerned, 60% can be identified with these tools (usually the most important species in the medical microbiological domain). The API® identification system is one example of a rapid identification technique, with results which have been evaluated by several authors (Logan & Berkeley, 1984). It is composed of two strips with a range of microtubes containing various dehydrated substrates. The medium is reconstituted by bacterial suspension. After 48 h incubation, biochemical reactions are revealed by colour indicator or addition of reagents. Identification is made using adapted software or by comparison with a reference library. The most important aspects of this method are rapid results, small space requirements and cost reduction of medium preparation. A new, unknown species cannot be identified using the API system. In this case it will be necessary to use other testing methods.

2. *Hybridization DNA–DNA*

Molecular analysis of the genome has improved understanding of phylogenetic relationships and placed taxonomy on a more rational basis. For 30

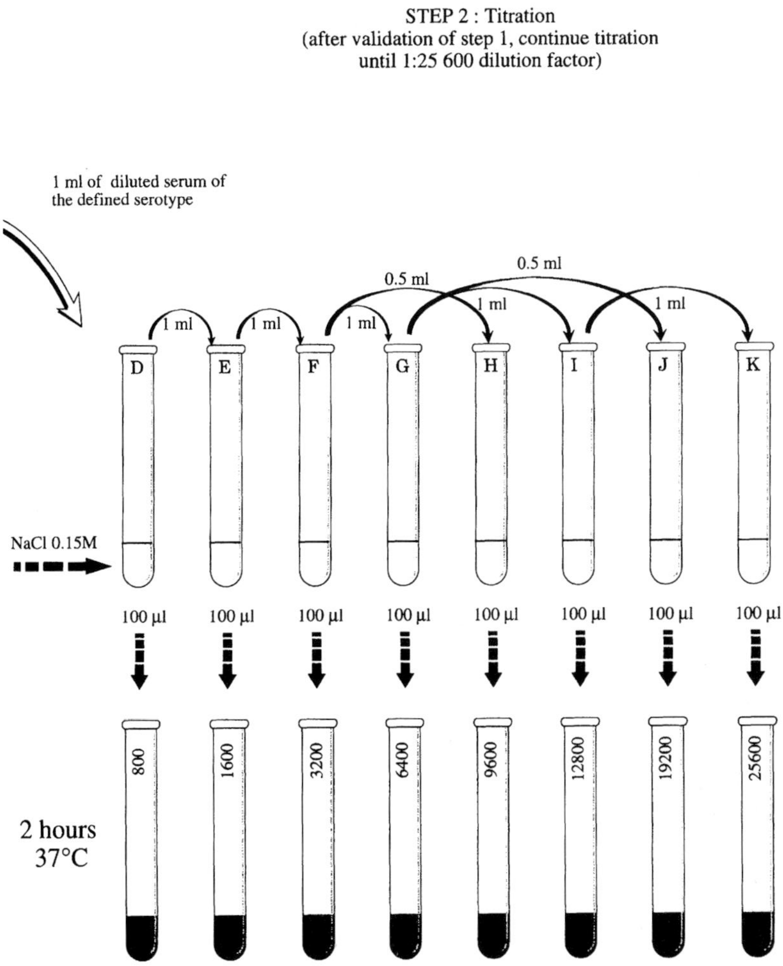

Figure 8 (continued)

years, this technique has allowed the definition of the various bacterial species. The species defined by this technique have not always agreed with those established earlier on phenotypic characteristics (Grimont, 1988). Numerous studies have been performed in order to distinguish *thuringiensis, cereus* and *anthracis* species as well as between serotypes of *B. thuringiensis* (Kaneko *et al.*, 1978; Nakamura, 1994; Carlson *et al.*, 1994) without challenging these species or using H-antigens as a basis of *B. thuringiensis* classification. The genetic relationship based on DNA homology has been worked out with *B. sphaericus* strains (Krych *et al.*, 1980). Six homology groups have been identified, one subgroup, IIA, contains all strains pathogenic to mosquito larvae.

3. Classification by bacteriophage typing

Sensitivity of bacteria to certain bacterial viruses called phages allows classification of *B. thuringiensis* and *B. sphaericus* according to this susceptibility. There are 11 bacteriophages that have been used for *B. sphaericus* and 14 for *B. thuringiensis*, but phage typing of the latter is inconsistent with serotyping and does not permit classification. Bacteriophages can be isolated from soil or occasionally from the bacteria themselves.

The response to the bacteriophages of *B. sphaericus* allows differentiation of *B. sphaericus* strains into five groups (Yousten, 1995 personal communication). In this species, there are some relationships

Table 3 Correlation between class of crystal protein, type of parasporal crystalline inclusion and entomopathogenic activity.

Crystal toxin class		*Mol. wt. (kDa)*[a]	*Structure*	*Serotype*[b] *example*	*Activity spectra*
Cry I	A	131 or 133	bipyramidal	1	Lepidoptera
	B	138	bipyramidal	3	Lepidoptera
	C	135	bipyramidal	6	Lepidoptera
	D	133	bipyramidal	7	Lepidoptera
	E	133	bipyramidal	7	Lepidoptera
Cry II	A	71	cuboidal	3	Lepidoptera/Diptera
	B	71	cuboidal	3	Lepidoptera
Cry III	A	73	bipyramidal	8	Coleoptera
	B	74	bipyramidal	8	Coleoptera
	C	129	bipyramidal	8	Coleoptera
	D	73	bipyramidal	8	Coleoptera
Cry IV	A	134	heterogeneous	14	Diptera
	B	128	heterogeneous	14	Diptera
	C	78	heterogeneous	14	Diptera
	D	72	heterogeneous	14	Diptera
Cry V		81	bipyramidal	3	Lepidoptera/Coleoptera
Cry VI		?	?	?	Nematodes
Cyt	A	28	heterogeneous	14	*cytolytic* (blood cells)

After Höfte & Whiteley (1989); Lereclus *et al.* (1989).
[a] Molecular weight in kDa.
[b] Crystal toxin type and serotype are not necessarily strictly linked; for example the PG-14 isolate of *B. thuringiensis morrisoni* (serotype 8a, 8b) produces CryIV toxins whereas the 256-82 isolate from the same serotype produces CryIII toxin.

between DNA homology group, H-antigen serology and bacteriophage typing classification (Yousten *et al.*, 1980).

4. Gas chromatography of cellular fatty acids

The Hewlett–Packard Microbial Identification System (MIS) is composed of a gas chromatograph, an automatic injector and a computer analysis system. The principle is to prepare methyl esters of cellular fatty acids which are then separated by gas chromatography on a 25 m capillary column and to send the results from the flame ionization detector (FID) to the computer for analysis. Identification software (Sherlock, Microbial ID, Inc, Newark, DE) compares the profiles obtained by an evolutive database that groups most of the profiles of known bacterial genus and species (as well as yeasts and fungi). The analysis of cellular fatty acids is now well recognized as a method for identification. It takes into account the true standardization of methods and culture conditions used (as profiles and proportion of fatty acids vary according to different growth stages).

Frachon *et al.* (1991) showed that the fatty acid composition of *B. sphaericus* strains can be linked to mosquitocidal activity and can thus differentiate pathogenic from non-pathogenic strains within two different groups. Although with *B. thuringiensis* strains no evidence of such a relationship was observed, nevertheless, this tool is very useful for distinguishing two isolates belonging to a same serotype of *B. thuringiensis*.

3. ISOLATION

A Prospecting

B. thuringiensis and *B. sphaericus* are soil bacteria, but they are also abundant in insects and, being cosmopolitan, can be found in any biotope.

A general search for *Bacillus* spp. consists of random sampling of soil, leaves, trees and dead insects (larvae or adults). When searching for a particular species pathogenic towards a specific insect, one must look for diseased insects in their natural biotope. For example, for *B. sphaericus* toxic to mosquito larvae, samples of water, mud and substrates from breeding sites should be examined. In fact the majority of mosquitocidal strains were isolated from the breeding site environment; therefore sampling must be relevant to the ecology of the organisms in the biotope.

Nevertheless, a pathogen can also be isolated from a non-susceptible insect to that pathogen. For example, *B. sphaericus* strain 2362 was isolated from an adult of *Simulium damnosum* in Nigeria, although it is strictly pathogenic to Culicidae larvae. So there are no hard and fast rules for prospecting; thoroughness, luck and good eye-sight are valuable!

B Isolation

It is important to record all information concerning sampling, especially when different people are involved in surveys. Insect or animal name and physiological state, geographic zone, type of biotope, and date must be noted as well as each important detail related to sampling. One must record whether earlier bacterial treatments were performed in the area of sampling, even years before (spores are very resistant). Check also to ensure that sampling vials are sterile.

Each sample must be collected from the field in a separate bag or tube to avoid contamination.

After arrival in the laboratory, each sample should be divided into 4 g lots and distributed into tubes containing 10 ml peptonated sterile water and vortexed.

Note: One must keep in mind that there might be some human pathogenic bacteria in each soil sample!

For water samples, a large volume can be concentrated through a 0.22 μm (Millipore type) filter. The filter is then placed in a sterile water peptonated tube and agitated. Insects should be crushed in a Potter tube. The samples should then be treated as follows:

1. Heat-shock the 10 ml samples at 80°C for 10 min to kill all vegetative forms. Spores of *Bacillus* species are heat-resistant whereas vegetative cells of bacilli and non-spore-forming bacteria are not.
2. Plate onto a Petri dish containing a sporulation medium such as MBS medium (Appendix medium no. 17) or normal (UG or HCT) medium (Appendix media nos. 13, 14 and 15) and incubate for 24 h at 30°C. The former medium allows the growth of most of *Bacillus* species, the latter is not as rich and can be used for *B. thuringiensis*. Of course these conditions do not allow for selection of all bacteria, just the most commonly known entomopathogenic species. If it is necessary to isolate all *Bacillus* species present in a sample, plate on to several different media and incubate at different temperatures as some *Bacillus* species are thermophilic whereas others are psychrophilic, and grow only at low temperatures.
3. Grow each colony in a 10 ml suspension (UG with glucose or MBS media) on an orbital shaker for 48 h at 30°C. Check microscopically for purity, and stage of culture.
4. Plate a loop of the suspension to isolate single pure colonies (Figure 3). Incubate for 24–48 h. Use single colonies for preparing bacterial suspensions for further identification, characterization and bioassay.

Colony morphology can help to distinguish a *B. thuringiensis* colony from, for example, a *B. sphaericus* colony. The former forms white, rough colonies which spread out and can expand over the plate very quickly whereas *B. sphaericus* presents white, small, round, brilliant and smooth colonies that do not spread on the plate (Plate 17).

For further characterization, colonies are grown in a sporulating medium. Prepare a glucose stock-solution (30% w/v), filtered on a membrane filter 0.22 μm and store at 4°C.

For *B. thuringiensis*, cultures are grown in tubes or in flasks containing a 1% final concentration of the glucose solution as a carbon source. Procedures for isolating *B. popilliae* and related species and *Serratia* spp. are presented in Chapter III-4.

4 CULTIVATION

A On artificial media

B. thuringiensis strains can use a carbohydrate source to grow whereas *B. sphaericus* is unable to metabolize carbohydrates and thus needs proteinaceous media. Amino acids are the preferred nitrogen source for *B. sphaericus,* which requires the vitamins biotin and thiamine to grow and calcium and manganese ions for sporulation (Lacey, 1984; Russell *et al.*, 1989).

Natural resistance of *Bacillus* strains to certain antibiotics such as chloramphenicol, streptomycin, bacitracin (*B. sphaericus*) and ampicillin (*B.t.i.*, for example) can be used to select isolates and follow the fate of these bacteria in the field (Yousten *et al.*, 1982; Kalfon *et al.*, 1986). Different media with or without antibiotics can be used: MBS medium (Kalfon *et al.*, 1983), BATS medium (Yousten *et al.*, 1985) and NYSM (Yousten & Wallis, 1984) to grow *B. sphaericus*. *B. thuringiensis* is usually grown on a UG medium (de Barjac & Lecadet, 1976) or HCT medium. Many different parameters can interfere with growth, good rate of sporulation and production of entomopathogenic toxins (Yousten *et al.*, 1984). In some developing countries, *Bacillus* sp. has been produced at a very low cost using local agricultural residues or waste (Obeta & Okafor, 1984).

In this chapter we give only one medium recipe for each of the two bacteria *B. sphaericus* and *B. thuringiensis*. These media are commonly used at the Pasteur Institute, Paris, France, and have been proven to encourage good growth, sporulation and production of parasporal bodies in both cases (see Appendix for bacterial culture media and reagents). Procedures for *in vivo* production of *B. popilliae* and related species are presented in Chapter III-4.

B Example of *B. thuringiensis* culture

1. Preparation of a preculture

From a stock-tube (see Section 5) or a colony inoculate a tube containing 10 ml UG medium to serve as a preculture (see Appendix media nos. 13 and 14).

After incubation on a shaker for 48 h at 30°C and observation under a microscope, the sporulated preculture is heat-shocked at 78–80°C for 10 min to kill all vegetative forms in order to obtain a better homogeneity of growth of the new culture to be inoculated with the preculture.

There should be a maximum 200 ml of culture medium in a 1 litre Erlenmeyer flask to allow for sufficient aeration.

A few drops of preculture (1 ml per 200 ml) is enough for inoculation. However, when the bacterial culture volume is higher using a fermentor, an intermediate preculture is inoculated from the heat-shocked culture (for example 200 ml for 4.5 litre of bacterial culture). After 5–6 h incubation at 30°C, the fermentor will be inoculated.

2. Bacterial growth

1. Prepare a 1 l Erlenmeyer flask containing 100 ml UG medium and close the flask with a cotton plug. After sterilization (121 °C 15 min), add 1% final concentration of glucose (which has been previously filtered or autoclaved 105 °C 10 min).
2. Inoculate the flask directly with the preculture and incubate with orbital agitation for 48–72 h at 30°C until cell lysis is complete.
3. Check under a phase-contrast microscope for lysis of the cells, the sporulation rate and presence of protein parasporal bodies.
4. Centrifuge the FWC (final whole culture) for 15–20 min at 7000 rpm.
5. Resuspend the pellet of spores–crystals in 0.5 M NaCl for 15 min to avoid exoprotease activity.
6. Centrifuge.
7. Resuspend the pellet in distilled or demineralized water twice.
8. Centrifuge.
9. Then either:
   (a) Resuspend the pellet in a water volume identical to the initial one. Dispatch in aliquots and freeze at –20°C. Bioassay on the insect target as other characterization of the strain can also be made from dilutions of one aliquot; or
   (b) Keep the pellet of spores–crystals frozen and use to prepare powder, or freeze dried samples using a lyophilizor. Coprecipitation with the lactose–acetone technique of Dulmage *et al.* (1970) is commonly used to prepare acetone powders, which are easy to produce and can be kept at 4 °C until use.

3. Cell and spore counting

Before centrifugation the number of cells and spores can be evaluated by counting the number of cells (before heat-shock) and number of spores (after heat-shock) present by plating series of bacterial dilutions (0.1 ml per plate) onto solid medium in Petri dishes.

Three Petri dishes are plated per dilution and incubated for 24 h at 30°C. Usually under those conditions, a *B. thuringiensis* culture produces ca 10^9–10^{10} cells/ml of original suspension with roughly 100%

sporulation, whereas *B. sphaericus* culture produces ca 10^8–10^9 cells/ml with a lower and variable sporulation rate. Plates that produce 30–150 colonies provide the most reliable counts.

Counting gives an idea of the growth of a culture and can be used, as can the optical density, to compare various cultures but it is not a valuable tool for evaluating the entomopathogenic potential of a strain as it does not reflect precisely the quantity of parasporal inclusions responsible for toxicity. In fact, the majority of bacterial cells produce parasporal inclusions but one cannot be sure whether each cell produces one, two or no crystals.

The larvicidal activity is sometimes expressed as a dilution of the final whole culture though this can be misleading if different strains have achieved different final populations. It may also be expressed as the quantity of protein (method of Bradford, 1976 or Lowry, 1951) or in grams of powder prepared from the strain.

5 PRESERVATION

Storage of bacterial strains is an important function of any laboratory. Preservation must maintain both culture viability and maintain cultural characteristics. Numerous problems may appear either during manipulation, or during storage (genetic mutation, plasmid loss, loss of characteristics, selection of resistant populations) until a strain is completely lost.

No method is 100% reliable and some are more or less adapted to certain species. Other parameters can influence the choice of a particular technique, such as cost, dispatching, mailing, number of strains to store, management of stocks, etc.

A Short-term storage

The simplest solution for low-cost equipment is to reinoculate strains on a new nutrient medium. The first advantage is the immediate availability of the strain, but the inconveniences are numerous, such as contamination and loss of certain characteristics.

Maintenance of final whole cultures (FWC) at 4 °C is satisfactory for one or two weeks. Most FWC are usually stored at –20°C for months before use. It is better to stock the culture under bacterial pellet form to avoid eventual degradation of protein by excreted proteases.

When strains are plated on nutrient agar, colonies on Petri dishes covered with Parafilm® can be kept at 4 °C. Each month, reinoculate a fresh agar plate as shown in Figure 5. This technique is compatible with long-term storage and convenient for internal laboratory use.

B Long-term storage

Freezing for several months can be used as a form of long-term storage for internal laboratory use. Isolate colonies on a nutrient agar plate under optimal growth conditions and sample a large quantity of colonies that have been homogenized in 5 ml of a 17% final concentration of sterile glycerol solution. Glycerol acts as a cryoprotectant. Place 500 µl aliquots into cryotubes and freeze immediately at –80°C. Cultures can also be frozen in glycerol at –20°C.

Always freeze several samples as the steps of freezing and thawing are harmful for bacteria.

1. Storage of spores of Bacillus *sp. on filter paper*

The storage of strains takes into account the ability of sporulated bacteria to resist dessication. The Pasteur Institute *Bacillus* collection is kept in spore form. Spores can germinate after more than 30 years of storage.

1. Heat the final whole cultures at 80 °C for 12 min to select for spores. Put a drop of each heated culture on to a piece of sterile filter paper previously placed in a long thin tube.
2. Let the filter paper dry in the tube for 2–3 weeks at 37 °C.
3. Seal the tube under sterile conditions by melting the glass shut.
4. Keep the stock-tube at 4 °C until use.
5. When used, file the tube and pour the filter paper into the appropriate medium for growth.

These stock-tubes are convenient for sending strains. Long-term storage of *B. popilliae* spores on microscope slides is presented in Chapter III-4.

2. Freeze drying

This technique is considered the most efficient for long-term storage and conservation of strain characteristics. Distribution of lyophilized material is practical for supplying strains to other workers as no special storage conditions are required.

It is particularly useful for badly sporulated *Bacillus* strains, for oligosporogenous strains or for strains with spores which are less time-resistant. Apart from collection and storage, it is also used for bacterial products in order to avoid the problem of hydrophobicity sometimes linked with bacterial powder. The equipment, composed of a vacuum pump with a quick freeze dryer system, is frequently found in laboratories. The steps are as follows.

1. Grow strains in optimum conditions, preferentially on an agar plate.
2. Harvest colonies and homogenize them in sterile physiological saline containing 20% horse serum.
3. Place 100 µl into 1–2 ml sterile lyophilized tubes with a pipette, avoid dropping any of the suspension on the outer edge of the tube. Close the tube.

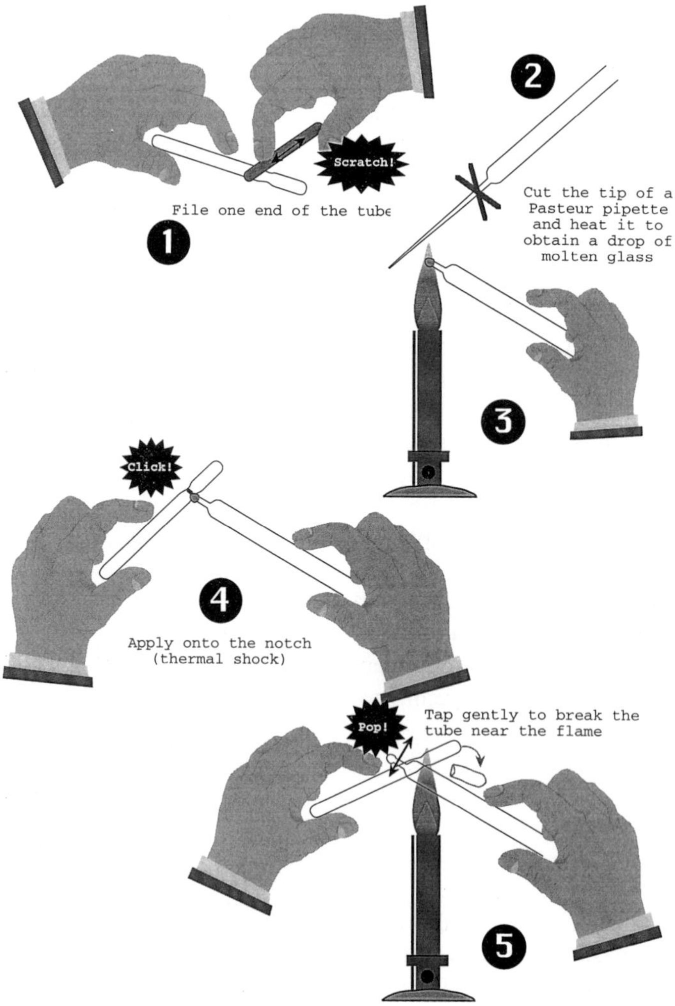

Figure 9 Method for safe opening of lyophilized tube.

Note: The freeze-dryer operator must be experienced with the equipment. It should not be done for the first time without supervised instructions.

4. Immerse the tubes deeply into a mixture of ethanol (100%) and solid CO_2 so that they freeze very quickly (use protective glasses and cryo-gloves).
5. Remove the cotton plug and attach the tubes to the freeze-dryer as explained in the user instructions.
6. After lyophilization, seal the tubes with flame under a vacuum.
7. Each lot of tubes must be checked systematically. Growth should be tested from one tube in order to check viability, purity and characterisitcs of the strain. To reduce the aerosol problems encountered when opening a lyophilized tube, file one end of the tube, apply a previously heated Pasteur pipette on it to obtain a circular split, and gently tap the end of the tube in front of a flame to open the tube completely. Add drops of nutrient broth and then use as a bacterial suspension (Figure 9). Lyophilized tubes can be stored at room temperature, 4 °C or below 20 °C for many years in an active stage. When growth is needed, the germination rate of spores of *B. thuringiensis* is enhanced by heat shock (65 °C for 30 min; Krieg, 1981).

3. *Production of primary powders and formulations*

When a strain has proved its entomopathogenic potential, experimental formulations can be produced in order to check its toxicity against a broad range of insect species under controlled conditions in the laboratory or in the field. Primary powders, such as acetone-precipitated powders (Dulmage *et al.*, 1970), can be made in the laboratory, or different formulations such as micronized suspensions, emulsifiable concentrates, wettable powders, granules and briquets can be made by commercial producers. All of these formulations are made according to the target insect and its biotope and must be adapted to the mode of application requested. Some of these formulations are quite stable in terms of toxicity when maintained in appropriate conditions. In fact, formulations must resist high temperatures (especially those used in tropical countries), degradation (addition of preservatives, UV protectants, adjuvants to increase adherence to foliage, baits, dispersants for water breeding site treatments, etc.). These bacterial products can be stored for months or years if the parameters are regularly controlled, most especially the toxicity of the preparation. More detailed information on the preservation, large-scale production, bioassay and formulation of *Bacillus* entomopathogens is available in a World Health Organization booklet (Dulmage *et al.*, 1990). Procedures for bioassay of primary powders and formulated bacteria against terrestrial, aquatic and soil-dwelling insects are presented in later chapters.

REFERENCES

Bradford, M. M. (1976) A rapid and sensitive method for the quantitation of microgram quantities of protein utilizing the principle of protein-dye binding. *Anal. Biochem.* **72**, 248–254.

Brenner, D. J. (1973) Deoxyribonucleic acid reassociation in the taxonomy of enteric bacteria. *Int. J. Syst. Bacteriol.* **23**, 298–307.

Carlson, C. R., Caugant, D. A. & Kolsto, A-B. (1994) Genotypic diversity among *Bacillus cereus* and *Bacillus thuringiensis* strains *Appl. Environ. Microbiol.* **60**, 1719–1725.

Charles, J-F., Nielsen-Leroux, C. & Delécluse, A. (1996) *Bacillus sphaericus* toxins: Molecular biology and mode of action. *Annu. Rev. Entomol.* **41**, 451–472.

Crickmore, N., Zeigler, D. R., Feitelson, J., Schneft, E., Lambert, B., Lereclus, D., Gawron-Burke, C. & Dean, D. H. (1995) Revision of the nomenclature for the *Bacillus thuringiensis cry* genes. *Annual Meeting of the Society for Invertebrate Pathology*, Ithaca, NY, 16–21 July, pp. 14.

de Barjac, H. & Bonnefoi, A. (1962) Essai de classification biochimique et sérologique de 24 souches de *Bacillus* du type *B. thuringiensis*. *Entomophaga*, **7**, 5–31.

de Barjac, H. & Frachon, E. (1990) Classification of *Bacillus thuringiensis* strains. *Entomophaga* **35**, 233–240.

de Barjac, H. & Lecadet, M-M. (1976) Dosage biochimique de l'exotoxine thermostable de *B. thuringiensis* d'après l'inhibition d'ARN-polymérases bactériennes. *C. R. Acad. Sci. Paris* **282**, 2119–2122.

de Barjac, H. Larget-Thiery, I., Cosmao Dumanoir, V. & Ripouteau, H. (1985) Serological classification of *Bacillus sphaericus* strains in relation with toxicity to mosquito larvae. *Appl. Microbiol. Biotechnol.* **21**, 85–90.

Dulmage, H. T., Correa, J. A. & Martinez, A. J. (1970) Coprecipitation with lactose as a mean of recovering the spore-crystal complex of *Bacillus thuringiensis*. *J. Invertebr. Pathol.* **15**, 15–20.

Dulmage, H., Yousten, A., Singer, S. & Lacey, L. (1990)

Guidelines for production of *Bacillus thuringiensis* H14 and *Bacillus sphaericus*. UNDP/World Bank/WHO booklet TDR/BCV/90.1 Geneva.

Favret, M. E. & Yousten, A. A. (1985) Insecticidal activity of *Bacillus laterosporus* *J. Invertebr. Pathol.* **45**, 195–203.

Frachon, E., Hamon, S., Nicolas, L. & de Barjac, H. (1991) Cellular fatty acid analysis as a potential tool for predicting mosquitocidal activity of *Bacillus sphaericus* strains. *Appl. Environ. Microbiol.* **57**, 3394–3398.

Goldberg, L. J. & Margalit, J. (1977) A bacterial spore demonstrating rapid larvicidal activity against *Anopheles sergentii, Uranotaenia unguiculata, Culex univitattus, Aedes aegypti* and *Culex pipiens*. *Mosq. News* **37**, 355–358.

Gordon, R., Haynes, W. & Pang, C. (1973) *The genus Bacillus*. US Dept. of Agriculture Handbook no. 427.

Grimont, P. A. (1988) Use of DNA reassociation in bacterial classification. *Can. J. Microbiol.* **34**, 541–546.

Höfte, H. & Whiteley, H. R. (1989) Insecticidal crystal proteins of *Bacillus thuringiensis*. *Microbiol. Rev.* **53**, 242–255.

Kalfon, A., Larget-Thiéry, I., Charles, J-F. & de Barjac, H. (1983) Growth, sporulation and larvicidal activity of *Bacillus sphaericus*. *Eur. J. Appl. Microbiol. Biotechnol.* **18**, 168–173.

Kalfon, A., Lugten, M. & Margalit, J. (1986) Development of selective media for *Bacillus sphaericus* and *Bacillus thuringiensis* var. *israelensis*. *Appl. Microbiol. Biotechnol.* **24**, 240–243.

Kaneko, T., Nozaki, R. & Aizawa, K. (1978) Deoxyribonucleic acid relatedness between *Bacillus anthracis*, *Bacillus cereus* and *Bacillus thuringiensis*. *Microbiol. Immunol.* **22**, 639–641.

Krieg, A. (1981) The genus *Bacillus* insect pathogens. In *The Prokaryotes, a handbook on habitats, isolation, and identification of bacteria* (eds M.P. Starr, H. Stolp, H. G. Trüper, A. Balows & H. G. Schlegel) Vol II, pp. 1742–1755. Springer-Verlag, Berlin, Heidelberg, New York.

Krieg, A., Huger, A. M., Langenbruch, G. A. & Schnetter, W. (1983) *Bacillus thuringiensis* subsp. *tenebrionis*: ein neuer, gegenüber Larven von coleopteran wirksamer Pathotyp. *Z. Ang. Entomol.* **96**, 500–508.

Krych, V. K., Johnson, J. L. & Yousten, A. A. (1980) Deoxyribonucleic acid homologies among strains of *Bacillus sphaericus*. *Int. J. Syst. Bacteriol.* **30**, 476–484.

Lacey, L. A. (1984) Production and formulation of *Bacillus sphaericus*. *Mosq. News* **44**, 153–159.

Lawrence, D., Heitefuss, S. & Seifert, H. (1991) Differentiation of *Bacillus anthracis* from *Bacillus cereus* by gaz chromatographic whole-cell fatty acid analysis. *J. Clin. Microbiol.* **29**, 1508–1512.

Lereclus, D., Bourgouin, C., Lecadet, M-M., Klier, A. & Rapoport, G. (1989) Role, structure, and molecular organization of the genes coding for the parasporal δ-endotoxins of *Bacillus thuringiensis*. In *Regulation of procaryotic development* (eds I. Smith, R. A. Slepecky & P. Setlow), pp. 255–276. American Society for Microbiology, Washington, DC.

Logan, N. A. & Berkeley, R. C. W. (1984) Identification of *Bacillus* strains using the API system. *J. Gen. Microbiol.* **130**, 1871–1882.

Lowry, O. H., Rosebrough, N. J., Farr, A. L. & Randall, R. J. (1951). Protein measurement with the Folin phenol reagent. *J. Biol. Chem.* **193,** 265–275.

Miteva, V., Abadjieva, A. & Grigorova, R. (1991) Differentiation among strains and serotypes of *Bacillus thuringiensis* by M13 DNA fingerprinting. *J. Gen. Microbiol.* **137**, 593–600.

Nakamura, L. K. (1994) DNA relatedness among *Bacillus thuringiensis* serovars. *Int. J. Syst. Bacteriol.* **44**, 125–129.

Norris, J. R., Berkeley, R. C. W., Logan, N. A. & O'Donnell, A.G. (1981) The genera *Bacillus* and *Sporolactobacillus*. In *The Prokaryotes, a handbook on habitats, isolation, and identification of bacteria*, (eds M. P. Starr, H. Stolp, H. G. Trüper, A. Balows & H. G. Schlegel), Vol II, pp. 1711–1742. Springer-Verlag, Berlin, Heidelberg, New York.

Obeta, J. A. & Okafor, N. (1984) Medium for production of primary powder of *Bacillus thuringiensis* subsp. *israelensis*. *Appl. Environ. Microbiol.* **47**, 863–867.

O'Donnell, A. G., Berkeley, R. C. W., Claus, D., Kaneko, T., Logan, N. A. & Nozaki, R. (1980) Characterization of *Bacillus subtilis, Bacillus pumilus, Bacillus licheniformis*, and *Bacillus amyloliquefaciens* by pyrolysis gas-liquid chromatography, deoxyribonucleic acid-deoxyribonucleic acid hybridization, biochemical tests and API systems. *Int. J. Syst. Bacteriol.* **30**, 448–459.

Russell, B. L., Jelley, S. C. & Yousten, A. A. (1989) Carbohydrate metabolism in the mosquito pathogen *Bacillus sphaericus* 2362. *Appl. Environ. Microbiol.* **55**, 294–297.

Singer, S. (1973) Insecticidal activity of recent bacterial isolates and theirs toxins against mosquito larvae. *Nature* **244**, 110–111.

Singer, S. (1974) Entomogenous bacilli against mosquito larvae In *Developments in industrial microbiology*, vol. 15, pp. 187–194. American institute of Biological Sciences, Washington, DC.

Smibert, R. & Krieg, N. (1994) Phenotypic testing. In *Methods for general and molecular bacteriology* (eds P. Gerhardt, R. G. E. Murray, W. Wood & N. Krieg). American Society for Microbiology, Washington, DC.

Sneath, P. H. A. (1986) Endospore-forming gram-positive rods and cocci. In *Bergey's manual of systematic bacteriology* (eds P. H. A. Sneath, N. S. Mair, M. E. Sharpe & J. G. Holt), pp 1104–1140. Williams and Wilkins, Baltimore, London, Los Angeles, Sydney.

Stager, C. E. & Davis, J. R. (1992) Automated systems for identification of microorganisms. *Clin. Microbiol. Rev.* **5**, 302–327.

Stahly, D., Andrews, R. & Yousten, A. (1991) The genus *Bacillus* : insect pathogens. In *The Prokaryotes, a handbook on the biology of bacteria* (eds A. Ballows,

H. Truper, M. Dworkin, W. Harder & K. Schliefer). Springer-Verlag, New-York.
Stead, D. E., Sellwood, J. E., Wilson, J. & Viney, I. (1992) Evaluation of a commercial microbial identification system based on fatty acid profiles for rapid, accurate identification of plant pathogenic bacteria. *J. Appl. Bacteriol.* **72**, 315–321.
Thiery, I. & de Barjac, H. (1989) Selection of the most potent *Bacillus sphaericus* strains based on activity ratios determined on three mosquito species. *Appl. Microbiol. Biotechnol.* **31**, 577–581.
White, P. J. & Lotay, H. K. (1980) Minimal nutritional requirements of *Bacillus sphaericus* NCTC 9602 and 26 other strains of this species : the majority grow and sporulate with acetate as sole major source of carbon. *J. Gen. Microbiol.* **118**, 13–19.
Wistreich, G. A. & Lechtman, M. D. (1988) *Microbiology*, 5th edn. Macmillan, New York, 913 pp.
Yousten, A. A. & Wallis, D. A. (1984) Batch and continuous culture production of the mosquito larval toxin of *Bacillus sphaericus* 2362. *J. Indust. Microbiol.* **2**, 277–283.
Yousten, A. A., de Barjac, H., Hedrick, J., Cosmao Dumanoir, V. & Myers, P. (1980) Comparison between bacteriophage typing and serotyping for the differenciation of *Bacillus sphaericus* strains. *Ann. Microbiol. (Inst. Pasteur)* **131B**, 297–308.
Yousten, A. A., Jones, M. E. & Benoit, R. E. (1982) Development of selective/differential bacteriological media for the enumeration of *Bacillus thuringiensis* serovar. *israelensis* (H14) and *Bacillus sphaericus* 1593. *World Health Org. mimeo.* doc. WHO/VBC/82844, 7 pp.
Yousten, A. A., Wallis, D. A. & Singer, S. (1984) Effect of oxygen on growth, sporulation and mosquito larval toxin formation by *Bacillus sphaericus* 1593. *Curr. Microbiol.* **11**, 175–178.
Yousten, A. A., Fretz, S. B. & Jelley, S. A. (1985) Selective medium for mosquito-pathogenic strains of Bacillus sphaericus. Appl. Environ. Microbiol. 49, 1532–1533.

APPENDIX: BACTERIAL CULTURE MEDIA AND REAGENTS

1. Nutrient broth

13 g nutrient broth in 1 litre distilled water

Transfer 8 ml per screw-cap tube (17 × 150 mm) or 10 ml per Pyrex tube (22 mm diameter) Sterilize at 120°C for 15 min.

2. BNO3 (nitrate nutrient broth)

Nutrient broth	1 litre
Potassium nitrate (KNO_3)	10 g

Adjust to pH 7.6. Dispatch 8 ml per screw-cap tube. Sterilize at 120°C for 15 min.

3. Nutrient agar

Nutrient agar	28 g
Distilled water	1 l

Dispatch ca 200 ml into 250 ml-flasks or 8 ml into screw-cap tube. Sterilize at 120°C for 15 min. Let solidify at a sloping position after autoclaving

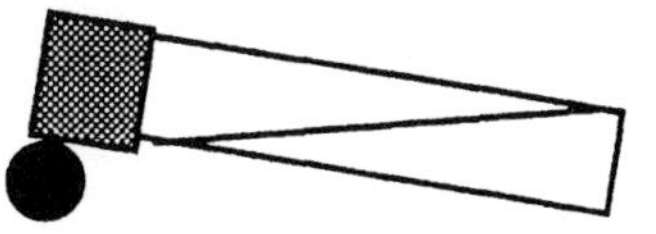

4. Nutrient agar pH 6

As nutrient agar but adjust to pH 6.

Dispatch 9 ml per screw-cap tube and allow to cool on a slope after sterilization

5. Gram stain

Crystal violet	1 g
Distilled water	100 ml
Potassium iodide	2 g
Iodine	1 g
Distilled water	20 ml

Adjust to 100 ml after complete dissolution

Basic Fuchsin	0.1 g
Distilled water	100 ml
Ethanol 95%	50 ml
Acetone	50 ml

6. Craigies

Nutrient broth	13 g
Bacto Agar	2 g
Distilled water	1 l

Adjust to pH 7.2. Put a small glass cylinder into each screw-cap tube and dispatch 11 ml per tube

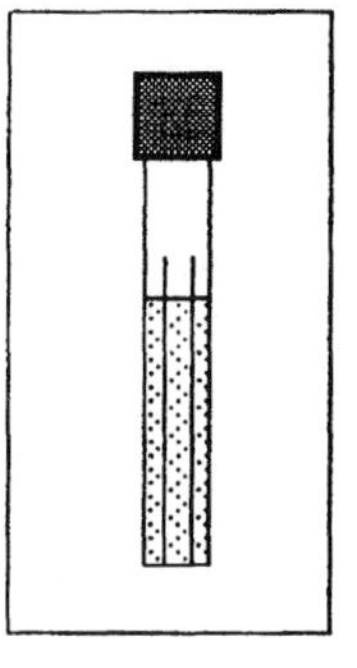

7. Meat extract medium

Meat extract	4 g
Sodium chloride	5 g
Pancreatic peptone or Bacto peptone	10 g
Distilled water	1 l

Precipitate at 120°C for 30 min. Adjust to pH 7.3–7.4.
Filter then sterilize at 110°C for 30 min.

8. Nutrient gelatin

Meat extract medium (double concentrated)	500 ml
Peptone	10 g
Sodium chloride	5 g
Gelatin	150 g
Distilled water	500 ml

Dissolve the peptone and sodium chloride in boiling meat extract medium. Add water and gelatin while stirring. Let it boil until dissolution then cool to 40°C. Adjust to pH 7.4–7.6. Add either an egg white (previously mixed) or 75 ml horse-serum. Let precipitate for 30 min at 112/115°C. Filter on humid filter paper. Adjust to pH 7.0–7.2. Dispatch 10 ml per screw-tube and sterilize for 20 min at 120°C.

9. Nutrient gelatin without nitrate

Meat extract medium (see no. 7)	2 l
Gelatin	75 g
Nutrient agar	12 g
Trypsic peptone or Bacto peptone	20 g
Potassium chloride (KCl)	10 g

Adjust to pH 7.6–7.8 and cool to 50°C.
Add:

Horse serum	100 ml

Precipitate at 120°C for 15 min then filter when hot (heating funnel).
Add :

Glucose	20 g

Dispatch 15 ml per screw-tube
Sterilize at 110°C for 30 min.

10. Arginine dihydrolase medium

L-Arginine hydrochloride	5 g
Yeast extract	3 g
Glucose	1 g
Bromocresol purple solution (1.6%)	1 ml
Distilled water	1 l

Adjust to pH 6.3–6.4. Distribute 5 ml per tube
Autoclave at 120°C for 15 min. Make a control tube without arginine.

11. 'Ammonium salt sugars' base

Ammonium dibasic phosphate ($(NH_4)_2HPO_4$)	1.0 g
Potassium chloride (KCl)	0.2 g
Magnesium sulphate heptahydrate ($MgSO_4,7H_2O$)	0.2 g
Yeast extract	0.2 g
Agar	15.0 g
Distilled water	1 l

Boil to dissolve, adjust to pH 7 then add

Bromocresol purple	0.05 g

Distribute 10 ml into tubes and sterilize at 120°C for 15 min.
For use: melt, cool to 50–60°C and add sterile carbohydrate solution to 1% final concentration.

12. Christensen

Bacto peptone	1 g
Sodium chloride	5 g
Monopotassium phosphate (KH_2PO_4)	2 g
Agar	20 g
Distilled water	1 l

Adjust to pH 6.8–7.0 then add

Glucose	1 g
Phenol red	0.12 g

Distribute in 10 ml amounts into screw-cap tubes (17 mm × 145 mm) and sterilize at 105°C for 30 min.
Prepare a stock solution of 20% urea in distilled water and sterilize on a 0.22 μm filter. Store at 4°C.

13. Stock solutions

Stock solution 1:

Magnesium sulphate heptahydrate ($MgSO_4$, $7H_2O$)	12.3 g
Manganese monohydrate sulphate ($MnSO_4$, $1H_2O$)	0.17 g
Zinc heptahydrate sulphate ($ZnSO_4$, $7H_2O$)	1.4 g

Mix into 1 litre-vial, dissolve gently by heating then adjust to one litre with distilled water.

Stock solution 2:

Ferric sulphate ($Fe_2(SO_4)_3$)	2 g
Distilled water	105 ml
Sulphuric acid (H_2SO_4)	3 ml

Heat for 4 min then filter.
Adjust to one litre with distilled water.

Stock solution 3:

Calcium chloride ($CaCl_2$, $2H_2O$)	14.7 g
Distilled water	1 l

The above stock solutions can be kept for several weeks at room temperature.

14. UG Usual medium

Bacto peptone	7.5 g
Potassium phosphate solution	100 ml
Stock solution 1	10 ml
Stock solution 2	10 ml
Stock solution 3	10 ml
Distilled water	870 ml

Potassium phosphate solution:

Monopotassium phosphate (KH_2PO_4)	68 g
Distilled water	1 l

Adjust to pH 7.4. Dispatch 110 ml per 1 litre Erlenmeyer flask or 10 ml into tubes (22 mm diameter).
Sterilize at 120 °C for 15 min.

15. Usual medium with nutrient agar

Usual medium + Bacto agar 15 g per 1 litre

Bring to boil while stirring.
Dispatch in flasks and sterilize at 120 °C for 15 min. These flasks can be kept at room temperature, before use boil then cool to 55 °C.

16. Starch nutrient agar

Potassium phosphate solution	100 ml
Stock solution 1	10 ml
Stock solution 2	10 ml
Stock solution 3	10 ml
Peptone	7.5 g
Distilled water	870 ml

Potassium phosphate solution:

Mono potassium phosphate (KH_2PO_4)	68 g
Distilled water	1 l

Adjust to pH 7.4 and add:

Nutrient agar	20 g

Filter hot after autoclaving. Prepare 150 ml starch suspension (commercial starch) 7% in hot distilled water and add it to the filtered nutrient agar. Dispatch 100 ml into 125 ml flasks

17. M.B.S.

Mono potassium phosphate (KH_2PO_4)	6.8 g
Bacto-tryptose	10.0 g
Yeast extract	2.0 g
Magnesium sulphate heptahydrate ($MgSO_4,7H_2O$)	0.3 g
Calcium chloride dihydrate ($CaCl_2,2H_2O$)	0.2 g
Stock solution	10.0 ml
Distilled water	1 l

Stock solution:

Manganese sulphate ($MnSO_4, 1H_2O$)	2.0 g
Ferric sulphate ($Fe_2(SO_4)_3$)	2.0 g
Zinc sulphate ($ZnSO_4, 7H_2O$)	2.0 g
Distilled water	1 l

Adjust to pH 7.2. Dispatch 110 ml per 1 litre Erlenmeyer flask or 10 ml into tubes (22 mm diameter). Sterilize at 120 °C for 15 min. For plating in Petri dish, add 1.5% nutrient agar to M.B.S medium.

18. MRVP medium

Polypeptone or trypsic peptone	5 g
Glucose	5 g
Sodium chloride	5 g
Distilled water	1 l

Dissolve by gently heating. Adjust to pH 7 and dispatch 5 ml in screw cap tube. Sterilize for 30 min at 105 °C.

19. Kovacs reagent

Para-dimethylaminobenzaldehyde	5 g
Isoamyl alcohol	75 ml

Dissolve by gently heating in a water-bath at 50 °C, then add slowly while stirring (use gloves, protective glasses and chemical hood):

Hydrochloric acid 37%	25 ml

Store at 4 °C for less than 3 months

20. Nitrate test reagents

Reagent NO3 A

Parasulphanilic acid	0.8 g
Acetic acid (5N solution in distilled water)	100 ml

Reagent NO3 B

Alpha-naphthylamine	0.5 g
Acetic acid (5N solution in distilled water)	100 ml

Warning: Alpha-naphthylamine is carcinogenic. Use gloves for handling and disposing of the product.

21. VP reagents

Reagent VP A

Potassium hydroxide	40 g
Distilled water	100 ml

Reagent VP B

Alpha-naphthol	6 g
Absolute ethanol	100 ml

Warning: alpha-naphthol is harmful.

CHAPTER III-2

Bacteria: Laboratory bioassay of bacteria against aquatic insects with emphasis on larvae of mosquitoes and black flies

LAWRENCE A. LACEY

Yakima Agricultural Research Laboratory, USDA-ARS, 5230 Konnowac Pass Road, Wapato, WA 98951, USA.

1 INTRODUCTION

The insect families Culicidae (mosquitoes) and Simuliidae (black flies) contain some of the most medically important insect species. Interest in the development and use of *Bacillus* entomopathogens as microbial control agents of black fly and mosquito larvae has been greatly increased with the discovery of varieties of *Bacillus thuringiensis* with elevated larvicidal activity against these insects (de Barjac, 1978; Lacey & Undeen 1986). Their activity against certain families in the suborder Nematocera is principally due to the production of Cry IVA-D and Cry IIA toxins. These are predominantly found in *B. thuringiensis* var. *israelensis* and the PG-14 isolate of *B. thuringiensis* var. *morrisoni*. Additionally, several isolates of certain varieties of *Bacillus sphaericus* and isolates of *Clostridium bifermentans* also have good potential for the microbial control of pest and vector mosquitoes (Lacey & Undeen 1986; Thiery *et al.*, 1992).

In order to assess and compare the efficacy of any bacterium against both mosquito and black fly larvae, the bacterial toxins responsible for larvicidal activity must be ingested by the larvae. Comparative efficacy testing must be conducted under repeatable conditions which permit normal or close to normal feeding rates and do not produce abnormally high mortality (above 10%) in control larvae. Although there are other aquatic Nematocera that are susceptible to bacteria and/or their toxins, I will emphasize methods for the testing of bacteria against larvae of mosquitoes and black flies. Readers interested in the bioassay of *Bacillus thuringiensis* var. *israelensis* against chironomid larvae can refer to procedures used by Ali *et al.* (1981) and other authors cited by Dejoux & Elouard (1990) and Lacey & Mulla (1990). A multitude of aquatic non-target organisms are not directly affected by ingesting bacteria or bacterial toxins active against mosquitoes and/or black flies, but they may be affected by formulation components or the consequences of removing target species from the ecosystem. Results of testing and a review of the literature on the effects of entomopathogenic bacteria on non-target aquatic insects are

MANUAL OF TECHNIQUES IN INSECT PATHOLOGY
ISBN 0–12–432555–6

presented by Dejoux & Elouard (1990) and Lacey & Mulla (1990).

2 BIOASSAY OF *BACILLUS* PATHOGENS AGAINST MOSQUITO LARVAE

The bioassay of bacteria against larvae of most mosquito species is a relatively straightforward procedure. Larvae of most pest and vector mosquitoes are filter feeders and readily ingest particulates in the size range of bacteria and bacterial inclusions (Pucat, 1965; Merritt & Craig, 1987). Some notable exceptions are predatory mosquitoes that eat aquatic insects including the larvae of other mosquito species.

A Factors affecting activity

A number of factors, such as mosquito species and age, number of mosquito larvae per bioassay container, water quality and temperature, volume and depth of water, presence or absence of food and other particulates, shape of the bioassay container and factors related to the bacterial preparation (size of particles, affect of adjuvants) can significantly influence toxin activity and the results of bioassays. Many of these factors also influence feeding rate and hence the amount of toxin ingested. The following sections present information on various bioassay parameters and how they might influence bioassay results.

1. Effect of species

All filter feeding mosquito species thus far tested are susceptible to *B. thuringiensis* var. *israelensis*. There is considerably more variability among mosquito species in susceptibility to *B. sphaericus*. *Culex, Psorophora, Anopheles* and *Mansonia* species are highly susceptible to *B. sphaericus*, but many *Aedes* species, especially *Aedes aegypti* are not susceptible or considerably less susceptible. Recently, high levels of resistance to this bacterium have been reported in certain populations of *Culex quinquefasciatus*, a species that is normally very susceptible.

2. Feeding behaviour

In addition to other susceptibility factors related to species, the feeding behaviour of a particular species will influence the amount of inoculum with which the larvae will come into contact. Some species may feed predominantly at the surface of the water (*Anopheles* species), within the water column (some *Culex* spp.) or predominantly off the substratum (some *Aedes* spp.) or some species may combine column and substrate feeding. Species that feed mostly at the surface usually consume less toxin due to settling of toxic moieties from their feeding zone and thus appear to be less susceptible to entomopathogenic bacteria. Thiery (personal communication) has observed that *Anopheles* larvae exposed to *B. thuringiensis* var. *israelensis* or *B. sphaericus* in very shallow water (i.e. where settling of toxin has less influence) are still less susceptible to the toxin than are *Culex* or *Aedes* species.

The major difference in standardized bioassays for isolates of *B. thuringiensis* and *B. sphaericus* will be the target species. Because *Ae. aegypti* is relatively easy to rear and its eggs can be stored dry between rearings, it is an ideal test animal for bioassays of *B. thuringiensis* varieties. Due to the lack of susceptibility in *Ae. aegypti* to *B. sphaericus, Cx. quinquefasciatus* is the preferred test animal for this bacterium.

3. Number of larvae/container, volume and quality of water, and features of the bioassay container

Because mosquito larvae, especially those feeding within the water column and off the substratum can be very efficient in removing particulates, the number of larvae per container will influence the amount of toxic moieties that will be available to each individual (Sinègre *et al.*, 1981a). Crowding may also negatively influence normal feeding rates. The amount of toxin per larva will also be a function of the volume of water used for the bioassay. Five ml of water per larva has enabled consistent and repeatable bioassays in studies conducted at the USDA-ARS Medical and Veterinary Entomology Research Laboratory (MAVERL; Gainesville, FL, USA) on both *B. sphaericus* and *B. thuringiensis* against several mosquito species (Lacey & Singer, 1982; Lacey *et al.*, 1988). The shape of the container and water depth may also affect the availability of bacterial

Figure 1 Bioassay of bacteria against mosquito larvae at the USDA-ARS Medical and Veterinary Entomology Laboratory, Gainesville, FL. (Courtesy of Al Undeen).

toxin. Small disposable cups (depth ca. 4 cm) filled with 100 ml of chlorine-free water (well water or aged tap water), are used for bioassays at MAVERL (Figure 1). The quantity of water used for bioassay at the Pasteur Institute (Paris, France) is 150 ml of water. The presence of chlorine can significantly reduce the larvicidal activity of *B. thuringiensis* varieties (Sinègre *et al.*, 1981b) and be detrimental to larvae. Twenty larvae per container are used at MAVERL and 25 larvae per container are used at Pasteur Institute.

4. Larval age

The larval age group that enables good survival in controls and consistent results are late third and early fourth instars. Younger larvae (first and second instars) do provide more sensitive targets, but tend to be less hardy and unable to survive much more than 24 h without food (Lacey & Singer, 1982). Older fourth instars may pupate before the end of the exposure period. With the exception of first instars, mosquito larvae imbibe very little water. If solubilized toxins are to be assayed, it will be necessary to use first instars or to encapsulate the toxin(s).

5. Temperature

Within the range of temperatures tolerated for a given species, the activity of bacterial toxins in mosquito larvae is, for the most part, positively correlated with temperature (Sinègre *et al.*, 1981b; Wraight *et al.*, 1987; Lacey *et al.*, 1988). Many of the species most utilized for laboratory bioassay (e.g. *Ae. aegypti, Cx. quinquefasciatus, An. albimanus* and *An. stephensi*) are tropical in origin and thrive at 27 °C. Lower temperatures may be required for other species from more temperate climes.

6. Effect of food and particulates on activity of bacterial toxins

Depending on its nature and quantity, particulate matter can reduce or enhance the feeding rate of mosquito larvae (Dadd, 1970) and hence the amount of toxic inclusions that are ingested by larvae (Ramoska & Pacey, 1979; Sinègre *et al.*, 1981a; Aly, 1988). Even the addition of excess food that is normally desirable could inhibit feeding, dilute the toxins in the gut or interfere with the activity of toxins. During the exposure/incubation period, the absence of food will enable an accurate and repeatable bioassay of bacterial pathogens. Late third and early fourth instars can withstand starvation with little or no effect on control mortality. In assays involving younger larvae that require more than 48 h exposure and incubation, a small amount of food (5 mg^{-1}), such as finely ground lab chow, could be added to the assay cups in the second 24 h of exposure. In assays at Pasteur Institute, food is provided to older instars of *Anopheles* and *Culex* species during bioassays (Thiery, personal communication).

B Selection of dosage

Almost invariably one observes a dose-dependent mortality response in mosquito larvae to *B. thuringiensis* and *B. sphaericus* varieties. The exception being when bacterial preparations contain such low amounts of Nematocera-active toxin that the particulate load required to induce mortality also inhibits feeding. Skovmand (personal communication) suggests that hypersensitivity to toxin might also inhibit feeding. A range of 5–7 concentrations of promising candidate bacteria should be selected, which produce mortality of 10–90% with two above and two below the 50% mortality level.

Dosage is usually expressed in milligrams of

spore powder or formulation per litre of water or in parts per million (ppm). The use of international toxicity units (ITU) to compensate for the day to day fluctuations in larval response to *B. thuringiensis* var. *israelensis* has been proposed (de Barjac & Larget, 1979, 1984). ITUs are calculated by comparing bacterial preparations to a standard and using the ratio of the LC_{50} values of the standard and test sample times the assigned potency (in ITUs) of the standard. For example:

$$\text{Potency (ITU) of sample} = \frac{LC_{50}\text{ for standard} \times \text{potency (ITU) of standard}}{LC_{50}\text{ for sample}}$$

The potency of IPS-78, the first standard prepared by de Barjac, for example, was arbitrarily assigned a potency value of 1000 ITU mg^{-1} (de Barjac & Larget, 1979). It has been replaced by IPS-82 since 1982 (de Barjac & Larget, 1984). Primary powders of *B. thuringiensis* var. *israelensis* were also proposed by Dulmage *et al.* (1985) as US standards. Various parameters that can influence potency of products based on *B. thuringiensis* var. *israelensis* are presented by Skovmand (1996).

Several reasons for avoiding the use of spore count as a measure of dosage are presented by Dulmage *et al.* (1990). Nevertheless, if suspensions of fresh cultures (48–72 h growth on agar) are used in assays, such as in preliminary screening of recently isolated strains, spore count (after heat shock at 80 °C for 12 min) may be the only indication of dosage available. Cultures should also be checked for the presence of parasporal inclusion bodies (where endotoxins of *B. thuringiensis* and some other *Bacillus* species are found). Autoclaved suspensions that produce mortality in larvae indicate the production of exotoxins, some of which may be harmful to vertebrates (Melin & Cozzi, 1990). As in the other assays, tests with whole cultures of cells should include a range of dilutions. Ultimately, preparation of primary powders would enable comparative bioassays against a standard.

C Number of replicates and tests

For each concentration and control, there should be a minimum of four sets (i.e separate containers) per test. Although a test conducted on a single date with several sets of each dosage will usually provide a homogeneous set of data, it is advantageous to run replicate tests over time to account for the variation in susceptibility that is observed in different cohorts of mosquito larvae obtained from the same colony. Three replicate tests on separate test dates are the usual minimum. In general, the higher the coefficient of variation (standard deviation divided by the mean) for mortality data, the greater the need to do more replicate tests.

D Bioassay protocol

In order to have repeatable results the various bioassay parameters, as well as rearing conditions, should be standardized. Suggested parameters for standardized bioassay of varieties of *B. thuringiensis* and *B. sphaericus* against mosquito larvae are presented in Table 1. Several similar protocols have been recommended for the repeatable bioassay of *B. thuringiensis* varieties (de Barjac & Larget, 1979, 1984, McLaughlin *et al.*, 1984; Dulmage *et al.*, 1990) against mosquito larvae. The following is an eclectic combination of procedures for the preparation of inoculum and bioassay against mosquito larvae.

Table 1 Suggested parameters for bioassay of varieties of *Bacillus thuringiensis* and *Bacillus sphaericus* against mosquito larvae.

Dosage of bacteria	mg/l (ppm)[a]
No. concentrations	5–7[b]
No. of cups/control & concentration/test	4
Replicate tests (separate dates)	3
Age of larvae	late 3rd early 4th
Duration of test	24–48 h[c]
Volume of water	100 ml
Source of water	non-chlorinated[d]
Food	none[e]

[a] Weight of primary powder or formulation; with fresh cultures of spores, spore count (after heat shock; 80 °C for 12 min) may be the only indication of dosage available; several reasons for avoiding the use of spore count as a measure of dosage are presented by Dulmage *et al.* (1990).

[b] Concentration should be chosen such that at least two produce mortality between 10–50% and at least two produce mortality between 50–90%.

[c] 24 h for *B. thuringiensis* var. *israelensis*; 48 h for *B. sphaericus*.

[d] Well water or dechlorinated (aged and/or aerated) tap water.

[e] If very young larvae are used, addition of food may be necessary. Finely ground and suspended food (e.g. Purina Lab Chow, Tetramin) will be required for larval diet.

1. Preparation of inoculum

Suspensions of inoculum for bioassay should be freshly prepared from primary powders, formulated product or fresh cultures just prior to each bioassay. Initial homogenates are prepared with standards and primary powders by suspending 50 mg of powder in 10 ml of deionized or distilled water and agitating for 10 min using a bead mill (20 ml penicillin flask with several 6 mm glass beads) or similar method. Addition of a wetting agent such as (0.1%) Tween 80 will improve wetting of primary powders. Subsequent dilutions can then be made from the homogenate using routine dilution procedures (see Chapter III-1). Dulmage *et al.* (1990) describe making a 'stock' suspension from the above homogenate by adding 0.1 ml of the homogenate to 9.9 ml of water. The stock is then resuspended using a Vortex agitator. Following the procedures of de Barjac & Thiery (1984) cited by Dulmage *et al.* (1990), subsequent dilutions are made in the container in which the assay will take place by adding 15–120 µl of stock suspension directly into assay cups containing 150 ml of water. The use of more dilute suspensions is advantageous because accuracy is increased when larger volumes are measured. McLaughlin *et al.* (1984) recommended making a series of dilutions to be used for treatments using a minimum of 10 ml of bacterial suspensions for making subsequent dilutions and for applying the diluted suspensions to the bioassay cups. Skovmand (personal communication) points out that commercial products are not highly homogeneous and suggests larger quantities (1 g) be used for preparing initial suspensions.

2. Handling of insects and application of inoculum

Materials and methods for rearing various species of mosquitoes used for bioassays are presented by Gerberg *et al.* (1994) and are not covered here. When larvae are in the late third to early fourth instar they are removed from rearing trays by sieving and placed in chlorine-free water without food. If a chlorine-free source of water is not available, tap water that has been aerated for 48 h will suffice. Either 20 or 25 late third or early fourth instars are transferred to each bioassay container using a Pasteur pipette with the narrow tip removed or an eye dropper with a wide opening. Filing or fire polishing the tip may be required to avoid larval injury.

Small (ca. 4–5 cm deep) containers holding 100 ml of water provide an adequate arena in which to bioassay bacteria against most species of mosquitoes. Four cups per concentration for each isolate and control per test date are recommended. It is easier to first add larvae and then fill with the balance of the 100 ml of water, rather than adding the water first and then subtracting the amount that was added along with the larvae. If 10 ml of bacterial suspension will be added to each container rather than µ l amounts, the containers should only be filled to 90 ml prior to adding treatments. Total amount of liquid based on weight is another method of determining when the appropriate amount of water has been added to assay containers. One method of adding larvae without the addition of more water, is to use small sieves or nets for their transfer.

3. Incubation and assessment of mortality

After the appropriate treatments are added to each container, the larvae are then incubated for 24 h at 25–27 °C in bioassays of *B. thuringiensis* var. *israelensis* and 48 h with *B. sphaericus*. Temperature is evenly maintained at MAVERL by placing the cups with larvae in large trays containing 2–3 cm of water. The trays are then set on thermostatically controlled heat tapes within a closed cabinet. The sensor for the thermostat is placed in one of the trays under water. Incubators or climate cabinets can also be used to maintain constant temperature.

After incubation, the larvae are assessed for mortality. If larvae fail to respond to tapping the side of the assay cup, they are probed with a needle. Lack of response or only very weak response (i.e. the larva is moribund) is scored as dead. It is best to base calculations of percentage mortality on the number of larvae originally placed in the containers and number of living larvae after the exposure/incubation period due to the fact that dead larvae are often consumed by the survivors.

4. Data processing

If control mortality does not exceed 10% the data are kept and analysed by probit analysis after adjusting for control mortality using Abbott's (1925) formula:

$$\text{adjusted \% mortality} = \frac{\text{observed \% mort.} - \text{ave. control \% mort.}}{100\% - \text{ave. control mort.}}$$

A bioassay is the use of a living organism to assay, or measure the amount of a substance, such as a toxin, in an unknown sample. A single discriminating concentration of bacterial suspensions may also be used to study the effect of abiotic and/or biotic factors on larvicidal activity of candidate isolates or formulations. In such cases the amount of toxin is already known and some other biological effect, not the amount of toxin is being tested. Strictly speaking, these tests are not bioassays. Various statistical tests for analysis of variance and mean separation and greater detail on probit analysis are presented in Chapters II and IV-3.

E Considerations for predatory larvae

The testing of efficacy of bacteria against predatory species of mosquitoes, notably *Toxorhynchites* species and certain species of *Psorophora* as well as others, will necessitate first feeding the bacteria to a prey species, such as *Aedes aegypti*, and then adding the predator. Procedures used for testing of *B. sphaericus* and *B. thuringiensis* var. *israelensis* against *Toxorhynchites* spp. are presented by Larget & Charles (1982) and Lacey (1983).

3 BIOASSAY OF *BACILLUS THURINGIENSIS* ISOLATES AGAINST BLACK FLY LARVAE

Like mosquitoes, the larvae of black flies are aquatic. Their habitats include running water ranging from small creeks to huge rivers. Although they are normally thought of as filter feeders, a range of feeding strategies is observed for the family that also include grazing (scrapers), deposit feeders and predators (Currie & Craig, 1987). The larvae of most vector or pestiferous species (i.e. those that warrant control) are filter feeders for a significant amount of the time. Using their labral fans and mucous secretions, black fly larvae are capable of filtering particles ranging from 0.09 to 350 μm from the water column (Currie & Craig, 1987). The combination of their filtering efficiency and susceptibility make black fly larvae ideal targets for *B. thuringiensis* var. *israelensis*, even though the period of exposure is relatively brief.

A Factors affecting activity

Factors that affect the susceptibility of black fly larvae to *B. thuringiensis* var. *israelensis* are, for the most part, the same as those that affect susceptibility of mosquito larvae (species, larval age, temperature, presence of food and other particulates, etc.; Lacey *et al.*, 1978; Molloy *et al.*, 1981; Coupland, 1993). The major difference between bioassays of bacteria against black flies and mosquitoes is the requirement for running water for black fly larvae. A number of bioassay systems have been proposed for the laboratory evaluation of bacteria against larval black flies (Lacey *et al.*, 1982). The current required for bioassay can be artificially created by using a magnetic stirrer and stir bar in a beaker of water (Undeen & Nagel, 1978), with air bubbles (Lacey & Mulla, 1977) (Figure 2), using orbital shakers or other artificial methods (Lacey *et al.*, 1982) or more naturally in

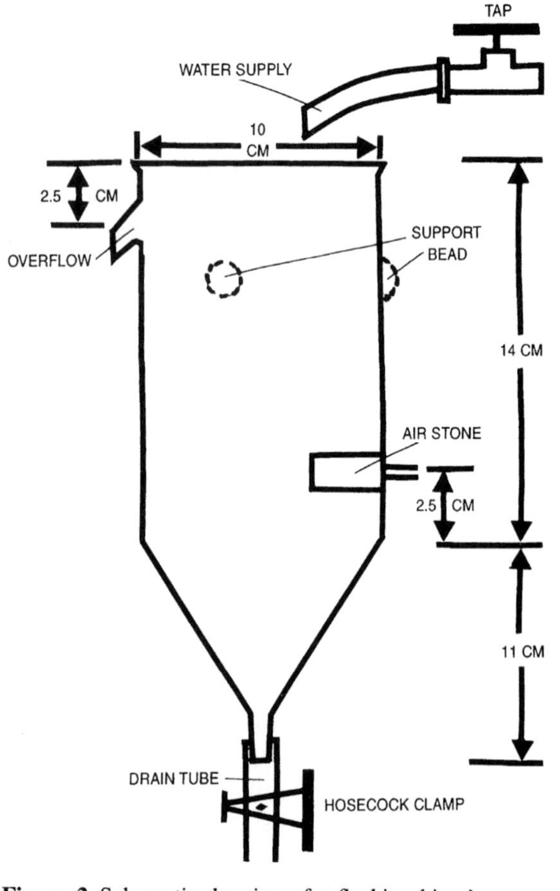

Figure 2 Schematic drawing of a flushing bioassay system. (From Lacey & Mulla, 1977, *J. Econ. Entomol.* **70**, 453–456, with permission.)

systems which utilize flowing water (Gaugler *et al.*, 1980; Guillet *et al.*, 1985; Coupland, 1993) (Figures 3 and 4). As with mosquito larvae, the most important factor is to enable normal larval feeding rates and avoid excess control mortality. Systems that create a current using magnetic stirrers, aquarium air pumps and the like will be the least expensive, but the current is not as laminar as that obtained with

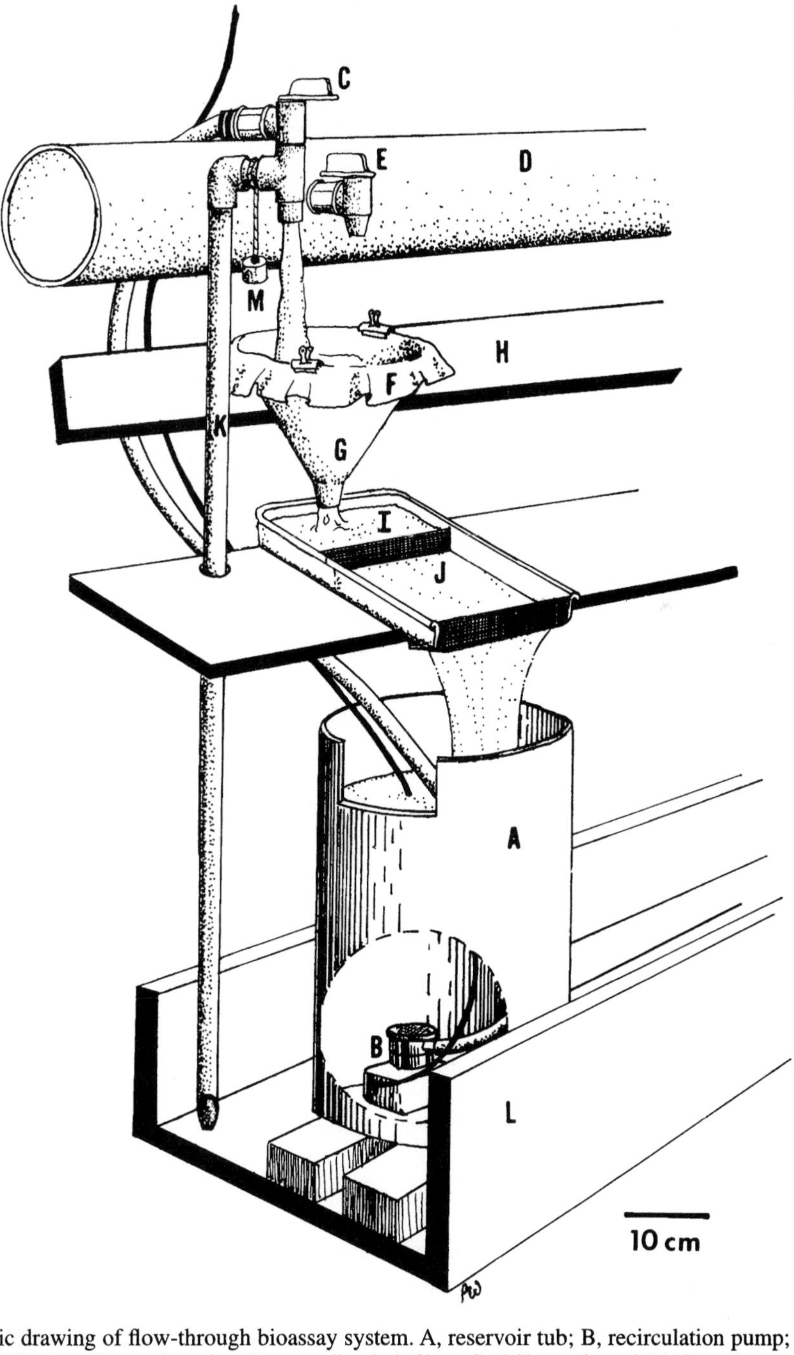

Figure 3 Schematic drawing of flow-through bioassay system. A, reservoir tub; B, recirculation pump; C, recirculation valve; D, water supply pipe; E, supply valve; F, organdie cloth filter; G, delivery funnel; H, funnel support; I, tray reservoir section; J, tray attachment section; K, standing waste pipe; L, waste trough; M, cap. (From Gaugler *et al.*, 1980, *Can. Entomol.* **112**, 1271–1276, with permission.)

Figure 4 Flow-through system utilized by Guillet *et al.* for stream side bioassay of candidate *B. thuringiensis* var. *israelensis* formulations. (Courtesy of Pierre Guillet.)

flowing water. The systems used by Gaugler *et al.* (1980) and Coupland (1993) can be operated in both recirculation and flow-through modes.

In addition to current, other environmental conditions may also have to be met. Species that are normally found in pristine, cold, fast flowing streams may be difficult to use for laboratory bioassays unless their environmental conditions are adequately simulated.

Larval feeding rates can be highly influenced by particle concentration in the water. Feeding efficiency increases at lower particulate concentrations (Kurtak, 1978) and increasing particle concentrations up to an optimum will increase the quantity of food consumed, but an excessive amount of particulates will actually inhibit feeding (Gaugler & Molloy, 1980). If feeding is inhibited before or during exposure to *B. thuringiensis*, fewer toxic inclusions will be consumed. If inhibition occurs following ingestion of toxin, mortality can be increased due to the fact that ingested particulates remain stationary in the midgut and hence prolong contact time with receptors on the target epithelium (Gaugler & Molloy, 1980).

B Exposure period

Length of exposure is another factor that will differ from mosquito bioassays. Under most operational conditions, the time interval during which *B. thuringiensis* var. *israelensis* formulations are in contact with black fly larvae is fairly brief because stream flow carries the material downstream quickly. Initial application time can be short as in the case of aerial application, or extended to 10–30 min when applied at ground level. To approximate natural conditions more realistically, exposure of the larvae to the bacterium in the laboratory should not be too protracted. For a standard bioassay in closed systems, I suggest 30 min of exposure. For screening purposes only, varieties of *B. thuringiensis* that produce low amounts of toxins active against black flies will require a higher dosage and longer exposure time to produce significant mortality (Lacey *et al.*, 1978).

C Bioassay protocol

The preparation of bacterial suspensions and number of replicates is as described above for mosquito bioassays. Table 2 presents some suggested standardized parameters for bioassay of *B. thuringiensis* varieties against black fly larvae.

1. Handling of insects

Except in the few laboratories with a black fly colony, larvae have to be collected from a breeding site. Getting from the field to the lab with larvae of some species may require transport of the larvae in aerated containers of water under cool or cold conditions. Several species, including *Simulium damnosum* and *S. vittatum*, can be transported for short periods of time on the damp vegetation on which they were collected. After removal of excess water, the plants are placed loosely in plastic bags in a cool

Table 2 Parameters for bioassay of *Bacillus thuringiensis* varieties against black fly larvae.

Concentration of bacteria	mg/l (ppm)[a]
Duration of exposure (min)	30
No. concentrations	5–7[b]
No. containers/control and concentration/test	4
Replicate tests (separate dates)	3
Age of larvae	penultimate and early last instar
Duration of test	24 h
Volume of water	1000 ml
Source of water	non-chlorinated[c]
Food	5 mg/l after exposure[d]

[a] Weight of primary powder or formulation; with fresh cultures of spores, spore count (after heat shock; 80 °C for 12 min) may be the only indication of dosage available; several reasons for avoiding the use of spore count as a measure of dosage are presented by Dulmage *et al.* (1990).

[b] Concentration should be chosen such that at least two produce mortality between 10–50% and at least two produce mortality between 50–90%.

[c] Well water or dechlorinated (aged and/or aerated) tap water.

[d] Following the 30 min exposure period and finely ground and suspended (5 mg^{-1}) Purina Lab Chow, Hog Chow, Tetramin, and the like have been used for larval diet.

place for transport to the lab. An insulated box, such as an ice chest, works well. A layer of ice separated from the bags of larvae by a sheet of styrofoam is ideal for ensuring cool, but not excessively cold temperatures.

In the lab, the larvae should be removed from the bags and placed in trays in chlorine-free water. Penultimate instar larvae or young last instars (those which do not have melanized 'gill spots' (organs that will become the pupal respiratory filaments)) can be gently removed using a soft latex eye dropper by nudging the larvae with the dropper near the anal gills at the same time as drawing water into the dropper. A camel's hair brush can also be used. Immediately transfer the larva to the container or system in which bioassay will be conducted. Twenty larvae per replicate container provides sufficient test animals and will avoid the entanglement of larvae in their own silk that occurs when too large a number of larvae are in a container.

The larvae should be allowed to acclimate for approximately 3 h before exposure to candidate bacteria. Before adding bacterial suspensions, dead, detached and pupating larvae should be removed. Usually if healthy penultimate instars are used, the number of larvae that are not attached and actively feeding will be low or zero. If larvae are removed, additional larvae should be added to bring the count to the desired number for each bioassay container. After acclimation, and when larvae are feeding (as evidenced by the opening of cephalic fans), add bacterial suspensions. In closed bioassay systems, including recirculating systems, the desired concentrations can be added all at once. If a flow-through system is used the suspension will have to be dripped continuously in the flowing water for the desired exposure period. Figure 4 shows the plastic bottles used by Guillet *et al.* (1985) to meter in bacterial suspensions into small gutters through which stream water flows. With isolates containing high amounts of Nematocera-active toxin(s) an exposure period of 30 min is recommended.

2. *Termination of exposure*

After exposure in closed non flow-through systems, it is necessary to change the water in which the larvae are exposed. This usually causes some of the larvae to detach and care must be taken not to lose them when the water is decanted. Detachment will also result in temporary cessation of feeding, which can result in increased mortality. In the system used by Lacey & Mulla (1977) (Figure 2), the inoculum is flushed from the bioassay container by adding water to each container from individual taps while it drains from the bottom and overflows. Larvae are not exposed to air or cessation of current and hence, do not detach. Flushing of inoculum without larval detachment is ideal when using flow-through systems such as that of Gaugler *et al.* (1980).

3. *Incubation and post-exposure care of insects*

After changing the water, it will be important to add food to enable as close to normal feeding rates as possible. Five ppm of finely ground lab chow or similar food added as a suspension should enable normal feeding rates. If feeding inhibition is noted (labral fans are held closed for longer intervals than normal) a lower concentration of food should be used.

Larvae are then incubated for at least 24 h before assessment of mortality. A lower temperature (20 °C), than that used for mosquitoes is recommended to approximate stream conditions more realistically. After incubation, larvae are observed for signs of life. When in doubt regarding the condition of the larvae, use a probe. Data are treated as described previously for mosquito larvae. One exception may be necessary. Because black fly larvae do not always respond favourably to collection, handling and the laboratory environment, it may be necessary to accept a higher level of control mortality; 20% control mortality may have to be the cut off point for acceptable test results. Because of lower temperatures and shorter exposure times, it may be necessary to take a second reading 24 h after the first.

Readers interested in field assessment of bacterial pathogens against simuliid larvae are referred to Undeen & Lacey (1982). A combination of field exposure and laboratory assessment of *B. thuringiensis* var. *israelensis* formulations against black fly larvae is presented by Undeen & Colbo (1980) and Lacey & Undeen (1984). The assay of bacteria against other aquatic organisms, especially those found in running water can be quite complex depending on their requirements. The bioassay system and methods that are chosen should simulate the aquatic system in which the organism is normally found as closely as possible.

4 CONCLUSIONS

Data from laboratory bioassays can provide useful information on species susceptibility, including development of resistance, and for comparing isolates of bacteria, particularly when primary powders of bacterial strains are compared to a standard. However, care must be exercised when using laboratory-derived data to make predictions of relative activity under field conditions. For example, ITU ratings based on bioassays against *Ae. aegypti* do not always correspond to relative efficacy of *B. thuringiensis* var. *israelensis* formulations against field populations of mosquito larvae (Dame *et al.*, 1981). Similarly, the ITU ratings for *B. thuringiensis* var. *israelensis* formulations as determined with mosquito larvae are often not correlated with activity against black fly larvae in laboratory bioassays (Molloy *et al.*, 1984) nor under field conditions (Lacey & Undeen, 1984). Even laboratory bioassays of formulated *B. thuringiensis* var. *israelensis* conducted with black fly larvae could provide misleading information regarding relative efficacy under field conditions. Molloy *et al.* (1984) observed a positive correlation between particle size and efficacy in lab bioassays, yet under field conditions formulations with smaller particle sizes provide greater effective carry (i.e. kill black fly larvae for a greater distance downstream), probably due to more rapid settling of the larger particles.

Once effective isolates are identified in laboratory bioassays, the most realistic evaluation of formulated bacterial microbial control agents will be under field conditions.

ACKNOWLEDGEMENTS

I thank Ole Skovmand, Isabelle Thiery, Al Undeen and James Coupland for their review of the manuscript. Photographs supplied by Randy Gaugler, Al Undeen, Pierre Guillet and Mir Mulla are gratefully appreciated.

REFERENCES

Abbott, W. S. (1925) A method of computing the effectiveness of an insecticide. *J. Econ. Entomol.* **18**, 265–267.

Ali, A., Baggs, R. D. & Stewart, J. P. (1981) Susceptibility of some Florida chironomids and mosquitoes to various formulations of *Bacillus thuringiensis* serovar. *israelensis*. *J. Econ. Entomol.* **74**, 672–677.

Aly, C. (1988) Filter feeding of mosquito larvae (Diptera: Culicidae) in the presence of the bacterial pathogen *Bacillus thuringiensis* var. *israelensis*. *J. Appl. Entomol.* **105**, 160–166.

Coupland, J. B. (1993) Factors affecting the efficacy of three commercial formulations of *Bacillus thuringiensis* var. *israelensis* against species of European black flies. *Biocontrol Sci. Tech.* **3**, 199–210.

Currie, D. C. & Craig, D. A. (1987) Feeding strategies of larval black flies. In: *International symposium on ecology and population management of black flies* (eds R. Merritt & K. C. Kim), Pennsylvania State University Press, pp. 155–170.

Dadd, R. H. (1970) Comparison of rates of ingestion of particulate solids by *Culex pipiens* larvae: phagostimulant effect of water-soluble yeast extract. *Entomol. Exp. Appl.* **13**, 407–419.

Dame, D., Savage, K., Meisch, M. & Oldacre, S. (1981) Assessment of industrial formulations of *Bacillus thuringiensis* var. *israelensis*. *Mosq. News* **41**, 540–546.

de Barjac, H. (1978) Un nouveau candidat à la lutte biologique contre les moustiques: *Bacillus thuringiensis* var. *israelensis*. *Entomophaga* **23**, 309–319.

de Barjac, H. & Larget, I. (1979) Proposals for the adoption of a standardized bioassay method for the evaluation of insecticidal formulations derived from serotype H.14 of *Bacillus thuringiensis*. World Health Organisation mimeo. doc. WHO/VBC/79.744, 16 pp.

de Barjac, H. & Larget, I. (1984) Characteristics of IPS 82 as standard for biological assay of *Bacillus thuringiensis* H-14 preparations. World Health Organisation mimeo. doc. WHO/VBC/84.892, 10 pp.

Dejoux, C. & Elouard, J.-M. (1990) Potential impact of microbial insecticides on the freshwater environment, with special reference to the WHO\UNDP\World Bank, Onchocerciasis Control Programme. In *Safety of Microbial Insecticides* (eds M. Laird, L. A. Lacey & E. W. Davidson), pp. 65–83. CRC Press, Boca Raton.

Dulmage, H. T., McLaughlin, R. E., Lacey, L. A., Couch, T. L., Alls, R. T. & Rose, R. I. (1985) HD-968-S-1983, a proposed U.S. standard for bioassays of preparations of *Bacillus thuringiensis* subsp. *israelensis*-H-14. *Bull. Entomol. Soc. Am.* **31**, 31–34.

Dulmage, H. T., Correa, J. A. & Gallegos-Morales, G. (1990) Potential for improved formulations of *Bacillus thuringiensis israelensis* through standardization and fermentation development. In *Bacterial control of mosquitoes and black flies: biochemistry, genetics, and applications of Bacillus thuringiensis israelensis and Bacillus sphaericus* (eds H. de Barjac & D. Sutherland), pp. 110–133. Rutgers University Press, New Brunswick.

Gaugler, R. & Molloy, D. (1980) Feeding inhibition in black fly larvae (Diptera: Simuliidae) and its effects on the pathogenicity of *Bacillus thuringiensis* var. *israelensis*. *Environ. Entomol.* **9**, 704–708.

Gaugler, R., Molloy, D., Haskins, T. & Rider, G. (1980) A bioassay system for the evaluation of black fly (Diptera: Simuliidae) control agents under simulated stream conditions. *Can. Entomol.* **112**, 1271–1276.

Gerberg, E. J., Barnard, D. R. & Ward, R. A. (1994) Manual for mosquito rearing and experimental technique. *Am. Mosq. Control Assoc. Bull.* **5** (revised). 98 pp.

Guillet, P., Hougard, J.-M., Doannio, J., Escaffre, H. & Duval, J. (1985) Evaluation de la sensibilité des larves du complexe Simulium damnosum à la toxine de *Bacillus thuringiensis* H 14. 1. Méthodologie. *Cah. ORSTOM, sér. Ent. méd. Parasitol.* **23**, 241–250.

Kurtak, D. C. (1978) Efficiency of filter feeding of black fly larvae (Diptera: Simuliidae). *Can. J. Zool.* **56**, 1608–1623.

Lacey, L.A. (1983) Larvicidal activity of *Bacillus* pathogens against *Toxorhynchites* mosquitoes (Diptera: Culicidae). *J. Med. Entomol.* **20**, 620–624.

Lacey, L. A. & Mulla, M. S. (1977) A new bioassay unit for evaluating larvicides against blackflies. *J. Econ. Entomol.* **70**, 453–456.

Lacey, L. A. & Mulla, M. S. (1990) The safety of *Bacillus* pathogens of mosquitoes and black flies for nontarget organisms in the aquatic environment. In *Safety of microbial insecticides* (eds M. Laird, L. A. Lacey & E. W. Davidson), pp. 169–188. CRC Press, Boca Raton.

Lacey, L. A. & Singer, S. (1982) Larvicidal activity of new isolates of *Bacillus sphaericus* and *Bacillus thuringiensis* (H-14) against anopheline and culicine mosquitoes. *Mosq. News* **42**, 537–543.

Lacey L. A. & Undeen, A. H. (1984) The effect of formulation, concentration, and application time on the efficacy of *Bacillus thuringiensis* (H-14) against black fly larvae under natural conditions. *J. Econ. Entomol.* **77**, 412–418.

Lacey, L. A. & Undeen, A. H. (1986) Microbial control of black flies and mosquitoes. *Annu. Rev. Entomol.* **31**, 265–296.

Lacey, L. A., Mulla, M. S. & Dulmage, H. T. (1978) Some factors affecting the pathogenicity of *Bacillus thuringiensis* Berliner against blackflies. *Environ. Entomol.* **7**, 583–588.

Lacey, L. A., Undeen, A. H. & Chance, M. M. (1982) Laboratory procedures for the bioassay and comparative efficacy evaluation of *Bacillus thuringiensis* var. *israelensis* (serotype 14) against black flies (Simuliidae). *Misc. Pub. Entomol. Soc. Am.* **12**, 19–23.

Lacey, L. A., Lacey, C. M. & Padua, L. E. (1988) Host range and selected factors influencing the mosquito larvicidal activity of the PG-14 isolate of *Bacillus thuringiensis* var. *morrisoni*. *J. Am. Mosq. Control Assoc.* **4**, 39–43.

Larget, I. & Charles, J. F. (1982) Etude de l'activité larvicide de *Bacillus thuringiensis* variété *israelensis* sur les larves de Toxorhynchitinae. *Bull. Soc. Pathol. Exp.* **75**, 121–130.

McLaughlin, R. E., Dulmage, H. T., Alls, R., Couch, T. L., Dame, D. A., Hall, I. M., Rose, R. I. & Versoi, P. L. (1984). U.S. standard bioassay for the potency assessment of *Bacillus thuringiensis* serotype H-14 against mosquito larvae. *Bull. Entomol. Soc. Am.* **30**, 26–29.

Melin, B. E. & Cozzi, E. (1990) Safety to nontarget invertebrates of lepidopteran strains of *Bacillus thuringiensis* and their β-exotoxins. In *Safety of microbial insecticides* (eds M. Laird, L. A. Lacey & E. W. Davidson), pp. 149–167. CRC Press, Boca Raton.

Merritt, R. W. & Craig, D. A. (1987) Larval mosquito (Diptera: Culicidae) feeding mechanisms: mucosubstance production for capture of fine particles. *J. Med. Entomol.* **24**, 275–278.

Molloy, D., Gaugler, R. & Jamnback, H. (1981) Factors influencing efficacy of *Bacillus thuringiensis* var. *israelensis* as a biological control agent of black fly larvae. *J. Econ. Entomol.* **74**, 61–64.

Molloy, D., Wraight, S. P., Kaplan, B., Gerardi, J. & Peterson, P. (1984) Laboratory evaluation of commercial formulations of *Bacillus thuringiensis* var.

israelensis against mosquito and black fly larvae. *J. Agric. Entomol.* **1**, 161–168.

Pucat, A.M. (1965) The functional morphology of the mouthparts of some mosquito larvae. *Quaestiones Entomol.* **1**, 41–86.

Ramoska, W. A. & Pacey, C. (1979) Food availability and period of exposure as factors of *Bacillus sphaericus* efficacy on mosquito larvae. *J. Econ. Entomol.* **72**, 523–525.

Sinègre, G., Gaven, G. & Jullien, J. L. (1981a) Contribution à la normalisation des épreuves de laboratoire concernant des formulations expérimentales et commerciales du sérotype H-14 de *Bacillus thuringiensis*. III. Influence séparée ou conjointe de la densité larvaire, du volume ou profondeur de l'eau et de la présence de terre sur l'efficacité et l'action larvicide résiduelle d'une poudre primaire. *Cah. ORSTOM, sér. Entomol. méd. Parasitol.* **19**, 157–163.

Sinègre, G., Gaven, B. & Vigo, G. (1981b) Contribution à la normalisation des épreuves de laboratoire concernant des formulations expérimentales et commerciales du sérotype H-14 de *Bacillus thuringiensis*. II. Influence de la température, du chlore résiduel, du pH et de la profondeur de l'eau sur l'activité biologique d'une poudre primaire. *Cah. ORSTOM, sér. Entomol. méd. Parasitol.* **19**, 149–155.

Skovmand, O. (1996) Parameters influencing the potency of products based on *Bacillus thuringiensis* var *israelensis*. *J. Econ. Entomol.* in press.

Thiery, I, Hamon, S., Gaven, B. & de Barjac, H. (1992) Host range of *Clostridium bifermentans* serovar. *malaysia*, a mosquitocidal anaerobic bacterium. *J. Am. Mosq. Control. Assoc.* **8**, 272–277.

Undeen, A. H. & Colbo, M. H. (1980) The efficacy of *Bacillus thuringinesis* var. *israelensis* against blackfly larvae (Diptera: Simuliidae) in their natural habitat. *Mosq. News* **40**, 181–184.

Undeen, A. H. & Lacey, L. A. (1982) Field procedures for the evaluation of *Bacillus thuringiensis* var. *israelensis* (serotype 14) against black flies (Simuliidae) and nontarget organisms in streams. *Misc. Pub. Entomol. Soc. Am.* **12**, 25–30.

Undeen, A. H. & Nagel, W. L. (1978) The effect of *Bacillus thuringiensis* ONR-60A strain (Goldberg) on *Simulium* larvae in the laboratory. *Mosq. News* **38**, 524–527.

Wraight, S. P., Molloy, D. & Singer, S. (1987) Studies on the Culicine mosquito host range of *Bacillus sphaericus* and *Bacillus thuringiensis* var. *israelensis* with notes on the effects of temperature and instar on bacterial efficacy. *J. Invertebr. Pathol.* **49**, 291–302.

CHAPTER III-3

Bacteria: Bioassay of *Bacillus thuringiensis* against lepidopteran larvae

MICHAEL R. McGUIRE*, LUIS J. GALAN-WONG† & PATRICIA TAMEZ-GUERRA†

* USDA-ARS, National Center for Agricultural Utilization Research, 1815 N University St, Peoria, IL 61604, USA
† Departamento de Microbiología, Facultad de Ciencias Biológicas, Universidad Autónoma de Nuevo León, S Nicolas de los Garza, Nuevo León AP 270 64450, México

1 INTRODUCTION

The order Lepidoptera contains some of the most damaging pests known to humans. Control of these pests has relied almost exclusively on chemical pesticides. The discovery that *B. thuringiensis* was capable of killing certain species of lepidopteran larvae provided a true breakthrough in the management of these pests because *B. thuringiensis* is relatively easy to produce in large quantities and can be made commercially. However, as previously described, there is a wide diversity of *B. thuringiensis* isolates and new ones are continually discovered. The use of bioassay to ascertain activity of new isolates or formulations of bacteria is a necessary task if information related to pesticidal activity is desired. Although bioassays are principally used to assess the insecticidal activity of the protein toxins found in the parasporal inclusions of *B. thuringiensis*, bioassays may also be employed to determine the role of spores and spore/toxin interactions in the activity of a particular isolate or fermentation run. Bioassays can be used to determine host range, relative activity, speed of kill, and other factors that are important in the selection of strains, fermentation conditions and formulation ingredients used to produce this very diverse species. Even within Lepidoptera, susceptibility can vary greatly depending on which type of toxin or combination of toxins is present (see Chapter III-1, Table 3). Because the interaction between spore and crystal is not well understood and insects vary in their susceptibility to different spore/crystal combinations, it is important that spore counts are done in the initial phases of assessing activity. Ultimately, however, fermented bacteria must be assayed for insecticidal activity against the insect species of interest as well as non-target insects.

A clear understanding of what criteria will be used to assess activity (e.g. mortality, morbidity, growth reduction, etc.) must be delineated prior to starting bioassays. Also, statistical analysis of data recovered from the bioassays must be preplanned. There is a large number of statistical software packages available but statistical tests must be used judiciously and care must be taken to select

MANUAL OF TECHNIQUES IN INSECT PATHOLOGY
ISBN 0-12-432555-6

the appropriate test. See Chapters II and V-3 for a full discussion of the use of appropriate tests. In addition, Robertson & Priesler (1992) give an excellent overview of the relation of probit analysis to bioassay.

Most of the main parameters that must be considered in the design and implementation of bioassays for the evaluation of *B. thuringiensis* against lepidopteran larvae are similar to those presented elsewhere in this manual for a broad range of entomopathogens. There can be an infinite variety of procedures for considering the myriad potential targets and their feeding habits. One standard bioassay that has been adopted by industry to ensure consistency within and among products was developed by Dulmage *et al.* (1971) and should be read by all those interested in developing a bioassay screening protocol. However, this assay may not be appropriate for all phases of screening new isolates or formulations. Below, we present a framework of selected assays successfully used against a few insect species. It is up to the individual conducting the research to modify these approaches to fit the needs of the particular species, target host plant and application conditions. In all cases, the test insects must be vigorous, free of pathogens, genetically representative of the natural population, and similarly aged.

2 DIET-BASED BIOASSAYS

Diet incorporation assays are an excellent way to determine activity and host range of bacteria, viruses and protozoa when a direct measure of activity is desired. This method can also be valuable when small quantities of bacteria are produced using less than commercial-scale fermentation equipment (e.g. shake flask cultures, solid phase agar, etc.) that is commonly available in most laboratories. In addition, diet-based assays remove the need to maintain a constant supply of plants in the greenhouse or field. However, insects that cannot be reared on artificial media cannot easily be tested with an artificial diet assay.

The choice of diet incorporation or surface contamination really relates to the behaviour of the test insect. *Heliothis virescens*, for example, feeds by grazing across the surface of the diet and surface contamination may be preferable. An insect

Table 1 Diet used for rearing and bioassay of lepidopteran insects.

Group	*Ingredient*[a]	*Diet quantity*	*Bioassay quantity*
I	Distilled H_2O (heated to boiling)	1200 ml	1200 ml
	4M KOH	18 ml	4N KOH 6 ml or 22% w/v
	Casein	126 g	0
	Alfalfa meal (entomological grade)	54 g	Soybean flour, 241.8 g
	Sucrose	126 g	43.80 g
	Wesson salts mixture	36 g	36 g
	Alphacel	18 g	0
	Wheat germ	108 g	108 g
	Ascorbic acid	14.5 g	1.92 g
	Aureomycin (250 mg/capsule)	0.5 g	Chlortetracycline-HCl 0.17 g
	Sorbic acid	4 g	3.4 g
	15% Methyl-*p*-hydroxybenzoate solution	35 ml	5.0 g
	10% choline chloride w/v solution	36 ml	15%, 24.9 ml
	10% formaldehyde w/v solution	13 ml	13.3 or 15.0
II	Agar dissolved in 2500 ml boiling DW[b]	90 g	34.8 in 2000 ml boiling DW
III	Vitamin solution A[c]	12 ml	4 ml
IV	Vitamin solution B[d]	12 ml	0

[a] Dulmage *et al.* (1971). Prepare group I in a 1-gal Waring blender equipped with a variable speed transformer; add the components in the order listed with the blender operating at very slow speed. Add group II, cool to about 60–65 °C, and add groups III and IV. Adjust transformer to maximum output and continue blending at slow speed for 2 min. Final pH of diet will be about 5.2.

[b] DW = distilled water.

[c] Vitamin solution A: nicotinic acid amide, 12.0 g; calcium pantothenate, 12.0 g; thiamine hydrochloride, 3.0 g; pyridoxine hydrochloride, 3.0 g; biotin, 0.24 g; vitamin B_{12}, 0.024 g; distilled water 1000 ml.

[d] Vitamin solution B: riboflavin, 60 g; folic acid, 3.0 g; dissolved in a solution of 2.24 g KoH in 1000 ml distilled water.

like *Ostrinia nubilalis*, however, often spends little time on the surface and will bore into the medium. Thus, diet incorporation is preferable for this type of insect. One factor that is consistent across both tests is that the active agent must be stable in the diet. Many recipes for diets contain antimicrobial agents that can suppress the activity of bacteria. Conversely, the diet should not support the germination and growth of bacterial spores over time. Sometimes this can be a dilemma because ingredients in the diet that suppress growth of *B. thuringiensis* may also inhibit activity after ingestion. Careful screening of diet ingredients may be necessary to determine the role of diet ingredients in the activity of *B. thuringiensis*. A suggested diet for rearing and for bioassay of *H. virescens, Spodoptera exigua, Helicoverpa zea, Trichoplusia ni* and others is presented in Table 1.

Chapter II gives an excellent graphical interpretation of both surface contamination and diet incorporation assays. Below, we present some of the more technical information that is essential to obtaining reliable and consistent results from diet-incorporation bioassay of *B. thuringiensis.*

A Cup preparation

Containers for larvae should be uniform, clean, and able to hold larvae individually. Disposable 9 g, clear plastic cups are preferable but care must be taken to ensure lids are fastened securely enough to prevent escapes. Often, when a larva feeds on *B. thuringiensis*, it begins wandering and will find a way of the cup if possible. Plastic jelly trays with sealing Mylar lids also work well. For development of data suitable for LC_{50} determination, a minimum of 50 cups per dilution is required.

B Sample preparation

All suspensions should be prepared and diluted in sterile distilled water or in a buffered saline solution containing 0.85% NaCl, 0.6% K_2HPO_4, and buffered to pH 6.5 with citrate buffer solution. If the test sample is difficult to wet, 2.5 ml of a 1% Tween 80 solution may be added to each 100 ml of the saline solution.

1. Concentrations

Selection of suitable concentrations for accurate LC_{50} determination may require some preliminary experimentation with high and low concentrations of the test sample. Once a suitable range of concentrations is established, six or seven serial dilutions of test material should be made. As has been previously discussed, concentrations at the extreme ends of the dose–response curve are less informative to the probit model than data points in the linear phase. In general at least two concentrations above and two concentrations below the LC_{50} concentration should be used. Control samples using water or saline are essential for correct interpretation of any bioassay. In general, no corrections to the final mortality of each dilution are necessary if control mortality is less than 5%. However, treatment data may be corrected for control mortality above 5% by using the formula developed by Abbott (1925). Experiments where control mortality exceeds 20% should be repeated.

2. Making the dilutions

It is very easy to make errors in dilution and in keeping track of dilutions while setting up an experiment. The following set of procedures is offered as guidance to minimize these errors.

1. Tubes with the diluted sample should be labelled with masking tape with the final concentration in the diet.
2. Each suspension is homogenized with a vortex mixer just prior to making the next dilution in the series and again just before adding it to the diet.
3. The diet should be cooled to 55 °C and maintained at that temperature until used. Samples are diluted 1 : 10 into aliquots of diet and mixed thoroughly by blending for 4 min at high speed in a Waring blender (a variable speed transformer should be used to bring the blender up to speed to avoid splashing) or blending for 2 min at high speed in a malted milk mixer. While mixing, the tape is transferred to the mixing container. Timers should be used to precisely control the amount of mixing that occurs.
4. The diet/dilution can then be poured into cups and allowed to solidify and dry before addition of larvae. The amount of diet in each cup should be sufficient to maintain the life of the insect throughout the incubation phase but is not

otherwise critical because the concentration of *B. thuringiensis* is constant throughout the diet. An amount of approximately 5 ml is usually sufficient for most species. If plastic jelly trays are used, diet is added to each cavity and allowed to set for about an hour. After larvae are added, a sheet of Mylar is then pressed onto the surface and heated with an iron to bond the surfaces. Small holes can then be punched in the Mylar over each cavity by using a small board with small brads.

5. The masking tape can then be transferred to the tray containing the cups.

C Infestation, incubation and evaluation

Larvae should be similarly aged, free from pathogens or other contaminants and highly vigorous. Larvae must be carefully transferred individually to cups using a camel hair brush. If possible, the larva should be allowed to 'silk' off the brush and onto the diet surface. Damage to the insect from this transfer process is difficult to observe but will affect the results. While infesting, larvae should be transferred to the most dilute sample first. If the brush touches the diet surface, it should be cleaned to avoid contaminating subsequent larvae. The cups containing larvae should be incubated at constant temperature and humidity and no movement of larvae among cups should be allowed. Temperature and humidity levels will vary for different species but, in general, 25–30 °C and 40–60% RH are acceptable.

Assays are evaluated after 4–7 days by examining each cup. Because the plastic covers are clear, there is usually no need to open the cup to assess mortality. If there is any question, the larva should be prodded to ascertain mortality. If any movement is observed, the larva must be scored as alive.

3 BIOASSAY OF GRANULAR FORMULATIONS OF *B. THURINGIENSIS*

A Laboratory bioassays

Granular formulations present a challenging case for determining the activity of *B. thuringiensis* on or in the granules. In our experience, *B. thuringiensis* granules (even those formulated with feeding deterrents) placed in an enclosed dish with neonate larvae will kill all larvae, probably because larvae sample their environment by tasting. Therefore the *B. thuringiensis* must be extracted from the granule to determine its activity in a laboratory setting.

1. B. thuringiensis *extraction*

If *B. thuringiensis* is simply sprayed onto the outside of the granule, as is the case with many corn grit or sand-based formulations, soaking the granules in water with a detergent such as Tween followed by vigorous agitation (e.g. a Waring blender) is suitable. The suspension that is recovered by filtering can then be used for bioassay. If, however, the granules contain the *B. thuringiensis* within a matrix such as cornstarch or other polymer, the granule must be digested to liberate the *B. thuringiensis*. In the case of cornstarch granules, α-amylase will digest the material sufficiently to retrieve the *B. thuringiensis*. Two ml of a 1 mg/ml suspension of α-amylase can be added to 100 mg granules and incubated at 37 °C for 1 h. This mixture is then ground up with a tissue homogenizer for 10 s.

2. Bioassay of extracted B. thuringiens

Once the *B. thuringiensis* is extracted from the granules, it can be tested either through diet incorporation as previously described or it can be tested with the droplet feeding method (Figure 1) described in Chapter II (also, see McGuire *et al.*, 1994).

B Greenhouse bioassays

Although data gathered from the tests above will indicate survival of *B. thuringiensis* on or within the granules after exposure to specified conditions (e.g. sunlight), diet incorporation or droplet assays (Chapter II) will not indicate acceptance of the granule or the *B. thuringiensis* itself by target insects. Therefore, bioassay on foliar surfaces is a necessary part of any screening programme dealing with the activity or acceptability of new *B. thuringiensis* isolates and formulations. The anti-feedant effects caused by *B. thuringiensis* are well known and formulations designed to stimulate feeding may help to alleviate this problem (Bartelt *et al.*, 1990). In

Figure 1 Droplet feeding bioassay. Neonate larvae are allowed to feed on droplets of *Bacillus thuringiensis* suspensions. Those larvae that feed, as evidenced by the colour in the gut, are removed to diet cups for incubation.

general, *B. thuringiensis* granules are used to control insect pests that attack plants with 'holding' structures such as the whorl or leaf axils of a corn plant. One example is the European corn borer, *O. nubilalis* in corn. Currently, the use of *B. thuringiensis* is increasing in corn and is starting to replace some of the more toxic synthetic chemical compounds.

1. Application and infestation

In the greenhouse, formulations can be tested on corn plants simply by placing an appropriate amount of granules in the whorl of V6–V7 (6–7 leaves present on the corn plant) stage plants and then introducing 20–25 neonate *O. nubilalis*. 75 mg granules per whorl will approximate actual field application rates of 11.2 kg/ha (10 lbs/acre) in a typical field.

2. Incubation and evaluation

A week later, the live corn borers can be easily counted by unrolling the whorl of the plant. Most larvae will be within the whorl but a few may be in the leaf midrib or within the axils of the lower leaves (Figure 2). Although about half the corn borers in the controls will crawl off the plants, this test still gives a good approximation of formulation acceptance by insects in relation to the surrounding corn tissue and has been shown to yield results similar to those seen under field conditions (Gillespie *et al*., 1994). Because of the high control mortality, at least six replications per treatment should be done. To analyse these data, analysis of variance followed by a multiple range test to examine differences among means (e.g. protected least significance difference test

Figure 2 Corn plants infested with European corn borer larvae. The plant on the right has been treated with granules containing *Bacillus thuringiensis* while the plant on the left is an untreated control. Note the amount of damage that has occurred in the control plant over a one-week period.

available on many computer statistical software packages) will indicate differences in activity of new preparations.

4 BIOASSAY OF SPRAYABLE FORMULATIONS OF *B. THURINGIENSIS* ON LEAF SURFACES

In Chapter II, Evans and Shapiro outlined a leaf disk method of bioassay wherein a larva was allowed to consume an entire leaf disk which had been contaminated with virus. This assay works well for larger instar insects, but for neonates, the disk is rarely eaten entirely and the anti-feedant effects of *B. thuringiensis* may inhibit full ingestion leading to variability in the amount of toxin obtained by each larva. Another example of a leaf disk test is to allow insects to live on the contaminated disk for a specified period of time and then assess mortality. One advantage to this method is the presence of uncontaminated leaf tissue (i.e. the underside of the leaf disk) which will give some information of feeding preference by insects. Cotton is an excellent choice, assuming the test insect will feed on and survive on the foliage, because it lasts well in sealed dishes, provides plenty of food and moisture, and will support insects such as *O. nubilalis* for up to a week. If other test insects will not survive on cotton, other plants can be used. However, some screening must occur prior to bioassay to ensure that the quality of the leaf tissue does not deteriorate to the point of weakening insects over the anticipated course of the study. It may be necessary to immerse the stem of the leaf in a sealed tube containing water, agar or nutrient medium to maintain the necessary leaf quality. Another aspect that must be considered in choice of leaf disk size and number of insects per leaf disk is the cannibalistic nature of the larvae. Insects such as *O. nubilalis* and *T. ni* will live in large numbers together without eating each other. Insects such as *H. virescens*, however, are cannabilisitic and must be held individually.

A Application of *B. thuringiensis*

There are three main ways of treating plants: spraying from overhead (e.g. track sprayer, Figure 3), spreading a known amount across a marked area of a leaf, or leaf dipping. Each offers different advantages.

Figure 3 A Devries research track spray chamber. The control panel on the right allows modification of nozzle speed and direction while pressure is regulated externally in the air supply. To modify the chamber for rain, the air supply line is replaced with a water supply line, causing water to be expelled from the nozzle. A simple electronic addition to the unit creates a continually traversing nozzle and allows for rain simulation.

1. Track sprayer

The track sprayer (Devries Research Track spray Booth, Hollandale, MN) can be used to simulate actual field applications. The nozzle is interchangeable, the spray pressure adjustable, and the track assembly has speed adjustment. Volumes of water applied can range from 46 l/ha (5 gal/acre) up to 460 l/ha (50 gal/acre) or higher depending on the above adjustments. Also, spray splashing, dripping, etc., is more likely to imitate field situations. This type of application is ideal for screening large numbers of *B. thuringiensis* isolates or formulations when absolute accuracy of dose response is not essential. This method will also give valuable data related to formulation performance coming through an agricultural-type nozzle. Care must be taken to ensure that all areas of each leaf in the chamber were contacted by the spray suspension. Areas not touched by the spray can be marked out with a permanent marker and not used in subsequent assays. Disadvantages to this method centre around the amount of dripping and splashing that are involved that may lead to inconsistent deposition of toxin on the leaf surfaces. However, once a series of formulations has been tested under more precise conditions, this method can be used to more accurately reflect actual field conditions.

2. Spreading

The spreading method involves marking an area on a leaf surface, say 33 cm^2, with a permanent marker, and then applying a carefully measured amount of solution (100 μl to simulate 233 l/ha (25 gal/acre)) to that area via pipette. The droplet is then carefully spread across the leaf surface with a glass rod and allowed to dry. If necessary to achieve even spreading, a small amount (0.1% v/v) of Tween 80 can be added. This method gives a carefully controlled dosage of active ingredient and can be used to do LC_{50} determinations, or for examining subtle differences in activity and acceptance of *B. thuringiensis* or formulations. Disadvantages of this method include the labour intensiveness of the application and the use of only one disc per leaf to avoid contamination.

3. Leaf dipping

This method requires much larger volumes of material than either of the two previously described methods. An excised leaf or entire plant is immersed in a suspension of *B. thuringiensis* and then allowed to dry. All surfaces of the plant are covered but there can be a very large difference in the amount of material adhering to the treated surfaces. Plants such as cabbage have very waxy leaves that may repel liquids and lead to a very low accumulation of toxin. Also, leaves on the same plant may have quite different characteristics based on growing conditions and age. This method should only be used if other methods are not available.

B Concentration of *B. thuringiensis*

Concentration of *B. thuringiensis* should be tested to determine the optimal dosage for discerning differences among treatments. For example, field rates of *B. thuringiensis* will kill 100% of test *O. nubilalis* neonates when applied under laboratory conditions. To determine if a particular isolate or formulation is more or less effective than a commercial product, it is best to have a dosage of the commercial product that will not kill 100%. Ideally, a concentration that will kill insects in the straight line portion of the log dose curve will allow for optimal estimation of loss or gain of activity due to formulation or exposure to environmental factors. For *O. nubilalis*, a dosage of 20 mg Dipel 2X for every 50 ml spray solution applied at a rate of 233 l/ha will kill approximately 80% of the test insects. However, for other insect species, this dosage may need to be changed depending on the susceptibility of the particular insect.

C Disk bioassay

After application and any subsequent treatments (e.g. solar exposure, rain test, see below), leaf disks can be removed and placed in plastic dishes.

1. Removal of leaf disk

Tools can be made by sharpening the edge of metal pipe to quickly cut the disc from the leaf. The size of the disk is not critical, because a certain amount of material has been applied per unit area, but the disk must be large enough to support the number of insects that will be placed in the dish. A piece of filter paper should be added to the dish and can be left dry to soak up moisture from a succulent leaf such as cotton, or water can be added to the paper to provide moisture to a thin or dry leaf.

2. Addition of insects

For *O. nubilalis*, ten neonates will survive well for the three-day duration of the test on a leaf disk of 33 cm^2. Larvae should be added following the techniques described above. For cannibalistic insects, small disks and small dishes may be used to hold insects individually.

3. Incubation and assessment

After application of larvae, the dish is sealed and incubated for a time sufficient for mortality to occur (this will vary depending on species, size, *B. thuringiensis* dosage, etc. and must be ascertained for each situation). Self sealing dishes can be used, or seal the dish with two wraps of stretched parafilm. Dishes should be incubated at constant temperature amenable to insect development. After a specified time, the dishes are unsealed and the larvae counted. For *O. nubilalis* neonates, 3 days is sufficient for mortality to occur. Other, less susceptible insects may require longer incubation times.

D Measuring solar stability of foliar deposits of *B. thuringiensis*

Inactivation of *B. thuringiensis* by sunlight is a well-known phenomenon and has hindered acceptance of *B. thuringiensis* products. The effect of different strains or formulations of *B. thuringiensis* on sunlight stability can be measured in the laboratory by using equipment that is becoming increasingly available.

1. Solar simulators

Solar simulators such as the Suntest CPS (Heraeus, Hanau, Germany) shown in Figure 4, will produce sunlight in both quality and quantity as seen in nature (McGuire *et al.* 1996). Generally, a xenon-based light source is used and then filtered through special glass to provide an adequate representation of sunlight.

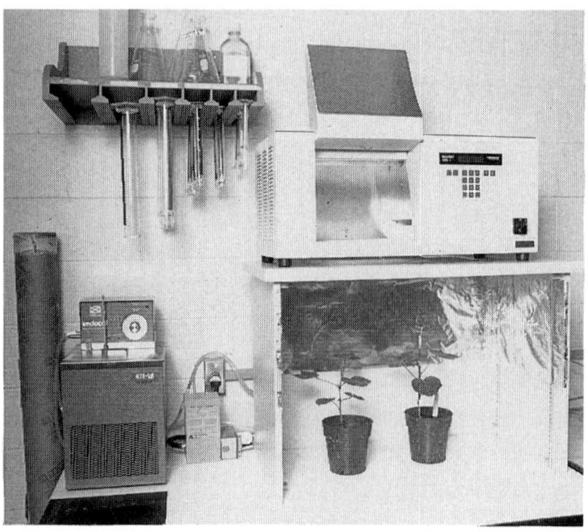

Figure 4 A Suntest CPS solar simulator showing plants placed beneath the light source. The door is open to show the reflective sides of the chamber. In this modification, the floor to the chamber has been removed to allow light to pass onto the plants. A piece of Tefcel plastic is placed across the opening of the floor to prevent burning of the leaves. In normal operation, the plants are raised to just below the plastic.

2. Treatment of leaves

Cotton plants should be treated with the glass rod technique described above and then placed under the light source and exposed for designated periods of time. Plants must be staked up such that all leaves that are exposed are equidistant from the light source.

3. Exposure

Care must be taken to ensure the plant does not get burned. This can be avoided by placing a piece of plastic called Tefcel (DuPont Corp., Wilmington, DE) between the plant and light source. There is some loss of energy as the light passes through the plastic but the representation of wavelengths remains unchanged. For 50% loss of *B. thuringiensis* activity, 8 h exposure to the light at a distance of approximately 20 cm, is generally necessary but may vary depending on host plant, formulation and distance from the source. After exposure, the leaf disks can be removed and bioassayed as described above. Untreated plants and plants treated with *B. thuringiensis* but no solar exposure must be included within the same assay to account for variability that occurs between assays.

E Measuring rainfastness of foliar deposits of *B. thuringiensis*

Rainfall also washes *B. thuringiensis* deposits off foliar surfaces. Fermentation conditions and formulation may greatly affect the rainfastness of *B. thuringiensis* so procedures to measure this characteristic could be useful.

1. Rain simulator

The spray chamber described above can be converted into a rain chamber. Water is fed to the line that provided air pressure to the spray nozzle, and, when turned on, water comes through the nozzle. An electronic switch can be installed that keeps the head traversing back and forth over the plants. Again, by changing speed, nozzle, pressure, etc., the rate of rainfall can be adjusted to mimic a light rain or a heavy downfall.

2. Preparation of plants

After plants are treated with *B. thuringiensis* and allowed to dry thoroughly, they can be placed in the rain simulator. It is important that the plants be placed in the centre of the simulator at a distance of about 1 m from the spray nozzle, and 0.5 m from the ends of the chamber. Plants placed too close to the ends of the chamber may receive more water because the nozzle assembly briefly pauses at each end.

3. Rain application

A good test of rainfastness involves application of 5 cm rain over a 50-min period. This simulates a relatively heavy rain and gives a good indication of the ability of a formulation to resist rain. Accumulation may be measured simply by placing a rain gauge in the simulator at the same height as the leaves. After drying, leaf disks can be cut and assayed for retention of activity as above.

5 CONCLUSIONS

Laboratory bioassay can be an extremely useful tool to screen isolates of *B. thuringiensis* for activity against multiple insect species and to determine the effect on insects provided with a choice of potential feeding substrates. Ultimately, field tests which are often time consuming and expensive will have to be conducted. By using the full range of laboratory and greenhouse bioassays available, many potential treatments can be eliminated before the field work begins.

REFERENCES

Abbott, W. S. (1925) A method of computing the effectiveness of an insecticide. *J. Econ. Entomol.* **18**, 265–267.

Bartelt, R. J., McGuire, M. R. & Black, D. A. (1990) Feeding stimulants for the European corn borer (Lepidoptera: Pyralidae): additives to a starch-based formulation for *Bacillus thuringiensis* Berliner. *Environ. Entomol.* **19**, 182–189.

Dulmage, H. D., Boening, O. P., Rehnborg, C. S. & Hansen, D. G. (1971) A proposed standardized bioassay for formulations of *Bacillus thuringiensis* based on the International Unit. *J. Invertebr. Pathol.* **81**, 240–245.

Gillespie, R. L. McGuire, M. R. & Shasha, B. S. (1994) Palatability of flour granular formulations to European corn borer (Lepidoptera: Pyralidae). *J. Econ. Entomol.* **87**, 452–457.

McGuire, M. R., Shasha, B. S., Eastman, C. E. & Oloumi-Sadeghi, H. (1996) Effect of starch and flour based formulations on rainfastness and solar stability of *Bacillus thuringiensis*. *J. Econ. Entomol.* **89**, 863–869.

McGuire, M. R., Shasha, B. S., Lewis, L. C. & Nelsen, T. C. (1994) Residual activity of granular starch-encapsulated *Bacillus thuringiensis*. *J. Econ. Entomol.* **87**, 631–637.

Robertson, J. L. & Priesler, H. K. (1992) *Pesticide bioassays with arthropods*, CRC Press, Boca Raton.

CHAPTER III-4

Bacteria of soil-inhabiting insects

MICHAEL G. KLEIN

* USDA-Agricultural Research Service, Application Technology Research Unit, OSU-OARDC, 1680 Madison Ave., Wooster, OH 44691, USA

1 INTRODUCTION

Adults and/or larvae of many scarabs are serious pests of crops, ornamentals and turf throughout the world, and are likely to increase in significance (Jackson, 1992). In addition, scarabs such as the Japanese beetle, *Popillia japonica*, are quarantined to prevent introduction and damage as has occurred in the United States and Terceira Island, Azores (Lacey *et al.*, 1994). Jackson & Glare (1992) surveyed the general status of pathogens against scarabs, and recent reviews (Klein & Jackson, 1992; Klein, 1995) examined the status of bacteria against turf and soil pests in general. Despite their importance, and concerns about conventional pesticides, there are few microorganisms available for use against soil-inhabiting pests.

This chapter concentrates on three groups of bacteria, two *Bacillus* species and the genus *Serratia*, and describes methods for their isolation, identification, propagation, bioassay and preservation. *Bacillus thuringiensis* is a naturally occurring soil bacterium which produces protein crystals toxic to many insects. It shows considerable promise as an alternative to conventional insecticides for suppressing white grubs. Many of the techniques for dealing with scarab active *B. thuringiensis* differ little from those for other *B. thuringiensis* strains and are covered in detail in Chapter III-1. The history of the milky disease bacterium, *B. popilliae*, has been reviewed by Klein (1992) and will not be covered in detail here. It was the first microbial agent registered in the United States and has been used for suppressing Japanese beetle populations since the early 1940s. *Serratia* species have been associated with insect diseases (Klein & Jackson, 1992), but commercial production of one of these bacteria was achieved only recently. A product containing *S. entomophila* is now being used against the grass grub, *Costelytra zealandica*, in New Zealand pastures (Jackson *et al.*, 1992). Following ingestion, these bacteria adhere to the foregut intima and are especially numerous around the cardiac valve. After adhesion, the insect ceases feeding and clears the gut leading to a depletion of fat bodies and the

MANUAL OF TECHNIQUES IN INSECT PATHOLOGY
ISBN 0–12–432555–6

characteristic amber coloration. Bacteria continue to grow in the foregut and, in advanced stages, may give the appearance of a bacterial plug at the cardiac valve. The diseased larvae may live for more than one month (Jackson *et al.*, 1993). The bacteria are commercially produced by fermentation and applied to pastures with subsurface application equipment.

2 ISOLATION

Most bacterial pathogens can be isolated from dead and diseased larvae using standard bacteriological techniques (see Chapter III-1). Haemolymph, gut sections, or other infected tissue can normally be streaked under aseptic conditions onto nutrient agar (see Appendix) or similar media. Single colonies can then be isolated and purified. Certain fastidious bacteria, such as the milky disease organisms (*B. popilliae* and *B. lentimorbus*), require more specialized media discussed below. Preliminary differentiation of bacteria can be made on the basis of cell morphology, results of Gram-staining (Chapters III-1 and VIII-1), and utilizing *Bergey's Manual of Systematic Bacteriology* (Krieg & Holt, 1984; Sneath *et al.*, 1986). Once a particular pathogen is suspected, selective agars can be employed.

Bacteria can be categorized and identified according to their growth requirements on the standard or specialized media listed in Bergey's Manual. Alternatively, identification systems such as the Enterotube® II or Oxi/Ferm Tube® II may be purchased from a number of companies. These systems have 12 or more biochemical and carbohydrate utilization tests to help in the separation of Gram-negative bacteria. A coded response to the various tests is used to identify the species of bacteria involved. The Biolog MicroStation™ System (Biolog, Hayward, CA) utilizes 95 biochemical reactions and a computer program to identify either Gram-negative or -positive bacteria. However, both systems have limitations and will not identify all bacteria.

A *Bacillus popilliae*

Milky disease bacteria fit the classic pattern of obligate pathogens. In nature, they are only found associated with diseased scarabs, or as spores in the surrounding soil, and require specialized conditions (bodies of their hosts) for growth and sporulation.

New isolates of *B. popilliae* can be obtained from the haemolymph of infected scarab larvae (Plate 1). Stahly *et al.* (1992a) recently reviewed procedures used in isolating these bacteria.

1. Surface disinfect grub with water and 0.5% sodium hypochlorite (NaClO, diluted chlorine bleach).
2. Puncture the cuticle with a needle (or snip off leg).
3. Collect spores in sterile water and rinse free of haemolymph.

Spores from haemolymph can also be collected onto glass slides for later isolation (see Section, 6A). Spores or vegetative rods of *B. popilliae* can then be diluted and directly plated on J-medium or MYPGP agar (Appendix). St Julian *et al.* (1963) recommend 0.1% tryptone as a standard diluent, particularly if vegetative rods are involved. Problems arising from poor germination and/or outgrowth of spores have been noted and can inhibit strain isolation (Stahly *et al.*, 1992a). Only 1–5% of the spores are likely to produce colonies on agar. Numerous attempts to increase germination with heat treatments or chemical additives have given variable results. The most successful procedures involved heating spores at 60°C for 15 min in calcium chloride ($CaCl_2$, 1 mM) at pH 7, and suspending them in cabbage looper haemolymph and tyrosine at an alkaline pH (Stahly *et al.*, 1992a).

Ironically, poor spore germination may be beneficial for strain isolation. Milner (1977) has taken advantage of the poor spore germination in isolating *B. popilliae* var. *rhopaea* from soil.

- Soil suspension (2 g wet weight of soil in 20 ml water) mixed with a germinating medium (0.5% yeast extract, 0.1% glucose, at pH 6.5) and alternately subjected to heat shock 7 times at 70 °C for 20 min.
- Following that procedure, which killed vegetative cells and germinated spores of contaminant bacteria, an aliquot was spread on J-medium and incubated anaerobically at 28 °C for 7 days.

There is no information on how this procedure works with other varieties of *B. popilliae*. Stahly *et al.* (1992b) note that this is a very time consuming procedure and is not an accurate method for quantifying spores in soil. An alternative procedure that they suggested utilized natural resistance to vancomycin which is shown by many strains of *B. popilliae* to isolate the bacteria from spore powder or from inoculated soil.

- 1 g of spore powder Doom® (Fairfax Biological Lab., Clinton Corners, NY), or 1 g of each, Doom and soil, placed in 5 ml of water and heated at 60 °C for 15 min to kill vegetative cells.
- Dilutions made in water and 0.1 ml plated on MYPGP agar containing vancomycin (150 µg/ml).
- Mould growth, a particular problem when soil samples are analysed, prevented by the addition of cycloheximide (1 mg/ml).
- Plates incubated at 30 °C for up to 3 weeks. Maximum *B. popilliae* colony counts occurred in about 9 days.

Vancomycin did not interfere with the colony forming ability of *B. popilliae*. Although not all *B. popilliae* strains are vancomycin resistant, this is a valuable tool for isolation of those that are.

B *Serratia* spp.

A rapid typing system has been developed for isolating amber disease bacteria based on the use of selective agars (O'Callaghan & Jackson, 1993).

- Soil or larval extracts plated on caprylate-thallous agar (CTA) (Appendix) to allow *Serratia* spp. to grow, while repressing other micro-organisms (Starr *et al.*, 1976).
- Gut tract from larvae with signs of amber disease extracted by holding each end of a larva with a forceps and pulling in opposite directions.
- The gut, which normally stays attached to the anterior end, crushed in water and spread on CTA, or directly streaked onto the agar.
- Cream coloured colonies, 2 mm in diameter, form after 4–6 days at 27 °C and can be transferred to selective media to complete the identification process (see Section 3B).

B Bacillus thuringiensis

Bacillus cereus and *B. thuringiensis* are commonly isolated from soil and insects. Claus & Berkeley (1986) suggest the use of 50% (v/v) ethanol for at least 1 h to clean up *Bacillus* spores before isolating the bacterium. This procedure will kill vegetative cells but not increase the chance for mutations as found after heating the culture above 70 °C for 10 min. They also described two media for selective isolation of *B. cereus*. The first relies on polymyxin to suppress Gram-negative organisms, and an egg yolk reaction to identify the *B. cereus*.

- Commercially available medium, polymyxin pyruvate, egg yolk, mannitol, bromothymol blue agar (PEMBA) (Appendix) can be used.
- On this mixture, *B. cereus* colonies are slightly rhizoid, have a distinct turquoise blue colour, and are usually surrounded by an egg yolk precipitate of similar colour.
- Alternatively, a liquid enrichment broth (peptone, 5.0 g; meat extract, 5.0 g; KNO_3, 10 g; distilled water to 1000 ml) can be used.
- After the medium has been sterilized at 121 °C for 15 min, pasteurized (10 min, 80 °C) soil samples (5 g/100 ml) are added to the enrichment broth.
- After 24 h incubation at 30 °C, the suspension is streaked on to nutrient agar plates and held at 37 °C for colony formation.

B. thuringiensis can be isolated using the procedures above. The main difference between *B. cereus* and *B. thuringiensis* is the presence of parasporal bodies, or crystals, in the latter. Other isolation procedures for *B. thuringiensis* are presented in Chapter III-1. A series of biochemical tests outlined by Claus & Berkeley (1986) and Thiery & Franchon (Chapter III-1), and the keys presented in Chapter III-1 can help identify *Bacillus* species isolated from scarabs.

3 IDENTIFICATION

A *Bacillus popilliae*

Dutky (1940) first described two obligate pathogens, *B. popilliae* and *B. lentimorbus*, as the causative agents of Type A and B milky disease in the Japanese beetle. The distinguishing feature separating the two bacteria is the presence of a crystal, or parasporal body, in *B. popilliae* (Figure 1). The sporangium has a 'footprint' shape with the spore forming the sole,

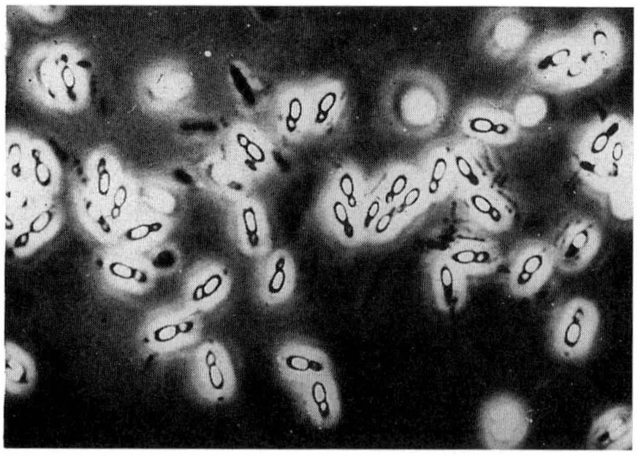

Figure 1 *Bacillus popilliae*, the causal agent of milky disease. Note footprint-shaped sporangium with a large terminal spore and smaller parasporal body (crystal).

and the parasporal body the heel. Keys for distinguishing these two bacteria are presented in Chapter III-I. Numerous species, varieties and strains of milky disease bacteria have been reported. Although it has been suggested that milky disease bacteria should be placed in the genus *Clostridium*, it is generally agreed that on the basis of their aerobic growth they are properly placed in the genus *Bacillus* (Milner, 1981). The general inability of the milky disease bacteria to grow and sporulate on standard microbiological media has resulted in confusion of the taxonomy of the group. Standard biochemical reactions used in the identification of most bacteria (Krieg & Holt, 1984; Sneath *et al.*, 1986) have not been very useful with the *B. popilliae* group. Milner (1981) found insignificant differences among four varieties of *B. popilliae* (var. *popilliae*, *lentimorbus*, *melolonthae*, and *rhopaea*) after 20 standard tests. The physiology of one milky disease bacterium, *B. popilliae* var. *popilliae*, has been extensively examined. Those interested in exploring this subject are directed to the review of Stahly *et al.* (1992a) to obtain the extensive list of references available.

Several investigators have treated strains of the milky disease bacteria as varieties of *B. popilliae* (Milner, 1981; Ellis *et al.*, 1989; Klein & Jackson, 1992). Others (Gordon *et al.*, 1973; Claus & Berkeley, 1986; Stahly *et al.*, 1992a) separate *B. popilliae* and *B. lentimorbus* as valid species. The two bacteria differ not only in the presence of a parasporal body, but also in their lipid composition, spore surface topography, and their antigen antibody reactions (Claus & Berkeley, 1986; Stahly *et al.*, 1992a). In addition, their DNA GC (guanine, cytosine) ratios suggest separate species. Stahly *et al.* (1992a) also suggested species status for the Australian isolate originally described as *B. euloomarahae*.

To help clear up the taxonomic confusion, Milner (1981), Claus & Berkeley (1986), and Stahly *et al.* (1992a) have suggested obtaining information on host specificity, spore size and morphology, parasporal body structure, type of peptidoglycan, major type of quinone, DNA GC ratios, nucleic acid hybridization, fatty acid composition, polar lipids, serology, menaquinones, and electrophoresis of enzymes.

Milner (1981) expanded on Dutky's division of milky disease into Type A and B to place the milky disease bacteria into four groups. He retained the basic premise that Type A contains a parasporal body and Type B lacks a parasporal body. Type A_1 has a large spore with the parasporal body overlapping the spore and often being small. In Type A_2, the parasporal body is often large and is separated from the large spore. Type B_1 differs from B_2 in that the former has a large central spore, and the latter has a small, often eccentric, spore. Only isolate 'RM 12' from *Anoplognathus porosus* in Australia fits the Type A_2 profile. Most varieties isolated to date are similar to var. *popilliae* (Type A_1) (see table in Klein & Jackson, 1992). This includes bacteria from important hosts around the world.

B *Serratia* spp.

After isolation on CTA, bacteria conforming in appearance to *Serratia* spp. are transferred to three other media to provide the final determination of the amber disease bacteria (Jackson *et al.*, 1993).

- Isolates first streaked on DNase–toluidine blue agar (Appendix) where *Serratia* spp. develop a pink halo around the growing colonies after 24 h incubation at 30 °C.
- Final identification takes place by spotting individual colonies onto adonitol agar and itaconate agar media (Appendix).
- On adonitol agar, *S. entomophila* produces yellow colonies, and *S. proteamaculans* produces blue/green colonies after incubation at 30 °C for 24 h. Growth on itaconate agar after 96 hours incubation at 30 °C is another characteristic of *S. entomophila*.

In addition, *S. entomophila* can be differentiated from other *Serratia* species by serotyping (Allardyce *et al.*, 1991). Bacterial adhesion to the grass grub foregut has been shown to be a primary determinant of pathogenicity. Pathogenic and non-pathogenic strains can be differentiated by the presence of a plasmid (Glare *et al.*, 1993).

C *Bacillus thuringiensis*

Identification of *B. thuringiensis* is covered by Larget & Frachon (Chapter III-1) and will not be detailed here. There has been some activity against scarab larvae from strains of *B. thuringiensis* that are active against other Coleoptera. The Cry III protein from those strains appears to be responsible for the activity. The most potent isolate against scarab larvae is the Buibui strain (Ohba *et al.*, 1992). Buibui is the Japanese name for the main target of this bacterium, the cupreous chafer, *Anomala cuprea*. The Buibui strain is further characterized as being *B. thuringiensis* serovar *japonensis* (flagellar antigen 23). The parasporal inclusion, or protein crystal, is spherical to ovoid.

4 PROPAGATION

A *Bacillus popilliae*

The inability to produce highly infective *B. popilliae* spores on artificial media has severely restricted research on this bacterium and the use of milky disease spore powder as a biopesticide. Until now, milky disease products have been made by finding naturally infected larvae or by collecting white grubs, particularly Japanese beetle larvae, infecting them and extracting the spores.

1. In vitro method

As noted earlier in this chapter, numerous varieties and strains of the bacterium *B. popilliae* are responsible for the milky disease infections of scarab larvae. Unlike *B. cereus* and *B. thuringiensis*, *B. popilliae* is extremely fastidious in its growth requirements and does not sporulate well in liquid media in fermentors or on agar-solidified nutrient media. A number of more complex media have been used to support its growth. St Julian et al. (1963) developed what became the standard medium (J-medium, see Appendix) for obtaining growth of *B. popilliae* cells. The final pH was 7.3–7.5, and as with all the media discussed here, the carbohydrate source was sterilized by filtration and added after the rest of the medium had been autoclaved. Other workers have used modifications of the basic J-medium. Lingg & McMahon (1969) observed better growth when the proportions of tryptone and yeast extract were reversed to provide 1.5% tryptone, 0.5% yeast extract, 0.6% K_2PO_4, 0.2% glucose and 1.5% agar with a final pH of 7.0–7.2. Costilow & Coulter (1971) suggested the more complex MYPGP medium (Appendix). More recently, Ellis *et al.* (1989) replaced the glucose in the J-medium with trehalose (the major sugar in haemolymph) resulting in a medium with 1% tryptone, 0.5% yeast extract, 0.3% K_2HPO_4, 0.2% trehalose and 1.5% agar. *B. popilliae* has also been found to grow on Difco® Brain Heart Infusion and Difco® Todd-Hewitt Broth (Appendix) media (D. Dingman, personal communication).

The *in vitro* method of spore production patented by Ellis *et al.* (1989) for milky disease bacteria is not being utilized. Bacteria reported to be *B. popilliae* produced on artificial media were identified as *B. polymyxa* and *B. amylolyticus* (Stahly & Klein, 1992). The bacterial make-up of the commercial products and a lack of infectivity from anything other than *B. popilliae* (Stahly & Klein, 1992) caused withdrawal of all commercially *in vitro*-produced milky disease spore powder.

2. In vivo *method*

Dutky (1942) provided the following procedure for production of milky disease bacterial spores in larvae. Films of dried haemolymph from diseased larvae on glass microscope slides (see Section 6A for details) may be kept as stock cultures of *B. popilliae* spores for over 30 years.

a. Preparation of inoculum

1. A standardized spore suspension is obtained by removing the spores from the slides with 0.5 ml of distilled water. This is facilitated by stroking the moistened film with a sterile 1 ml pipette to bring the spores into suspension.

2. The suspension is allowed to run into a sterile test tube by holding the tip of the pipette against one corner of the slide and tilting the slide toward the pipette tip.
3. Fresh distilled water is flooded over the slide and the procedure is repeated.
4. Spore counts on the suspension are made using a counting chamber, and the number of spores in the suspension is adjusted by the addition of sterile distilled water to ca. 3×10^8 ml^{-1}.

To prepare sufficient spore suspension to inoculate 500–1000 larvae, the following will be helpful.

1. Prepare a test tube (A) containing 1 ml of sterile distilled water and a second tube (B) containing 2 ml of sterile distilled water.
2. Wash spores from two culture slides into the inoculum tube (B) with an additional 1 ml portion of sterile distilled water so that the total volume of the suspension is 3 ml.
3. Fill a capillary pipette to the 0.01 ml mark with a loopful of this suspension and discharge the capillary pipette into the sample tube (A).
4. Shake the tube, withdraw a loopful of the suspension from the sample tube and fill a counting chamber.
5. Place the counting chamber on the mechanical stage of a microscope and allow the chamber to stand for 2 min to permit settling of the spores in the chamber.
6. Using a 65× objective, count the number of spores in 5 large squares of a Levy, or similar, counting chamber (80 small squares).

The number of spores in 5 large squares multiplied by 5 equals the number of spores $\times 10^6$ per ml of the suspension in tube (B). This figure divided by 100 equals the volume to be made in order to have the desired concentration of 3×10^8 spores per ml of suspension. Subtract 3.00 to determine the volume of sterile distilled water which must be added. Additional details on the use of counting chambers are presented in Chapter V-3.

b. Inoculating and incubating larvae

1. A hypodermic syringe with a 27 gauge needle is filled with the adjusted suspension and the entrapped air is expelled from the needle. Care must be exercised to ensure that no air bubble is present in the needle. Such a bubble will result in the failure of the small volumes to be injected into the grub.
2. The loaded syringe is then put into the microinjector (Figure 2), and the micrometer screw is turned forward until a droplet is forced from the end of the needle. This droplet is removed with a piece of absorbent cotton.
3. To quantify the spore dosage, an inoculating dosage is discharged into a dry, sterile test tube and sterile distilled water (1 ml) is then pipetted into the tube.
4. Counts are made on this sample with a haemocytometer to confirm the spore dosage.

Larvae are injected as follows.

1. The grub is held firmly, but lightly, between the thumb and forefinger, the dorsal posterior is positioned outward and the grub is forced onto the needle point so that the needle enters the dorsal portion of the suture between the second and third posterior abdominal segments. Care must be taken that the needle enters horizontally so as not to puncture the intestine. In case of accidental puncture of the intestine of a grub during injection, the hypodermic needle should be sterilized immediately by swabbing with 0.5% sodium hypochlorite solution before proceeding with further inoculation. Otherwise, larvae may be infected with septicaemia-producing bacteria contaminating the needle. Care must be also be taken to ensure that the site of puncture is free from soil particles.
2. The larva is then allowed to hang suspended on the needle during the injection (Figure 2).
3. The injection is made by depressing the pressure bar of the ratchet mechanism of the microinjector which forces an inoculating dosage into the body of the grub.

It is preferable to use full-grown larvae, although small third-instar and even second-instar larvae may be used if larger larvae are not available. By volume, the dosage approximates 3.3 μl. Since the inoculum is adjusted to 3×10^6 spores ml^{-1}, the resulting spore dosage is approximately 10^6 spores. Diseased grubs can also be produced by microinjection of vegetative cells of *B. popilliae* into grubs. The procedure is similar except that only 10–100 cells are injected in each larva.

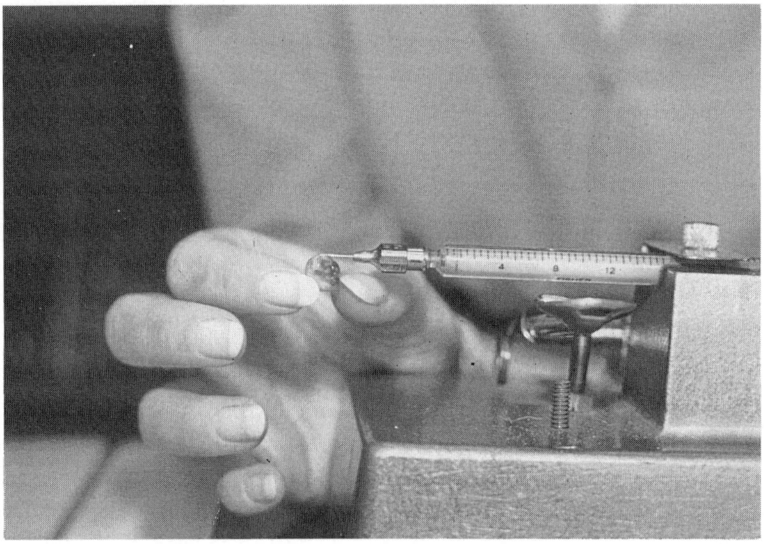

Figure 2 Dutky-Fest microinjector used to infect scarab larvae with milky disease bacteria. Note larva in position on the needle.

c. Preparation of soil for production of diseased grubs

- Boxes having a capacity of 500 larvae are used to hold the injected larvae during the period of incubation. These are equipped with metal cross-section separators that divide each of 5 layers into 100 compartments.
- After injection, the inoculated larvae are each dropped into a separate compartment. As soon as all compartments in a layer are occupied, soil (heated at 100 °C for 24 h) and brought up to 60% of the ball point) is added to fill the compartments.
- A flat, metal separator is placed on top of the filled layer, and a new cross section is put in the box. The compartments in the new layer are partially filled with soil, injected larvae are placed in these compartments, and the whole process is repeated until all 5 layers have been filled.
- The soil should contain 225 g of grass seed for each 40 kg of soil which, when sprouted, will serve as food for the larvae during incubation.
- The boxes are incubated at 30 °C for 10–12 days. High humidity should be maintained in the incubation chamber to prevent excessive drying out of the soil.

The use of soil at the proper moisture level is essential for satisfactory survival of larvae and spore production. Too low soil moisture will cause insufficient germination of the seed used for food and will prevent larvae from reaching maximum weight and hence, will reduce the spore yield. Excessive moisture will increase larval mortality by interfering with the normal gas exchange in the soil. The moisture value which is optimum for spore production and larval survival is dependent on the soil type and has been found, for a large number of soil types, to be that which just prevents adherence of soil particles to the cuticle of the larvae. This is approximately 60% of the 'ball point' of the soil.

To determine the ball point of the soil, 100 g soil is weighed into containers of approximately 250 ml capacity, and water is added from a burette in 2 ml increments.

The soil is mixed thoroughly after each addition until it forms a plastic mass with just an excess of free moisture. When this point is reached, the soil ball formed will re-form when the ball is broken up and the mass agitated by rotating the container.

The amount of water added to bring the 100 g sample to this state is recorded as the 'ball point' value of the soil. Between 60 and 65% of this value is the moisture content which the soil should contain for use in the incubation boxes. These values are determined and expressed on the basis of ml of water per 100 g air-dried soil.

d. Preparation of spore powder

1. After incubation, boxes are dismantled, and the grubs with milky disease signs are screened out of the soil and dropped into a battery jar of ice water. The ice water inactivates the larvae, permitting thousands of them to be placed in the jar

without danger of loss of spores due to larvae nipping one another.
2. The larvae are then washed in a colander to remove adhering soil particles, returned to the glass jar, packed with ice and held in a refrigerator at 1–2 °C until used.
3. When sufficient numbers of diseased larvae have accumulated, the grubs are crushed by running them through a meat chopper or blender. After all the larvae have been run through the machine, it is washed out with a small quantity of water to remove the adhering grub tissue.
4. The resulting suspension is placed in a graduated cylinder, brought up to an even volume, and spore counts are made on this suspension.
5. The suspension is then added to the carrier ($CaCO_3$, calcium carbonate, precipitated, USP) so that the mixture will contain 10^9 spores per gram of dry material.
6. The moist paste ($CaCO_3$ plus spore suspension) is mixed thoroughly by running it through a mixing device, such as a blade cutter or trowel mixer.
7. After thorough mixing has been accomplished, the moist paste is passed through a high-speed impeller-type blower which shears the agglomerated particles.
8. Drying of the dust is accomplished either by drawing heated air through the blower and exposing the finely divided particles to the warm air blast, or the paste can be spread out in a thin layer in front of a dehumidifier and pulverized when dry.

   Ellis *et al.* (1989) suggested that freeze-drying of the concentrate is preferable to the original air-drying procedure. When dry, this material is the stable concentrated spore-dust preparation.
9. The concentrate is then mixed with 9 parts of a dry carrier (talcum powder, marble flour, etc.) and stored until used. The final mixture contains 10^8 spores per gram.

The spore content of the dry spore-dust concentrate may be checked as follows:

1. To 10 g of the powder in a volumetric flask (200 ml capacity), is added approximately 100 ml of distilled water, the flask is shaken to wet the particles, and 20 ml of concentrated hydrochloric acid is added to the suspension.
2. The flask is gently rotated with the stopper removed until gas evolution ceases.
3. Distilled water is added until the volume is 200 ml, the flask is stoppered and shaken vigorously.
4. A loopful of this suspension is then used to fill a cell counting chamber. The count on the suspension should be ca. 5×10^7 spores/ml. The actual count in millions is then multiplied by 0.02 yielding spore counts in 10^9 per g of the concentrated dust mixture.

Syringes should be flushed with distilled water and then with alcohol immediately after use to avoid a plugged needle or 'freezing' of the plunger in the syringe. Careful records should be made of the number of larvae injected, spore and volume dosages, the number of larvae living at the end of the incubation period, the number diseased or not infected, and the yields. From such records, it is easy to control the process fully and correct any serious faults in procedure as they occur.

B *Serratia* spp.

Jackson *et al.* (1992) have discussed the commercial development of the amber disease bacterium against the grass grub in New Zealand in general, and of *S. entomophila* in particular. In contrast with other grass grub pathogens, *S. entomophila* can be mass produced *in vitro*. Strains of *S. entomophila* were selected on the basis of their virulence, persistence in the environment, ease of production and storage qualities.

The amber disease bacteria are produced on an industrial scale by fermentation similar to that of *B. thuringiensis* discussed in Chapter III-1. However, fermentation of *S. entomophila* produces a non-spore-forming bacterium that is applied as a living organism. Production of a high yield product is needed to economically provide sufficient bacteria to treat pastures at 4×10^{13} cells/ha to promote an epizootic. Improvements in fermentation technology have resulted in a yield of 4×10^{10} bacteria/ml, and a product volume of 1 l/ha. Shelf-life of the product is more than three months. Phage-resistant strains of *S. entomophila* provide a bacterium that can be fermented without a loss in pathogenicity or decrease in field performance.

For small-scale field work, nutrient broth can be used. For laboratory assays, Wilson *et al.* (1992)

reported that *S. entomophila* is grown overnight in Luria Bertani (LB) broth (Appendix) at 25 °C. Numbers of bacteria are estimated by dilution plate counting on LB agar (Appendix).

C *Bacillus thuringiensis*

The mass propagation and fermentation of scarab active strains of *B. thuringiensis* is the same as for other *B. thuringiensis* products and is presented by Thiery and Frachon (Chapter III-1). *B. thuringiensis* will grow and sporulate readily on nutrient agar (Appendix). Suzuki *et al.* (1992) outlined their procedure for producing spores and crystals for laboratory tests. Bacteria which had been purified on NYS agar (Appendix) for 24 h at 30 °C were cultured in NYS broth medium overnight on a rotary shaker at 350 rpm. NYS medium (200 ml) was placed in a 1 l baffled flask and inoculated with 2 ml of the seed culture. Cells were cultured for 3 days, and the spore/crystal complex was collected by centrifugation for 15 min at 15 000 ***g***, and rinsed 3 times with distilled water. In later tests, Suzuki *et al.* (1994) used 10 000 ***g*** for 10 min and added potassium sorbate (0.01%) to prevent bacterial growth.

5 BIOASSAY

A *Bacillus popilliae*

There is considerable interest in determining the virulence and host range of the many varieties of *B. popilliae* discussed above. Dutky (1941) outlined a series of tests of milky disease organisms (by injection and feeding) that have been used and slightly modified by many researchers, but is still generally valid today. As the first indication of susceptibility to *B. popilliae* infection he recommended injection of spores as follows:

- Use an inoculating dose of $1\text{–}2 \times 10^6$ spores per larva, as outlined in Section 4A.2.
- Use uninjected larvae and those injected with sterile water as controls.
- Hold larvae in individual tins or vials, in soil at 60% of the ball point with sprouted grass for food (Section 4A.2), and examine daily for 15 days.

Propawski & Yule (1990) replaced the soil–grass mixture with an artificial diet. Milner (1981) noted that the age of spores used in infectivity tests should be standardized (one month or more) since fresh spores may germinate poorly. St Julian *et al.* (1963) found that vegetative cells of var. *popilliae* could be tested after dilution in 0.01% tryptone. However, injections should be completed within 1 h of dilution. The cause of larval mortality and all macroscopic symptoms must be confirmed by microscopic examination. The recent demonstration of low infectivity with injected free spores of *B. popilliae* produced *in vitro* (Ellis *et al.*, 1989) suggest that feeding tests are a more reliable indication of bacterial virulence than is injection. With a feeding test, the bacteria must cross the intestinal lining and invade the haemocoel in order to establish the disease.

For his feeding tests, Dutky (1941) used a standard spore–talc preparation as described above.

- The spore–talc preparation was mixed with soil at 2×10^9 spores kg^{-1} air-dried soil, and the mixture was brought up to 60% of the ball point (Section 4A.2).
- A rate of 2×10^9 or more spores per kg of soil should result in about 85% milky larvae after one month (Dutky, 1963).
- Rates of 0.25 and 0.5×10^9 kg^{-1} soil have resulted in 50% milky Japanese beetle larvae (Dutky 1963). However, it should be remembered that those values were obtained by pooling the results of many tests and that several factors including age of spores, nutritional status of the host and incubation temperature can influence the results (Milner 1981).
- To obtain a better idea of the infectivity of the spore preparation, a range of 4–5 rates between 0.25×10^9 and $1\text{–}2 \times 10^{10}$ spores per kg soil should be used. Spore concentrate must be used for rates above 8×10^9 spores per kg soil.
- Soil for these tests should be at about 60% of the ball point of the soil.
- Larvae are incubated at 30 °C in the soil–spore powder medium with 10 g of seed per kg of soil added for food.
- Twenty or more larvae should be used for each concentration and rates should be replicated a minimum of three times.
- Examination of larvae twice a week for one

month is needed to establish the infectivity of the milky disease variety or strain being tested.
- Inclusion of an untreated control, as well as a standardized spore preparation of known virulence, is essential for interpreting the results.

Other methods of feeding *B. popilliae* spores have been used. Poprawski & Yule (1990) used a pipetting gun fitted with a blunt glass tip to force-feed 0.1 ml of bacteria in 0.1% tryptone at rates of 2–3 × 10^7 spores per ml to individual *Phyllophaga* larva. The placement of the spores on pieces of carrot or grass roots (Milner, 1981), and sugarcane roots (David & Alexander, 1975) has also been employed. Milner (1981) reported a LD_{50} of about 1 × 10^7 spores per larva in his direct feeding studies. Dingman (1996) noted that natural feeding of spores to grubs does not quantitatively or, consistently, introduce the pathogen into the host. A technique to force feed *B. popilliae* spores to Japanese beetle and European chafer larvae has been used to induce and study milky disease. Per oral injection was best done with a blunt 33 gauge needle and volumes of about 5 μl. Larvae were anaesthetized with a stream of CO_2 gas, held ventral side up, and by wiggling the mandibles against the needle, the needle could be inserted into the oral aperture and slid about 0.75 cm into the ventriculus. The procedure was done using a stereomicroscope at 6× magnification.

Results of these bioassay studies have shown that, in general, there is very little cross-infectivity of the strains of *B. popilliae* isolated from one grub species for grubs of another species. However, as noted above, the host range for varieties of *B. popilliae* is much greater when spores are injected rather than being fed. Milner (1981) notes that spore germination, host defence reactions and host nutrition, all contribute to the specificity of milky disease bacteria.

B *Serratia* spp.

Bioassays of amber disease bacteria are based on the ability of pathogenic bacteria to cause a cessation of feeding by infected larvae.

- Before testing starts, grass grub larvae are collected from the field and apparently healthy larvae are allowed to feed on carrot for 24 h in individual compartments.
- Those feeding after 24 h are selected for tests.
- Larvae are confined individually in 2 cm^3 compartments in covered trays and each is fed with a 20 mg piece of carrot.
- For tests, the carrot is impregnated with about 10^5 bacteria in 5 μl of suspension.
- Larvae are assessed after 7 days at 27 °C in the dark for appearance of the amber coloration and feeding.
- Infected larvae will not feed on another piece of carrot if it is offered.

C *Bacillus thuringiensis*

Sharpe (1976) found that a preparation of *B. thuringiensis* var. *galleriae* applied with a small spatula directly to the mouthparts of the Japanese beetle was able to invade the haemocoel and the parasporal crystal was found to be toxic. About 20% of diseased larvae contained sporulating *B. thuringiensis* as the only septicaemic agent. Later, Sharpe & Detroy (1979) showed that *B. thuringiensis* preparations were only effective when given to actively feeding Japanese beetle larvae.

Suzuki *et al.* (1992) used laboratory reared larvae. They collected eggs by allowing adult scarabs to oviposit in sand, and used the follow-up procedure.

- Eggs are transferred to compost medium and allowed to hatch.
- First to third instars of various scarabs are fed in individual cups containing 1 g of sterilized compost with various amounts of *B. thuringiensis* crystals for up to 21 days at 25 °C in the dark.
- Mortality is assessed at weekly intervals.
- Ten larvae are used for each dosage and each experiment is repeated three times.

Later, Suzuki *et al.* (1994) suggested rearing scarab larvae at 25 °C, 60% relative humidity, and 12/12 dark/light.

Spores and/or crystals of *B. thuringiensis* can be directly added to agar medium to which 5% sucrose has been added as a feeding stimulus (Ladd, 1988). Small larvae can be placed directly in holes made in the agar, or agar plugs can be removed and fed to larvae (A. Slaney, personal communication).

6 PRESERVATION

A *Bacillus popilliae*

Both Type A and B *B. popilliae* spores may be collected on glass slides and retained for future use in infectivity tests (Dutky, 1963). This method for preparing milky disease stock culture slides involves the following:

1. Wash about 20 milky larvae in running water, followed by thorough rinsing in 40–50% alcohol;
2. Transfer to 45–50% alcohol for 5 min.
3. Transfer grubs to water at 45–50°C for 5 min.
4. Remove grubs and place on clean cheesecloth or blotting paper until used.
5. Bleed each grub by puncturing behind the head capsule with a needle, on to a clean slide (do not puncture or rupture gut).
6. Place a second slide over the first leaving about 1.25 cm at the ends and separate the slides with a quick pulling movement in opposite directions;
7. Allow slides to dry and store in a slide box for future use.

This procedure is designed to keep the blood from coagulating and to allow the maximum recovery of spores from a grub. Longer conditioning times may be required for larger larvae. In addition, the blood from larger infected grubs can be placed on more than one set of slides, so that 4–6 slides may be made from a single grub. Spores isolated on microscope slides by Dr S. Dutky in 1945 using the above procedure, and subsequently stored at room temperature, were as infective as spores from any other source when tested against Japanese beetle larvae in 1983 (Klein & Jackson, 1992).

For vegetative cells, lyophilization of bacteria in the early stationary phase in a medium with less than 0.2% fermentable carbohydrate has been recommended. Vegetative cells of *B. popilliae* var. *popilliae* and var. *lentimorbus* have also been maintained by transferring to fresh J or MYPGP agar every 7–10 days (Stahly *et al.*, 1992a). Gordon *et al.* (1973) used monthly transfers of *B. popilliae* in tubes of J-medium without glucose, but with 0.1% agar. Milky disease bacteria can also be preserved by adding glycerol (to 15%) to mid-log phase cells in MYPGP broth and storing the cell suspensions at –70°C (Dingman, personal communication). Lyophilization of vegetative cells, in either bovine serum or a solution of 5% sodium glutamate and 0.5% gum tragacanth, has also been successful (Stahly *et al.*, 1992a). Spores can be washed and added to dry, sterile soil, placed on microscope slides and dried, or kept as suspensions in water or alcohol (Milner, 1981; Stahly *et al.*, 1992a).

B *Serratia* spp.

Detailed suggestions for the maintenance of bacterial cultures from different genera are presented in Volumes I and II of Bergey's Manual (Krieg & Holt, 1984; Sneath *et al.*, 1986). In general, once purified, bacterial isolates should be stored for short-term preservation on agar slants or stabs at room temperature or under refrigeration. As with all microorganisms, species should not be subcultured at short intervals so as to prevent loss of virulence and the selection of mutant strains. For long-term storage, bacteria should either be kept in glycerol at –70°C, under liquid nitrogen or be lyophilized. Representative strains of insect pathogenic bacteria should be deposited in a recognized reference collection so the isolates will be properly maintained and available for future use.

C *Bacillus thuringiensis*

As with non-spore-forming bacteria, the spore-forming species should not be maintained by subculturing at short intervals (Claus & Berkeley, 1986). Rather, sporulated cultures can be kept for at least one year at room temperature on agar slants protected from drying. Both sporulating and vegetative strains can be preserved by freeze-drying in a medium of skim milk powder (20%, w/v) supplemented with 5% (w/v) meso-inositol, and subsequently stored under vacuum. As an alternative, cells can be preserved in liquid nitrogen using nutrient broth supplemented with 10% (v/v) glycerol or 5% (v/v) dimethyl sulphoxide as a cryoprotective. This latter method is recommended if vegetative cells are to be protected (Claus & Berkeley, 1986). Other methods for long-term storage of *Bacillus* spp. can be found in Chapter III-1.

REFERENCES

Allardyce, R. A., Keenan, J. I., O'Callaghan, M. & Jackson, T. A. (1991) Serological identification of *Serratia entomophila*, a bacterial pathogen of the New Zealand grass grub (*Costelytra zealandica*) *J. Invertebr. Pathol.* **57**, 250–254.

Claus, D. & Berkeley, R. C. W. (1986) Genus *Bacillus* Cohn 1872. In *Bergey's manual of systematic bacteriology*, vol. II (eds P. H. A. Sneath, N. S. Mair, M. E. Sharpe & J. G. Holt), pp. 965–1599. Williams & Wilkins, Baltimore.

Costilow, R. N. & Coulter, W. H. (1971) Physiological studies of an oligosporogenous strain of *Bacillus popilliae*. *Appl. Microbiol.* **22**, 1076–1084.

David, H. & Alexander, K. C. (1975) Natural occurrence of milky disease bacterium *Bacillus popilliae* Dutky on white grubs in India. *Current Sci.* **44**, 819–820.

Dingman, D. W. (1996) Description and use of a peroral injection technique for studying milky disease. *J. Invertebr. Pathol.* **67**, 102–104.

Dutky, S. R. (1940) Two new spore-forming bacteria causing milky disease of the Japanese beetle. *J. Agric. Res.* **61**, 57–68.

Dutky, S. R. (1941) Testing the possible value of milky diseases for control of soil-inhabiting larvae. *J. Econ. Entomol.* **34**, 217–218.

Dutky, S. R. (1942) Method for the preparation of spore-dust mixtures of type A milky disease of Japanese beetle larvae for field inoculation. *USDA Bur. Entomol. Pl. Quar.*, ET-192, 15 pp.

Dutky, S. R. (1963) The milky diseases. In *Insect pathology: an advanced treatise*, vol. II (ed. E. A. Steinhaus), pp. 75–115. Academic Press, New York.

Ellis, B-J., Obenchain, F. & Mehta, R. (1989) *In vitro* method for producing infective bacterial spores and spore-containing insecticidal compositions. US Patent, 4,824,671.

Glare, T. R., Corbett, G. E. & Sadler, A. J. (1993) Association of a large plasmid with amber disease of the New Zealand grass grub, *Costelytra zealandica*, caused by *Serratia entomophila* and *S. proteamaculans*. *J. Invertebr. Pathol.* **62**, 165–170.

Gordon, R. E., Haynes, W. C. & Pang, C. H-N. (1973) *The genus Bacillus*. USDA Agric. Handbook 427, 283 pp.

Jackson, T. A. (1992) Scarabs – pests of the past or future? In *Use of pathogens in scarab pest management* (eds T. A. Jackson & T. R. Glare), pp. 1–10. Intercept Ltd, Andover.

Jackson, T. A. & Glare, T. R. (1992) *Use of pathogens in scarab pest management*, Intercept Ltd, Andover, 298 pp.

Jackson, T. A., Pearson, J. F., O'Callaghan, M. O., Mahanty, H. K. & Willocks, M. J. (1992) Pathogen to product-development of *Serratia entomophila* (Enterobacteriaceae) as a commercial biological control agent for the New Zealand grass grub (*Costelytra zealandica*). In *Use of pathogens in scarab pest management* (eds T. A. Jackson & T. R. Glare), pp. 191–198. Intercept Ltd, Andover.

Jackson, T. A., Huger, A. M. & Glare, T. R. (1993) Pathology of amber disease in the New Zealand grass grub *Costelytra zealandica* (Coleoptera: Scarabaeidae). *J. Invertebr. Pathol.* **61**, 1–8.

Klein, M. G. (1992) Use of *Bacillus popilliae* in Japanese beetle control. In *Use of pathogens in scarab pest management* (eds T. A. Jackson & T. R. Glare), pp. 179–189. Intercept Ltd, Andover.

Klein, M. G. (1995) Microbial control of turfgrass insects. In *ESA handbook of turfgrass insect pests* (eds R. L. Brandenburg & M. G. Villani), pp. 95–100. ESA Publications, Lanham.

Klein, M. G. & Jackson, T. A. (1992) Bacterial diseases of scarabs. In *Use of pathogens in scarab pest management* (eds T. A. Jackson & T. R. Glare), pp. 43–61. Intercept Ltd, Andover.

Klein, M. G. & Kaya, H. K. (1995) *Bacillus* and *Serratia* species for scarab control. *Mem. Inst. Oswaldo Cruz* **90**, 87–95.

Krieg, N. R. & Holt, J. G. (1984) *Bergey's manual of systematic bacteriology*, vol. I. Williams & Wilkins, Baltimore, 964 pp.

Lacey, L. A., Amaral, J. J., Coupland, J. & Klein, M. G. (1994) The influence of climatic factors on the flight activity of the Japanese beetle (Coleoptera: Scarabaeidae): implications for use of a microbial control agent. *Biol. Control* **4**, 298–303.

Ladd, T. L. Jr (1988) Japanese beetle (Coleoptera: Scarabaeidae): influence of sugars on feeding response of larvae. *J. Econ. Entomol.* **81**, 1390–1393.

Lingg, A. J. & McMahon, K. J. (1969) Survival of lyophilized *Bacillus popilliae* in soil. *Appl. Microbiol.* **17**, 718–720.

Milner, R. J. (1977) A method for isolating milky disease, *Bacillus popilliae* var. *rhopaea*, spores from the soil. *J. Invertebr. Pathol.* **30**, 283–287.

Milner, R. J. (1981) Identification of the *Bacillus popilliae* group of insect pathogens. In *Microbial control of pests and plant diseases 1970–1980* (ed. H. D. Burges), pp. 45–59. Academic Press, London.

O'Callaghan, M. & Jackson, T. A. (1993) Isolation and enumeration of *Serratia entomophila* – a bacterial pathogen of the New Zealand grass grub, *Costelytra zealandica*. *J. Appl. Bacteriol.* **75**, 307–314.

Ohba, M., Iwahana, H., Asano, S., Suzuki, N., Sato, R. & Hori, H. (1992) A unique isolate of *Bacillus thuringiensis* serovar *japonensis* with a high larvicidal activity specific for scarabaeid beetles. Letters in *Appl. Microbiol.* **14**, 54–57.

Poprawski, T. J. & Yule, W. N. (1990) Bacterial pathogens of *Phyllophaga* spp. (Col., Scarabaeidae) in Southern Quebec, Canada. *J. Appl. Entomol.* **109**, 414–422.

St Julian, G., Pridham, T. G. & Hall, H. H. (1963) Effect of diluents on viability of *Popillia japonica* Newman larvae, *Bacillus popilliae* Dutky, and *Bacillus lentimorbus* Dutky. *J. Insect Pathol.* **5**, 440–450.

Sharpe, E. S. (1976) Toxicity of the parasporal crystal of

Bacillus thuringiensis to Japanese beetle larvae. *J. Invertebr. Pathol.* **27**, 421–422.

Sharpe, E. S. & Detroy, R. W. (1979) Susceptibility of Japanese beetle larvae to *Bacillus thuringiensis*: associated effects of diapause, midgut pH, and milky disease. *J. Invertebr. Pathol.* **34**, 90–91.

Sneath, P. H. A., Mair, N. S., Sharpe, M. E. & Holt, J. G. (1986) *Bergey's manual of systematic bacteriology*, vol II. Williams & Wilkins, Baltimore, 1599 pp.

Stahly, D. P. & Klein, M. G. (1992) Problems with *in vitro* production of spores of *Bacillus popilliae* for use in biological control of the Japanese beetle. *J. Invertebr. Pathol.* **60**, 283–291.

Stahly, D. P., Andrews, R. E. & Yousten. A. A. (1992a) The genus *Bacillus*: insect pathogens. In *The procaryotes*, vol. 2 (eds A. Balows, H. G. Truper, M. Dworkin, W. Harder & K. H. Schliefer), pp. 1697–1745. Springer-Verlag, New York.

Stahly, D. P., Takefman, D. M., Livasy, C. A. & Dingman, D. W. (1992b) Selective medium for quantitation of *Bacillus popilliae* in soil and in commercial spore powders. *Appl. Environ. Entomol.* **58**, 740–743.

Starr, M. P., Grimont, P. D. A., Grimont, F. & Starr, P. B. (1976) Caprylate-thallous agar medium for selectively isolating *Serratia* and its utility in the clinical laboratory. *J. Clin. Microbiol.* **4**, 270–276.

Suzuki, N., Hori, H., Ogiwara, K., Asano, S., Sato, R., Obha, M. & Iwahana, H. (1992) Insecticidal spectrum of a novel isolate of *Bacillus thuringiensis* serovar *japonensis*. *Biol. Control* **2**, 138–142.

Suzuki, N., Hori, H., Tachubabam M. & Asano, S. (1994) *Bacillus thuringiensis* strain Buibui for control of cupreous chafer, *Anomala cuprea* (Coleoptera: Scarabaeidae), in turfgrass and sweet potato. *Biol. Control* **4**, 361–365.

Wilson, C. J., Mahanty, H. K. & Jackson, T. A. (1992) Adhesion of bacteria (*Serratia* spp.) to the foregut of grass grub (*Costelytra zealandica* (White)) larvae and its relationship to the development of amber disease. *Biocontrol Sci. Technol.* **2**, 59–64.

APPENDIX

Caprylate-thallous agar (CTA)

Method: Makes 1l (45–50 plates)

Solution A:

0.15 g $MgSO_4.7H_2O$	Magnesium sulphate heptahydrate
0.68 g KH_2PO_4	Potassium dihydrogen orthophosphate
2.61 g K_2HPO_4	Dipotassium hydrogen orthophosphate anhydrous (Potassium phosphate)
1.0 ml $CaCl_2$	Calcium chloride solution (1%)
10.0 ml tion	Trace element solu- (see below)
1.1 ml $CH_3(CH_2)_6.COOH$	*n*-octanoic acid (caprylic acid)
0.25 g Tl_2SO_4	**Thallous sulphate**
0.1 g	Yeast extract

Warning: Care must be taken with thallous sulphate. It is extremely poisonous if inhaled. Avoid contact with skin, eyes and clothing.

Add to 500 ml distilled water, in the order given, dissolving each ingredient *completely* before adding the next. Adjust pH to 7.20. (Use K_2HPO_4 to raise pH; KH_2PO_4 to lower.)

Solution A can be made up to a week in advance, sterilized and stored at 4 °C.

Solution B:

7.0 g NaCl	Sodium chloride
1.0 g $(NH_4)_2SO_4$	Ammonium sulphate

Dissolve in 500 ml distilled water. Adjust pH to 7.20. (Use K_2HPO_4 to raise pH; KH_2PO_4 to lower.) Add 15 g of Difco agar. Add magnetic stirbar; heat to boiling.

Sterilize Solution A and Solution B separately for 15 min at 121 °C, 15psi. Add Solution A to Solution B aseptically, stir vigorously. To prevent precipitation from occurring, pour agar while still hot.

Trace element solution for CTA

1.96 g H_3PO_4	Trihydrogen phosphate
0.055 g $FeSO_4.7H_2O$	Ferrous sulphate hepta hydrated
0.0287 g $ZnSO_4.7H_2O$	Zinc sulphate hepta hydrated
0.0223 g $MnSO_4.H_2O$	Manganous sulphate hydrated
0.0025 g $CuSO_4.5H_2O$	Cupric sulphate hydrated
0.003 g $Co(NO_3)_2.6H_2O$	Cobaltous nitrate hexa hydrated
0.0062 g H_3BO_3	Boric acid powder

Dissolve in 1 l distilled water. Store at 4 °C.

Deoxyribonuclease agar (DNAse)

Method: Makes 1 l (45–50 plates)

Solution A:

900.0 ml	distilled water
37.8 g	DNAse agar

Add magnetic stirbar; dissolve agar in water; heat to boiling.

Solution B:

90.0 ml	distilled water
0.09 g	Toluidine blue O

Dissolve indicator in water; mix thoroughly.

Sterilize both solutions separately for 15 min at 121 °C, 15 psi. Add Solution B to Solution A; stir before pouring.

Adonitol agar

Method: Makes 1 l (45–50 plates)

8.33 g Peptone
4.17 g NaCl (sodium chloride)

Dissolve completely in distilled water. Adjust pH to 7.4.

Add:

10.0 ml	Bromothymol blue solution (see below)
5.0 g	Adonitol (dissolved in 20ml distilled water)

Bring the volume of the solution to 1 l using distilled water.

Add:

12.5 g	Bacto agar

Add a magnetic stirbar; heat to boiling. Sterilize for 30 min at 121 °C, 15 psi. Pour while still hot.

Bromothymol blue solution

0.2 g	Bromothymol blue
5.0 ml	0.1M NaOH Sodium hydroxide
95.0 ml	Distilled water

Sterilize for 15 min at 121 °C, 15 psi. Store at room temperature.

Itaconate agar

Method: Makes 1 l (45–50 plates)

6.0 g Na_2HPO_4	Sodium phosphate (dibasic)
3.0 g KH_2PO_4	Potassium phosphate (monobasic)
0.5 g NaCl	Sodium chloride
1.0 g NH_4Cl	Ammonium chloride

Dissolve ingredients in 1 l distilled water in the order given. Adjust pH to 7.0.

Add:

15 g	Difco agar

Add magnetic stirbar; heat to boiling. Sterilize for 30 min at 121 °C, 15 psi. Allow to cool.

Before pouring add:

10 ml	0.01M $CaCl_2$ solution (sterile)
1 ml	1M $MgSO_4.7H_2O$ solution (sterile)
10 ml	20% itaconate solution (filter sterilized)

J-medium

Method: Makes 1 l (45–50 plates)

5.0 g	Tryptone
15.0 g	Yeast extract
3.0 g	K_2HPO_4 Dipotassium hydrogen othophosphate anhydrous (potassium phosphate)
2.0 g	Glucose (filter sterilized)
1000 ml	Distilled water

Dissolve ingredients in distilled water. Adjust pH to 7.3–7.5. If a solidified medium is desired, add 20 g agar and a magnetic stirbar; heat to boiling. Sterilize for 15 min at 121 °C, 15 psi. Add glucose after mixture cools.

Luria Bertani agar

Method: Makes 1 l (45–50 plates)

15.0 g	Agar
10.0 g	Tryptone
10.0 g	NaCl Sodium chloride
5.0 g	Yeast extract

Dissolve ingredients in distilled water. Add magnetic stirbar; heat to boiling. Adjust pH to 7.5. Sterilize for 15 min at 121 °C, 15 psi.

Luria Bertani broth

Method: Makes 1 l

10.0 g	Pancreatic digest of casein
10.0 g	NaCl Sodium chloride
5.0 g	Yeast extract

Dissolve ingredients in distilled water. Add magnetic stirbar; heat to boiling. Adjust pH to 7.2. Sterilize for 15 min at 121 °C, 15 psi.

MYPGP medium

Method: Makes 1 l (50 plates)

10.0 g	Mueller–Hinton broth
10.0 g	Yeast extract
3.0 g	K_2HPO_4 Di-potassium hydrogen orthophosphate anhydrous (Potassium phosphate)
1.0 g	$C_3H_3O_3Na$ Sodium pyruvate
0.5 g	Glucose (filter sterilized)
1000 ml	Distilled water

Dissolve ingredients in distilled water. Adjust pH to 7.1. If a solidified medium is desired, add 20 g of agar and a magnetic stirbar; heat to boiling. Sterilize for 15 min at 121 °C, 15 psi. Add glucose after mixture cools.

Nutrient agar

Method: Makes 1 l (45–50 plates)

5.0 g	Peptone
3.0 g	Beef extract
15.0 g	Agar
1000 ml	Distilled water

Dissolve ingredients in distilled water. Add magnetic stirbar; heat to boiling. Adjust pH to 6.8. Sterilize for 15 min at 121 °C, 15 psi.

NYS agar

Method: Makes 1 l (45–50 plates)

1.0 g	Nutrient broth
1.0 g	Tryptone
2.0 g	Casamino acid
0.5 g	Yeast extract
15.0 g	Agar

Dissolve ingredients in distilled water. Add magnetic stirbar; heat to boiling. Adjust pH to 7.2. Sterilize for 15 min at 121 °C, 15 psi.

PEMBA

Method: Makes 1 l (45–50 plates)

1.0 g	Peptone	
10.0 g	D-Mannitol	
0.1 g	$MgSO_4.7H_2O$	Magnesium sulphate heptahydrate
2.0 g	NaCl	Sodium chloride
2.5 g	Na_2HPO_4	Sodium phosphate (dibasic)
0.25 g	KH_2PO_4	Potassium dihydrogen orthophosphate
0.12 g	Bromothymol blue (water soluble)	
18.0 g	Agar	
1000 ml	Distilled water	

Dissolve ingredients in distilled water. Adjust pH to 7.4. Dispense 90 ml amounts into bottles; sterilize for 15 min at 121 °C, 15 psi.

Before using, add the following solutions to *each* bottle of molten and cooled (50 °C) agar:

5.0 ml 20% w/v $C_3H_3O_3Na$ Sodium pyruvate, filter sterilized
100 units/ml Polymyxin, filter sterilized
5.0 ml Egg yolk emulsion (Oxoid SR 47)

If samples suspected of containing large numbers of fungi are to be examined, 1 ml 0.4% (w/v) actidione, filter sterilized, may also be added.

Brain heart infusion

Method: Makes 1 l

16.4 g	Calf brains, infusion from
20.5 g	Beef heart, infusion from
0.8 g	Proteose peptone
0.16 g	Dextrose
0.41 g	NaCl Sodium chloride
0.2 g	Na_2HPO_4 Sodium diphosphate (dibasic)

Dissolve in distilled water. Sterilize for 15 min at 15 psi.

Todd Hewitt broth

Method: Makes 1 l

33.3 g	Beef heart, infusion from
1.33 g	Neopeptone
0.13 g	Dextrose
0.13 g	NaCl Sodium chloride
0.03 g	Na_2HPO_4 Sodium phosphate (dibasic)
0.17 g	Na_2CO_3 Sodium carbonate

CHAPTER IV

Research methods for entomopathogenic Protozoa

ALBERT H. UNDEEN* & JIŘÍ VÁVRA†

* Medical and Veterinary Entomology Research Laboratory, USDA/ARS, Gainesville, Florida 32604, USA

† Department of Parasitology and Hydrobiology, Faculty of Sciences, Charles University, Vinicna 7, 128 44 Prague 2, Czech Republic

1 INTRODUCTION

'Protozoa', defined here as Protists (eukaryotic, single cell organisms) with 'animal' affinities, are currently divided to two kingdoms and 34 phyla (Corliss, 1994). However, for the sake of simplicity the modified classification of Levine *et al.* (1980), considering the Protozoa as a subkingdom of the kingdom Protista and dividing the Protozoa into seven phyla is used in this review. Representatives of the phyla Sarcomastigophora (flagellates and amoebae), Apicomplexa (gregarines, neogregarines, coccidia), Microspora (microsporidia) and Ciliophora, (ciliates) commonly occur in insects, together with some organisms of protistan nature of which the taxonomic position is uncertain.

Although any tissue or organ of insects may be infected, the various groups of protozoa characteristically infect particular sites in insect larval stages or adults. Amoebae and flagellates have a preference for various parts of the digestive tract (pharynx, salivary glands, proventriculus, midgut, Malpighian tubules). Apicomplexa, are intracellular or/and epicellular, and usually infect gut tissues, Malpighian tubules and fat bodies, but also have forms occupying the haemocoel. Microsporidia are strictly intracellular and their representatives can be found in any type of host tissues. The few ciliates that occur in insects appear to be restricted to the hemocoel.

The protozoa–insects symbiotic relationships vary from those in which both partners are indifferent to each other, the insect simply providing the spatial niche for the protozoan (commensalism), to those in which the protozoa living inside the insect host are essential for its survival (mutualistic relationship) or in which the protozoan is truly pathogenic to the insect host (parasitism). The focus here will be on those groups that are pathogenic to their hosts, providing only basic data necessary for the identification of non-pathogenic taxa. Blood-sucking insects serve as vectors of many protozoa, especially the flagellates (trypanosomes and their relatives) and Apicomplexa (*Plasmodium* and other blood parasites). Although extremely important from medical and veterinary aspects, these pathogens of man and

MANUAL OF TECHNIQUES IN INSECT PATHOLOGY
ISBN 0–12–432555–6

animals are beyond the scope of this chapter and will be mentioned only briefly.

2 IDENTIFICATION

This section provides a brief account of the Protozoa that are commonly found in insects and the insects in which they occur. Additional sources, such as Jahn's brief, but very informative booklet *How to know Protozoa*, Kudo's *Protozoology* (Kudo, 1966), Weiser's *Atlas of insect diseases* (1977), *Laboratory guide to insect pathogens and parasites* (Poinar & Thomas, 1984), *An illustrated guide to the protozoa* (Lee *et al.*, 1985), *Biological control of vectors* (Weiser, 1991) and *Insect pathology* by Tanada & Kaya (1993) need to be sought for further information. The variety of the protozoa to be identified preclude any detailed account of their life cycles, some of which are quite complex. As many protozoan pathogens are host specific, the host groups can be a useful starting point for identifying the pathogen. Table 1 is a short key to the phyla of protozoan pathogens in insects. Methods for viewing and identifying these various insect pathogens are found in Chapter VIII-1.

Table 1 Key to the major groups of protozoa pathogenic to insects.

1a.	Active stages containing cilia, flagella or pseudopodia; non-motile resting states not containing spores	2
1b.	Active stages absent, resting stage spores or cysts containing spores	3
2a.	Cilia or sucking tentacles present	Ciliophora
2b.	No cilia: motile stages with flagella or amoeboid movement; resting stages thick-walled cysts, frequently in the intestine or Malpighian tubules of the host	Sarcomastigophora
3a.	Spores with a coiled filament	4
3b.	Spores without a coiled filament	5
4a.	Spores discoidal; filamentous cell 3–4 coils around the circumference, three cells stacked in the centre, filamentous cell straightens upon hatching	*Helicosporidium*
4b.	Spores variously shaped but most often ovoid, 1–2 nuclei, filament forming one to many coils inside the spore and everts rapidly upon germination	Microspora
5a.	Meronts with an apical complex of organelles for penetration; spores mostly round or navicular containing sporozoites	Apicomplexa
5b.	Small spore formed from three cells	*Nephridiophaga periplanetae* and relatives

A Phylum: Sarcomastigophora

Diagnostic features Flagellates and amoebae move by flagella, pseudopodia, or both types of organelles (Levine *et al.*, 1980). Resistant cysts are formed by some forms. Representatives of the following orders might be encountered in insects:

1. Order: Kinetoplastida

Diagnostic features Uninucleate flagellates (insect representatives have a single flagellum), identifiable by the presence of a kinetoplast, a body that stains like the nucleus, but represents actually a part of the mitochondrion in which a large amount of extranuclear DNA is accumulated. Size range: 4–15 μm. Generally referred to as Trypanosomatids.

Occurrence Trypanosomatids are the most commonly occurring flagellate protozoa in insects being found primarily in the orders Heteroptera and Diptera, in the gut lumen or attached to epithelial cells of the oesophagus. They are more rarely seen in salivary glands or insect mouthparts.

Methods They are observed in fresh microscopic preparations of the gut contents, haemolymph or salivary glands in host tissue squashed in insect saline. Giemsa staining on dry or wet smears reveals the nucleus and kinetoplast.

Identification Fresh preparation shows the flagellates as rapidly moving, either rod-shaped or undulating bodies. Staining reveals a relatively large nucleus and the kinetoplast stained as a dense body located close to the basal body of the single flagellum. Genera are identified by the spatial relationship between the kinetoplast and the nucleus, presence or absence of the flagellum, its form, and the position of its emergence from the cell (Lee *et al.*, 1985).

Pathogenicity The species occurring in the gut are seldom pathogenic and usually act as subpathogenic stressors, unless the multiplication of the flagellates is so extensive that it will block the digestive tract. Those representatives which penetrate to the salivary gland via hemolymph tend to be more virulent to the insect host.

Representatives The following genera occur in insects:

- *Leptomonas*, characterized by the presence of promastigote stages, sometimes occurring together with cyst-like stages. Hosts: Hemiptera, Diptera, Hymenoptera, Blattoidea, Lepidoptera, Siphonaptera, Anoplura, digestive tract lumen. Non-pathogenic, e.g. *Leptomonas ctenocephali* from the dog flea *Ctenocephalides canis*, *Leptomonas pyrrhocoris* a ubiquitous parasite of the hemipteran *Pyrrhocoris apterus*.
- *Herpetomonas*, characterized by the occurrence of opisthomastigotes sometimes occurring also with promastigotes. Hosts: Diptera, digestive tube lumen. Non-pathogenic, e.g. *Herpetomonas muscarum* from *Musca domestica* and many other flies.
- *Crithidia*, have choanomastigote stages in the lumen of the digestive tracts of insects in the orders Diptera, Hemiptera, Trichoptera, and Hymenoptera. Non-pathogenic, e.g. *Crithidia fasciculata* from various culicine mosquitoes.
- *Blastocrithidia* shows epimastigote stages and cyst-like stages. Hosts: Diptera, Hemiptera, Siphonaptera with the digestive tube lumen as the typical location. They are pathogenic in some species, e.g. *Blastocrithidia triatomae* from kissing bugs *Triatoma* spp.
- *Phytomonas* has promastigote stages. These are plant pathogens (living in plant vascular systems) with insects serving as vectors. In order for transmission to be accomplished, the flagellate must invade not only the gut lumen, but also the salivary glands of the insect hosts. This is why these flagellates are expected to have some degree of virulence to their hosts (Hemiptera and some Diptera), e.g. *Phytomonas davidi* from various heteroptera feeding on milkweeds (*Euphorbia*).
- *Leishmania* has promastigote stages in the digestive tube of sandfly adults and amastigote stages in vertebrate macrophages, causing human and animal leishmaniasis. Although they live seemingly as non-pathogenic commensals in the digestive tube of their insect hosts, their adherence to the stomodeal valve may hinder, to some extent, the feeding of the insect, e.g. *Leishmania* spp. in the gut of *Phlebotomus* spp.

  Trypanosoma has a life cycle with trypomastigote stages as the most typical form in vertebrate tissues (typically in blood), whereas trypomastigote, epimastigote and, sometimes, also amastigote stages occur in the digestive tract, salivary glands and mouth parts of insects. Diptera, Hemiptera and Siphonaptera serve as vectors of these organisms that are sometimes serious pests of man and animals, causing trypanosomiasis. Pathogenicity to insects is various and depends on the life cycle of the flagellate in the host: some of these flagellates live as non-pathogenic commensals in the digestive tube of their insect hosts (e.g. *Trypanosoma cruzi* in the gut lumen of reduviid bugs), some penetrate from the gut to the haemocoel and the salivary glands (e.g. *Trypanosoma rangeli* in reduviid bugs) and are thus pathogenic to the insect vector. In some members of the genus *Trypanosoma*, Diptera of the genus *Glossina*, *Tabanus*, *Stomoxys*, serve as insect hosts and vectors. Their development in the insect host might be either very complex, involving different parts of the digestive system and salivary glands (e.g. *T. brucei* in *Glossina* spp.), or might be limited to a simple survival in the proboscis of the blood-sucking insect, enabling mechanical transmission (e.g. *T. evansi* in horse flies and *Stomoxys* flies). No reliable data are available on the pathogenicity of these flagellates for their insect hosts.

2. *Order: Retortamonadida*

Diagnostic features Small, uninucleate flagellates with 2 or 4 flagella. One flagellum is directed backwards and moves in a groove-like depression situated in the anterior part of the body and representing a cytosome. Cysts have the form of a very broad pear. Size range of the flagellate: 6–14 μm.

Occurrence It can readily be found in the hindguts of insects such as *Tipula* larvae, *Phyllophaga*, *Gryllotalpa*, *Blatta orientalis* and *Amitermes*.

Methods Fresh preparation of the gut contents, dry smear stained with Giemsa, wet smear stained with Heidenhain's iron haematoxylin, or Protargol silver staining.

Identification Two flagella are present in those species inhabiting insect gut. The wet smear staining mentioned above will reveal the flagella and fibrils bordering the cytosome. The nucleus, both flagella and cytostomal fibrils are stained in cysts.

Pathogenicity Non-pathogenic to their hosts.

Representatives Insect-inhabiting species belong to the genus *Retortamonas*, e.g. *Retortamonas blattae* from the gut of the oriental cockroach.

3. Order: Oxymonadida

Diagnostic features Small or medium size, uninucleate to exceptionally multinucleate, flagellates, usually with four flagella arranged in two pairs. Either three of these flagella are oriented forward, the fourth one serving as trailing flagellum and partly adhering to the cell, or all four flagella are directed backwards adhering for some length along the cell surface. A microtubular rod, called an axostyle, runs lengthwise through the body of the flagellate. Some forms have a prominent extensible rostellum, which is the holdfast organelle attaching the protozoan to the host gut.

Occurrence All representatives of this order occur in guts of wood-eating insects (termites and woodroaches), *Melolontha* and *Tipula* larvae.

Methods Fresh preparation of gut contents, wet smear stained with Heidenhain's iron haematoxylin, Klein's or Protargol silver staining.

Identification See Lee *et al.* (1985).

Pathogenicity Harmless mutualists and endocommensals.

Representative Oxymonas grandis

4. Order: Trichomonadida

Diagnostic features Medium size protozoa, usually with single nucleus, 4–6 flagella, one of which is a trailing flagellum attached to the cell surface and forming an undulating membrane. The cell contains an axostyle. Some insect-inhabiting members of this order have cells with many nuclei, many flagella and many axostyles.

Occurrence Most representatives of this order are parasites of vertebrates, but several genera occur also in the gut of termites.

Identification See Lee *et al.* (1985)

Methods Fresh preparation of gut contents, wet smear of gut contents stained by Heidenhain's iron haematoxylin or Protargol silver staining. Identification: see Lee *et al.* (1985).

Pathogenicity Harmless commensals.

Representatives Pseudotrypanosoma, Calonympha.

5. Order: Hypermastigida

Diagnostic features Highly organized multiflagellated but uninucleate protozoa. Their flagella are distributed as either an anterior tuft, a plate(s) or in spiral rows. Staining might reveal highly developed and organized 'parabasal bodies' (actually Golgi apparati) arranged in proximity of basal bodies of flagella.

Occurrence Mutualistic symbionts living in the guts of termites, woodroaches and cockroaches.

Methods See order Oxymonadida above

Identification See Lee *et al.* (1985).

Pathogenicity None; these flagellates are mutualistic symbionts, facilitating the digestion of cellulose for their hosts, some of which cannot live without the flagellates.

Representatives Lophomonas, Trichonympha.

6. Order: Amoebida

Diagnostic features Protozoa characterized by movement using pseudopodia. They are polyphyletic in origin, some having only amoebic forms, others being able to form resistant cysts and still others having flagellated forms during parts of their life cycles.

Occurrence Most amoebae are free-living but a few live in the gut or Malpighian tubules of insects.

Methods Fresh preparation of gut contents, squash of Malpighian tubules in insect saline. Dry smears stained by Giemsa, wet smears stained by haematoxylin stains.

Identification Identification is very difficult, involving knowledge of the type of pseudopodia formation, the structure of the nucleus after staining on wet smear, ability to form cysts and flagellate stages and the host. See Lee *et al.* (1985).

Pathogenicity Some amoebae are harmless commensals in the gut of insects (e.g. *Endamoeaba blattae* in cockroaches and several species of the same genus in the gut of termites). Only three genera, *Malameba* from Orthoptera, *Malphigamoeba* from Hymenoptera (*Apis mellifera*) and *Malpighiella* from the dog flea (*Ctenocephalides canis*) are known to be pathogenic to insects. They have amoebic forms and cysts in the lumen of the gut and Malpighian tubules, an example being *Malameba locustae* from the Malpighian tubules of grasshoppers of several genera.

Representatives As noted under pathogenicity.

B Phylum: Apicomplexa

All members of this phylum are parasitic. The diagnostic feature of this phylum is the 'apical complex', an organelle present in the sporozoite (the motile infective stage). The apical complex consists of a conoid (a cone of spirally wound microtubules), rhoptries and micronemes (tear-shaped secretory gland-like formations) that aid in penetration of the host cell – all apicomplexa are totally or partly intracellular during at least a part of their life cycle. Apicomplexan parasites are, however, most often and most easily identified by their resistant spores (properly called oocysts) which contain a specific number of sporozoites. The oocysts are oval or lemon-shaped refractile bodies with thick walls, sometimes containing another cyst-like body (or bodies), the sporocyst. Filiform or banana shaped, uninucleate sporozoites are present in different, often diagnostic numbers inside the sporocyst/oocyst. A new host is infected by ingesting the oocyst from which the sporozoites escape in its digestive tract.

The phylum contains many representatives occurring in insects, the classes Gregarina and Coccidia being the most important. The distinguishing characteristic between these two classes is the size of gamonts (stages ready to engage in part of the life cycle involving sexuality) – large in gregarines, small in coccidia – and the modification of the apical complex in trophic stages. In gregarines the apical complex is changed into a feeding and holdfast organelle (epimerite or mucron), in coccidia the apical complex is not modified in stages where it exists.

1. Class: Gregarina

This class has two orders that infect insects, Eugregarinida and Neogregarinida.

a. Order: Eugregarinida

Diagnostic features Most eugregarines are epicellular parasites divided into two regions, the nucleated deutomerite and the anterior, non-nucleated protomerite. The eugregarine lives embedded in a host cell by the structure, either a mucron or an epimerite, at the tip of the protomerite. The epimerite , usually a feature of the cephaline gregarines, is divided by a septum from the rest of the protomerite. The mucron performs the same function in non-segmented eugregarines. Some gregarines are, however, fully intracellular, at least in a part of their life cycle. The epimerite or mucron mediates the uptake of nutrients from the host cell and anchors the organism. This is why they have a wide range of forms from simple, button-like mucrons to very complicated epimerites with protrusions in the form of hairs, hooks, etc. The form of the epimerite, the body size of the eugregarine, its location within the host and the host, itself are major taxonomic characters used in eugregarine classification. An investigator usually finds a gregarine at the stage of the life cycle when it is a fully grown trophozoite, already detached from the host gut epithelium and moving freely in the gut lumen. Such a gregarine is a long oval or filiform cell, usually in the range of several tens or even hundreds of microns, moving by slow, gliding movements without any change in body shape. The cell might or not might be divided by a transverse septum. Frequently one encounters gregarines in syzygy, a prelude to copulation in which two individuals are attached either tailwise or sidewise (Figure 1). Another characteristic stage of eugregarines is the gametocyst, a large spherical formation in which two gregarines encyst and form gametes. A mature gametocyst

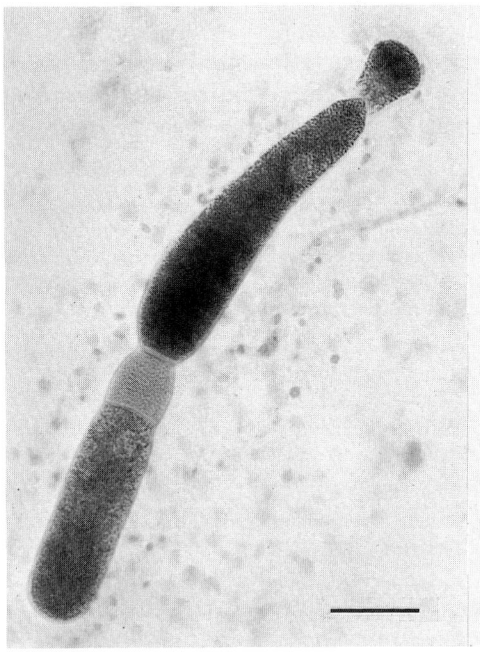

Figure 1 Syzygy of gamonts of *Gregarina cuneata*, a eugregarine from the gut of the mealworm *Tenebrio molitor* (bar = 50 μm). (Micrograph courtesy of Jaroslav Weiser.)

contains several hundred lemon-shaped oocysts, the shape also being diagnostic. These are released from the gametocyst by rupture of its wall or are evacuated by means of evaginable tubes, the sporoducts.

Occurrence Eugregarines are parasites of invertebrates and their presence might be expected in all insect orders. They are usually restricted to the gut although some species live in Malpighian tubules or in coelomic cavities.

Methods Fresh preparation of gut contents and body cavity fluids. A wet smear stained with haematoxylin will reveal a single nucleus in the posterior segment of the gregarine body and the septum, when present. Epimerites or mucrons are rarely seen in permanent preparation as they frequently remain embedded in host tissues during preparation. Staining also sometimes reveals fine longitudinal ribs on the gregarine pellicle. These are epicytar folds thought to be involved in movement or feeding. Sporozoites within oocysts can be revealed by Giemsa staining after an acid hydrolysis or by staining with haematoxylin stains (e.g. Heidenhain's).

Identification See Lee *et al.* (1985). The following simple key that allows the determination of main eugregarine families from insects.

Pathogenicity Because eugregarines do not divide as trophozoites and the only divisions occur during formation of oocysts (which often takes place outside of the host), their pathogenic effect on the host is minimal. Some pathogenic effect is, however, present in the forms living in Malpighian tubules and in the coelomic cavity.

Representatives Gregarines of the mealworm and *Ascocystis* spp. in the gut and Malpighian tubules of tree hole mosquitoes.

b. Order: Neogregarinida

Diagnostic features Neogregarines are morphologically dissimilar to eugregarines, usually being smaller and having, with the exception of a single genus, a non-segmented body. Reproduction is more complex than in the Eugregarines, involving schizogony during the vegetative part of the life cycle and again during gamogony and sporogony (Figure 2).

Occurrence They occur as epicellular but frequently as intracellular parasites in various organs in the haemocoel, Malpighian tubules, intestines, and the fat body of insects, which are their usual hosts.

Methods Dry smears are stained with Giemsa and wet smears with haematoxylin stains. These reveal uninucleate trophonts, multinucleate meronts, merozoites with single nuclei, gametocysts and oocysts. Sporozoites within the oocyst can be revealed by Giemsa staining after an acid hydrolysis or by staining with haematoxylin stains (e.g. Heidenhain's).

Identification Table 3 is provided to aid in the identification of these Gregarines.

Table 2 Key to the families of Eugregarinidae (modified from Weiser, 1966).

1a.	Round gametocysts in body cavity	Diplocystidae
1b.	Trophozoites and gametocysts in the gut	2
2a.	Epimerite conspicuous, sometimes structurally modified	3
2b.	Epimerite of the trophozoite is structurally simple	4
3a.	Gametocysts with one or several sporoducts	Gregarinidae
3b.	Gametocyst without sporoducts, open by rupture	Hirmocystidae
4a.	Gametocyst with a 2-layer envelope, with large, cup-shaped residual body. Oocysts evacuated first as a cyst-like structure, later being dispersed in chains	Stylocephalidae
4b.	Gametocyst with one layer envelope, open by rupture, oocysts not in chains	Actinocephalidae

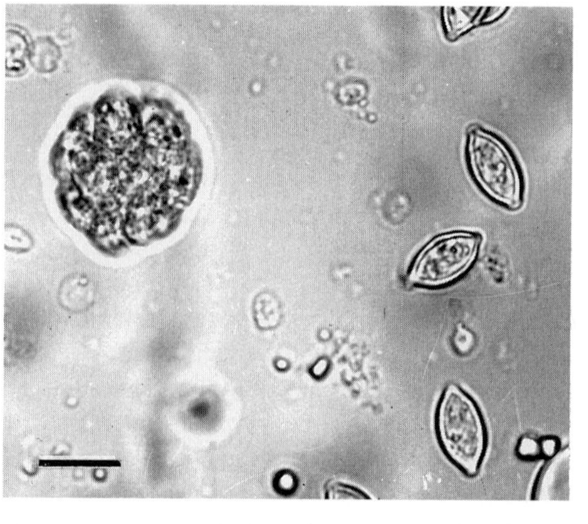

Figure 2 Maturing oocysts and free sporocysts of a neogregarine (*Farinocystis tribolii*) from the fat body of *Tribolium* sp. (bar = 10 μm). (Micrograph courtesy of Jaroslav Weiser.)

Table 3 Key to the most common genera of Neogregarines infecting insects.

1a.	Forms having, in the life cycle, two merogonies with different size of nuclei ('micronuclear' and 'macronuclear' merogony). Ophryocystidae	2
1b.	Only one merogony, with relatively large nuclei present, trophozoites band-like and wide. Schizocystidae	7
2a.	Meronts of the first and second merogonies are morphologically different. Gametocytes at the early syzygy are uninucleate. Ophryocystinae	3
2b.	Meronts of the first and second merogony are morphologically similar, but are multinucleate. Gametocytes at early syzygy multinucleate. Machadoellinae	6
3a.	Trophozoites and merozoites with root-like extensions attaching the parasite to host cells, gametocyst with single oocyst	*Ophryocystis*
3b.	Trophozoites and merozoites without root-like extensions gametocyst with several oocysts	4
4a.	One or two oocysts in a gametocyst	*Mattesia*
4b.	Gametocyst with more than two oocysts .	5
5a.	Maximum 40 oocysts per gametocyst	*Menzbieria*
5b.	More than 40, but usually 200–300 oocysts per gametocyst	*Lipocystis*
6a.	Trophozoites worm-like, gametocytes in syzygy usually with 4 nuclei, gametocyst with 4 oocysts	*Machadoella*
6b.	Trophozoites round in shape, gametocytes in syzygy with 8 nuclei, about 30 oocysts in a gametocyst	*Farinocystis*
7a.	Oocysts egg-like. Caulleryellinae	8
7b.	Oocysts naviculate	9
8a.	One oocyst per gametocyst	*Tipulocystis*
8b.	Eight oocysts per gametocyst	*Caulleryella*
9a.	Trophozoites round or oval in shape	10
9b.	Trophozoites long and band-like, disintegrating into groups of merozoites	*Schizocystis*
10a.	Oocysts naviculate, with 4 spines at each pole, more than 100 oocysts per gametocyst	*Syncystis*
10b.	Oocysts naviculate to egg-shaped, without spines, 16 oocysts per gametocyst	*Lipotropha*

Pathogenicity Because there is extensive multiplication of the parasite in the host, neogregarines are more virulent than eugregarines and infection is often lethal.

Representatives Mattesia dispora, Farinocystis tribolii

2. Class: Coccidia

This is a group of apicomplexans which are all intracellular parasites and multiply extensively during both the vegetative and sporogonic phase of the life cycle. All but a few coccidian species are host specific. Almost any tissue can be infected but the infection sites of individual species can be quite specific. Insect-infecting coccidia are members of the order Eucoccidiida and belong to the suborders Adeleina and Haemosporina.

a. Suborder: Adeleina

Diagnostic features Intracellular parasites multiply in host cells by merogony, displaying multinucleate meronts and uninucleate merozoites inside infected cells. Sexual stages (gamonts) are formed in relatively small numbers and may be found in long-lasting syzygy before copulation, the male gamont being much smaller and adhering to the much larger female gamont. Round or oval oocysts with a specific number of sporocysts and sporozoites are formed as the final stage of development (Figure 3).

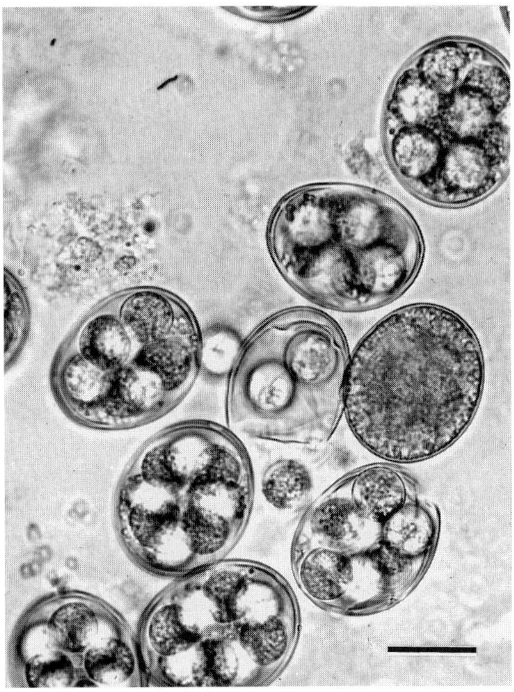

Figure 3 The coccidian *Adelina tribolii* showing young oocysts and sporocysts forming within the oocysts (bar = 20 μm). (Micrograph courtesy of Jaroslav Weiser.)

Occurrence Various invertebrates, most representatives in insects. Most genera use insects as a single host, several genera are parasites of vertebrates and use an insect for the sexual part of development and the formation of sporozoites.

Methods Dry smear stained with Giemsa and wet smear with haematoxylin. Sporozoites within the oocyst can be revealed by Giemsa staining after an acid hydrolysis or with haematoxylin. (e.g. Heidenhain's).

Identification The five genera that are entirely entomoxenous are *Adelina*, *Chagasella*, *Legerella*, *Ithania* and *Barrouxia*. In all these parasites the diagnostic stage is an oocyst, a spherical or subspherical thick-walled body, approximately 10–50 μm in size, containing spherical or ellipsoidal sporocysts which, in turn, contain sporozoites. The shape and size of the oocyst, the number of sporocysts it contains and the number of sporozoites within the sporocysts are diagnostic (Table 4).

Pathogenicity Pathogenic effect is more pronounced as more multiplication cycles take place in the insect host, resulting in increased mortality of the host or its greater sensitivity to insecticides. No reliable data exist on the influence of the representatives of the Haemogregarina and Hematozoon on insects which serve as their vectors.

Representative Adelina melolontha.

b. Suborder: Haemosporina

Haemosporina are coccidian parasites causing serious human and animal diseases, using insects as vectors. For example, mosquitoes transmit human and animal malaria caused by the genus *Plasmodium*. Hippoboscids, midges and *Chrysops* are vectors of bird parasites of the genus *Haemoproteus* and black flies transmit the bird parasite *Leucocytozoon*.

Diagnostic features Only the sexual part of the life cycle takes place in the insect host which, in contrast to the adeleid coccidia, does not involve the syzygy of gamonts. Thin walled oocysts are formed in the gut tissues and no sporocyst is formed. When mature, the oocyst breaks, releasing the filiform sporozoites, which migrate to the salivary glands of the insect.

Occurrence Gut tissues, haemolymph and salivary glands of blood sucking insects of the order Diptera.

Methods Dry smears stained with Giemsa, wet smears with Giemsa or haematoxylin stains.

Identification Upon dissection, oocysts can be seen on the surface of the gut and filamentous sporozoites in the salivary gland.

Pathogenicity Very mild if any, shortening the life span of the female insect.

Representative Plasmodium falciparum.

C Phylum: Microspora

This phylum encompasses the most numerous and most important protozoan pathogens of insects. Their classification is based on life-cycle studies and electron microscopy.

Diagnostic features All are obligatorily intracellular organisms, multiplying in the host cell in the form of small paucinucleate meronts or plasmodia. Excessive mortality, reduced longevity, or reduced fecundity are sometimes the indications of a microsporidium in laboratory-reared insects. Heavy infections in lightly pigmented hosts can be seen through the cuticle as a white or off-white coloration, frequently accompanied by swellings caused by hypertrophy of infected cells. Sometimes the body of the insect will be deformed, often swollen by masses of developing spores. However, in insects with opaque and heavily scleratized cuticles, spores can occur in numbers appearing to entirely fill the host without producing obvious signs. Immune responses

Table 4 Genera of coccidia affecting insects.

Genus	*Oocyst shape*	*No. sporocysts/oocyst*	*No. sporozoites/sporocyst*
Adelina	subspherical – spherical	3–30	2
Chagasella	ovoid	3	4–6 or more
Legerella	spherical	0	15–40
Ithania	spherical	1–4	9–33
Barrouxia	spherical	many	1

to microsporidia will sometimes show up as dark spots of melanized parasites or infected tissues. Infected tissues are identifiable during dissection of larger hosts by swelling and altered colour due to masses of spores or spore-filled cysts.

The nuclear component is a single nucleus or two tightly adhering nuclei in the form of a 'coffee bean', called a diplokaryon. At the end of the life cycle, spores are formed that contain an eversible tube ('polar tube') through which the spore contents ('germ', or 'sporoplasm') are inoculated into a host cell when the spore germinates. Spores will eventually be released into the habitat, usually after the host dies but sometimes before death, in faeces or silk. These spores infect the next host and are the most frequently encountered stage of microsporidia. The spores are generally small, several micrometres in size. Most are ovoid to tubular in shape (Figure 4) but some have external ornamentation. They are refringent under phase-contrast microscopy and often a vacuolar space can be seen at the posterior pole of the spore.

Transmission electron microscopy reveals the very complex internal organization of the spore. The internally coiled polar filament serves to inject the spore contents into a host cell, and is the most diagnostic feature of a microsporidium (Figure 5).

In the light microscope the presence of the polar filament can be confirmed by inducing the spore to germinate mechanically (by crushing the spores slightly under the coverslip) or by chemical means. Spores are quite resistant and can still be found in insects that have been pinned, alcohol-preserved or which are badly decayed.

Occurrence Microsporidia are parasites of animals, especially arthropods, and probably occur in all insect orders. Development is entirely intracellular in a wide range of host cells, with the gut epithelial cells and fat body cells being the most common. Examination of insects in laboratory rearings, especially those failing to pupate frequently yield a microsporidium.

Methods Dry smears are stained with Giemsa or Gram stain (or its modification). The Robinow–Piekarski method – Giemsa stain following acid hydrolysis – is used for determining the number of nuclei (either 1 or 2) in microsporidian spores on dry smears. Wet smears are stained with

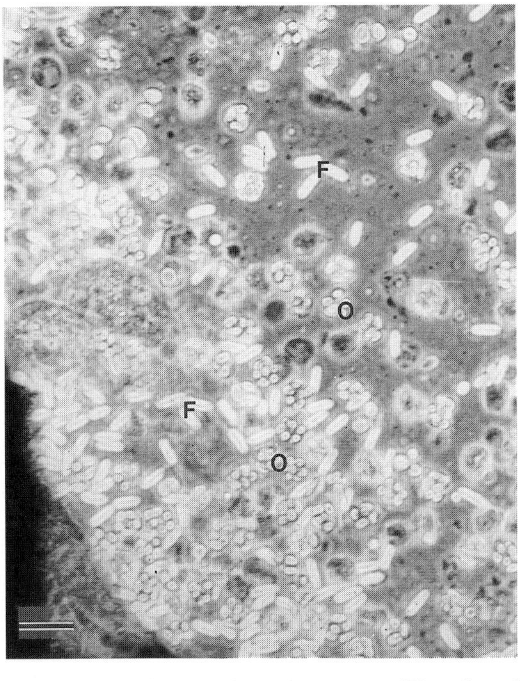

Figure 4 Elongate, binucleate free spores (F) and small spores in groups of eight (octospores) (O) of *Vairimorpha* sp. in a fresh smear of fire ant (*Solenopsis* sp.) fat body, viewed with a phase-contrast microscope (bar = 10 μm).

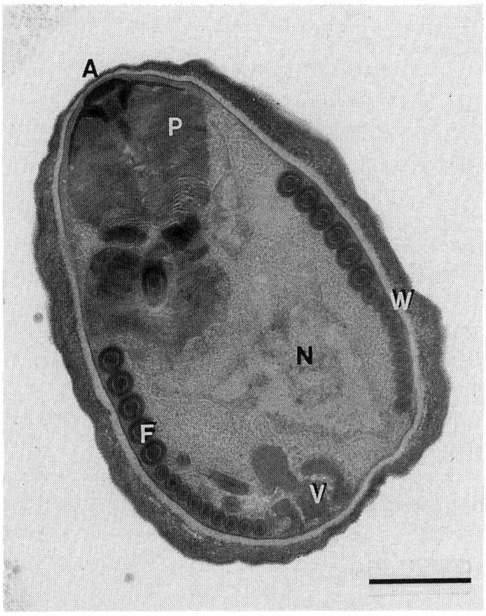

Figure 5 Electron micrograph of a longitudinal section through an immature microsporidian (*Amblyospora* sp.) spore showing the polar filament (F), nucleus (N), spore wall (W), polaroplast (P), posterior vacuole (V) and the polar tube anchoring region (A) (bar = 1 μm).

Table 5 Identification table for microsporidia of insects.

Genus	*ST*	*1*	*2*	*3*	*4*	*5*	*6*	*7*	*8*	*9*	*Type host (Order: Family), other hosts*
Amblyospora	1	D	S	S	+	8	B	6 × 4	a 9	f	*Culex* (Diptera: Culicidae), Simuliidae, Trichoptera
	2	D	D	D	0	0	C	11 × 3	i	o,ov	
Bohuslavia		D	S	S	+	8–16	P	6 × 3	i 10	f	*Endochironomus* (Diptera: Chironomidae)
Burenella	1	D	D	D	0	0	Oe	6 × 3	i 26	f	*Solenopsis* (Hymenoptera: Formicidae), Coleoptera
	2	D	S	S	[+]	8	OLe	7 × 4	i 57	f	
Buxtehudea		?	S	S	0	0	S	3	i 10	m	*Petrobius* (Thysanura: Machilidae)
Campanulospora		D	D	D	0	0	Oe	5 × 3	i 12	m	*Delia* (Diptera: Muscidae)
Canningia		S	S	S	0	0	Oe	2 × 1	i 6		*Pityokteines* (Coleoptera: Scolytidae)
Caudospora		D	D	D	0	0	Oc	6 × 5	i 12	f	*Eusimulium* (Diptera: Simuliidae)
Chapmanium		D	S	S	+	8	P	3 × 3	a 8	f	*Corethrella* (Diptera: Chaoboridae), Heteroptera, Chironomidae
Chytridiopsis			S	S	+	many	S	1–2	i 2	m	*Blaps* (Coleoptera: Tenebrionidae), Trichoptera
Coccospora		D	S	S	[+]	8	S	2–3	a 3	f	*Tanypus* (Diptera: Chironomidae), Arctopeliopia
Cougourdella[a]		S	S	S	[+]	4	L	7 × 2	i 6	f	*Megacyclops* (Crustacea), Trichoptera
Culicospora	1	D	S	S	+	2–8	L	14 × 7	i 12	f	*Culex* (Diptera: Culicidae)
	2	D	D	D	0	0	C	11 × 4	i 2	o	
Culicosporella	1	D	S	S	+	2–8	B	4 × 5		f	
	2	D	D	D	+	8	L	7 × 4	i 10	ov	*Culex* (Diptera: Culicidae)
	3	D	D	D	0	0	Ce	2 × 1			
Cylindrospora		D	S	S	[+]	8	C	10 × 7	i 0	f	*Chironomus* (Diptera: Chironomidae)
Cystosporogenes		S	S	S			Ee	2 × 1	i 10	s	*Operophtera* (Lepidoptera: Geometridae)
Duboscqia		S	S	S	+	16	O,E	5 × 3		f	*Reticulotermes* (Isoptera: Rhinotermitidae) Diptera, Chaoboridae
Edhazardia	1	D	D	D	0	0	Cb	9 × 3	i 6	m,g	*Aedes* (Diptera: Culicidae)
	2	D	S	S	+	1	L	8 × 4	i 10	f	
	3	D	S	S	+	8	B	8 × 6		f	
Endoreticulatus		S	S	S	0	0	OC-R	3 × 2	i 6	m	*Leptinotarsa* (Coleoptera: Chrysomelidae), Lepidoptera
Episeptum		S	S	S	[+]	4	P	2 × 1	a 6	f	*Holocentropus* (Trichoptera: Polycentropodidae)
Flabelliforma		S	S	S	+		O	3 × 2	4	m	*Phlebotomus* (Diptera: Psychodidae)
Golbergia		D	D	D	0	0	Pc	6 × 3		f,sg	*Culex* (Diptera: Culicidae)
Gurleya[a]		S	S	S	+	4	O	5–6	i 7	h	*Daphnia* (Crustacea), Diptera, Lepidoptera, Ephemeroptera, Odonata, Isoptera, Trichoptera
Hazardia	1	D	S	S	0	0	P	5 × 2	i 10	f	*Culex* (Diptera: Culicidae)
	2	D	D	D	0	0	P	5 × 2	i 10	f,h	
Helmichia		D	S	S	+	8	Ceb	3 × 1	i 0	f	*Endochironomus* (Diptera: Chironomidae)
Hessea		D	D	SD	+	N	SE	2	i 3	m	*Sciara (*Diptera: Sciaridae)
Heterovesicula	1	D	D	D	[+]	N	Oe	5 × 2	i 11	f	*Anabrus (Orthoptera: Tettigoniidae)*
	2	D	S	S	+	8	P	4 × 2	i 9	f	
Hirsutusporos		D	D	D	0	0	OLf	7 × 4	i 20	f	*Austrosimulium* (Diptera: Simuliidae)
Hyalinocysta		D	S	S	[+]	8	O	4 × 5	a 8	f	*Culiseta* (Diptera: Culicidae)
Issia		D	D	D	+	2 joined	OP	4 × 2		f	*Plectrocnemia* (Trichoptera), Culicidae
Janacekia		D	S	S	+	1	OP	7 × 4	i 20	f	*Simulium* (Diptera: Simuliidae)
Merocinta	1	D	S	S	0	0	EB	3 × 2	1	m	*Mansonia* (Diptera: Culicidae)
	2	D	D	D	0	0	Ren	3 × 2			
Napamichium		D	S	S	+	8	P	6 × 4	a 14	f	*Endochironomus* (Diptera: Chironomidae)
Neoperezia		D	S	S	+	2	Oe	6 × 3	i 25	f	*Chironomus* (Diptera: Chironomidae)
Nolleria			S	S	[+]	200	S	2	i 2	m	*Ctenocephalides* (Siphonaptera: Pulicidae)
Nosema	1	D	D	D	0	0	O	4 × 2	i 10	s	*Bombyx* (Lepidoptera: Bombicidae), most other insects
	2	D	D	D	0	0	O	4 × 2	i 4	m	
Nudispora		D	S	S	+	8	O	2 × 1	a 6	f	*Coenagrion* (Odonata: Coenagrionidae), Trichoptera
Octosporea		D	D	D	[+]	8	OC	7 × 3	i 12	m	*Musca* (Diptera: Muscadae)

Table 5 *contd.*

Genus	*ST*	*1*	*2*	*3*	*4*	*5*	*6*	*7*	*8*	*9*	*Type host (Order: Family), other hosts*
Orthosomella		S	S	S	0	0	Ce	4 × 1	? 6	s	*Operophtera* (Lepidoptera: Geometridae)
Ovavesicula		D	S	S	+	32	S,0	3 × 2	i 6	mt	*Popillia* (Coleoptera: Scarabeidae)
Parathelohania	1	D	D	D	0	0	O	7 × 3	i 5	o	*Anopheles* (Diptera: Culicidae)
	2	D	S	S	+	8	Oco	5 × 3	a 6	f	
Pegmatheca		D	S	S	+	8	E	3 × 2	i 5	f	*Simulium* (Diptera: Simuliidae), Trichoptera
Pernicivesicula		D	S	S	+	24–64	Cee	14 × 1	i 0	f	*Pentaneurella* (Diptera: Chironomidae)
Pilosporella	1	D	S	S	+	8	OV	3	i 4	f	
	2	D	D	D	0	0	Ce	5 × 3	i 6		*Wyeomia* (Diptera: Culicidae)
Polydispyrenia		D	S	S	+	n	O	5 × 3	i 6	f	*Simulium* (Diptera: Simuliidae), other simuliids
Pyrotheca[a]		S	S	S	[+]	4	L	7 × 2	i 5	f	*Cyclops* (Copepoda: Cyclopoidae)
Resiomeria		D	S	S	+	8	Ceb	7 × 2	a 10	f	*Aeshna* (Odonata: Aeshnidae)
Ringueletium		D	D	D	0	0	OL	8 × 5	i 16	f	*Gigantodax* (Diptera: Simuliidae)
Scipionospora		D	D	D	+	8–32	Ce	6 × 1	i 0	f	*Endochironomus* (Diptera: Chironomidae)
Spherospora		D	S	S	+	8–32	S	2	i 0	?	*Gigantodox* (Diptera: Simuliidae)
Stempellia	1			S	+	4	O	6 × 2	a 10	f	*Ephemera* (Ephemeroptera: Ephemeridae)
	2			S	0	0	P	3 × 1	? 6	f	
Striatospora			S	S	+	8	C	small	i 0	f	*Chironomus* (Diptera : Chironomidae)
Systenostrema		.	S	S	+	8	P	3 × 5	a 16	f	*Tabanus* (Diptera : Tabanidae), Odonata, Orthoptera
Tardivesicula		S	S	S	+	16–32	Ce	4 × 1	i 10	f	*Limnephilus* (Trichoptera: Limnephilidae)
Telomyxa		S	S	S	+	2 joined	O	4 × 2	i 12	f	*Ephemera* (Ephemeroptera: Ephemeridae), Coleoptera
Thelohania[a]		D	S	S	+	8	O	5 × 4	i 18	m	*Crangon* (Crustacea: Decopoda), many insect orders
Toxoglugea		D	S	S	+	8	Cebb	4 × 2	i 4	f	*Bezzia* (Diptera: Ceratopogonidae), Chironomidae, Chaoboridae, Plecoptera, Odonata, Heteroptera
Trichooctosporea		D	S	S	+	8	Ofc	7 × 4	a 8	f	*Aedes* (Diptera: Culicidae)
Trichoduboscqia		D	S	S	+	16–32	Pe	4 × ?	i 6	f	*Epeorus* (Ephemeroptera: Heptageniidae)
Tricornia		D	S	S	+	8	Op	3 × 2	a 5	f	*Mansonia* (Diptera: Culicidae)
Tuzetia[a]		S	S	S	+	1	P	6 × 3	i 15	o	*Cyclops* (Crustacea, Copepoda), Ephemeroptera, Coleoptera, Odonata.
Vairimorpha	1	D	D	D	0	0	O	6 × 3	i 13	f	*Pseudaletia* (Lepidoptera: Noctuidae), other
	2	D	S	S	+	8	E	4 × 2	i 12	f	Lepidoptera, Hymenoptera
Vavraia		S	S	S	+	8–64	Oe	4 × 2	i 15	s	*Culex* (Diptera: Culicidae), other culicids, Trichoptera
Weiseria		D	D	D	0	0	Ppc	4 ×	i	f	*Prosimulium* (Diptera: Simuliidae), Culicidae, Ephemeridae

Column headings:

ST = Spore type (1, 2, or 3). For those species that have more than one type of spore in the host, each spore is listed separately (the spores in the non-insect intermediate hosts are not included).

1 = Nuclei in meronts (S, single; or D, diplokarya)

2 = Nuclei in sporonts (S or D)

3 = Nuclei in spores (S or D)

4 = Sporophorous vesicle (+, Present; 0, absent; [+], present but not persistent)

5 = Number of spores in sporophorous vessicle;

6 = Spore shape (B, barrel; C, cylindrical; E, ellipsoid; L, lanceolate; O, ovoid; OL, oval; P, pyriform; R, reniform; S, spherical) and other features (b, bent; c, caudal spine; co, other caudal ornamentation; e, elongate; f, filamentous spore wall)

7 = Spore length and width to the nearest micrometre

8 = Polar filament type (i, isofilar – uniform diameter throughout; a, anisofilar – non-uniform diameter; number of coils)

9 = Tissues infected (f, fat body; g, gastric caeca; h, haemocytes; hy, hypodermis; m, midgut; mt, Malphigian tubules; o, oenocytes; ov, ovary; s, several; sg, salivary glands)

[a] Type host is not an insect and those species found in the insects might be misnamed.

Heidenhain's haematoxylin. Lacto-aceto-orcine (Hazard & Brookbank, 1984) is used to reveal chromosomes in wet smears. Heidenhain's haematoxylin, Giemsa-colophonium and a modified Gormoi triple stain can all be used for revealing spores in tissue sections (Hazard & Brookbank, 1984). These stains are found in Chapter VIII-1. For special techniques see Vávra & Maddox (1976) and Hazard *et al.* (1981).

Identification Microsporidia are easily identified by their characteristic spores but their classification requires an understanding of the developmental cycles of the organism and of its structural characters, including the fine structure of developmental stages and spores. The classification is based mainly on: (1) character of nuclei during the initial multiplicative cycle of the organism (merogony), in the presporal stages (sporogony) and in spores. The nuclei can be either single or joined in pairs (diplokarya); (2) whether spores produced by the parasite are in direct contact with the host cell cytoplasm or are enclosed in another membrane-like formation, the sporophorous vesicle (SPV). Under light microscopy, the SPV-forming microsporidia spores are seen in packets containing a characteristic number of spores (one to many). The SPV, however, is a fragile structure in some microsporidia or encloses only a single spore. The definite presence or absence of the SPV must be determined by transmission electron microscopy. As microsporidia are relatively host specific, the host is an excellent aid in species identification. However, relatively recently it has been demonstrated that microsporidia often have complex life cycles with more than one type of spore formed in either the same or different hosts.

Table 5 provides some orientation in the identification of microsporidia in insects. More detailed information on individual genera can be found in Sprague *et al.* (1992). Because many microsporidia are host specific organisms, located in specific tissues, this information is also presented. Classification is made more difficult by the fact that some microsporidia have complicated life cycles forming more than one kind of spore. In some cases different sporulation cycles occur in different stages of the host (e.g. in larvae and imagos), in other cases they might occur within a single individual. Some genera of microsporidia infecting mosquitoes form yet another type of spore in a non-insect (copepod) intermediate host.

Pathogenicity Phylogenetic studies using molecular biology methods show that microsporidia have existed for a very long time. As such, they are well adapted to the parasitic way of life and are slow to kill the host, taking advantage of the entire host life span for maximizing spore production. It is only at the latest stage of infection that microsporidian spores are produced in massive numbers and kill the host. Many microsporidia are transmitted vertically, from the female to the eggs of the next generation. This is also called transovarial (within the eggs) or transovum (spores contaminating the surface of the eggs). The developmental stages and spores responsible for this are not produced in high numbers, a thorough search often being required to determine whether the insect is infected. These infections tend to be benign.

Representatives Nosema apis, *Vairimorpha necatrix.*

D Phylum: Ciliophora

Diagnostic features Ciliates are characterized as organisms propelled by rows of cilia and possessing two different types of nuclei: a larger macronucleus, involved in vegetative functions of the organism and a small micronucleus, involved in sexuality. Under its plasmalemma, the cell of a ciliate contains flattened, membrane-bound alveoli and a complex system of microtubules and fibres called the infraciliature.

Occurrence Most ciliates are free-living organisms, but many parasitic forms exist. Only two genera infect insects as endoparasites: *Lambornella* and *Tetrahymena*, both of which occur in Diptera (Culicidae, Chironomidae and Simuliidae). An insect infected by ciliates is identified by the presence of actively swimming organisms in the haemolymph. These parasitic ciliates also have free-swimming infectious stages, actively seeking hosts in the larval habitat. A much wider range of ciliate genera and species occur as epibionts on insects larvae and adults in aquatic environments.

Methods Ciliate nuclei are revealed by staining of the cell with a wide variety of stains (haematoxylin, Giemsa, etc.). Infraciliature is revealed by silver impregnation. Ciliate identity can be established by nigrosine staining.

Identification Classification of ciliates is based on infraciliature organization, mostly in the area around the cytosome as revealed by silver staining. The cells of ciliates occurring as endoparasites of insects are covered with regular longitudinal rows of cilia and have a well-defined buccal cavity containing several rows of cilia arranged as membranelles (order Hymenostomatida). The basal bodies of these oral cilia are revealed in silver preparation and their shape is used as a generic characteristic.

Epibiotic ciliates usually belong to the order Peritrichida, have bell- or goblet-shaped bodies with a ciliary belt at the apical end of the body and are attached to the substratum at the posterior end of the body, usually by a stalk. These ciliates use the insect body merely as an attachment site as they are feeding on bacteria and small particulate food brought to the cytosome by whirling of their cilia. Although many of these ciliates are specifically adapted to live on certain parts of insect bodies, they are not pathogenic. However, in certain situations some peritrichous ciliates can cover an aquatic insect (e.g. mosquito larva) by a dense mat of cells and effectively hinder its movement. This is an abnormal situation indicating poor water quality (high bacterial count). Frequently, dying aquatic insect larvae are covered by a dense layer of ciliates feeding on bacteria growing on the insect cuticle.

Pathogenicity *Lambornella clarki* causes decreased survivorship of mosquito adults and parasitic castration of its female hosts. *Tetrahymena* is pathogenic when ciliate numbers in haemolymph are high (Egerter & Anderson, 1985).

Sometimes an investigator is confronted with a dead insect larva full of actively moving ciliates. Usually these ciliates have not caused the death of the host, but are feeding as scavengers on decaying tissues and the bacteria growing in the dead larva. Ciliates of the genus *Tetrahymena* and *Ophryoglena* are often found in dead aquatic insect larvae.

Representative Lambornella clarki.

E Parasites of uncertain protozoan affinities

Two pathogens of insects exist which are spore-forming organisms but have no clear affinity to any protozoan group previously mentioned. In fact, the spores of both groups develop from several cells rather than just one as is typical for Protozoa.

1. Nephridiophaga

In earlier literature, this organism was considered to be a haplosporidium, assigned to the phylum Haplosporidia. The most modern protozoan classification schemes have excluded them from the new phylum, Ascetospora, to which true Haplosporidia (parasites of molluscs) now belong.

Diagnostic features Organisms occur as small multinuclear plasmodia, later undergoing differentiation of nuclei into generative nuclei and somatic nuclei, the latter remaining unchanged in the plasmodium. A thick spore wall, without a lid (an apical lid is characteristic of *Haplosporidium*), is formed around each generative nucleus. These spores also lack the polar filament that is typical for microsporidia. The spores are 5–8 μm in size and have a mid-longitudinal concavity.

Occurrence Roaches, Coleoptera, Dermaptera, Hymenoptera (*Apis mellifera*).

Methods Examination of fresh squash preparation of Malpighian tubules with a phase contrast microscope will reveal the spores. Giemsa-stained smears of the same material will reveal small multinucleate plasmodia.

Identification See Lange (1993).

Pathogenicity None observed.

Representatives *Nephridiophaga periplanetae* from *Periplaneta americana*.

2. Helicosporidium

Helicosporidium is a monotypic genus of unknown affinity that has spores originating from more than one cell.

Diagnostic features Discoidal spores are composed of three centrally stacked cells surrounded by three to four coils of another cell in the form of a long filament (Figure 6) considerably different in nature from the polar filament in microsporidian spores (Figure 5). All four cells are packaged in a

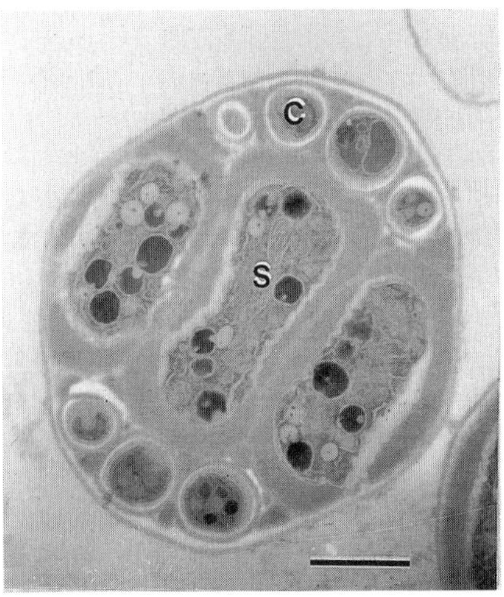

Figure 6 Electron micrograph of *Helicosporidium* sp. in cross section revealing the coiled cell (C), the three stacked cells (S) and the spore wall (bar = 1 μm).

smooth, somewhat transparent shell. Spores range in diameter between 5 and 9 µm, with a thickness of approximately one-half to three-quarters the diameter. Upon hatching, the coiled cell straightens and appears to break the outer shell (Kellen & Lindegren, 1974). Whether it is the three central cells or the coiled filamentous one that penetrates and infects the host is still uncertain. *Helicosporidium* from mosquito larvae are are also infectious to lepidopterous larvae, suggesting a low degree of host specificity (Avery & Undeen, 1987). Common sites of infection are muscle and fat body cells.

Occurrence Culicidae, Lepidoptera, Collembola, Coleoptera.

Methods *Helicosporidium* is readily identified by phase contrast microscopy of squash preparations from living or dead hosts. In the spore, Giemsa staining differentiates the nuclei and cytoplasm of developmental stages and the three stacked cells but not the filamentous cell. Spores can be germinated by alternately drying and wetting them. The inner cells quickly lyse in water but the coiled cell persists, looking much like a small roundworm. The empty shell is oval and the site of breakage has inward turning edges.

Identification The spores are characteristic and not easily mistaken for any other protozoan, particularly after they have been germinated or crushed under a coverslip, revealing the coiled filamentous cell.

Pathogenicity Virulence is variable. One *Helicosporidium* isolate was lethal to both mosquito and lepidopteran larvae.

F Describing a species

Even if the reader's intention is not necessarily to describe a new species, the list presented below offers a guideline for examination of the pathogen. Some of the information is easy to obtain whereas other characteristics require the colonization of the host and pathogen, experimental infections and timed examinations – impossible tasks in the case of some hosts or pathogens, especially if only available from field collections.

1. *Information required*

- Host, including any obligate intermediate hosts
- Site of infection
- Transmission
- Interface (membranes between parasite and host cell contents)
- Any other parasite–host relationship
- Development of the pathogen – the whole cycle if known
- Spore or spores (there may be more than one kind)
  Number of nuclei
  Shape and size (preferably fresh and stained)
  Spore wall structure – endospore and exospore; surface ornamentations or projections
  Internal structures as appropriate and diagnostic within each protozoan group
- Locality
  Type of habitat
  Geographic location – latitude and longitude or proximity to a durable well-known landmark.
- Where the type specimens were deposited
- Nucleic acid sequences. In recent years the use of DNA sequencing, primarily of the ribosomal RNA gene, is adding a new dimension to microsporidian identification and taxonomy.
- Other available information, such as:
  Experimental host range
  Comparison with similar species

2. *Deposition of reference slides*

Whenever a new species is described, a type specimen should be deposited with the International Protozoan Type Collection at the National Museum of Natural History of the Smithsonian Institution. Correspondence should be addressed to:

Museum Specialist, Division of Echinoderms and Lower Invertebrates, Department of Invertebrate Zoology, National Museum of Natural History, Smithsonian Institution, Washington, DC 20560, USA.

The material must be permanently mounted on glass microscope slides, affixed with labels listing genus, species, author and year; material type (paratype, lectotype, etc.); collection locality; genus, species and class of the host. Slides must be arranged in order and be accompanied by a listing of this information: stain, collector, number of specimens, original field number, identifier, pertinent remarks, higher classification. A short letter should be sent with the specimens stating that they are being donated. Wet specimens (specimens in alcohol or formalin) must contain labels with the same information (except stain) or be accompanied by papers containing this information.

Type material (specimens from which a new species was described) will form a separate collection and must be accompanied by separate data sheets. The descriptive paper of a type material must be accepted for publication or be in press before type specimens are submitted to the Museum. At least one copy of each publication should be supplied to the Museum's library.

The museum will affix an accession number to each slide. This number comes from the Registrar's office and all pertinent papers can be located by this number at any time. All material will be assigned a catalogue number in addition to the accession number. Slides will be stored numerically by catalogue number in 'Technicon' slide cabinets. Wet specimens will have a special section in the collection area. Slides requiring catalogue numbers for forthcoming publications will rapidly be accommodated.

Collections will be loaned in full or in part to investigators or institutions; the duration of the loan being arrived at by negotiation, but not to exceed one year. Loans are renewable annually upon request with justification. Utmost care of materials loaned is assumed.

G General techniques useful for identification of protozoa

1. Examination

a. Fresh smears

It is always best to examine live insects. Those already dead will often have secondary microbial activity, obscuring the cause of death. Presporal stages are not easily visible in fresh smears. Larger insects must be macerated separately and a small part of the fluid, without pieces of cuticle, can then be placed on a slide for examination. Small insects can be crushed under a coverslip whole. To obtain a clean preparation, it is best to remove the gut contents before macerating the insect. In order to determine the site of infection, the insect can be dissected and pieces of tissue carefully removed and examined individually. Phase contrast is usually the best optical system to use for examining these fresh smears.

b. Preserved material

Preserved material can also be examined for infection but not as easily as live tissues. However, spores can still be found and identified in methanol preserved or even dried, pinned insects. Many identification features will be lost in spores that have been preserved in these ways. Formalin is a better preservative for infected insects. It is far preferable to fix the infected insect or tissues by the method appropriate for the technique to be used later in viewing the parasite.

2. Measuring

The size and shape of the spore is a fairly consistent feature of most protozoan spores or cysts, although it is sometimes variable between hosts or influenced by the temperature prevailing during development. Spore size is reported in terms of its anterior – posterior length and its width at the plane of maximum diameter. Published measurements are usually (and preferably) from live spores. Spores and developmental stages can also be measured in stained preparations but their dimensions vary widely according to osmotic conditions prevailing when the smear was made. Greater precision is obtained by measuring more spores, however, a point of diminishing return sets in at approximately 30.

a. Ocular micrometer

The most frequently available device for measuring objects in a microscopic field is an ocular micrometer. This is simply a ruled grid placed in one of the oculars of the microscope and calibrated with a stage micrometer. This device lacks precision at the small sizes of most protozoan spores.

b. Image-splitting eyepiece

For more precise measurements, the 'Image-splitting eyepiece' (Vickers Instruments, Malden, MA) is used. This instrument contains a mirror system that splits the object into two images. Coloured filters are incorporated so that each image is differently coloured (green or red). A calibrated dial controls the distance between the two images. The two images are moved apart with the dial until they just touch. The distance between the superimposed images (0 on a well-aligned instrument) and the point at which the two images touch is read from the dial. The instrument is rotated 90° and the procedure repeated for the other axis. Measurements in µm are obtained by multiplying the number from the dial by a calibration factor obtained by measuring a stage micrometer. Good measurements depend on clear and consistent focus to provide a clear edge to the image of the spore. For this reason, phase contrast should not be used. Under bright field lighting, the substage diaphragm should be adjusted to obtain the lowest degree of contrast practical.

3. *Immobilization*

Apart from their small size, the major problem in the measurement of spores is their immobilization while it is being done. Spores move with the flow of the suspending medium under the coverslip and are subjected to Brownian motion. A few methods have been used to immobilize the spores.

a. Agar immobilization method

- Cover a large area of a microscope slide with hot, molten 1.5% agar and let cool. The objective is to obtain a very even, flat surface on the agar.
- A very small drop of concentrated spores (less than 5 µl) is placed in the middle of a coverslip and inverted on top of the solidified agar. The spores will be immobilized between the coverslip and the agar.
- Only those spores that are in clear focus at both ends can be measured.

b. Polylysine
Coating the slide with a film of polylysine first will immobilize some of the spores (Mazia *et al.*, 1975).

c. Oil
Paraffin (mineral) oil can also serve as a trap to immobilize spores (Vávra, 1964). A drop of the oil is placed on a slide and a coverslip with a small drop of dense spore suspension is applied on top of the oil. Water, having a better affinity for glass, spreads out on the surface of the coverslip, leaving spores individually trapped in 'holes' in the oil. Vávra (1976) cautioned that phase contrast can not be used to examine water–oil slides. Desiccation of any of these preparations can be slowed by sealing the edges of the coverslip with molten petroleum jelly or paraffin.

3 BIOASSAY AND EXPERIMENTAL INFECTIONS

It is frequently necessary to determine the viability or relative infectivity of a protozoan. In the case of non-motile dormant stages, cysts or spores, there is often no easy way to determine viability other than by an experimental infection. Inviable microsporidian spores will sometimes be less refractive under phase-contrast lighting but, otherwise, there are no reliable visual cues or staining methods to differentiate live from dead spores. Although a bioassay, or experimental infection is the only sure method to determine spore viability, there are other correlates. Spores that are capable of germination are almost always viable. Microsporidian spores contain a high concentration of trehalose. This sugar is gradually lost from dead spores and these sugar-depleted spores are less dense than viable spores (Undeen & Solter, 1996). Therefore, measurements of intrasporal sugars or density can sometimes be used when no other methods are possible.

A Bioassay

Bioassays are used to test the presence, viability, or quantity of infective forms of a protozoan. To conduct a quantitative bioassay, spores must be counted and a dilution series made so that a known concentration of spores can be fed to the host. A control, fed only the suspension medium but otherwise handled in the same way as those in the test groups, must always be run as a check for mortality unrelated to the protozoan. The test insects must be healthy and of uniform age and size. All other environmental factors subject to control must be as consistent as possible within and between tests. The test should be replicated to assure consistency and to provide sufficient numbers to demonstrate a statistical significance. Details of statistical analyses of bioassay data are found in Chapters II and III.

1. Counting

To calculate a dosage or concentration for any bioassay, the spores, or other infectious stages must be accurately counted. The most common counting chamber used for infective stages is the haemocytometer, designed for counting blood cells. The counting grid is divided into 25 large squares, each divided into 16 smaller squares. The procedure described below is for a Brite-line® phase-contrast haemocytometer.

a. Use of the haemocytometer

1. Thoroughly mix a reasonably clean spore suspension (at least filtered of the larger contaminants) to ensure even distribution of spores.
2. Place the special, thick haemocytometer coverslip on a clean, dry haemocytometer.
3. Remove a sample of the suspension with a pipette. Carefully deposit about 10 µl of spore suspension in the notch and watch the sample be drawn by capillary action into the chamber. Fill – but do not overfill – both sides of the haemocytometer chamber.
4. Place the haemocytometer on the microscope stage. **Note:** The haemocytometer is thicker than a normal slide, therefore, the objective lens must be raised before inserting the haemocytometer.
5. Allow the spores to settle for a minute or so.
6. Count the cells in the large squares. Count cells lying on the left and upper double lines. Do not count cells lying on the right and lower double lines.
7. Count both chambers and use the mean of the two counts.
8. If there are more than one cell per large square, count five large squares (four corners and centre). Spores/ml = $5 \times 10^4 \times$ the number of spores in the count.
9. If there are fewer than 1 cell per large square, count all 25 large squares. Spores/ml = $10^4 \times$ the number of spores in the count.
10. If there are more than 100 cells per large square, dilute a sample of the suspension and count the diluted sample.
11. If the stock suspension was diluted before counting, divide the concentration by the dilution factor (a fraction) to obtain the concentration in the stock solution.
12. Clean the haemocytometer and coverslip with distilled water and blot dry. Avoid using materials that will scratch the haemocytometer's surface.

2. Application of dose

Bioassays differ primarily by the method of feeding spores to the host. A *dose* is the known number of spores or other infectious stages fed to the host. The dose can be fed to the test host by applying it to a small amount of its natural food or an artificial diet. The amount of food carrying the spores must be standardized within the test and sufficiently small that the entire amount can be consumed within a short period. Any test organism failing to consume the entire amount must be eliminated from the test. Another method is to offer the starved host a small (1–2 µl) droplet of spore suspension on a calibrated bacteriological inoculating loop. The loop is dipped into a suspension of spores by a standard technique to ensure that an equal volume is picked up each time, and then the droplet is touched to the mouthparts of the test insect. Some lepidopterous larvae will imbibe the spore suspension as soon as the loop touches. Here too, only those hosts that consume the entire quantity should be included in the assay.

Results of these procedures are usually presented as an LD (lethal dose), ID (infective dosage) followed by a subscript percentage, without the % sign, most often 10, 50 or 90. For example, LD_{50} is the dose that, on average, killed 50% of the larvae; ID_{90} is the dose expected to infect 90% of the test insects (for further explanation see references in Chapters II and III).

Lethal or infective concentrations are used when the exact numbers of spores ingested by the test organisms cannot be determined. A concentration series of spores applied this way, either by per volume or unit area, forms the basis of an assay in which results are reported as lethal or infective concentrations instead of dosages. Uniformity in application is the key to consistency for these techniques. For filter-feeding organisms, such as mosquito larvae, spores are mixed with the water in a known concentration. The volume of water in the exposure vessel and the time that the larvae remain in it are important factors. With the same concentration of spores in the water, larvae in a large volume might receive a higher dose than larvae in a small volume with the same concentration. Likewise, larger numbers of larvae may receive a smaller dose than fewer larvae in an equal volume of water. Therefore, standardization of variables such as temperature, volume of water, duration of exposure and amount of food in the water must be rigidly controlled. The notation used in this case is 'LC or IC', 'lethal concentration' or 'infective concentration', with subscripts as described above for dose.

'Infection' is usually a more useful parameter than 'lethal' for Protozoa. Mortality due to the protozoan is extended over such a long time that mortality by natural causes can be excessively high. The chronic nature of so many protozoan diseases is expressed in reduced longevity and fecundity. The effect of these diseases on insect populations are better assessed in terms of reproductive potential than outright mortality, requiring studies beyond bioassays. More details of bioassays are found in Chapters II and III.

3. *Bioassay set-up*

The following bioassay procedure is typical but many variations are possible and even necessary. The correct dosage or concentration cannot be established without a preliminary or 'range-finding' assay. A tenfold dilution series is generally used for the preliminary assay. This and more precise bioassays are performed as described below.

a. *Dilution series*

A dilution series of the pathogen infectious stages is made within the dosage range of the preliminary assay where between ca. 5 and 95% infection occurred, a double dilution series is usually satisfactory but at least five dosage groups within the above percentage infection range should be included in the assay. Dilutions are preferable to measuring material directly from the stock solution because accuracy is better when larger volumes are being measured and each test group receives the same volume.

b. *Selection of test organisms*

All the test organisms must be healthy and of the same age and size. Higher numbers are of course better but as few as 10 per dosage group can be used.

c. *Replicates and controls*

To guard against the loss of a dosage group and to assess the variability within the assay, each group should be set up in triplicate. Always include a control group that is treated in an identical manner except that they receive no spores. Further details can be found in references in Chapters II and III.

d. *Feed or expose the test organisms*

Feed the spores to the test organisms by whatever means appropriate using the same volume for each dosage group. Include 'sham controls' which subject identical groups of insects to the same handling and dosage procedure except that no pathogen is applied.

e. *Development time*

Allow time for the infections to develop or for mortality to occur, maintaining groups under uniform conditions optimum for development of the host and the protozoan.

4. *Scoring the bioassay*

a *Mortality*

- Count the number of dead and live test organisms in each group, including the control (uninfected) group. As the dead insects are sometimes missing, the usual procedure is to count the live insects. The number of dead = the number originally set minus the live insects.
- Combine the numbers from all three repetitions of each dosage group.
- Calculate the percentage of dead insects in each group: 100 × (number dead/number originally set).
- Using Abbot's formula, correct the percentage mortality:

  Corrected % mortality = $100 \times (T\% - C\%)/(100\% - C\%)$

  where T% = the percentage of dead test organisms and C% = the percentage of dead control organisms.

b. *Infection*

- Count the number of infected and uninfected organisms in each dosage group and in the control group. Examine and score both dead and live animals.
- Calculate % infection: 100 × (number infected/total number scored). In this case the missing organisms are omitted from consideration.

5. Calculation of lethal or infective concentrations

a. Computer

The analysis can be done using probit analysis with PC SAS software (SAS Systems for Windows, Release 6.08, SAS Institute, Inc., Cary, NC 27513). (See references in Chapters II and V-3.)

b. Graph paper

The percentage infection can also be graphed as a function of dosage on log-probit graph paper (this graph paper converts percentages into probits). This plot facilitates the fitting of a straight line to the sigmoid dosage–response curve.

- Enter % infection or corrected % mortality as points on log-probit paper. Test groups with 0 and 100% infection are excluded from the calculation of LC, IC, LD, or LC values because they are open ended – any lower dosage will continue to cause no infections and any higher dosage will continue to produce 100% infection.
- 'Eye fit' a straight line through the points such that there are an equal number of equally spaced points above and below the line.
- The IC_{50} is the concentration straight down from the intersection of the eye fitted line and the horizontal 50% infection line (centre of the graph). IC_{10} and IC_{90} values are obtained from the intersection of the line with the 10% and 90% lines on the graph.

B Other viability tests

Viability can be assessed without infecting host animals. Viable microsporidian spores must be capable of germination. This is a rapid process, easily observable with a phase-contrast microscope. Usually within a few minutes of stimulation the spores lose their refringency and a discharged polar tube can be seen (Figure 7). Spores treated with gamma radiation or ultraviolet light were the only demonstrated exceptions to this generality. Irradiated spores were capable of germination for several days after they were no longer capable of infecting the host (Undeen & Vander Meer, 1990). Viability is probably indicated by germination *in vitro* even when the germination stimuli are obviously abnormal, that is, different from those prevailing in the gut of the host.

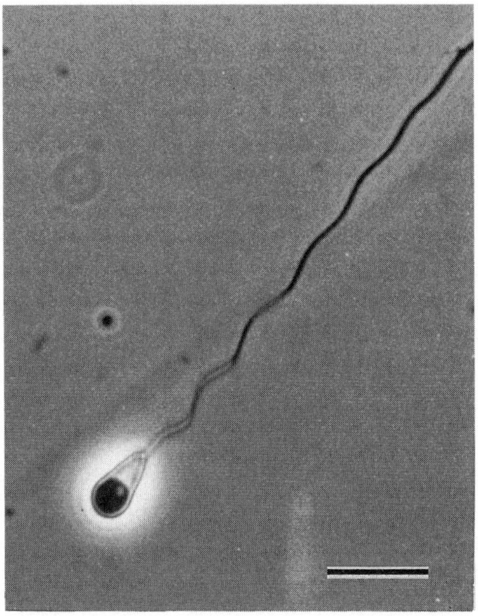

Figure 7 Germinated microsporidian (*Edhazardia aedis*) spore (bar = 10 μm).

1. Spore germination

a. In vitro *stimuli*

The spores of many microsporidia germinate in alkaline (pH 9.0–11.0) solutions of monovalent salts (KCl, CsCl, NaCl, RbCl, KI, KBr and many others). *Nosema bombycis, Vairimorpha* spp. and probably many others germinate in a neutral buffered salt solution after a few minute's pretreatment in a strongly alkaline (>pH 11) solution (Ohshima, 1964). *Spraguea lophii* spores were germinated in phosphate buffered saline at pH 8.5–9.0 with 0.1–0.5% porcine mucin (Sigma) (Weidner *et al.*, 1984). The most effective ionic and pH conditions for a particular species can be determined by experimentation. *Glugea hertwigi* spores, stored in pH 7.0 phosphate buffer, germinated after transfer to 1–5 mM calcium ionophore A23187 in a 0.1 M, pH 9.0–9.5 carbonate buffer (Weidner *et al.*, 1984).

Dried spores of *Nosema whitei* germinate immediately upon rehydration in distilled water, an apparently natural stimulus for a microsporidium of a host that lives in a dry environment. Spores of *Nosema apis* (Olsen *et al.*, 1986) and *Nosema locustae* (Undeen & Epsky, 1990) and many other microsporidia germinate after desiccation only if they are rehydrated in a solution with the right pH and ion content. Prolonged exposure to low concentrations of polyethylene glycol greatly enhanced *in vitro* germination of *N. locustae* in pH 9–10, 0.05 M NaCl (Undeen & Epsky, 1990).

The spores of many species can be induced to germinate when rehydrated in a buffered salt solution after they have been subjected to partial dehydration (Undeen & Avery, 1984). This can be observed by placing a thin drop of spore suspension on a slide and permitting it to dry until only a small wet spot remains in the centre. The area is then flooded with a drop of the putative germination solution and a coverslip applied. A ring of germinated spores, marking the zone where the correct degree of drying has occurred, will be seen a few minutes later. Partial desiccation can be obtained in a more quantitative manner by placing the spores in a hyperosmotic sucrose solution (1.6–1.8 M) for about 15 min, then flooding with the test solution. Although sometimes producing high percentages of germination, drying – rehydration is not necessarily the physiologically normal stimulus.

Spores of many species will germinate in the presence of 2–5% hydrogen peroxide (H_2O_2), although sometimes at very low percentages. Contrary to other germination stimuli, germination in H_2O_2 is often better at low temperatures.

The effect of calcium in the germination medium is variable among species. Calcium influx was thought to stimulate germination (Pleshinger & Weidner, 1985), and calcium antagonists such as EGTA (a calcium chelator) block germination of *Spraguea lophii* spores germinated in glycylglycine or carbonate buffers in the presence of mucin at pH 9. The addition of $CaCl_2$ to the germination medium generally inhibits germination (Ishihara, 1967; Undeen, 1978, 1983). In such cases, EGTA enhances germination (Malone, 1984). Weidner has demonstrated the inhibitory activity of the calcium antagonists, lanthanum, and verapamil and the calmodulin inhibitors, chlorpromazine and trifluoperazine, on the germination of *S. lophii* spores (Weidner & Halonen, 1993).

b. Scoring germination

Germinated spores appear dark with an obviously empty spore case. If examined within a few minutes after germination, the sporoplasm can be seen at the end of the thin polar tube in the form of a minute drop of cytoplasmic material. A quantitative measure of viability can be determined by the percentage of spores that germinate and a number of means are at hand to determine this.

(i) *Microscopically.* The concentration of spores in the test solution must be about 10^6 spores/ml so that there will be sufficient spores under the coverslip. Too many spores in the test solution will cause them to clump by entanglement of the polar tubes, making counting difficult or impossible. The test solution containing the spores is mixed briefly and a sample (10–20 μl for an 18 × 18 mm coverslip) is placed on a slide and covered. A phase-contrast microscope at 400× magnification is best for counting the spores. Begin near one corner of the coverslip and count all spores in the microscope field, scoring those that turned black or have an attached polar tube as germinated and those unchanged (refringent, white) as ungerminated. Without looking into the microscope, move the slide to another area and again count all the spores in that field. In addition to removing a source of bias, the motion of the field sometimes causes 'seasickness'. Continue in this manner, systematically choosing fields in every section of the coverslip until the desired number of spores is reached. The number of spores to be counted depends on the precision required. For most purposes a total of 200 seems to be adequate. If it is necessary to differentiate within a few percentage points, more spores will need to be counted.

(ii) *Spectrophotometry.* Rather than counting spores to obtain a percentage germination, a suspension of spores can be germinated in a cuvette, while in the specimen chamber of a recording spectrophotometer (Undeen & Avery, 1988a,b; Undeen & Frixione, 1990). The phase-contrast darkening of the spores during germination results in a decrease of about 50% in optical density (wavelength set at 625 ηm) during complete germination, progressing along a sigmoid curve. The time before germination begins, the rate of germination (slope) and the final percentage germination (maximum OD reduction) can all be determined, providing information on germination kinetics. Spores for this procedure must be well purified. The cuvette chamber must be equipped with a temperature-regulating device; spore germination is quite temperature sensitive (Ishihara, 1967; Undeen, 1978). The germination solution is placed in the cuvette and brought to the temperature set in the chamber. The spores are added to the cuvette, mixed quickly, the cuvette placed in the chamber, and recording begun immediately. At 30 °C in 0.1 M NaCl, pH 9.5, germination of *Nosema algerae* spores begins after a lag time of 1 min and is complete 3–4 min later.

(iii) *Plate assay*. An indirect measure of germination can be made using a haemoglutination plate. Spores are added to each well in a quantity that will, after they settle out, form a visible pellet on the bottom of the well. Test germination solutions are placed in the wells and the spores mixed with it. If the spores do not germinate, they will form the small pellet on the bottom. Due to the entanglement of the polar tubes, germinated spores will settle over a wide area of the bottom, forming no visible pellet.

c. In vivo *germination*

If an *in vitro* stimulus cannot be found, the spores can be fed to the host organism for evaluation of *in vivo* germination. After feeding spores to the host for a period of time roughly equal to the filling time of the gut, the food plug is dissected out and examined for germinated spores. Germination usually occurs in a specific area of the gut (Undeen, 1976), therefore, if quantitation of germination is required, only the spores beyond that region should be scored. The polar tube is quickly digested and, therefore, seldom seen in the gut contents and germinated spores can be identified only by their black empty cases. It is helpful to use purified spores for this procedure so that germinated spores will not be confused with immature spores that are also black in phase contrast. In some instances, viable spores do not all germinate in one passage through the gut (Kramer, 1973) and a low percentage germination might be the normal course of events.

2. *Sugar assay*

Microsporidian spores contain high concentrations of trehalose, a disaccharide of glucose, which appears to be essential for germination (Undeen & Vander Meer, 1994). Microsporidian spores lose this sugar during long-term storage. Recent tests have shown that spores lacking this sugar are inviable (Undeen & Solter, 1996). After injury, the trehalose concentration decreased and the concentration of reducing sugar increased. When the percentage of reducing sugar exceeded 20% of the total sugars, few spores in the sample were still viable. Therefore, the viability of the spores can be estimated by the results of two sugar assays (Undeen & Solter, 1996), the hot anthrone test for total carbohydrates (Van Handel, 1985) and the Nelson's test for reducing sugars (Clark, 1964). Gas chromatography has also been used to measure the amounts of sugar in extracts from microsporidian spores (Undeen & Vander Meer, 1994). These sources can be accessed for detection and measurement procedures. Methods for extraction of sugars from microsporidian spores are presented in Section 5.

3. *Buoyant density*

Trehalose-depleted spores are less dense than viable spores. In a continuous Ludox density gradient, these dead spores are found in a band about 1 cm above the viable spores of the same species (Undeen & Solter, 1996). As stated above, spores can also be inviable if much of the trehalose has been converted into glucose, in such a case there is no detectable change in density. This situation is, however, temporary. Thus, light spores are dead but those of normal density are not necessarily viable. The best way to compare the test spores with the fresh, viable spores is to run a separate gradient with each of the two samples and a third gradient with a mixture of both (a fourth gradient will have to be prepared to balance the centrifuge). The mixture will unequivocally resolve the issue of density differences; the single sample gradients will reveal which is the heavy one and which the light. This test is especially useful because it is nondestructive and can be conducted during spore purification. Procedures for the density gradient are given in Section 5B.

4 PROPAGATION AND PRODUCTION

Large numbers of protozoan spores can sometimes be obtained from field-collected insects. However, for greater certainty that the spores are all of the same species, production in the laboratory is preferred as a source of material for experimental work. Laboratory propagation of a protozoan is generally limited by two factors, the amenability of its host to laboratory culture and our ability to transmit the pathogen to the host. The latter is a problem with many microsporidian spores that are produced in one host but infect an unknown intermediate host. A more convenient, alternate host can often be used for production of microsporidia that do not have a high degree of physiological host specificity.

Spores can be fed to the host on artificial diet or their natural food. Experimentation is necessary in order to find the optimum dosage for best spore production. Feeding too many spores can kill the host early, reducing spore production – too few spores will result in a low percentage infection. However, the infectious and lethal dosage ranges are generally quite broad, with 10 times the 100% infectious dosage frequently not causing excessive mortality. Outlined below are some methods that have been routinely used for the production of some Protozoa. A more thorough review of mass production methodologies can be found in Brooks (1980, 1988).

A Microsporidia *in vivo*

1. Amblyospora californica

Amblyospora californica is a highly host specific parasite of the mosquito *Culex tarsalis* with an obligate intermediate copepod host. The life cycle of *A. californica* includes a cycle in which the female larvae survive to adults that lay infected eggs, carrying the infection through an apparently endless number of generations. Large numbers of spores that are infectious only to the intermediate host (a copepod) are produced in the male larvae. Larvae with this developmental sequence usually fail to pupate and die from the infection. Spores of this microsporidium are easily produced in large quantities simply by maintaining a colony of infected mosquitoes, harvesting spores from fourth instar male larvae. Two precautions must be taken. The colony needs frequent augmentation of males from an uninfected colony for breeding purposes and the colony has occasionally to be purged of uninfected individuals. Purging is done by obtaining eggs from isolated, individual females. Only the females from cohorts that produced infected males are retained for the new breeding population.

2. Edhazardia aedis

Edhazardia aedis spores are produced in vertically infected (transmitted from the female, through the egg, to the next generation) *Aedes aegypti* larvae. Thus spore production requires two host generations. Second instar larvae are infected by placing them in deionized water contaminated with $1 \times 10^3 - 1 \times 10^4$ spores/ml with larval density at about 1 larva/ml. A small amount of larval food (about 0.4 g/l of an equal mixture of powdered alfalfa pellets and a hog chow supplement without animal fat) is added to insure normal feeding. After 24 h, the larvae are transferred to a larger volume of water, suitable for optimal larval development. Pupae are collected and transferred to cages. The adults are supplied with cotton soaked in sugar solution and provided a blood meal so that eggs will be produced. Eggs are collected and hatched and the larvae are reared for 5–7 days, until pupation begins. Most of the infected larvae are delayed in development and fail to pupate. Spores are harvested from fourth instar larvae.

3. Nosema algerae

Nosema algerae has been described as a parasite of mosquitoes but has an extremely broad host range. Moderate numbers of spores are produced by feeding spores to early instar mosquito larvae and then harvesting the spores from the adults. For highest production, harvest from the adults should be delayed until mortality from the parasite has begun. Depending on temperature and age, the larvae should be heavily infected approximately 5 days after pupal eclosion. Approximately 10^6 spores are obtained from each mosquito. As spores of *Nosema algerae* are intolerant of desiccation, they must be harvested from living insects.

4. Nosema locustae

Nosema locustae spores have been produced in quantities sufficient to treat thousands of acres of rangeland in efforts to suppress grasshoppers (Henry & Oma, 1981). The grasshoppers were infected with *N. locustae* by feeding them lettuce on which spores had been sprayed. Spores were applied to the lettuce at a rate of 10^6 spores per 2000 fifth instar nymphs for two consecutive days and then again on the fourth day. *Nosema locustae* infections developed slowly, the numbers of spores per insect increasing until, at 32 days after the first spore feeding, there were nearly 5×10^9 spores per male and about twice that in females (Henry & Oma, 1981). For this microsporidium and others that have a wide host range, there is considerable latitude in choice of a production host.

5. *Vairimorpha necatrix*

Vairimorpha necatrix is a microsporidium of Lepidoptera with a moderately broad host range. The best spore production is accomplished by feeding fourth instar larva approximately 10^5 spores each. This may be done by placing the spores in or on a piece of diet small enough to ensure that all of it is eaten within 24 h. For a less quantitative but still reliable feeding, the spores can be layered on the surface of the artificial diet. Spores develop in the fat bodies and are ready for harvest in about 12–14 days at 24 °C. Pupation of infected larvae is either delayed or does not occur at all. On the order of 10^9 spores will be produced in each larva, most of them of the elongated, binucleate type. The shorter octospores (meiospores) develop later on.

6. *Nosema* and *Endoreticulatus* spp.

Nosema and *Endoreticulatus* spp. in Lepidoptera develop more slowly than *V. necatrix*, dictating that infections be initiated earlier in larval life (second instar) than was recommended for *V. necatrix*. Larval development is slowed by the disease and spores can be harvested 14–20 days post-infection.

7. *Microsporidia with broad host ranges*

Nosema algerae and *Vavraia culicis* (and other microsporidia with broad host ranges) can often be produced more efficiently in a larger alternate host. *Helicoverpa zea* larvae are reared to the third or fourth instar and then starved overnight, individually (they are cannibalistic) in small containers. A small drop (ca. 10 µl) is then added to each container and the larvae are held again for several hours or overnight. The larvae are then returned to individual containers of diet and reared to adults. Approximately 10^9 spores are produced in each *H. zea*, a thousand times the yield from a mosquito. Spores are harvested when the adults begin to die from the disease.

8. *Injection of spores*

Microsporidia with broad host and tissue ranges can also be propagated in alternate hosts by intrahaemocoelomic injection of an aseptic spore suspension. With a steady hand, larger insects can be injected with a disposable 'tuberculin syringe' (1 ml, 25 gauge, ⅜ in. needle). A smaller needle and a microapplicator produces more consistent results because it causes less damage and provides for better control of the dose (Undeen & Maddox, 1973; Pilley *et al.*, 1978; Weiser, 1978). Lepidopteran larvae are best injected through the planta (or base) of a proleg, with the needle entry at a shallow angle to prevent puncturing the gut. Microbes released from the gut cause a fatal septicaemia. Injection through the proleg base also helps to reduce bleeding. It is desirable to anaesthetize the larva with CO_2 or to immobilize them and decrease the turgor of the haemolymph to minimize bleeding. Chilling also makes the insect easier to handle but the hosts must be warmed immediately after injection because cold temperatures can seriously reduce spore germination. The stage of the host injected and the stage from which spores are harvested depend on the development time of the microsporidium and the host as well as the minimum size and age of organism that can be injected (Pilley *et al.*, 1978). Spore production of *N. algerae* in *H. zea* is maximized by injecting third instar larvae and harvesting spores from the adults a few days after emergence. It is unlikely that all species of microsporidia will infect insects by injection as easily as *N. algerae*. The spore must be able to germinate in the blood in order to infect the susceptible tissues. Those microsporidia that germinate well in haemolymph and have the lowest host and tissue specificity, have the best chance of infecting by this route.

B *In vitro*

1. *Cell-free media*

So far, obligate insect pathogens can not be grown *in vitro*. The mosquito parasitic ciliates, *Tetrahymena pyriformis* and *Lambornella clarki* replicate both in the larval habitat and in the host. They can be cultured *in vitro* in a vitamin-supplemented, septic cerophyl (powdered wheat leaf) extract (Washburn *et al.*, 1988). To prepare the medium, 0.25–0.5 g cerophyl is boiled in 100 ml water and then filtered through glass wool to remove undissolved material. The extract can be stored at 5 °C until needed. The medium must be inoculated with a bacterium before use to provide a food source for the ciliates.

2. Cell culture

Mass production of microsporidian spores in cell culture is not yet practical. Jaronski (1984) is an excellent source of information on production in cell culture. Some microsporidia are easily inoculated into cell culture (Jaronski, 1984; Kurtti *et al.*, 1990; Hayasaka *et al.*, 1993) and are even capable of infecting cells derived from organisms as distantly related to their natural insect hosts as mammals (Ishihara, 1968; Undeen, 1975). A cell culture can be inoculated with a microsporidium by the addition of explanted tissues obtained by sterile dissection from an infected host (Sohi, 1971; Sohi & Wilson, 1976). More often, established cell lines are infected by inoculating cultures with aseptic spores. The first problem to be surmounted is inducing the spores to germinate in the culture medium. Inoculation of cultured cells with microsporidia, such as *N. algerae*, that germinate in simple salt solutions, is straightforward. Otherwise, the culture medium must be altered to temporarily meet the germination requirements of the spores without damaging the cells. Alternatively, the spores must be pretreated so that they germinate in a medium which is normally unstimulatory. Several methods have been used.

a. Modification of the culture medium

Infection of the cells occurs while the spores are actively germinating, a process that is usually complete within a few minutes. Since the spores are adapted to germinate in the gut, the culture medium might need to be altered for spore germination and infection to occur. The time cells have to be in the germination medium is minimized by changing the culture medium as soon as germination is complete. During the germination period, the temperature should be near the upper limits for the survival of the microsporidium or the cell cultures so that the maximum percentage germination will be obtained.

b. Conditioning the spores

Spores of some species of microsporidia require pretreatment, or 'priming', and then germinate in a second solution, the culture medium in this instance. There is no one method that is always successful. Some techniques that might be tried are described below.

(i) *High pH – neutralization.* Many microsporidia germinate in a near-neutral solution after residing for 10–30 min in an alkaline solution (pH 11.0 or above), usually a low concentration (ca. 0.01 M) of NaOH or KOH. In addition to an alkaline pretreatment, some spores require chelation of bivalent metal ions for optimum germination. The following is a method used by Kurtti *et al.* (1990) to inoculate cell cultures with spores of *V. necatrix*. An aseptic suspension of spores (sufficient spores for 5–10 spores/cell) were suspended in 5 ml of 5 mM EDTA in 0.5 mM Tris-HCl, pH 7.5 for 30 min at room temperature. The spores were pelleted by centrifugation at 260 ***g***, resuspended and held for 30 min in a priming solution of 0.01 M KOH in 0.17 M KCl. Cells were suspended in their culture medium and both the cells and the spores were centrifuged at 260 ***g*** for 5 min at room temperature. The cell pellet was resuspended in 1 ml of 0.17 M KCl in 1 mM Tris-HCl with 10 mM EDTA at pH 8 (the germination solution) and immediately mixed with the spore pellet. After 3 min the suspension of cells and spores was poured into 30 ml of fresh culture medium (plus 50 mg/ml gentamycin to prevent bacterial growth) and placed in the appropriate culture vessels. It should be noted that the cells were suspended in the germination medium first and the spores added last because germination proceeds so rapidly after stimulation that, if the spores were added first, many would have germinated before they came into contact with the cells.

(ii) *Other priming systems used.* Spores of *Nosema michaelis* from the blue crab were primed for 90–120 min in Michaelis veronal-acetate buffer (9.7 g sodium acetate and 2.9% (wv) sodium barbiturate in 500 ml CO_2-free distilled water); they discharged in cell culture medium (199) with glutamine and Hank's salts (Weidner, 1976).

(iii) *Timing.* There is a stimulating period between the addition of the stimulant and expulsion of the microsporidian polar filament. With proper control of germination conditions, spores can be stimulated and then quickly transferred to another, normally unfavourable, solution where they will complete the germination process. Once they have been stimulated, *N. algerae* spores will complete germination over the next few minutes in any solution, even distilled water or in the presence of substances that are normally inhibitory (Undeen & Frixione, 1990). Stimulation takes 20–60 s at 30 °C in 0.1 M NaCl at pH 9.5. Eversion of the polar tubes begins about the time all spores are stimulated (60 s) and is complete 4–5 min later. Therefore, with careful timing, spores can be stimulated in solutions that are unfavourable to cells, then transferred to the cell cultures in a volume of germination solution too small to affect the culture medium.

The time between stimulation and germination (eversion of the polar tube) can by extended by stimulating the spores in a high concentration of sucrose (about 1.7 M for *N. algerae*) or polyethylene glycol. The process of germination will continue after dilution in the culture medium (Undeen & Frixione, 1990).

C Controlling infections in insect colonies

Protozoa, especially microsporidia, can become a serious problem in insect colonies. An organism, later determined to be a microsporidium (*Nosema bombycis*), was found by Louis Pasteur to be the causative agent of 'pebrine disease' in silkworms. He determined that the disease was transmitted from mother to progeny through the eggs. Using this knowledge, he was able to initiate new, uninfected colonies from offspring of moths that were isolated for oviposition and then found afterwards to be uninfected. Today, this is still the most reliable way to rid a laboratory colony of unwanted infection by microsporidia.

Selection of uninfected individuals is even easier when the contaminating microsporidium is not transmitted transovarially. A colony of *Anatis efformata* was freed of infection by *Pleistophora schubergi* by only one generation of individual matings (Briese & Milner, 1986). Removal of dead adults from oviposition containers and rinsing the eggs with water was sufficient to keep colonies of anopheline mosquitoes free of infection by *Nosema algerae* (Alger & Undeen, 1970). Sodium hypochlorite (bleach) is effective in destroying microsporidian spores on the surface of insect eggs. There are some drugs that can be fed to insects for control of microsporidia but none are known to eliminate the disease.

1. Selecting uninfected progeny

In order to free an insect colony from contamination by a microsporidium, the following procedure should be used.

1. Isolate gravid females individually in sterilized containers.
2. After oviposition is completed, dissect the female and examine for signs of infection.
3. Rear only the offspring of uninfected females and check them for infection.
4. Destroy all cohorts where infection is found.
5. If the new colony started in this way is uninfected, destroy the contaminated colony and sterilize everything to be used in the new colony.

2. Sanitation

If the contaminating microsporidium is one that is not transmitted within the egg, the disease can be controlled by sterilization of all equipment used to rear the animal. All dead adults or other insect material must be separated from the eggs and the eggs rinsed in distilled water. If the spores are deposited along with the eggs and adhere to them (transovum transmission), 0.25–5% of a commercial bleach product (which is usually about 5% sodium hypochlorite) can be used for sterilization. In most cases a concentration and treatment time can be found that will kill the microsporidian spore without damaging the eggs. Sanitation measures used in bacteriology labs are good guidelines for avoiding contamination problems with microsporidian spores. All counter tops, glassware, dissection instruments, and even pens, pencils, chemical jars – anything that is used in proximity to the microsporidia and the insects – must be routinely sterilized by heat or wiped down with bleach or another antimicrobial. Some spores, particularly those that tolerate desiccation, are hard to kill and some agents might not be completely effective. Ethanol, for example, can evaporate before all the spores are destroyed. These measures should remain a part of routine colony maintenance.

5 MANAGING SPORES

A Extraction from the host

Whole infected organisms or just the excised infected tissues can be triturated by a number of means including a tissue grinder, mortar and pestle or a blender. The choice depends on the particular organism and the volume of material being processed. For example, large fragile spores, such as those of the microsporidium *Edhazardia aedis* are damaged by harsh disruption methods such as the blender. Cleaning of the spores can be facilitated by harvesting spores only from the infected tissues dissected from the host instead of the whole organism. In some instances only the fat body is infected and, because the organ is almost entirely replaced by spores, little further purification is necessary. Because the quality of tap water varies by time and locale it is best to use deionized or distilled water for all spore-extraction procedures.

Spores will sometimes be lost through germination, possibly stimulated by solutes released into the water from the triturated hosts. Keeping the volume of water high in relation to the amount of tissue to be triturated helps avoid this problem. Germination of microsporidian spores can be prevented by extracting spores in a pH 9.0-buffered, 0.001–0.05 M ammonium chloride solution. Maintaining cool temperatures throughout the process also limits adventitious germination.

B Purification

Purity facilitates counting and measuring of spores and is also essential for obtaining pure extracts for biochemical studies. Freeing the spores of host tissues and microbial contaminants is often necessary for storage. Bacterial and fungal growth quickly destroys stored batches of protozoan spores. Therefore, the time spores remain in the triturated insects or any other medium that will support microbial growth should be minimized. Although this is not particularly important if they are to be frozen or dried, there is still opportunity for microbial growth after they are thawed.

1. Filtration and centrifugation

Purification schemes usually rely on differences in density and size between protozoan spores and contaminants. Filtration and centrifugation are, therefore, the methods used. Small quantities of triturated host material can be filtered through about 2–5 mm of wet cotton packed in a syringe (Undeen & Becnel, 1992). Larger quantities can be vacuum-filtered through a cotton pad in a Buchner funnel. Laboratory tissues, cheesecloth or other coarse fabric also serve as filtration beds for removal of large pieces of host material. Some microsporidian spores, such as those of *Caudospora* spp., have external ornamentation that increases their loss during filtration. *Until experienced with a particular organism, microscopically examine all filtrates, supernatants and residues for spores before discarding them.*

Because of their high density, a series of washes and centrifugations will free most microsporidian spores of dissolved and most particulate contaminants. Spores are usually more dense than most host tissues and centrifugation concentrates them near the bottom of the residue. The supernatant can be decanted and the detritus from the top of the residue carefully resuspended in a small volume of water and discarded. The spores can then be resuspended in water and the process repeated. If a considerable loss of spores is acceptable, a fairly clean suspension can be obtained by a few such cycles. The process of 'triangulation', expands on this process and, although time consuming, yields a fair harvest of clean spores using only low-speed centrifugation (Cole, 1970).

2. Density gradient centrifugation

Density gradient centrifugation is fast and the yield is high. Sucrose gradients can be used for terrestrial-host microsporidia that tolerate desiccation, but some microsporidia, especially those of aquatic hosts, may be desiccated and killed by sucrose concentrations high enough to suspend them. Colloidal silica does not have this disadvantage. Two silica colloids, Ludox® HS40 (duPont), with a density of 1.303 g/ml (Undeen & Alger, 1971; Undeen & Avery, 1983) and Percoll®, a Sigma product (Jouvenaz, 1981), with a density of 1.130 g/ml, have come into common usage for this purpose. Both are alkaline (pH 9.7) and high pH causes some spores to germinate, a problem that can be solved by the addition of ca. 0.01 M ammonium chloride to the gradient components before mixing (Undeen & Avery, 1983). Percoll can be neutralized but its density is too low to suspend most microsporidian spores which have density values in the range 1.180–2.200 (Undeen & Solter, 1996).

a. Continuous Ludox gradients

Materials

- Centrifuge capable of at least 10 000 ***g***
- Magnetic stirrer
- Small magnetic stirring bar
- Density gradient mixer
- Centrifuge tubes – round bottom
- Ludox HS-40 (or Percoll)
- 1.0 M NH_4Cl (optional)

Procedure

1. Determine the volume of the centrifuge tube and subtract the volume of the crude spore suspension planned to go onto the gradient. Place Ludox, in the amount of one half of this remaining volume, in the front chamber, the one with the outflow tube, along with a magnetic stirring bar. An equal volume of water is placed in the back chamber.
2. Adding 0.01– 0.05 M NH_4Cl to each chamber of the gradient mixer reduces the risk of spores germinating in the gradient.
3. Fill the U-shaped siphon tube with water and place it across the wall between the two chambers. (Density gradient mixers can be purchased with a valved conduit built into the bottom, connecting the two chambers.)
4. Fix the discharge tubing to a point against the wall of the centrifuge tube near the top so that the fluid will flow slowly down the side of the centrifuge tube and air can escape. (A cork or rubber stopper with a notch cut into each side works well for this.) Begin stirring and then start the flow through the tubing into the centrifuge tube. As the Ludox flows into the centrifuge tube, water passes from the distal chamber to the proximal one, mixing with the more dense Ludox and producing a progressively lower concentration. This diminishing concentration of Ludox runs slowly down the side of the centrifuge tube, layering on top. Allow it to flow until both chambers are empty.
5. The gradient is complete when both chambers are empty and the flow through the outflow tube stops. This produces a continuous gradient with a density of 1.303 g/ml at the bottom grading to 1.000 at the top.
6. An aliquot of the filtered crude spore suspension is layered on top of the gradient. For high purity do not overload the gradient. If the spore suspension is too concentrated, considerable detritus that otherwise might remain above the spore band, will be carried down with the spores.
7. The gradients have commonly been centrifuged at about 16 000 ***g***. for 30 min but optimum centrifugation has never been experimentally determined. Allow the centrifuge to decelerate slowly; braking will create a vortex and mix the upper region of the gradient.
8. After centrifugation, the mature spores are usually concentrated in a white band, 2–3 mm wide, 60–70% down the gradient, below most of the contaminants (Figure 8). Immature

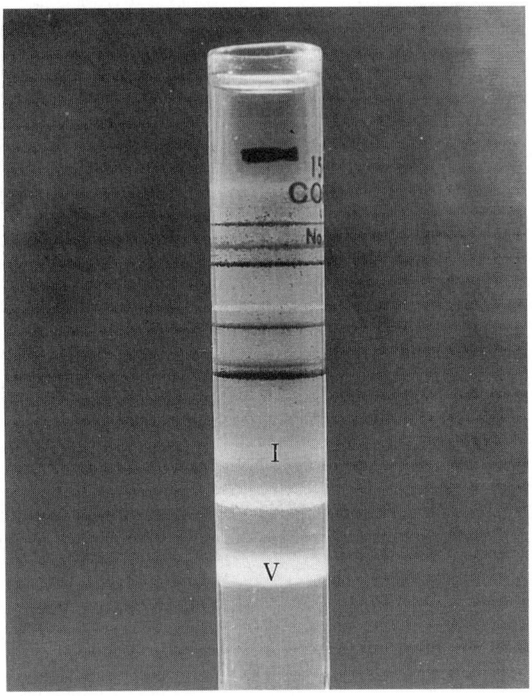

Figure 8 Viable (V) and inviable (I) microsporidian (*Nosema locustae*) spores in a Ludox density gradient. The other bands are density standards.

spores, soft insect tissues and bacteria will be above the spores. Insect viral polyhedra and fungal spores – but little else – are found with or below the spores. There is no guarantee that spores of all species will be in this zone. *Edhazardia aedis* has a lower and more variable density, demonstrating the need to determine, individually, the density of each species.

9. If there are too few spores to form a visible band, the gradient can be fractionated and each fraction examined for spores. Even if spores are present in small numbers they can be found by diluting the fractions with water (so that the spores will not remain suspended) and centrifuged to concentrate any spores that might be present. Alternatively, a pipette can be carefully inserted into the gradient and small samples taken at various levels to be examined microscopically for spores.
10. As soon as centrifugation is complete and the spores are located, they should be removed from the gradient and washed two or three times in water to remove the silica. Spores can be damaged by leaving them for several hours in Ludox.

Percoll® can be autoclaved to provide sterility and neutralized to enhance the survival of cells. In a procedure used by Iwano & Kurtti (1995), spores were layered on the top of neutralized 100% Percoll. After centrifugation at 39 000 ***g*** for 40 min, a 'shelf' of silica upon which the spores layered, formed near the bottom of the gradient.

b. Discontinuous gradients

If a density gradient mixer is not available, discontinuous gradients can be constructed by carefully layering a concentration series of the gradient material, starting with the most concentrated at the bottom. With a little experimentation, a discontinuous gradient of only two phases can be used to purify spores. The lower concentration is made sufficiently dense to just pass the spores and the higher one just sufficient to suspend them (Undeen *et al.*, 1993).

c. Continuous flow density gradient centrifugation

Litre – quantities of spore suspension have been purified by feeding it slowly into density gradients in a continuous-flow centrifuge (Undeen & Avery, 1983). Because the diluted crude spore suspension is fed slowly onto the gradient, the same quantity of gradient material will purify many times more spores than could be accommodated in a single batch.

Procedure

1. Filter the suspension to prevent blockage of the passageways in the distribution head.
2. If necessary, add a small amount of detergent (0.1% v/v, sodium dodecyl sulphate) to the filtered spore suspension to prevent formation of a layer of fat that tends to trap spores near the inlet tube.
3. Make the density gradients in the specialized tubes for the continuous flow centrifuge as described above.
4. Stir the crude spore suspension continuously to prevent settling of the spores and feed it slowly into the centrifuge which is running at the relative centrifugal force described above for density gradients. Save the outflow and check for the presence of spores before discarding.
5. After all the spore suspension has passed through the centrifuge, water is fed through the system to clear the tubing of spores.
6. The centrifuge is stopped without braking, the tubes are removed and the bands containing the spores are withdrawn. If the tube is opaque, the contents of the gradient can be aspirated from it through a small diameter tubing (ca. 1–2 mm). The suction tube is placed at the bottom of the gradient, fluid passing through the tube first will be clear, high concentration Ludox and then become cloudy as the spores begins to pass through. Start collecting the fluid at this point and stop collecting when the fluid once again becomes clear.
7. Rinse the spores free of the Ludox as described above.

C Obtaining aseptic spores

Aseptic spores are needed to inoculate cell cultures or inject into a host (Undeen & Maddox, 1973; Undeen & Alger, 1975; Pilley *et al.*, 1978).

1. Density gradient centrifugation

Spores that are sufficiently heavy to settle below the bacterial contaminants can be cleaned with a Ludox density gradient. Spores such as *E. aedis* that are exceptionally light cannot be cleaned in this way. To limit the numbers of bacteria present at the outset, the spores must be extracted from living hosts and cleaned immediately thereafter. The spores must be well dispersed and the gradients not overloaded, otherwise bacteria can be carried down with the spores. Fungal contamination can be a problem because some fungal spores are similar in density to microsporidian spores. The band of spores is extracted from the gradient, cleaned by several rinses in sterile water and treated with antibiotics. This method does not provide absolute sterility; therefore, the spores must be used immediately, before the bacterial levels increase.

2. Antibiotics

Pilley (1978) found the antibiotics, tetracycline, streptomycin and kanamycin to be better than several others tested at suppressing the growth of microbes in *V. necatrix* spores harvested from lepidopteran larvae. These spores were stored at 4 °C (for as long as 3 years) in 100 mg/ml tetracycline hydrochloride and 500 mg/ml neomycin sulphate with no ill effect.

3. Sterile dissection

Sterile spores can be obtained from live hosts by sterile dissection of infected tissues (Weiser, 1978). The host is surface-sterilized with bleach (sodium hypochlorite) or ethanol, pinned and the integument cut longitudinally on the ventral side, taking care not to puncture the gut. Infected tissues are removed with sterile implements, transferred to sterile water and rinsed two or three times in sterile water to remove host contaminants. These spores can be inoculated into cell cultures or injected into insects without fear of septicaemia.

D Transporting

It will frequently be necessary to ship a microsporidium to another laboratory for an expert opinion on its identity. Shipment time for unfixed material should be kept to a minimum to limit microbial activity that can destroy the sample.

1. Living infected hosts

Caution: a permit may be required to ship live organisms into other states or countries.

Live material is always preferred because developmental stages deteriorate along with the host cells after death. Live cells will be available for microscopic examination and staining according to the practices preferred at the collaborator's laboratory. A living host is particularly important for preparation of specimens for electron microscopy. Live hosts can be shipped, Federal Express, in labelled plastic boxes or vials supplied with host plant material or crushed paper towelling to reduce trauma.

2. Host cadavers

Spores remain identifiable and viable in insect cadavers for an indefinite period. Protozoan spores in terrestrial hosts usually withstand desiccation of the host without immediate loss of viability. However, microsporidian spores from aquatic insects do not appear to tolerate desiccation. Putrefaction is likely to occur during shipping but, if they are not held in these conditions too long, the spores may still be viable. They should be shipped by the most rapid mail service available.

3. Live spores

In order to avoid the possibility of desiccation or putrefaction, spores can be purified and shipped in deionized or distilled water. If the suspensions still contain host material or other contaminants, the addition of streptomycin and fungizone will help to protect the spores from microbial activity. Live spores permit transmission of the microsporidium in the collaborator's lab to obtain early developmental stages and to evaluate the host range, two important factors in its identification. Even if the microsporidium can not be transmitted, the live spores are needed for the evaluation of spore size and other morphological features.

4. Fixed host tissues

Small pieces of infected tissues can be fixed in 1–2% glutaraldehyde and shipped for later preparation for electron microscopy (see Chapter VIII-1). Tissues that have been fixed in formalin, ethanol, or other common histological fixatives are useful for diagnosis of microsporidioses but provide limited information about the identity of the microsporidium. Ethanol is probably the worst fixative for subsequent morphological study of microsporidian spores. Ethanol-fixed microsporidia have been processed for electron microscopy but the results have been poor. However, fixation in 70% ethanol appears to be an adequate means for preserving nucleic acids for sequencing.

5. Dried smears

Dried tissue smears on microscope slides can be easily mailed. These samples can be shipped unfixed or fixed with absolute methanol before mailing. Dried and, especially, unfixed smears need protection from household insects and humidity.

E Extraction of substances from spores

In order to liberate their contents, spores are most commonly disrupted by agitating them vigorously with glass beads. Cell disrupters such as the French press, Parr Bomb, X-press and sonicators are unable to break most microsporidian spores. The method of choice is dictated by the substance to be recovered. For proteins, heating and foaming must be avoided. Nuclear DNA needs to be treated as gently as possible to prevent excessive shearing of the long molecules.

1. Nucleic acids

a. Germination

Ribosomal RNA is easily extracted by agitation of the spores with glass beads but the same procedure can cause excessive shearing of nuclear DNA. The most gentle extraction procedure, when possible, is to germinate the spores. In this procedure, the primary problem is the protection of the nucleic acids from nucleases. A reasonably successful procedure is described in Undeen & Cockburn (1989). When germinating high concentrations of spores, the polar tubes entangle, forming a solid mass of spores, a problem that can be ameliorated by the addition of 2-mercaptoethanol which dissolves the polar tubes.

b. Agitation with glass beads

For those spores that can not be germinated *in vitro*, carefully timed disruption with glass beads provides reasonable results.

Procedure

1. Approximately 50 mg of spores are suspended in 0.2 ml Bead Beater Solution (BBS), 0.4 ml Tris-saturated phenol and combined with 0.4 g glass beads (0.5 mm diameter) in a 1.5 ml Eppendorf tube.
2. The tube is capped, covered with parafilm® and shaken with a bead beater for 1 min at high speed.
3. The tube is centrifuged in an Eppendorf centrifuge for 1 min at 10 000 rpm.
4. The aqueous phase (top) is withdrawn and transferred to another tube.
5. A 0.2-ml aliquot of BBS was added to the phenolic phase; this mixture was vortexed, centrifuged, and the aqueous phase was again extracted and combined with the first aqueous supernatant.
6. At 4 °C, the aqueous supernatant is extracted with tris-saturated phenol, centrifuged for 5 min, and;
7. Extracted with an equal volume of a 1 : 1 solution of phenol and chloroform.
8. A final extraction is made with an equal volume of chloroform.
9. The nucleic acids remaining in the aqueous phase are precipitated by adding 1 part sodium acetate to 9 parts final DNA extract and 2.5 parts cold absolute ethanol followed by chilling at –80 °C for 15 min.
10. The nucleic acids are pelleted by centrifugation, dried to remove the alcohol, resuspended in 20–30 µl of 10 mM, pH 8.5 Tris buffer, and stored at –80°C.

Alternatively: A more gentle agitation was used to extract nuclear DNA from spores (Undeen & Cockburn, 1989), using procedures similar to those described for the bead beater. Small volumes (0.1–0.5 ml) of spore suspension (10^7–10^9 spores/ml are combined with equal volumes of 0.5 µm glass beads in 10 × 75 mm glass culture tubes and shaken at high speed on a vortex mixer for 30–60 s. About 60–80% of the spores are disrupted and the DNA was not severely sheared.

2. Enzymes

Foaming and heating must be avoided to prevent denaturing of the protein that will inactivate the enzyme. Two methods of spore disruption are suggested below.

a. Braun homogenizer
Procedure

1. Two to four ml aliquots of spore suspensions containing approximately 10^9 spores per ml, are combined in a Braun homogenizer flask and shaken for 50 s in a Braun homogenizer.
2. In order to prevent denaturation of proteins by the heat generated by the friction of the beads, the flask is cooled by a spray of liquid CO_2 while being shaken. The flow rate of CO_2 must be carefully controlled; if too slow the flask will overheat; if too fast the contents of the flask will freeze and the spores will not be disrupted.
3. The homogenates are removed as described for the bead beater and centrifuged in a refrigerated centrifuge for 30 min at 4000 ***g***.
4. The supernatant are used immediately or frozen at –20 °C until use. Glycerol can be added before freezing for those enzymes that are sensitive to freezing.

b. Freeze – grinding
Spores were frozen in a mortar and then ground them with a pestle (Conner, 1970). The material was allowed to thaw, pool in the bottom of the mortar, and was then refrozen. Several cycles of freezing and grinding disrupted about 95% of the spores. The materials are not subjected to any heating during this procedure and loss of material on the glass beads is avoided. This material was said to be usable for immunological studies without further extraction. This procedure has also been followed with spores that were frozen at –196 °C in liquid nitrogen (Strick, 1993).

3. Carbohydrates

Spores were disrupted for extraction of sugars by grinding with glass beads for about one minute using either a bead beater or a vortex mixer without need for cooling.

Procedure

1. A 200 μl sample of spores suspended in deionized water was combined with an equal volume of 0.45 mm glass beads (Braun) in a 10 × 75 mm borosilicate culture tube.
2. The tube was shaken for 1 min at the highest speed on a vortex mixer (SP S8220) or in a Mini Beadbeater (Biospec Products) for 50 s.
3. The homogenate (ca. 100 μl) was withdrawn from the beads and an additional 100 μl aliquot of deionized water was added and shaken briefly, withdrawn and combined with the homogenate in a 1.5 ml Eppendorf tube.
4. The homogenates were immediately placed in boiling water for 5 min to stop enzymatic activity, then centrifuged at high speed (Eppendorf, model no. 5415) for 5 min.
5. The supernatant is retained for sugar assays.

6 STORAGE

Optimal storage conditions need to be experimentally determined for each species. The environmental conditions into which the spores are normally released are useful guidelines. Generally speaking, microsporidia and perhaps other Protozoa from terrestrial hosts tolerate desiccation and freezing; those from aquatic hosts do not. Microsporidia from most aquatic hosts must be stored in distilled or deionized water. When in doubt, the safest course of action is to purify the spores and hold them in an aqueous suspension in a refrigerator. Experience with *E. aedis* spores, however, has shown that not all microsporidian spores are tolerant to refrigerator temperatures. Brooks (1988) presents an excellent review of spore storage.

A Refrigeration

Highly purified spores of most species survive well in cold (ca. 5 °C) deionized or distilled water. Refrigerated spores should be held in tightly capped vials and checked frequently for evaporation. Longevity varies considerably among species with some remaining viable longer than 2 years (Brooks, 1988). *Nosema algerae* spores have retained their viability after 10 years in the refrigerator (Undeen, personal observation). Microbial activity in the suspension medium appears to be deleterious to microsporidian spores. Contamination by fungal spores is difficult to avoid because many of them are of the same density as microsporidian spores, making them impossible to remove by even the best of purification schemes. Massive fungal overgrowth is often responsible for destroying vials of spores in the refrigerator, even when the spores appear to be quite clean otherwise. A mixture of 100 mg penicillin, 100 units streptomycin and 0.25 mg fungizone per ml of suspension medium is a combination that is routinely used to retard microbial growth. Purity is the critical factor for optimizing refrigerated storage; antibiotics help but will not substitute for purification of the spores.

Edhazardia aedis is the only microsporidium known to be adversely affected by temperatures just above freezing (0–5 °C), a trait that might be shared by other microsporidia of tropical origin. It survives for approximately one month at temperatures between 10 and 30 °C but less than 24 h in the refrigerator (Undeen *et al.*, 1993). The developmental stages within *Aedes aegypti* eggs and larvae are more tolerant to chilling than the spores.

B Frozen

1. Household-type freezer

a. Spores

Most terrestrial species can be stored frozen at –20 to –30 °C in a household freezer. The inclusion of 50% glycerol in the suspension as a cryoprotectant is frequently required. Even under the best of conditions, repeated freezing and thawing causes spores to lose viability; therefore, spores should be stored in small aliquots so that they need to be thawed only once.

b. Cadavers

Nosema locustae spores are routinely stored in frozen grasshopper cadavers (Henry & Oma, 1981). This method appears to be equal to storage in water, without the necessity of cleaning the spores before storage. *Nosema locustae* spores, formulated on a bran bait, have been applied to large tracts of range land for the control of grasshoppers. Formulated this way, the dry spores have a relatively short shelf life. Therefore, the cadavers are removed from frozen storage, the spores extracted in water and then sprayed on the bran shortly before field application is anticipated.

2. Liquid nitrogen

For all but the microsporidia of aquatic hosts, liquid nitrogen is probably the most reliable method for long-term storage.

a. Spores

Spores that are tolerant of near 0 °C freezing conditions can also be stored for prolonged periods in liquid nitrogen. Cryoprotectants such as glycerol, dimethylsulphoxide or sucrose are frequently required (Maddox & Solter, 1996). The high density of sucrose and the slight toxicity of dimethylsulphoxide leave glycerol as the preferred cryoprotectant. A little experimentation with cryoprotectant concentration might be necessary. Fifty percent sucrose or glycerol and about 10% dimethylsulphoxide are good starting points.

The simplest procedure for preparing spores for storage in liquid nitrogen is to dissect out heavily infected tissues, homogenize them with a tissue grinder, filter the homogenate through cotton or other fine-mesh material and rinse once or twice with deionized or distilled water. Density gradient centrifugation can also be used. To avoid freezing and thawing spores a number of times, spores should be stored in several small aliquots. In one routine procedure, a 0.5 ml each of spore suspension and glycerol are placed in a cryopreservation vial and then plunged directly into liquid nitrogen, without precooling. Crude homogenates of spores can be frozen directly but antimicrobial agents should be added for protection of the spores after the vials are thawed.

b. Cadavers

Preservation of spores can sometimes be accomplished by freezing a small, intact host in a cryovial.

c. Cell cultures

Stocks of cultured cells are commonly stored in liquid nitrogen. Whenever tested, the microsporidium infecting the cells also survived under cryopreservation. In one study (Sohi & Wilson, 1976) the infected cells were mixed with 10% dimethylsulphoxide and cooled at 1 °C/min from room temperature to –40 °C in an ethanol – solid CO_2 bath then plunged into liquid nitrogen. Spores were viable after rapid thawing in a 30 °C water bath. According to Jaronski (1984) developmental stages of *N. algerae* and *Nosema eurytremae* also survived freezing in the host cell.

C Dry

1. Desiccated

The spores of some microsporidia such as *Nosema whitei*, survive for extended periods in the dried host cadaver and then germinate as soon as they come into contact with water. Most terrestrial host microsporidia are to some degree tolerant to desiccation but they usually survive longer in refrigerated aqueous suspensions or frozen.

2. Lyophilized

Spores that can be dried or frozen might also survive in a freeze-dried state. Vacuum drying, without prior freezing was shown to be preferable to freeze drying in some instances (Lewis & Lynch, 1974). Microsporidian spores can be lyophilized and stored in flame-sealed vials under vacuum (Bailey, 1972; Lewis & Lynch, 1974; Pilley, 1976).

D *In situ*

Some species, such as *E. aedis*, are best 'stored' within the living host. A pathogen of the mosquito *Aedes aegypti*, it is vertically transmitted within the egg and will retain viability as long as the host eggs remain viable, a period of several months.

REFERENCES

Alger, N. E. & Undeen, A. H. (1970) The control of a microsporidian, *Nosema* sp. in an anopheline colony by an egg-rinsing technique. *J. Invertebr. Pathol.* **15**, 321–337.

Avery, S. W. & Undeen, A. H. (1987) Some characteristics of a new isolate of *Helicosporidium* and its effect upon mosquitoes. *J. Invertebr. Pathol.* **49**, 246–251.

Bailey, L. (1972) The preservation of infective microsporidan spores. *J. Invertebr. Pathol.* **20**, 252–254.

Baker, M. D. (1987) Phylogenetic relationships of five microsporidian genera based on ribosomal RNA sequence data. PhD Thesis, University of Illinois.

Briese, D. T. & Milner, R. J. (1986) Effect of the microsporidian *Pleistophora schubergi* on *Anaitis efformata* (Lepidoptera: Geometridae) and its elimination from a laboratory colony. *J. Invertebr. Pathol.* **48**, 107–116.

Brooks, W. M. (1980) Production and efficacy of protozoa. *Biotech. Bioeng.* **22**, 1415–1440.

Brooks, W. M. (1988) Entomogenous protozoa. In *CRC Handbook of natural pesticides*, vol. V, *Microbial insecticides*, Part A, *Entomogenous protozoa and fungi* (ed. C. M. Ignoffo), pp. 1–149. CRC Press, Boca Raton.

Clark, J. M. (1964) *Experimental Biochemistry*. W. H. Freeman, San Francisco, 266 pp.

Cole, R. J. (1970) The application of the 'triangulation' method to the purification of *Nosema* spores from insect tissues. *J. Invertebr. Pathol.* **15**, 193–195.

Conner, R. M. (1970) Disruption of microsporidian spores for serological studies. *J. Invertebr. Pathol.* **15**, 138.

Corliss, J. O. (1994) An interim utilitarian ('user friendly') hierarchical classification and characterization of the Protists. *Acta Protozool.* **33**, 1–51.

Egerter, D. E. & Anderson, J. R. (1985) Infection of the western treehole mosquito, *Aedes sierrensis* (Diptera: Culicidae), with *Lambornella clarki* (Ciliophora: Tetrahymenidae). *J. Invertebr. Pathol.* **46**, 296–304.

Hayasaka, S., Sato, T. & Inoue, H. (1993) Infection and proliferation of microsporidians pathogenic to the silkworm *Bombyx mori* L. and the Chinese oak silkworm *Antheraea pernyi* in lepidopteran cell lines. *Bull. Natl. Inst. Seric. Entomol. Sci.* **7**, 47–63.

Hazard, E. I. & Brookbank, J. W. (1984) Karyogamy and meiosis in an *Amblyospora* sp. (Microspora) in the mosquito *Culex salinarius*. *J. Invertebr. Pathol.* **44**, 3–11.

Hazard, E. I., Ellis, E. A. & Joslyn, D. J. (1981) Identification of microsporidia. In *Microbial control of plant pests and plant diseases 1970–1980* (ed. H. D. Burges), pp. 163–182. Academic Press, New York.

Henry, J. E. & Oma, E. A. (1981) Pest control by *Nosema locustae*, a pathogen of grasshoppers and crickets. In *Microbial control of pests and plant diseases 1970–1980* (ed. H. D. Burges), pp. 573–586. Academic Press, New York.

Ishihara, R. (1967) Stimuli causing extrusion of polar filaments of *Glugea fumiferanae* spores. *Can. J. Microbiol.* **13**, 1321–1332.

Ishihara, R. (1968) Growth of *Nosema bombycis* in primary cell cultures of mammalian and chicken embryos. *J. Invertebr. Pathol.* **11**, 328–329.

Iwano, H. & Kurtti, T. J. (1995) Identification and isolation of dimorphic spores from *Nosema furnacalis* (Microspora: Nosematidae). *J. Invertebr. Pathol.* **65**, 230–236.

Jaronski, S. T. (1984) Microsporida in cell culture. *Adv. Cell Culture* **3**, 183–299.

Jouvenaz, D. P. (1981) Percoll: An effective medium for

cleaning microsporidian spores. *J. Invertebr. Pathol.* **37**, 319.

Kellen, W. R. & Lindegren, J. E. (1974) Life cycle of *Helicosporidium parasiticum* in the naval orangeworm, *Paramyelois transitella. J. Invertebr. Pathol.* **23**, 202–208.

Kramer, J. P. (1973) Differential germination among spores of the microsporidian *Octosporea muscaedomesticae. Z. Parasitenk.* **41**, 61–64.

Kudo, R. (1966) *Protozoology*, C. C. Thomas, Springfield. 1174 pp.

Kurtti, T. J., Munderloh, U. G. & Noda, H. (1990) *Vairimorpha necatrix*: Infectivity for and development in a lepidopteran cell line. *J. Invertebr. Pathol.* **55**, 61–68.

Lange, C.E. (1993) Unclassified protists of arthropods: The ultrastructure of *Nephridiophaga periplanetae* (Lutz and Splendore, 1903) n. comb., and the affinities of the Nephridiophagidae to other protists. *J. Euk. Microbiol.* **40**, 689–700.

Lee, J. L., Hutner, S. H. & Bovee, E. C. (1985) *An Illustrated Guide to the Protozoa.* Society of Protozoologists. Allen Press, Lawerence, KS, 629 pp.

Levine, N. D., Corliss, J. O., Cox, F. E. G., Deroux, G., Grain, J., Honigberg, B. M., Leedale, G. F., Loeblich III, A. R., Lom, J., Lynn, D., Merinfeld, E. G., Page, F. C., Poljansky, G., Sprague, V., Vávra, J. & Wallace, F. G. (1980) A newly revised classification of the protozoa. *J. Protozool.* **27**, 37–58.

Lewis, L. C. & Lynch, R. E. (1974) Lyophilization, vacuum drying, and subsequent storage of *Nosema pyrausta* spores. *J. Invertebr. Pathol.* **24**, 149–153.

Maddox, J. V. & Solter, L. (1996) Long term storage of viable microsporidian spores in liquid nitrogen. *J. Invertebr. Pathol.* **43**, 221–225.

Malone, L. A. (1984) Factors controlling in vitro hatching of *Vairimorpha plodiae* (Microspora) spores and their infectivity to *Plodia interpunctella, Heliothis virescens*, and *Pieris brassicae. J. Invertebr. Pathol.* **44**, 192–197.

Mazia, D., Schatten, G. & Sale, W. (1975) Adhesion of cells to surfaces coated with polylysine. *J. Cell Biol.* **66**, 198–200.

Ohshima, K. (1964) Effect of potassium ion on filament evagination of spores of *Nosema bombycis* as studied by the neutralization method. *Annot. Zool. Jpn* **37**, 102–103.

Olsen, P. E., Rice, W. A. & Liu, T. P. (1986) In vitro germination of *Nosema apis* spores under conditions favorable for the generation and maintenance of the sporoplasms. *J. Invertebr. Pathol.* **47**, 65–73.

Pilley, B. M. (1976) The preservation of infective spores of *Nosema necatrix* (Protozoa: Microsporida) in *Spodoptera exempta* (Lepidoptera: Noctuidae) by lyophilization. *J. Invertebr. Pathol.* **27**, 349–350.

Pilley, B. M. (1978) The storage of infective spores of *Vairimorphia necatrix* (Protozoa; Microspora) in antibiotic solution at 4°C. *J. Invertebr. Pathol.* **31**, 341–344.

Pilley, B. M., Canning E. U. & Hammond, J. C. (1978) The use of a microinjection procedure for large-scale production of the microsporidian *Nosema eurytremae* in *Pieris brassicae. J. Invertebr. Pathol.* **32**, 355–358.

Pleshinger, J. & Weidner, E. (1985) The microsporidian spore invasion tube. IV. Discharge activation begins with pH-triggered Ca^{2+} influx. *J. Cell Biol.* **100**, 1834–1838.

Poinar, G. O. & Thomas, G. M. (1984) *Laboratory guide to insect pathogens and parasites*, Plenum Press, New York, 392pp.

Sohi, S. S. (1971) *In vitro* cultivation of hemocytes of *Malacosoma disstria* Hubner (Lepidoptera, Lasiocampidae). *Can. J. Zool.* **49**, 1355–1358.

Sohi, S. S. & Wilson, G. G. (1976) Persistent infection of *Malacosoma disstria* (Lepidoptera, Lasiocampidae) cell culture with *Nosema disstriae* (Microsporida, Nosematidae). *Can. J. Zool.* **54**, 336–342.

Sprague, V. (1977) *Comparative Pathobiology*, Vol. 2, *Systematics of the Microspoiridia* (eds. L. A. Bulla and T. C. Cheng). Plenum Press, New York.

Sprague, V., Becnel, J. J. & Hazard, E. I. (1992) Taxonomy of phylum Microspora. *Crit. Rev. Microbiol.* **18**, 285–295.

Streett, D. A. & Briggs, J. D. (1982) An evaluation of sodium dodecyl sulphate – polyacrylamide gel electrophoresis for the identification of microsporidia. *J. Invertebr. Pathol.* **40**, 159–165.

Strick, H. (1993) Disruption of microsporidian spores for biochemical analysis. *Z. Angew. Zool.* **77**, 3–4.

Tanada, Y. & Kaya, H. K. (1993) *Insect Pathology.* Academic Press, San Diego, 666 pp.

Undeen, A. H. (1975) Growth of *Nosema algerae* in pig kidney cell cultures. *J. Protozool.* **22**, 107–110.

Undeen, A. H. (1976) *In vivo* germination and host specificity of *Nosema algerae* in mosquitoes. *J. Invertebr. Pathol.* **27**, 343–347.

Undeen, A. H. (1978) Spore hatching processes in some *Nosema* species with particular reference to *Nosema algerae* Vávra and Undeen. *Misc. Publ. Entomol. Soc. Am.* **11**, 29–50.

Undeen, A. H. (1983) The germination of *Vavraia culicis* spores. *J. Protozool.* **30**, 274–277.

Undeen, A. H. & Alger, N. E. (1971) A density gradient method for fractionating microsporidian spores. *J. Invertebr. Pathol.* **18**, 419–420.

Undeen, A. H. & Alger, N. E. (1975) The effect of the microsporidian, *Nosema algerae*, on *Anopheles stephensi. J. Invertebr. Pathol.* **25**, 19–24.

Undeen, A. H. & Avery, S. W. (1983) Continuous flow-density gradient centrifugation for purification of microsporidia spores. *J. Invertebr. Pathol.* **42**, 405–406.

Undeen, A. H. & Avery, S. W. (1984) Germination of experimentally non-transmissible microsporidia. *J. Invertebr. Pathol.* **43**, 299–301.

Undeen, A. H. & Avery, S. W. (1988a) Spectrophotometric measurement of *Nosema algerae* (Microspora: Nosematidae) spore germination rate. *J. Invertebr. Pathol.* **52**, 253–258.

Undeen, A. H. & Avery, S. W. (1988b) Ammonium chloride inhibition of the germination of spores of *Nosema algerae* (Microspora: Nosematidae). *J. Invertebr. Pathol.* **52**, 326–334.

Undeen, A. H. & Becnel, J. J. (1992) Longevity and germination of *Edhazardia aedis* (Microspora: Amblyosporidae). *Biocontrol. Sci. Technol.* **2**, 247–256.

Undeen, A. H. & Cockburn, A. F. (1989) The extraction of DNA from microsporidia spores. *J. Invertebr. Pathol.* **54**, 132–133.

Undeen, A. H. & Epsky, N. D. (1990) *In vitro* and *in vivo* germination of *Nosema locustae* (Microspora: Nosematidae) spores. *J. Invertebr. Pathol.* **56**, 372–379.

Undeen, A. H. & Frixione, E. (1990) The role of osmotic pressure in the germination of *Nosema algerae* spores. *J. Protozool.* **37**, 561–567.

Undeen, A. H. & Maddox, J. V. (1973) The infection of nonmosquito hosts by injection with spores of the microsporidian *Nosema algerae. J. Invertebr. Pathol.* **22**, 258–265.

Undeen, A. H. & Solter, L. F. (1996) The sugar content and density of living and dead microsporidian (Protozoa: Microspora) spores. *J. Invertebr. Pathol.* **67**, 80–91.

Undeen, A. H. & Vander Meer, R. K. (1990) The effect of ultraviolet radiation on the germination of *Nosema algerae* Vávra and Undeen (Microsporida: Nosematidae) spores. *J. Protozool.* **37**, 194–199.

Undeen, A. H. & Vander Meer, R. K. (1994) Conversion of intrasporal trehalose into reducing sugars during germination of *Nosema algerae* (Protista: Microspora) spores: a quantitative study. *J. Euk. Microbiol.* **41**, 129–132.

Undeen, A. H., Johnson, M. A. & Becnel, J. J. (1993) The effects of temperature on the survival of *Edhazardia aedis* (Microspora: Amblyosporidae), a pathogen of *Aedes aegypti. J. Invertebr. Pathol.* **61**, 303–307.

Van Handel, E. (1985) Rapid determination of glycogen and sugars in mosquitoes. *J. Am. Mosq. Control Assoc.* **1**, 299–300.

Vávra, J. (1964) Recording microsporidian spores. *J. Insect Pathol.* **6**, 258–260.

Vávra, J. (1976) Structure of the Microsporidia. In *Comparative pathobiology, Vol. 1. The biology of the microsporidia* (eds L. A. Bulla & T. C. Cheng), pp. 2–85. Plenum Press, New York.

Vávra, J. & Maddox, J. V. (1976) Methods in microsporidiology. In *Comparative pathobiology, Vol. 1. The biology of the microsporidia* (eds. L. A. Bulla & T. C. Cheng), pp. 298–313. Plenum Press, New York.

Washburn J. O., Gross, M. E., Mercer, D. R. & Anderson, J. E. (1988) Predator-induced trophic shift of a free-living ciliate: Parasitism of a mosquito larva by their prey. *Science* **240**, 1193–1195.

Weidner, E. (1976) The microsporidian spore invasion tube. The ultrastructure, isolation, and characterization of the protein comprising the tube. *J. Cell Biol.* **71**, 23–34.

Weidner, E. & Halonen, S. K. (1993) Microsporidian spore envelope keratins phosphorylate and disassemble during spore activation. *J. Euk. Microbiol.* **40**, 783–788.

Weidner, E., Byrd, W., Scarborough, A., Pleshinger, J. & Sibley, D. (1984). Microsporidian spore discharge and the transfer of polaroplast organelle membrane into plasma membrane. *J. Protozool.* **31**, 195–198.

Weiser, J. (1966) *Nemoci Hmyzu* (*Insect Diseases*) (in Czech). Academia, Prague. 554 pp.

Weiser, J. (1977) *An atlas of insect diseases*. Academia, Prague. 240 pp.

Weiser, J. (1978) Transmission of microsporidia to insects via injection. *Spol. Zool.* **42**, 311–317.

Weiser, J. (1991) *Biological control of vectors* (*Manual for collecting, field determination and handling of biofactors for control of vectors*). Wiley, New York, 189 pp.

CHAPTER V-1

Fungi: Identification

RICHARD A. HUMBER

USDA-ARS Plant Protection Research Unit, US Plant, Soil & Nutrition Laboratory, Tower Road, Ithaca, New York 14853-2901, USA

1 INTRODUCTION

Most scientists who find and try to identify entomopathogenic fungi have little mycological background. This chapter presents the basic skills and information needed to allow non-mycologists to identify the major genera and, in some instances, most common species of fungal entomopathogens to the generic or, in many instances, to the specific level with a degree of confidence.

Although many major species of fungal entomopathogens have basic diagnostic characters making them quickly identifiable, it must be remembered that species such as *Beauveria bassiana* (Bals.) Vuill., *Metarhizium anisopliae* (Sorok.) Metsch, and *Verticillium lecanii* (Zimm.) Viégas are widely agreed to be species complexes whose resolutions will depend on correlating molecular, morphological, pathobiological and other characters (Soper *et al.*, 1988; Humber, 1996). The keys in this chapter cannot treat the total variation known for these common genera and species, but the information given is a detailed guide to the diagnostic characters of many important fungal entomopathogens.

This chapter also discusses the preparation of mounts for microscopic examination. Similar points are covered in other chapters, but good slide mounts and simple issues of microscopy are indispensable skills for facilitating the observation of key taxonomic characters. Many publications discuss the principles of microscopy, but a manual by Smith (1994) is easy to understand and notable for its many micrographs showing the practical effects of the proper and improper use of a light microscope.

The recording of images presents a wholly new set of options and challenges in increasingly computerized laboratories. Until this century, the only visual means to record microscopic observations was with drawings; such artwork, whether rendered freehand or with the aid of a camera lucida, still remains an important means of illustrating many characters. The photographs in this chapter were acquired directly as digital files and then adjusted, composed into plates and labelled with photographic software, and printed with a dye sublimation printer. Such a

MANUAL OF TECHNIQUES IN INSECT PATHOLOGY
ISBN 0–12–432555–6

non-traditional approach to scientific illustration will, undoubtedly, become much more common in the near future.

2 PREPARATION AND OBSERVATION OF MICROSCOPE SLIDES

The identification of most entomopathogenic fungi necessarily depends on the observation of microscopic characters. Fortunately, however, many common entomopathogens can, with relatively little experience, be easily identified to the genus or, in some instances, the species by observation with either the unaided eye or low magnifications from hand lenses or stereo microscopes. Species identifications usually require confirmation of essential microscopic characters.

The ease with which key microscopic characters can be seen is directly affected by the quality of one's microscopy and slide preparative techniques. The following sections outline the few major skills needed to use a microscope well or to make good slide preparations.

A Köhler illumination: the first and most important step

The key to observing fine details in a microscope is not magnification; it is optical resolution, the ability to distinguish two adjacent objects. Many factors can affect image resolution, but the first and most important is to maintain Köhler illumination when using bright field or differential interference optics. Phase-contrast images are much less sensitive to the physical settings of a microscope, but it is always a good idea to maintain Köhler illumination at all times.

The following steps to achieve Köhler illumination should be repeated for each objective used. Focus sharply on any object in a slide and then:

1. Close down the field diaphragm (at the light source) and adjust the height of the condenser so that both the inner edge of this iris diaphragm and the object in the slide are sharply focused when seen through the eyepieces.
2. Open the field diaphragm until its image nearly fills the field of view and then centre the field diaphragm image in the field of view with the condenser's centring screws.
3. Adjust the opening of the condenser diaphragm. The image of this diaphragm is seen by removing an eyepiece and looking down the inside of the microscope body; a focusing telescope can be useful but is not truly necessary for this step. The condenser diaphragm should be adjusted so that its opening fills some 80–90% of the diameter of the image in this back focal plane.

The condenser diaphragm should never be opened wider than the full diameter of the back focal plane; the resulting 'glare' of too much uncollimated light in the system severely degrades the image resolution. A frequent error in light microscopy is to close down the condenser diaphragm too far to increase the image contrast, but the resulting interference effects (seen as increasing graininess and darkening of object edges) also dramatically reduces image resolution.

B Coverslips

Microscopic image resolution is also affected by the type and thickness of coverslips used in slide preparations. The optics of microscope lenses are calculated to allow maximal resolution with no. 1½ coverslips (0.16–0.19 mm thick); maximal resolution is lower with either no. 1 and no. 2 coverslips (with thicknesses of 0.13–0.17 and 0.17–0.25 mm, respectively). Use glass coverslips for diagnostic work. Plastic coverslips are too thick and cause intolerable image degradation; they should be reserved for specialized experiments and avoided for general microscopic observations.

Full-sized 18 or 22 mm square or round coverslips may not be the most practical size for diagnostic purposes or whenever one must make large numbers of mounts in a short time. The total amount of glass and mounting medium to be used can be greatly reduced by scribing square coverslips into quarters with a diamond or carbide pencil and a slide edge as a straightedge, and then gently breaking those coverslips along the scratches if they do not break during the scribing. Ten or twelve such miniature coverslips can fit on a standard slide. Not only is less material consumed in this process, but the smaller area under each coverslip makes it easier to locate the fungus to be observed.

C Mounting media

Regardless of the mounting medium used, it is important to use no more than is needed to fill the volume under the coverslip. It is alright to use too little mounting medium, but using too *much* floats the coverslip, does not flatten the material to be examined, and prevents any later sealing with nail polish or other slide sealants. Mounting medium can be removed and a preparation further flattened without spreading mounting medium all over the slide (or microscope) and without lateral movement on the specimen by placing the slide into a pad of bibulous paper and applying whatever pressure is needed.

The choice of mounting medium and the means of preparing slide mounts can profoundly affect the apparent sizes of taxonomically important structures (Humber, 1976).

Recipes for some useful mounting media are given in the Appendix to this chapter. These include pure lactic acid (to which acidic stains such as aniline blue or aceto-orcein may be added), lactophenol (which is more useful for semi-permanent mounts than is lactic acid, and is also compatible with acidic stains), and aceto-orcein (a very useful general mount for diagnostic purposes that can hydrate even dried specimens and is nearly required for identifying entomophthoralean fungi).

D Handling of the material to be observed

Novice slide-makers often include too much material in a slide with the mistaken belief that 'more is better'. In fact, the most useful slides usually include the very little amount of material that has been carefully teased apart and spread in the mounting medium. Using only small amounts of material in mounts may force repeated preparations to see specific structures, but the effort required is often distinctly rewarded by the results. In all practicality, most preparations for diagnostic uses can be prepared fairly quickly since the most critical characters may be readily seen regardless of the care in preparation. Mounts intended for photography and or archival preservation, however, do benefit greatly from the most fastidious possible preparative attention.

The most useful tools for preparing slides of many fungi are not standard dissecting needle probes. The points of such probes are much too large to tease apart delicate fungal structures. The best tools may be '0' and 'minuten' insect pins mounted in soft wood sticks (e.g. the thick wooden match sticks available in the US or wooden chopsticks). The blunt ends of stainless steel '0' (whose heads have been cut off) or 'minuten' insect pins should be pushed into the sticks. The points of both of these types of pins remain small and distinctly pointed even when viewed at high magnification (see Figure 1). The '0' pins are superb for coarse operations or teasing apart leathery or hard structures; 'minuten' pins are excellent for manipulating hyphae, conidiophores, or other delicate structures. These insect pins are also versatile tools for manipulating cultures. The points of '0' pins can be pounded out into very useful microspatulas. Standard or flattened points of '0' needles can be flame sterilized but the points of 'minuten' pins may melt and even burn if flamed; autoclaving in glass Petri dishes or in groups in folded foil packets is a convenient way to sterilize these pins.

The art of making good slides consistently is, once again, mostly a matter of practice and common sense. Most taxonomically important structures can be detected well enough at magnifications of 50–75× to know if a slide merits examination on the compound microscope. Virtually all microscopic examination of entomopathogenic fungi for diagnostic purposes can be done at a magnification of 400–450×; oil immersion is only rarely needed.

E Semi-permanent slide mounts

Most slide mounts are made strictly for immediate observation rather than for long-term storage for

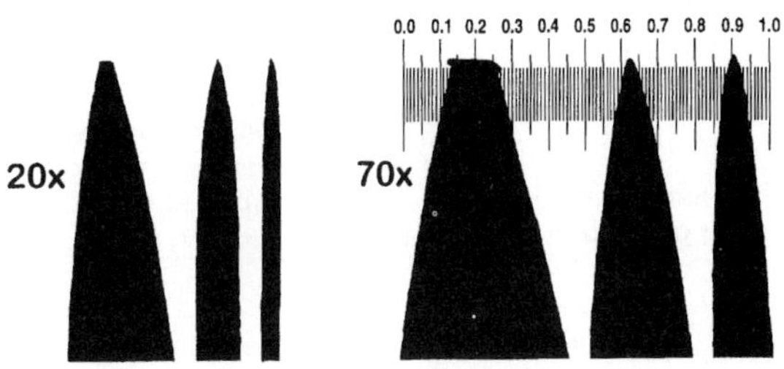

Figure 1 Comparative appearances at low magnification of dissecting needle tips (left to right: standard dissecting needle, 0' insect pin, and 'minuten' insect pin). The higher magnification set is superimposed over an ocular micrometer scale (total length, 1.0 mm).

later reference. Many differing techniques can be used to make semi-permanent slides, but those most useful for invertebrate pathogens involve means to seal slides prepared with aqueous mounting media.

A very short-term seal may be obtained by painting a melted mixture of roughly equal amounts of paraffin and petroleum jelly around the coverslip. Extreme caution must be used if melting this mixture over an open flame (alcohol burner, etc.) since paraffin vapour is highly flammable.

Coverslips are most often sealed by ringing them with fingernail polish, Canada balsam or another slide-making resin. Apply a relatively narrow and thin first layer; once the sealant is dried, a thicker and more secure seal can be built up by later applications of the sealant, but always be sure that the edges of the subsequent layer(s) cover the inner and outer edges of earlier layer(s). Such preparations may remain sealed for several months but should not be relied on to last for years. No sealing method is likely to work unless only a *minimal* amount of mounting medium is included under the coverslip; slides on which any amount of mounting medium protrudes from under the coverslip will probably fail to seal.

More secure, longer-lasting aqueous mounts can be prepared with methods using two coverslips of unequal sizes (Kohlmeyer & Kohlmeyer, 1972). The basic method shown in Figure 2 is simple: The material is spread in a minimal drop of mounting medium on the small coverslip; the large coverslip is then lowered onto the small one; the smaller coverslip of this sandwich is then attached to the standard microscope slide by a drop of glycerol, immersion oil or resin; and the space under the edge of the large coverslip is filled with a permanent sealant. Kohlmeyer & Kohlmeyer (1972) modified this basic procedure with a preliminary sealing of the small coverslip onto the large one and allowing this first ring to dry before attaching the sandwich to the slide. Such a procedure is easier to describe than to execute flawlessly. Several points should be heeded to increase the likelihood of success:

- The relative size differences of the coverslips should be small. Pairing 18 mm and 22 mm square coverslips is suitable; mixing square and round coverslips should be avoided.
- It takes practice to get the sizes of the drops of fluids small enough.
- It is easiest to use a small paint brush to apply the sealant.
- Adjusting the viscosity and solvent concentration in the sealant is the most difficult problem in this technique. Too much solvent tends to create bubbles in the sealing ring and may destroy the longevity of the mount. Inadequately thinned sealant may be too viscous to fill the space under the large coverslip.
- Excess (hardened) sealant can be cut away with a razor blade to improve the cosmetic appearance of the preparation.

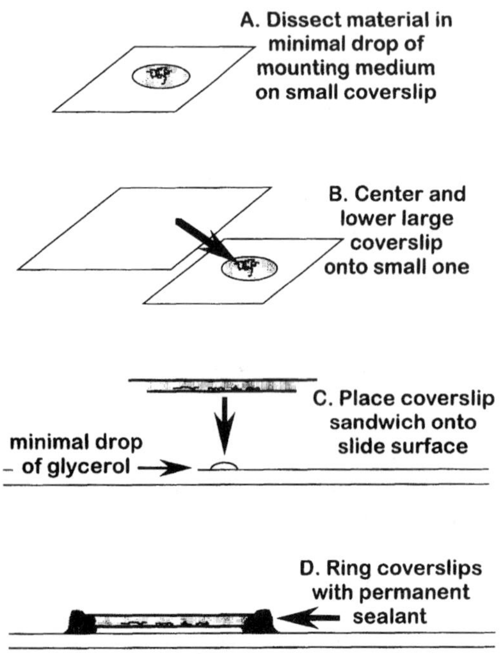

Figure 2 Outline of the procedure to make coverslip 'sandwiches' and semi-permanent slides.

3 KEY TO MAJOR GENERA OF FUNGAL ENTOMOPATHOGENS

This key should be used together with the taxonomic treatments and photos in Section 4. The key includes all fertile (spore-bearing) states most likely to be found for the genera treated. A greater number of entomopathogenic fungal genera are illustrated and keyed (although in less detail) by Samson *et al.* (1988).

Those with access to the World Wide Web may find a glimpse of the possible future of taxonomic mycology there in the form of an interactive key to

Fusarium species (Seifert, 1995; <http://res.agr.ca/brd/fusarium/>). Few species of this complex genus affect insects but this interactive key offers a significant model for future similar on-line keys to pathogens of invertebrates that could become important and highly accessible tools for a broad spectrum of scientists, regardless of their academic backgrounds and specialties.

Vegetative states of most fungi have little taxonomic value and are not characterized in the key. If no spores are seen in a collection, specimens (or cultures) should be incubated for a further time in room conditions of temperature, humidity and light and, if reasonable, part of any fresh collection of infected specimens should be incubated in a humid chamber at 100% RH for 24–48 h but watch closely for fast-growing fungal and bacterial saprobes that may soon overwhelm a real pathogen.

It is assumed that this key will be used primarily with infected specimens but most of the included fungi should also be identifiable from sporulating cultures so long as the user is aware of the host's identity and has a general idea about the appearance of the fungus on that host.

A brief glossary of terms used in the key and generic discussions is presented at the end of this chapter and should help to clarify many potential questions. More detailed definitions of terms can be found in many mycological textbooks or in *Ainsworth & Bisby's Dictionary of the Fungi* (Hawksworth *et al.*, 1995).

1\. Spores and hyphae or other fungal structures visible on exterior of host or host body is obscured by fungus; few or no spores form inside host cadaver 2

1a. Fungal growth and sporulation wholly (or nearly wholly) confined to *interior* of host body 30

2\. Elongated macroscopic structures (synnemata or club-like to columnar stromata) project from host 3

2a. Fungal growth may cover all or part of the host and may spread onto the substrate but large, projecting structures are absent 10

3\. Conidia form on synnemata and/or on mycelium on the host body 4

3a. Flask-like to laterally flattened fruiting structures (perithecia) present whether on or submersed in an erect, dense to fleshy, club-like to columnar stroma or on body of host; if mature, containing elongated asci with thickened apical caps 9

4\. Conidia formed in short to long chains 5

4a. Conidia produced singly on many separate denticles on each conidiogenous cell or, if in some sort of slime, singly (slime sometimes not evident) or in small groups in a slime droplet 7

5\. Conidiogenous cells flask-like, with swollen base and a distinct neck, borne singly or in loose clusters; chains of conidia often long and divergent (when borne on clusters of conidiogenous cells) . . . .*Paecilomyces*

5a. Conidiogenous cells short, with rounded to broadly conical apices (not having a distinctly narrowed and extended neck) 6

6\. Conidiogenous cells clustered on more or less swollen vesicle on short to long, conidiophores projecting laterally from synnemata and/or the hyphal mat covering the host; conidia pale to yellow or violet in mass; affecting spiders . . . . *Gibellula*

6a. Conidiogenous cells borne at apices of broadly branched, densely intertwined conidiophores that form a compact hymenium; conidia borne in parallel chains and usually green in mass . . . . *Metarhizium*

7\. Conidiogenous cell with swollen base and elongated, narrow to spine-like neck; conidia formed singly (usually with a distinct slime coating) or small groups in a slime droplet . . . . *Hirsutella*

7a. Conidiogenous cells producing several to many conidia, each formed singly on separate denticles 8

8. Conidiogenous cell with an extended, denticulate apex (growing apex repeatedly forms a conidium and regrows [rebranches] just below the new conidium) .... *Beauveria*
8a. Conidiogenous cell short and compact, cylindrical to broadly clavate, with apex studded by many denticles, each of which bears a single conidium .... *Hymenostilbe*

9. Erect stroma bears perithecia superficial to partially or fully immersed (with only small circular opening raised above stromatic surface); perithecia scattered or aggregated into more or less differentiated, apical or lateral fertile part; asci (if present) with thickened apical cap perforated by narrow canal and filiform ascospores (that usually dissociate into one-celled part spores); conidia, if simultaneously present, being formed on host body, on lower portion of stroma, or on separate synnemata .... *Cordyceps*
9a. Perithecia occur *only* on or partially immersed in a cottony to woolly hyphal layer covering host ...... *Torrubiella*

10. Fungus covering host is a stroma (fleshy to hard mass of intertwined hyphae); sporulation occurs in cavities below the stromatic surface . 11
10a. Host partially to completely covered by wispy, cottony, woolly, or felt-like growth or by a dark-coloured, extensive patch having columns and chambers below its surface but *not* forming a dense stroma 12

11. Spores are fusoid, one-celled conidia discharged in a slime mass from fertile chambers immersed in the stroma but not set off by a differentiated wall .... *Aschersonia*
11a. Globose to flask-like perithecia delimited by a distinct wall are immersed in stroma and contain elongated asci with thickened apices or, at maturity, a (non-slimy) mass of globose, ovoid or rod-like spores formed by dissociation of multiseptate ascospores; *Aschersonia* conidial state often present on same stroma .... *Hypocrella*

12. Fungus a dark brown to black, sometimes extensive patch on woody plant parts; upper surface dense to felt-like, with elongated or clavate thick-walled cells (teleutospores) remaining attached; open chambers and vertical fungal columns underlie the more or less solid upper surface and shelter living scale insects, some of which contain prominently coiled haustorial hyphae .... *Septobasidium* (see Couch, 1938; not treated here)
12a. Fungal hyphae emerging from or covering host are colourless to light coloured, wispy to cottony, woolly, felt-like or waxy-looking mat 13

13. Flask-like to laterally compressed perithecia present, superficial to partially immersed in fungus covering the host; asci elongate, with thickened apex; when mature, filiform multiseptate ascospores tend to dissociate into 1-celled part-spores; conidial state(s) may occur simultaneously on host body or synnemata; especially on spiders or homopterans .... *Torrubiella*
13a. Spores form on external surfaces of the fungus; no sexual structures (perithecia) are present 14

14. Conidia form on cells with elongated denticulate necks bearing multiple conidia on awl- to flask-shaped or short blocky conidiogenous cells; conidia form singly or successively in dry chains or slime drops (Hyphomycetes) 15
14a. Conidia forcibly discharged and may rapidly form forcibly or passively dispersed secondary conidia (Entomophthorales) 22

15. Conidiogenous cell with an extended,

denticulate apex (growing apex repeatedly forms an conidium and regrows (rebranches) just below the new conidium) . . . . *Beauveria*

15a. Conidiogenous cells are awl- to flask-shaped, with or without an obvious neck; conidia borne singly, in chains, or in slime drops 16

16. Conidia single or in chains on apices of conidiogenous cells 17

16a. Conidia aggregate in slime drops at apices of conidiogenous cells 20

17. Conidia borne singly on conidiogenous cell with swollen base and one or more narrow, elongated necks; conidia globose or, if not, usually having an obvious slime coat; especially on mites . . . . *Hirsutella*

17a. Conidia borne in chains, not covered by any obvious slime 18

18. Conidiophores much branched in a candelabrum-like manner but very densely intertwined, and forming nearly wax-like fertile areas; conidiogenous cells short, blocky, without apical necks; conidial chains long and, usually, laterally adherent in prismatic columns or continuous plates . . . . *Metarhizium*

18a. Conidiophores individually distinct and unbranched or with a main axis and short side branches bearing single or clustered conidiogenous cells 19

19. Conidiogenous cells flask-like, with swollen base and a distinct neck, borne singly or in loose clusters; chains of conidia often long and divergent (when borne on clusters of conidiogenous cells) . . . . *Paecilomyces*

19a. Conidiogenous cells short and blocky with little obvious neck, borne in small clusters on short branches grouped in dense whorls on (otherwise unbranched) conidiophores; conidial chains short; especially on Noctuidae (Lepidoptera) . . . . *Nomuraea*

20. Conidia aggregating in slime droplets with morphology either (1) macroconidia, elongated, gently to strongly curved with somewhat pointed ends, one or more transverse septa and usually a short (basal) bulge or bend ('foot') and/or (2) microconidia aseptate, with variable morphology; conidiogenous cells often distinctly thicker than vegetative hyphae; hyphae often with terminal or intercalary chlamydospores (thick-walled spore-like swellings of vegetative cells; surface smooth or decorated) . . . . *Fusarium*

20a. Conidiogenous cells little thicker than hyphae, occurring singly or grouped into regular clusters and/or whorls; conidia one-celled; mycelium highly uniform in diameter 21

21. Conidiogenous cells usually tapering uniformly from base to truncate apex, usually without a swollen base or distinct neck; occurring singly, in pairs or whorled along hyphae or in terminal clusters . . . . *Verticillium*

21a. Conidiogenous cells with a swollen to flask-like base and a (usually short) neck often bent out of axis of the conidiogenous cell; conidiogenous cells borne singly, clustered, or in whorls aggregating in loose 'heads' on erect apically branching conidiophores poorly differentiated from vegetative hyphae . . . . *Tolypocladium*

22. In aceto-orcein, primary conidia obviously uninucleate and sometimes seen to be bitunicate (with outer wall layer lifting partially off of spores in liquid mounts) 23

22a. In aceto-orcein, primary conidia obviously multinucleate or nuclei not readily seen 26

23. Conidia long clavate to obviously elongated (length/width ratio usually ≥2.5), papilla broadly conical, often

with a slight flaring or ridge at junction with basal papilla 24

23a. Conidia ovoid to clavate; papilla rounded and frequently laterally displaced from axis of conidium ... 25

24. Conidia readily forming elongate secondary capilliconidia attached laterally to and passively dispersed from capillary conidiophores; rhizoids and cystidia not thicker than hyphae; rhizoids numerous, often fasciculate or in columns ...... *Zoophthora*

24a. Conidia never forming secondary capilliconidia; conidia often strongly curved and/or markedly elongated; rhizoids and/or cystidia 2–3× thicker than hyphae; especially on dipterans (or other insects) in wet habitats (on wetted rocks, in or near streams, etc.) *Erynia*

25. Conidia never producing secondary capilliconidia; rhizoids 2–3× thicker than hyphae, terminating with prominent discoid holdfast; cystidia at base 2–3× thicker than hyphae, tapering towards apex .... *Pandora*

25a. Conidia never producing secondary capilliconidia; rhizoids not thicker than hyphae, numerous, solitary to fasciculate, with weak terminal branching system or sucker-like holdfasts; cystidia as thick as hyphae, often only weakly tapered .... *Furia*

26. In aceto-orcein, nuclei staining readily, with obviously granular contents 27

26a. In aceto-orcein, nuclei not readily visible or not staining 29

27. Conidia with apical point and broad flat papilla; discharged by cannon-like expulsion of fluid from conidiogenous cell forming halo-like zone around conidia after discharge .... *Entomophthora*

27a. Conidia without apical projection and discharged by eversion of a rounded (not flat) papilla 28

28. Conidia pyriform with papilla merging smoothly into spore outline; formed by direct expansion of tip of conidiogenous cell (with no narrower connection between conidiogenous cell and conidium); rhizoids never formed .... *Entomophaga*

28a. Conidia globose with papilla emerging abruptly from spore outline; formed on conidiogenous cells with a narrowed neck below the conidium; if present, rhizoids 2–3× thicker than hyphae, with discoid terminal holdfast .... *Batkoa*

29. Conidia globose to pyriform, papilla rounded, with many (inconspicuous) nuclei; secondary conidia: (a) single, forcibly discharged and resembling primaries; (b) single, passively dispersed capilliconidia formed in axis of capillary conidiophore or (c) numerous on a primary conidium, small, forcibly discharged (microconidia) .... *Conidiobolus*

29a. Conidia globose to pyriform, papilla flattened, usually 4-nucleate; secondary conidia (a) forcibly discharged, resembling primary or (b) almond- to drop-shaped, laterally attached to a capillary conidiophore with a sharp subapical bend; especially on aphids or mites .... *Neozygites*

30. Affecting larval bees (Apidae and Megachilidae), causing chalkbrood; fungus in cadavers is white or black, organized as large spheres (spore cysts) containing smaller-walled spherical groups (asci) of (asco)spores .... *Ascosphaera*

30a. Affecting insects other than bees; spores formed individually rather than in spherical groups of inside larger spheres 31

31. Spores formed *inside* a fungal cell, in a more or less loosely fitted outer (sporangial) wall 32

31a. Spores forming directly at apices of hyphae or hyphal bodies by budding or intercalary (thick-walled but not confined loosely inside remnant of another cell) 33

32. Spores (oospores) thick-walled, smooth walled, colourless; formed inside irregularly shaped cell (oogonia); some cells in thick mycelium producing narrow tube through cuticle with evanescent terminal vesicle from which motile, biflagellate zoospores are released; affecting mosquitoes .... *Lagenidium*

32a. Spores (resistant sporangia) globose or subglobose, golden-brown with hexagonally reticulated surface; formed inside close fitting thin (but evanescent) outer wall .... *Myiophagus*

33. Affecting gregarious cicadas (Homoptera: Cicadidae); terminal segments of abdominal exoskeleton drop off to expose loose to compact, colourless to coloured fungal mass; spores thin-walled or, if thick-walled, with strongly sculptured surface *Massospora*

33a. Not affecting cicadas, with spores occurring throughout body (not confined to terminal abdominal segments) 34

34. Spores (zygospores or azygospores) with outer surfaces smooth or with surface irregularly roughened, warted, or spinose; colourless to pale or deeply coloured (various colours possible), brown, grey, or black 35

34a. Spores (thick-walled resistant sporangia) with surface regularly decorated with ridges, pits, punctations, striations, reticulations; yellow–brown to golden-brown 37

35. Resting spores grey, brown or black (outer wall is coloured; inner wall is hyaline), with smooth or rough surface; binucleate but nuclei often not staining strongly in aceto-orcein if spore wall is cracked; infected hosts from which conidia were discharged and then produced almond- to drop-shaped secondary capilliconidia should be evident in the infected population; affecting aphids, scales, or mites .... *Neozygites*

35a. Resting spores colourless, coloured, or dark, surfaces smooth or rough; infected host population may or may not include cadavers producing conidia but, if present, conidia not as above 36

36. When spores are gently crushed in aceto-orcein (to crack walls and partially extrude cytoplasm), nuclei are poorly stained (or unstained) and, if seen, do not have obviously granular contents (Ancylistaceae) *Conidiobolus*

36a. When spores are gently crushed in aceto-orcein (to crack walls and partially extrude cytoplasm), nuclei stain well and have obviously granular contents ...... *Entomophthoraceae* (genus undetermined)

37. Sporangia ellipsoid (not globose), with a preformed dehiscence slit (may not be obvious); wall very thick, golden-brown, pitted to elaborately sculptured; affecting larvae/pupae of mosquitoes (or midges) .... *Coelomomyces*

37a. Sporangia globose or subglobose, with no visible dehiscence slit; wall relatively thin; surface with low (hexagonally) reticulated ridges; affecting terrestrial insects ..... *Myiophagus*

4 DIAGNOSES AND CRITICAL CHARACTERS OF MAJOR ENTOMOPATHOGENS

This section is organized by fungal classes, starting with the conidial fungi that are the most commonly encountered fungal entomopathogens and moving through the ascomycetes and basidiomycetes, zygomycetes, oomycetes and chytridiomycetes that are progressively less common and may have narrower host ranges. Generic treatments include a brief diagnosis, and lists of major (but not all) diagnostic characters, characterizations of some common and important species, references to taxonomic literature useful for species identification, and, in some instances, further comments.

Labels on the figures correspond to the lettered diagnostic characters of the genera and species.

A Deuteromycota: Hyphomycetes

These conidial fungi produce their spores on exposed hyphae rather than in some sort of closed fruiting structure; *Aschersonia* is the only major entomopathogenic genus *seeming* to be an exception to this generalization. Even on the relatively uncommon occasions when conidial fungi (anamorphs) occur together with their sexual states (teleomorphs), both morphs have different scientific names. The hyphae of Hyphomycetes and their teleomorphs are frequently septate. Most entomopathogenic Hyphomycetes grow readily on many common cutlure media; surprisingly few of these fungi are difficult to grow *in vitro* or have specialized nutritional requirements.

Species in nearly every genus of entomopathogenic Hyphomycetes are distinguished by the morphologies of their conidia and conidiogenous cells and by the identity of their hosts. Other distinctive characters used in these genera are specifically noted in the generic treatments.

Important general reference works for identifying many more genera of Hyphomycetes than treated here are Carmichael *et al.* (1980), Samson (1981) and Samson *et al.* (1988).

1. Aschersonia *Montagne and* Hypocrella *Saccardo (Figure 3)*

Conidial state: *Aschersonia*, with stroma hemispherical or cushion-shaped (sometimes indistinct), superficial, usually light to brightly coloured (yellow, orange, red, etc.), covering host insect, with one or more conidia-forming zones (locules) sunken into stroma and opening by wide pore or irregular crack; conidia hyaline, one-celled, spindle-shaped, extruded onto stromatic surface from locules in slime masses. **Sexual state:** *Hypocrella*, with perithecia (walled structures containing asci and ascospores) globose to pyriform, immersed in stroma with opening protruding from stroma; asci cylindrical, with prominent hemispherical apical thickening penetrated by a narrow canal; ascospores filiform, with numerous transverse septa, dissociating at maturity to produce numerous cylindrical part-spores but sometimes remaining intact. **Hosts:** coccids and aleyrodids.

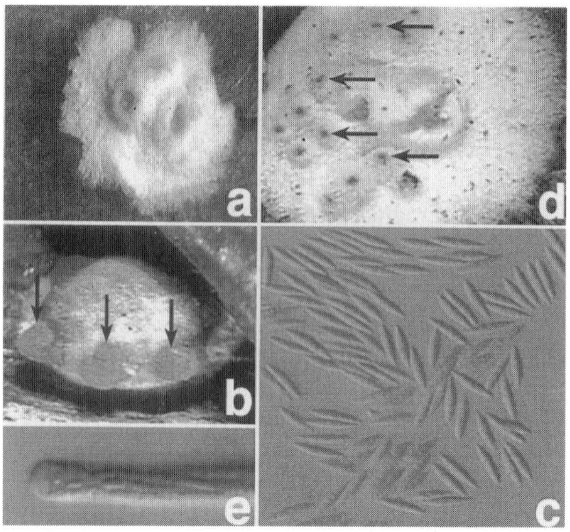

Figure 3 *Aschersonia* (a–c) and *Hypocrella* (d–e). (a) Stroma with three depressed conidiogenous areas. (b) Slimy masses of spores (arrows) on stromatic surface. (c) Conidia. (d) Stroma bearing *Hypocrella* perithecia (arrows indicate perithecial ostioles) and two conidiogenous zones of the *Aschersonia* state. (e) Thickened apex of ascus.

a. Key diagnostic characters

(a) Stroma: presence, size, cross-sectional profile, colour. **(b)** (*Aschersonia*) Conidiogenous locules: sunken in stroma, arrangement on stroma, release of conidia in slime. **(c)** (*Aschersonia*) Conidia elongated, fusoid, aseptate. **(d)** (*Hypocrella*) Perithecia: embedded in stroma. **(e)** (*Hypocrella*) Asci: long, with apical thickening penetrated by a narrow channel.

b. Major species

Aschersonia aleyrodis Webber – stromata ca. 2 mm diam. × 2 mm high, orange to pink or cream-coloured, surrounded by thin halo of hyphae spreading on leaf surface. Conidia bright orange in mass, 9–12 × 2 μm.

c. Main taxonomic literature

Petch (1914, 1921); Mains (1959a,b).

d. General comments

Aschersonia species are the conidial states of the less frequently found *Hypocrella* states; both genera are widespread in the tropics and subtropics. *Hypocrella* perithecia are immersed in the surface of stromata on which *Aschersonia* state may also occur. The taxon-

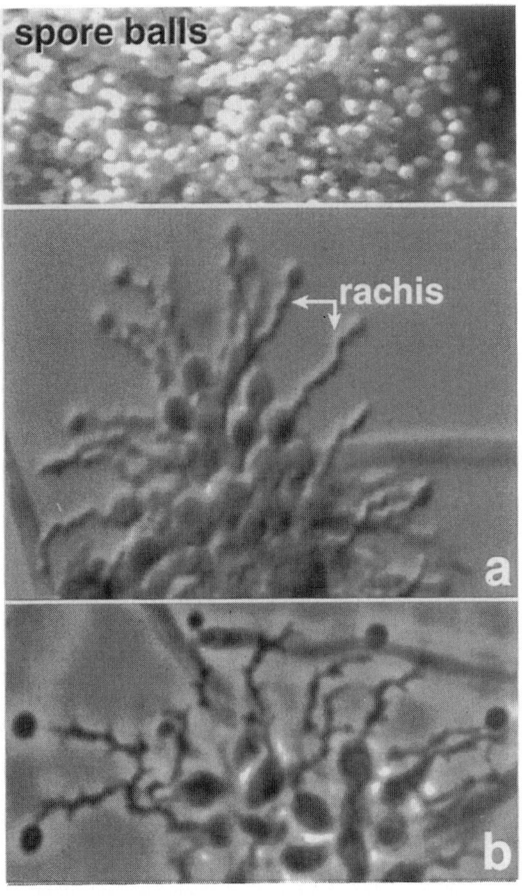

Figure 4 *Beauveria bassiana.* (Upper) Spore balls representing dense clusters of large numbers of conidiogenous cells and conidia. (a,b) Conidiogenous cells with globose bases and extended, denticulate raches.

omy for both genera is that of Petch (1914, 1921) but needs thorough revision.

2. Beauveria *Vuillemin (Figure 4)*

Forming a dense white covering on host exoskeleton, occasionally synnematous (forming erect fascicles of hyphae); conidiogenous cells usually densely clustered (or whorled or solitary), colourless, with globose or flask-like base and denticulate (toothed) apical extension (rachis) bearing one conidium per denticle; conidia aseptate. **Sexual state:** *Cordyceps* (for *B. brongniartii*; Shimazu *et al.*, 1988). **Hosts:** extremely numerous and diverse.

a. Key diagnostic characters

(a) Conidiogenous cells: extending apically into sympodial rachis. **(b)** Rachis: denticulate, with one conidium per denticle. (Note: rachis ***must*** be denticulate to be identified as *Beauveria*). **(c)** Conidia: size, shape, and surface characteristics.

b. Major species

B. bassiana (Balsamo) Vuillemin: conidia nearly globose, ≤3.5 μm diam. *B. brongniartii* (Saccardo) Petch: conidia long ovoid to cylindrical, 2.5–4.5 (6) μm long; mostly on Scarabaeidae (Coleoptera). *B. amorpha* Samson & Evans: conidia short cylindrical, flattened on one side or curved, 3.5–5 × 1.5–2.0 μm.

c. Main taxonomic literature

Hoog (1972); Samson & Evans (1982).

3. Fusarium *Link (Figure 5)*

Fruiting body (if present) a stromatic pad (sporodochium) with conidial hymenium on its surface, pale tan to yellow to orange or red. Conidiophores solitary or aggregated, simple or branched, bearing apical conidiogenous cells. Conidiogenous cells (phialides) short, cylindrical to much elongated, awl-like; forming one or two conidial types: *macroconidia* curved to canoe-shaped with sometimes prominent foot-like appendage on basal cell, with one or more transverse septa, usually released in slime heads or spore masses, and/or *microconidia* aseptate, small, ovoid to cylindrical, produced in slime or dry and in chains from elongate,

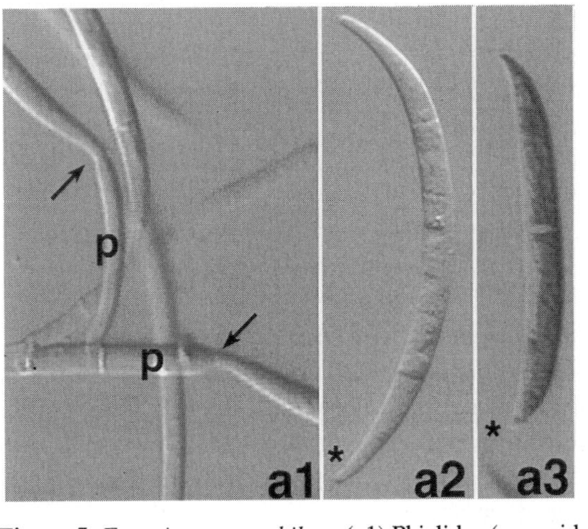

Figure 5 *Fusarium coccophilum.* (a1) Phialides (p; conidiogenous cells) bearing developing macroconidia; arrows indicate the slight bend ('foot') at the base of a new conidium. (a2, a3) Macroconidia with transverse septa and poorly differentiated feet (*).

awl-shaped conidiogenous cells. **Sexual state:** *Nectria.* **Hosts:** scales and many other insects.

a. Key diagnostic characters
(a) Macroconidia are the most important diagnostic character for the genus. These spores are highly variable in morphology, but usually are elongated, more or less curved, frequently described as boat- or canoe-shaped, and include one to several transverse septa. **(b)** Macroconidia: morphology, number of septa, colour. **(c)** Colours of mycelium and/or exudations into culture media.

b. Main taxonomic literature
Booth (1971); Nelson *et al.* (1983).

c. General comments
Few of the many *Fusarium* species seem to be entomopathogens (Booth, 1971; Hajek *et al.*, 1993). Specialists in *Fusarium* taxonomy disagree strongly about how to define or identify taxa, and no authoritative revision of *Fusarium* taxonomy synthesizing molecular and more traditional characters is available. Seifert (1995) provides an illustrated key to *Fusarium* species (FusKey: *Fusarium* Interactive Key) on the World Wide Web at <http://res.agr.ca/brd/fusarium/>.

4. Gibellula *Cavara (Figure 6)*

Synnemata (if present) white, yellowish, greyish to distinctly violet when fresh; becoming brownish with age, with conidial heads in a compact hymenium or more or less isolated on synnema; conidiophores septate, rough-walled, with a small apical swelling (vesicle); conidiogenous cells (phialides) clustered on short swollen cells on vesicle, without obvious necks, with apical wall becoming progressively thickened; conidia aseptate, smooth, single or in short chains. **Sexual state:** *Torrubiella.* **Hosts:** spiders.

a. Key diagnostic characters
(a) Synnemata: presence, colour, arrangement of conidiophores on synnema. **(b)** Wall texture on hyphae or conidiophores. **(c)** Conidiophores: size, surface texture, presence/absence of narrow isthmus below terminal vesicle.

b. Major species
G. pulchra (Saccardo) Cavara: conidiophores long, projecting from surface of white, lilac, yellow or orange synnemata. *G. leiopus* (Vuillemin) Mains: conidiophores short, crowded, forming dense fertile areas on lilac to purple synnemata.

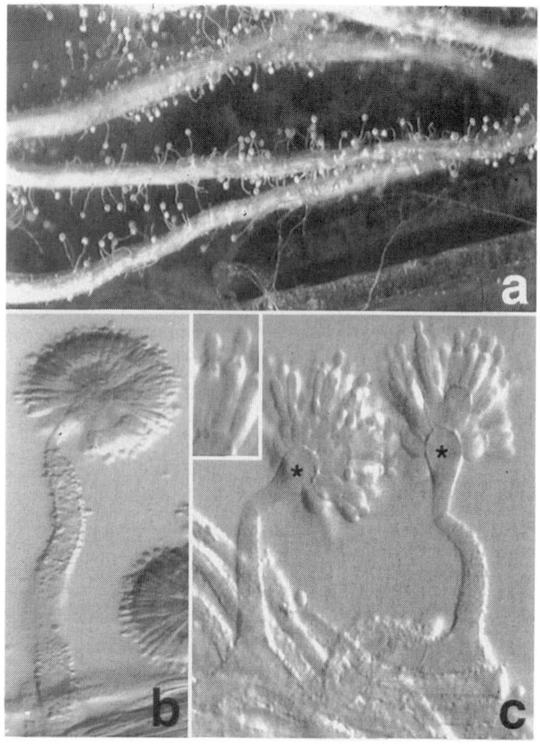

Figure 6 *Gibellula pulchra.* (a) Synnemata with laterally projecting conidiophores. (b) Rough-surfaced conidiophore narrowing apically to a vesicle bearing densely clustered conidiogenous cells. (c) Apical vesicles (*) of conidiophores bearing layer of 'metulae' and clusters of conidiogenous cells (phialides) whose apices thicken progressively (inset) with each conidium formed.

c. Main taxonomic literature
Samson & Evans (1992).

d. General comments
The apical vesicles of *Gibellula* conidiophores resemble those of *Aspergillus* but *Aspergillus* has (usually) globose conidia borne on phialides with a short neck. *Pseudogibellula formicarum* (Mains) Samson & Evans closely resembles *Gibellula pulchra* but is an insect rather than spider pathogen and has conidiogenous cells whose apices multiply denticulate (Samson & Evans, 1973). A second conidial state of *Gibellula/Torrubiella* species, *Granulomanus* sp. (Hoog, 1978), forms elongated

cylindrical conidia on polyphialidic conidiogenous cells.

5. Hirsutella *Patouillard (Figure 7)*

Synnemata (if present) erect, often prominent, thin, compact, hard or leathery; conidiogenous cells (phialides) scattered to crowded, projecting laterally from synnema or from hyphae on host body, swollen basally and narrowing into one or more slender necks; conidia aseptate (rarely 2-celled), hyaline, round, rhombic, elongate or like segments of citrus fruit, covered by persistent mucus, borne singly or 2 or more in droplets of mucus. **Sexual state:** *Cordyceps* or *Torrubiella*. **Hosts:** many diverse insects (one species affecting nematodes).

a. Key diagnostic characters
(a) Conidiogenous cell (generic character): with swollen to flask-like base and one or more elongated, narrow necks. **(b)** Conidia: borne singly or in small groups, covered by persistent slime (slime absent from some species). **(c)** Synnemata: present in most species but some species not forming synnemata. **(d)** Conidiogenous cell (specific characters): shape, size, with a single neck or polyphialidic.

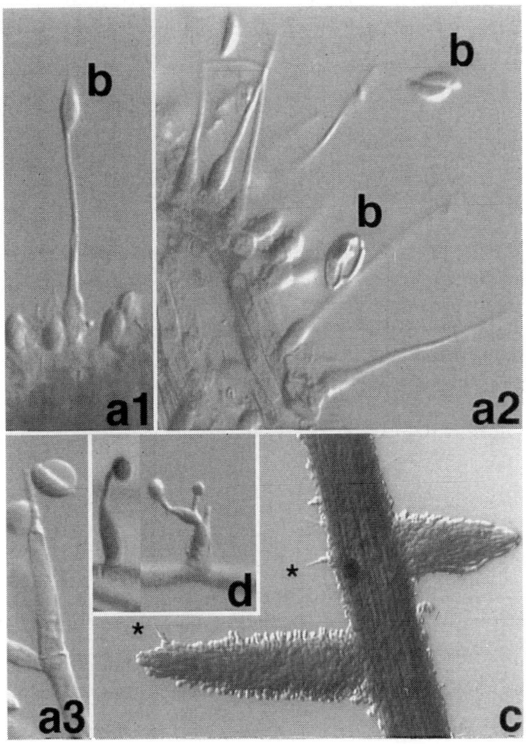

Figure 7 *Hirsutella*. (a1–a3) Conidiogenous cells with swollen bases and narrow, extended necks forming conidia with a slime coating (esp. the conidial clusters, (b) in a2 and the conidium in a3). (c) Synnema with lateral branches and conidiogenous cells (*). (d) Mono- (left) and polyphialidic (right) conidiogenous cells.

b. Major species
H. citriformis Speare: on leaf- and planthoppers (Homoptera: Cercopidae, Delphacidae); synnemata long, numerous, grey or brown, with many short lateral branches. *H. rhossiliensis* Minter & Brady: on nematodes and mites; not forming synnemata; conidiogenous cells with a single short, narrow neck; conidia resemble orange segments (straight on one side, curved on the other) or ellipsoid. *H. thompsonii* Fisher: on mites; conidiogenous cells with a distinctly swollen base and one short, narrow neck or polyphialidic; conidia globose with a smooth or wrinkled surface and no obvious slime layer.

c. Main taxonomic literature
Mains (1951); Minter & Brady (1980); Samson *et al.* (1980); Minter *et al.* (1983); Evans & Samson (1982); Rombach & Roberts (1989).

d. General comments
Despite the importance of this genus, *Hirsutella* has never been monographed, and its literature is dispersed. Identifying *Hirsutella* species can be difficult due to this lack of a monograph and also because the morphologies of *Hirsutella* species intergrade with species of *Verticillium*, *Tolypocladium*, and other genera (Humber & Rombach, 1987).

6. Hymenostilbe *Petch emend. Samson & Evans (Figure 8)*

Synnemata cylindrical or slightly tapered apically, covered by compact layer of conidiogenous cells; conidiogenous cells polyblastic, bearing solitary conidia on short denticles; conidia aseptate, hyaline, smooth or roughened. **Sexual state:** *Cordyceps*. **Hosts:** diverse insects.

a. Key diagnostic characters
(a) Synnemata present. **(b)** Conidiogenous cells form compact hymenium. **(c)** Conidiogenous cells polyblastic. **(d)** Teleomorphic connections with

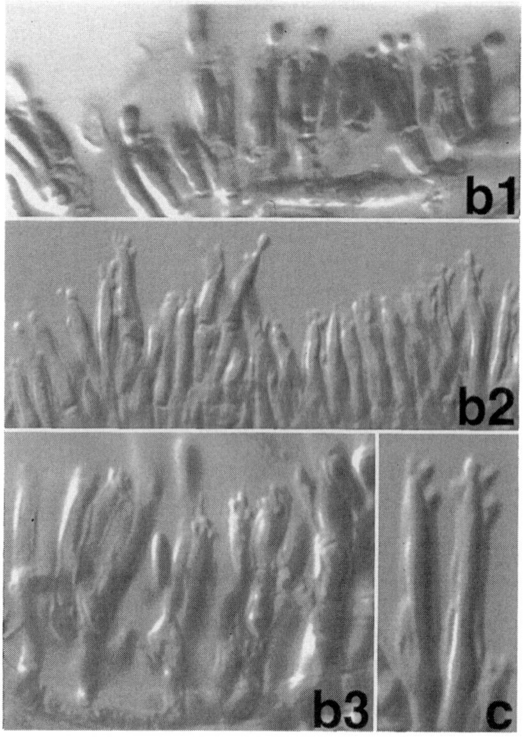

Figure 8 *Hymenostilbe*. (b1–b3) Hyphae bearing lateral conidiogenous cells with denticulate apices. (c) Apical denticles (conidia have been dislodged).

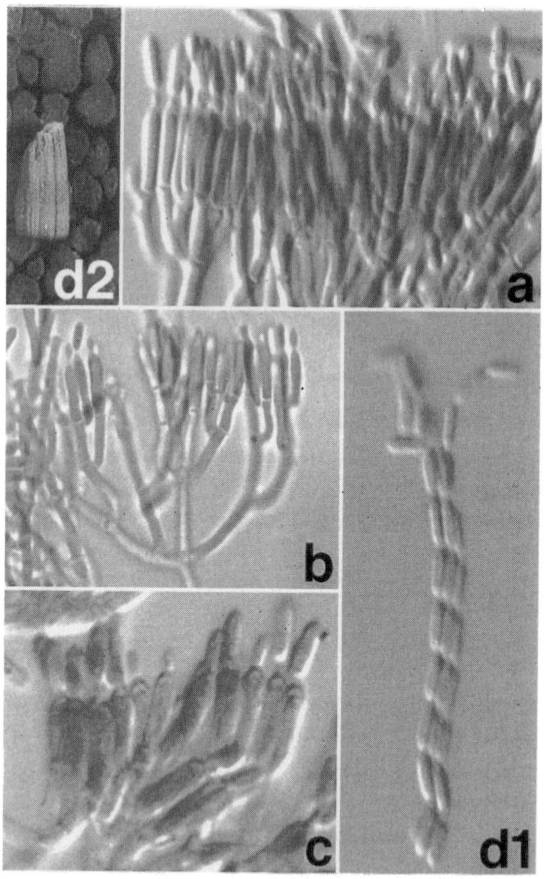

Figure 9 *Metarhizium anisopliae*. (a) Conidiogenous cells (phialides) forming a dense layer (hymenium). (b) Branched conidiophore; note that conidiogenous cells develop in a common plane. (c) Thickened, blunt tips of conidiogenous cells. (d1) Laterally adherent conidial chains. (d2) Lateral and top views of conidial columns.

Cordyceps spp. (which may occur together with *Hymenostilbe* synnemata).

b. Major species

H. dipterigena Petch: on flies; associated with *Cordyceps dipterigena;* synnemata brown, long.

c. Main taxonomic literature

Samson & Evans (1975).

7. Metarhizium *Sorokin (Figure 9)*

Mycelium often wholly covering affected hosts; conidiophores in compact patches; individual conidiophores broadly branched (candelabrum-like), densely intertwined; conidiogenous cells with rounded to conical apices, arranged in dense hymenium; conidia aseptate, cylindrical or ovoid, forming chains usually aggregated into prismatic or cylindrical columns or a solid mass of parallel chains, pale to bright green to yellow–green, olivaceous, sepia or white in mass. **Sexual state:** *Cordyceps* (Liang *et al.*, 1991). **Hosts:** extremely numerous and diverse.

a. Key diagnostic characters

(a) Conidiogenesis occurring in dense hymenia. **(b)** Conidiophores branching repeatedly at broad angles and resembling a candleholder (although observation of individual conidiophores is difficult). **(c)** Conidiogenous cells clavate or cylindrical, with a rounded to conical apex, no obvious neck; apical wall progressively thickening as conidia are produced. **(d)** Conidia produced in long chains; chains often adhere laterally to form prismatic columns or solid plates.

b. Major species

M. anisopliae (Metschnikoff) Sorokin var. *anisopliae*: conidiogenous cell cylindrical; conidia ≥9 μm long, cylindrical and often with a slight central narrowing, forming very long, laterally adherent chains, usually some shade of green.

M. anisopliae (Metsch.) Sorok. var. *majus* (Johnston) Tulloch: morphology as for *M.a.* var. *anisopliae* but conidia ≤11 μm; host usually a scarabaeid. *M. flavoviride* Gams & Rozsypal: conidiogenous cells clavate to broadly ellipsoid; conidia light grey–green in mass, ovoid (not cylindrical), 7–11 μm long; relatively slow to develop. (two varieties; see Rombach *et al.*, 1986.)

c. Main taxonomic literature

Rombach *et al.* (1986, 1987).

8. Nomuraea *Maublanc (Figure 10)*

Mycelium septate, white, with flocculent overgrowth, sparse in culture to dense on insects (often completely covering the host), usually becoming green, or purple–grey to purple as sporulation proceeds; conidiophores single or (rarely) synnematous (if synnematous, with a sterile base and distal fertile zone), erect, bearing whorls of short and blocky branches (metulae) with clusters of short phialides on metulae; conidiogenous cells short, with blunt apices and little if any distinct neck; conidia aseptate, smooth, round to ovoid or elongate and slightly curved, in short, divergent chains, pale to dark green, purple–grey to purple, or (rarely) white in mass. **Sexual state:** *Cordyceps* (for *N. atypicola*). **Hosts:** Noctuidae (Lepidoptera) or spiders.

a. Key diagnostic characters

(a) Conidiophores: with conigenous cells in dense, individually distinct whorls. **(b)** Conidiogenous cells: short, blocky, with no distinct neck (but seemingly papillate at apex). **(c)** Conidia: one-celled, in short, divergent chains.

b. Major species

N. rileyi (Farlow) Samson: on Noctuidae (especially larvae); conidial mass light (grey–green), covering host; conidia ovoid, in short chains. *N. atypicola* Yasuda: on spiders; conidial mass lavender-grey to purple.

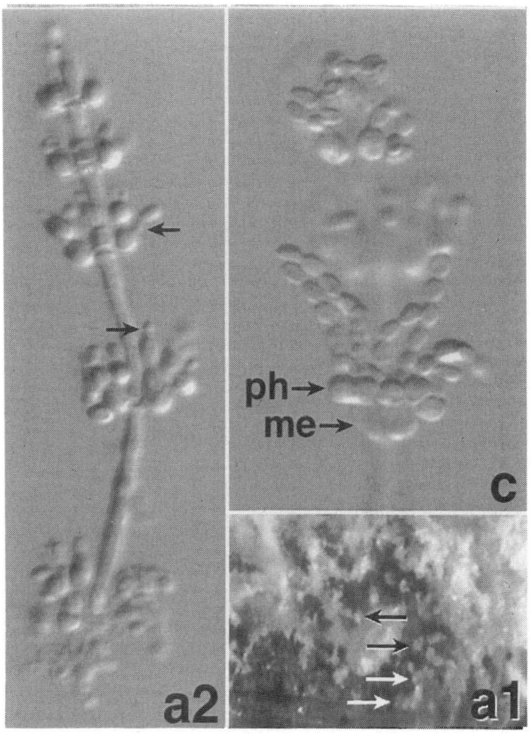

Figure 10 *Nomuraea rileyi.* (a1) Note beaded appearance of conidiogenous whorls (arrows) on conidiophores. (a2) Whorls of blocky conidiogenous cells with newly forming conidia (arrow). (c) Short, divergent chains of conidia produced on phialides (ph) clustered on short sterile cells (metulae; me).

c. Main taxonomic literature

Samson (1974).

9. Paecilomyces *Bainier (Figure 11)*

Conidiophores usually well developed, synnematous in many species, bearing whorls of divergent branches and conidiogenous cells (phialides), colourless to pigmented (but not black, brown or olive); conidiogenous cells with a distinct neck and base flask- to narrowly awl-shaped or nearly globose, borne singly or in groups in whorls on conidiophores, on short side branches or in apical whorls; conidia aseptate, hyaline to coloured, in dry divergent chains. **Sexual state:** *Cordyceps* or *Torrubiella*. **Hosts:** numerous, diverse insects.

a. Key diagnostic characters

(a) Conidiophores often synnematous. **(b)** Conidiogenous cells (phialides) single or whorled, with

Figure 11 *Paecilomyces*. (a) Synnemata. (b1) Clusters of divergent conidial chains. (b2–b4) Conidiogenous cells (phialides) with swollen bases and prominent necks. (c) Conidial chains elongate from the base as new conidia form. (a, b1, b2, c = *P. farinosus*; b3 = *P. javanicus*; b4 = *P. fumosoroseus*.)

swollen (flask-like, clavate to globose) base and a distinct neck; orientation of phialides gives cluster of spore chains a feathery to cottony appearance. **(c)** Conidia: in long chains, one-celled, ovoid to elongate (rarely globose).

b. Major species

P. farinosus (Holm ex S.F. Gray) Brown & Smith – synnemata often present; conidia short fusoid to lemon-shaped, ≤3 μm long, smooth walled, white to cream-coloured in mass. *P. fumosoroseus* (Wize) Brown & Smith: synnemata usually present; conidiophores and phialides with smooth uncoloured walls; conidia long ovoid, ≤4 μm long, rosy-tan to smoky-pink (or grey) in mass. *P. lilacinus* (Thom) Samson: conidial mass grey to tan with indistinct lavender shading; conidiophores slightly coloured (in comparison to conidia), with roughened walls; conidia ellipsoid to fusoid, 2–3 μm long.

c. Main taxonomic literature

Samson (1974).

d. General comments

There are many entomopathogenic species of *Paecilomyces* but few have been connected to sexual states (which are always *Cordyceps* spp.). Some *Paecilomyces* species are difficult to identify; such common and important species (for biocontrol) as *P. fumosoroseus* and *P. lilacinus* can be difficult to distinguish unless one keeps the morphologies of the conidia and conidiophores firmly in mind. As with many entomopathogens, the morphologies of critical characters may vary depending on whether the observations are made from specimens or cultures.

10. Tolypocladium *W. Gams (Figure 12)*

Conidiophores irregularly branched; conidiogenous cells (phialides) single or clustered, often forming terminal 'heads' with aggregated clusters of conidiogenous cells; conidiogenous cells with globose to flask-like base, narrowing abruptly to distinct neck that often bends away from axis of conidiogenous cell; conidia globose to cylindrical, aseptate, colourless, in slime heads; especially from nematoceran dipteran hosts. **Sexual state:** *Cordyceps* (for *T. inflatum* [=*T. niveum*]; Hodge *et al.*, 1996). **Hosts:** mostly small dipterans.

a. Key diagnostic characters

(a) Conidiogenous cells (phialides): with flask-like to subglobose base, short narrow necks often bent out of the axis of the base; occurring singly or in whorls on vegetative cells. **(b)** Conidia: aseptate, released in slime drops.

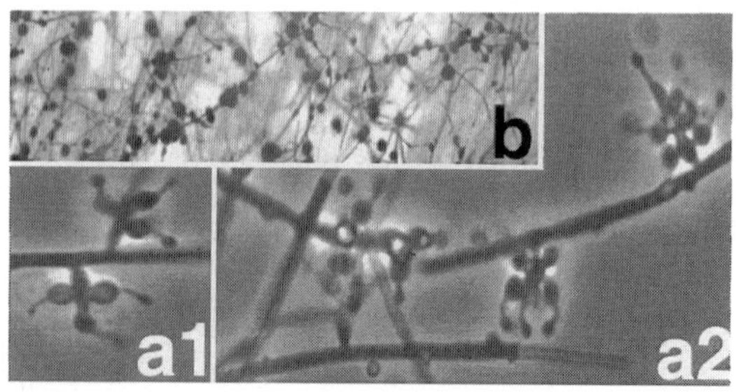

Figure 12 *Tolypocladium inflatum*. (a1,a2) Conidiogenous cells with swollen bases and thin, often bent necks. (b) Mucoid conidial balls formed atop clusters of conidiogenous cells.

b. Major species

T. cylindrosporum W. Gams: conidia cylindrical, straight or slightly curved. *T. extinguens* Samson & Soares: conidia subglobose, short ovoid or slightly curved (bean-shaped).

c. Main taxonomic literature

Bissett (1983); Samson & Soares (1984).

d. General comments

This genus has conidiogenous cells (phialides) that, may resemble those of species of *Verticillium*, *Hirsutella* and other genera forming conidia in slime drops. The shapes of *Tolypocladium* conidia tend to be distinct from those usually seen in *Hirsutella* but strongly resemble those of *Verticillium*; *Tolypocladium* phialides differ from those in *Verticillium* by having a distinct, occasionally bent neck. Except for releasing conidia into slime droplets, *Tolypocladium* phialides could be mistaken for young conidiogenous cells of *Beauveria* that had not yet formed an elongated, denticulate rachis. *Tolypocladium* remains a poorly circumscribed genus despite its use for the commercial production of the antibiotic cyclosporin.

11. Verticillium *Nees per Link (Figure 13)*

Conidiophores little differentiated from vegetative hyphae; conidiogenous cells (phialides) in whorls (verticils) of 2–6, paired, or solitary on hyphae or apically on short side branches; conidia hyaline, aseptate, borne in slime droplets or dry chains. **Sexual state:** *Cordyceps*, *Torrubiella* for entomopathogens. **Hosts:** scales, aphids, other insects or nematodes.

a. Key diagnostic characters

(a) Conidiogenous cells (phialides): occurring in whorls or pairs (or singly); elongated and usually tapering uniformly from the base. **(b)** Conidia: released into slime drops at apices of phialides.

b. Major species

V. lecanii (Zimmermann) Viégas [species complex]: conidia short ovoid to obviously elongate and cylindrical with rounded apices, 2–10 μm long, usually 1–1.7 μm wide. *V. fusisporum* W. Gams: conidia fusoid (spindle-shaped), 4–6 μm long.

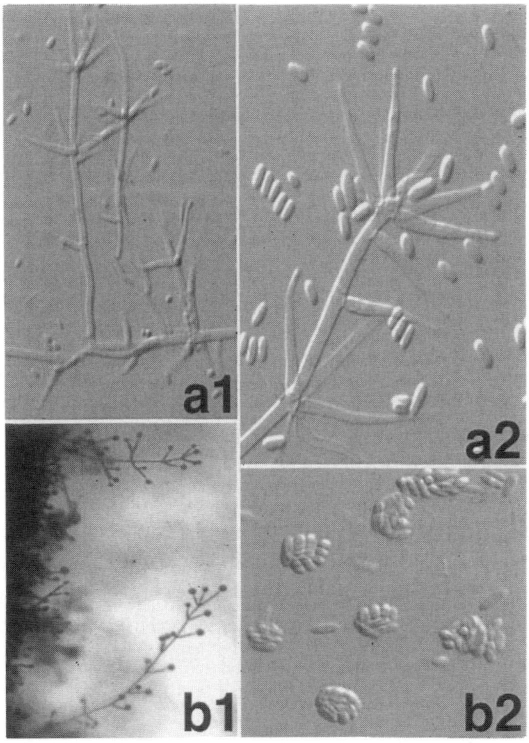

Figure 13 *Verticillium lecanii.* (a1, a2) Tapered conidiogenous cells (phialides) occur singly or in whorls. (b1) Mucoid conidial balls formed apically on individual phialides. (b2) Conidial balls and individual conidia.

c. Main taxonomic literature

Gams (1971, 1988).

B Ascomycota and Basidiomycota

The ascomycetous sexual states (teleomorphs) of the conidial states (anamorphs) discussed above (and, indeed, the great majority of entomopathogenic hyphomycetes in general) belong to a very narrow range of genera in the Clavicipitales (Pyrenomycetes). The Clavicipitales – the order containing *Claviceps purpurea*, which causes ergot in rye and other grasses – is characterized by having long asci with a prominent apical thickening (cap) traversed by a thin channel through which the ascospores are discharged. Clavicipitalean ascospores tend to be thread-like and multiseptate; while still in the asci, however, these ascospores frequently dissociate at the septa to form one-celled 'part spores'. The teleomorphs associated with entomopathogenic hyphomycetes are relatively rare or difficult to connect unambiguously with anamorphic taxa.

Apart from the Clavicipitales, there are relatively limited examples of insect-associated ascomycetes. A few *Nectria* species (Pyrenomycetes: Hypocreales) with *Fusarium* conidial states affecting mostly scales. *Ascosphaera* spp. are the remarkable causative agents of chalk-brood, a serious disease affecting bee larvae. Entomogenous Loculoascomycetes belong mostly to the genera *Myriangium* and *Podonectria* and are not treated here. The most mystifying and, perhaps diverse entomogenous ascomycetes may be the minute ectoparasites of the order Laboulbeniales (Tavares, 1985).

The Septobasidiales (Teliomycetes), mainly *Septobasidium* spp., comprise the only entomogenous basidiomycetes. These fungi are, effectively, zoophilic rusts whose nourishment derives wholly from partial parasitism of scale insect populations underlying crust-like fungal thalli. The global knowledge of these fungi depends heavily on a classic monograph by Couch (1938). *Septobasidium* is included in the key in Section 3 but is not discussed separately.

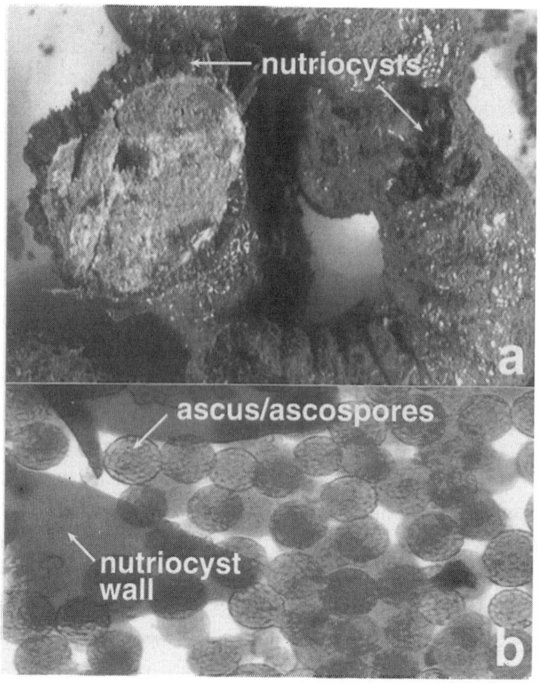

Figure 14 *Ascosphaera aggregata.* (a) Melanized nutriocysts under cuticle of bee larva. (b) Ruptured nutriocyst walls and globose asci containing large numbers of ovoid ascospores.

1. Ascosphaera *Spiltoir & Olive (Figure 14)*

[Ascosphaeromycetes: Ascosphaerales] Affecting larval bees (Hymenoptera: Apoidea); mycelium septate, with sex organs (globose ascogonia and papillate trichogynes) formed on separate mycelia; fertilized ascogonia swell to form large 'nutriocyst' (spore cyst) containing many individual asci; each ascus internally producing numerous ascospores. Appearance of mature infection is a blackened (or white) larva filled with a dense mass of balls (nutriocysts) containing balls (asci) containing ovoid to elongate spores.

a. Key diagnostic characters
(a) Pathogenic to bee larvae. **(b)** Mature infection is notable for the presence of balls (ascospores) within balls (asci) within balls (nutriocysts).

b. Major species
A. aggregata Skou: affecting leaf-cutting bees (Megachilidae) in North America; nutriocysts 140–550 × 100–400 μm; ascospores long ovoid to cylindrical, 4–7 μm long. *A. apis* (Maassen ex Claussen) Olive & Spiltoir: affecting honeybees (*Apis mellifera*) worldwide; nutriocysts 50–120 μm diam.; ascospores short ovoid or allantoid (bean-shaped), 2–3.5 μm long.

c. Main taxonomic literature
Skou (1972, 1988); Rose *et al.* (1984).

2. Cordyceps *Fries (Figure 15)*

[Pyrenomycetes: Clavicipitales] Forming one or more erect stromata on a host, with perithecia confined to an apical (or subapical) fertile portion or with scattered on stromatic surface; perithecia flask-shaped, superficial to fully immersed in stroma; asci elongated, with thickened apical cap penetrated by a fine pore, with eight filiform, multiseptate ascospores which usually fragment to form 1-celled part-spores. **Asexual states:** *Beauveria, Hirsutella, Hymenostilbe, Metarhizium, Nomuraea, Paecilomyces, Verticillium* and other genera. **Hosts:** numerous, diverse insects.

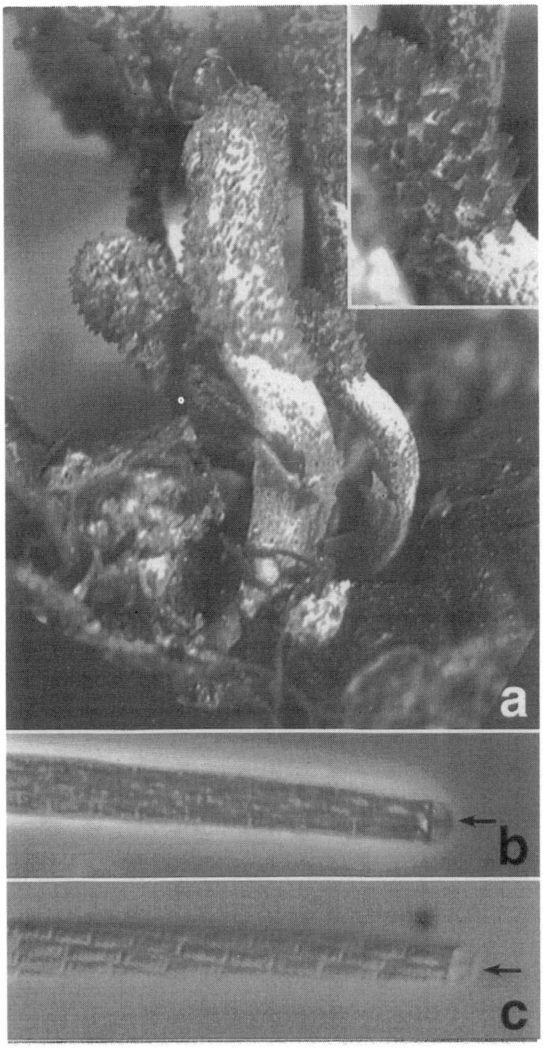

Figure 15 *Cordyceps*. (a) Stromata bearing superficial perithecia (esp. in inset). (b,c) Asci with thickened tips and a distinct central canal (arrows); in (c), threadlike ascospores that, at maturity. dissociate to form part-spores. (a, inset = *Cordyceps* sp. from lepidopteran; b,c = *C. corallomyces*.)

a. Key diagnostic characters

(a) Erect stroma(ta) bearing asci in perithecia. **(b)** Asci: filiform, with a thickened apical cap having a central channel. **(c)** Ascospores: at first 8 filiform spores per ascus with many transverse septa; in most species, dissociating at maturity to yield 1-celled part-spores.

b. Major species

C. lloydii Fawcett: on ants; stromata white to cream-coloured, ≤1 cm high, with discoid apical fertile part; perithecia partially immersed. *C. militaris* Link:Fries: on lepidopterans; stromata single (rarely multiple) per host, <10 cm high, thickly clavate and unbranched, orange, with swollen fertile part at apex. *C. sinensis* (Berkeley) Saccardo: on larval lepidopterans; stromata dark brown to black, <8 cm high, with elongated and little swollen apical fertile portion; used in Chinese herbal medicine. *C. tuberculata* (Lebert) Mains: on lepidopterans; stromata off-white, several per host, with partially immersed, sulphur to bright yellow perithecia scattered toward apices. *C. unilateralis* (Tulasne) Saccardo: on ants; fertile portion a swollen pad borne below apex of stroma; ascospores remain filiform (not dissociating to part-spores).

c. Main taxonomic literature

Kobayasi (1941, 1982); Mains (1958); Kobayasi & Shimizu (1983).

d. General comments

The taxonomy of *Cordyceps* and *Torrubiella* is problematic because so many of the species (ca. 280 in *Cordyceps* and more than 50 in *Torrubiella*) are poorly described, little is known of their anamorphs, and the types for many of these fungi cannot be borrowed from the National Science Museum in Tokyo, the major repository for these types. Kobayasi's (1941, 1982) extensive keys to *Cordyceps* and *Torrubiella* are unusable unless one is already completely familiar with the full spectrum of their species. The easiest way to identify many of these fungi is to compare specimens with the exquisite watercolour illustrations in Kobayasi & Shimizu (1983, whose text is in Japanese) and then to work backwards from a tentative identification using the (English) keys and descriptions in Kobayasi (1941, 1982). Mains (1958) provides a good key to North American *Cordyceps* species.

3. Torrubiella *Boudier (Figure 16)*

[Pyrenomycetes: Clavicipitales] Stroma absent or poorly developed as a light- to brightly-coloured mycelial mat (subiculum) on the host; perithecia elongate, white to yellow to orange or red, superficial to immersed; asci elongated, with thickened apical cap penetrated by a fine pore, with 8 filiform, multiseptate ascospores filiform, multiseptate, fragmenting at maturity to form 1-celled part-spores;

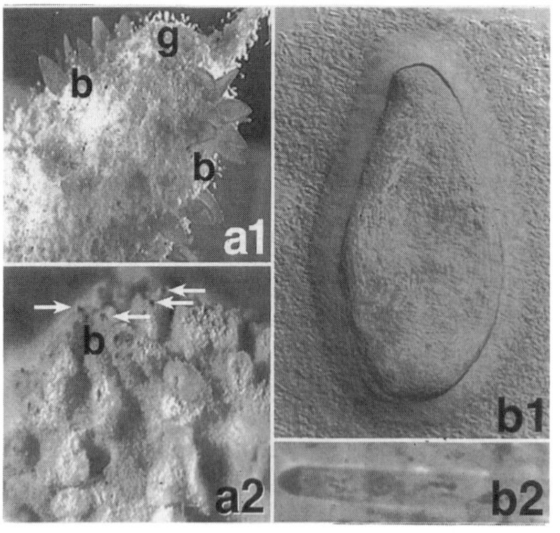

Figure 16 *Torrubiella.* (a1,a2) Perithecia on flocculent (a1) to stromatic (a2) mass covering host; perithecia (b) either superficial (a1) or clad by mycelium (a2); conidiophores of a *Gibellula* anamorph (g) are visible in a1. (b1) Perithecium (from a2) showing thick covering of loose hyphae; interior is filled by a sharply delineated bubble. (b2) Thickened apical cap of developing ascus from (b1).

especially on spiders or scale insects. **Asexual states:** *Gibellula, Granulomanus, Hirsutella, Verticillium* and other genera. **Hosts:** spiders, Homoptera.

a. Key diagnostic characters

(a) Subiculum (mycelium covering host) compact to woolly, *not* organized into distinct stroma (either erect as in *Cordyceps* or covering host). **(b)** Perithecia: immersed to superficial. **(c)** Host: generally spiders (or Homoptera).

b. Major species

None of the more than 50 species and varieties in this genus is especially common. See the general comments above for *Cordyceps*. Most species are found on spiders and are often occur together with an anamorphic (conidial) state; most non-araneous species of *Torrubiella* affect scale insects but are easily distinguished from *Hypocrella* by the absence of a compact, dense stroma.

c. Main taxonomic literature

Kobayasi (1982); Kobayasi & Shimizu (1982, 1983); Humber & Rombach (1987).

C Zygomycota, Entomophthorales

The most significant entomopathogens in the Zygomycota belong in the order Entomophthorales (Roberts & Humber, 1981). These fungi produce thick-walled resting spores (zygospores or azygospores) as their overwintering states but are more often found producing numerous forcibly dispersed conidia that serve as dispersive and infective units. Conidia that do not contact a suitable host upon discharge have a nearly unique ability to produce one or more types of secondary conidia which may also forcibly dispersed.

Outside the Entomophthorales, the only other significant associations with arthropods are found in the Trichomycetes, a diverse group of fungi that are mainly endocommensal in the guts of insects or crustaceans (Lichtwardt, 1986).

Entomopathogenic entomophthoraleans generally develop vegetatively in a host haemocoel by producing small, readily circulated, rod-like to irregularly hyphal bodies rather than thread-like hyphae. In many genera, these vegetative states are naturally wall-less and may be distinctly mobile and ever-changing in shape; many of these cells closely resemble the host's haemocytes. The thick-walled spores of these fungi can be referred to as 'resting spores' although in strict mycological terms they are zygospores or azygospores (terms that denote *only* the presence or absence of gametangial conjugations prior to their formation).

'Rhizoids' and 'cystidia' are terms applied to many different sorts of structures in a wide spectrum of cryptogamic organisms but whose meanings for the Entomophthorales are noted in the brief glossary (184*f*). Cystidia but not rhizoids may be produced *in vitro*. Rhizoids and cystidia are taxonomically significant structures for several of the genera discussed below.

Despite the widely varying taxonomies used in them, the best general works for identifying a broad range of entomopathogenic entomophthoralean species include MacLeod & Müller-Kögler (1970, 1973), Waterhouse & Brady (1982), Keller (1987, 1991) and Balazy (1993). Several approaches to the familial and generic taxonomy of this order have appeared since the 1960s when the use of *Entomophthora* as the sole genus for nearly all the order's entomopathogens came under question. A revised phenetic classification considering only these entomopathogens (Remaudière & Hennebert, 1980; Remaudière & Keller, 1980) provoked detailed reanalyses of the taxonomic characters of the Entomophthorales (Humber, 1981) and the counter-proposition of a comprehensive, evolutionarily based classification (Ben-Ze'ev & Kenneth, 1982a,b) that, as further refined by Humber (1989), is used below.

1. Batkoa *Humber (Figure 17)*

[Entomophthoraceae] Hyphal bodies elongated, walled (not protoplastic); conidiophores simple with a narrow 'neck' between conidium and conidiogenous cell; primary conidia globose, multinucleate, discharged by papillar eversion; rhizoids (if present) 2–3 times diameter of vegetative hyphae or conidiogenous cells, with prominent terminal discoid holdfast; resting spores bud laterally from parental hypha; unfixed nuclei have granular contents staining in aceto-orcein.

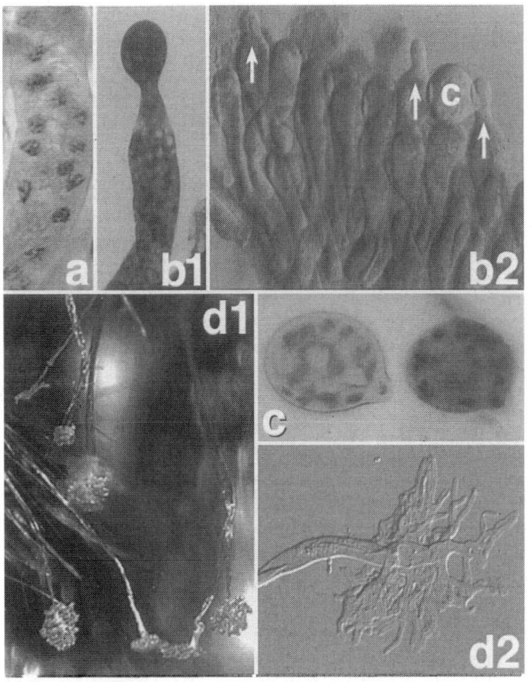

Figure 17 *Batkoa.* (a) Nuclei in hypha stained by aceto-orcein. (b1) Developing conidium with narrow neck between conidiogenous cell and conidium. (b2) Extended tips of conidiogenous cells (arrows) before conidia develop, and a discharged conidium (c). (c) Multinucleate conidia stained in aceto-orcein. (d1,d2) Rhizoids with discoid terminal holdfasts.

a. Key diagnostic characters

(a) Nuclei: contents granular and stain in aceto-orcein. **(b)** Conidiogenous cells: with distinct apical narrowing below conidium. **(c)** Conidia: globose, multinucleate. **(d)** Rhizoids (if present): thick, with discoid terminal holdfast

b. Major species

B. apiculata (Thaxter) Humber: conidia ca. 30–40 μm diam.; papilla often with pointed extension (apiculus); on homopterans and flies. *B. major* (Thaxter) Humber: conidia ca. 40–50 μm diam.; papilla often with pointed extension; on diverse insects.

c. Main taxonomic literature

MacLeod & Müller-Kögler (1973) (as *Entomophthora* spp); Humber (1989); Keller (1987), 1991 (as *Entomophaga* spp.).

2. Conidiobolus *Brefeld (Figure 18)*

[Ancylistaceae] Mycelium initially coenocytic but becoming septate, often forming walled, elongate hyphal bodies; conidiophores unbranched; primary conidia globose to pyriform with rounded apex and prominent papilla, multinucleate, forcibly discharged by papillar eversion; secondary conidia on primary conidia form: (1) singly, resembling primaries, forcibly discharged; (2) multiply, forcibly discharged (microconidia; in subgenus *Delacroixia*) or (3) as cylindrical capilliconidia passively dispersed from capillary conidiophore (in subgenus *Capillidium*); resting spores (zygospores, rarely azygospores) forming in or little displaced from hyphal axis (not budding laterally), conjugations are usually between adjacent cells in a hypha; unfixed nuclei unstained (or poorly stained and without coarsely granular contents) in aceto-orcein.

a. Key diagnostic characters

(a) Nuclei: not staining in aceto-orcein; contents not obviously granular. **(b)** Conidiophores: simple (or rarely, basally bifurcate). **(c)** Conidia: globose to pyriform, multinucleate. **(d)** Resting spores: formed in axis of host hypha, mostly as zygospores. **(e)** Vegetative cells: walled (not protoplastic). **(f)** Subgenera are defined by types of secondary conidia formed (Ben-Ze'ev & Kenneth, 1982a).

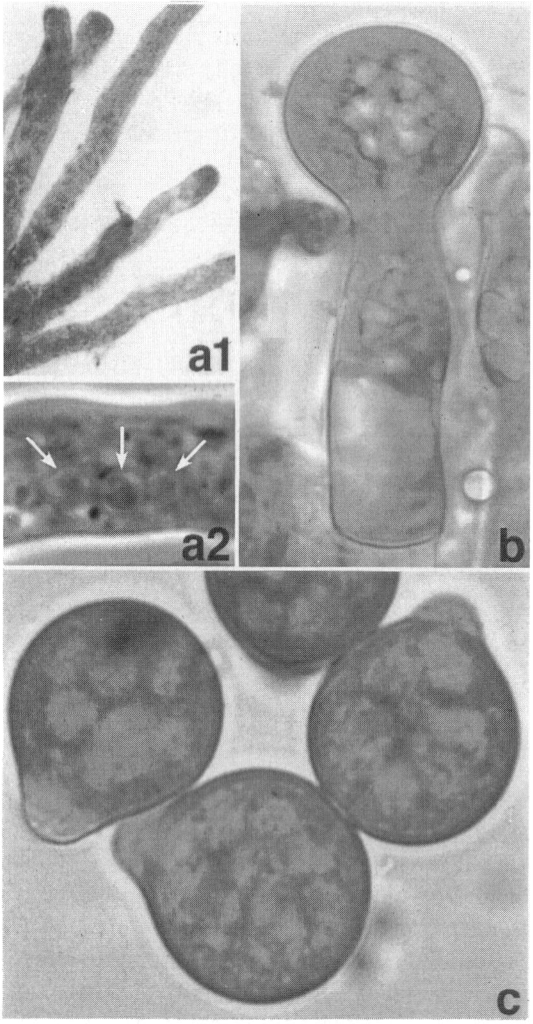

Figure 18 *Conidiobolus.* (a1) Hypha in aceto-orcein showing lack of nuclear differentiation. (a2) Nuclei (arrows) with clear nucleoplasm and dense central nucleolus (living hypha; phase contrast). (b) Conidium developing at apex of conidiogenous cell. (c) Globose conidia with rounded papillae. (a1, a2, b = *C. thromboides*; c = *C. obscurus*.)

b. Major species

C. coronatus (Costantin) Batko: older conidia become covered with villose spines (1–2 up to many μm long); often forming secondary microconidia; a weak pathogen of insects or vertebrates. *[Note: positive identification* requires *presence of villose conidia]. C. obscurus* (Hall & Dunn) Remaudière & Keller: conidia globose, hemispherical papilla emerges abruptly from spore outline, 30–40 μm diam.; nuclei may showing faint (finely granular) peripheral staining in aceto-orcein; no capilliconidia or microconidia formed; especially on aphids. *C. thromboides* Drechsler: conidia mostly pyriform, papilla merges gradually into spore outline, 17–30 μm diam.; no capilliconidia or microconidia formed; especially on aphids.

c. Main taxonomic literature

King (1976, 1977); Keller (1987).

3. Entomophaga *Batko (Figure 19)*

[Entomophthoraceae] Hyphal bodies fusoid to beaded, amoeboid protoplasts, later rod-like to spherical; conidiophores simple; primary conidia pyriform to ovoid, multinucleate, discharged by papillar eversion; rhizoids and cystidia not formed; resting spores bud laterally from parental hypha; unfixed nuclei have granular contents staining in aceto-orcein.

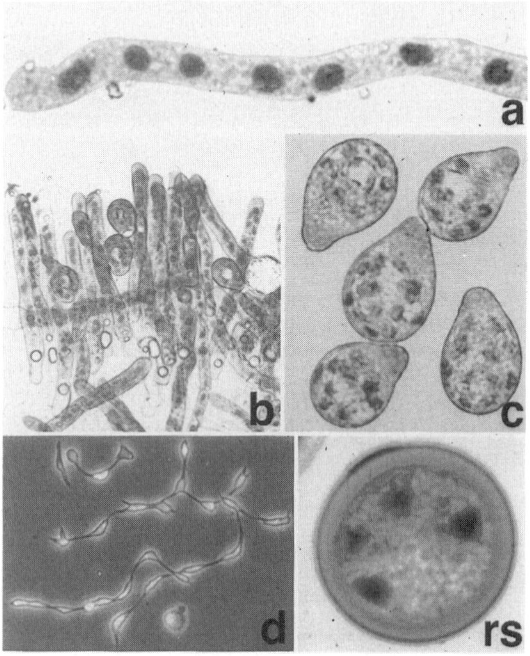

Figure 19 *Entomophaga.* (a) Nuclei stained by aceto-orcein. (b) Unbranched conidiogenous cells and conidia. (c) Pear-shaped, multinucleate conidia. (d) Amoeboid vegetative protoplasts (in culture). (e) Immature, 4-nucleate resting spore; resting spores of many *Entomophaga* species are binucleate when fully mature.

a. Key diagnostic characters
(a) Nuclei: with coarsely granular contents stained in aceto-orcein. **(b)** Conidiogenous cells: no elongated, narrow neck subtending conidium. **(c)** Conidia: pyriform, multinucleate. **(d)** Vegetative growth: protoplastic, bead-like to fusoid or irregularly hyphoid; walled only late in development in host.

b. Major species
E. aulicae (Hoffman in Bail) Batko [species complex]: affecting many lepidopterans; *E. maimaiga* Humber, Soper & Shimazu (Soper *et al*., 1988) specifically affects gypsy moths, *Lymantria dispar*, in Japan in North America. *E. grylli* (Fresenius) Batko [species complex]: affecting diverse acridids (Orthopteran); *E. calopteni* (Bessey) Humber (Humber, 1989) specifically affects melanopline (spur-throated) grasshoppers and forms resting spores but not primary conidia.

c. Main taxonomic literature
Keller (1987); Soper *et al*. (1988); Balazy (1993).

4. Entomophthora *Fresenius (Figure 20)*

[Entomophthoraceae] Vegetative cells short, rod-like (with or without cell walls); conidiophores simple; conidiogenous cells club-shaped; primary conidia with prominent apical point and broad, flat basal papilla, with 2–12 (to ca. 40) nuclei, forcibly discharged by cannon-like mechanism with discharged conidia attached to substrate in a droplet of discharged cytoplasm; rhizoids (if present) ca. diameter of hyphae, numerous, isolated or fasciculate; resting spores bud laterally from parental hypha; unfixed nuclei have granular contents staining in aceto-orcein.

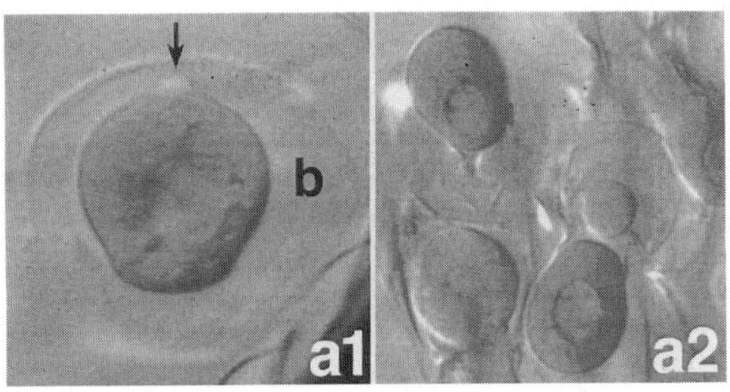

Figure 20 *Entomophthora*. (a1) Discharged primary conidium with apiculus (arrow) and a broad, nearly flat basal papilla, surrounded by a halo-like droplet of cytoplasm (b). (a2) Secondary conidia are broadly obovoid, not prominently apiculate, and form rapidly on discharged primary conidia (ghosts of which are visible below the secondary conidia).

a. Key diagnostic characters
(a) Primary conidia: with a broad, flat papilla and a pointed apical projection; at specific level: conidial size, number and size of nuclei. **(b)** Primary conidial discharge: cannon-like; discharged conidia surrounded by halo-like droplet of cellular material. **(c)** Secondary conidia more broadly clavate, non-apiculate. **(d)** Rhizoids: present or absent; location, number, morphology if present.

b. Major species
E. culicis (Braun) Fresenius: affecting mosquitoes and blackflies; conidia binucleate. *E. muscae* (Cohn) Fresenius [species complex]: affecting muscoid flies; individual species in complex distinguished by size, conidial nuclei (number and size), and hosts affected. *E. planchoniana* Cornu – affecting aphids (especially in Europe).

c. Main taxonomic literature
MacLeod *et al*. (1976); Keller (1987).

5. Erynia *Nowakowski [sensu Humber (1989)] (Figure 21)*

[Entomophthoraceae] Hyphal bodies rod-like to filamentous, walled; conidiophores digitately branched (rarely simple); primary conidia pyriform (clavate) to elongate, curved or straight with rounded to acute apices, uninucleate, basal papilla conical (often with slight flare at junction with spore) or rounded, bitunicate (outer wall layer *may* separate in liquid mounts), discharged by papillar eversion; secondary conidia resemble primaries or more globose, discharged by papillar eversion; rhizoids 2–3× diameter of hyphae, without differentiated terminal holdfast; cystidia columnar (some may branch), 2–3× diameter of hyphae; resting spores bud laterally from parental hypha; unfixed nuclei have granular contents staining in aceto-orcein.

a. Key diagnostic characters
(a) Conidiophores: branching close to conidiogenous cells; it is often difficult to separate individual

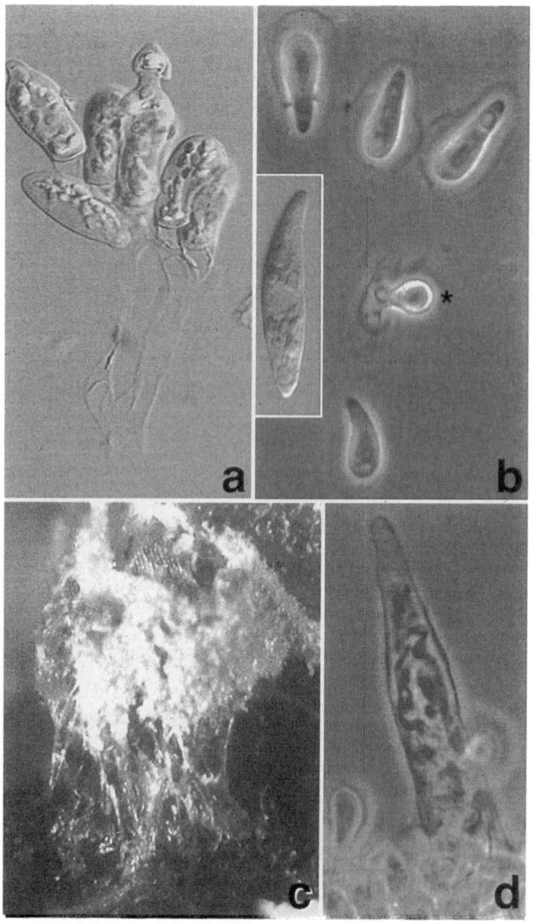

Figure 21 *Erynia*. (a) Branched conidiophore with short, blocky conidiogenous cells. (b) Bitunicate primary conidia with outer wall layer completely separated; nearly globose secondary conidium (*) is forming one spore; inset: conidium with conical papilla. (c) Rhizoids emerging from head of simuliid fly. (d) Cystidium with primary conidium (out of focus at left indicating relative size). (a, b, d = *E. aquatica*; b inset, c = *E. conica*.)

conidiophores. **(b)** Conidia: uninucleate, bitunicate, clavate to elongate, apices rounded or tapered to a blunt point; papillae broadly conical (and flared at junction with conidium) or rounded. **(c)** Rhizoids: 2–3× thicker than hyphae, without discoid terminal holdfast. **(d)** Cystidia: 2–3× thicker than hyphae, not strongly tapered.

b. Major species

E. aquatica (Anderson & Anagnostakis) Humber: on Culicidae and Chironomidae (Diptera), esp. in temporary, cold snow-melt pools; conidia long clavate 30–40 × 15–18 μm. *E. conica* (Nowakowski) Remaudière & Hennebert: on Culicidae and Chironomidae (Diptera); conidia 30–80 × 12–15 μm, often gently curved, tapering to subacute apex. *E. ovispora* (Nowakowski) Remaudière & Hennebert: on nematoceran dipterans; conidia broadly ovoid to ellipsoid, 23–30 × 12–14 μm. *E. rhizospora* (Thaxter) Remaudière & Hennebert: on Trichoptera; conidia lunate to straight, 30–40 × 8–10 μm, broad in middle; tapering strongly to a blunt apex; resting spores wrapped within layer of fine brown hyphae, often on surface of host rather than within.

c. Main taxonomic literature

Keller (1991) (*Erynia* as defined by Remaudiere & Hennebert, 1980); Balazy, 1993 (as *Zoophthora* subgenus *Erynia*).

6. Furia *(Batko) Humber (Figure 22)*

[Entomophthoraceae] Hyphal bodies hypha-like, walled; conidiophores branched (rarely simple); primary conidia clavate to obovoid, uninucleate, basal papilla rounded, bitunicate (outer wall layer *may* separate in liquid mounts), discharged by papillar eversion; rhizoids numerous, may be fasciculate, with diameter of vegetative hyphae or conidiogenous cells, no discoid terminal holdfast; cystidia with diameter of hyphae or conidiogenous cells; resting spores bud laterally from parental hypha; unfixed nuclei have granular contents staining in aceto-orcein.

a. Key diagnostic characters

(a) Conidiophores: branching close to conidiogenous cells. **(b)** Conidia: uninucleate, bitunicate, obovoid to clavate; apices and papillae rounded. **(c)** Rhizoids: as thick as hyphae, with sucker-like attachments or weak terminal branching systems but no discoid terminal holdfast; numerous, single, fasciculate, or in pseudorhizomorphs. **(d)** Cystidia: as thick as hyphae.

b. Major species

F. americana (Thaxter) Humber: conidia obovoid, 28–35 × 14–16 μm; on cyclorrhaphan (muscoid) flies. *F. virescens* (Thaxter) Humber: conidia broadly clavate to obovoid, 20–30 × 9–13 μm; especially on Noctuidae (Lepidoptera).

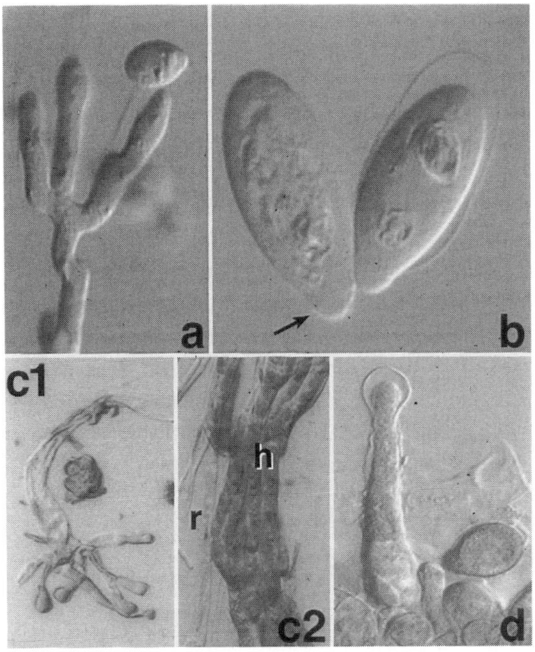

Figure 22 *Furia*. (a) Branched conidiophore; most are more richly branched than this. (b) Bitunicate primary conidia showing line of separation (arrow) between papilla and spore body, with outer wall layer separating (at right). (c1) Rhizoid with sparsely branched apical holdfast. (c2) Comparative diameters of rhizoids (r, devoid of cytoplasm) and hyphae (h). (d) Cystidium projecting from hymenium (note relative size of conidium at right).

c. Main taxonomic literature

Li & Humber (1984); Keller (1991) (as *Erynia* spp.); Balazy (1993) (as *Zoophthora* subgenus *Furia* spp.).

7. Massospora *Peck (Figure 23)*

[Entomophthoraceae] Hyphal bodies spherical to elongate, initially wall-less; development confined to and filling terminal abdominal segments of host; conidiophores lining small cavities in host abdomen; conidia 1–6 (mostly 2) nucleate, passively dispersed upon disarticulation of abdominal exoskeleton of living cicada; resting spores thick-walled, surface deeply reticulate, budded off from parental hyphal bodies; conidia and resting spores not usually formed within same host individual; restricted to emergences of gregarious cicadas (Homoptera: Cicadidae), especially in North and South America; unfixed nuclei have granular contents staining in aceto-orcein.

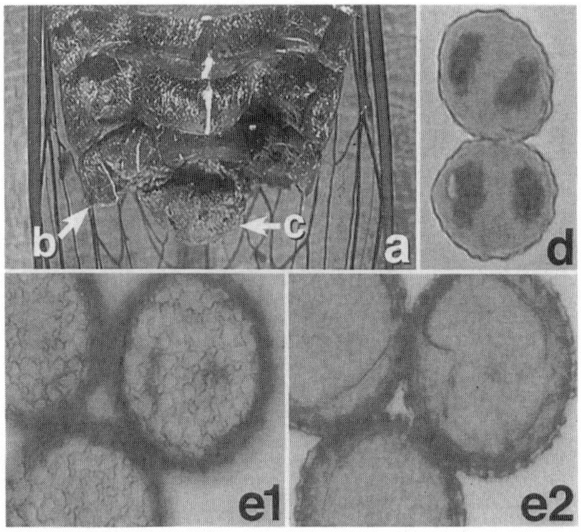

Figure 23 *Massospora*. (a) Abdomen of cicada with terminal segments fallen away (at b) to expose a dense mass of conidia (c) below. (d) Primary conidia of *M. cicadina* have two nuclei and a warted surface. (e1, e2). Surface (e1) and optical section (e2) of *M. cicadina* resting spores showing reticulate, deeply sculptured surface.

a. Key diagnostic characters

(a) Affecting gregarious cicadas (with a high specificity for individual cicada species). **(b)** Fungal growth largely confined to interior of terminal 3–4 abdominal segments. **(c)** Spore dispersal from living cicadas with internal spore mass exposed by disarticulation and sloughing away of terminal abdominal exoskeleton. **(d)** Conidia: size, shape, number and position of nuclei. **(e)** Resting spores: size, morphology of decorated surface.

b. Major species

M. cicadina Peck: affecting 17-year cicada (*Magicicada septemdecim*).

c. Main taxonomic literature

Soper (1974, 1981).

8. Neozygites *Witlaczil (Figure 24)*

[Neozygitaceae] Hyphal bodies irregularly shaped, rod-shaped or spherical, usually 3–5 nucleate; conidiophores simple; primary conidia round, ovoid or broadly fusoid, with relatively flattened basal papilla, mostly 4-nucleate, forcibly discharged a short distance by papillar eversion; secondary conidia usually (more or less almond-shaped) capilliconidia

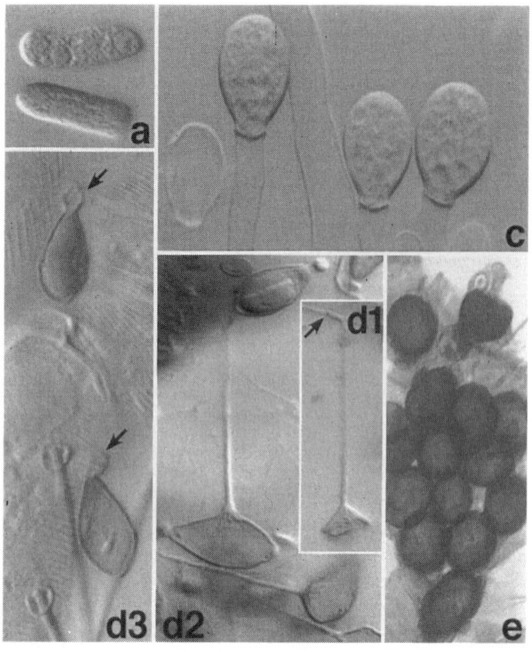

Figure 24 *Neozygites* (from mites). (a) Rod-like hyphal bodies. (c) Pyriform/ovoid conidia; note the nearly flat papilla. (d1–d3). Secondary capilliconidia produced atop capillary conidiophores (d1, arrow notes usual apical bend of conidiophore; d2). (d3) Capilliconidia attached to mite leg by apical slime drop (haptor; arrows). (e) Darkly melanized resting spores.

passively dispersed from capillary conidiophores; resting spores bud laterally from conjugation bridge between gametangia (hyphal bodies), black to smoke-grey, binucleate; nuclei in unfixed material staining poorly in aceto-orcein (except during mitosis); especially on Homoptera, thrips, and mites.

a. Key diagnostic characters

(a) Hyphal bodies: rod-like to spherical, routinely 4-nucleate. **(b)** Nuclei: not staining well in aceto-orcein during interphase, but with vermiform chromosomes and central metaphase plate staining well during mitosis. **(c)** Conidia: mostly 4-nucleate (3–5 overall), with flattened papilla. **(d)** Secondary conidia: like primaries (and forcibly discharged) or passively dispersed, drop- to almond-shaped capilliconidia, on capillary conidiophores, with a mucoid apical droplet (haptor). **(e)** Resting spores: zygospores with conspicuously darkened outer wall layer, arising from conjugation bridge between gametangia, ovoid and smooth or round (and usually roughened), binucleate.

b. Major species

N. fresenii (Nowakowski) Remaudière & Keller: conidia subglobose, 17–20 μm diam.; zygospores ovoid, 25–50 × 15–30 μm; on aphids. *N. floridana* (Weiser & Muma) Remaudière & Keller: conidia 10–14 μm diam.; zygospores subglobose, dark brown, outer wall roughened, 14–25 μm in long dimension; on tetranychid mites. [*Note: N. floridana is the taxon most often identified from tetranychid mites, but descriptions of Neozygites spp. from mites are incomplete and overlap too much to distinguish meaningfully among these species.*] *N. parvispora* (MacLeod, Tyrrell & Carl) Remaudière & Keller: conidia small, 9–15 μm diam.; zygospores black, spherical to flattened, 15–20 μm diam.; especially on thrips.

c. Main taxonomic literature

Keller (1991); Balazy (1993).

9. Pandora *Humber (Figure 25)*

[Entomophthoraceae] Hyphal bodies filamentous, protoplastic or walled; conidiophores digitately branched; primary conidia clavate to obovoid, uninucleate, basal papilla rounded, bitunicate (outer wall layer *may* separate in liquid mounts), discharged by papillar eversion; secondary conidia similar to primary or more nearly globose; rhizoids 2–3× diameter of hyphae or conidiogenous cells, with discoid terminal holdfast; cystidia taper, at base, 2–3× diameter of hyphae or conidiogenous cells; resting spores bud laterally from parental hypha; unfixed nuclei have granular contents staining in aceto-orcein.

a. Key diagnostic characters

(a) Conidiophores: branching digitately. **(b)** Conidia: uninucleate, bitunicate, obovoid to clavate; apices and papillae rounded. **(c)** Rhizoids: 2–3× thicker than hyphae, relatively sparse, with prominent terminal discoid holdfast. **(d)** Cystidia: 2–3× thicker than hyphae at base, tapering toward bluntly pointed apex. **(e)** Vegetative growth: hyphoid, protoplastic or walled in host.

b. Major species

P. blunckii (Bose & Mehta) Batko: on lepidopterans (esp. diamondback moth, *Plutella xylostella*); conidia pyriform, 15–20 × 7–11 μm. *P. delphacis* (Hori) Humber: especially on planthoppers (Homoptera: Delphacidae); conidia broadly clavate, 30–35 ×

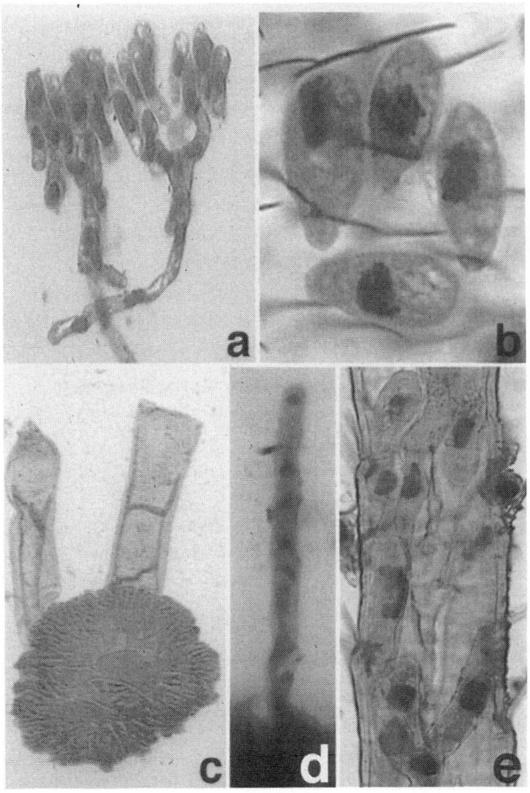

Figure 25 *Pandora*. (a) Branched conidiophores. (b) Uninucleate primary conidia (not showing bitunicate nature). (c) Discoid terminal holdfasts of rhizoids. (d) Cystidium projecting from hymenium, with individual nuclei stained in aceto-orcein. (e) Hyphal bodies with deeply stained nuclei in leg of aphid.

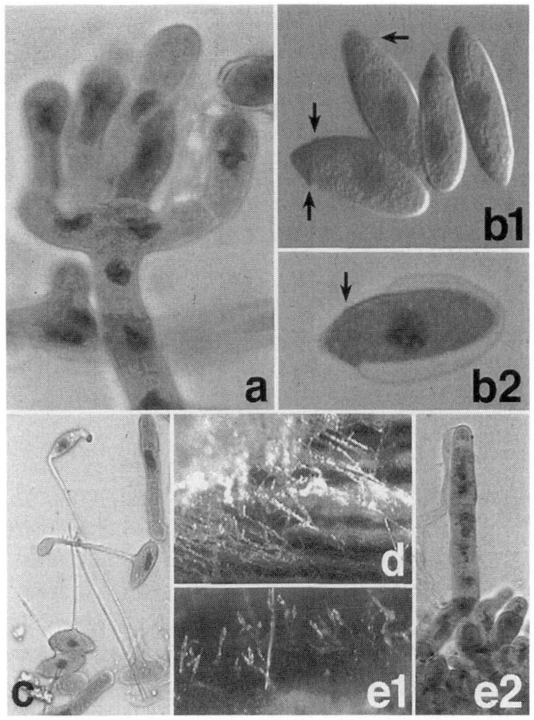

Figure 26 *Zoophthora*. (a) Branched conidiophore with prominently stained nuclei. (b1) Primary conidia showing conidal papilla and slight flaring at junctions (arrows) between papilla and spore body. (b2) Conidium showing flared junction and bitunicate nature. (c) Secondary capilliconidia. (d) Numerous thin rhizoids attaching host to substrate. (e1) Low magnification of rhizoids projecting from hymenium. (e2) Cystidium is about as thick as conidiogenous cells.

12–18 μm; growing and sporulating well *in vitro* on diverse media. *P. neoaphidis* (Remaudière & Hennebert) Humber: on aphids; conidia broadly clavate, 15–40 × 9–16 μm, often with papilla laterally displaced; often difficult to isolate or reluctant to grow and sporulate well *in vitro*.

c. Main taxonomic literature

Humber (1989); Keller (1991) (as *Erynia* spp.); Balazy (1993) (as *Zoophthora* subgenus *Neopandora* spp.).

10. Zoophthora *Batko [sensu Humber (1989)] (Figure 26)*

[Entomophthoraceae] Hyphal bodies rod-like to hyphoid, walled; conidiophores digitately branched (rarely simple); primary conidia clavate to obovoid, uninucleate, basal papilla rounded, bitunicate (outer wall layer *may* separate in liquid mounts), discharged by papillar eversion; secondary conidia (1) resembling primaries or more globose and discharged by papillar eversion or (2) elongate capilliconidia passively dispersed from capillary conidiophore; rhizoids as thick as vegetative hyphae, numerous, individual or fasciculate, discoid terminal holdfast absent; cystidia as thick as hyphae; resting spores bud laterally from parental hypha; unfixed nuclei have granular contents staining in aceto-orcein.

a. Key diagnostic characters

(a) Conidiophores: branching digitately close to conidiogenous cells. **(b)** Conidia: papilla broadly conical (with slight flare at junction with spore), rarely rounded, elongate (L/D usually 2.5), straight or curved, apex rounded or tapering to point. **(c)** Secondary conidia: frequently forming elongated capilliconidia passively dispersed from capillary conidiophore or forming forcibly discharged conidia generally resembling primaries. **(d)** Rhizoids: as thick as hyphae, holdfasts (if present) weakly differentiated, sparse terminal branching systems, and/or small and sucker-like; usually numerous, single or fasciculate or forming thick pseudorhizomorphs. **(e)** Cystidia: as thick as hyphae, untapered or weakly tapered, toward apex

b. Major species

Z. phalloides Batko: conidia elongated, curved, with bluntly rounded apices; on aphids. *Z. phytonomi* (Arthur) Batko: conidia cylindrical, straight; resting spores colourless (or, possibly, darkened with roughened outer wall); especially affecting *Hypera* spp. (Coleoptera: Curculionidae) on alfalfa. *Z. radicans* (Brefeld) Batko [species complex; see Balazy, 1993]: conidia more or less bullet-shaped, tapering to bluntly pointed apex, 15–30 μm long, with L/D ratio of ca. 2.5–3.5.

c. Main taxonomic literature

Keller (1991); Balazy (1993).

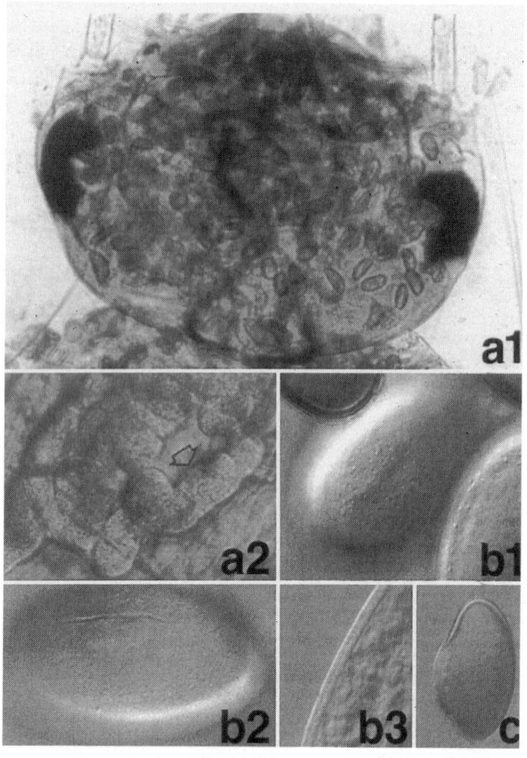

Figure 27 *Coelomomyces.* (a1) Resistant sporangia in mosquito head (photo: D.W. Roberts). (a2) Wall-less hyphal bodies in mosquito haemocoel (from Padua *et al.*, 1986, *J. Invertebr. Pathol.* **48**, 286). (b1–b3) Resistant sporangia with punctate surface texture (in b1), preformed germination slit (b2), and bilayered wall structure (b3). (c) Early stage of resistant sporangium germination with gelatinous plug bulging through the germination slit. (a2–c = *C. stegomyiae.*)

D Watermolds: Chytridiomycetes and Oomycetes

These fungi produce uni- or biflagellate zoospores that are both dispersive and infective units. Flagellate zoospores may be released from two possible sorts of sporangia, those with either thin or thick walls (usually propagative sporangia versus meiosporangia in which meiosis occurs, respectively). It would be unusual to detect hosts infected by these fungi during their vegetative states; these fungi are usually only detected when sporangia have been formed or are releasing zoospores.

1. Coelomomyces *Keilin (Figure 27)*

[Chytridiomycetes: Blastocladiales] Mycelium parasitic in haemocoel of aquatic dipteran larvae, coarse, wall-less in early development, budding off thick-walled resistant meiosporangia with thick deep golden or yellow-brown walls decorated by folds, ridges, warts, pits, etc.; sporangia releasing posteriorly uniflagellate zoospores on germination; zoospores infecting copepods or other aquatic crustaceans; haploid mycelium in crustacean haemocoel is wall-less, cleaving off posteriorly uniflagellate gametes that fuse in pairs; biflagellate zygotes encysting on and infecting dipteran hosts. (See life-cycle diagram and other figures in Chapter V-4.)

a. Key diagnostic characters

(a) Mosquito (or other aquatic dipteran) larvae becoming filled with golden-brown resistant sporangia; before sporangiogenesis, body filled with wall-less hyphae. **(b)** Resistant sporangia: ovoid, thick-walled, golden-brown, usually with obvious sculpturing or punctation on surface and a preformed germination slit. **(c)** Zoospores: posteriorly uniflagellate when released from germinating resistant sporangia or (as gametes) from haploid thalli in copepod or cladocerans. Note: Gametes fuse in pairs; biflagellate zygote swims to and encysts on dipteran host.

b. Major species

C. indicus Iyengar: resistant sporangia 25–65 × 30–40 μm, with longitudinal ridges frequently anastomosing; affecting *Anopheles* spp. in Africa, India, Australasia and Philippines. *C. dodgei* Couch & Dodge: resistant sporangia 37–60 × 27–42 μm with longitudinal slits (striae) 3–4 μm apart (giving banded appearance to sporangia); affecting *Anopheles* spp. in North America. *C. psorophorae* Couch: resistant sporangia 55–165 × 40–80 μm, surface with closely spaced punctation (seen in optical section as vertical channels through wall 2–10 μm thick); affecting aedine and culicine mosquito larvae in the Northern Hemisphere.

c. Main taxonomic literature

Bland *et al.* (1981); Couch & Bland (1985).

2. Myiophagus Thaxter (Figure 28)

[Chytridiomycetes: Blastocladiales] Monotypic genus. Vegetative thallus endozoic, coenocytic, branched with frequent constrictions becoming devoid of cytoplasm during formation of sporangia, and dissociating to produce free sporangia at maturity; sporangia globose (oval, ellipsoidal, or fusiform), with slightly thickened wall, forming 1–5 exit papillae; zoospores posteriorly uniflagellate, uninucleate, oval to elongated (4–5 × 7–7.5 μm), with yellow to orange granules in anterior; resting sporangia mostly spherical, with golden-coloured outer wall decorated with polygonal reticulation, cracking open upon germination to extrude globose sporangium and releasing zoospores; affecting scales, weevils and lepidopterans (*not* known from mosquitoes or insects with aquatic stages).

a. Key diagnostic characters

(a) Resistant sporangia: globose, thick-walled, golden-brown, with prominent hexagonal reticulation of surface. **(b)** Zoospores (*rarely* observed): posteriorly uniflagellate.

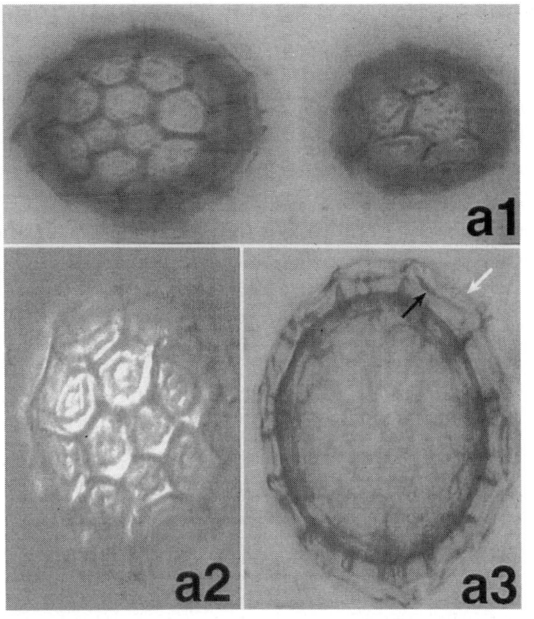

Figure 28 *Myiophagus ucrainicus.* (a1–a3) Reticulate surface of resistant sporangia. (a3) The reticulate wall is distinctly bilayered; the outer (white arrow) is the parental sporangium and the inner (black arrow) is the spore.

b. Only species

M. ucrainicus (Wize) Sparrow: resistant sporangia golden-brown with hexagonally reticulated surface, 20–30 μm diameter; zoospores (rarely observed) posteriorly uniflagellate, released on germination of resistant sporangia or from globose thin-walled sporangia.

c. Main taxonomic literature

Sparrow (1939); Karling (1948).

3. Lagenidium *Schenk (See figures in Chapter V-4)*

[Oomycetes: Lagenidiales] Mycelium parasitic in haemocoel of mosquito larvae, coarse and thick, coenocytic at first, becoming septate and forming oval to spherical segments that serve as zoosporangia or sex organs; partially differentiated contents of zoosporangia extruding through thin evacuation tube (7–10 × 50–300 μm) to thin-walled vesicle, formed outside host body; zoospores laterally biflagellate and reniform, cleaving in vesicle and dispersing on dissolution of vesicle wall; oospores (zygotes) thick-walled forming between segments of same or adjacent mycelial strand. (*See life-cycle diagram and other figures in Chapter V-4.*)

a. Key diagnostic characters

(a) Mycelium initially coenocytic, becoming cellular with each cell then acting as a zoosporangium or gametangium. **(b)** Zoosporogenesis: individual cells (zoosporangia) develop a thin exit tube penetrating host cuticle; partially cleaved cytoplasmic blocks are transferred through this tube into a growing vesicle at apex of the tube; zoospores complete differentiation in the vesicle and are dispersed on the disappearance of the evanescent gelatinous vesicle wall. **(c)** Zoospores: laterally biflagellate (with anterior tinsel flagellum and posterior whiplash flagellum). **(d)** Resistant spores: thick-walled oospores, globose, formed by conjugations of adjacent cells (gametangia).

b. Major species

L. giganteum Couch: affecting mosquitoes (only entomopathogenic species).

c. Main taxonomic literature

Sparrow (1960); Bland *et al.* (1981).

d. General comments

Pythium species (Oomycetes: Peronosporales), rarely affect insects, but when they do resemble *L. giganteu* since the zoospores of both genera are dispersed from external vesicles in the same manner. *Pythium*, however, produces only thin hyphae (≤3–4 μm diam.) in which cytoplasm remains readily visible after the production and release of zoospores whereas the hyphae of *L. giganteum* are coarse (generally ≥5 μm diam.) and their contents convert entirely into zoosporangia or oospores.

REFERENCES

Balazy, S. (1993) *Flora of Poland (Flora Polska), Fungi (Mycota)* **24**, 1–356. Polish Acad. Sci., W. Szafer Inst. Botany, Kraków.

Ben-Ze'ev, I. & Kenneth, R. G. (1982a) Features-criteria of taxonomic value in the Entomophthorales: I. A revision of the Batkoan classification. *Mycotaxon* **14**, 393–455.

Ben-Ze'ev, I. & Kenneth, R. G. (1982b) Features-criteria of taxonomic value in the Entomophthorales: II. A revision of the genus *Erynia* Nowakowski 1881 (=*Zoophthora* Batko 1964). *Mycotaxon* **14**, 456–475.

Bissett, J. (1983) Notes on *Tolypocladium* and related genera. *Can. J. Bot.* **61**, 1311–1329.

Bland, C. E., Couch, J. N. & Newell, S. Y. (1981) Identification of *Coelomomyces*, Saprolegniales, and Lagenidiales. In *Microbial Control of Pests and Plant Diseases 1970–1980* (ed. H. D. Burges), pp. 129–162. Academic Press, London.

Booth, C. (1971) *The genus* Fusarium. Commonwealth Mycological Institute, Kew, Surrey, UK. 237 pp.

Carmichael, J. W. Kendrick, W. B., Connors, I. L. & Sigler, L. (1980) *Genera of Hyphomycetes.* University of Alberta Press, Edmonton. 386 pp.

Couch, J. N. (1938) *The genus* Septobasidium. University of North Carolina Press, Chapel Hill. 480 pp.

Couch, J. N. & Bland, C. E. (eds) (1985) *The genus* Coelomomyces. Academic Press, Orlando. 399 pp.

Evans, H. C. & Samson, R. A. (1982) Entomogenous fungi from the Galapagos Islands. *Can. J. Bot.* **60**, 2325–2333.

Gams, W. (1971) Cephalosporium-*artige Schimmelpilze (Hyphomycetes)*. Gustav Fischer Verlag, Stuttgart. 262 pp.

Gams, W. (1988) A contribution to the knowledge of nematophagous species of *Verticillium*. *Neth. J. Plant Path.* **94**, 123–148.

Hajek, A. E., Nelson, P. E., Humber, R. A. & Perry, J. L. (1993) Two *Fusarium* species pathogenic to gypsy moth, *Lymantria dispar. Mycologia* **85**, 937–940.

Hawksworth, D. L., Kirk, P. M., Sutton, B. C. & Pegler, D. N. (1995) *Ainsworth & Bisby's Dictionary of the Fungi*, 8th edn. International Mycological Institute, CAB International, Wallingford. 616 pp.

Hodge, K. T., Krasnoff S. B. & Humber, R. A. (1996) *Tolypocladium inflatum* is the anamorph of *Cordyceps subsessilis. Mycologia* **88**, 715–719.

Hoog, G. S. de (1972) The genera *Beauveria, Isaria, Tritirachium*, and *Acrodontium* gen. nov. *Stud. Mycol.* **1**, 1–41.

Hoog, G. S. de (1978) Notes on some fungicolous Hyphomycetes and their relatives. *Persoonia (Leiden)* **10**, 33–81.

Humber, R. A. (1976) The systematics of the genus *Strongwellsea* (Zygomycetes: Entomophthorales). *Mycologia* **68**, 1042–1060.

Humber, R. A. (1981) An alternative view of certain taxonomic criteria used in the Entomophthorales (Zygomycetes). *Mycotaxon* **13**, 191–240.

Humber, R. A. (1989) Synopsis of a revised classification for the Entomophthorales (Zygomycotina). *Mycotaxon* **34**, 441–460.

Humber, R. A. (1996) Fungal pathogens of the Chrysomelidae and prospects for their use in biological control. In *Biology of Chrysomelidae* (eds P. H. Jolivet, M. L. Cox & T. H. Hsiao), Vol. IV, in press. SPB Academic Publishing, The Hague.

Humber, R. A. & Rombach, M. C. (1987) *Torrubiella ratticaudata* sp. nov. (Pyrenomycetes, Clavicipitales) and other fungi from spiders on the Solomon Islands. *Mycologia* **79**, 375–382.

Karling, J. S. (1948) Chytridiosis of scale insects. *Am. J. Bot.* **35**, 246–254.

Keller, S. (1987) Arthropod-pathogenic Entomophthorales of Switzerland. I. *Conidiobolus*, *Entomophaga*, and *Entomophthora*. *Sydowia* **40**, 122–167.

Keller, S. (1991) Arthropod-pathogenic Entomophthorales of Switzerland. II. *Erynia*, *Eryniopsis*, *Neozygites*, *Zoophthora*, and *Tarichium*. *Sydowia* **43**, 39–122.

King, D. S. (1976) Systematics of *Conidiobolus* (Entomophthorales) using numerical taxonomy. II. Taxonomic considerations. *Can. J. Bot.* **54**, 1285–1296.

King, D. S. (1977) Systematics of *Conidiobolus* (Entomophthorales) using numerical taxonomy. III. Descriptions of recognized species. *Can. J. Bot.* **55**, 718–729.

Kobayasi, Y. (1941) The genus *Cordyceps* and its allies. *Sci. Rep. Tokyo Bunrika Daig., Sect. B* **5**, 53–260.

Kobayasi, Y. (1982) Keys to the taxa of the genera *Cordyceps* and *Torrubiella*. *Trans. Mycol. Soc. Jpn* **23**, 329–364.

Kobayasi, Y. & Shimizu, D. (1982) Monograph of the genus *Torrubiella*. *Bull. Nat. Sci. Museum, Ser. B (Botany)* **8**, 43–78.

Kobayasi, Y. & Shimizu, D. (1983) *Iconography of vegetable wasps and plant worms*. Hoikusha Publ. Co. Ltd, Osaka. 280 pp.

Kohlmeyer, J. & Kohlmeyer, E. (1972) Permanent microscopic mounts. *Mycologia* **64**, 666–669.

Li, Z. & Humber, R. A. (1984) *Erynia pieris* (Zygomycetes: Entomoph-thoraceae): description, host range and notes on *Erynia virescens*. *Can. J. Bot.* **62**, 653–663.

Liang, Z.-Q., Liu, A.-Y. & Liu, J.-L. (1991) A new species of the genus *Cordyceps* and its *Metarhizium* anamorph. *Acta Mycol. Sinica* **10**, 257–262.

Lichtwardt, R. W. (1986) *The Trichomycetes, fungal associates of arthropods*. Springer-Verlag, New York. 343 pp.

MacLeod, D. M. & Müller-Kögler, E. (1970) Insect pathogens: Species described originally from their resting spores mostly as *Tarichium* species (Entomophthorales: Entomophthoraceae). *Mycologia* **62**, 33–66.

MacLeod, D. M. & Müller-Kögler, E. (1973) Entomogenous fungi: *Entomophthora* species with pear-shaped to almost spherical conidia (Entomophthorales: Entomophthoraceae). *Mycologia* **65**, 823–893.

MacLeod, D. M., Müller-Kögler, E. & Wilding, N. (1976) *Entomophthora* species with *E. muscae*-like conidia. *Mycologia* **68**, 1–29.

Mains, E. B. (1951) Entomogenous species of *Hirsutella, Tilachlidium,* and *Synnematium*. *Mycologia* **43**, 691–718.

Mains, E. B. (1958) North American entomogenous species of *Cordyceps*. *Mycologia* **50**, 169–222.

Mains, E. B. (1959a) North American species of *Aschersonia* parasitic on Aleyrodidae. *J. Insect Pathol.* **1**, 43–47.

Mains, E. B. (1959b) Species of *Hypocrella*. *Mycopathol. Mycol. Appl.* **11**, 311–326.

Minter, D. W. & Brady, B. L. (1980) Mononematous species of *Hirsutella*. *Trans. Br. Mycol. Soc.* **74**, 271–282.

Minter, D. W., Brady, B. L. & Hall, R. A. (1983) Five hyphomycetes isolated from eriophyid mites. *Trans. Br. Mycol. Soc.* **81**, 455–471.

Nelson, P. E., Toussoun, T. A. & Marasas, W. F. O. (1983) *Fusarium species: an illustrated manual for identification*. Pennsylvania State University Press, University Park. 193 pp.

Petch, T. (1914) The genera *Hypocrella* and *Aschersonia*. *Ann. R. Bot. Gdns Peradeniya* **5**, 521–537.

Petch, T. (1921) Studies in entomogenous fungi. II. The genera of *Hypocrella* and *Aschersonia*. *Ann. R. Bot. Gdns Peradeniya* **7**, 167–278.

Remaudière, G. & Hennebert, G. L. (1980) Révision systématique de *Entomophthora aphidis* Hoffm. in Fres. Description de deux nouveaux pathogènes d'aphides. *Mycotaxon* **11**, 269–321.

Remaudière, G. & Keller, S. (1980) Révision systématique des genres d'Entomophthoraceae à potentialité entomopathogène. *Mycotaxon* **11**, 323–338.

Roberts, D. W. & Humber, R. A. (1981) Entomogenous fungi. In *Biology of conidial fungi* (eds G. T. Cole & W. B. Kendrick), vol. 2, pp. 201–236. Academic Press, New York.

Rombach, M. C. & Roberts, D. W. (1989) *Hirsutella* species (Deuteromycotina; Hyphomycetes) on Philippine insects. *Philip. Entomol.* **7**, 491–518.

Rombach, M. C., Humber, R. A. & Roberts, D. W. (1986) *Metarhizium flavoviride* var. *minus* var. nov., a pathogen of plant- and leafhoppers of rice in the Philippines and Solomon Islands. *Mycotaxon* **27**, 87–92.

Rombach, M. C., Humber, R. A. & Evans, H. C. (1987) *Metarhizium album* Petch, a fungal pathogen of leaf- and planthoppers of rice. *Trans. Br. Mycol. Soc.* **88**, 451–459.

Rose, J. B., Christensen, M. & Wilson, W. T. (1984) *Ascosphaera* species inciting chalkbrood in North America and a taxonomic key. *Mycotaxon* **19**, 41–55.

Samson, R. A. (1974) *Paecilomyces* and some allied Hyphomycetes. *Stud. Mycol.* **6**, 1–119.

Samson, R. A. (1981) Identification: entomopathogenic Deuteromycetes. In *Microbial control of pests and plant diseases 1970–1980* (ed. H. D. Burges), pp. 93–106. Academic Press, London.

Samson, R. A. & Evans, H. C. (1973) Notes on entomogenous fungi from Ghana. I. The genera *Gibellula* and *Pseudogibellula*. *Acta Bot. Neerl.* **22**, 522–528.

Samson, R. A. & Evans, H. C. (1975) Notes on entomogenous fungi from Ghana. III. The genus *Hymenostilbe*. *Proc. Koninkl. Nederl. Akad. Wetensch., Ser. C* **78**, 73–80.

Samson, R. A. & Evans, H. C. (1982) Two new *Beauveria* spp. from South America. *J. Invertebr. Pathol.* **39**, 93–97.

Samson, R. A. & Evans, H. C. (1992) New species of *Gibellula* on spiders (Araneida) from South America. *Mycologia* **84**, 300–314.
Samson, R. A. & Soares, G. G., Jr (1984) Entomopathogenic species of the hyphomycete genus *Tolypocladium*. *J. Invertebr. Pathol.* **43**, 133–139.
Samson, R. A., McCoy, C. W. & O'Donnell, K. L. (1980) Taxonomy of the acarine parasite *Hirsutella thompsonii*. *Mycologia* **72**, 359–377.
Samson, R. A., Evans, H. C. & Latgé, J.-P. (1988) *Atlas of entomopathogenic fungi*. Springer-Verlag, Berlin. 187 pp.
Seifert, K. (1995) FusKey: *Fusarium* interactive key. [World Wide Web page: <http://res.agr.ca/brd/fusarium/>]. Agriculture and Agri-Food Canada, Ottawa.
Shimazu, M., Mitsuhashi, W. & Hashimoto, H. (1988) *Cordyceps brongniartii* sp. nov., the teleomorph of *Beauveria brongniartii*. *Trans. Mycol. Soc. Jpn* **29**, 323–330.
Skou, J. P. (1972) Ascosphaerales. *Friesia* **10**, 1–24.
Skou, J. P. (1988) Japanese species of *Ascosphaera*. *Mycotaxon* **31**, 173–190.
Smith, R. F. (1994) *Microscopy and photomicrography: a working manual*, 2nd edn. CRC Press, Boca Raton. 162 pp.
Soper, R. S. (1974) The genus *Massospora*, entomopathogenic for cicadas, Part, I, Taxonomy of the genus. *Mycotaxon* **1**, 13–40.
Soper, R. S. (1981) New cicada pathogens: *Massospora cicadettae* from Australia and *Massospora pahariae* from Afghanistan. *Mycotaxon* **13**, 50–58.
Soper, R. S., Shimazu, M., Humber, R. A., Ramos, M. E. & Hajek, A. E. (1988) Isolation and characterization of *Entomophaga maimaiga* sp. nov., a fungal pathogen of gypsy moth, *Lymantria dispar*, from Japan. *J. Invertebr. Pathol.* **51**, 229–241.
Sparrow, F. K., Jr (1939) The entomogenous chytrid *Myrophagus* Thaxter. *Mycologia* **31**, 439–444.
Sparrow, F. K., Jr (1960) *Aquatic Phycomycetes*, 2nd edn. University of Michigan Press, Ann Arbor. 1187 pp.
Tavares, I. (1985) Laboulbeniales (Fungi, Ascomycetes). *Mycologia Memoir* no. 9. J. Cramer, Lehre. 627 pp.
Waterhouse, G. M. & B. L. Brady. (1982) Key to the species of *Entomophthora* sensu lato. *Bull. Br. Mycol. Soc.* **16**, 113–143.

BRIEF GLOSSARY OF MYCOLOGICAL TERMS

Italicized terms appearing in the definitions are separately defined in this section. Irregular plurals of terms appear in parentheses at the start of definitions. Terms are applicable within the context of this chapter to fungi listed in brackets at the end of definitions.

Ascus (*asci*). Cell in which a single nucleus undergoes meiosis, after which one or more (usually eight) *ascospores* are cleaved out of the cytoplasm. [Ascomycota]

Capilliconidium (*capilliconidia*). A passively dispersed conidium produced apically on a long, slender (capillary) *conidiophore* arising from another conidium. [Entomophthorales, e.g. *Neozygites*, *Zoophthora*]

Conidiogenous cell. The cell on which a conidium forms, usually with only a single place (locus) on which a conidium forms; some conidiogenous cells have two or more conidiogenous loci. [Hyphomycetes, Entomophthorales]

Conidiophore. A simple or branched hypha or hyphal system bearing conidiogenous cells and their conidia. [Hyphomycetes, Entomophthorales]

Conidium (*conidia*). Fungal mitospore formed externally on a *conidiogenous cell*; conidia are not formed wholly inside any other cell (*ascus*, *sporangium*, etc.) nor as external meiospores (basidiospores on a basidium, the cell in basidiomycetes in which both karyogamy and meiosis occurs prior to basidiosporogenesis). [Hyphomycetes, Entomophthorales]

Cystidium (*cystidia*). In the Entomophthorales, more or less differentiated hyphae that precede and facilitate the emergence of the developing *conidiophores* through the host cuticle; cystidia usually project above the *hymenium*, but soon lose their turgor and collapse. Cystidia are rarely seen on any but very fresh specimens. [Entomophthorales; e.g. *Pandora* spp.]

Denticle. One of several to many small, conical to spike- or thorn-like or truncate projections on a conidiogenous cell, each of which bears a single conidium. [Hyphomycetes; e.g. *Beauveria* or *Hymenostilbe* spp.]

Hymenium (*hymenia*). A compact palisade layer of sporulating cells (conidiogenous cells, asci, etc.). [Hyphomycetes; Ascomycota; Entomophthorales]

Papilla (*papillae*). The basal portion of an entomophthoralean conidium by which conidia are attached to conidiogenous cells and which is usually involved in forcible discharge of conidia. [Entomophthorales]

Perithecium (*perithecia*). A globose, ovoid or pear-shaped walled structure in which *asci* and *ascospores* form; perithecia may be superficial or partially to fully immersed in the fruiting body. Each perithecium has an apical opening (ostiole) through which the *ascospores* are discharged. [Ascomycota: Pyrenomycetes]

Polyphialide. A conidiogenous cell having more than one neck, each of which produces one or more conidia; relatively common in *Hirsutella* species that do not form synnemata. [Hyphomycetes]

Rachis (*raches*). A geniculate (or sometimes zigzag) apical extension of a conidiogenous cell produced by sympodial branching of the elongating extension beneath each successive conidium formed. [*Beauveria* spp.]

Rhizoid. In the Entomophthorales, more or less differentiated hyphae that contact and anchor a host to the substrate; they may or may not have differentiated terminal holdfasts. [Entomophthorales]

Sporangium (*sporangia*). A cell or 'spore sac' inside of which (mitotic or meiotic) spores form; this is a very general term that can be correctly applied to diverse structures in nearly every class of fungi.

Stroma (*stromata*). A loose to fleshy or dense mass of vegetative hyphae on or in which spores (conidia or ascospores) are produced. Conidial stromata are usually very dense and compact, not extending very far above the host or substrate) (e.g. *Aschersonia* spp.); ascomycetous stromata bearing *perithecia* may be either low and compact (e.g. *Hypocrella* spp.) or upright and club- to column-like (e.g. *Cordyceps* spp.). [Hyphomycetes; Ascomycota]

Synnema (*synnemata*). An erect, branched or simple (unbranched) aggregation of hyphae; loosely fasciculate to compact, leathery or brittle in consistency, bearing conidiogenous cells and conidia. [Hyphomycetes; e.g. *Hirsutella*]

Zoospore. A uni- or biflagellate motile spore produced in a *zoosporangium*. [Chytridiomycetes; Oomycetes]

Zoosporangium (*zoosporangia*). The *sporangium* in which flagellate *zoospores* develop; *zoospores* and *zoosporangia* are formed only by watermolds. [Chytridiomycetes; Oomycetes]

APPENDIX: RECIPES OF STAINS AND REAGENTS

Aceto-orcein (nuclear stain/mounting medium)

Orcein (natural or synthetic)	1.0 g
Acetic acid, glacial	45.0 ml

Dissolve the orcein in hot glacial acetic acid, dilute 1 : 1 with distilled water and reflux or boil for at least 5 min. If boiled, replace lost volume with 50% glacial acetic acid. Filter at least twice to remove undissolved particulates. This stain continues to throw a precipitate over time and requires periodic clarification by filtration. There are many other ways of preparing aceto-orcein, most of which recommend refluxing; this simplified procedure yields a very satisfactory stain.

Lactophenol (mounting medium)

Phenol (crystals)	20 g
Lactic acid	20 g
Glycerol	40 g
Distilled water	20 ml

Lactic acid (mounting medium)

Anhydrous lactic acid with or without the addition of stains such as acid fuchsin, aniline blue or other acidic dyes can be used as an outstanding temporary to semi-permanent mounting medium.

CHAPTER V-2

Fungi: Entomophthorales

BERNARD PAPIEROK* AND ANN E. HAJEK†

*Institut Pasteur, 25, rue du Dr Roux, 75724 Paris Cedex 15, France
†Cornell University, Department of Entomology, Ithaca, New York 14853-0901, USA

1 INTRODUCTION

The order Entomophthorales (Zygomycotina) includes about 200 species that are pathogenic to insects and mites. These entomopathogens are notable for the epizootics they induce in populations of many insects, principally Homoptera, Lepidoptera, Orthoptera and Diptera. Their most noteworthy characteristic is that conidia are forcibly discharged from the conidiophores, which develop at the host surface from hyphal bodies penetrating and elongating through the cuticle. This is the case for most entomopathogenic Entomophthorales, but not for species of the genus *Massospora*, for which the conidia detach passively, or for one member of the *Entomophaga grylli* species complex which produces cryptoconidia inside of cadavers (Humber & Ramoska, 1986). Moreover, conidia in some *Neozygites* species infecting mealybugs are not really discharged.

Most entomopathogenic Entomophthorales possess two further biological characteristics: (i) the fungus multiplies as protoplasts and/or hyphal bodies (having a cell wall), after having invaded the host, and (ii), hyphal bodies can develop into thick-walled spores inside or more rarely, outside the cadaver. These resting spores, which are azygospores or zygospores according to the species, allow the fungus to survive adverse conditions. In most cases, the resting spore is a dormant stage and requires environmental stimulation for germination.

These characteristics should be kept in mind when studying entomopathogenic Entomophthorales. They indeed influence the design of methods and techniques for isolating these fungi, keeping them in culture, and bioassaying them for infectivity.

As regards generic names for Entomophthorales, there is presently a disagreement between specialists. In particular, the subdivision of the genus *Erynia* into *Erynia, Furia* and *Pandora,* as proposed by Humber (1989, and Chapter V-1) is accepted by the junior author, but rejected by Keller (1991) and the senior author. This is why two generic names are cited together for the same species in this chapter in the

MANUAL OF TECHNIQUES IN INSECT PATHOLOGY
ISBN 0-12-432555-6

case of *Erynia neoaphidis* (= *Pandora neoaphidis*), and *Erynia crustosa* (= *Furia crustosa*).

2 ISOLATION

In many cases, except for species sporulating on living insects, entomophthoralean fungi are isolated by directly inoculating various media with conidia produced from cadavers. Once the host is filled with fungal cells and killed by entomopathogenic Entomophthorales, conidiophores are produced that will bear actively discharged conidia. However, there are a few exceptions. For instance, *Strongwellsea* species develop only in the abdomen of dipteran hosts and form circular holes in the ventral surface (Batko & Weiser, 1965; Humber, 1976). Conidia are discharged as the host continues to live. Conidial sporulation on living insects also occurs in the genus *Massospora*, a pathogen specific for cicadas (Homoptera, Cicadidae) (Soper, 1974, 1981), and in *Entomophthora erupta*, a pathogen specific for Miridae (Heteroptera) (Ben-Ze'ev *et al.*, 1985; B. Papierok, unpublished data).

In several instances, direct isolation through conidia is not possible and other methods have been designed which allow isolation of the fungus using other fungal stages (i.e. resting spores, hyphal bodies, or protoplasts). However, regardless of the fungal stage used for isolation, it is a rule that isolation must be performed as quickly as possible after collection.

A Collecting infected insects and cadavers

1. Collecting cadavers

Insects dying from entomophthoralean infections can frequently be recognized by conspicuous positioning on the substrate (for instance, Orthoptera killed by *E. grylli* systematically adhere to the top of grasses), a peculiar colour, or the presence of sporulating structures and possibly rhizoids, which allow the cadaver to be fixed to the substrate.

According to the pathogen species as well as the host species, cadavers can be found in various characteristic locations:

- on grasses and small plants, in open fields and in forest areas;
- on trees, on the underside of leaves or on bark, especially at the border of forests;
- on substrates adjacent to aquatic ecosystems (banks of lakes, ponds, streams and rivers), very close to the water level on mosses, stones, and plants;
- on inner walls of structures, e.g. houses, cellars, cowsheds and henhouses.

Once cadavers are discovered, the collecting procedure is as follows;

- Carefully remove cadavers from the substrate, using fine forceps. If possible, the cadaver should be removed with a piece of the adhering substrate.
- Place each specimen in a polystyrene box or a glass tube. The container should not be completely closed, in order to prevent the atmosphere inside from becoming saturated, which generally allows fungi to develop too rapidly. For this reason, it is advisable to use ventilated boxes or glass tubes (e.g. with the opening covered by a piece of gauze or with a plastic lid riddled with small holes).
- Transport boxes or tubes to the laboratory, keeping them cool to arrest fungal development.

2. Collecting living infected specimens

Collecting living infected specimens is suitable for species sporulating on living insects. This is also appropriate when one needs to collect living infected insects that will soon die. Insects are sampled using appropriate devices and procedures. Once in the laboratory, they are kept in incubators for several days. Individuals should be monitored daily, and living or dead insects exhibiting symptoms of infection should be removed for fungal isolation. For those species requiring isolation from protoplasts, haemolymph samples can be taken from surface-sterilized, living insects that are suspected of being infected (see Section 2D).

B Isolation using conidia

Isolation of entomopathogenic Entomophthorales can be undertaken using two basically similar methods: (i) the showering method which could also be named the method of 'descending conidia', as

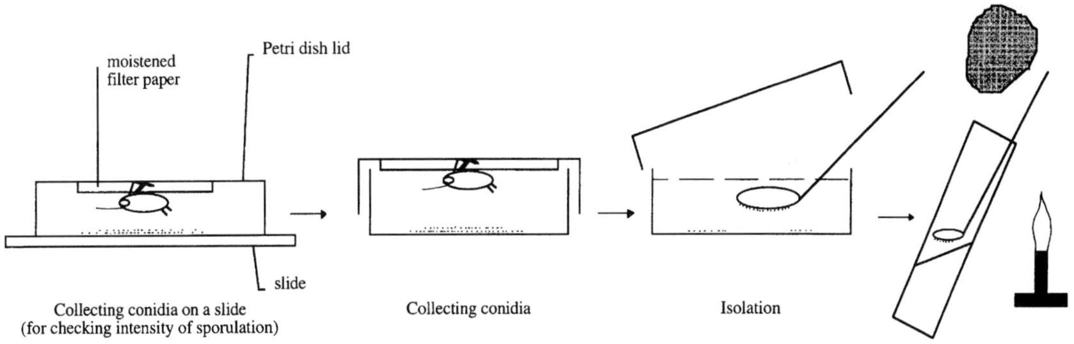

Figure 1 Procedure for isolation of Entomophthorales using the showering method (after Papierok, 1989, Silvie & Papierok, 1991).

illustrated by Papierok (1989), and Silvie and Papierok (1991), or (ii) the method of 'ascending conidia' (Keller, 1987, 1994). Because they require conidial discharge, these methods are not suitable for isolation of *Massospora* species or some *Neozygites* species (i.e. those infecting mealybugs).

1. The 'descending conidia' showering method

Cadavers that are filled with hyphal bodies but from which conidiophores have not yet emerged should be surface sterilized by dipping them successively in 65–70% ethanol (10–15 s), 2% sodium hyplochlorite solution (2–3 min), and sterile water (a few seconds in two successive baths) (see Chapter I).

Insects bearing conidiophores are placed on a moistened piece of tissue, paper towel or filter paper that is attached to the inside of a small sterile Petri dish lid (e.g. 35 × 10 mm). For large cadavers (more than 10–15 mm long) larger Petri dishes can be used but the risk of contamination is then greater than with smaller dishes. To prevent contamination, large cadavers should be cut into several sections (e.g. the whole abdomen, pieces of a few abdominal segments, thorax), and then prepared as smaller specimens.

After having prepared the infected specimen, the Petri dish lid is inverted over the base of a sterile Petri dish, which can contain culture medium. Conidial production begins within a few hours to a day, depending on whether conidiophores were already formed. Conidia are collected for periods of varying lengths (from a few minutes to a few hours), depending on the intensity of sporulation. In the case of a Petri dish containing culture medium, the lid with the specimen attached is replaced by a sterile lid. If a Petri dish with an empty base is used, conidia which land on the bottom are transferred to media by rubbing a piece of solid medium across the spores and subsequently using this spore-bearing medium to inoculate a culture tube (Figure 1).

2. The 'ascending conidia' showering method

Either insects bearing conidiophores already, or insects with no emerging conidiophores that have been surface sterilized (as described earlier in this chapter), are placed in the base of small Petri dishes (e.g. 35 × 10 mm) on water or on a moistened piece of tissue or filter paper. A sterile slide or cover glass is placed above the dish in order to collect discharged conidia. This device is kept in a sterile chamber (e.g. a Petri dish of larger size) until sufficient conidia have been collected (Figure 2). The spores are then removed by rubbing a piece of solid medium across the showered spores and subsequently transferring these spores to solid medium in a culture tube.

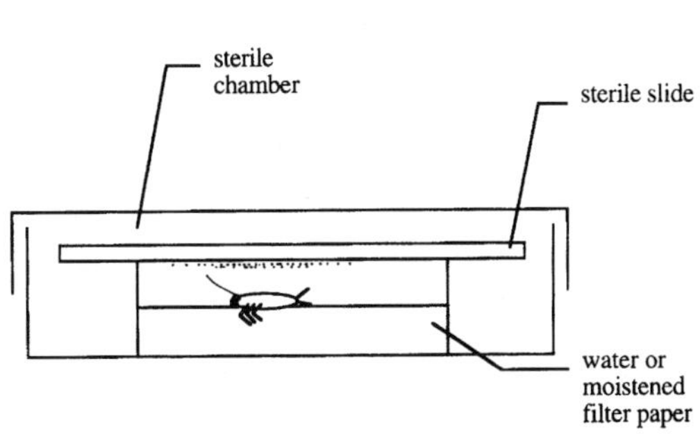

Figure 2 Procedure for collecting ascending conidia (after Keller, 1994).

C Isolation using resting spores

Resting spores formed inside the cadaver should be prepared for isolation as follows (after Tyrrell & MacLeod, 1975):

- place the cadavers filled with resting spores in sterile distilled water;
- gently break up specimens with forceps or a glass rod;
- sonicate in order to create a suspension;
- surface sterilize spores in 2–3% sodium hypochlorite solution (2–3 min);
- wash spores in sterile distilled water, at least 2 times;
- spread the spore suspension on culture medium.

In the case of resting spores formed outside the cadaver, preparation should be as follows:

- remove clumps of spores with fine forceps and place them in sterile distilled water;
- gently break up spore clumps with fine forceps or a glass rod and possibly sonicate in order to create a suspension;
- surface sterilize, wash, and inoculate in the same way as specified above.

It is frequently very difficult to separate resting spores from other micro-organisms in the environment, especially if resting spores are collected in the soil or on plants. In these cases, attempting to start cultures on media and even agar can lead to overgrowth of the entomophthoralean by saprophytes. As an alternative, test insects can be caged over the location of the resting spores for variable lengths of time. The insects are then reared and, several days later, the fungal pathogen can be isolated either from protoplasts and hyphal bodies (see Section 2D) or, after host death, from conidia (see Section 2B).

D Isolation using vegetative stages: hyphal bodies and protoplasts

It is occasionally useful to attempt isolation of the fungus from infected, living insects. The pathogen is then developing as protoplasts and/or hyphal bodies.

The procedure to be used for large specimens (e.g. caterpillars) consists of:

- superficially cleaning the insect cuticle with a 65–70% alcohol solution (or 2–3% sodium hypochlorite solution);
- collecting a small amount of haemolymph either directly from the haemocoel with a syringe and an appropriate needle, or as it is released after removal of a proleg.
- inoculating the medium.

For very hairy caterpillars, bacterial contamination has been a problem in isolating fungal pathogens. In these cases, cadavers are used to shower conidia onto larvae of a permissive species from a laboratory colony. It is important to repeat this procedure, varying the inoculation concentration, because excessive inoculation may result in septicaemia and not mycosis. Just before inoculated larvae die, haemolymph samples are collected by removing a proleg. The haemolymph, which contains fungal cells, is then introduced into Grace's insect tissue culture medium plus 5% fetal bovine serum plus gentamicin (see Section 2F.3). Fungi isolated using this procedure will grow as hyphal bodies or protoplasts according to the species. Initiation of growth can be very slow (frequently one to several weeks) and cultures growing with cell walls can eventually change to growth as protoplasts, so patience is required.

For small arthropods, the whole insect should be prepared according to the procedure described in the following section.

E Isolation using whole specimens (living or dead)

This method, suitable for small arthropods like aphids, was extensively used by Remaudière *et al.* (1976 a,b). It involves simply placing cadavers on solid media. Prior to inoculation of the medium with cadavers, insects should be surface sterilized as described earlier. Because this operation can be harmful to the pathogen when the host's cuticle is broken, this method should be used only with insects without emergent fungal structures. Furthermore, this method of isolation does not allow subsequent study of any original *in vivo* fungal material.

Disruption of cadavers with the help of fine forceps appears not to be necessary for cadavers that are filled with fungal mycelium. However disruption of the cuticle is essential with living insects, so that the fungal material contained in the body will contact the medium.

F Media for isolation

Classical mycological media, such as potato dextrose agar or malt extract agar, are unsuitable for the great majority of Entomophthorales; these species simply will not grow or will grow poorly. Entomophthoralean species generally need richer nutrients to grow and develop. The most common media for successfully growing Entomophthorales contain egg yolk. However, requirements of several species are more elaborate and, consequently, media, at times, have to be supplemented, e.g. with vitamins and amino acids. In addition, some species can only be isolated using insect tissue culture media.

1. Media containing egg yolk

A diversity of media containing egg yolk are mentioned in the literature. See the following references for variations in ingredients and preparation, both of which can be critical: coagulated egg yolk (Müller-Kögler, 1959; Gustafsson, 1965), egg yolk/Sabouraud maltose agar (Soper *et al.*, 1975), egg yolk with milk/Sabouraud dextrose agar (Milner & Soper, 1981), egg yolk/Sabouraud dextrose agar (Keller, 1987), coagulated egg yolk with milk (Keller, 1987), egg yolk/peptone broth agar (Bałazy, 1993). Because all these media are basically similar, complete descriptions of all of them would be redundant. We provide recipes for two variations routinely used in the Institut Pasteur: (for routine laboratory isolation Sabouraud dextrose agar supplemented with egg yolk and milk (Papierok, 1978) and for field isolation, coagulated egg yolk with milk (Papierok & Charpentié, 1982)) and the USDA Agricultural Research Service Collection of Entomopathogenic Fungi in Ithaca, NY (egg yolk/Sabouraud maltose agar: EYSMA) (see Appendix, recipes 1–3).

2. Compound media

Because of poor growth of several Entomophthorales species on media containing egg yolk, compound media supplemented with additional components (vitamins, salts, lipids, proteins, etc.) have occasionally been developed.

For instance, Ben Ze'ev (1980; in Keller, 1994) developed '*Entomophthora* complete medium', containing a carbon source (e.g. dextrose, maltose), a solution of 11 salts, tryptophan, yeast extract, casein hydrolysate, a solution of nine vitamins and agar. This complex medium has the advantage of being transparent, which allows microscopic observation of cultures. Furthermore, it also allows isolation of a few species which cannot be isolated using media containing egg yolk, e.g. *E. grylli* or *Entomophthora muscae* (Keller, 1987, 1994).

3. Insect cell culture-type media and substitutes

Tyrrell and MacLeod (1972) first observed the formation and proliferation of protoplasts originating from conidia of *Entomophaga aulicae* in Grace's medium (Grace, 1962). Since then, this insect tissue culture medium and modifications of it have been used not only for growing entomophthoralean protoplasts, but also for isolating species of this fungal group which exhibit high nutritional requirements. Other complex media have also been tested, as well as a simpler medium (Latgé & Beauvais, 1987; Beauvais & Latgé, 1988).

The insect tissue culture medium developed by Grace is available commercially (Gibco-BRL). Fungal isolations in Grace's medium are preferably attempted in 35 mm diam. Petri dishes containing the medium supplemented with 5% or 10% (v/v) heat-inactivated fetal calf serum (Dunphy *et al.*, 1978; Latgé & Beauvais, 1987; Soper *et al.*, 1988). In the peculiar case of *E. grylli*, MacLeod *et al.* (1980) found that a grasshopper thorax and abdominal aqueous extract (without digestive tract, wings and other appendages) added to the Grace's tissue culture medium plus 5% fetal bovine serum was necessary for successful isolation.

Modifications of Grace's medium for growth of *E. aulicae* have been investigated extensively by Dunphy and Nolan. In one study it was found that *E. aulicae* protoplasts did not need vitamins and required heat-inactivated fetal calf serum only when shaken (Dunphy & Nolan, 1982). *E. aulicae* could be grown in synthetic eastern hemlock looper haemolymph (Dunphy & Nolan, 1989) and a simplified defined medium with only eight amino acids (Nolan, 1988). Other tissue culture media, e.g. Mitsuhashi and Maramorosch's (Gibco-BRL), MGM-443 (Mitsuhashi, 1982) have also been successfully tested (Soper *et al.*, 1988; Eilenberg *et al.*, 1992).

A medium based on the need for fetal calf serum was developed by Holdom (1983) for isolation of *Entomophthora planchoniana* (see Appendix).

Previous attempts at isolating this aphid-pathogenic species on egg yolk-based media and even on the complex medium developed by Ben-Ze'ev (1980) had been unsuccessful. A medium developed by Beauvais & Latgé (1988) has been used for isolation of *Strongwellsea castrans* as protoplasts (Eilenberg *et al.*, 1992) (see Appendix).

G Tricks of the trade

As is the case with any biological material, experience is invaluable when attempting to isolate Entomophthorales.

1. Choice of isolation media

Choosing an isolation medium depends mainly on what one knows about the fungus expected to be isolated. Whether the fungus may be a new species, is not well known, or is suspected to be a well-known species, the first isolation method that should be tried is inoculating Sabouraud dextrose agar supplemented with egg yolk and milk, EYSMA, or coagulated egg yolk with milk (see Appendix) with conidia. The latter medium may be the most suitable because it is harder than the former two and therefore allows easier transfer of conidia from slides or Petri dish bases. Isolation from conidia also allows for both collection of conidia on slides to aid in identification, and retention of other fungal structures (e.g. conidiophores, hyphal bodies, etc.) and cadavers for records and subsequent taxonomic study. If, after initial efforts, the pathogen in question will not grow on egg yolk-based media, e.g. the *E. grylli* species complex, richer media will have to be used, supposedly due to requirements for complex nutrients and/or increased amounts of nutrients.

According to the host and fungal species and one's objectives, it might be appropriate to select a method based on the time it requires to prepare the media and/or conduct the isolation. For instance, isolation through whole specimens (see Section 2E) is recommended during extensive surveys for known species of Entomophthorales infecting aphids.

2. Specific considerations in preparing media

When preparing and pouring media, it is advised:

- to autoclave the different ingredients separately;
- to avoid running egg yolk-based medium down along the inner wall of the tubes or Petri dishes. This can disturb observations of initial fungal growth, and can even favour contamination if the medium is in contact with the cotton plug or opening of the Petri dish.

3. Antibiotics

Most of the media listed do not include antibiotics. For fungal isolation, Grace's insect tissue culture media (9.5 ml) plus fetal bovine serum (5 ml) is regularly amended with 45 μl gentamicin for fungal isolation (A. E. Hajek, unpublished data). This is possible because gentamicin has no effect on entomophthoralean protoplasts. The medium developed by Holdom (1983) for isolation of *E. planchoniana* (see Appendix) contains streptomycin and penicillin. These two antibiotics have no effect on growth of *Zoophthora radicans* isolates, unlike tetracycline, chloramphenicol and actidione (Ben-Ze'ev, 1980). Because of entomophthoralean sensitivity to many antibiotics, antibiotics are generally not added to isolation media for entomopathogenic Entomophthorales by the senior author.

4. How to avoid contamination

Essential rules are to operate using sterile methods as rapidly as possible. Avoiding contamination is essentially a matter of correct handling, but this can be very difficult with some host insects and/or under field conditions.

a. Maintaining sterility

As far as possible, preparation of specimens and fungal isolation should be carried out in a laminar flow hood. When such equipment is not available, a portable plexiglass box (25 × 40 × 40 cm) with holes for access on one side and a hinged lid has been used successfully. Under field conditions, fungal isolation can be conducted on any surface, preferably covered with a plastic tablecloth. It is recommended that the surface be disinfected with 65–80% alcohol and drafts be reduced before attempting isolation. A portable Bunsen burner, e.g. Labogaz® (Camping Gaz International) is suitable for isolation under sterile conditions.

All materials used should be sterile. Special materials such as filter paper or paper towels, water for moistening the substrate for cadavers, and cotton plugs for small isolation tubes (see Section 2), should be sterilized in appropriate packaging. If necessary, bottled water can be used.

b. Working rapidly

Isolations should be attempted as soon after insect death as possible, optimally within a few hours after collecting the material. This is especially important for cadavers which already bear conidiophores and which have to be prepared as specified in Section 2B. Sporulating structures of most Entomophthorales are indeed functional for less than 24 h at temperatures of 15–20 °C. Moreover these structures are rapidly contaminated, leading to contamination of isolates by bacteria or saprophytic fungi. Therefore, the period of time during which conidia are collected should be as short as possible. The intensity of spore discharge should be regularly monitored. If necessary, development of conidiophores can be slowed by cooling (e.g. 4 °C), although this should be avoided, if possible. If cooling must be used, cadavers should not be stored in a saturated atmosphere. The ability to slow down sporulation by cooling appears to be species dependent. For instance, this technique does not work well with the fly pathogens, *E. muscae* and *Entomophthora schizophorae*.

Cadavers with conidiophores not yet emerging through the cuticle can be stored for 24 h or more at 4 °C, provided they are kept under dry conditions. However, the longer the cadaver is kept in cool conditions, the longer the time required for subsequent appearance of conidiophores once cadavers are returned to higher temperatures and high humidity.

If conidiophores have not formed within 24–48 h of insect death, cadavers should be dissected and observed under the light microscope. Lack of conidiation can result from poorly growing hyphal bodies and/or internal production of resting spores instead of conidia.

5. Incubation conditions

As a general rule, inoculated tubes or Petri dishes should be incubated at an average temperature of 20 °C (18–23 °C).

6. Checking isolation containers

a. Monitoring fungal development

Once incubated, isolation tubes or Petri dishes can be observed daily in order to monitor germination of the inoculated spores. Opaque media (e.g. Sabouraud dextrose agar supplemented with egg yolk and milk, EYSMA, and coagulated egg yolk and milk) should be observed at a dissecting microscope. Details of fungal development are often difficult to see until mycelium is nearing sporulation. Transparent media (e.g. '*Entomophthora* complete medium' by Ben-Ze'ev, insect tissue culture-based liquid media) can be examined with the aid of an inverted microscope.

The fungal developmental rate varies according to species. On solid media, discharge of conidia onto inner walls of tubes or Petri dishes should be considered a sign of successful isolation. When resting spores represent the inoculative material, the absence of development can simply result from spore dormancy.

b. Checking the identity of the isolated entomophthoralean

The form and size of conidia produced on solid isolation media should be compared to those of conidia from the original infected specimen. One must keep in mind that, as a general rule, conidia produced *in vitro* are slightly larger than conidia from cadavers. In practice, conidia are obtained from solid media in a similar way as from cadavers (see Section 2B). A piece of mycelial mat is removed aseptically from the isolation container and placed on a moistened piece of filter paper (or paper towel) which is then attached to the inside of a Petri dish lid. The lid is inverted over a slide, where conidia will be collected.

Conidia are generally not produced on compound media (e.g. insect tissue culture-type media). To induce sporulation, the fungus can be transferred to Sabouraud dextrose agar supplemented with egg yolk and milk (after Holdom, 1983). Alternatively, host insects can be infected by injecting protoplasts from cultures, which will lead to development of mycosis and subsequent production of conidia (see Section 4A).

c. Overcoming contamination

Fungal contaminants (e.g. *Penicillium, Mucor, Cladosporium,* etc.) can grow rapidly in isolation tubes. Very little can be done to overcome this situation once it has started. A few saprophytic species of Entomophthorales, mostly of the genus *Conidiobolus* are an exception. These species can be contaminants on arthropods infected by pathogenic entomophthoraleans (Remaudière *et al.*, 1976b; Papierok, 1985). In order to prevent contamination, isolation attempts should be made rapidly, i.e. within a few hours after collecting the cadavers, and conidia should be collected for as brief a time period as possible (see Section 2G.4).

Bacterial contamination can occasionally be overcome because, initially, bacteria often grow slowly. Once a bacterial colony is detected, a piece of the mycelial mat should be removed sufficiently far from the bacteria and transferred to fresh medium. In some cases, the surface of the culture can be contaminated, while conidia are stuck to the inner wall of the container. Attempts to save the culture can be made by scraping these conidia with a piece of sterile solid medium and transferring them to fresh medium.

3 CULTURE AND PRODUCTION

A Routine culture maintenance

Once an entomophthoralean is successfully growing in the isolation container, the culture has to be maintained. When an entomopathogenic fungus is unable to grow *in vitro*, it can be kept in culture on the host. The culture should be initiated using infected insects that recently died, which serve as the source of inoculum to infect healthy ones, following procedures described in Section 4A. Infections to maintain the culture should be regularly carried out, according to the developmental rate and survival of the fungus.

Routine *in vitro* subculturing procedures depend on the isolation method and on the fungal stage occurring in the isolation medium: hyphal bodies grow on solid media while hyphal bodies and protoplasts both grow in insect tissue culture-type liquid media. These routine procedures are also generally used in the laboratory for production of selected stages of the fungus. Methods have been developed for production of only a few species on an industrial scale.

Entomophthoralean species that are only cultivated *in vitro* as protoplasts are necessarily kept in liquid media. Other species, which represent the majority, can be maintained on solid media, which is simpler and less costly.

1. Solid media

a. The most convenient media

Two convenient media for keeping most Entomophthorales in culture are Sabouraud dextrose agar supplemented with egg yolk and milk and EYSMA (see Appendix). Coagulated egg yolk (see Appendix) and '*Entomophthora* complete medium' should be tried for species growing poorly on the previous two media.

b. Techniques for transfer

Techniques for fungal transfer are as follows:

- carefully remove a small piece of the mycelial mat (3–4 × 3–4 mm) from the medium;
- transfer it in a new tube or Petri dish by first making a small depression in the middle of the surface of the medium, and then laying the mat down on the medium.

The growth of the fungus will occur partly from direct development of mycelium, and partly by germination and subsequent formation of hyphal bodies from conidia that have been discharged from the piece of culture and have landed on the new medium. We suggest keeping the tubes in a slightly tilted position for a few days after inoculation. This favours landing of discharged conidia on the medium rather than on the tube walls (Gustafsson, 1965).

c. Maintenance conditions

Freshly inoculated media should be kept at 18–20 °C for two weeks in order to secure consistent growth of the initial fungal inoculum. Afterwards, cultures can be stored at lower temperatures, for instance 12 °C and even 8 °C, so that the fungus grows more slowly and transfers are needed less frequently. Cultures should be kept in darkness or under a 12h photophase.

d. Frequency of subculture

For many Entomophthorales, transfers should be made every 2 or 3 months at most, when maintaining the fungus at 12 or 8 °C, respectively. However, some very fast growing Entomophthorales, especially *Conidiobolus coronatus*, may need more frequent transfers, e.g. every 4–6 weeks. Nevertheless, it is highly recommended that cultures are checked weekly or biweekly once the fungus has invaded the medium surface. Subculturing should be undertaken when fresh conidiophores are no longer detectable and/or the mycelial mat starts to liquefy.

e. Possible loss of culture capability

Entomophthoralean cultures can deteriorate (diminution of growth rate and/or loss of conidial production) if successively transferred too many times. However, this phenomenon is species-or even strain-specific. Such a decline has been observed with cultures of *Erynia neoaphidis* (= *Pandora neoaphidis*) (Rockwood, 1950), *Zoophthora* spp., *Erynia* spp. and strains of *Entomophthora culicis*. Conversely, cultures of almost all strains of *Conidiobolus obscurus* maintained over a period of more than 20 years have not changed (B. Papierok, unpublished data).

2. Liquid media

Liquid media are used for *in vitro* maintenance of Entomophthorales in the protoplast stage. The most common medium used is Grace's insect tissue culture medium supplemented with fetal calf serum (see Section 2F.*3*). Protoplast cultures are easily maintained in 25 cm^2 canted neck tissue culture flasks at 18–20 °C in total darkness. In this size flask, 9.5 ml Grace's insect tissue culture medium and 0.5 ml fetal calf serum are inoculated with 1 ml of protoplast culture. Cultures must be transferred to fresh media based on the density and speed of growth of the inoculum, e.g. *Entomophaga maimaiga* is transferred every 3–5 days. Transfer is necessary once protoplasts become swollen, spherical, and begin to appear granular internally. However, great care must be taken with the number of subcultures; experience with *E. maimaiga* has shown that subculturing can result in a loss of virulence (Hajek *et al.*, 1990c) although changes in morphology or growth rate are not always evident. To prevent loss of virulence of *E. maimaiga*, we routinely maintain abundant protoplast samples from the same original culture under liquid nitrogen and regularly thaw samples so that individual fungal lines are never subcultured more than 10 times.

As an alternative, cultures can be maintained at 4 °C and then subculturing is necessary only once a month. However, this method of maintenance is more suitable for temporary storage and not regular use in the laboratory for bioassays.

B Laboratory-scale production

1. Hyphal bodies

Hyphal bodies represent the basic developmental stage of Entomophthorales. They are, of course, easily found in solid media but cannot readily be separated from medium ingredients, especially the fatty ones (e.g. egg yolk and milk). Hyphal bodies are therefore best produced in liquid culture.

Dextrose (as a carbon source) and hydrolysates of proteins (as sources of other nutrients, basically nitrogen) are especially convenient substances for growth of various entomophthoralean species (Latgé, 1975, 1981; Latgé & Remaudière, 1975; Latgé *et al.*, 1978). Accordingly, a simple liquid medium using these basic components has been developed for growth of entomophthoralean hyphal bodies (see Appendix).

Once prepared, the medium is distributed in Erlenmeyer flasks, the volume per flask depending on the flask size. For instance, 150 ml flasks are filled with 50 ml medium. Flasks containing media are autoclaved for 30 min at 120 °C.

Flasks are initially inoculated with a few pieces of mycelial mat removed from cultures on solid medium. This mycelium should be in a fast-growing stage, e.g. taken from the periphery of fungal growth of 2–5-day-old cultures incubated at 20 °C. Flasks should be placed on a reciprocating shaker (100 oscillations/min) or a rotary shaker (150 rev./min) in darkness at 20–25 °C. Depending on the fungal species, optimal production of hyphal bodies should take 2–5 days.

For collecting hyphal bodies, liquid cultures at the appropriate stage should be filtered, possibly using a vacuum pump. Hyphal bodies collected on the paper should then be washed with sterile distilled water. Subcultures should be made by inoculating flasks with a suspension of hyphal bodies, 5% (v/v), from 2–5-day-old liquid cultures.

When cultivated in liquid media, hyphal bodies of many species aggregate. In order to prevent, or at least reduce, this phenomenon it is useful to add magnetic rods to flasks, and, immediately after inoculation, to place flasks on a magnetic stirrer for a few minutes.

Repeated subculturing in liquid media seems to be deleterious for many entomophthoralean species: formation of hyphal bodies is quantitatively and qualitatively affected. Consequently, many species should not be maintained in successive liquid culture for more than 2 months. After this period, mycelium should be transferred from liquid culture to a solid medium and then back to liquid culture. However, it is more convenient to restart liquid cultures from stock cultures that are continuously kept on solid media.

2. Conidia

Entomophthoralean conidia are produced in the aerial environment. Methods for obtaining conidia from cadavers have been described earlier (see Section 2B). With cultures on solid media, spontaneous conidiation is observed at the surface for many species, especially *C. coronatus*. On the other hand, there are practically no spontaneously formed conidia in cultures of *Conidiobolus osmodes* or *E. culicis*, for example.

Conidia can also be produced spontaneously from liquid cultures grown in flasks on a shaker. Under these conditions, conidiophores appear on mycelial aggregates either floating on the medium or stuck to the inner walls of the flasks.

The best way to obtain spores from *in vitro* hyphal bodies is as follows:

- detach the culture from the medium (solid medium) or filter the culture as specified in Section 3B.*1* (liquid medium);
- place the mycelium, or the masses of hyphal bodies, on very moist filter paper;
- attach the filter paper bearing the fungus to the inner wall of a Petri dish lid;
- invert the lid above the base.

For most species, a few hours' delay will occur before conidial discharge starts. A good trick is to carry out these preparations in the late afternoon and hold the plates overnight at 12–14 °C. This chilling generally results in an adequate cover of conidiophores on the fungal mass the following morning.

For some fungal species, a completely saturated atmosphere can be unfavourable to conidiophore formation and subsequent conidial discharge, so Petri dishes with mycelium should be kept slightly open.

3. Resting spores

A few entomophthoralean species are known to produce resting spores *in vitro*. For those species, production can occur in solid media, especially those supplemented with egg yolk. However, as with hyphal bodies (see Section 3B), it is most practical to obtain this developmental stage in liquid culture. Several authors have made significant progress toward mass producing the resting spores of *Conidiobolus* spp. for use in biological control and their techniques (see Section 3C) can be downscaled for laboratory production.

As with many fungi, Entomophthorales form spores in culture when starved. Induction of sporulation occurs when carbon or nitrogen sources are lacking and the resulting numbers of resting spores produced are in direct proportion to the quantity of nutrients. This is also affected by the type of nitrogen source; yeast extract is optimal although results appear to vary between different batches of this product. A suitable and simple basic liquid medium is the one specified in Section 3B which contains dextrose and yeast extract (see Appendix). Dextrose can be replaced by a vegetable oil, e.g. sunflower oil (30 ml/l). In order to maximize yields, respective amounts of carbon and nitrogen sources should most likely be slightly adapted to species, and even to fungal strains.

To remove variability due to the complexity of yeast extract, attempts have been made to develop chemically defined media. A simple medium containing dextrose, L-arginine, L-leucine, glycine and mineral salts allowed growth and sporulation of *Conidiobolus thromboides*. However, higher sporulation rates were consistently obtained on a medium with yeast extract (Perry & Latgé, 1980). According to Latgé & Sanglier (1985), the optimal defined medium for sporulation of *C. obscurus* contains dextrose, 11 amino acids, four vitamins (thiamin, biotin, folic acid and panthotenic acid), and four salts (phosphates, magnesium and zinc sulphates, manganese chloride).

Basically, culture conditions are similar to those for production of hyphal bodies (as described earlier in this chapter). However, sporulation seems to be more sensitive to physical conditions than fungal growth stage. A suitable combination of factors includes an initial pH of 6.5, a temperature of 20 °C, and complete darkness (Latgé *et al.*, 1978; Latgé & Sanglier, 1985). Time for complete resting spore formation should take 8–10 days, according to species.

Once completely formed, resting spores can be recovered by filtration and washed with sterile distilled water. The sporulation rate can be estimated by determining the number of spores per ml with a haemocytometer. Use of the haemocytometer is described in Chapter V-3, although we suggest counting five 1-mm^2 squares in five different haemocytometer chambers for increased accuracy (A. E. Hajek, unpublished data).

Two resting spore types can be distinguished in the Entomophthorales: quiescent spores that are able to germinate immediately after being produced (the less common type), and dormant spores that are unable to germinate without cessation of dormancy. Germinability can be evaluated by keeping batches of resting spores in sterile water in a humid chamber at 15°C. Contact with free water seems necessary. Germination occurs slowly and asynchronously and can last for a few weeks. If germination does not occur, spores should be considered dormant. Batches of spores should then be kept for several months at 4°C and 100% RH. Such a cold period should stimulate spore germination, as demonstrated with several species, including *C. obscurus* (Latgé *et al.*, 1978; Perry & Latgé, 1982), *Z. radicans* (Perry *et al.*, 1982) or *Erynia crustosa* (= *Furia crustosa*) (Perry & Fleming, 1989). However, this simulated winterization does not always result in germination in all species.

4. Protoplasts

For laboratory-scale production of entomophthoralean protoplasts, procedures for strain maintenance are used (see Section 3A).

C Large-scale production

Several authors have made significant contributions toward mass producing resting spores of *Conidiobolus* spp. for use in biological control (Gröner, 1975; Soper *et al.*, 1975; Matanmi & Libby, 1976; Latgé *et al.*, 1977; Latgé & Perry, 1980). For large-scale production of Entomophthorales, the fungus is produced on industrial-type liquid media in fermenters. Such a procedure involves successive inoculations of increasingly larger fermentation chambers. Production in fermenters was first achieved only for resting spores of *C. thromboides* and *C. obscurus* (Latgé *et al.*, 1977; Latgé 1980). Attempts have also been made to mass produce hyphal bodies of *E. neoaphidis* (= *P. neoaphidis*) (Latgé, 1982). Nolan (1990, 1993) successfully developed methodology for mass production of hyphal bodies of *E. aulicae*.

Zoophthora radicans can also be grown in large quantities (Soper, 1985). For this system, the fungus can be dried (termed 'marcescent process') and milled, and, once distributed in the environment, individual pieces of mycelium produce conidia on conidiophores.

1. Media

Suitable industrial-type media for mass production could contain the following ingredients:

- for *Conidiobolus thromboides* resting spores: 80 g/l dextrose (industrial grade product), and 20 g/l soybean flour (and/or cottonseed flour) (industrial grade product);
- for *Conidiobolus obscurus* resting spores: 60 ml/l unrefined corn oil, and 40 ml corn steep liquid;
- for *Erynia neoaphidis* (= *Pandora neoaphidis*) hyphal bodies: 60 g/l dextrose, and 20 g/l yeast extract.
- for *Zoophthora radicans* hyphal bodies: 40 ml/1 corn syrup, 10 g/l yeast extract, and 10 g/l peptone (pH 6.8).

These media should be supplemented with up to 1% of an antifoam agent.

In the case of *E. aulicae*, medium for mass production of hyphal bodies is modified Grace's medium, a complex mixture including 13 amino acids (Nolan, 1988, 1990, 1993).

2. Culture conditions

The fungus should first be grown in flasks on a shaker, following the procedure described in 3B. Flasks should contain at most 30% (v/v) of medium. After 24–96 h of growth, according to species and flask size, the mycelia can be used for inoculating a 2-litre fermenter. The volume of inoculum should amount to 8–10% of the volume of the medium. After a second 24–96 h, the mycelia thus produced serve as inoculum for a larger fermenter (i.e. 6, 10 or up to 25 litres).

Physical conditions for fermentation include complete darkness, 20° C, agitation of 700 rpm, pH of 6.5, and an aeration of 1 (or 0.5) volume of air per volume of medium per minute (v/v/m). Alternatively, *Z. radicans* grown in 25-litre carboys is agitated by constantly bubbling air through it (see Chapter V-3). Such conditions allow a maximum yield of *E. neoaphidis* (= *P. neoaphidis*) hyphal bodies or *Z. radicans* mycelium within three days. Production of mature resting spores of *C. thromboides* and *C. obscurus* takes longer (up to 10 days with the latter species).

3. *Fermentation difficulties*

Fermentation of Entomophthorales has been successful in containers of up to 25 litres. Results with larger containers (e.g. 100 litres and more) have mostly been unsuccessful, due to contamination. This contamination occurred both at the beginning and end of fermentation. Late contaminations are likely to result from the length of this process, e.g. resting spores produced by *Conidiobolus* spp. require 7–10 days. In order to circumvent this problem, one could possibly apply the process developed on a laboratory scale by Latgé and Perry (1980). These authors took into consideration two facts (i) resting spore formation is induced within 4 days, and (ii) after this time, the pre-resting spores (prespores) need no further nutrients. Accordingly, fermentation was stopped after 4 days to gather the prespores, wash them with water, formulate in humid clay, and maintain at 20 °C. Under these conditions, prespores matured within 4–5 days and contamination of culture media was prevented.

4. *Harvesting and storing*

For *Z. radicans*, after large-scale production, the mycelium is collected on cheesecloth and the liquid is removed under vacuum. The mycelium is washed several times with deionized water and mats are placed on wire racks to dry and incubate for 2 hours. Then, mycelial mats are sprayed with sterile 10% maltose until saturated, and dried for 4–5 h. Mats are incubated at 4 °C overnight then dried until 'crunchy' under continuous air flow (e.g. in a fume hood). Mats are then ground into a powder and flash frozen by immersion of containers holding mycelial powder into liquid nitrogen for storage (McCabe & Soper, 1982) (see Chapter V-5).

4 BIOASSAYS

When studying any pathogen, it is essential to experimentally infest healthy insects. This represents one of Koch's postulates (see Chapter I). Furthermore, such a procedure is the basis of possible *in vivo* culture of the pathogen. Experimental infections allow study of the mode of action of the pathogen, adding to our fundamental knowledge of pathogenesis. From the applied point of view, bioassays with entomopathogenic fungi help to determine their potential in biological control of pests.

Given the normal route of arthropod infection by fungi, conidia (the infective units) must come into contact with the host cuticle. It is using this basic information that the first attempts to infect insects with Entomophthorales were successfully undertaken. In fact, most infection experiments are still carried out in this way. As an alternative, for some experiments, fungal material is injected into the insect haemocoel.

Originally, bioassays with Entomophthorales were strictly qualitative, evaluating only whether or not a species was able to infect an insect. During the last 20 years, quantitative bioassay methods have been designed in order to estimate, among other things, differences in infectivity within and between fungal species and differing insect hosts.

A Infection methodology

1. *Conidial inoculation*

There are three ways of exposing insects to conidia discharged either from cadavers or cultures:

- placing test insects in a shower of conidia;
- bringing test insects and a surface covered with conidia into contact;
- exposing test insects to suspensions of known concentrations of conidia.

Once conidia are on the insect cuticle, test insects must be maintained for a limited time in a saturated atmosphere (100% RH) in order to ensure germination of conidia and subsequent penetration by the fungus.

a. *Conidial showers*

Conidia can be obtained from cadavers filled with hyphal bodies or from mycelia produced in either solid or liquid media. These sources of inoculum should be initially prepared as specified in Section 2B (cadavers) or 3B (cultures). Germinating resting spores can also be used as sources of conidia, as first explained by Valovage *et al.* (1984), although, at least for *E. maimaiga*, infections initiated by germ conidia produced by resting spores are fundamentally different from infections initiated by conidia (A. E. Hajek, unpublished data). All the previous sources of inoculum should be inverted over hosts, so that only primary conidia will be showered. Coverslips or slides placed beneath showers allow quantification as conidia/mm^2.

Groups of insects can be exposed to the conidial inoculum in tubes, the size of the tubes depending on the size of insects. For instance, 55 mm high × 25 mm diameter tubes are suitable for adult stages of aphids (Papierok & Wilding, 1979). Moist substrates bearing the sources of inoculum can be cut to the diameter of the tube and inverted above it. A fine gauze or grid can be placed between the 'discharging' disc and the insects, preventing the insects from making direct contact with the culture but allowing the conidia to pass through. The tubes should be placed in a saturated atmosphere (100% RH) for the duration of exposure.

Groups of insects can also be exposed in Petri dishes with cadavers or mycelial mats inverted over the hosts. Mycelial mats rarely produce conidia synchronously over the entire mat. Therefore, depending on the size of insects and Petri dishes, the mats should be rotated or the dish containing the insects placed on a revolving turntable so that all insects are showered with equivalent numbers of conidia.

b. Surface contamination

Insects can be infected with Entomophthorales by being brought into contact with a surface covered with conidia. Such a procedure is especially suitable for larvae and apterous adult forms, e.g. ant or termite workers which would move on such a surface. Two different approaches include:

- direct contact with sporulating cultures: this is appropriate for cultures which spontaneously produce numerous conidia in a uniform way, e.g. *C. coronatus*;
- contact with a surface covered with conidia: As early as 1871, Brefeld succeeded in infecting caterpillars of *Pieris brassicae* (Lepidoptera, Pieridae) with *Z. radicans* by exposing test insects to cabbage leaves dusted with conidia of the fungus. However, other substrates can be used, e.g. moist filter paper, plaster of Paris, plastic or glass Petri dishes. In practice, these surfaces are covered with conidia by placement above sporulating cadavers or cultures (as described earlier in this chapter). This latter procedure is the only one suitable for passively detached secondary conidia, such as the capilliconidia of *Zoophthora* (Glare *et al.*, 1985) and *Neozygites*.

c. Conidial suspensions

Conidia, originating from cadavers or mycelium cultured in solid or liquid media, are collected in sterile distilled water supplemented with a few drops of wetting agent (e.g. Atmos 300, Tween 40 or 80). Suspensions are brought into contact with test insects by means of: (i) direct spraying (a hand sprayer or a chromatography sprayer are convenient); (ii) topically applying a microvolume of highly concentrated suspension; or (iii) dipping insects into the suspension.

As an example of the steps involved in dipping insects into a conidial suspension, *E. maimaiga* conidia are showered into 0.25% Atmos 300 and 0.10% Tween-80 (Shimazu & Soper, 1986). Conidia are then collected by centrifugation every hour and stored at 4 °C (preferably on that same day) so that none germinate before bioassays are begun. Once an adequate number of conidia have been collected, the solution is decanted and replaced with 0.025% Atmos 300 and 0.01% Tween 80 to minimize effects of the detergent on test insects. Concentrations of suspensions should be estimated using a haemocytometer. Use of the haemocytometer is described in Chapter V-3, but we suggest counting five 1-mm^2 squares in five different haemocytometer chambers for greater accuracy (A. E. Hajek, unpublished data).

d. Post-inoculation incubation

Immediately after being exposed to the conidial inoculum, insects should be kept in a saturated atmosphere and possibly allowed to feed before being replaced in normal rearing conditions. The length of the period in very high humidity depends on each entomophthoralean/test insect system and on the temperature. At an average temperature of 18–20 °C for instance, the period can range from 12 to 24 h.

A saturated atmosphere can be produced by keeping the containers with test insects in a humid chamber. With winged insects, e.g. aphid alates or mosquitoes, it is convenient to apply small pieces of moistened filter paper to the interior of containers (Dumas & Papierok, 1989). Alternatively, the presence of free water in containers (e.g. a few drops of sterile distilled water) has been used to increase humidity and thus promote infection of non-winged insects, such as caterpillars (Ullyett & Schonken, 1940; Krejzová, 1971). For insects that can be cannibalistic, individuals are placed separately in plastic Petri dishes containing slightly moistened filter paper. Petri dishes are then sealed for the required amount of time.

e. Special considerations

Throughout all types of bioassays, conidial age and state must be considered. In a humid environment, entomophthoralean conidia will begin to germinate in several hours. These conidia frequently die when dried for very long. Therefore, conidial showering or collection must be conducted rapidly. For conidial collection, if showering is slow, conidia can be collected on an hourly basis and refrigerated for at least several hours (but with reduced survival if kept overnight). This arrests development so that conidia used for bioassays are at approximately the same stage.

Directly exposing insects to conidial showers is the best method for small (winged or apterous) insects like aphids or mosquitoes. Such insects can be confined in small containers for several hours (up to 24 h) without damage. This method is the most similar to natural conditions. However, it is very difficult in practice to expose insects to the same dose during replicated experiments.

In the case of larger insects, especially caterpillars, infection using surfaces covered with conidia are appropriate. However, it is important to try to expose insects to conidia of the same range of ages (and not yet germinating) in replicated experiments. Mobile insects can be maintained in showers of conidia by immobilizing them using thread and modelling clay (Gabriel, 1968).

Although convenient, the method involving application of a water suspension of conidia can give inconsistent results. Care must be given to maintain accurate homogeneous suspensions of conidia. Moreover, contact with free water may alter the adherent coat of the conidia, making them more sensitive to external conditions. For example, Baird (1957) noticed that the protoplasm coating the primary conidia of *E. muscae* dissolves in water. In addition, entomophthoralean conidia are relatively short-lived when submerged in water so they cannot be stored to conduct studies the day after showering.

Length of exposure of test insects to conidial showers or surfaces covered with conidia depends on the intensity of showering or on the number of spores on the surface, respectively. In these cases, the use of coverslips or slides for quantifying conidia over time or the use of a dissecting microscope, to look for conidia directly on insects, is highly recommended. Using any of the conidial inoculation methods, exposure of insects to excessive doses can result in death due to bacteriosis instead of mycosis.

2. Injection of fungal material

Experimental infection can also be achieved by injection of hyphal bodies or protoplasts into the haemocoel. Because of the size of these fungal cells, this procedure should be considered only for rather large insects. Until recently this procedure was used only on the last two instars of caterpillars such as *G. mellonella*, *Choristoneura fumiferana*, and *Bombyx mori* (Krejzová, 1971; Tyrrell & MacLeod, 1972). However, microinjection has more recently been regularly used with third instar *L. dispar* (A. E. Hajek, unpublished data) and grasshoppers (Ramoska *et al.*, 1988).

Hyphal bodies and protoplasts should be cultivated in liquid media, as described earlier in this chapter. The fungal material to be injected should come from 1- or 2-day-old cultures. The culture can be injected without special preparation or after being gently centrifuge-washed (e.g. for protoplasts, 4000 ***g***, 4 °C, 5 min; cf. Dunphy & Chadwick, 1985). With hyphal bodies, cultures can be filtered and the mycelium washed with sterile distilled water before being placed into a final volume of sterile distilled water.

Suspensions of hyphal bodies or protoplasts should be introduced into insects by intrahaemocoelic injection (in 10–50 μl amounts according to the size of the insects) via the footpad or base of a proleg. It is most important to insert cells into the haemocoel, with special care not to puncture the gut during this process. Microsyringes can be fitted with a drawn-glass needle or with a standard metal needle, e.g. 23 ***g*** for *E. maimaiga* protoplasts. The needle size used must be large enough that it does not clog and protoplasts must be agitated regularly within the syringe (e.g. by shaking the syringe) because they settle. Once injected, insects can be maintained under normal growth conditions; when injection is used, there is, of course, no need for maintenance in a saturated atmosphere after infection.

B Designing and analysing bioassays

1. Principles and difficulties

To measure the infectivity of an entomophthoralean, as with other pathogens, groups of test insects are exposed to various doses (or concentrations) of infective units. Results are conventionally expressed in terms of regression of probit-transformed percentage mortality on log dose and the LD_{50} (dose killing 50% of the test-insects) or LC_{50} (pathogen concentration killing 50% of the test-insects) is estimated (Finney, 1962, 1971) (but see Huber & Hughes, 1984, for a discussion of the use of the exponential or independent action model as an alternative to probit analysis). The time from infection to death should also be calculated either as mean time to death or LT_{50} (time from pathogen exposure until 50% of inoculated insects die) preferably using the non-truncated method of Allaway & Payne (1984).

Because of the route of fungal infection in insects, bioassays with fungi can be based on spraying test insects with suspensions of infective spores at given concentrations. However, such a procedure does not always work well with entomophthoralean conidia, due to their relatively brief survival time and occasional alteration when suspended in water. Consequently many classical bioassays with these fungal stages involve the exposure of insects to showers of conidia. Therefore, actual doses cannot be standardized before application, rendering the experimental results rather uncertain, and the procedures more time and labour consuming. On the other hand, bioassays using injection of protoplasts do not face these problems; they can be conducted using a range of concentrations of cells for injection.

2. Bioassays using conidia

Further discussion of bioassays using conidia will be presented in terms of direct conidial showers. The procedures for estimating the infectivity of entomophthoralean conidia are based on the general methodology for infecting insects with these fungi. Successive steps are involved: (i) available inoculum sources, either *in vivo* or *in vitro* (as described earlier in this chapter); (ii) exposure of groups of insects to various estimated quantities of inoculum; (iii) establishing the most favourable conditions for the infection process (as described earlier in this chapter); (iv) regularly monitoring infected specimens to record time of death; and (v) analysis of data. Given that some of these stages are specified in previous parts of this chapter, the following text will be devoted only to those stages not yet presented.

The basic principle of this bioassay method is that the number of conidia landing on an insect is regulated by the amount of time the insect is exposed to a conidial shower. Groups of test insects can be immobilized under conidial showers, if necessary. Immobilization allows more accurate determination of the doses received by the insects. The doses are estimated by counting the conidia (in number mm^{-2}) in sample areas (e.g. glass cover slips) on the same level as insects under the conidia shower. This procedure was first used by Wilding (1976) for apterous adults of the pea aphid, *Acyrthosiphon pisum* (Homoptera, Aphididae).

In most cases, however, it is definitely more practical not to immobilize insects under conidial showers (Figure 3). Groups of insects are submitted to conidial showers as specified in Section 4A. Under these conditions, the actual doses received by insects cannot be accurately estimated, but can be estimated for a fixed time period (t). The number of conidia landing on the bottom of the container during the period of exposure can be counted. Therefore, dose is estimated by the conidia (in number mm^{-2}) on the sample areas. This method was shown to be suitable for insects of small size, such as aphids (Papierok & Wilding, 1979, 1981; Milner & Soper, 1981; Papierok *et al.*, 1984; Oger & Latteur, 1985) and mosquitoes (Dumas & Papierok, 1989).

The production of conidia can be considered uniform during the experiment, which is acceptable for short exposures (up to 60 min). The quantity of inoculum is then estimated by averaging the numbers of conidia projected in equal amounts of time before and after the insect exposure. This method is suitable for any insects, but especially for larger ones, e.g. caterpillars, which prevent accurate

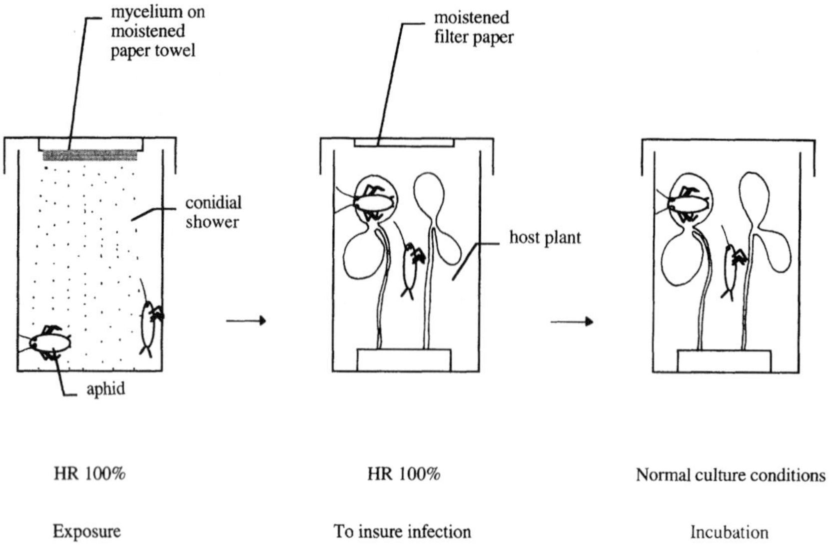

Figure 3 Successive steps for assaying infectivity of Entomophthorales against aphids (after Latgé & Papierok, 1988).

collection of conidia on the bottom of the container (Vandenberg & Soper, 1979).

In practice, test insects should be exposed to a given quantity of inoculum in groups of 5–10 individuals per container. For example, a bioassay designed to accurately estimate fungal infectivity could involve four exposure periods with five repetitions per period, which would provide up to 20 different dose levels.

3. *Bioassays using protoplasts*

Bioassays with hyphal bodies or protoplasts basically involve injection of suspensions of fungal cells of specific concentrations into test insects. Dose is, therefore, easier to control and repeat compared with bioassays using conidia.

4. *Data analysis*

Bioassays will consist of a number of insects exposed to the inoculum, a number of subsequently infected specimens, and a corresponding estimate of inoculum concentration for each container. Mortality percentages are probit-transformed and inoculum concentrations are log-transformed. If there is any mortality in the controls, treatment data should be altered using Abbott's correction (see Chapter V-3). The subsequent regression line of probit versus dose allows the estimation of LC_{50}, which characterizes the infectivity of a strain at the time of the experiment.

Computer programs have been developed to facilitate probit analysis. These programs test for significance of the regression and calculate the slope of the regression line, its standard error, the heterogeneity (chi-squared), the LC_{50}, and 95% fiducial limits.

5. *Tricks of the trade*

As with any bioassay, test insects should be as homogeneous as possible. The developmental stage, the age, and the physiological state of hosts should be standardized. Moreover, the origin of the laboratory colony should be well established. In order to reduce possible heterogeneity in results when the purpose of the experiment is to compare fungal lines, it is advisable to use only the offspring from two parents.

The fact that there are less than 20 papers presenting results from bioassays of Entomophthorales where LD or LC_{50} was calculated emphasizes the difficulties in applying this methodology to these fungi. However, the general bioassay procedures described above have proven accurate for a few fungal species (*Conidiobolus* spp., *E. neoaphidis* (= *P. neoaphidis*), *Z. radicans*) and different insect hosts (aphids, caterpillars, mosquitoes). Practical details (e.g. size of containers, time of exposure, etc.) have to be adapted to every situation, i.e. every pathogen/insect association. For the purposes of accurate data analysis, it is preferable to carry out preliminary infection experiments in order to determine the quantities of inoculum allowing mycosis rates of about 25 and 75% (Finney, 1971). Then, the inoculum concentrations applied in subsequent bioassays should be kept within the range of the corresponding values.

The most frequently published values for the slopes of regression lines range from 1 to 2. These values appear relatively low compared with chemical insecticides. Such values could result from rather high variability among the groups of insects submitted to given amounts of inoculum. As mentioned above, the requirement for very homogeneous test insects is therefore essential.

On the other hand, variation can be observed between various values of LD or LC_{50} for one strain with one test insect species. A weighted mean log LD or LC_{50}, and a weighted mean slope, when slope estimates are not significantly different from each other, should then be calculated. These problems have been discussed by Vanderberg & Soper (1979) and Oger & Latteur (1985). The latter authors demonstrated that the following experimental conditions: (five doses, ten *A. pisum* apterous adults, eight repetitions, i.e. 400 aphids in total)), appear to be the most efficient for statistically distinguishing two values of LC_{50} with a ratio of 1–2.

In addition to dose–response analysis, the time required by a pathogen to kill a host is an important aspect of its pathogenicity. The duration of an infection cycle can be estimated by regularly checking for dead infected insects in experimental containers. At a specific dose, according to the interval between two observations, mean time to death or LT_{50} can be calculated (Allaway & Payne, 1984). Researchers have conducted these types of studies to calculate LT_{50} across a range of doses (e.g. Van Beek *et al.*, 1988 – presented as time for 50% survival (ST50)) although such studies are difficult due to the necessity of maintaining independence among observations (see Chapter V-3).

5 SAMPLING ENTOMOPHTHORALES

Species in the Entomophthorales differ radically from Deuteromycete entomopathogens due to the active discharge of conidia by most species as well as production of resting spores within cadavers by many species. Studies of the biology and ecology of these pathogens have utilized techniques appropriate for sampling infected insects and/or pathogen propagules of this group.

A Sampling living insects and cadavers

Quantitative sampling to evaluate disease prevalence in a host population is fundamental to studies of all host/pathogen systems. Although it is important to evaluate levels of disease, it is also very important to simultaneously evaluate the density of the host population. It is beyond the scope of this chapter to review sampling procedures extensively, but some comments are appropriate specific to species of Entomophthorales.

For some species of Entomophthorales, hosts exhibit specific taxes prior to death, e.g. climbing by grasshoppers infected by *E. grylli*. Therefore, if sampling is not purposefully conducted in a variety of microhabitats, estimates of disease prevalence could be erroneous due to the ease (e.g. tops of grasses) or difficulty (e.g. tops of trees) in collecting infected insects at specific locations.

To estimate disease prevalence, living insects can be collected and reared to detect disease mortality. Reared insects must be checked on a regular basis to detect time of death and then to detect potential conidial production. After host death, cadavers should be transferred to humid chambers to promote conidial production thereby confirming the cause of death. It is very important to monitor such cadavers for several days after host death because conidia and conidiophores can degrade quite rapidly after production, even at 4 °C. These cadavers should be dissected and a smear observed under a compound microscope to detect potential conidial and resting spore production. In some species of Entomophthorales conidia and resting spores can be made in the same cadaver whereas in other species these different spore types are produced in different cadavers. In addition, dissection and microscopic examination will detect the potential coexistence of other pathogens.

An alternative to microscopic examination of individual cadavers is to use immunoassays. An enzyme-linked immunosorbent assay (ELISA) using polyclonal antibodies has been developed for accurate detection of *E. maimaiga* in living, infected larvae (Hajek *et al*., 1991). This assay was relatively straightforward to develop although antibodies required cross-absorption with *L. dispar* haemolymph. It is much less expensive than microscopic examination of individual cadavers. Unfortunately, this technique can only be used to detect conidia and hyphal bodies and will only begin detecting them 3 days after infection when holding insects at 20°C; accuracy at 3 days was 71.1% whereas 4 days after infection accuracy increased to 97.8%. Since *E. maimaiga* kills hosts relatively rapidly (5–7 days at 20°C), this leaves only a small window for use of ELISA. This assay will not detect resting spores so it cannot be used for cadavers and the antibody cross-reacts with the sympatric entomophthoralean pathogen of Lepidoptera *E. aulicae*. Although immunoassays hold promise for evaluation of disease prevalence, there are certain aspects of their use that must be taken into consideration.

As an alternative, DNA probes have been used for evaluation of field samples of grasshoppers potentially infected with members of the *E. grylli* species complex (Bidochka *et al*., 1996) and caterpillars potentially infected with members of the *E. aulicae* species complex (Hajek *et al*., 1996). These probes are extremely specific and have readily differentiated between the morphologically identical, closely related members of the *E. grylli* species complex and the *E. aulicae* species complex. Chapter VIII-3 presents a more in-depth treatment of DNA probe development and application.

Cadavers of insects killed by species of Entomophthorales can remain in the environment for prolonged periods, e.g. a small percentage of cadavers of *L. dispar* killed by *E. maimaiga* can remain on tree trunks for at least 9 months (A. E. Hajek, unpublished data). Therefore, while sampling to estimate disease prevalence, cadavers that are collected could represent mortality that occurred over a large time span. For this reason, the more accurate method for calculating disease prevalence is to rear living insects collected in the field.

If disease prevalence is being calculated using data from rearing sampled larvae (i.e. as for a life table analysis), it is important not to overestimate disease prevalence by counting equivalent disease mortality multiple times. For example, insects collected on day 1 and dying on day 20 represent equivalent mortality to insects collected on day 14 and dying on day 6 and, if both of these instances of disease mortality were counted, percentage infection would be inflated. van Driesche & Bellows (1996) present a more in-depth discussion of life tables and Hajek *et al*. (1990b) present a method of determining marginal attack rate of an entomophthoralean infecting a univoltine host.

As previously mentioned, microscopic examination can be used to quantify disease mortality.

B Conidia discharged by cadavers

As regards the number of conidia discharged from cadavers, this parameter can be estimated by collection of conidia and direct counting or by a spectrophotometric method. Estimation by direct counting can be accomplished according to methods developed for species infecting aphids, flies and caterpillars. Dead, infected specimens should be prepared as specified in Section 2B. Wilding (1971) and Mullens & Rodriguez (1985) have used spore trains, which allow calculation of the approximate number of conidia produced per hour, and consequently, the total amount of conidia discharged per cadaver. Shimazu & Soper (1986) suspended cadavers of caterpillars over beakers containing a known volume of 1% Triton-X plus 0.2% maleic acid. The conidia that were collected were counted to estimate total spore discharge. Milner (1981) and Papierok & Wilding (1981) allowed dead, infected insects to discharge conidia onto slides, which were replaced at regular intervals. Spores were counted within sample areas on each slide and the total numbers were then estimated.

To evaluate spore discharge by cadavers under field conditions, a 'sporometer' has been used (Hajek & Soper, 1992). This instrument consisted of a cylinder that rotated once every 24 h (originally from a hygrothermograph) around which a strip of Melinex tape (Burkard, Rickmansworth, UK) was attached. A cadaver just beginning to sporulate was attached over a specific location on the tape and was collected 24 h later. The numbers of spores on the tape was associated with hourly weather conditions recorded by weather recording equipment adjacent to the sporometer.

An original photometric technique for estimating conidial production from infected aphids was developed by Courtois *et al*. (1983). The principle of this technique, utilizing a precision luxmeter and a dark room, is based on the partial absorption by conidia of a luminous flux shining through. This method involves the preliminary calculation of the linear relation between observed photometric values and given numbers of conidia (estimated by direct counts). A similar technique, but using a spectrophotometer, was developed by Dumas (1986) for conidia produced from cultures.

C Evaluating spore viability and activity

1. Germination tests

Germination tests can be used to evaluate whether conidia or resting spores are alive or dead or to study the impact of abiotic or biotic factors on germination.

For conidia, this frequently involves showering conidia onto water agar, covering them with a coverslip and incubating at an appropriate temperature. Entomophthoralean conidia do not need exogenous nutrients to germinate. If a coverslip is not used, care must be taken not to count secondary conidia produced by germinating primary conidia as ungerminated primaries. This problem can be avoided by inverting plates with conidia so that secondaries produced are not deposited on plates. Under a coverslip, entomophthoraleans will grow as germ tubes. A germ tube greater than the spore diameter is generally considered as indication of germination.

It is critical to determine the optimal length of time after showering at which to make germination counts. For *E. maimaiga*, 50% germination in constant dark is reached 3.41 h after showering at 20 °C; in this system, results in constant light were dramatically different (8.99 h) so lighting conditions are an important variable to control (Hajek *et al.*, 1990a). Trying to make counts after germ tubes have grown extensively can potentially be very difficult due to interweaving of long germ tubes.

For adequate replication, germination quantification should not be based on several replicates using the same showering culture on the same day. Optimally, germination tests should be conducted on several different days using different source cultures of the same isolate for conidial production and counting at least 100 conidia per replicate.

Resting spore germination tests are actually not indications of whether resting spores are alive or dead because these spores can stay dormant for several years before germinating. However, germination tests demonstrate activity of resting spores during the period in question. Resting spore germination can be evaluated by placing resting spores on water agar and incubating at an appropriate temperature. As with conidia, nutrient agars are not required. Resting spores are generally obtained from cadavers and samples are frequently contaminated with bacteria and saprophytic fungi. To avoid contamination, 30 ppm mercuric chloride has been added to spores for 5 min, spores are washed and then germination is tested (Perry & Fleming, 1989). However, this treatment should be used with caution because it can kill resting spores of some species (A. E. Hajek, unpublished data).

When germinating, the interior of resting spores becomes granular, the double wall thins to a single wall, and a relatively thick germ tube emerges. Resting spore germination occurs much more slowly than conidial germination and, with *E. crustosa* (= *F. crustosa*) at 24 °C, resting spores stored for 6 months at 4 °C began germinating by 3 days and finished by 30 days, with a maximum of germination of 33% (Perry & Fleming, 1989). Therefore, observations of resting spore germination should be on an appropriate schedule.

2. Stains

Whether conidia are alive can also be determined by using fluorescent stains (see Chapter VIII-2). With fluorescein diacetate and propidium iodide, conidia of *Z. radicans* and *E. maimaiga* stained green if alive and red if dead (Firstencel *et al.*, 1990). This technique cannot be used for resting spores because they autofluoresce.

D Airborne conidia

Entomophthoralean conidia in the air can be sampled by several different types of samplers that use impaction to collect spores (Dhingra and Sinclair, 1985). Visual examination is then used to quantify collected spores. Such samplers are frequently the same aerial samplers used to take pollen counts. The characteristic shape and size of many entomophthoralean conidia aids in their identification in air samples. However, difficulties arise because spore shapes are sometimes not unique, e.g. the pear-shaped conidia of *E. grylli* and *E. aulicae* cannot be differentiated in aerial samples. The techniques most commonly used with airborne entomophthoralean conidia will be discussed, but for a more comprehensive treatment of techniques see Dhingra & Sinclair (1985).

The most basic method for examining airborne conidia is to place microscope slides at various locations in the environment to be sampled. Microscope slides are covered with a thin layer of silicon or double-sided clear adhesive tape to trap spores. This method is advantageous because it is inexpensive, allowing for abundant sampling, and easy to manipulate. The length of time that microscope slides are left in the field is important. If in the field too long, slides can become completely covered and thereafter will collect no more material during the period of time they are in the field.

To sample spores at low densities and in different locations in the environment, rotorods have been employed. Rotorods are portable and battery powered and therefore can be used in remote locations. They are efficient at collecting spores in the 20 μm range and are therefore suitable for most entomophthoralean species. These units consist of a small motor that rapidly rotates small plexiglass rods through the air to sample a specific numbers of litres of air per minute (e.g. 120 l air/min, although exact volume of air sampled varies by manufacturer). The plexiglass rods are covered with a light layer of silicon and impacted spores readily adhere to them. Rods are examined visually under a microscope to quantify conidia per volume of air. A major problem with rotorods is that they can only be used for short time periods because they readily become overloaded.

To sample over a time course, the Burkard 7-day volumetric spore trap (Burkard, Rickmansworth, Hertfordshire, UK) is regularly used. This is a bulky piece of equipment powered by a battery. This sampler consists of a drum that rotates once every 7 days. A piece of tape lightly covered with silicon is attached around the drum. A vacuum pulls ambient air through an orifice so that the air impacts the rotating tape. The orifice is approximately 0.5 m above the base of this unit and the unit swings so that the orifice samples into the wind. The tape on the drum must be replaced every 7 days. Numbers of spores are visually quantified on the tape, providing a relative measure of airborne conidial density across a time course. This is the principal instrument that has been used to evaluate diurnal periodicity in abundance of airborne spores in the field. Conidial densities can be associated with weather conditions to evaluate relationships with abiotic conditions. Major disadvantages with the use of volumetric spore samplers are their high cost and the lack of flexibility in locations where it can be used due to its large size.

E Resting spores or conidia in soil

Entomopathogenic Entomophthorales are likely to be able to survive in the soil as resting spores for as long as several years and as mycelia or conidia for at most several months. Detection of such inoculum is useful for ecological purposes as well as for isolation of new strains.

1. Use of living insects

Latteur (1977) first demonstrated that aphids could be used to reveal the presence of *C. obscurus* and *E. neoaphidis* (= *P. neoaphidis*) in soil. A crop hosted an aphid population infected by these two pathogens. The next year, sentinel aphids were used to demonstrate the presence of these two pathogens in the soil.

The '*Galleria* bait method', originally described for trapping entomoparasitic nematodes in soil, has proven efficient for the detection of entomopathogenic fungi, including Entomophthorales (Zimmermann, 1986; G. Zimmermann and H. Hokkanen, personal communication). Caging *Lymantria dispar* (Lepidoptera, Lymantriidae) larvae over soil containing resting spores or conidia of *E. maimaiga* has repeatedly been successfully used to study germination seasonality, spore survival and infection events (A. E. Hajek, unpublished data).

Whereas the previous methods can be used in the field or laboratory, the following procedures are designed for laboratory bioassays. For larvae of the gypsy moth (*L. dispar*), the greater wax moth, *Galleria mellonella* (Lepidoptera, Pyralidae), and possibly for other eruciform larvae (e.g. Coleoptera), the procedure is as follows:

- take soil samples from a depth of 5–10 cm;
- fill sterile containers with samples after sieving them but maintaining the soil moisture level;

- place several test insects into or on top of soil samples;
- store the sealed containers at room temperature and check for diseased or mummified larvae after 3 days to 2 weeks.

For aphids, the procedure is as follows:

- collect soil samples to a depth of 1.5 cm, without disturbing the soil surface;
- place these samples in sealed sterile Petri dishes and store at 4 °C, if necessary;
- confine batches of aphids for several hours on soil samples in Petri dishes, beneath a gauze covered by moistened paper towelling;
- remove aphids and hold them on plants for approximately 24 h in a saturated atmosphere;
- return the aphids to normal rearing conditions;
- check for diseased specimens (up to 5–6 days following exposure at 20 °C).

When the above procedures are used for detection of a pathogen in the resting spore stage, studies must frequently be conducted only when it is known that resting spores are able to germinate, i.e. usually during the field season for specific systems, or after dormancy requirements have been satisfied.

2. *Quantification of resting spores*

Another approach to studying resting spores, is to evaluate their densities in the soil. A method using a discontinuous density gradient of Percoll (Pharmacia) was first developed by MacDonald & Spokes (1981) for *C. obscurus*. First, 5 g soil is added to 0.1% (v/v) Triton-X 100 in 0.25 ml/l and then sonicated for 1 min. The resulting slurry was then added to the top of a stack of sieves (500, 250, 125, 63, and 20 μm openings) and washed through. The fraction collected on the 20 μm sieve is added to a discontinuous gradient of Percoll diluted with 0.15 M NaCl to 1.05, 110, and 1.13 g/ml. The sample is then centrifuged at 3800 *g* for 10 min. The 1.10 g/ml layer and its interfaces are then counted at 50× under a dissecting microscope.

This method was further used to quantify resting spores of *E. maimaiga* in soil, incorporating information on the accuracy of this method at different densities (from 10 to 100 000 resting spores/g dry soil) to arrive at a more exact estimate of density (Hajek & Wheeler, 1994).

As an alternative, a method using a discontinuous density gradient of sucrose plus $CaCl_2$ has been described (Li *et al.*, 1988). Although the materials for this method are less expensive, it requires much more time (Hajek & Wheeler, 1994).

Unfortunately, for all of the above methods, it is often impossible to tell exactly which species is present because resting spores of many species are morphologically identical. At present, researchers are left to assume the identity of spores based on their location and the past history of that area, e.g. resting spores in soil samples taken directly from the bases of tree trunks that previously were covered by cadavers of insects killed by *E. maimaiga* are assumed to be *E. maimaiga*.

ACKNOWLEDGEMENTS

We would like to thank the Institut Pasteur for its constant support of research on entomopathogenic fungi. We are also grateful to the USDA, ARS group in Ithaca for sharing information about techniques.

REFERENCES

Allaway, G. P. & Payne, C. C. (1984) Host range and virulence of five baculoviruses from lepidopterous hosts. *Ann. Appl. Biol.* **105**, 29–37.

Baird, R. (1957) Notes on a laboratory infection of Diptera caused by the fungus *Empusa muscae* Cohn. *Can. Entomol.* **89**, 432–435.

Bałazy, S. (1993) *Fungi* (*Mycota*). vol. XXIV. *Entomophthorales*. Polska Akademia Nauk, Instytut Botaniki im. W. Szafera, Kraków. 356 pp.

Batko, A. & Weiser, J. (1965) On the taxonomic position of the fungus discovered by Strong, Wells & Apple: *Strongwellsea castrans* gen. et sp. nov. (Phycomycetes, Entomophthoraceae). *J. Invertebr. Pathol.* **7**, 455–463.

Beauvais, A. & Latgé, J.-P. (1988) A simple medium for growing entomophthoralean protoplasts. *J. Invertebr. Pathol.* **51**, 175–178.

Ben-Ze'ev, I. (1980) Systematics of entomopathogenic fungi of the '*sphaerosperma*' group (Zygomycetes: Entomophthoraceae) and their prospects for use in biological pest control. PhD Thesis, Hebrew University of Jerusalem.

Ben-Ze'ev, I., Keller, S. & Ewen, A. B. (1985) *Entomophthora erupta* and *Entomophthora helvetica* sp. nov. (Zygomycetes: Entomophthorales), two pathogens of Miridae (Heteroptera) distinguished by pathobiological and nuclear features. *Can. J. Bot.* **63**, 1469–1475.

Bidochka, M. J., Walsh, S. R. A., Ramos, M. E., St. Leger, R. J., Silver, J. C. & Roberts, D. W. (1996) Fate of biological control introductions: monitoring an Australian fungal pathogen of grasshoppers in North America. *Proc. Natl. Acad. Sci. USA* **93**, 918–921.

Brefeld, O. (1871) Untersuchungen über die Entwicklung der *Empusa muscae* und *Empusa radicans* und die durch sie verursachten Epidemien der Stubenfliegen und Raupen. *Abh. Naturforsch. Ges. Halle* **12**, 1–50.

Courtois, P., Latteur, G. & Oger, R. (1983) Méthode rapide d'estimation de la quantité de conidies primaires de *Erynia neoaphidis* (Remaud. and Henn.) (Zygomycètes, Entomophthoraceae) produites par des momies de *Acyrthosiphon pisum* (Harris) (Hom.: Aphididae). *Parasitica* **39**, 173–182.

Dhingra, O. D. & Sinclair, J. B. (1985) *Basic plant pathology methods.* CRC Press, Boca Raton, Florida, 355 pp.

Dumas, J-L. (1986) Utilisation de la spectrophotométrie pour l'estimation des concentrations de conidies d'Entomophthorales dans les essais d'infection d'insectes. *Parasitica* **42**, 127–136.

Dumas, J.-L. & Papierok, B. (1989) Virulence de l'Entomophthorale *Zoophthora radicans* (Zygomycètes) à l'égard des adultes de *Aedes aegypti* (Dipt.: Culicidae). *Entomophaga* **34**, 321–330.

Dunphy, G. B. & Chadwick, J. M. (1985) Strains of protoplasts of *Entomophthora egressa* in spruce budworm larvae. *J. Invertebr. Pathol.* **45**, 255–259.

Dunphy, G. B. & Nolan, R. A. (1982) Simplified growth media for *Entomophthora egressa* protoplasts. *Can. J. Microbiol.* **28**, 815–821.

Dunphy, G. B. & Nolan, R. A. (1989) Development of *Entomophaga aulicae* protoplasts in synthetic eastern hemlock looper hemolymph. *Can. J. Microbiol.* **35**, 304–308.

Dunphy, G. B., Nolan, R. A. & MacLeod, D. M. (1978) Comparative growth and development of two protoplast isolates of *Entomophthora egressa. J. Invertebr. Pathol.* **31**, 267–269.

Eilenberg, J., Wilding, N. & Bresciani, J. (1992) Isolation *in vitro* of *Strongwellsea castrans* (Fungi: Entomophthorales) a pathogen of adult cabbage root flies, *Delia radicum* (Dipt.: Anthomyiidae). *Entomophaga* **37**, 65–77.

Finney, D. J. (1962) *An introduction to statistical science in agriculture*, 2nd edn. Munksgaard, Copenhagen, Denmark, 216 pp.

Finney, D. J. (1971) *Probit analysis,* 3rd edn. Cambridge University Press, Cambridge, 333 pp.

Firstencel, H., Butt, T. M. & Carruthers, R. I. (1990) A fluorescence microscopy method for determining the viability of entomophthoralean fungal spores. *J. Invertebr. Pathol.* **55**, 258–264.

Gabriel, B. P. (1968) Histochemical study of the insect cuticle infected by the fungus *Entomophthora coronata. J. Invertebr. Pathol.* **11**, 82–89.

Glare, T. R., Chilvers, G. A. & Milner, R. J. (1985) A simple method for inoculating aphids with capilliconidia. *Trans. Br. Mycol. Soc.* **85**, 353–354.

Grace, T. D. C. (1962) Establishment of four strains of cells from insect tissues grown *in vitro. Nature (London)* **195**, 788–789.

Gröner, A. (1975) Production of resting spores of *Entomophthora thaxteriana. J. Invertebr. Pathol.* **26**, 393–394.

Gustafsson, M. (1965) On species of the genus *Entomophthora* Fres. in Sweden. II. Cultivation and physiology. *Lantbrukshšgskolans Ann.* **31**, 405–457.

Hajek, A. E. & Soper, R. S. (1992) Temporal dynamics of *Entomophaga maimaiga* after death of gypsy moth (Lepidoptera: Lymantriidae) larval hosts. *Environ. Entomol.* **21**, 129–135.

Hajek, A. E. & Wheeler, M. M. (1994) Application of techniques for quantification of soil-borne entomophthoralean resting spores. *J. Invertebr. Pathol.* **64**, 71–73.

Hajek, A. E., Carruthers, R. I. & Soper, R. S. (1990a) Temperature and moisture relations of sporulation and germination by *Entomophaga maimaiga* (Zygomycetes: Entomophthoraceae), a fungal pathogen of *Lymantria dispar* (Lepidoptera: Lymantriidae). *Environ. Entomol.* **19**, 85–90.

Hajek, A. E., Humber, R. A., Elkinton, J. S., May, B., Walsh, S. R. A. & Silver, J. C. (1990b) Allozyme and RFLP analyses confirm *Entomophaga maimaiga* responsible for 1989 epizootics in North American gypsy moth populations. *Proc. Natl. Acad. Sci. USA* **87**, 6979–6982.

Hajek, A. E., Humber, R. A. & Griggs, M. H. (1990c) Decline in virulence of *Entomophaga maimaiga* (Zygomycetes: Entomophthorales) with repeated *in vitro* subculture. *J. Invertebr. Pathol.* **56**, 91–97.

Hajek, A. E., Butt, T. M., Strelow, L. I. & Gray, S. M. (1991) Detection of *Entomophaga maimaiga* (Zygomycetes: Entomophthorales) using enzyme-linked immunosorbent assay. *J. Invertebr. Pathol.* **58**, 1–9.

Hajek, A. E., Butler, L., Walsh, S. R. A., Silver, J. C., Hain, P. T., Hastings, F. L., ODell, T. M. & Smitley, D. R. (1996) Host range of the gypsy moth (Lepidoptera: Lymantriidae) pathogen *Entomophaga maimaiga* (Zygomycetes: Entomophthorales) in the field versus laboratory. *Environ. Entomol.* **25**, 709–721.

Holdom, D. G. (1983) *In vitro* culture of the aphid pathogenic fungus, *Entomophthora planchoniana* (Zygomycetes: Entomophthorales). *J. Aust. Entomol. Soc.* **22**, 188.

Huber, J. & Hughes, P. R. (1984) Quantitative bioassay in insect pathology. *Bull. Entomol. Soc. Am.* **30**, 31–34.

Humber, R. A. (1976) The systematics of the genus *Strongwellsea* (Zygomycetes: Entomophthorales). *Mycologia* **68**, 1042–1060.

Humber, R. A. (1989) Synopsis of a revised classification for the Entomophthorales (Zygomycotina). *Mycotaxon* **34**, 441–460.

Humber, R. A. & Ramoska, W. A. (1986) Variations in entomophthoralean life cycles: Practical implications. In *Fundamental and applied aspects of invertebrate pathology* (eds R. A. Samson, J. M. Vlak & D.

Peters), pp. 190–193. Found. 4th International Colloqium Invertebrate Patholology, Wageningen.

Keller, S. (1987) Arthropod-pathogenic Entomophthorales of Switzerland. I. *Conidiobolus, Entomophaga* and *Entomophthora. Sydowia* **40**, 122–167.

Keller, S. (1991) Arthropod-pathogenic Entomophthorales of Switzerland. II. *Erynia, Eryniopsis, Neozygites, Zoophthora and Tarichium. Sydowia* **43**, 39–122.

Keller, S. (1994) Working with arthropod-pathogenic Entomophthorales. *Bull. OILB/SROP* **17**, 287–307.

Krejzová, R. (1971) Versuchsinfektionen der Raupen von *Galleria mellonella* L. und *Antheraea pernyi* L. durch Vertreter der *Entomophthora*-Gattung. *I. Vest. Cs. Spol. Zool.* **35**, 107–113.

Latgé, J.-P. (1975) Croissance et sporulation de 6 espèces d'Entomophthorales. 1. Influence de la nutrition carbonée. *Entomophaga* **20**, 201–207.

Latgé, J.-P. (1980) Sporulation d'*Entomophthora obscura* Hall & Dunn en culture liquide. *Can J. Microbiol.* **26**, 1038–1048.

Latgé, J.-P. (1981) Comparaison des exigences nutritionnelles des Entomophthorales. *Ann. Microbiol. (Inst. Pasteur)* **132B**, 299–306.

Latgé, J.-P. (1982) Production of Entomophthorales. *Proc. IIIrd Int. Coll. Invertebrate Pathol.*, Brighton, pp. 164–169.

Latgé, J.-P. & Beauvais, A. (1987) Wall composition of the protoplastic Entomophthorales. *J. Invertebr. Pathol.* **50**, 53–57.

Latgé, J.-P. & Papierok, B. (1988) Aphid pathogens. In *Aphids. Their biology, natural enemies and control* (eds A. K. Minks & P. Harrewijn), vol. 2B, pp. 323–335. Elsevier, Amsterdam.

Latgé, J.-P. & Perry, D. (1980) The utilization of an *Entomophthora obscura* resting spore preparation in biological control experiments against cereal aphids. *Bull. OILB/SROP III* **4**, 19–25.

Latgé, J.-P. & Remaudière, G. (1975) Croissance et sporulation de 6 espèces d'Entomophthorales. III. Influence des concentrations de carbone et d'azote et du rapport C/N. *Rev. Mycol.* **39**, 239–250.

Latgé, J.-P. & Sanglier, J. J. (1985) Optimisation de la croissance et de la sporulation de *Conidiobolus obscurus* en milieu défini. *Can. J. Bot.* **63**, 68–85.

Latgé, J.-P., Soper, R. S. & Madore, C. D. (1977) Media suitable for industrial production of *Entomophthora virulenta* zygospores. *Biotechnol. Bioengineer.* **19**, 1269–1284.

Latgé, J.-P., Remaudière, G. & Papierok, B. (1978) Un exemple de recherche en lutte biologique: les Champignons *Entomophthora* pathogènes de Pucerons. *Bull. Soc. Pathol. Exot.* **71**, 196–203.

Latteur, G. (1977) Sur la possibilité d'infection directe d'aphides par *Entomophthora* à partir de sols hébergeant un inoculum naturel. *C.R. Acad. Sci. Paris* 284, série D, 2253–2256.

Li, Z., Soper, R. S. & Hajek, A. E. (1988) A method for recovering resting spores of Entomophthorales (Zygomycetes) from soil. *J. Invertebr. Pathol.* **52**, 18–26.

McCabe, D. E. & Soper, R. S. (1982) Preparation of an entomopathogenic fungal insect control agent. US Patent Pending no. 06/419637.

MacDonald, R. M. & Spokes, J. R. (1981) *Conidiobolus obscurus* in arable soil: A method for extracting and counting azygospores. *Soil Biol. Biochem.* **13**, 551–553.

MacLeod, D. M., Tyrrell, D. & Welton, M. A. (1980) Isolation and growth of the grasshopper pathogen *Entomophthora grylli. J. Invertebr. Pathol.* **36**, 85–89.

Matanmi, B. A. & Libby, J. L. (1976) The production and germination of resting spores of *Entomophthora virulenta* (Entomophthorales, Entomophthoraceae). *J. Invertebr. Pathol.* **27**, 279–285.

Milner, R. J. (1981) Patterns of primary spore discharge of *Entomophthora* spp. from the blue green aphid, *Acyrthosiphon kondoi. J. Invertebr. Pathol.* **38**, 419–425.

Milner, R. J. & Soper, R. S. (1981) Bioassay of *Entomophthora* against the spotted alfalfa aphid *Therioaphis trifolii* f. *maculata. J. Invertebr. Pathol.* **37**, 168–173.

Mitsuhashi, J. (1982) Media for insect cell cultures. In *Advances in cell cultures* (ed. K. Maramorosch), vol. 2, pp. 133–196. Academic Press, New York.

Mullens, B. A. & Rodriguez, J. L. (1985) Dynamics of *Entomophthora muscae* (Entomophthorales: Entomophthoraceae) conidial discharge from *Musca domestica* (Diptera: Muscidae) cadavers. *Environ. Entomol.* **14**, 317–322.

Müller-Kögler, E. (1959) Zur Isolierung und Kultur insektenpathogener Entomophthoraceen. *Entomophaga* **4**, 261–267.

Nolan, R. A. (1988) A simplified, defined medium for growth of *Entomophaga aulicae* protoplasts. *Can. J. Microbiol.* **34**, 45–51.

Nolan, R. A. (1990) Enhanced hyphal body production by *Entomophaga aulicae* protoplasts in the presence of a neutral and a positively charged surface under mass fermentation conditions. *Can. J. Bot.* **68**, 2708–2713.

Nolan, R. A. (1993) An inexpensive medium for mass fermentation production of *Entomophaga aulicae* hyphal bodies competent to form conidia. *Can. J. Microbiol.* **39**, 588–593.

Oger, R. & Latteur, G. (1985) Description et précision d'une nouvelle méthode d'estimation de la virulence d'une Entomophthorale pathogène de pucerons. *Parasitica* **41**, 135–150.

Papierok, B. (1978) Obtention *in vivo* des azygospores d'*Entomophthora thaxteriana* Petch, champignon pathogène de pucerons (Homoptères, Aphididae). *C.R. Acad. Sci. Paris* **286**, série D, 1503–1506.

Papierok, B. (1985) Données écologiques et expérimentales sur les potentialités entomopathogènes de l'Entomophthorale *Conidiobolus coronatus* (Costantin) Batko. *Entomophaga* **30**, 303–312.

Papierok, B. (1989) On the occurrence of Entomophthorales in Finland. I. Species attacking aphids (Homoptera, Aphididae). *Ann. Entomol. Fennici* **55**, 63–69.

Papierok, B. & Charpentié, M. J. (1982) Les champignons se développant en Côte-d'Ivoire sur la fourmi *Paltothyreus tarsatus* F. Relation entre l'Hyphomycète *Tilachlidiopsis catenulata* sp. nov. et l'Ascomycète *Cordyceps myrmecophila* Cesati 1846. *Mycotaxon* **14**, 351–368.

Papierok, B. & Wilding, N. (1979) Mise en évidence d'une différence de sensibilité entre 2 clones du puceron du pois *Acyrthosiphon pisum* Harr. (Homoptères: Aphididae), exposés à 2 souches du champignon Phycomycète *Entomophthora obscura* Hall & Dunn. *C.R. Séances Acad. Sci. Paris* **288**, série D, 93–95.

Papierok, B. & Wilding, N. (1981) Etude du comportement de plusieurs souches de *Conidiobolus obscurus* (Zygomycètes, Entomophthoraceae) vis-à-vis des pucerons *Acyrthosiphon pisum* et *Sitobion avenae* (Hom. Aphididae). *Entomophaga* **26**, 241–249.

Papierok, B., Valadão L., Tôrres, B. & Arnault, M. (1984) Contribution à l'étude de la spécificité parasitaire du champignon entomopathogène *Zoophthora radicans* (Zygomycètes, Entomophthorales). *Entomophaga* **29**, 109–119.

Perry, D. F. & Fleming, R. A. (1989) *Erynia crustosa* zygospore germination. *Mycologia* **81**, 154–158.

Perry, D. F. & Latgé, J.-P. (1980) Chemically defined media for growth and sporulation of *Entomophthora virulenta*. *J. Invertebr. Pathol.* **35**, 43–48.

Perry, D. F. & Latgé, J.-P. (1982) Dormancy and germination of *Conidiobolus obscurus* azygospores. *Trans. Br. Mycol. Soc.* **78**, 221–225.

Perry, D. F., Tyrrell, D. & Delyzer, A. J. (1982) The mode of germination of *Zoophthora radicans* zygospores. *Mycologia* **74**, 549–555.

Ramoska, W. A., Hajek, A. E., Ramos, M. E. & Soper, R. S. (1988) Infection of grasshoppers (Orthoptera: Acrididae) by members of the *Entomophaga grylli* species complex (Zygomycetes: Entomophthorales). *J. Invertebr. Pathol.* **52**, 309–313.

Remaudière, G., Keller, S., Papierok, B. & Latgé, J.-P. (1976a) Considérations systématiques et biologiques sur quelques espèces d'*Entomophthora* du groupe *sphaerosperma* pathogènes d'insectes (Phycomycètes, Entomophthoraceae). *Entomophaga* **21**, 163–177.

Remaudière, G., Latgé, J.-P., Papierok, B. & Coremans-Pelseneer, J. (1976b) Sur le pouvoir pathogène de quatre espèces d'Entomophthorales occasionnellement isolées d'aphides en France. *C.R. Acad. Sci. Paris 283, série D*, 1065–1068.

Rockwood, L. P. (1950) Entomogenous fungi of the family Entomophthoraceae in the Pacific Northwest. *J. Econ. Entomol.* **43**, 704– 707.

Shimazu, M. & Soper, R. S. (1986) Pathogenicity and sporulation of *Entomophaga maimaiga* Humber, Shimazu, Soper and Hajek (Entomophthorales: Entomophthoraceae) on larvae of the gypsy moth, *Lymantria dispar* L. (Lepidoptera: Lymantriidae). *Appl. Entomol. Zool.* **21**, 589–596.

Silvie, P. & Papierok, B. (1991) Les ennemis naturels d'insectes du cotonnier au Tchad. Premières données sur les champignons de l'ordre des Entomophthorales. *Coton Fibres Trop.* **46**, 293–308.

Soper, R. S. (1974) The genus *Massospora*, entomopathogenic for cicadas, part I, taxonomy of the genus. *Mycotaxon* **1**, 13–40.

Soper, R. S. (1981) New cicada pathogens: *Massospora cicadettae* from Australia and *Massospora pahariae* from Afghanistan. *Mycotaxon* **13**, 50–58.

Soper, R. S. (1985) *Erynia radicans* as a mycoinsecticide for spruce budworm control. *Proc. Symp. Microbial Control of Spruce Budworms and Gypsy Moths. USDA, Forest Service GTR-NE-100*, 69–76.

Soper, R. S., Hollbrook, F. R., Majchrowicz, I. & Gordon, C. C. (1975) Production of *Entomophthora* resting spores for biological control. *Maine Life Sci. Agric. Exp. Sta., Tech. Bull.* **76**, 15 pp.

Soper, R. S., Shimazu, M., Humber, R. A., Ramos, M. E. & Hajek, A. E. (1988) Isolation and characterization of *Entomophaga maimaiga* sp. nov., a fungal pathogen of gypsy moth, *Lymantria dispar*, from Japan. *J. Invertebr. Pathol.* **51**, 229–241.

Tyrrell, D. & MacLeod, D. M. (1972) Spontaneous formation of protoplasts by a species of *Entomophthora*. *J. Invertebr. Pathol.* **19**, 354–360.

Tyrrell, D. & MacLeod, D. M. (1975) *In vitro* germination of *Entomophthora aphidis* resting spores. *Can. J. Bot.* **53**, 1188–1191.

Ullyett, G. C. & Schonken, D. B. (1940) A fungus disease of *Plutella maculipennis* Curt. in South Africa, with notes on the use of entomogenous fungi in insect control. *Union Sth Africa Dept. Agr. Forestry Sci. Bull.* **218**, 1–24.

Valovage, W. D., Nelson, D. R. & Frye, R. D. (1984) Infection of grasshoppers with *Entomophaga grylli* by exposure to resting spores and germ conidia. *J. Invertebr. Pathol.* **43**, 274–275.

Van Beek, N. A. M., Wood, H. A. & Hughes, P. R. (1988) Quantitative aspects of nuclear polyhedrosis virus infections in lepidopterous larvae: The dose-survival time relationship. *J. Invertebr. Pathol.* **51**, 58–63.

Vandenberg, J. S. & Soper, R. S. (1979) A bioassay technique for *Entomophthora sphaerosperma* on the spruce budworm, *Choristoneura fumiferana*. *J. Invertebr. Pathol.* **33**, 148–154.

Van Driesche, R. G. & T. S. Bellows, Jr (1996) *Biological Control*. Chapman & Hall, New York, 539 pp.

Wilding, N. (1971) Discharge of conidia of *Entomophthora thaxteriana* Petch from the Pea aphid *Acyrthosiphon pisum* Harris. *J. Gen. Microbiol.* **69**, 417–422.

Wilding, N. (1976) Determination of the infectivity of

Entomophthora spp. *Proc. Ist. Int. Coll. Invertebr. Pathol.*, Kingston, pp. 296–300.

Zimmermann, G. (1986) The '*Galleria* bait method' for detection of entomopathogenic fungi in soil. *J. Appl. Entomol.* **102**, 213–215.

APPENDIX: MEDIA

Sabouraud dextrose agar supplemented with egg yolk and milk

80% Sabouraud dextrose agar
20% of a mixture of egg yolk (40%) and milk (60%)

The respective amounts of each ingredient are calculated according to the final volume of medium wanted. The volume of a medium-sized egg yolk is about 15–18 ml. The different ingredients are prepared separately as follows (Figure 4):

- for Sabouraud agar: 'Gelose de Sabouraud' (Diagnostics Pasteur) (45 g/l), is placed in a flask with distilled water and sterilized at 120°C for 30 min, and then maintained at 50–60°C;
- for dextrose–yeast extract–agar: dextrose (20 g/l), yeast extract (10 g/l), agar (20 g/l), and distilled water are placed in a flask and sterilized at 120°C for 30 min, and then kept at 50–60°C. It is most efficient to sterilize the above two flasks simultaneously;
- for milk: sterilize whole milk at 120°C for 30 min, and keep at room temperature;
- for egg yolk: place fresh eggs in a mixture of 90% alcohol (200 ml) and 2% sodium hypochlorite (800 ml) for surface sterilization. Break egg shells with forceps near the flame of a Bunsen burner. Gently remove the egg yolk, and pour it into a sterile graduated cylinder;
- add milk to egg yolks in the graduated cylinder, and stir the contents with a sterile glass rod, until the mixture is homogeneous;
- pour the egg yolk–milk mixture into the flask containing dextrose–yeast extract-agar, and shake;
- pour the accumulated mixture into the flask containing Sabouraud agar, and shake;

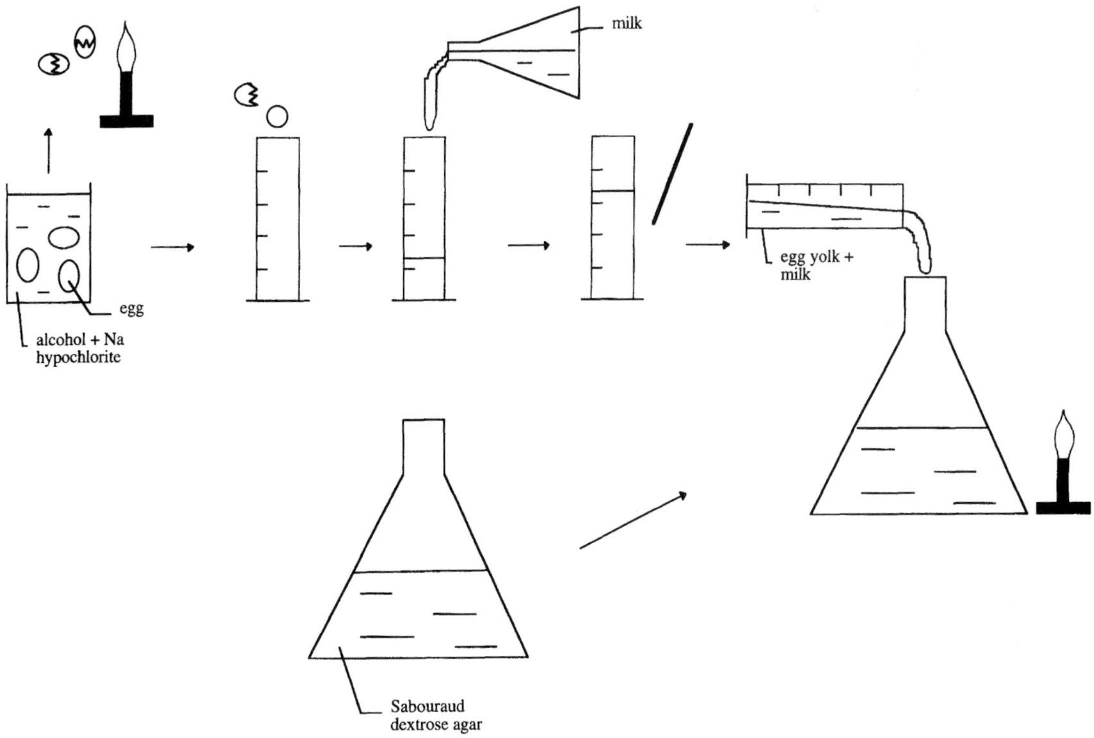

Figure 4 Procedure for supplementing Sabouraud dextrose agar with egg yolk and milk.

- immediately pour the medium into sterile Petri dishes or tubes. An automatic dispenser, such as a multiple pipetter (e.g. Multipette, Eppendorf) fitted with a 50 ml sterile reservoir, (e.g. Combitips, Eppendorf) is efficient;
- if tubes are being used, immediately push a cotton plug in after filling, and lightly tilt until the medium hardens.

Coagulated egg yolk and milk

60% egg yolk.
40% whole milk.

Because this medium is harder than the previous one, it is quite useful for isolation by rubbing showered conidia off of the surfaces of Petri dishes (see 'Isolation using conidia'). For this reason, and for work away from a laboratory, this medium is frequently prepared in small tubes (60 × 12 mm). The procedure is as follows:

- prepare the two ingredients separately, and then mix them together as explained above for the preparation of Sabouraud–dextrose–agar supplemented with egg yolk and milk;
- distribute the medium in small tubes, each tube being plugged with cotton and then lightly tilted immediately after filling;
- sterilize in the oven at 100 °C for 30 min;
- let rest at room temperature for 24 h;
- replace cotton plugs with rubber ones, taking extreme care to plug securely.

These tubes can be stored for several months or even a few years, as long as they are kept in the dark. After inoculation, tubes must be plugged with cotton.

Egg yolk/sabouraud maltose agar (EYSMA)

24 g maltose
6 g peptone
6 g yeast extract
9 g agar
12 eggs
600 ml distilled water

- To each of six 500 ml Fleakers® (18 cm tall beakers that are convenient to use; Corning Inc.) add 100 ml distilled water;
- To each of three of the Fleakers, add 8 g maltose and to each of the remaining three Fleakers, add 2 g peptone, 2 g yeast extract, and 3 g agar;
- Cover and autoclave all Fleakers; for 15 min
- Soak 12 hen's eggs in 50% ethanol for approximately 30 min;
- In a sterile hood, add two broken egg yolks to each Fleaker. After egg yolks are added, procedures must be undertaken quickly due to rapid coagulation as ingredients cool. Therefore, it is easiest to add egg yolks to only one pair of Fleakers at a time;
- Thoroughly mix the contents of pairs of Fleakers containing different ingredients by pouring back and forth, and dispense into Petri dishes.

Holdom's protoplast medium

2% neopeptone (Difco)
5% dextrose
10% fetal calf serum
25 units/ml penicillin G
50 μg/ml streptomycin sulphate

Neopeptone and dextrose are autoclaved separately at 109 °C for 25 min, and supplements, i.e. fetal calf serum plus 25 units/ml penicillin G plus 50 μg/ml streptomycin sulphate, are added immediately before use.

Beauvais and Latgé protoplast medium

0.4% dextrose
0.5% yeast extract (Difco)
0.65% lactalbumin-hydrolysate (Difco)
0.77% NaCl
10% fetal calf serum

Technical considerations are as follows: osmotic pressure 400 ± 20 mOsm, pH 6.5, autoclaved 30 min at 115 °C, distributed into 35 mm diam. Petri dishes. Lactalbumin hydrolysate can be replaced v/v with yeast extract.

Liquid medium for laboratory-scale hyphal body production

40 g/l dextrose
10 g/l yeast extract (Difco)

[optional] 1% wetting agent (e.g. Tween 40 or 80, Rhodorsil® 426 R, Prolabo).

CHAPTER V-3

Fungi: Hyphomycetes

MARK S. GOETTEL AND G. DOUGLAS INGLIS

Agriculture Canada Research Centre, PO Box Main, Lethbridge, Alberta T1J 4B1, Canada.

1 INTRODUCTION

Hyphomycetes are filamentous fungi that reproduce by conidia generally formed aerially on conidiophores arising from the substrate. Many genera of entomogenous fungi occur in the Hyphomycetes and are described in form genera on the basis of their morphological features. Therefore phylogenetic relationships cannot be implied based on these genera. The most common route of host invasion is through the external integument, although infection through the digestive tract is possible. Conidia attach to the cuticle, germinate, and penetrate the cuticle. Once in the haemocoel, the mycelium ramifies throughout the host, forming yeast-like hyphal bodies often referred to as blastospores. Host death is often due to a combination of the action of a fungal toxin, physical obstruction of blood circulation, nutrient depletion and invasion of organs. After host death, hyphae usually emerge from the cadaver and, under appropriate conditions of temperature and humidity, produce conidia on the exterior of the host. Conidia are then dispersed by wind or water. Under dry or cool conditions, insects often remain intact as fungus-filled 'mummies', the fungus only emerges and sporulates once the cadaver is brought under appropriate conditions for fungal growth and conidiation. Consequently, entomopathogenic Hyphomycetes are among the most commonly encountered insect pathogens, as cadavers often remain intact for long periods and the external mycelium is conspicuous (see Chapter I).

Most entomopathogenic Hyphomycetes are facultative pathogens and are relatively easily grown in pure culture on defined or semi-defined media. However, many non-pathogenic micro-organisms, including Hyphomycetes, quickly colonize insect cadavers, especially if they are not already colonized by a pathogen. Consequently, it is often a challenge to determine if a hyphomycete isolated from an insect cadaver was responsible for the insect's death.

Entomopathogenic Hyphomycetes have one of the widest spectra of host ranges among entomopathogens. A variety of factors may determine or influence the susceptibility of a host to infection by a

MANUAL OF TECHNIQUES IN INSECT PATHOLOGY
ISBN 0–12–432555–6

fungal pathogen. These include the fungal strain, the host's physiological state, nutrition, defence mechanisms, cuticular and epicuticular micro-organisms as well as a number of other diverse factors such as the environment. Prediction of the ecological host range based on laboratory bioassay results remains a challenge, as in most cases, insects are much more susceptible to infection under laboratory or caged situations than they are in their field environment. To be useful in predicting field virulence, laboratory bioassays must attempt to mimic those conditions most likely to be expected in nature.

This chapter provides techniques for the study of entomopathogenic Hyphomycete pathogens. Due to the diversity of entomopathogenic fungi and the very wide range of hosts that they infect, we provide information both in the form of generalizations and via specific examples. In many cases, readers will need to adapt these techniques to specific requirements.

The reader is referred to Feng *et al.* (1994), Hajek & St Leger (1993) and Tanada & Kaya (1993) for more recent reviews on entomopathogenic Hyphomycete biology, epizootiology, pathogenesis and development as microbial control agents.

2 ISOLATION AND IDENTIFICATION

Entomopathogenic Hyphomycetes can most often be isolated from insect cadavers or from soil. However, methods that differentially isolate entomopathogenic fungi may be required when isolating from soil or when contaminants are a problem.

A Isolation from cadavers

Entomopathogenic Hyphomycetes may be harvested directly from insect cadavers on which the fungus has already sporulated. If external sporulation has not occurred, cadavers may be placed in suitable environments and the fungus allowed to produce hyphae or conidia externally. Alternately, cadavers may be homogenized and the homogenate plated on an appropriate agar medium. In cases where isolation is difficult, it may be necessary to experimentally infect more insects, before finally obtaining a pure culture (see Section 5).

In many instances, it is desirable to surface disinfest cadavers to remove potential contaminants on their integument. However, this is possible only if the fungus has not yet emerged and sporulated on the host surface. The most common 'disinfectants' are sodium hypochlorite (1–5%) and/or ethanol (70%) (Chapter I). The efficacy of disinfestation can be confirmed by swirling the treated cadaver in a nutrient broth (with or without antibacterial agents) and checking for microbial growth after 1 to 3 days. This method may yield unsatisfactory results if the integrity of the cadaver integument is poor.

If the identity of the fungus is known, it is preferable to use a selective medium developed for the particular fungus in question (Section 2B). However, if no selective medium exists or if the identity of the fungus is unknown, virtually any medium used for propagation of entomopathogenic Hyphomycetes can be used (see Section 3). We routinely use Sabouraud dextrose agar with yeast extract (SDAY) (Appendix 2) supplemented with streptomycin sulphate (0.08%) and penicillin (0.03%).

It may be necessary to first induce external growth of entomopathogenic Hyphomycetes by placing the cadaver in an environment with high relative humidity (e.g. on a water agar medium amended with antibacterial agents, on moistened filter paper, or adjacent to moistened cotton batten in a sealed sterile container such as a Petri dish sealed with parafilm or plastic film). Once insects that have died of mycosis are placed under conditions of high humidity vegetative growth and/or sporulation on the surface of the integument usually occurs within a few days at 20–25 °C. Once adequate growth is observed on cadavers, the fungus can be transferred to an appropriate agar medium with a needle. If prolific conidiogenesis is observed on the cadaver, conidia can be streaked on to an agar medium directly or conidia can be suspended in buffer or water before being streaked. Addition of antibiotic such as chloromycetin (50 μg ml^{-1}) to the buffer or water may be useful. It may also be possible to obtain relatively pure colonies of entomopathogenic Hyphomycetes by attaching the cadaver to the lid of the Petri dish with an adhesive such as double-sided sticky tape. Conidia released from the cadaver which land on the medium may produce mycelial colonies that are relatively free of contaminants.

Another isolation method requires the homogenization of cadavers followed by dilution plating of the homogenate on an appropriate selective medium. The surface-disinfested cadaver can be homogenized using a variety of methods. Potter–Elvehjem tubes may be used for larger insects and micropestles in 1.5 ml microcentrifuge tubes for smaller insects. Polytron tissue grinders and blenders may also be used to homogenize insects, particularly those with hard exoskeletons, but it is more difficult to maintain a sterile environment using this equipment.

B Selective media

Selective media are frequently used for the isolation of entomopathogenic Hyphomycetes. Most bacteria and Actinomycetes are inhibited by low pH and, in general, fungal growth will be favoured on media where the pH is less than 5. In most instances, however, inhibition of bacteria is achieved by amending media with antibacterial agents. Frequently used wide-spectrum antibacterial agents include chloramphenicol, tetracycline and streptomycin. These agents may be used alone or in combination with agents with a more specific mode of action (e.g. penicillin, novobiocin and chloramphenicol). Crystal violet is commonly used as a colouring agent to provide contrast for visualizing light-coloured colonies, and it also inhibits the growth of Gram-positive bacteria. Antibiotics must be used with caution as they may be inhibitory to some species or strains of fungi.

Inhibition of contaminant fungi is more problematic than bacteria, and fungal contaminants are invariably a problem when attempting to isolate entomopathogenic Hyphomycetes from soil. Fungi in the genera *Trichoderma*, *Mucor* and *Rhizopus* are fast growing and can rapidly (*ca* 2–3 days) obscure colonies of entomopathogenic Hyphomycetes making isolation difficult or impossible. Species of *Penicillium* and *Aspergillus* can also be problematic because they are prolific sporulators and are common in soil. Undesirable fungi can be inhibited by amending media with fungicides but other materials such as Rose Bengal, oxgall, *O*-phenyl-phenol, and/or sodium desoxycholate have also been used. In addition to facilitating isolation of entomopathogenic Hyphomycetes, use of semi-selective media has advanced our knowledge of host targeting, population dynamics and their saprotrophic behaviour.

1. Beauveria *and* Metarhizium *spp.*

Media based on the differential activity of the fungicide, dodine (*N*-dodecylguanidine monoacetate) have been successfully used to isolate *Beauveria* and *Metarhizium* spp. from soil and insect cadavers. An oatmeal agar medium amended with 650 μg/ml dodine suppressed growth of *Penicillium* spp., *Trichoderma viride* and Mucorales species but supported growth of *B. bassiana* and *Metarhizium* spp. (Beilhartz *et al.*, 1982). Subsequently, Chase *et al.* (1986) confirmed that an oatmeal medium amended with 600 μg/ml of dodine resulted in good isolation of *B. bassiana*; crystal violet was added to the medium to enhance the contrast with fungal colonies (Appendix 1). However, this concentration of dodine was inhibitory to *M. anisopliae*. When the concentration was reduced to 500 μg/ml and 400 μg/ml of the fungicide benomyl (Benlate) was added to the medium, both *B. bassiana* and *M. anisopliae* were effectively isolated from soil (Chase *et al.*, 1986). Liu *et al.* (1993) observed that even 300 μg/ml of dodine inhibited isolation of some isolates of *M. anisopliae*. They found that a reduced concentration of dodine (10 μg/ml) in combination with 500 μg/ml of cycloheximide increased the efficacy of recovery of *Metarhizium* from soil (Appendix 1).

2. Culicinomyces clavisporus

Panter and Frances (unpublished) found that low concentrations of thiabendazole (2 μg/ml) inhibited growth of *Cladosporium*, *Penicillium* and *Mucor* species but permitted growth of *C. clavisporus* (Appendix 1).

3. Paecilomyces lilacinus

Paecilomyces lilacinus is more tolerant of high salt concentrations than many other fungi. The amendment of an agar medium with sodium chloride, pentachloronitrobenzene, benomyl and Tergitol was found to facilitate recovery of *P. lilacinus* from soil by inhibiting growth of contaminant fungi such as *Rhizopus* and *Trichoderma* (Mitchell *et al.*, 1987; Appendix 1).

4. Verticillium lecanii

A number of selective media have been devised for plant pathogenic taxa of *Verticillium* (*V. albo-atrum* and *V. dahliae*) (Tsao, 1970). The efficacy of these media for *V. lecanii* have yet to be established.

C Isolation from soil

Entomopathogenic fungi are generally considered poor competitors and are usually heterogeneously distributed in soil, putatively in or near insect cadavers. Pooling soil samples usually increases the frequency of isolation. Soil can be collected with any number of tools but soil corers are most frequently used since a specified volume can be removed. The depth is usually limited to the top 10–15 cm of the organic and/or A horizon. The collection tool should be surface-disinfested between samples to avoid cross-contamination. Upon collection, the soil samples are usually placed in a cool environment (approximately 5 °C). Although entomopathogenic fungi may remain viable in soil for relatively long periods, it is recommended that samples be processed as quickly as possible, usually within 5 days of collection.

The methods described below favour the isolation of propagules. To isolate fungi present in soil as actively growing or dormant hyphae, a number of methods have been devised (e.g. immersion methods, direct hyphal isolation, and washing methods). For a summary of these, see Parkinson (1994).

1. Soil dilution plating

The direct isolation of entomopathogenic fungi from soil usually relies on the use of a semi-selective medium. The most popular technique is the soil dilution plate method (see Section 4, for a description of the quantitative isolation methods). A commonly used procedure is to place 10 g of soil into 90 ml of sterile water or buffer (usually ≈ pH 6–7). The sample is then homogenized to release propagules from the soil matrix. The preferred technique is to use a commercial blender but stirring the slurry with a magnetic stir bar or on a mechanical shaker for 20–60 min may be sufficient.

After homogenization, aliquots of 100–200 μl are spread on to an appropriate medium using an L-shaped rod. In cases where the density of entomopathogenic fungi in soil is high, it may be necessary to dilute the homogenate further. However, densities of entomopathogenic fungi in soil are usually low, and larger volumes (≈ 1 ml) of the original sample may be spread on to the agar medium using an inclined rotary motion of the plate; an agar concentration of 2–3% may facilitate absorption of the solution into the medium. When propagules are associated with soil particles, it may be necessary to increase the viscosity of the dilution solution. Increasing the viscosity will prolong the sedimentation of particles which will reduce the variation in the number of particles transferred. A number of materials may be used, but the most common and easy to prepare is a low concentration of agar (≤0.2%).

Once the homogenate is spread on the agar medium, cultures are incubated at an appropriate temperature (20–25 °C for most taxa) for 3–7 days. Individual colonies can then be transferred to a suitable nutrient medium. For prolific sporulators, if possible, make the transfers before sporulation occurs.

2. Soil plating

In some instances, it may be desirable to plate soil directly which is faster than soil dilution plating. The simplest method is to sprinkle soil particles on to the surface of an agar medium and allow fungal hyphae to radiate out from the particles. Unless a selective medium is used, this usually provides unsatisfactory isolation of entomopathogenic fungi due to overgrowth of contaminant fungi.

A better strategy is to embed the particles in an agar medium (Warcup method). For this technique approximately 5–15 g of soil is placed in a sterile Petri dish. The soil may be sieved prior to its placement in the dish or the large soil aggregates can be disrupted once in the dish. Approximately 15 ml of an appropriate molten (≤50 °C) agar medium is added to the dish and the soil particles dispersed using a swirling motion. After incubation of the cultures (usually at 20–25 °C), isolates can be transferred to suitable medium. Disadvantages are that colonies embedded in the medium may be difficult to retrieve, and it is relatively arduous to dilute the original sample with sterile soil or sand if densities of colony forming units (cfu) in the soil are high.

3. Insect baiting

Insect baits may be used to indirectly isolate fungi from soil. In general, entomopathogenic Hyphomycetes are considered to be weak saprotrophs but since they possess the ability to infect living insects, they can gain access to a living insect relatively free of competitors. Although larvae of the greater wax moth (*Galleria mellonella*) are most commonly used, larvae of other insects such as the large flour beetle (*Tribolium destructor*) and the pine bark beetle (*Acanthocinus aedilis*) may also be used (Zimmermann, 1986). Soil samples are placed in containers, the soil is moistened, larvae are added to the soil and incubated for approximately 14 days in conditions that favour development of the target fungus. Although the quantity of soil may limit the larval density, in general, 5–15 larvae are used. The soil is agitated, or the containers are repositioned periodically to ensure that the larvae remain exposed to the soil. Cadavers are collected at intervals and processed for internal fungi (see Section 2A). Non-soil dwelling insects (i.e. *G. mellonella*) are very susceptible to infection by entomopathogens in soil which increases the sensitivity of this method.

D Purification

After isolation, it is usually necessary to ensure that the isolated fungus is free from contaminant microorganisms (excluding mycoviruses) and, if desired, that it represents a single genotype. Most entomopathogenic Hyphomycetes are relatively prolific sporulators and can be satisfactorily streaked on an agar medium containing an antibacterial agent(s) (see Section 2B). Individual colonies free of bacteria can then be collected.

Colony-forming units (cfu) do not necessarily originate from a single propagule and therefore additional steps may be desirable to ensure that a cfu represents a single genotype. The two most common methods used to achieve genotype purity are isolation of individual conidia or the isolation of hyphal tips. Entomopathogenic Hyphomycetes are heterokaryons, and heterokaryotic mycelium may give rise to uninucleate or homokaryotic conidia, or to homokaryotic hyphal tips (Webster, 1986). Therefore, both of the above methods do not necessarily ensure a unique karyotype.

Although a micromanipulator can be used, single conidia or germlings of entomopathogenic Hyphomycetes can be isolated without the aid of a manipulator. Conidia of entomopathogenic Hyphomycetes are generally too small to isolate with a dissecting microscope so a compound light microscope should be used. An inverted compound microscope provides ample working area. A number of methods have been described for the isolation of fungal propagules using compound microscopes, but the light location method requires no attachments to an objective lens (Tuite, 1969). A suspension of conidia is spread on to a relatively transparent agar medium (12–15 ml per dish); care must be taken to ensure that the density of conidia is low enough to allow isolation of individual propagules. For the isolation of non-germinated conidia, the cultures should be examined soon after the carrier has absorbed into the medium. For germlings, the cultures should be maintained at 20–25 °C for *c.* 6–12 h before being examined. The Petri dish is placed on the stage and the culture is examined under bright field illumination. The 10× objective lens is preferable to larger magnification lenses since it provides the largest field with an adequate level of resolution. All other objectives should be removed to allow for more working room. Once a segregated conidium or germling has been located, the stage is adjusted so that the target propagule is in the centre of the field. The objective lens is swung to one side, the microscope diaphragm is closed until only the target area is illuminated, and a disk of agar (4–5 mm in diameter) is excised and aseptically transferred on to an agar medium in a Petri dish or slant tube.

Hyphal tip isolations are, in general, easier to conduct than single spore isolations, but it is not always possible to isolate a single hypha. Mycelium or a suspension of conidia is placed centrally on an agar medium. To promote diffuse hyphal growth which will facilitate isolation, the medium used should be relatively low in available nutrients. The margin of the colony should be examined with a dissecting microscope or compound light microscope, a section of agar containing a hyphal tip is removed using a sterile knife or needle and aseptically transferred on to an agar medium.

E Identification

The identification of the vast majority of Hyphomycetes is based on conidiogenesis. Therefore, successful identification of entomopathogenic Hyphomycetes relies on adequate observation of both conidia and conidiogenous cells. A number of methods are used to prepare Hyphomycetes for microscopy. Whole mount slide preparations (see Chapter V-1) can be quickly prepared, and, if done correctly, are very useful for identifying Hyphomycetes; whole mounts are the method of choice for phialidic taxa that are prolific sporulators, such as *Aspergillus* and *Penicillium* species. However, considerable disruption to conidiogenous cells and dehiscence of conidia may occur during this type of preparation. Therefore, we use slide culture preparations for more critical observations of conidiogenesis (Figure 1). Small blocks of an appropriate agar medium (approximately 1-cm square) are placed on a sterile glass coverslip (22 × 30 mm) situated on 2% water agar medium in a Petri dish (Harris, 1986). The agar block is then inoculated and a second sterile coverslip is then placed on top of the block. The lid of the Petri dish is replaced and the culture is incubated at an appropriate temperature. Vegetative growth may be visible on the surface of the coverslip(s) within a few days, but the culture must be incubated long enough to allow conidiogenesis. Sporulation can usually be observed using a dissecting microscope. The top coverslip and, if adequate, the bottom coverslip, are removed from the agar block and prepared for mounting; it is recommended that replicate cultures be established to allow for situations where the coverslip is removed before adequate sporulation had occurred.

A number of methods are available for preparing semi-permanent and permanent microscope slides (see Chapter V-1). Polyvinyl alcohol mounting medium (PVA; Appendix 3) is easy to prepare and use, has a long shelf life and provides high quality and stable slide preparations. To mount coverslips from slide cultures, the material is first wetted by flooding the coverslip with a PVA wetting agent (Appendix 3). Excess wetting agent is removed by blotting and the coverslip is then mounted in a drop of PVA. The PVA is allowed to harden in an oven or on a slide warmer at 40 °C for *c*. 24–36 h. Slide preparations should be examined under phase-contrast, but if bright field illumination is preferred, acid fuchsin can be added to the PVA solution (Appendix 3).

See Chapter V-1 for a dichotomous key for the identification of the genera of entomopathogenic fungi.

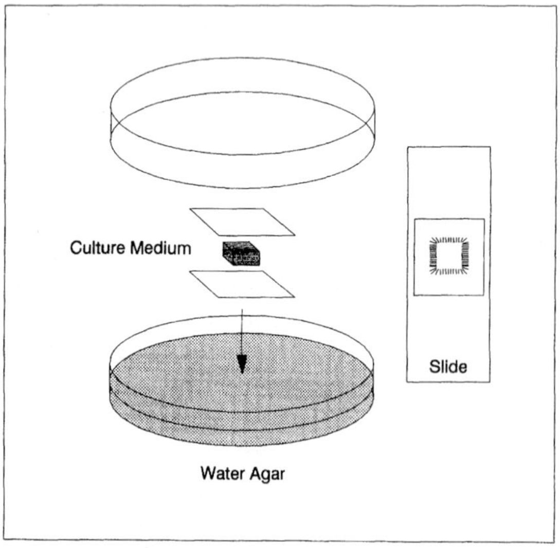

Figure 1. Slide culture. A small block of an appropriate agar medium is inoculated, sandwiched between two sterile cover slips, and incubated on 2% water agar in a Petri dish. Once the fungus has started to sporulate, the agar is removed, and the coverslips are mounted on a microscope slide.

3. PROPAGATION AND STORAGE

Detailed studies on fungal pathogenesis, virulence, physiology and evaluation as microbial control agents require methods for the propagation and storage of mycelia and infection propagules. Most entomopathogenic fungi are easily propagated on defined or semi-defined media containing suitable nitrogen and carbon sources. Conidia of most species also store well under refrigeration, although several simple precautions may extend their shelf life considerably. Mycelia and blastospores can also be refrigerated or dehydrated after treatment with a cryoprotectant.

Submerged culture is used for production of mycelia and blastospores. Some species of entomopathogenic Hyphomycetes may also produce conidia under submerged culture, but these conidia are usually short-lived. Conidia of entomopathogenic Hyphomycetes are most often produced aerially using surface culture techniques. In most cases, the choice of culture method will be dictated by the type of propagule desired.

Continuous serial subculturing on artificial media is to be avoided as this can result in a change in the phenotype (e.g. attenuation of virulence). In many cases, virulence can be restored after a single or several passages through an insect host. However, it is wise to preserve the isolate as soon as possible in order to preserve the original genotype. Methods for preservation of cultures are presented in Chapter V-5.

A Surface culture

Surface culture is used for the routine maintenance of isolates and for production of conidia. One of the most commonly used media by insect pathologists is Sabouraud's Dextrose agar supplemented with yeast extract (SDAY) (Appendix 2); however, many other media such as Cornmeal, Czapeck-Dox, malt extract, nutrient, potato dextrose agar (PDA), and Sabouraud's maltose agar are also often used. Some researchers advocate more complex media such as mixed cereal agar (Appendix 2) claiming that there is less chance of loss of vigour when such media are used. However, this has not been substantiated and this media has not been widely accepted in insect pathology, although most entomopathogenic Hyphomycetes including *Beauveria bassiana*, *Culicinomyces clavisporus*, *Metarhizium anisopliae*, *Tolypocladium cylindrosporum* and *Verticillium lecanii* will grow and sporulate well on this medium (Goettel, 1984).

At the laboratory level, surface culture is usually accomplished on agar media within glass bottles or disposable plastic Petri plates. The agar surface is inoculated under sterile conditions with a suspension of either conidia or blastospores; initial growth is faster when blastospores are used. The dishes are incubated normally at 25–27 °C under a variety of light conditions. To reduce dehydration of the media, Petri dishes can be sealed with Parafilm. Within 7–10 days, the cultures will normally have sporulated and the conidia are harvested either by direct scraping from the surface using a sterile rubber 'policeman' or by washing off with sterile distilled water.

For larger-scale production of conidia, cheaper nutritive substrates such as rice, bran or cereal grains are used. Moistened substrates are autoclaved in wide-mouthed jars, autoclavable plastic bags or tin trays. After cooling, the substrate is inoculated with a conidial or blastospore suspension and incubated at room temperature. The use of rice in polypropylene plastic bags is currently the most widely used method for production of *M. anisopliae* (Mendonça, 1992) and *Metarhizium flavoviride* (Jenkins & Goettel, 1997). Par-boiled autoclaved rice in the plastic bags is inoculated with a blastospore suspension. Sterile water is then added to give a 35–40% moisture content. After 10 days of fermentation, the bags are opened and contents are dried in a dehumidified room for at least 24 h. Conidia are then extracted by sifting.

Use of inert substrates as a base for the aerial growth and sporulation of a fungus enables the complete separation of the pathogen from the nutritive medium. Goettel (1984) used cellophane (i.e. cellulose film) as a semi-permeable membrane which allows harvest of the fungus virtually free of nutritive substrate contamination. Approximately 1 part bran (by wt.) and 10 parts distilled water (by vol.) are combined in autoclavable pans (e.g. tin cookware roasting pans). Sheets of presoaked cellophane are then layered over the bran mixture. Each pan is then placed into an autoclavable plastic bag and autoclaved for 1 h at 138 kPa (20 psi, approx. 125 °C). Extreme care is taken in cooling the autoclave as slowly as possible to prevent the bran mixture from boiling, thereby causing the cellophane to lift off the bran surface. The pans are allowed to cool and the surface of the cellophane is inoculated with the fungus using a hypodermic needle; inoculum is injected on to the surface of the cellophane in one corner of the pan and spread over the surface by moving the pan in a swirling motion. The amount of inoculum required must be adjusted according to the size of pans used. After incubation for 1–2 weeks, the pans are removed from the autoclave bags and the cellophane with adhering fungus is gently lifted off the bran surface. The fungal mat, which consists almost exclusively of conidia, is scraped off the cellophane surface using a spatula. Up to 2.3×10^8 conidia per cm^2 cellophane surface were obtained with several entomopathogenic Hyphomycetes after a 14-day incubation period at 20 °C (Goettel, 1984).

Inert substrates can also be used as a base for the aerial growth and sporulation of entomopathogenic Hyphomycetes. For instance, Bailey & Rath (1994) used a nutrient-impregnated membrane for production of *M. anisopliae* conidia. Strips of absorbent membrane (e.g. Superwipe, an absorbent fibrous material), are dipped into a conidial–nutrient solution (e.g. Ultra-High Temperature treated milk (1.4%), sucrose (0.2%) and 2.1×10^5 conidia/ml). The membranes are then incubated in bottles or steel boxes. Yields are in the order of 10^7 conidia/cm^2.

Aerial conidia can also be produced on the surface of still liquid medium. Kybal and Vlček (1976) used a polyethylene hose with a diameter of approximately 300 cm, sealed at both ends and partially filled with broth inoculated with the fungus. Sterile air is then pumped through the hose. While mycelial growth occurs within the liquid medium, sporulation occurs at the liquid–gas interface. After incubation, the liquid is drained and the conidia are harvested. Using this system, after 12 days incubation, Samšiňáková *et al.* (1981) obtained 10^9 conidia of *B. bassiana* cm^2 liquid surface (peptone (0.8%) and sorbitol (1%)).

B Submerged culture

Submerged culture is used for the production of mycelia, blastospores or, sometimes, conidia. The type of product obtained (i.e. mycelium, blastospore or conidium) can usually be controlled by selecting specific fungal strains and varying culture conditions. At the laboratory level, submerged culture is usually accomplished in flasks with magnetic stirrer bars, shake flasks on rotary or wrist shakers, or in larger plastic or glass containers as air-lift fermenters.

Although most entomopathogenic Hyphomycetes will produce blastospores in submerged culture, specific parameters need to be evaluated and adjusted for every strain studied for optimum blastospore production. For instance, Kleespies & Zimmermann (1992) found that each strain of *M. anisopliae* differed in its growth pattern and physiology according to the culture medium used. Some of the isolates formed discrete mycelial pellets and produced few blastospores, whereas other strains grew as mycelium and formed numerous blastospores. Production of blastospores was generally improved by the addition of 5% polyethylene 200 or up to 1.2% Tween 80 and by adjusting the carbon and nitrogen source and pH. Bidochka *et al.* (1987) obtained highest yields of blastospores of *B. bassiana* in a peptone (1%)–glucose (2%) medium, whereas more mycelial mass was produced in a glucose (2%)–peptone (1%)–yeast extract (0.2%) medium.

Under the proper conditions, certain species and/or strains produce conidia in submerged culture. For instance, out of 14 isolates of *Hirsutella thompsonii*, only one was found to be capable of producing submerged conidia, and only if corn steep liquor (1%) and Tween 80 (0.2%) were added to the culture medium (van Winkelhoff & McCoy, 1984). *M. flavoviride* can be induced to switch from vegetative growth to conidiation by varying the ratio of brewers' yeast to sucrose and quantity of yeast in the media; more than 1.5×10^9 conidia/ml were produced in a medium containing yeast (2–3%) and sucrose (2–3%) (Jenkins & Prior, 1993). Similarly, submerged conidia of a *B. bassiana* isolate were produced in a sucrose (2%)–yeast extract (0.5%)–basal salts medium (1.7×10^8 conidia/ml), whereas mostly blastospores were produced in a sucrose (2.5%)–yeast extract (2.5%) medium (7.4×10^8 blastospores/ml) (Rombach, 1989).

For production of larger quantities of biomass, air-lift fermentation in 10 or 20 l autoclavable plastic carboys is used (McCoy *et al.*, 1975; Pereira & Roberts, 1990). Autoclaved nutrients and an anti-foaming agent (V-30 (25 ppm), Dow Corning or vegetable oil (0.1%)) are added to the flasks in up to 1 l water. An antibiotic solution, fungal inoculum and filter-sterilized water (using an inline autoclavable filter with a 0.5 μm pore size) are then added until the desired volume is reached. Compressed filtered air (using the same inline filters described above) is injected into the flask through a glass pipette extending to the bottom and is allowed to escape through an opening at the top of the flask. A recent modification is to use a plastic tube that has its bottom plugged but which contains small perforations in the bottom 10 cm. The tube is extended to the bottom of the flask. This gives better aeration and mixing, and produces less foaming (R. Pereira, personal communununication). The air is first warmed and humidified by passing through water in an Erlenmeyer flask kept on a hot plate set at low heat or wrapped with heating tape. Mycelia are harvested by filtering after 3–4 days incubation. Using this method, yields of *H. thompsonii* were 30 g wet weight/l culture medium (dextrose (0.5%), yeast extract (0.5%), peptone (0.05%) supplemented with basal salts) (McCoy *et al.*, 1975). Air-lift fermentation is often used for the production of biomass which is used for inoculation of solid substrates such as rice. For instance, a simple medium of sucrose (1.5%) and nutritional yeast (1.5%) is used to produce inoculation cultures of *B. bassiana* and *M. anisopliae* using this method (R. Pereira, personal communication).

C Propagule storage

Once the fungus is cultured, it must be stored unless it is used immediately. Most of the same methods used for preservation of fungal cultures (Chapter V-5) can also be used for short- or long-term storage of fungal propagules. Conidia, blastospores and mycelia can be stored at 4 °C for up to several weeks or even months. However, this depends much on the species and/or strain in question. Conidia usually store best if they are kept under dry cool conditions. If cooling is not possible, then the moisture content can be critical. For instance, Hedgecock *et al.* (1995) obtained gradual decline of viability of *M. flavoviride* conidia with a 5% moisture content, after 4 months storage at 38 °C, whereas a rapid loss of viability occurred at a 15% moisture content. In contrast, Daoust & Roberts (1983) reported that conidia of *M. anisopliae* survived best when RH was high (97%) at moderate temperatures (19–27 °C) or low (0%) at low temperature (4 °C). For storage under dry conditions, conidia should be harvested, air dried for several days preferably in a biosafety cabinet and then maintained over anhydrous silica gel crystals.

Blastospores can be stored wet or dry at 4 °C. For instance, Gardner & Pillai (1987) stored blastospores of *T. cylindrosporum* in distilled water for 5 months at 4 °C without appreciable loss of viability. Blachère *et al.* (1973) developed a method for drying and storage of blastospores of *Beauveria brongniartii*. Blastospores are centrifuged, mixed with formulating ingredients (Appendix 4) and dried at 4 °C. Essentially no loss of viability occurred after 8 months of storage at 4 °C in vacuum-sealed plastic bags.

Mycelia can probably be stored wet for short periods at 4 °C but precise information on this method of storage is lacking. Mycelia can also be stored dry. Mycelia are harvested during their active growth phase of submerged cultivation, vacuum filtered, washed, sprayed with a cryoprotectant sugar solution, air dried and stored at 4 °C. Upon rehydration, the mycelia produce conidia. Mycelia treated with 10% solutions of maltose or sucrose produced more conidia upon rehydration than when mycelia were treated with a dextrose solution (Pereira & Roberts, 1990). In contrast, maltose and dextrose were superior with *B. bassiana* dried mycelia stored at room temperature. Roberts *et al.* (1987) used 10% sucrose for drying mycelia of *C. clavisporus*.

4 PROPAGULE ENUMERATION

It is usually necessary to quantify propagules in both laboratory and field experiments. Quantification of inoculum for bioassays or field application is usually accomplished through direct counts; enumeration of propagules from insects or infested substrates (e.g. leaves or soil) is based primarily on direct counts or indirect recovery methods (e.g. dilution plating). Direct counts should normally be adjusted to take into account the proportion of viable propagules.

A Direct enumeration

1. Inoculum

a. Preparation of suspension

Propagules are usually formulated in a water carrier but any number of carriers may be used (e.g. solid substrates). Although propagules with hydrophilic cell walls are readily suspended in water (e.g. *T. cylindrosporum* and *V. lecanii*), this is not the case for conidia possessing hydrophobic cell walls (e.g. *B. bassiana* and *M. anisopliae*). To uniformly suspend hydrophobic propagules in water, it is necessary to sonicate and/or use mechanical suspension methods. Mechanical suspension of propagules using micropestles (Figure 2) provides excellent suspension of *Beauveria bassiana* and *Metarhizium flavoviride* conidia in water without causing damage to cells. Although a surfactant may facilitate suspension of propagules, it is generally not necessary and may interfere in the adherence of the propagule to the host insect. To suspend hydrophobic conidia, harvested conidia are placed in a 1.5-ml microcentrifuge tube, ≈ 0.5 ml of sterile water is added to the tube, the micropestle is inserted into the tube, and the

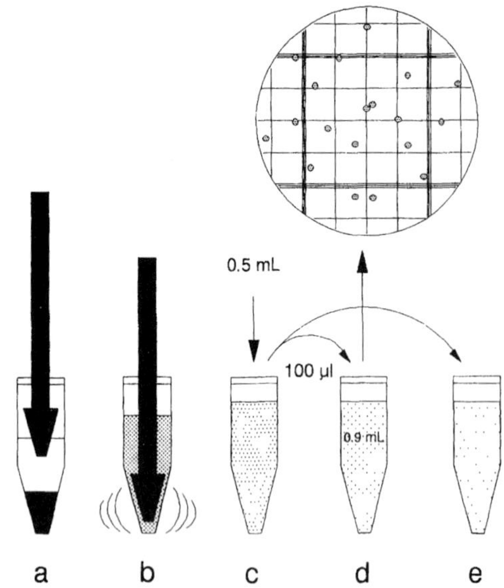

Figure 2 Suspension, enumeration and density adjustment of entomopathogenic Hyphomycetes propagules: (a) propagules at the bottom of a 1.5 ml microcentrifuge tube are gently agitated in ≈ 0.5 ml of carrier with a sterile micropestle; (b) clumps of propagules are separated by vigorous agitation of the suspension aided by the use of a pestle motor; (c) the volume of the carrier is increased to ≈ 1 ml by the addition of 0.5 ml of the carrier; (d) the suspension is diluted 10 ×, 10 µl is loaded into a haemocytometer, and the number of propagules per ml calculated; and (e) the original suspension is diluted accordingly to obtain the desired target concentration.

conidial mass is gently agitated with the micropestle by hand (prevents liberation of conidia into air). The micropestle is then attached to the motor (e.g. Kontes, Concept Inc, Clearwater, FL) and the suspension is vigorously agitated while moving the pestle in an up and down, and side to side motion (*c.* 30 s). Some evidence now suggests that oil is a more efficacious carrier than water for propagules (Bateman *et al.*, 1993; Inglis *et al.*, 1996a) and conidia with hydrophobic cell walls are readily suspended in oil using micropestles.

b. Quantification

A haemocytometer is commonly used to quantify numbers of propagules per unit volume or weight. The propagule suspension is vortexed and then loaded on to each side of the haemocytometer. Propagules must first be allowed to settle; for a water carrier, propagules settle rapidly (*c.* 5 min) but in oil, adequate settling may require more than 1 h. When using water, care must be taken that the water does not evaporate before measurements are taken. Consequently, measurements should be taken as soon as possible after the propagules settle. The cell type (be it 'A', 'B' or 'C') on the haemocytometer that contains a countable number of propagules (≈ 20–100) is selected, and the number of the propagules within each of five cells on each side of the haemocytometer are counted. For each type, the five cells should comprise the four corner cells and the middle cell; the fifth 'A' cell consists of all 25 'C' cells (Figure 3). If greater than 10% of the cells appear clustered, the entire procedure should be

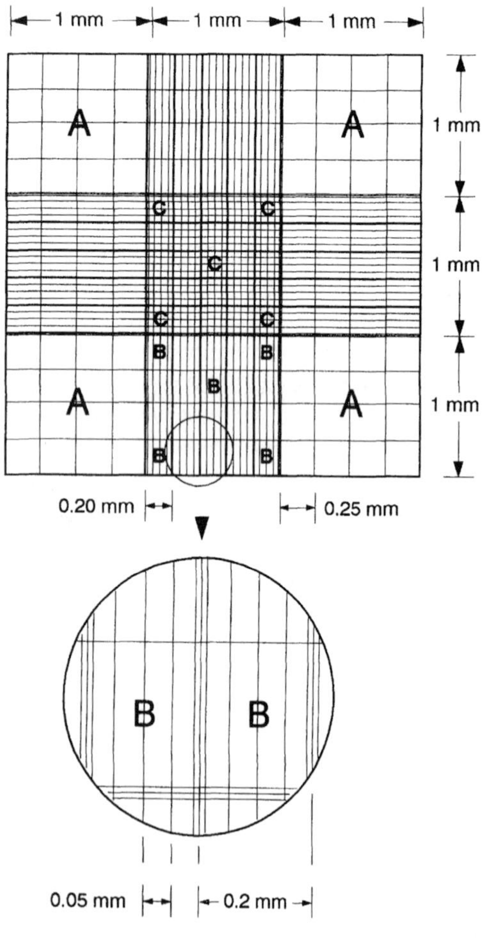

Figure 3 An Improved Neubauer haemocytometer used to enumerate propagule densities. The number of propagules in each of five cells (A, B or C) are counted; the fifth A cell consists of all 25 C cells. The average number of propagules per cell are multiplied by the volume conversion factor to obtain an estimation of the number of propagules per ml.

repeated, making sure the cells are dispersed by vigorous pipetting or vortexing of the original cell suspension.

The reference squares within each cell are used to facilitate counting; the boundary of each cell is determined by the centre line in the group of three. To avoid counting propagules twice, count those that are touching the centre line only when they are on the top and left centre line; those that touch the bottom and right side centre lines of the cell are not counted. The average number of propagules per cell is calculated ($n = 10$) and a good measure of count uniformity is when the standard deviation (SD) is within 10–15% of the mean. To obtain an estimation of the number of propagules per ml in the original suspension, the average number of propagules per cell is then multiplied by the volume conversion factor. For Improved Neubauer haemocytometers, the total volume of each 'C' cell is 0.02 cm × 0.02 cm × 0.01 cm = 4.0×10^{-6} cm^3 (Figure 3). The average number of propagules per 'C' cell is divided by this number, or multiplied by its inverse (2.5×10^5) to obtain the number of propagules per ml; for 'A' and 'B' cells, means are multiplied by 1.0×10^4 and 2.0×10^5, respectively. For example, a mean count of 50 propagules per 'C' cell with SD of ≤10 ($n = 10$) is considered to be a good count, and 50 is multiplied by the conversion factor, 2.5×10^5 to obtain a value of 1.25×10^7 propagules per ml.

The original suspension is then diluted to obtain the desired target concentration. If the desired number is 1×10^6 propagules per ml (target concentration), the target concentration is divided by the actual concentration (1.25×10^7 propagules per ml) to obtain a value of 0.08. Therefore, 0.08 ml of the original suspension added to 0.92 ml of an appropriate carrier will result in 1 ml of a suspension of 1×10^6 propagules per ml (target concentration). The total volume of inoculum required is determined and the appropriate dilution made. For instance, if 10 ml of inoculum at the target concentration of 1×10^6 propagules per ml is desired, then 0.8 ml of the original suspension is added to 9.2 ml of the appropriate carrier to obtain 10 ml of suspension with 1×10^6 propagules.

In instances where the concentration of propagules in the original suspension is too high to get an accurate count on the haemocytometer (e.g. >300/cell), it is necessary first to dilute the suspension prior to enumeration. For example, if an insect is to be treated with 10^5 propagules in 0.5 µl, the concentration of propagules per unit volume that is required is 2×10^8 propagules/ml. At this density, each 'C' cell of the haemocytometer would contain ≈ 800 propagules (2×10^8 propagules per ml/2.5×10^5) which is too numerous to count. Therefore, the original solution is diluted 10× (e.g. 100 µl of the propagule suspension into 900 µl of water) and the concentration is reduced to ≈ 80 propagules per 'C' cell. The mean number of propagules per ml in the diluted sample is then multiplied by the dilution factor, in this case by 10, to obtain the concentration of propagules in the original sample. When dilution of the original sample is necessary, it is desirable to obtain counts from two independent dilutions. For dose–mortality experiments suspensions for each dose can be prepared from independent samples or the original sample can be diluted. However, when dilution of the original sample is used, it is important that the accuracy of the dilutions be confirmed.

A number of other methods can be used to estimate concentrations of propagules. Turbidometric methods have been used to estimate propagule concentrations of entomopathogenic Hyphomycetes. This method relies on the transmittance of light through the propagule suspension the percentage light transmitted will diminish in proportion to the turbidity. Turbidity readings must be standardized in terms of numbers of propagules per unit volume. The relationship between turbidity and propagules per unit volume is usually determined using a haemocytometer. Although the turbidometric method is relatively simple and rapid, it is subject to errors due to variation in size, shape and clumping of propagules. Two additional methods used are dilution plating and the most-probable number method (Section 4B.1, below).

c. Viability

In contrast to haemocytometer counts and the turbidometric methods, plating techniques (see Section 4**B**2 below) provide a measure of viable propagules per unit volume. However, they usually take 3–7 days before data are obtained and generally provide a conservative estimate of propagule concentrations. Since haemocytometer and turbidometric methods do not distinguish between viable and non-viable propagules, it is necessary to determine spore viability so that doses can be prepared on the basis of viable propagules. Germination techniques provide results usually within 24 hours, while it may be possible to obtain results within an hour using vital dyes.

To calculate viable propagules per unit volume, total counts estimated with the haemocytometer are multiplied by the germination percentage obtained using either of the methods described below. For example, if the viability of propagules was 75% and the total count from the haemocytometer was 2×10^8 propagules per ml, the number of viable propagules per ml is $2 \times 10^8 \times 0.75 = 1.5 \times 10^8$ viable propagules per ml.

(i) *Germination.* Spore viability is most often determined by germinating propagules on a translucent agar medium such as SDAY or PDA. Propagules from the same batch to be used in the bioassay should be processed immediately prior to the preparation of inoculum (*c.* 1 or 2 days). A commonly used procedure is to prepare a suspension of propagules in water and to spread the suspension on to the surface of the medium at a density sufficiently high to permit rapid observation yet sparse enough to limit obscuration of propagules from overgrowth by hyphae.

A satisfactory density of spores is usually obtained by spreading approximately 10^6 propagules in 100 µl on to media in an 8.5 cm-diameter Petri dish. Propagules are incubated in the dark at an appropriate temperature for a specified period (usually 18–24 h) and the area to be observed is fixed (e.g. lactophenol, see Chapter VII); it is usually sufficient to place a couple of drops of the fixative on the surface of the medium and then to place a glass coverslip(s) over the area. Once the fixative has absorbed into the medium, the area can be examined microscopically. Phase-contrast microscopy is usually preferred but if bright field examination is used, the propagules should be stained (e.g. lacto-fuchsin; Appendix, 3 or lactophenol cotton blue; Chapter VII) after fixation and prior to placement of the coverslip. An inverted compound microscope may also be used.

Viability of conidia in an oil carrier may also be determined as outlined above. However, problems occur when attempting to enumerate propagules of species with small round conidia (e.g. *B. bassiana*) as the oil can form tiny emulsion droplets on the agar surface which are difficult to distinguish from ungerminated conidia. To circumvent this, propagules in an oil emulsion should be stained with a drop of lacto-fuchsin (Appendix 3) which is miscible in oil, prior to placement of the coverslip.

Propagules are usually considered viable if germtube lengths are two times the diameter of the propagule in question. For some taxa, conspicuous swelling (e.g. *B. bassiana*) or formation of a barbell shape (e.g. *T. cylindrosporum*) of germinating conidia may be used to indicate viability. Numbers of germinated and non-germinated propagules in arbitrarily-selected fields of view or in parallel transects, usually defined with an ocular micrometer, are counted. Although a minimum of 200 propagules may be sufficient, counting 500 or more can increase accuracy. It is desirable to determine the viability of propagules on replicate cultures and at various positions on the same plate.

A major limitation to this technique is that the rate of propagule germination is usually normally distributed and hyphae from early germinating propagules can obscure non-viable and later germinating propagules, thereby affecting the accuracy of the counts. To circumvent this problem, benomyl (Benlate) has been used to inhibit the hyphal development of *B. bassiana* and *M. anisopliae* (Milner *et al.*, 1991; Inglis *et al.*, 1996a). The mode of action of benomyl is by inhibition of spindle formation during mitosis as a result of binding to tubulin. Hyphomycetes in the series Phialosporae and Sympodulosporae (ascomycetous affinities) are highly sensitive (Edgington *et al.*, 1971) and at relatively low concentrations (ca. 0.005% Benlate 50 WP per ml); benomyl permits germ-tube formation but prevents further hyphal development thereby preventing overgrowth.

(ii) *Vital stains.* Vital stains can be used to rapidly determine conidial viabilities. Several fluorochromes have been used successfully to determine viability of conidia of entomopathogenic Hyphomycetes. The optical brightener Tinopal BOPT can be used to differentiate between viable and non-viable conidia of *M. anisopliae* and *B. bassiana* (Jimenez & Gillespie, 1990). Conidia are stained with a 0.05% Tinopal (bistriazinyl amino stilbene) solution in 0.05% Triton X-100 for 30 min and are viewed using a fluorescence microscope under UV light. Viable conidia fluoresce only weakly whereas non-viable conidia fluoresce brightly. This technique was unsuitable for conidia of *M. anisopliae* var. *majus*.

Fluorescein diacetate (FDA) can be used alone, or in conjunction with propidium iodide (PI), to determine viabilities of *Paecilomyces fumosoroseus* and *B. bassiana* conidia (Schading *et al.*, 1995). Aqueous suspensions of conidia (4 µl) are mixed with equal amounts of freshly prepared working solutions of FDA and PI (optional) (Appendix 3) in a dark room illuminated with a 40 W photographic safelight. The mixture is stirred with a pipette tip and covered with a cover slip. The slides are then viewed under epifluorescence using a 450–490 nm (blue light) exciter filter and a barrier filter in conjunction with a DM500 dichromatic mirror. Viable conidia fluoresce bright green, and, if PI is also used, nuclei of dead conidia fluoresce red. Only PI will fluoresce when green light (515–560 nm) is used. If PI is not used, after counting the green conidia using epifluoresence, just enough substage light (phase contrast) is added to identify and count all conidia present within the same field. Viable conidia maintain their fluorescence for only short periods (10–30 s) once normal substage light is applied. Concentrations of the fluorochromes used are critical; by experimenting with concentrations, these staining techniques can probably be adapted to assess conidial viability of most species of entomopathogenic Hyphomycetes. Further information on using fluorescent techniques is presented in Chapter VIII-2.

2. *On insect surfaces*

Direct quantification of propagules on the surface of insect integuments has been successfully accomplished in relatively few instances. Scanning electron microscopy can be used to obtain information on the spatial distribution of propagules but it is very labour intensive and is not conducive to estimation of propagule densities. The use of tissue specific and non-specific dyes offer enormous potential for direct enumeration of hyphomycetous propagules. Conidia of *V. lecanii* stained with the fluorescent dye Uvitex were enumerated on the integument of aphids using epifluorescence microscopy in conjunction with image analysis (Girard & Jackson, 1993) and FDA is used to observe conidia of *P. fumosoroseus* germinating directly on sweet potato whitefly nymphs or on foliage itself (R. L. Schading, unpublished observations cited in Schading *et al.*, 1995).

Stains can be rendered tissue-specific by conjugation to tissue-specific antibodies. The most common method is to produce polyclonal antibodies in rabbits or chickens to protein or carbohydrate antigens on the surface of cell walls of propagules and/or hyphae. Briefly, the antibodies are bound to cell wall antigens, unbound antibodies are removed, the tissue is exposed to anti-rabbit or anti-chicken IgG conjugated with the fluorochrome, fluorescein isothiocyanate (FITC), the tissue is re-washed and examined microscopically. However, the cuticle of many insects autofluoresces thereby obscuring the propagules stained with fluorescent dyes. *In situ* quantification of hyphomycetous fungi with fluorescent dyes has been most successful on small, soft-bodied insects (e.g. aphids). More details on the use of fluorochromes are presented in Chapter VIII-2.

B Indirect enumeration

While direct enumeration techniques may increase the sensitivity of enumeration and provide information on the spatial distribution of propagules, these methods are considerably more labour intensive than indirect methods. Furthermore, direct enumeration cannot be used in situations where physical or microbial contaminants obscure the propagules being enumerated. For indirect enumeration, the propagules must first be recovered. Recovery methods differ according to the substrates on which the propagules are present.

1. *Recovery*

a. *Insects*

Normally, living insects should be killed before they are processed; mechanical injury, exposure to CO_2, or freezing have all been used to kill insects, but the method selected should have a negligible effect on the fungus. Propagules must first be either washed from the surface of the insect or obtained through homogenization of the host. Sonication has been shown to efficaciously remove bacteria from leaf surfaces but the use of sonication for the recovery of entomopathogenic Hyphomycetes from insect integuments has not been studied extensively.

(i) *Washing*. Although propagules of entomopathogenic fungi may be strongly attached to the cuticle of insects, they can be dislodged by vigorous washing. To recover propagules by washing, insects are shaken in the wash solution using a mechanical shaker. The wash solution may be water or buffer amended with a surfactant (e.g. 0.01% Tween 80) but caution must be used in selecting a surfactant since they can be toxic to fungi. Both reciprocal and rotary shakers are used but the former is preferred since it provides a more vigorous agitation of insects. If a rotary shaker is employed it may be necessary to add coarse sterile sand or small glass beads to the wash solution to aid in the mechanical detachment of propagules from the integument. The speed at which the shakers are operated should be as high as possible (> 200 rpm), without causing disruption of the integument. The duration of the washing also varies, but times in excess of 1–2 h are generally used.

The volume of the wash solution and the type of vessel used also influence the efficacy of the wash procedure. Wash volumes in excess of 5 ml in relatively wide-diameter containers result in the greatest turbulence of the wash solution. Since relatively large wash volumes are required, it may be necessary to combine insects to increase the density of propagules released into the wash solution. The differential recovery of propagules from the external integument and the relatively low labour requirements are the primary advantages to the wash method. With insects that readily regurgitate their foregut contents, it may be necessary to coat the mouthparts (e.g. with molten paraffin) prior to washing (Inglis *et al.*, 1996a). The primary disadvantage to the wash method is the decreased sensitivity of the technique as a result of the large wash volumes required and recovered colonies usually represent a conservative estimate of inoculum densities due to clumping of propagules and/or inadequate detachment from the integument.

(ii) *Homogenization.* The homogenization method may be more efficacious than the wash method for recovery of propagules that are strongly attached to the cuticle or that are located in relatively inaccessible locations (e.g. cuticular folds). Since insects may be homogenized in relatively small volumes, the sensitivity of the homogenization method is generally greater than the wash method; the increased sensitivity facilitates measurements of inoculum densities between individual insects in the population. Furthermore, the homogenization method is conducive to the recovery of propagules from specific tissues (e.g. excised alimentary canals).

Homogenization of insects is usually accomplished using pestles or mechanical grinders (see Section 2A). To reduce a possible defence response, a chilled buffer should be used and the homogenate should be processed as rapidly as possible.

Disadvantages to the homogenization method include the inability to distinguish between propagules on the integument surface with those in the haemolymph and/or alimentary canal following the homogenization of whole insects; the possible inactivation of propagules due to encapsulation and/or melanization resulting from liberation of the haemolymph; conservative estimates of populations due to clumping and strong attachment of propagules to the integument; and the inability to distinguish between propagule types (e.g. conidia, blastospores or hyphal fragments).

b. Foliage and spore traps

Quantification of propagule density per unit area is often required in laboratory and field experiments, particularly following spray application. Propagules can be recovered from a number of substrates that possess a defined or measurable area. Two examples of substrates used to assess propagule densities include glass coverslips and leaves. Round coverslips (approximately 10- to 15-mm diameter) are suitable since they are relatively robust and readily fit into containers used for washing. One to several coverslips can be attached to an object such as a plastic Petri dish lid with double-sided tape or other adhesive. The lids with attached coverslips can be placed at the base of a spray tower in the laboratory or on the soil surface in the field. If desirable, the height of the coverslips can be varied (e.g. relative to the plant canopy) by attaching the lids to the tops of stakes with glue. The dishes should be removed as soon after application as possible and stored at low temperature (≈ 5 °C) until they can be processed.

To recover propagules, coverslips are either pooled or washed individually. The wash solution is diluted, spread on to an agar medium and the number of cfu counted at the appropriate dilution. Populations are calculated as: ($n \times$ dilution factor)/area; where n is the number of propagules at the desired dilution; area is usually reported in cm^2. In field settings, it is recommended that coverslips be placed at defined sites within each plot (subplots) to obtain a measure of spatial variation.

Measures of propagule density per unit area can also be obtained from foliage. Leaves are arbitrarily collected from a defined position in the canopy and stored at low temperatures prior to being processed. If a leaf area meter is available, leaves can be cut into pieces approximately the same size and shape. If a leaf area meter is unavailable, a cutting device of defined area (e.g. a cork borer) can be used to remove a disk of leaf tissue; this is considerably more time consuming and increases potential variation as a result of the smaller areas obtained from each leaf. The leaf pieces or disks are usually pooled prior to recovery. Propagules can be detached from the leaf segments using conventional washing, sonication, homogenization, or with a stomacher blender. Once in suspension, propagules are diluted, cfu recovered on an appropriate medium, and propagule densities per unit area calculated as described below.

To obtain a measure of the spatial distribution of propagules on leaves, it may be possible to use a leaf imprint technique. Leaves are uniformly pressed against the surface of an appropriate agar medium, the outline of the leaf is marked, and the position of cfu recorded after incubation of the cultures.

It is often useful to relate propagule densities to droplet deposition data. To obtain a measure of droplet deposition, water-or oil-sensitive papers or strips can be used. They are usually placed in proximity to the coverslips or leaves from which propagule populations are quantified. Water-sensitive cards cannot be used under conditions of high humidity, precipitation or where they may come into contact with water on vegetation or supports (e.g. dew). For qualitative assessments of size and density, cards can be visually compared to standards. For quantitative estimation of droplet density, the number of droplets per unit area (usually 1 cm^2) can be counted with the aid of a magnifying lens or stereomicroscope. In addition to droplet densities, image analysis systems can provide information on droplet area, size and percentage coverage.

c. Soil habitats

The soil dilution plate method is the most common technique used for quantifying propagules in soil. Soil should be collected to a defined depth with a soil corer. If possible (particularly in field experiments), it is recommended that samples be removed from a number of specific locations (e.g. subplots) within a sample area. Soil samples in plastic bags should be placed in a cool environment until they can be processed. The soil samples are usually mixed in bags and subsamples of soil are removed, weighed and homogenized in water or buffer (see Section 2C).

Although any soil weight can be used, it is most common to place 10 g of soil into 90 ml of the homogenization liquid. The use of blenders to homogenize soil is relatively labour intensive and it is more difficult to maintain sterility. In a method designed to expedite the homogenization procedure, particularly when the objective is to recover conidia from soil, 1 g samples are vigorously vortexed in 9 ml of water or buffer for a specific period of time (usually between 30 and 60 s). After homogenization, the slurry is diluted, aliquots are spread on to a selective agar medium (see Section 2), and cfu are counted at the appropriate dilution.

d. Aquatic habitats

Water samples and/or insects can be collected with a variety of sampling devices (Atlas & Bartha, 1981). The water samples can then be concentrated through centrifugation or diluted as required, and cfu enumerated using the spread plate method. Fresh water contains a number of potential fungal contaminants (i.e. Saprolegniales) and a semi-selective medium should be used if possible. The relationship between inoculum density (in water or soil) and pathogenicity can also be quantified.

2. Enumeration

a. Dilution plating

The most popular method for quantifying inoculum of entomopathogenic fungi on/in insect hosts or soil is based on dilution plating (Figure 4). After dislodgement of propagules from insect tissues or soil particles, the suspension is diluted as required (usually in a 4- to 10-fold dilution series). The dilution method usually provides a conservative estimate of populations in soil primarily due to the attachment of propagules to soil particles and/or to conidial aggregation. When densities of entomopathogenic hyphomycete propagules are low, the dilution method may lack sensitivity primarily due to the relatively large initial dilution (dilution factor = 50–100).

(i) *Spread plating*. For spread plates, aliquots of 100–200 μl are generally spread on to an appropriate medium; for most applications a semi-selective medium is recommended (see Section 2B). The

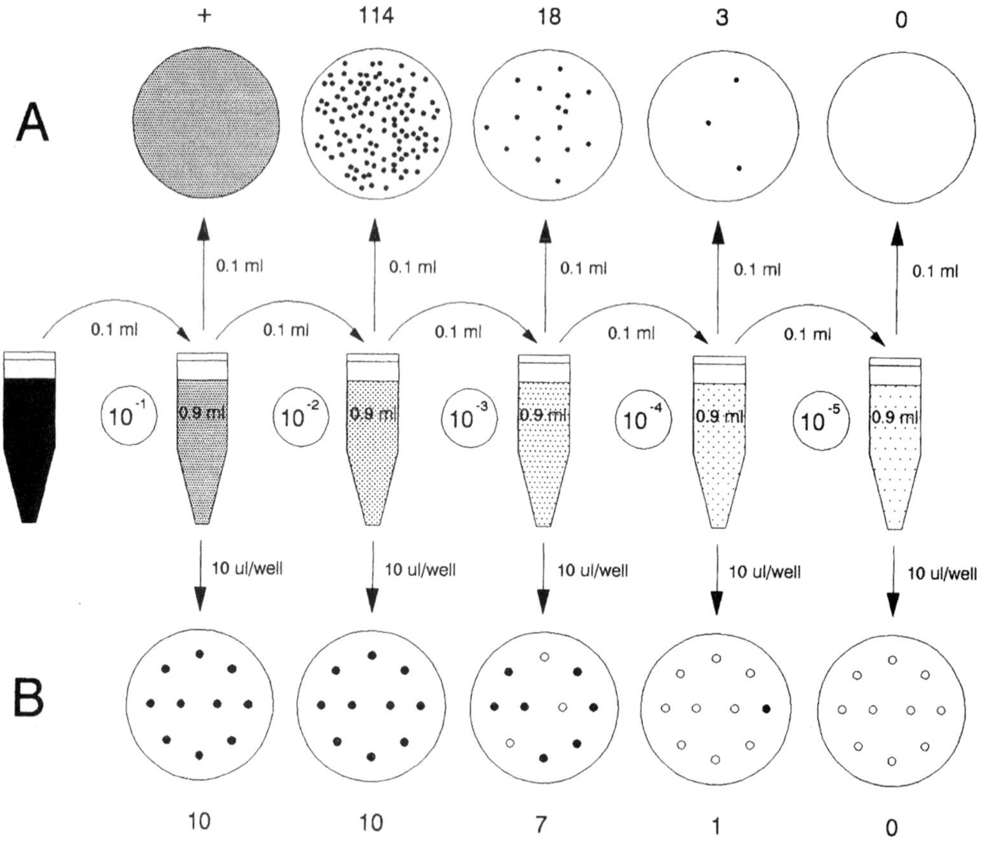

Figure 4 Enumeration of propagules using the dilution plating and most-probable number (MPN) methods. The original suspension is diluted in a 10-fold dilution series by placing 100 μl in 0.9 ml of water or buffer; each dilution is vortexed (≈ 10 s) before the transfer is made. For the dilution plating method (A), 100 μl of the suspension is uniformly spread on the surface of an agar medium and the culture is incubated. After an appropriate amount of time, the number of colony-forming units (cfu) are counted at the dilution yielding 30 to 300 cfu per dish. The culture from the 10^{-1} dilution contains too many colonies to obtain an accurate count; colonies were confluent and crowded each other. The 10^{-2} dilution culture contained 114, well-distributed and easily counted colonies. The culture from the 10^{-3} dilution contained 18 colonies, seven more than would be predicted from the 10^{-2} culture. The percentage error in each colony in such a high dilution makes such a count unreliable but duplicate or triplicate cultures may be used to minimize this error. The number of viable propagules per ml = number of colonies × tube dilution factor × plating dilution factor. Using the 10^{-2} culture, propagules per ml = 114 × 100 × 10 = 1.14×10^5 propagules per ml. For the MPN method (B), 10 replicate samples (10 μl) are placed on the agar medium from each dilution tube; wells are made with a cork borer to prevent spreading of the carrier. Three consecutive dilutions are selected at or immediately before propagule extinction and the number of positive wells are recorded. In this case the dilutions used may be 10^{-2} to 10^{-4} (10–7–1) or 10^{-3} to 10^{-5} (7–1–0). The MPN (per propagules in the first dilution used) are acquired from an MPN table (eg. Meynell & Meynell, 1970). If 10^{-2} is used as the minimum dilution, the MPN is 11.6. If 10^{-3} is used, the MPN is 1.16. The number of viable propagules per ml = MPN × inverse of the minimum dilution × plating dilution factor. Since 10 μl was placed in each well, the plating dilution factor is 1 ml/0.01 ml or 100. Using the 10^{-2} as the minimum dilution, propagules per ml = $11.6 \times 10^2 \times 100 = 1.16 \times 10^5$ propagules per ml. In this example, the dilution plating and MPN methods provide a similar estimation of propagule density.

solution should be allowed to absorb into the medium before cultures are transferred to an appropriate temperature (20–25°C for most taxa). To obtain a measure of variation due to plating, it is recommended that two or more dishes be prepared at each dilution. Dishes are incubated for 3–7 days depending on the medium or organism, and since there is evidence that light exposure can inhibit colony development in some taxa (Chase *et al.*, 1986), cultures should be maintained in the dark.

After an appropriate amount of time, colonies are counted at the dilution yielding 30 to 300 cfu per dish. To increase the accuracy of the counts (especially for smaller colonies) a stereo microscope at 2 or 3× may be used. Densities are usually presented as cfu/g of insect or per insect (standardized by weight) or as cfu/g of soil dry weight. To determine the moisture content of soil, a weighed subsample (10 g) is dried at approximately 110°C for a minimum of 12–24 h and the percentage of water in the sample is calculated. The density of conidia per unit fresh weight is then adjusted accordingly.

In an attempt to reduce error that may occur during dilution and/or plating, and to expedite the procedure, mechanized plating systems have been developed. Spiral platers, initially shown to be effective for bacteria (Gilchrist *et al.*, 1973) have recently been used for enumeration of entomopathogenic Hyphomycetes (Fargues *et al.*, 1996). Image analysis systems can expedite the enumeration of colonies and provide a permanent record of cultures. However, as a result of the hyphal growth characteristics and relatively rapid growth rate exhibited by many entomopathogenic Hyphomycetes, it is often necessary to make 'judgement calls' when colonies grow together or are found in close proximity to each other.

(ii) *Most probable number*. An alternative to the spread plate method is the most probable number technique (MPN) (Woomer, 1994). The MPN method was developed for the enumeration of bacteria in water and has been adapted for use with filamentous fungi (Figure 4). As with the spread plate method, it is first necessary to separate the fungal propagules from the substrate. Once in suspension, the sample is diluted to the point of propagule extinction. Dilutions are usually 4- or 10-fold, and three to ten replicate samples per dilution are required. Although a liquid medium may be used, a solid substrate is generally used for filamentous fungi. As with the spread plate method, it is desirable to use a selective medium (see Section 2B).

A physical barrier is usually required to prevent spreading of the suspension across the agar surface; this is accomplished by placing a glass or polypropylene ring in the medium, or by making a well in the agar surface with a cork borer. Volumes placed within the wells will vary according to the size of the well and the absorptive properties of the medium. The cultures are incubated and after an appropriate amount of time, the wells are scored as positive or negative for growth of the target hyphomycete. The number of positives and negatives at the dilutions before extinctions are used. From MPN tables based on a Poisson distribution, numbers of viable propagules can be obtained and density of propagules per unit volume or weight is then calculated (Meynell & Meynell, 1970). Although it is considered to be less accurate than the spread plate method, the MPN method may provide comparable results (Harris & Sommers, 1968). The salient advantages of the MPN technique are its greater overall flexibility and scope of application. For example, micro-organisms with specific attributes can be readily identified (e.g. fungi with chitinase activity if a chitin medium is used).

b. Host assays

Another approach for enumeration of fungal propagules is through use of bait host insects. This technique is relatively labour intensive and is used principally for fungi for which there is no selective medium available or for aquatic samples. Susceptible insects are exposed to the propagule infested substrate (e.g. leaf surface, soil homogenate or water sample) for a defined period in a controlled environment, and the incidence of mortality is recorded periodically. Concentrations of propagules per unit area or volume of infested substrate/water can then be estimated according to dose–mortality data obtained under similar conditions with known numbers of propagules.

Another approach which can be used to quantify infective units in both water and soil is the MPN method this technique is frequently used for assessing infective units of filamentous fungi that are obligate parasites (e.g. Ciafardini & Marotta, 1989). Although the MPN method is subject to many of the same limitations as the previously reported incidence of mortality method, it does not rely on a dose–mortality curve to estimate densities. Normally the insect pest of interest is used, but the sensitivity of the test is increased by using a highly susceptible insect host. For example, *G. mellonella* larvae are more sensitive to infection by entomopathogenic Hyphomycetes in soil than are soil-inhabiting insects. The methods used are analogous to those described previously for enumeration of fungal propagules on an agar medium. The substrate (i.e. water or soil) is serially diluted and insects are exposed to the diluted substrates for a defined period of time. The dilution where no mortality is observed is selected, and the number of infective units (IU) are determined from MPN tables, and IU per volume of water or per unit weight of soil are calculated.

C Molecular techniques

Molecular methods have been used to quantify biomass of fungi in a variety of habitats. However, use of these techniques for quantifying propagules of entomopathogenic Hyphomycetes has not been extensively applied. Enzyme-linked immunosorbent assay (ELISA) (e.g. Guy & Rath, 1990) and more recently, DNA and RNA probes are two techniques that could be used to quantify biomass of entomopathogenic Hyphomycetes. Molecular techniques are presented in Chapter VIII-3.

5 INFECTION AND BIOASSAY

Methods for the rapid, systematic infection of insects for use in other studies (e.g. sporulation of fungi on cadavers), for studies on the mode of infection, or for the *in vivo* maintenance of a pathogen, seldom require precise dosing. The simplest method for such pathogen transmission is to introduce cadavers on which the fungus is sporulating to cultures of healthy host insects. This may be the only method to maintain fastidious, obligate pathogens, but entomopathogenic Hyphomycetes are seldom obligate and fastidious.

In contrast dose–time– mortality responses are used to compare differences between pathogen strains or species, or to evaluate effects on infectivity of factors such as environment, methods of targeting, formulation, or propagule type. Therefore, bioassays must be repeatable and strict control of dose and dosing method (i.e. inoculation) must be exercised. A number of other factors such as host species, age and physiological condition, size of bioassay chamber and incubation conditions (e.g. temperature, humidity, photoperiod) may also affect bioassay results.

Although general bioassay designs may remain the same, no standardized bioassay systems are available for entomopathogenic Hyphomycetes. This is mostly due to the wide range of hosts which vary in their bioassay requirements. Therefore, specific bioassays must be developed for most host–pathogen combinations.

A Experimental design and analyses

Although bioassays can provide valuable information on the pathogen–insect–environment interaction, the value of the results obtained depends on the design, execution, analysis and interpretation of results. The objective of this section is to summarize the salient aspects of bioassay design and analysis. For more detailed descriptions of the statistical analyses presented, the reader should consult the biometrics literature.

1. *Definitions*

An *experimental unit* is the unit to which a treatment is applied (e.g. a group of insects); the term *plot* is synonymous with experimental unit. A *variable* is a measurable characteristic of an experimental unit and variables are either *dependent* (Y) or *independent* (X); independent variables are controlled by the experimenter. Dependent variables (e.g. measured response due to the effects of independent variables) are classed as *discrete* (e.g. real number values between specified limits) or *continuous* (i.e. any value between certain limits); discrete values are typically either counts or quantal (i.e. living or dead). A *treatment* is an agent or condition (independent variable) whose effect is to be measured and compared with other treatments. When a treatment is applied to more than one experimental unit, it is *replicated.* Replication is necessary to provide an estimate of experimental error (measure of variation among experimental units). The complexity of bioassay designs range from those that are relatively simple (e.g. one treatment is compared to another, usually a control treatment in a binary experiment) to those that are more complex (e.g. factorial designs). A *factor* is an independent variable and, in factorial experiments, each factor has at least two *levels* (several states within each factor). An *interaction* between two factors occurs if the level of one factor alters the impact of the other factor on the dependent variable.

2. Design and analysis

The design of a bioassay is of paramount importance to its successful outcome, and the researcher should consult a biometrician for advice whenever possible. The initial step of any bioassay is to establish the objective(s) of the experiment and to formulate hypotheses to test (e.g. H_0 hypotheses). It is important that the concomitant attributes of the bioassay (e.g. environment variables) be pertinent and should be considered relative to the objectives and design of the experiment. Once hypotheses have been formulated, treatments and an appropriate design are selected to answer the questions posed. The statistical design employed should be the simplest design that provides an acceptable level of precision without compromising the need for replication, power and appropriate tests.

Analysis of variance (ANOVA) is frequently used to evaluate the impacts of entomopathogenic Hyphomycetes on insects in bioassays. The two most common designs to which ANOVA is applied in bioassays, are the completely randomized design (CRD) and the randomized complete block design (RCBD). There are two sources of variation in the CRD; among experimental units within a treatment (experimental error), and among treatment means. Although the CRD may maximize degrees of freedom for estimating experimental error, it may be inefficient since the experimental error encompasses the entire variation among experimental units except that due to treatments. For this reason, the CRD is usually limited to bioassays that are conducted in uniform environments where experimental units are essentially homogeneous (e.g. controlled environment chambers). In most bioassays, the RCBD is preferred over the CRD since the RCBD incorporates a measure of variation among blocks (replicates). Variation accounted for by block differences can be removed from the total variation, the experimental error is typically the block by treatment interaction, which serves as a baseline against which variation among treatments can be assessed. Selection of the appropriate model is facilitated by listing all possible sources of variation and their associated degrees of freedom (e.g. ANOVA table), and identifying the tests to be conducted.

Most bioassays are influenced by a number of unmeasurable variables (e.g. dose estimation, the physiological status of the insects or of the fungus) and therefore, it is imperative that the bioassays be repeated at different times. One strategy is to repeat the entire experiment and compare results between trials. It is necessary to confirm homogeneity of variance (e.g. using Bartlett's test) before the trials are combined; variances associated with different treatment means should be homogeneous. Even when the variances between treatments are found to be homogeneous, the final decision on whether data from different trials should be combined rests with the researcher (i.e. whether the data between trials are consistent). In contrast to repeating the entire experiment, another strategy used in bioassays with entomopathogenic Hyphomycetes is to conduct each block of the experiment separately as a RCBD (single or multifactor experiment) with the blocks conducted at different times. Each of the blocks is carried out autonomously from each other and variation not attributable to the treatments can be removed by the model (e.g. block effect). However, if the experimental unit is not consistent, and this change influences the treatment effect, then this design may not be appropriate.

One of the assumptions of parametric statistics is that data are drawn from a normal distribution. To confirm this, data should be tested for normality, kurtosis and skewness (e.g. D'Agostino *et al.*, 1990; Univariate Procedure of SAS, SAS, 1991). In conjunction with normality testing, residual versus predicted (ANOVA model) values should be plotted to determine that variances are homogeneous. In cases where the data are not normally distributed or variances are heterogeneous, an appropriate transformation may be used. Log, square root and arcsine transformations are frequently used to normalize distributions and/or homogenize data (Little & Hills, 1978). Data obtained from bioassays are often based on counts (e.g. number of dead insects) that are expressed as percentages or proportions of the total sample. These type of data are typically binomially distributed and variances tend to be larger in the middle ($\approx$ 50%) relative to the two ends of the range. Application of an arcsine transformation to such data may normalize the variances. However, the arcsine transformation will only be effective if the range of percentages is greater than 40% (Little & Hills, 1978) and most of the means do not lie between 30 and 70% (Snedecor & Cochran, 1980).

When evaluating quantal mortality data (i.e. percentage dead) it is useful to know if an insect died of mycosis or other causes. To remove effects not due to the entomopathogen, mortality observed in the control treatment (i.e. treated with the carrier alone) is considered using Abbot's formula (Abbot, 1925). Since some death would be expected in control treatment insects, the proportion of insects killed by the entomopathogen alone is:

$$P = [(C - T)/C)] \times 100$$

where P is the estimated percentage of insects killed by the entomopathogen alone, C is the percentage of the control insects that are living, and T is the percentage of the treated insects that are living after the experimental period. Abbot's formula can also be applied to a variety of other types of data (e.g. propagule viability or germination tests). When applied to count data, Abbot's formula is also called the odds ratio or the cross products ratio (Johnson *et al.*, 1986; Schaalje *et al.*, 1986):

$$P = [1 - ((T_A/T_B)\,(C_A/C_B))] \times 100$$

where P is the estimated percentage of insects killed by the pathogen, T and C are count data from the treated and control treatments before (B) and after (A) treatment with the pathogen.

Researchers may want to determine if insects died of mycosis or other factors; colonization of cadavers by the hyphomycete can be checked to confirm fungal activity (see Section 5C). Although this may provide adequate results when the proportion of cadavers colonized is high, it generally provides a conservative estimate of mycosis; although non-colonized insects may have died from mycosis, the fungus could have been competitively excluded from the substrate by saprotrophic micro-organisms. Extreme care should be exercised in interpreting these types of data.

Once the data have been collected and the assumptions for analysis of variance have been satisfied, the data may be analysed as predetermined in the experimental design stage of the bioassay. Analysis of single factor experiments as either a CRD or RCBD is relatively straightforward. However, subsequent to a significant F-test for the factor in question (i.e. treatments), it may be necessary to determine which of the treatment means ($n \geq 3$) are significantly different. This is usually accomplished using a mean separation test at a selected α level (discrete variables) or with preplanned comparisons using orthogonal contrasts; the least square means function of SAS is frequently used for unbalanced designs or for interactions between factors (e.g. in a factorial experiment). A mean separation test that controls both type I error (rejecting the H_0 when it is true) and type II error (accepting the H_0 when it is false) should be used (see Jones (1984) for a comparison of means tests). Even if mean differences are statistically significant, these differences should be deemed biologically 'significant' by the researcher in order to be valid. Accurate interpretation of the experimental results is of paramount importance in the formulation of new hypotheses and in the implementation of subsequent experiments.

Factorial designs are used in many bioassays with entomopathogenic Hyphomycetes. This type of analysis allows the experimenter to determine the degree to which factors influence each other (i.e. whether an interaction exists between factors). Although the results obtained from factorial experiments may be more difficult to interpret than single factor experiments, the information obtained on the interaction between factors is often important. For example, when comparing the efficacy of two entomopathogens across a number of doses, the interaction between taxa and dose may be more important to the researcher than the response of the individual factors alone. In the simplest type of factorial experiment, the same error term (residual error term) is used to test all factors. In some instances it may be necessary to use different error terms (e.g. split-plot designs). Observations from repeated measure experiments (e.g. when the same group of insects is observed at different times) are not independent and thus are correlated. Split-plot models (in time) with a Box correction (Gomez & Gomez, 1984; Milliken & Johnson, 1984) can be applied to repeated measurement data (e.g. disease progress curves); the Box correction reduces the degrees of freedom for time, the time by treatment interaction and the residual error(time) by time−1.

Regression analysis is frequently used to analyse the efficacy of entomopathogenic Hyphomycetes, particularly in instances where the relationship between dose and mortality is of interest to the researcher. Besides providing a measure of efficacy (e.g. lethal dose), this type of analysis may also provide important information on the mechanism of pathogenesis. The discrete data required for this type of analysis are quantal. Although probit-, or logit-transformations may be used to linearize the response, which is typically sigmoidal in its untransformed plot, a number of other models (e.g. log-log) can also be used (Robertson & Preisler, 1992). There is no evidence to indicate the superiority of the probit versus logit models, and both methods provide similar median lethal dose (LD) results (Robertson & Preisler, 1992). How well the data fit the assumptions of the model is called the goodness-of-fit and this is usually tested using a χ^2 test; values predicted by the model are compared to actual values to derive this statistic. Additional information typically obtained from this type of analysis includes: LD_{50} and LD_{95} with 95% confidence intervals (95%); slope and standard error of the slope; and *y*-intercept of the regression. In bioassays with more than one treatment, the dose–response lines can be tested for parallelism and for a common *y*-intercept using log-likelihood ratio tests (Finney, 1971).

The two most important factors determining the power of dose–mortality analyses are dose selection and sample size. Selection of doses depends on the lethal dose of interest. For example, doses that provide a response between 25 and 75% are most useful for determinations of LD_{50}. The time at which data are collected is dependent on the researcher, and data collected at different times (e.g. day 5 and 6) can be analysed separately. Sample size also influences the precision of the analyses. Robertson *et al.* (1984) concluded that 240 insects were required for a reliable response in a typical bioassay, although a sample size of 120 insects was adequate in most instances. As indicated earlier, it is important that dose–mortality experiments be repeated. Results from replicate bioassays can be compared using a 'common-line' model (Finney, 1971). A number of software packages (e.g. POLO, GLIM, S108 Multiline Quantal Bioassay Program) that analyse dose–mortality data are available commercially or from non-profit organizations (Russell *et al.*, 1977; Payne, 1978; Morse *et al.*, 1987).

In time–dose response experiments (i.e. disease progress), dose is kept constant and time is varied. This contrasts with dose–mortality experiments where the reverse is true (i.e. dose is varied but time at which mortality is assessed is constant). Time course analysis provides a measure of lethal time, usually reported as the time at which 50% of the test insects have died. The use of probit- or logit-regression models to analyse time–mortality data is only valid if different groups of insects are used at each time. If the same group of insects is used, the data will be correlated and therefore analysis with standard probit techniques is invalid. In situations where it is not possible to obtain independent samples for each observation time (i.e. the number of insects is limited), methods that permit analysis of correlated response data must be used. Correlated data such as survivorship curves may be fitted to a Weibull function (Pinder *et al.*, 1978), and median lethal times with upper and lower 95% confidence limits estimated (i.e. using the SAS LIFEREG Procedure, SAS, 1991).

Throne *et al.* (1995) described a method for analysing correlated time–mortality data using log-log, logit, or probit transformation of proportion of insects killed (program available from James E. Throne, US Grain Marketing Research Laboratory, USDA-ARS, 1515 College Avenue, Manhattan, KS 66502). In many instances it is desirable to analyse both time– and dose–mortality data. Data from time–dose–response experiments are usually analysed by modelling time trends separately for each dose or by estimating dose trends separately for each time (Robertson & Preisler, 1992). However, Preisler & Robertson (1989) describe a method that estimates mortality over time in insects exposed to a series of increasing doses of insecticides (time–dose–mortality data); regression based on the complementary log-log model was used to analyse time trends for all dose levels simultaneously. Recently, Nowierski *et al.* (1996) applied the complementary log-log model to time–dose–mortality relationships for several entomopathogenic Hyphomycetes in grasshopper bioassays.

Analysis of covariance (ANCOVA) combines features of ANOVA and regression. Although ANCOVA is an extremely powerful technique, it has not been extensively applied to bioassays with entomopathogenic Hyphomycetes. Analysis of covariance can be used to remove variability associated with the dependent variable by including a concomitant variable in the model. The most common use of ANCOVA is to increase the precision in randomized bioassay experiments (Snedecor & Cochran, 1987). In such applications, the covariate (X) is a measurement (e.g. insect weight or dose) taken on each experimental unit before treatments are applied that predicts to some degree the final response of Y on the unit. By adjusting for the covariate, the experimental error is reduced and thus a more precise comparison among treatments is achieved. It is assumed that the slopes of the regression of Y and the concomitant variable or covariate, do not differ significantly among the treatments. Analysis of covariance may also be used to adjust for sources of bias in bioassay experiments (Steel & Torrie, 1960; Snedecor & Cochran, 1987). For example, in studying the relationship between food consumption and dose, a measure of food consumption from insects treated with varying doses of the entomopathogen as well as the initial size of each insect are recorded and differences between the mean size of the insects exposed to the different doses are noted. If food consumption is linearly related to size, differences found in consumption among different dose treatments may be due, in part, to insect size. Size is consequently included in the ANCOVA model to remove bias.

Analysis of covariance can also be used to test for differences in regression relationships (intercepts and slopes) among treatments. Application of ANCOVA to dose–mortality data can be used as an alternative to traditional probit-or logit-analysis with χ^2 tests. Using the general linear model (GLM) procedure of SAS, heterogeneity of slopes can be tested. The analyses can also test for differences in intercepts assuming a constant regression relationship among treatments (SAS Institute Inc., 1991).

A less well-known statistical method that may be useful for analysing bioassay data is the application of the competing risks theory. This theory deals with situations in which there is interest in the failure (or exit) times of individuals, where the subjects are susceptible to two or more causes of failure, and where the failure occurs over time (Johnson, 1992). Explanation of the models involved and several formulations of the theory with application to insect experimentation are provided by Schaalje *et al.* (1992).

B Inoculation

Presentation of a precise dose to the host is imperative for accurate, repeatable bioassay results. Whereas *per os* inoculation is required with most other pathogens, in contrast, most Hyphomycetes require some form of inoculation of the integument. This, at times, can be difficult to accomplish with precision depending on the size and type of insect and the requirement to inoculate large numbers of insects. Therefore, rapid methods have been developed that simplify inoculation, yet still provide repeatable results even though precise dose levels are not always known. Very indirect methods such as allowing the insect to walk on the surface of a sporulating culture have been used, however, such methods are crude and should be avoided, other than possibly for experimental transmission requirements.

1. Injection

Inoculation by injection is most often used when large numbers of infected insects are required, when studying internal immunological responses, or when attempting to maintain a pathogen in hosts. Most commonly, an aqueous suspension of propagules is injected using a 1-ml tuberculin syringe fitted with a fine needle (e.g. 30 gauge) using a motorized microinjector to drive the syringe plunger. Volume of inoculum injected depends on the size of host insect; small volumes should be used to avoid disruption of the insect's physiology as much as possible. The microinjector is calibrated to deliver the desired volume, most often by using oil; the oil is expelled on a preweighed filter paper and then the paper is reweighed. The weight of the oil is then divided by its specific gravity to determine the volume delivered.

Hand-held or otherwise immobilized insects (e.g. adhered to sticky tape) are inoculated by piercing the intersegmental membrane and injecting the propagules directly into the haemocoel. Insects can first be immobilized with CO_2 or chilling if required. Large numbers of insects can be injected, especially if the microinjector is equipped with a foot pedal. If a precise dose is required, care must be taken that the spores do not settle within the syringe during inoculation.

Because the cuticle is an important barrier to infection, inoculation by injection has seldomly been used in comparing virulence. However, Ignoffo *et al.* (1982) used inoculation through injection to demonstrate that resistance may not be solely at the integumental level; larvae of *Anticarsia gemmatalis*, a normally resistant species, injected with either blastospores or conidia of *Nomuraea rileyi*, were much more resistant that larvae of *Trichoplusia ni*, a susceptible insect. In addition, good dose–time–mortality results were obtained suggesting that this method of inoculation may be useful in bioassays of other insect–pathogen combinations.

2. Per os

Once again, since entomopathogenic Hyphomycetes generally enter the host's body via the integument, *per os*, or oral inoculation is seldom used, unless the objective is specifically to demonstrate infection via the alimentary tract. In such cases, microinjection devices have often been used; this method is virtually identical to the injection method described above except that the end of the needle is blunted with fine emery cloth and the inoculum is introduced directly into the mouth or gut. However, it is difficult to administer the dose without puncturing or damaging the gut wall.

An alternative method is to present the inoculum via a bait or food source. The easiest method is to incorporate infective propagules directly on to the surface or within a bait (e.g. spores can be mixed in a sugar solution for presentation to flies or applied on to leaves for presentation to leaf-eating insects). Further details on inoculating methods using baits are presented below (p. 236). Whatever the method of inoculation, it is virtually impossible to prevent surface contamination of the insect (see Chapter 1). It must be noted that even with proper disinfestation, external infection can occur from conidia excreted in the frass (Allee *et al.*, 1990). Consequently, any conclusions on *per os* infection based on *per os* inoculated insects should be tempered with histological results (see Chapter VIII-1).

3. *Topical*

Inoculum is most commonly administered by some form of topical application or contamination. The method adopted usually depends on the size of insect, the number of insects to be inoculated, the formulation used, and the precision required. Fluorescent dyes (0.1% w/w; Day-Glo Colour Corp., Cleveland, OH) in oil or aqueous formulations can be used to determine which insect body parts come into contact with formulated conidia (e.g. baits) (Inglis *et al.*, 1996a) after laboratory inoculation or field application.

a. Direct application

(i) *Immersion.* Although the dose cannot be measured precisely, immersion of insects into suspensions of propagules has been used successfully in bioassay of entomopathogenic Hyphomycetes. A series of aqueous suspensions are prepared with increasing concentrations of propagules. Insects are dipped singly into a suspension for a specified time. When using hydrophobic conidia, however, it would be expected that conidia would immediately adhere to the cuticle and much of the water would remain in the container (i.e. the concentration of conidia in the dipping suspension would change with every insect dipped). Consequently, in order to ensure that the dose received by each insect is as constant as possible, a separate suspension should be used for each insect dipped (i.e. insects should not be consecutively dipped into the same suspension). This is especially important if large insects are dipped into small volumes of aqueous inoculum. Insects can also be placed in small screened cages or bags that are dipped into propagule suspensions. Using dipping methods, dosages are usually expressed as the number of conidia ml^{-1} of suspension.

An alternative method used especially with small insects (e.g. aphids), is to flood propagule suspensions over the insect (Hall, 1976); the insects are placed on a filter paper in a Buchner funnel, and a spore suspension is gently poured in, immersing the insects. After a specified time (several seconds), the suspension is quickly drained off by suction. Alternatively, the insects are placed on detached leaf pieces or disks and flooded with the inoculum as described above.

A novel approach for bioassay of entomopathogenic fungi against fourth instar nymphs of the silverleaf whitefly has been described by Landa *et al.* (1994). Drops of conidial suspensions are placed singly on a sterile microscope slide. One fourth instar nymph is placed into each drop and the slides are then dried in a laminar flow hood. This assay system was successfully tested with *Paecilomyces fumosoroseus, Verticillium lecanii* and *B. bassiana.*

(ii) *Spraying.* One of the most common inoculation methods is to spray the propagules directly on to the host. Several experimental spray devices are available commercially. The ones most commonly used for application of entomopathogenic Hyphomycetes are stationary sprayers such as the Potter spray tower. Track sprayers, where the spray nozzles are moved over the host at a controlled speed can also be used (see Chapter III-3). Less expensive, yet very efficient systems can be easily developed using a plastic cylinder and an artist's air-brush. The equipment must be calibrated and care must be taken that the delivery of the inoculum at the host level is uniform. During calibration and host treatment, droplet size, density and distribution pattern and propagule deposition should be monitored as previously described (Section 4). During the spray operation, insects are often immobilized on a sticky surface such as double-sided sticky tape, by chilling or by treatment with CO_2. Insects that are not immobilized should be removed from the spray arena as soon as possible to prevent them from picking up additional inoculum from the sprayed substrate. Dosages are usually expressed as number of propagules cm^{-2} surface area.

(iii) *Droplet.* With larger insects, a precise droplet of the inoculum can be placed directly on to the insect's surface. Microinjectors (see above) or micropipettes can be used to apply volumes as little as 0.5 µl. However, this is usually not possible when applying aqueous suspensions, especially when using micropipettes, as the water droplet is difficult to deposit on the hydrophobic insect cuticle. Deposition of the droplet in an area where it is absorbed by capillary motion may be helpful (e.g. at the pronotal shield of locusts). Care must be taken when using oil formulations as the oil itself can be toxic. Doses are usually expressed as the number of propagules/insect.

b. Indirect application

Rather than presenting the inoculum directly on to the insect surface, indirect methods can be used to present inoculum via a secondary substrate. The most common method is to inoculate a substrate and then transfer the insects on to it; insects pick up the inoculum by contact with the substrate as they feed or move on it. A less common method is to present the inoculum within a food source; insects then surface contaminate themselves in the process of eating. The methods for deposition of the substrate are essentially the same as described above; inoculum is deposited on the surface of a substrate by either dipping, spraying or direct deposition.

Application of inoculum on to a leaf surface is commonly used with plant feeding insects such as caterpillars. Leaves or leaf disks are treated and presented to insects in bioassay containers such as Petri dishes. This method has been used successfully to bioassay many entomopathogenic fungus/host combinations including caterpillars and the Colorado potato beetle (Ignoffo *et al.*, 1983 and references therein).

Loss of inoculum and/or sticking of the leaf disk on to the surface of the bioassay container is often a problem, especially with oil formulations. To overcome this, Inglis *et al.* (1996c) presented leaf disks impaled on insect pins to grasshopper nymphs: 5 mm-diameter lettuce disks were inoculated with 0.5 µl of an oil/conidial suspension of *B. bassiana.* Each inoculated disk was pierced with a pin and suspended approx. 2 cm into the bioassay vial from a foam plug. Insects were allowed to feed for a certain period and only insects that had completely consumed the bait were used in the assay. Alternatively, the area of the leaflet or disk consumed can be recorded and the relative inoculum calculated accordingly; however, this is very time consuming.

For assay of *B. bassiana* against adult flies, Watson *et al.* (1995) treated 35 cm^2 sheets of plywood with either dry or wet formulations. Flies were exposed by placing CO_2 anaesthetized flies on the treated surface and covering them with an inverted Petri dish bottom for 3 h.

4. Aquatic

Inoculation of aquatic insects is usually accomplished by introducing the insects directly into suspensions containing the infective propagules. Dosages are expressed as number of propagules ml^{-1} of rearing medium. Such stationary systems are usually adequate when insects such as mosquito or chironomid larvae are assayed. However, systems using running water are necessary when assaying black fly larvae (see Chapter III-2). Simple stationary methods may be satisfactory with faster-acting pathogens such as *Culicinomyces clavisporus*. However, continuous exposure of larvae to various concentrations of conidia is not an ideal bioassay system with slower acting Hyphomycetes such as *Tolypocladium cylindrosporum* (Goettel, 1987). The effective dose can vary according to length of exposure, as mosquitoes are continually reingesting conidia that are still viable when excreted resulting in great variability between replicates. A limited exposure time may be more appropriate (Nadeau & Boisvert, 1994); larvae are placed in cups containing the inoculum suspension for a set period, harvested, rinsed and then placed into new containers containing water without inoculum.

5. Soil

Prior to commencing inoculation of soil, the classification, texture, cation exchange capacity, organic matter content, pH and moisture characteristics of the soil should be determined. Although soil can be stored moist at low temperatures for a period of time with little effect on the microflora, it is desirable to use soil as soon as possible after collection to minimize possible storage effects. For longer storage periods, the soil should first be weighed (to determine the percentage moisture content) and then stored dry.

Quantification of the influence of propagule density and distribution in soil on the efficacy of entomopathogens is important. Propagules may be applied to soil as a dry preparation, in aqueous suspension, or formulated in/on a solid carrier (e.g. wheat bran or alginate pellets). Propagules may be applied to the soil surface or uniformly mixed throughout the soil profile. The incorporation of dry propagules into soil can result in clumping of inoculum, either when the propagules are added to moistened soil or when they are added to dry soil which is subsequently moistened. A satisfactory distribution of propagules may be achieved by spraying conidia (e.g. with an airbrush) on to moistened soil while it is continuously mixed. Once incorporated, the soil moisture level (i.e. by weight) can then be increased to the desired level without affecting the distribution of propagules.

To test the distribution of propagules, soil cores should be removed from various locations, the samples weighed, propagules recovered on an agar medium, cfu counted and the number of cfu per unit weight of soil calculated (see p. 227). The number of cfu in soil at the different sample locations are compared with each other, and to theoretical populations, to obtain a measure of propagule uniformity.

Once propagules have been incorporated into the soil and the proper moisture level achieved, soil may be dispensed into containers that usually range in volume from 200 ml to 1 l, but volumes may be increased according to need. Insects are situated in the soil at a specific depth during soil placement or they are placed on the surface and permitted to move into the soil. Although most bioassays have focused on insects that inhabit soil for a portion of their life cycle (e.g. scarab beetle larvae), the influence of entomopathogenic Hyphomycetes on insects that are exposed to soil for a relatively short period of time (e.g. ovipositing grasshoppers) can also be tested using these protocols.

C Incubation and mortality assessments

After inoculation, the insects must be incubated, preferably under controlled environmental conditions. This is usually carried out in environmental chambers or cabinets controlling factors such as temperature, photoperiod and humidity. Methods for the study of effects of environmental factors are presented in the next section. Choice of bioassay chamber, feeding regime and incubation method are important in successful bioassay and will vary according to the needs of the host. Insects can be incubated singly or bulked in cages or assay chambers. In general, incubation conditions should be those that favour survival of non-inoculated insects. Control mortalities should be kept below 10%.

Mortality assessments are generally made daily and, due to the slow acting nature of most entomopathogenic fungi, may need to be carried out for up to 2 weeks post-inoculation. Cadavers must be removed before the fungus sporulates to prevent horizontal transmission. When evaluating mortality data, it is useful to know if an insect died of mycosis or other causes. To determine if insects died of mycosis, colonization of the cadaver by the hyphomycete is evaluated; cadavers are incubated in a high moisture environment (e.g. on moistened filter paper or water agar) and, if the cadavers are subsequently colonized by the hyphomycete, these insects are considered to have died from mycosis.

1. Insects in epigeal habitats

Larger insects are usually incubated singly in plastic containers, such as 500 ml food containers. They can also be pooled and incubated in small cages. Insects should be fed as required and conditions for proper growth and development provided.

Larvae are usually transferred to individual compartments in plastic trays and reared on artificial diets. Care should be taken that the diets do not contain antibiotics that could interfere with disease progression. For smaller insects inoculated directly on their host plant (e.g. detached leaf or leaf disk), incubation is often carried out in Petri dishes containing either water agar, moistened filter paper or a soaked piece of cotton batten. Under such conditions of high humidity, detached leaves or leaf disks usually remain viable and are able to provide nutrients for their host for many days. Care should be taken not to place too many insects on a leaf surface, as this accelerates leaf deterioration. Insects should be transferred to fresh leaf surfaces as required.

In the Landa *et al.* (1994) bioassay technique for whiteflies, the glass slides with inoculated nymphs are incubated under conditions of saturated humidity. Nymphs are assessed daily according to the degree of fungal development.

2. *Insects in aquatic habitats*

Insects in aquatic habitats, such as mosquito larvae, are most commonly incubated in 200 ml water in 500 ml plastic food containers or beakers. High variability in mortality between replicates often occurs. Goettel (1987) attributed this to differences in microbial flora and fauna that establish in the different replicate containers and suggested that this could possibly be overcome by inoculating each container with a standard suspension before introducing the larvae; however, this has not yet been tested.

3. *Insects in soil habitats*

Most bioassays with soil-inhabiting insects are conducted in soil placed in containers in a controlled environment chamber. Temperature is the easiest variable to control. Although soil temperatures may be similar to air temperatures in controlled environment chambers, where possible, it is recommended that the temperature in soil be recorded. There are numerous sensors or transducers that can be used to measure temperature (Livingston, 1993a). Water availability in soil is a more problematic parameter to control than temperature (see Section 6A.3).

Since in most soil assays the insects are cryptic, daily mortality assessment is often not possible. Consequently, mortality assessments are made at the termination of the bioassay period.

6 ASSESSMENT OF ENVIRONMENTAL PARAMETERS

The ultimate challenge in the study of entomopathogenic Hyphomycetes is to predict field efficacy based on laboratory-acquired data. *In vitro* assays can be used to determine pertinent variables affecting fungal growth, development and pathogenicity. Persistence of propagules can also be estimated through field studies. Ultimately, bioassays using target hosts and incorporating the pertinent environmental variables will provide information most suited for prediction of efficacy under field conditions.

A Fungal tolerances

There is great phenotypic and genotypic variability present among strains of entomopathogenic Hyphomycetes which affect, among other things, persistence in the field. To better predict efficacy under field conditions, intra-specific tolerances to environmental constraints should first be determined in laboratory assays. Such assays have been developed to study the three most important parameters: sunlight, temperature and humidity. Furthermore, persistence in the field must be verified.

1. *Sunlight*

Natural sunlight is one of the more important factors affecting survival of propagules under field conditions, the ultraviolet radiation -B (295–320 nm) component being the most detrimental. However, irradiation at different wavelengths may be beneficial by promoting photoreactivation, a phenomenon whereby the detrimental effects of UV irradiation are counteracted by the organism. Consequently, assays assessing tolerance to sunlight should preferably use polychromatic light at a temperature favourable to photoreactivation (Fargues *et al.*, 1996).

Conidia are uniformly deposited on a substrate (a variety of substrates can be used including glass slides, Petri plates, filter paper or foliage), air dried and then exposed to a source of simulated sunlight.

Natural sunlight is variable and unpredictable and therefore should be avoided, especially if replicates are to be made on different days. Several artificial sunlight devices are available commercially. Long-pass filters are used to block short wavelengths under 295 nm to simulate natural sunlight (Rougier *et al.*, 1994). The substrates containing the propagules are irradiated for the desired period. The irradiance received can be varied by changing the distance of the substrate from the light source. Intensity of UV-B radiation should be measured with a radiometer. Adequate ventilation is paramount as irradiation can produce significant levels of ozone which, in itself, can be toxic to conidia. A non-irradiated (i.e. shaded) control should be kept.

Although it is most desirable to simulate natural sunlight as closely as possible and include polychromatic light, much information can still be gained using much simpler and cheaper light sources. For instance, Inglis *et al.* (1995a) tested the effects of UV protectants using a UV-B fluorescent bulb (Ultra-Lum, Carson, CA) which emits radiation from 260 to 400 nm with a peak at 300–310 nm. However, wavelengths under 295 nm which normally do not reach the earth's surface should be filtered using long pass filters if possible.

Following exposure, conidia are harvested and viability is assessed using any of the methods described above (Section 4). If germination counts are used, it is preferable to use the Benlate method (see Section 4A.3), as UV irradiation may delay germination in a proportion of the conidia, thereby increasing the problems caused by obstruction of counts due to hyphal growth of the early germinated conidia. Conidial survival is estimated by comparing the viability of the irradiated conidia with the viability of the shaded, control conidia

$$\%\ \text{survival} = \frac{\text{no. viable conidia following irradiation}}{\text{no. viable conidia in control}} \times 100$$

2. *Temperature*

Temperature is an important factor that determines the rate of germination, growth, sporulation and survival of entomopathogenic Hyphomycetes. Studies of temperature effects on these factors are generally straightforward. Controlled environment chambers are used to keep temperatures constant, generally to within ±1 °C.

a. *Spore germination*

In determining the effect of temperature on spore germination, it is important to realize that even very short periods of changes in temperature can significantly affect responses. Therefore, destructive sampling should be used whenever possible. Adequate numbers of replicated inoculated plates, preferably containing Benlate to inhibit further growth of early germinated conidia (see Section 4A.3) must be set up at several temperatures, usually between 5 and 40 °C if upper and lower limits are sought. Periodically, several replicates are removed and evaluated. It is simplest if such plates are fixed with a drop of fixative (e.g. lactophenol cotton blue), covered with a coverslip and evaluated later.

The percentage germination is then calculated for each plate (i.e. temperature by time combination). Mean percentages for each temperature by time combination are transformed to their Logit or Probit values to obtain a straight line relationship between germination and time. Maximum-likelihood methods are used to estimate lag phase and germination rate. (Hywel-Jones & Gillespie, 1990).

b. *Vegetative growth*

Effects of temperature on growth of entomopathogenic fungi are most easily assessed on semi-synthetic media in Petri plates using colony diameters. Petri plates containing an adequate medium (see Section 3A) are inoculated centrally, either by placing a conidial suspension (e.g. 0.1 μl) or a small plug (e.g. 6 mm diameter) taken from a fresh, unsporulated culture, and incubated for a period of time at several temperatures. For most entomopathogenic Hyphomycetes, a range of temperatures between 4 and 40°C should be chosen. Plates are incubated in total darkness for approximately 2 weeks under conditions of saturated humidity. Three to five replicate dishes should be prepared for each temperature/isolate combination. Surface radial growth is recorded daily using two perpendicular measurements which can be drawn at the bottom of each dish at the commencement of the experiment. If radial measurements are done rapidly, destructive sampling is not required.

Because radial growth from day 3 to day 12 usually fits a linear model ($y = a + bx$) where a is the growth velocity, growth rates (velocity in mm/day) are used as the main parameter to evaluate the influence of temperature on fungal growth (Fargues *et al.*,

1992). In order to compare maximum growth rates between isolates, analyses (e.g. ANOVA) can be done on relative growth rates (%) calculated from the maximum growth rate for each isolate.

c. Moisture

Effects of moisture on germination, growth and sporulation of entomopathogenic Hyphomycetes are usually carried out using media adjusted to different water activities. For aerial studies, the relative humidity (RH) is usually maintained with glycerol or saturated salt solutions (Appendix 4). Manipulation of water potentials of liquid and solid media is accomplished by adjusting the solute concentration in the medium and equilibrating the medium or test material in a closed chamber with controlled RH. Solutes used include various salts, glucose, sucrose, glycerol, or ployethyleneglycol (PEG). Since the solute in the medium may have other effects, it is prudent to test several. The water potential, as affected by the solutes, can be easily calculated as follows:

$$\psi_{ts} = -4.46 \times 10^{-3}\ T\Delta T$$

Where ψ_{ts} is pressure in megapascals (MPa), T is absolute temperature K and ΔT is the freezing point depression (Griffin, 1994). The reader is referred to Rockland (1960) and Dhingra & Sinclair (1985) for more information on use of solutes to control humidity and water potential.

Since aerial humidities occur at equilibrium only at the solution/air interface, it is important to include a method for air circulation for optimum humidity control. Such a system has been developed for study of effects of humidity on entomopathogenic Hyphomycetes (Fargues and Goujet, personal communication). Air is constantly circulated with a membrane air pump over a saturated salt solution in one chamber (18 × 27 × 18 cm), containing 1 kg of salt in 0.5 l distilled water, into a second chamber (27 × 36 × 18 cm) in which the test materials are placed. Air exchange is approximately one complete air change in the test chamber per 4–5 min. Humidity is monitored within the test chamber with probes attached to data loggers.

To study effects of water on germination, growth and sporulation of fungi, a series of agar or aqueous media are prepared using different molar solutions (e.g. glycerol or PEG) to obtain media with a range of water activities (a_w) (Magan & Lacey, 1984). The media are inoculated with dry spores and placed in the humidity controlled chambers, each treatment in a chamber corresponding to the a_w of the medium. For growth in liquid media, flasks need not be included in a controlled humidity environment; however, water lost due to evaporation must be replaced daily. Germination, growth and sporulation are then evaluated as described on p. 234 and 239). Inch & Trinci (1987) demonstrated that there was good correlation on effects of water activity on growth of *P. farinosus* between measurements from shake flasks and those from solid medium.

It is often desirable to determine the effect of RH on sporulation on the surface of the insect cadaver. Insects are experimentally infected with the pathogen (see Section 5). Immediately after death, cadavers are transferred to the controlled humidity chamber and incubated for the desired time, usually 10–15 days for most entomopathogenic Hyphomycetes. The cadavers are then either washed or homogenized and the conidia are enumerated (see Section 4).

The availability of water to micro-organisms in soil is affected by a number of factors, the most significant of which is soil texture. For example, the availability of water will not be equal in two soils with different textures but the same percentage water content (v/w). By controlling the water potential of soil, it becomes possible to study the effects of water on the efficacy of entomopathogenic Hyphomycetes, independent of soil texture and vice versa. The total water potential of soil is the sum of the component water potentials so that:

$$\psi_t = \psi_g + \psi_m + \psi_o + \ldots$$

where ψ_t is the total water potential, ψ_g is a gravitational potential constant, ψ_m is the matric potential, ψ_o is the osmotic potential and other less significant potentials (e.g. pressure potential) are indicated by dots. Matric potential is the potential arising from the attraction of the soil matrix for water (adsorption and capillary). The osmotic potential of soil arises from dissolved solutes and lowered activity of water attributable to interaction with charged surfaces (Livingston, 1993b). Water potential possesses units of pressure, usually MPa or bars where 1 MPa is equal to 10 bars. Saturated soil has a water potential of ≈ 0 bars, and as the soil becomes drier, the water potential becomes increasingly more negative. Soil water desorption curves can be determined using a number of methods including pressure plates, resistance blocks, tensiometers, thermocouple psychrometers, neutron scattering, gamma-ray attenuation, ultrasonic energy, and/or filter paper methods (Livingston, 1993b; Topp, 1993; Topp *et al.*, 1993). There are advantages and disadvantages to each of these methods, and the reader should consult any number of references to obtain additional information on the quantification of soil water potentials.

Studdert & Kaya (1990) investigated the effect of water availability in two soils (organic and sandy loam texture) on the efficacy of *B. bassiana* against beet armyworm (*Spodoptera exigua*) pupae. Soil containing conidia was dispensed into 200 ml plastic containers containing pupae, and the containers were covered with polyethylene sheets leaving an air space of ≈1.5 cm. Containers with relatively moist soils were placed in plastic containers covered with damp towels and kept in the dark at a controlled temperature; the towels were wetted periodically. Less than 2% of the initial soil water was lost during the experiment (10 days). For drier soils (–37 and –200 bars), containers were maintained in desiccators over saturated salt solutions. Saturated salt solutions are used to control the water content of the atmosphere (Dhingra & Sinclair, 1985; Appendix, 4)

Although it is possible to maintain a relatively constant water potential in soils kept in closed containers, this is not the case when soils are exposed to the atmosphere. Inglis *et al.* (unpublished) studied the susceptibility of ovipositing grasshoppers to *B. bassiana* conidia in soil. Grasshoppers choose the depth at which they oviposit according to soil texture and moisture. They will readily oviposit into soil at or near field capacity. However, moisture is rapidly lost from soil in cups, particularly under conditions of low ambient humidity typical of arid agroecosystems. The daily addition of water to the soil surface is unsatisfactory due to the abnormal placement of egg pods near the top of the soil profile (≈1–2 cm) by females. To reduce saturation at the surface but permit the addition of water to soil, containers were fitted with a central watering tube (Figure 5). The porous gravel base acted as a water reservoir from which water could rapidly spread across the soil bottom, and then move upward in the profile by capillary action. Capillary and absorption forces associated with soil matrix determine the field capacity of soil; field capacity is the point at which the

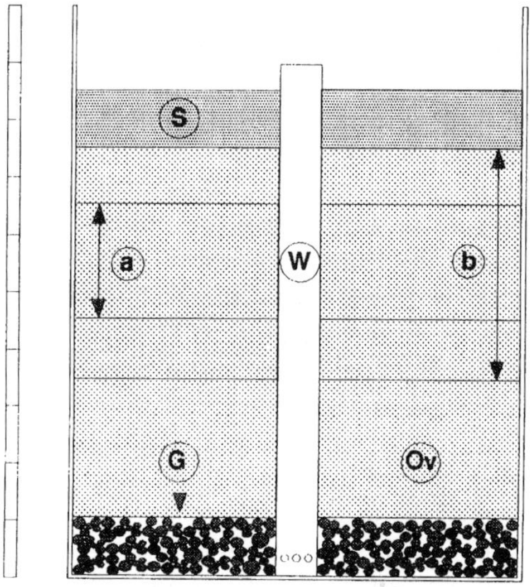

Figure 5 Soil container used to test the efficacy of *Beauveria bassiana* conidia in soil (Ov) against ovipositing grasshoppers. The watering tube (W) allowed the addition of water to the bottom of the soil profile. The porous gravel (G) facilitated lateral movement of water and acted as a reservoir from which water moved upward into the profile by capillary action. Soil moisture was maintained near field capacity at the depth where eggs were deposited. A layer of sterile soil (S) was placed on the surface to prevent liberation of conidia due to oviposition activity. The bar adjacent to the container is 10 cm in length in 1 cm increments. Most egg pods were deposited between the lines indicated by the arrow marked 'a', but ranged between the lines indicated by the arrow marked 'b'.

macropores are filled with air but water remains in micropores or capillary pores. Although the maintenance of soil at field capacity (or relative to field capacity) is an imprecise measure of matric potential, in some instances it is used as an estimation of water availability.

Effects of rain on persistence of propagules on leaf or insect surfaces can be evaluated using rainfall simulators (e.g. Tossell *et al.*, 1987). The substrate in question (e.g. leaf or insect surface) is inoculated with a known quantity of propagules (Section 5), and then the propagules are enumerated following exposure to simulated rain. The effects of the rain on conidial removal from leaves can be assessed using analysis of covariance with conidial populations on leaves before exposure to rain used as the covariate. Percentage reduction in *B. bassiana* conidia due to rain exposure was determined as: (number of propagules (e.g. cfu) prior to exposure – number of propagules after exposure)/number of propagules prior to exposure (Inglis *et al.*, 1995b).

B Inoculum persistence

It is often necessary to quantify rates of change in propagule populations over time, particularly when studying the effects of environmental parameters on the efficacy of entomopathogenic Hyphomycetes. To estimate persistence of entomopathogenic hyphomycete propagules in epigeal habitats, a number of substrates can be used (see Section 4B). Most often, conidial survival on leaves is measured, since a measure of area can be obtained.

Fransen (1995) used a leaf imprint technique to directly assess survival of *Aschersonia aleyrodis* spores on leaf surfaces; spores were inoculated on to leaf surfaces and incubated. Leaf imprints were made on water agar plates periodically up to 20 days after spore application. Immediately after each impression was made, the numbers of germinated and ungerminated spores were recorded microscopically. The agar plates were incubated under conditions favourable for spore germination (25 °C) for 24 h and the numbers of germinated and ungerminated spores were recorded again and reduction in germination was assessed.

However, enumeration of most propagules associated with a number of substrates including leaves, other plant organs (e.g. flowers or fruit), soil, and insects must be assessed through indirect methods (see Section 4B). Using such methods, it is often convenient to quantify populations per unit weight or insect (e.g. at a specific stadium), especially when it is difficult to accurately measure surface areas (e.g. insects). When significant variations in weight or size occur (e.g. individual insects or leaves), it is usually necessary to standardize weights. For example, cfu per mg can be calculated and then multiplied by the mean weight of the insects processed to obtain a measure of cfu per average insect. Although weight measurements are useful for comparing populations within a specific substrate, they are less accurate than area for comparing propagule densities between substrates. Although population densities or IU per unit area, weight or volume are most commonly used for assessing the persistence of propagules, the incidence of insect mortality or of propagule survival over time (e.g. effect of environmental parameters on conidial germinability *in vitro*) have been used as measures of persistence.

For statistical comparisons of propagule persistence between treatments, it is usually necessary to normalize the data before analysis; a logarithmic transformation is usually required with data from dilution plate counts. Although it is desirable to sample from different populations at each collection date, in most instances it is logistically difficult to do so, and samples are often obtained from the same population (e.g. same plot or plant) at each time. Repeated measure data have been analysed as a split-plot in time (Gomez & Gomez, 1984; Milliken & Johnson, 1984 see Section 5A.2).

C Host–fungus interactions

Bioassays using static conditions may be useful in comparing activity of different isolates, but they usually provide little information on the performance of the pathogen under field conditions. Consequently, bioassays must be developed that incorporate as many pertinent environmental parameters as possible. Inoculation techniques and the environmental conditions chosen should mimic as much as possible the natural situation and, more specifically, conditions at the level of the host microhabitat. By varying single or multiple factors, and by using information obtained on fungal tolerances, it should be possible to develop predictive models that would be indispensable in the development of entomopathogenic Hyphomycetes as microbial control agents. For instance, to study effects of temperature on a thermoregulating host such as the grasshopper, Inglis *et al.* (1996b) used bioassay cages fitted with incandescent bulbs to allow for behavioural thermoregulation by the host (Figure 6). Cages were placed in a large controlled temperature chamber, and the periods of 'simulated sunshine' were varied by limiting the time the bulbs, which provided a heat gradient, were turned on. Inoculated hosts were introduced into the system and monitored twice daily. Results were compared to those obtained in static temperature bioassays. Studies on the effects of ambient humidity on infection of hosts can be accomplished using controlled humidity chambers (see Section 6C.3). Preferably diurnal humidities and temperatures should be used to better mimic natural conditions.

Soil is very complex and the efficacy of ento-

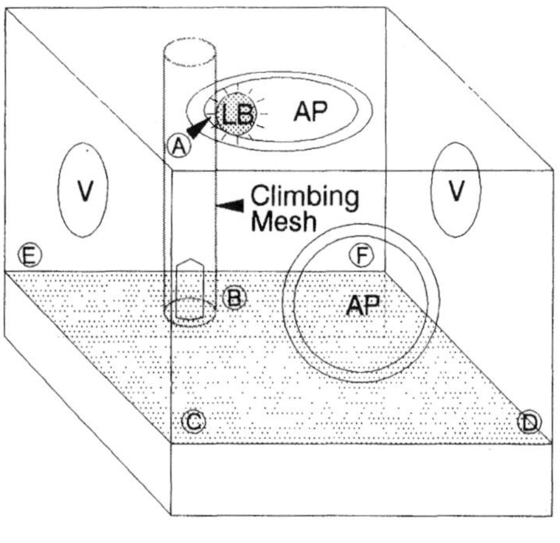

Figure 6 A bioassay cage used to test the effect of grasshopper body temperature on efficacy of *Beauveria bassiana*. To elevate their body temperature by basking, nymphs had to climb a plastic mesh (8 cm diameter × 28 cm high) towards a heat source (25 W incandescent light bulb; LB) in an aluminum cage (40 × 40 × 30 cm). The cage was equipped with a perforated metal floor, and a clear plexiglass top and front. Air flow was enhanced by mesh vents (V) and access ports (AP), and temperatures were monitored at points A to F.

mopathogens is affected by a number of biotic and abiotic variables. Soil parameters that influence the efficacy of entomopathogenic Hyphomycetes include temperature, texture, water activity, pH, exposure to agrochemicals and the presence of microflora (McCoy *et al.*, 1992). However, other soil characteristics (e.g. compaction and physical structure) likely affect entomopathogenic Hyphomycetes as well, but have not been extensively tested. It is important to consider the abiotic and biotic characteristics of soil in the experimental design phase of the bioassay.

The presence of a microflora can also influence the efficacy of entomopathogenic Hyphomycetes, especially in soil (e.g. Groden & Lockwood, 1991; Pereira *et al.*, 1993). The two most common methods used to study the effects of soil micro-organisms on the mycosis of insects are to compare efficacy in sterile or pasteurized soil versus non-sterilized soil, or to stimulate microbial activity by amending soil with nutrients and subsequently comparing disease development in insects in the amended and unamended soils. Sterilization of soil is accomplished using heat (with or without steam), chemical sterilants or irradiation (Wolf & Skipper, 1994). All sterilization methods result in chemical alterations, primarily to the organic component.

Heat is most commonly used to sterilize soil but high temperatures cause hydrolysation of carbohydrates, coagulation of proteins, and influence a variety of other chemical properties of soil (Wolf & Skipper, 1994). Heat is more effective when used in combination with high pressure steam (121 °C > 100 kPa = 15 PSI). Although steam enhances conduction and penetration, the sterilization efficacy is further facilitated if soil is dry and exposed in relatively thin layers. Pasteurization of soil is accomplished by exposing soil to relatively low temperatures (≈ 60–75 °C) for short periods of time. Although less severe (i.e. reduced chemical modifications to organic molecules), pasteurization generally does not affect thermophilic micro-organisms, bacterial endospores or cells in a desiccated state. *Bacillus* species and other spore-forming taxa (e.g. *Clostridium* spp.) are cosmopolitan in soil, and in soils where endospores are prevalent, tyndalization may be used. Tyndalization involves the repeated (2 to 3 times), short-term exposure of soil to high temperatures (e.g. 100 °C) at intervals (usually 24 h). The rationale for this method is that high temperature exposure stimulates endospore germination and vegetative growth then occurs. These vegetative cells are vulnerable to high temperature. The primary limitations to the use of sterilized or pasteurized soils include altered soil chemistry, a modified microbial community due to chemical changes, and the rapid colonization by pioneer colonizers following sterilization, which can result in the presence of large populations of a relatively small number of taxa.

The other strategy used to elucidate the effect of soil micro-organisms on entomopathogenic Hyphomycetes is to add nutrients (carbon and nitrogen) to soil. At periods following the amendment of soil, the possible effects of fungistasis are determined on vegetative growth (e.g. conidial germination and hyphal growth) and/or on disease development. Although the soil amendment method may be useful for establishing the potential effect of microbial antagonists on entomopathogenic Hyphomycetes, the selection for certain types of micro-organisms that results from the treatment may not be indicative of soils in nature. For example, the incorporation of simple carbohydrates (e.g. glucose) may favour fast-growing fungi with relatively simple nutritional requirements (e.g. Mucorales) to the detriment of taxa with more complex carbohydrate requirements (e.g. cellulolytic organisms). For both the sterilization and soil amendment methods, it is important that both total microbial biomass and microbial diversity be monitored.

In an effort to duplicate natural conditions, field-cage experiments can be used. However, results from field-cage assays must not be used as conclusive evidence of the fungus–host interactions that would normally take place in free-living insects. Cages provide a microclimate that is very different from that of the natural environment. For instance, cage screening provides shading and protection from wind. However, field-cages provide an excellent method to study host–pathogen interactions, especially if many of the factors have already been determined under laboratory conditions. For instance, Inglis *et al.* (1996d) investigated the influence of environmental conditions on the efficacy of *B. bassiana* against grasshoppers in field environments using caged insects. Testing laboratory-acquired bioassay results that showed that grasshoppers respond behaviourally to infection by thermoregulating and are able to recover from disease, it was demonstrated that a higher incidence and more rapid development of disease occurred in grasshoppers placed in shaded than in cages exposed to direct sunlight or protected from UV-B radiation. Methods for field-cage trials must be adapted for the different conditions and parameters being tested.

Once microclimatic constraints are better quantified and understood, it may be possible to overcome some of these inhibitory situations through improved formulation, strain selection, genotypic or phenotypic manipulation, and inoculum targeting. Identification of microlimatic constraints would also allow development of predictive models which would identify windows of opportunity, thereby optimizing efficacious use of these microbial control agents.

ACKNOWLEDGEMENTS

We thank Grant Duke for useful suggestions and help with compilation of literature; Toby Entz and Dan Johnson for their suggestions on experimental design and analyses, Jacques Fargues and Nathalie Smits for comments and suggestions on assessment of environmental parameters and John Vandenberg and Ann Hajek for critically reviewing the manuscript. This chapter was written while MSG was on work study leave at the Unité de Recherche en Lutte Biologique, Campus International de Baillarguet, INRA-Montpellier with fellowship support from the Institut National de Recherche Agronomique, the French Ministry of Education and Research, and the OECD Co-operative Research Programme: Biological Resource Management for Sustainable Agricultural Systems. This is LRC contribution 3879622.

REFERENCES

Abbott, W. S. (1925) A method for computing the effectiveness of an insecticide. *J. Econ. Entomol.* **18**, 265–267.

Allee, L. A., Goettel, M. S., Gol'berg, A., Whitney, H. S. & Roberts, D. W. (1990). Infection by *Beauveria bassiana* of *Leptinotarsa decemlineata* larvae as a consequence of fecal contamination of the integument following *per os* inoculation. *Mycopathologia* **111**, 17–24.

Atlas, R. M., and Bartha, R. (1981) *Microbial ecology: Fundamentals and Applications*, Addison-Wesley, Reading, 560 pp.

Bailey, L. A. & Rath, A. C. (1994) Production of *Metarhizium anisopliae* spores using nutrient-impregnated membranes and its economic analysis. *Biocontrol Sci. Technol.* **4**, 297–307.

Bateman, R. P., Carey, M., Moore, D. and Prior, C. (1993) The enhanced infectivity of *Metarhizium flavoviride* in oil formulations to desert locusts at low humidities. *Ann. Appl. Biol.* **122**, 145–152.

Beilhartz, V. C., Parbery, D. G. & Swart, H. J. (1982) Dodine: A selective agent for certain soil fungi. *Trans. Br. Mycol. Soc.* **79**, 507–511.

Bidochka, M. J., Pfeifer, T. A. & Khachatourians, G. G. (1987) Development of the entomopathogenic fungus *Beauveria bassiana* in liquid cultures. *Mycopathologia* **99**, 77–83.

Blachère, H., Calvez, J., Ferron, P., Corrieu, G. & Peringer, P. (1973) Etude de la formulation et de la conservation d'une préparation entomopathogène a base de blastospores de *Beauveria tenella* (Delacr. Siemaszko). *Ann. Zool. Ecol. Anim.* **5**, 69–79.

Carmichael, J. W. (1955) Lacto-fuchsin: A new medium for mounting fungi. *Mycologia* **47**, 611.

Chase, A. R., Osborne, L. S. & Ferguson, V. M. (1986) Selective isolation of the entomopathogenic fungi *Beauveria bassiana* and *Metarhizium anisopliae* from an artificial potting medium. *Florida Entomol.* **69**, 285–292.

Ciafardini, G. & Marotta, B. (1989) Use of the most-probable-number technique to detect *Polymyxa beta* (Plasmodiophoromycetes) in soil. *Appl. Environ. Microbiol.* **55**, 1273–1278.

D'Agostino, R. B., Belanger, A. & D'Agostino, R. B. (1990) A suggestion for using powerful and informative tests for normality. *Am. Statist.* **44**, 316–321.

Daoust, R. A. & Roberts, D. W. (1983) Studies on the prolonged storage of *Metarhizium anisopliae* conidia: Effect of temperature and relative humidity on conidial viability and virulence against moquitoes. *J. Invertebr. Pathol.* **41**, 143–150.

Dhingra, O. D. & Sinclair, J. B. (1985) *Basic Plant Pathology Methods.* CRC Press, Boca Raton, 355 pp.

Edgington, L. V., Khew, K. L. & Barron, G. L. (1971) Fungitoxic spectrum of Benzimidazole compounds. *Phytopathology* **61**, 42–44.

Fargues, J., Goettel, M. S., Smits, N., Ouedraogo, A., Vidal, C., Lacey, L. A. & Rougier, M. (1996) Variability in susceptibility to simulated sunlight of conidia among isolates of entomopathogenic Hyphomycetes. *Mycopathologia* (submitted).

Fargues, J., Maniania, N. K., Delmas, J. C. & Smits, N. (1992) Influence de la température sur la croissance in vitro d'hyphomycètes entomopathogènes. *Agronomie* **12**, 557–564.

Feng, M. G., Poprawski, T. J. & Khachatourians, G. G. (1994) Production, formulation and application of the entomopathogenic fungus *Beauveria bassiana* for insect control: Current status. *Biocontrol Sci. Technol.* **4**, 3–34.

Finney, D. J. (1971) *Probit Analysis.* Cambridge University Press, Cambridge.

Fransen, J. J. (1995) Survival of spores of the entomopathogenic fungus *Aschersonia aleyrodis* (Deuteromycotina: Coelomycetes) on leaf surfaces. *J. Invertebr. Pathol.* **65**, 73–75.

Gardner, J. M. & Pillai, J. S. (1987) *Tolypocladium cylindrosporum* (Deuteromycotina: Moniliales), a fungal pathogen of the mosquito *Aedes australis* II. Methods of spore propagation and storage. *Mycopathologia* **97**, 77–82.

Gilchrist, J. E., Campbell, J. E., Donnely, C. B., Peeler, J. T. & Delanay, J. M. (1973) Spiral plate method for bacterial determination. *Appl. Microbiol.* **43**, 149–157.

Girard, K. & Jackson, C. W. (1993) Using fluorescent microscopy and image analysis to assess distribution of *Verticillium lecanii* spores on *Rhopalosiphum padi. Proc. Soc. Invertebr. Pathol.* p. 55 (abstract).

Goettel, M. S. (1984) A simple method for mass culturing entomopathogenic Hyphomycete fungi. *J. Microbiol. Methods* **3**, 15–20.

Goettel, M. S. (1987) Studies on bioassay of the entomopathogenic hyphomycete fungus *Tolypocladium cylindrosporum* in mosquitoes. *J. Am. Mosq. Control Assoc.* **3**, 561–567.

Gomez, K. A. & Gomez, A. A. (1984) *Statistical Procedures for Agricultural Research.* Wiley, New York, 680 pp.

Griffin, D. H. (1994) *Fungal Physiology*, 2nd edn. Wiley-Liss, New York, 458 pp.

Groden, E. & Lockwood, J. L. (1991) Effects of soil fungistasis on *Beauveria bassiana* and its relationship to disease incidence in the Colorado potato beetle, *Leptinotarsa decemlineata*, in Michigan and Rhode Island soils. *J. Invertebr. Pathol.* **57**, 7–16.

Guy, P. L. & Rath, A. C. (1990) Enzyme-linked immunosorbent assay (ELISA) to detect spore surface antigens of *Metarhizium anisopliae. J. Invertebr. Pathol.* **55**, 435–436.

Hajek, A. & St Leger, R. J. (1993) Interactions between fungal pathogens and insect hosts. *Annu. Rev. Entomol.* **39**, 293–322.

Hall, R. A. (1976) A bioassay of the pathogenicity of *Verticillium lecanii* conidiospores on the aphid, *Macrosiphoniella sanborni. J. Invertebr. Pathol.* **27**, 41–48.

Harris, J. L. (1986) Modified method for fungal slide culture. *J. Clin. Microbiol.* **24**, 460–461.

Harris, R. F. & Sommers, L. E. (1968) Plate-dilution frequency technique for assay of microbial ecology. *Appl. Microbiol.* **16**, 330–334.

Hedgecock, S., Moore, D., Higgins, P. M. & Prior, C. (1995) Influence of moisture content on temperature tolerance and storage of *Metarhizium flavoviride* conidia in an oil formulation. *Biocontrol Sci. Technol.* **5**, 371–377.

Hywell-Jones, N. L. & Gillespie, A. T. (1990) Effect of temperature on spore germination in *Metarhizium anisopliae* and *Beauveria bassiana. Mycol. Res.* **94**, 389–392.

Ignoffo, C. M., Garcia, C. & Kroha, M. J. (1982) Susceptibility of larvae of *Trichoplusia ni* and *Anticarsia gemmatalis* to intrahemocoelic injections of conidia and blastospores of *Nomuraea rileyi. J. Invertebr. Pathol.* **39**, 198–202.

Ignoffo, C. M., Garcia, C., Kroha, M. J., Samšiňáková, A. & Kálalová, S. (1983) A leaf surface treatment bioassay for determining the activity of conidia of *Beauveria bassiana* against *Leptinotarsa decemlineata. J. Invertebr. Pathol.* **41**, 385–386.

Inch, J. M. M. & Trinci, P. J. (1987) Effects of water activity on growth and sporulation of *Paecilomyces farinosus* in liquid and solid media. *J. Gen. Microbiol.* **133**, 247–252.

Inglis, G. D., Goettel, M. S. & Johnson, D. L. (1995a) Influence of ultraviolet light protectants on persistence of the entomopathogenic fungus, *Beauveria bassiana. Biol. Control* **5**, 581–590.

Inglis, G. D., Johnson, D. L. & Goettel, M. S. (1995b) Effects of simulated rain on the persistence of *Beauveria bassiana* conidia on leaves of alfalfa and wheat. *Biocontrol Sci. Technol.* **5**, 365–369.

Inglis, G. D., Johnson, D. L. & Goettel, M. S. (1996a) Effect of bait substrate and formulation on infection of grasshopper nymphs by *Beauveria bassiana. Biocontrol Sci. Technol.* **6**, 35–50.

Inglis, G. D., Johnson, D. L. & Goettel, M. S. (1996b) Effects of temperature and thermoregulation on mycosis by *Beauveria bassiana* in grasshoppers. *Biol. Control* **7**, 131–139.

Inglis, G. D., Johnson, D. L. & Goettel, M. S. (1996c) An oil-bait bioassay method used to test the efficacy of

Beauveria bassiana against grasshoppers. *J. Invertebr. Pathol.* **67**, 312–315.

Inglis, G. D., Johnson, D. L. & Goettel, M. S. (1996d) Effects of temperature and sunlight on mycosis (*Beauveria bassiana*) of grasshoppers under field conditions. *Environ. Entomol.* (in press).

Jenkins, N. E. & Goettel, M. S. (1997) Methods for mass production of microbial control agents of grasshoppers and locusts. *Microbial Control of Grasshoppers and Locusts* (eds M. S. Goettel & D. L. Johnson) Memoirs Entomological Society of Canada (in press).

Jenkins, N. E. & Prior, C. (1993) Growth and formation of true conidia by *Metarhizium flavoviride* in a simple liquid medium. *Mycol. Res.* **97**, 1489–1494.

Jimenez, J. & Gillespie, A. T. (1990) Use of the optical brightener Tinopal BOPT for the rapid determination of conidial viabilities in entomogenous deuteromycetes. *Mycol. Res.* **94**, 279–283.

Johnson, D. L. (1992) Introduction: biology, ecology, field experimentation and environmental impact. In *Biological Control of Locusts and Grasshoppers* (eds C. J. Lomer & C. Prior), pp. 267–278. CAB International, Wallingford.

Johnson, D. L., Hill, B. D., Hinks, C. F. & Schaalje, G. B. (1986) Aerial application of the pyrethroid deltamethrin for grasshopper (Orthoptera: Acrididae) control. *J. Econ. Entomol.* **79**, 181–188.

Jones, D. (1984) Use, misuse, and role of multiple-comparison procedures in ecological and agricultural entomology. *Environ. Entomol.* **13**, 635–649.

Kleespies, R. G. & Zimmermann, G. (1992) Production of blastospores by three strains of *Metarhizium anisopliae* (Metch.) Sorokin in submerged culture. *Biocontrol Sci. Technol.* **2**, 127–135.

Kybal, J. & Vlček, V. (1976) A simple device for stationary cultivation of microorganisms. *Biotechnol. Bioengineer.* **18**, 1713–1718.

Landa, Z., Osborne, L., Lopez, F. & Eyal, J. (1994) A bioassay for determining pathogenicity of entomogenous fungi on whiteflies. *Biol. Control* **4**, 341–350.

Little, T. M. & Hills, F. J. (1978) *Agricultural experimentation.* Wiley, New York.

Liu, Z. Y., Milner, R. J., McRae, C. F. & Lutton, G. G. (1993) The use of dodine in selective media for isolation of *Metarhizium* spp. from soil. *J. Invertebr. Pathol.* **62**, 248–251.

Livingston, N. J. (1993a). In *Soil sampling and methods of analysis* (ed. M. R. Carter), pp. 673–682. Lewis Publishing, Boca Raton.

Livingston, N. J. (1993b) In *Soil sampling and methods of analysis* (ed. M. R. Carter), pp. 559–567. Lewis Publishing, Boca Raton.

Magan, N. & Lacey, J. (1984) Effect of temperature and pH on water relations of field and storage fungi. *Trans. Br. Mycol. Soc.* **82**, 71– 81.

McCoy, C. W., Hill, A. J. & Kanavel, R. F. (1975) Large-scale production of the fungal pathogen *Hirsutella thompsonii* in submerged culture and its formulation for application in the field. *Entomophaga* **20**, 229–240.

McCoy, C. W., Storey, G. K. & Tigano-Milani, M. S. (1992) Environmental factors affecting entomopathogenic fungi in soil. *Pesqui. Agropecu. Bras.* **27**, 107–111.

Mendonça, A. F. (1992) Mass production, application and formulation of *Metarhizium anisopliae* for control of sugarcane froghopper; *Mahanarva posticata*, in Brazil. In *Biological control of locusts and grasshoppers* (eds C. J. Lomer & C. Prior) pp 239–244. CAB International, Wallingford.

Meynell, G. G. & Meynell, E. (1970) *Theory and practice in experimental bacteriology.* 2nd edn. Cambridge University Press, Cambridge, 347 pp.

Milliken, G. A. & Johnson, D. E. (1984) *Analysis of messy data.* Vol. 1. *Designed experiments.* Van Nostrand Reinhold, New York, 473 pp.

Milner, R. J., Huppatz, R. J. & Swaris, S. C. (1991) A new method for assessment of germination of *Metarhizium* conidia. *J. Invertebr. Pathol.* **57**, 121–123.

Mitchell, D. J., Kannwischer-Mitchell, M. E. & Dickson, D. W. (1987) A semi-selective medium for the isolation of *Paecilomyces lilacinus. J. Nematol.* **19**, 255–256.

Morse, P., Hall, I. & Ludwig, K. (1987) S108 Multiline Quantal Bioassay Program. Fortran program available from, Agriculture and Agri-Food Canada, Sir John Carling Building, 930 Carling Ave, Ottawa, Ontario, K1A 0C5.

Nadeau, M. P. & Boisvert, J. L., (1994) Larvicidal activity of the entomopathogenic fungus *Tolypocladium cylindrosporum* (Deuteromycotina: Hyphomycetes) on the mosquito *Aedes triseriatus* and the black fly *Simulium vittatum* (Diptera: Simulidae) *J. Am. Mosq. Control Assoc.* **10**, 487–491.

Nowierski, R. M., Zeng, Z., Jaronski, S., Delgado, F. & Swearingen, W. (1996) Analysis and modeling of time-dose-mortality of *Melanoplus sanguinipes*, *Locusta migratoria migratorioides*, and *Schistocerca gregaria* (Orthoptera: Acrididae) from *Beauveria, Metarhizium*, and *Paecilomyces* isolates from Madagascar. *J. Invertebr. Pathol.* **67**, 236–252.

Padhye, A. A., Sekhon, A. S. & Carmichael, J. W. (1973) Ascocarp production by *Nannizzia* and *Arthroderma* on keritinous and non-keritinous media. *Sabouraudia* **11**, 109–114.

Parkinson, D. (1994) Filamentous fungi. In *Methods of soil analysis.* Part 2, *Microbiological and biochemical properties* (ed. S. H. Mickelson) pp 329–350. Soil Science Society of America, Madison, WI.

Payne, C. D. (ed.) (1978) *The GLIM System Release 3.77 Manual*, Numerical Algorithms Group.

Peirera, R. M. & Roberts, D. W. (1990) Dry mycelium preparations of entomopathogenic fungi, *Metarhizium anisopliae* and *Beauveria bassiana. J. Invertebr. Pathol.* **56**, 39–46.

Pereira, R. M., Stimac, J. L. & Alves, S. B. (1993) Soil antagonism affecting the dose-response of workers of the red imported fire ant, *Solenopsis invicta*, to *Beauveria bassiana* conidia. *J. Invertebr. Pathol.* **61**, 156–161.

Pinder, J. E., Wiener, J. G. & Smith, M. H. (1978) The Weibull distribution: a new method of summarizing survivorship data. *Ecology* **59**, 175–179.

Preisler, H. K. & Robertson, J. L. (1989) Analysis of time-dose-mortality data. *J. Econ. Entomol.* **82**, 1534–1542.

Roberts, D. W., Dunn, H. M., Ramsay, G., Sweeney, A. W. & Dunn, N. W. (1987) A procedure for preservation of the mosquito pathogen *Culicinomyces clavisporus*. *Appl. Microbiol. Biotechnol.* **26**, 186–188.

Robertson, J. L. & Preisler, H. K. (1992) *Pesticide bioassays with arthropods*. CRC Press, Boca Raton.

Robertson, J. L., Smith, K. C., Savin, N. E. & Lavigne, R. J. (1984) Effects of dose selection and sample size on the precision of lethal dose estimates in dose-mortality regression. *J. Econ. Entomol.* **77**, 833–837.

Rockland, L. (1960) Saturated salt solutions for static control of relative humidity between 5° and 40°C. *Anal. Chem.* **32**, 1375–1375.

Rombach, M. C. (1989) Production of *Beauveria bassiana* (Deuteromycotina. Hyphomycetes) sympodulo-conidia in submerged culture. *Entomophaga* **34**, 45–52.

Rougier, M., Fargues, J., Goujet, R., Itier, B. & Benateau, S. (1994) Mise au point d'un dispositif d'étude des effets du rayonnement sur la persistance des micro-organismes pathogènes. *Agronomie* **14**, 673– 681.

Russell, R. M., Robertson, J. L. & Savin, N. E. (1977) POLO: a new computer program for probit analysis. *Bull. Entomol. Soc. Am.* **23**, 209.

Samšiňáková, A., Kálalová, S., Vlček, V. & Kybal, J. (1981) Mass production of *Beauveria bassiana* for regulation of *Leptinotarsa decemlineata* populations. *J. Invertebr. Pathol.* **38**, 169–174.

SAS Institute Inc. (1991) *SAS System for Linear Models*. SAS Institute, Cary.

Schaalje, G. B., Charnetski, W. A. & Johnson, D. L. (1986) A comparison of estimators of the degree of insect control. *Commun. Statis.: Simul. Comput.* **15**, 1065–1086.

Schaalje, G. B., Johnson, D. L. & Van der Vaart, H. R. (1992) Application of competing risks theory to the analysis of effects of *Nosema locustae* and *N. cuneatum* on development and mortality of migratory locusts. *Environ. Entomol.* **21**, 939–948.

Schading, R. L., Carruthers, R. I. & Mullin-Schading, B. A. (1995) Rapid determination of conidia viability for entomopathogenic Hyphomycetes using fluorescence microscopy techniques. *Biocontrol Sci. Technol.* **5**, 201–208.

Snedecor, G. W. & Cochran, W. G. (1987) *Statistical methods*, The Iowa State University Press, Ames, IA.

Steel, R. G. D. & Torrie, J. H. (1960) *Principles and procedures of statistics. McGraw-Hill, New York.*

Studdert, J. P. & Kaya, H. K. (1990) Water potential, temperature, and clay-coating of *Beauveria bassiana* conidia: effect on *Spodoptera exigua* pupal mortality in two soil types. *J. Invertebr. Pathol.* **56**, 327–336.

Tanada, Y. & Kaya, H. K. (1993) *Insect Pathology*. Academic Press, London, 666 pp.

Throne, J. E., Weaver, D. K., Chew, V. & Baker, J. E. (1995) Probit analysis of correlated data: multiple observations over time at one concentration. *J. Econ. Entomol.* **88**, 1510–1512.

Topp, G. C. (1993) Soil water content. In, *Soil sampling and methods of analysis* (ed. M. R. Carter), pp. 541–557. Lewis Publishing, Boca Raton.

Topp, G. C., Galganov, Y. T., Ball, B. C. & Carter, M. R. (1993) Soil water desorption curves. In, *Soil Sampling and Methods of Analysis* (ed. M. R. Carter), pp. 569–579. Lewis Publishing, Boca Raton.

Tossell, R. W., Dickson, W. T., Rudra, R. P. & Wall, G. J. (1987) A portable rainfall simulator. *Canad. Agric. Engineer*. **29**, 155–162.

Tsao, P. H. (1970) Selective media for the isolation of pathogenic fungi. *Annu. Rev. Phytopathol.* **8**,157–186.

Tuite, J. (1969) *Plant Pathological Methods: Fungi and Bacteria*. Burgess, Minneapolis.

van Winkelhoff, A. J. & McCoy, C. W. (1984) Conidiation of *Hirsutella thompsonii* var. *synnematosa* in submerged culture. *J. Invertebr. Pathol.* **43**, 59–68.

Veen, K. H. & Ferron, P. (1966) A selective medium for the isolation of *Beauveria tenella* and of *Metarrhizium anisopliae*. *J. Insect Pathol.* **8**, 268–269.

Warcup, J. H. (1950) The soil-plate method for isolation of fungi from soil. *Nature* **166**, 117–118.

Watson, D. W., Geden, C. J., Long, S. J. & Rutz, D. A. (1995) Efficacy of *Beauveria bassiana* for controlling the house fly and stable fly (Diptera: Muscidae). *Biol. Control* **5**, 405–411.

Webster, J. (1986) *Introduction to fungi*. Cambridge University Press, Cambridge.

Wolf, C. & H. D. Skipper (1994) Soil sterilization. In *Methods of soil analysis*. Part 2. *Microbiological and biochemical properties* (ed. S. H. Mickelson) pp 41–51. Soil Science Society of America, Madison, WI.

Woomer, P. L. (1994) Most probable number counts. In *Methods of soil analysis. Part 2. Microbiological and biochemical properties* (ed. S. H. Mickelson) pp 59–79. Soil Science Society of America, Madison, WI.

Zimmermann, G. (1986) The '*Galleria* bait method' for detection of entomopathogenic fungi in soil. *J. Appl. Entomol.* **102**, 213–215.

APPENDIX

1. Selective media

Beauveria medium (Chase *et al.*, 1986)

- 2% oatmeal infusion
- 2% agar
- 550 μg/ml dodine (*N*-dodecylguanidine monoacetate)
- 5 μg/ml chlortetracycline
- 10 μg/ml crystal violet

Metarhizium medium (Veen & Ferron, 1966; Liu *et al.*, 1993)

- 1% glucose
- 1% peptone
- 1.5% oxgall
- 3.5% agar
- 10 μg/ml dodine (*N*-dodecylguanidine monoacetate)
- 250 μg/ml cycloheximide (actidione)
- 500 μg/ml chloramphenicol

Culicinomyces medium (Panter & Frances, unpublished)

- 1.5% nutrient agar
- 0.27% Lab-Lemco broth
- 500 μg/ml chloramphenicol
- 2 μg/ml thiabendazole

Paecilomyces lilacinus medium (Mitchell *et al.*, 1987)

- 3.9% potato dextrose agar
- 1–3% NaCl
- 0.1% Tergitol
- 500 μg/ml pentachloronitrobenzene
- 500 μg/ml benomyl
- 100 μg/ml streptomycin sulphate
- 50 μg/ml chlortetracycline hydrochloride

2. General culture media

Sabouraud Dextrose Agar + Yeast (SDAY)

- 10 g peptone
- 40 g dextrose
- 2 g yeast extract
- 15 g agar
- 1 l distilled water

Mixed cereal agar (Padhye *et al.*, 1973)

- 25 g Pablum Baby Mixed Cereal
- 5 g agar
- 250 ml water

Mix ingredients and boil in a sealed container as the baby cereal contains *Bacillus* spore formers. Let cool and autoclave in small amounts as this medium will boil over easily.

Liquid culture medium (Pereira & Roberts, 1990)

- 1% dextrose
- 1% yeast extract
- 0.05% antibiotic (200 000 units Penicillin, 250 mg streptomycin/ml)
- 0.1% sunflower oil

Liquid culture medium (Samšiňáková *et al.*, 1981)

- 2.5% glucose
- 2.5% soluble starch
- 2% corn-steep
- 0.5% NaCl
- 0.5% $CaCO_3$

pH adjusted to 5

3. Stains and Mounting Media

Polyvinyl alcohol (PVA) mounting medium

- dissolve 8.3 g PVA in 50 ml deionized water
- add 50 ml lactic acid
- add 5 ml glycerine and filter if necessary
- add 0.1 g acid fuchsin if desired
- keep at room temperature for 24 h before using

Polyvinyl alcohol wetting agent

- 50 ml 95% ethanol
- 25 ml acetone
- 25 ml 85% lactic acid

Lacto-fuchsin mounting medium and stain (Carmichael, 1955)

- 0.1 g acid fuchsin
- 100 ml lactic acid

Fluorescein diacetate (FDA) (Schading *et al.*, 1995)

Prepared by mixing 35 μl of a stock solution of FDA (4 mg FDA/ml acetone) in 4 ml deionized water, kept on ice protected from light and used within 1 h of preparation.

Propidium iodide (PI) (Schading *et al.*, 1995)

PI prepared by mixing 60 µl stock solution of PI (3 mg/ml deionized water) /5 ml deionized water and stored as for FDA.

4. Miscellaneous

Blastospore storage formulation (Blachère *et al.*, 1973)

- 1 kg blastospores (22% wet moisture)
- 1 kg silica powder
- 250 ml 200 g sucrose and 5 g sodium glutamate in water
- 250 ml liquid paraffin containing 10% polyoxyethylene glycerol oleate

Germination medium (Milner *et al.*, 1991; Inglis *et al.*, 1996b)

- 0.1% yeast extract
- 0.1% chloramphenicol
- 0.01% tween 80
- 0.001–0.005% Benlate (wp).

(c) Saturated salts used for regulation of relative humidities

Humidities vary from 1 to 2% from those previously published (J. Virolleaud, unpublished).

Saturated salt solutions	Relative humidity (%) at different temperatures (°C)								Solubility at 20°C (change)[a]
	5	10	15	20	25	30	35	40	
Lithium chloride $LiCl,H_2O$	14	14	13	12	12	12	12	11	81% (+)
Magnesium chloride $MgCl_2,6H_2O$	35	34	34	33	33	33	32	32	40% (=)
Potassium carbonate $K_2CO_3,2H_2O$	–	47	44	44	43	43	43	42	52% (+)
Magnesium nitrate $Mg(NO_3)_2,6H_2O$	58	57	56	55	53	52	50	49	43% (+)
Sodium chloride NaCl	76	76	76	76	75	75	75	75	36% (=)
Potassium chloride KCl	88	88	87	86	85	85	84	82	37% (+)
Potassium sulphate K_2SO_4	98	98	97	97	97	96	96	96	11% (+)

[a] Change in solubility at temperatures above 20°C.

CHAPTER V-4

Fungi: Oomycetes and Chytridiomycetes

JAMES L. KERWIN & ERIN E. PETERSEN

Botany Department, University of Washington, Seattle, Washington 98195 USA.

1 INTRODUCTION – PHYLOGENY AND LIFE CYCLES

Aquatic fungi, formerly grouped together as Phycomycetes, are a diverse group of organisms characterized by a motile, flagellated zoospore at some stage of their life cycle (Sparrow, 1960). It is now recognized that there are phylogenetically distinct groups within this artificial assemblage. This chapter will deal primarily with two genera of entomopathogenic organisms, *Lagenidium* (Oomycetes: Lagenidiales) and *Coelomomyces* spp. (Chytridiomycetes: Blastocladiales). These are two very different groups of organisms, with one common feature: *Lagenidium giganteum*, the species discussed in detail here, and *Coelomomyces* spp. are primarily parasites of mosquito larvae. Because they are parasites of these medically important arthropods, there has been much interest for the last two decades in developing them for use in operational mosquito control.

A *Lagenidium giganteum*

The Oomycetes are a group of organisms whose phylogenetic relationship to fungi and other taxonomic groups has been the subject of controversy for many years (Copeland, 1956; Barr, 1992). Although there are dissenting opinions, the prevailing view is that Oomycetes are related to heterokont algae, and are placed in the kingdom Chromista, which includes diatoms, the brown algae and all protists with chloroplast endoplasmic reticulum or tubular ciliary mastigonemes. Monographs describing the relationship of the Lagenidiales to related taxa by Karling (1981) and a revision of earlier views (Dick, 1996) are available.

Lagenidium giganteum is a robust, fast-growing facultative parasite which can be grown on a variety of undefined media (Domnas *et al.*, 1982; Kerwin *et al.*, 1986). Its most characteristic physiological feature is its inability to synthesize sterols. In the early 1960s several laboratories found that Oomycetes in the Pythiaceae, which includes important plant pathogenic members of *Pythium* and *Phytophthora*, lack

MANUAL OF TECHNIQUES IN INSECT PATHOLOGY
ISBN 0–12–432555–6

the ability to synthesize sterols, but require these compounds to produce oospores, their sexual spore (Hendrix, 1964; Elliot *et al.*, 1966). Subsequently Domnas *et al.* (1977) demonstrated that *L. giganteum*, also a sterol auxotrophic organism, required an exogenous source of sterol to reproduce asexually. This requirement was extended to sexual reproduction (oosporogenesis) by Kerwin & Washino (1983). The pertinent fact for this review is that in order to infect mosquito larvae, zoospores must be formed, either by an asexual or sexual process. The availability to the growing fungus of sterols structurally related to cholesterol, therefore, will determine whether the parasite only proliferates vegetatively, or produces zoospores with the potential to infect larval mosquitoes.

Infection by *L. giganteum* is initiated by laterally biflagellate zoospores (Figure 1A) that selectively attach to and encyst on the cuticle of mosquito larvae. The parasite then proliferates throughout the host (Figure 1B), with only the earliest stage of development characterized by filamentous growth. Individual

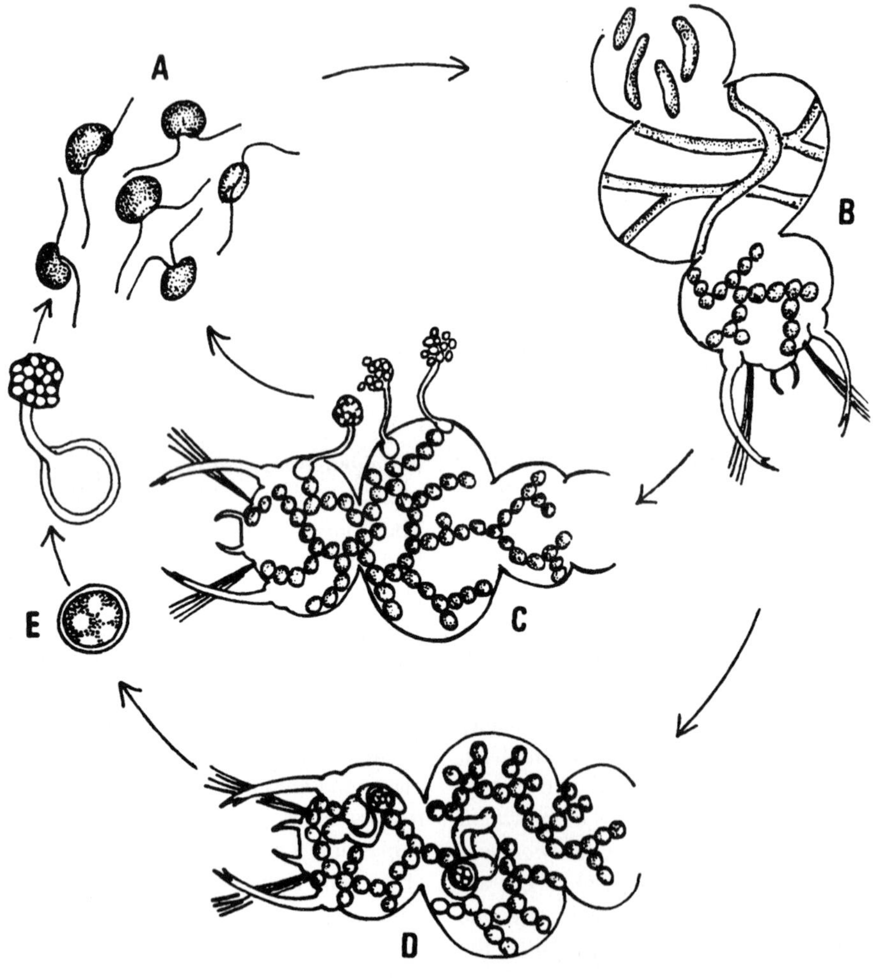

Figure 1 Life cycle of *Lagenidium giganteum.* Biflagellate zoospores (A) attach to and invade mosquito larvae (B), leading to development of mycelium and, ultimately, to septate hyphae. Each hyphal segment may become an asexual sporangium (C) from which zoospores are released at the tip of an exit tube. Depending on environmental conditions, the cycle from initiation of infection to asexual reproduction is usually completed in 24–72 h. Alternately, two adjacent hyphal segments may fuse (D) resulting in the production of dormant oospores (E). Under appropriate conditions, these dormant sexual spores will become activated, and germinate by production of an exit tube and subsequent zoospore release.

cells enlarge and usually within 2–3 days after initiation of infection the larva dies. At this point cells can enter either an asexual (Figure 1C) or sexual (Figure 1D) cycle. Asexual zoosporogenesis proceeds with development of an exit tube at the tip of which all cytoplasm migrates, followed by differentiation of 15–50 zoospores. Upon maturation these spores are released to infect new larval hosts. The sexual option (Figure 1D) involves fusion of two cells, one of which, the male cell or antheridium, produces a thin mycelial extension that fuses with the female cell or oogonium. All of the cytoplasm in the antheridium migrates into the oogonium, and a thick-walled dormant cell, the oospore, matures in 2–3 days. Depending upon environmental conditions, this spore can germinate within several weeks, or remain dormant for months or even years. Whereas other stages in the *L. giganteum* life cycle are relatively fragile and cannot persist without moisture, oospores can survive abrasion, desiccation and temperature extremes for at least 7 years (Kerwin *et al.*, 1986). Oospores germinate by dissolution of a series of layers making up the thick outer cell wall, followed by zoospore maturation at the tip of an exit tube in a process that is morphologically similar to asexual zoosporogenesis (Figure 1E). Selected stages of the parasite are shown in Figure 2. Scanning micrographs of sexual reproduction can be found in Brey (1985).

A second Oomycete, *Leptolegnia chapmanii*, has received some interest as a parasite with a degree of selectivity toward mosquito larvae (Seymour, 1984; McInnis *et al.*, 1985). The same techniques described for *L. giganteum* can be used for this species, and *L. chapmanii* is not further discussed.

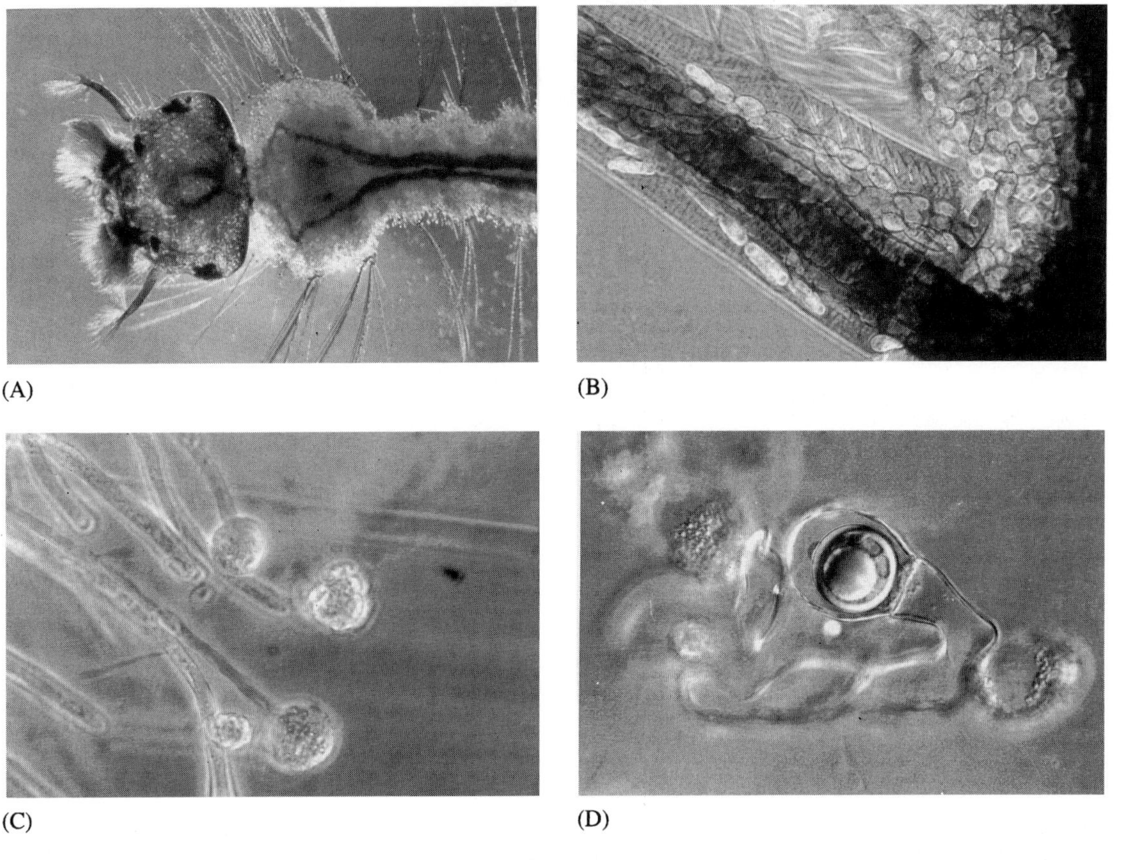

Figure 2 Micrographs of selected stages in the life cycle of *Lagenidium giganteum*. (A) *Culex tarsalis* larva infected with *L. giganteum*. The parasite kills its host by starvation. Upon completion of infection, all that remains of the larva is its cuticle. (B) Mycelia in the anal papilla of *Cx. tarsalis*, showing the characteristic oval to spherical hyphal cells. (C) Zoospores maturing at the tips of exit tubes. Upon maturation the zoospores swim off in search of a new host. (D) Mature oospore retained within the female oogonium, showing the point of fusion of the now empty male antheridium.

B *Coelomomyces* spp.

Chytridiomycetes are characterized by the production of posteriorly uniflagellate zoospores. Monographs on the genus (Couch & Bland, 1985), and on chytrids in general (Karling, 1977, 1981), have been published and can be referred to for detailed descriptions of taxonomy, life history and physiology. *Coelomomyces* spp. are all obligate parasites, and usually alternate between mosquitoes and microcrustacean hosts, with a concomitant alternation of sporophytic and gametophytic generations. *Coelomomyces* infections of chironomids have also been documented (Weiser & McCauley, 1971). Many species of this fungus have a very limited host range for both the mosquito and the crustacean.

As first described for *Coelomomyces psorophorae* (Whisler *et al.*, 1974), mosquito infection is initiated by biflagellate (2N) zygotes that selectively encyst on larval cuticle (Figure 3A). This sporophytic phase grows in the mosquito, and in approximately 7–10 days the cells mature into thick-walled, ovoid, resting sporangia (Figure 3B). These are roughly analagous to *L. giganteum* oospores in that they are the stage which can persist under adverse environmental conditions for months or years. Resting sporangia can be triggered to germinate by a variety of treatments, during which several hundred posteriorly

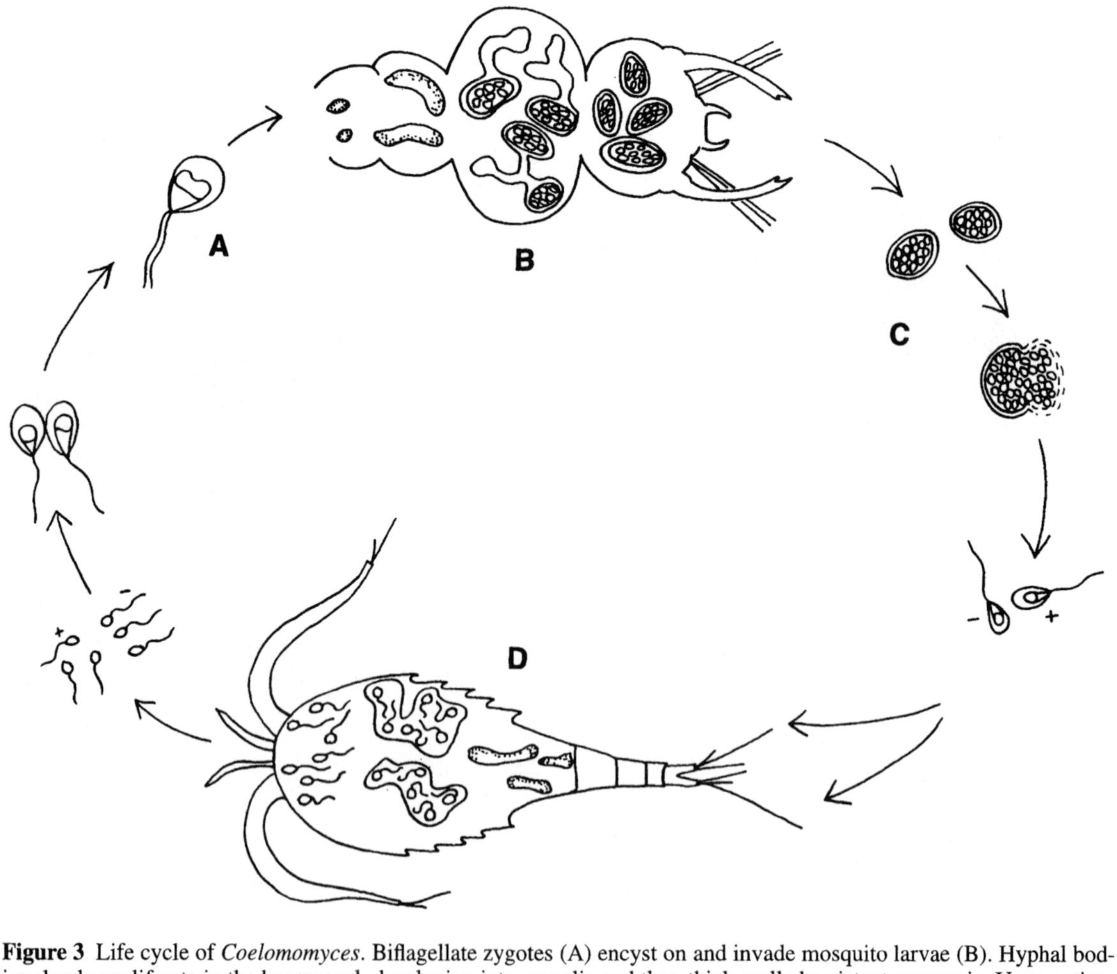

Figure 3 Life cycle of *Coelomomyces*. Biflagellate zygotes (A) encyst on and invade mosquito larvae (B). Hyphal bodies slowly proliferate in the haemocoel, developing into mycelia and then thick-walled resistant sporangia. Upon activation, resistant sporangia (C) release (+) and (–) meiospores, which specifically recognize, attach to and invade an appropriate crustacean host (D), in this case a copepod. Individual zoospores mature from a thallus to a gametangium, which upon maturation ruptures to release gametes. Opposite mating types fuse to form the zygotes which continue the mosquito infection cycle. The entire life cycle will take a minimum of 2 weeks. Adapted from Whisler *et al* (1974).

uniflagellate meiospores (1 N), are released (Figure 3C). These spores subsequently infect microcrustaceans, usually copepods or ostracods. Meiospores attach to, encyst on and then penetrate a crustacean host (Figure 3D), in the case of *C. psorophorae* the copepod *Cyclops vernalis*. Within 7–10 days the small, sparsely branched mycelium ramifies throughout the haemocoel. Following an appropriate signal, often associated with photoperiod (Federici, 1983; Lucarotti & Federici, 1984), there is very rapid differentiation of plus and/or minus uniflagellate gametes within the copepod. Fusion of gametes may occur within the host if both mating types are present. After a period of intense swarming within the copepod, a combination of enzymatic activity and mechanical pressure allows the spores to burst through the copepod cuticle. Following fusion of opposite mating types, the zygotes can invade a mosquito larva, completing the life cycle. Despite much conjecture and several questionable reports in the literature, there is no evidence of *Coelomomyces* cycling directly from mosquito to mosquito. Selected stages in the life cycle of *Coelomomyces* are shown in Figure 4.

2 ISOLATION

A Field isolation

For isolation of aquatic fungi in general, basic methods developed by pioneers in the field (Emerson, 1958; Sparrow, 1960) have been refined by Fuller

(A) (B) (C) (D)

Figure 4 Micrographs of selected stages in the life cycle of *Coelomomyces* spp. (A) Mycelia of *C. dodgei* in *Acanthocyclops vernalis*. (B) Entire copepod from (A). (C) Zygotes of *C. dodgei* encysted on the larval cuticle of *Anopheles quadrimaculatus*. (D) Resistant sporangia of *C. psorophorae* in *Culiseta inornata*. All micrographs are the courtesy of B. Federici.

(1978) and Fuller & Jaworski (1987). A common approach is to use a variety of baits, which, depending on the species of interest, can include crab shell, hair, boiled hemp seed, twigs, cellophane, apples and snake skin. These are suspended in an appropriate aquatic habitat for several days, and returned to the laboratory for microscopic examination or bioassay. An alternate approach is to collect water and naturally occurring substrates in these habitats for subsequent laboratory examination. Zoospores of these organisms are delimited only by a cell membrane. Since they lack a cell wall, it is important to take precautions that the samples are not exposed to extremes of temperature, pH, salinity or osmolarity.

Coelomomyces spp. are obligate parasites, so the only baits that can be used for these fungi are susceptible species of mosquitoes and/or crustacea.

Although saprophytic strains of *L. giganteum* have been isolated using chitin (Willoughby, 1969), we have found that it is much more effective to use mosquito larvae for this purpose, as described below. *Lagenidium giganteum* has also been isolated from chironomids, ceratopogonids (Frances *et al.*, 1989), copepods and daphnids (Couch, 1935), although most isolates will infect these other hosts only at very high zoospore densities (Nestrud & Anderson, 1994).

Isolation of mosquito parasitic organisms from larval breeding habitats is facilitated by the use of sentinel cages (Case & Washino, 1979). These cages are usually constructed from 2–4-litre plastic buckets, from which a portion of the sides and bottom have been removed and replaced with fine mesh (150–400 μm) nylon screen (Figure 5). Fishing bobbers are then attached at three equidistant points so that the top of the sentinel cage floats at least several centimetres above the water line. Mosquito larvae are added to the bucket and periodically removed to the laboratory for observation. For *L. giganteum*, sentinel mosquito larvae should be examined every 1–3 days since this parasite will usually complete its *in vivo* development within this period of time, depending upon larval age, density, species composition and environmental conditions. For *Coelomomyces*, larvae can remain in the sentinel cages for one week or even longer due to the slow development of these parasites. The time that larvae remain in the cages also depends on the age and developmental rate of the sentinel mosquito species. Since neither parasite will infect mosquito pupae, if late instar larvae are used as sentinels the sampling period will have to be shortened. For monitoring *Coelomomyces*, the same basic techniques can use the crustacean host as sentinels, but their much smaller size renders collection and monitoring of infection more difficult.

Figure 5 A sentinel cage used for field isolation of mosquito parasites, showing the nylon mesh sides and the fishing bobbers used to keep the top of the cage above water. A nylon mesh top (not shown) is used to minimize predation on sentinel mosquito larvae by water fowl.

The second option is to collect indigenous populations of mosquitoes or appropriate crustaceans for laboratory examination or bioassay. The standard method for field collection of mosquito larvae uses a 1 pint (0.47 l) long-handled dipper. A small mesh cotton or nylon net is used to concentrate the dip samples for transfer to the laboratory. The same collection method can be used for copepods, or aquatic light traps can be used for some species of copepods that are phototactic.

B Laboratory isolation

Upon return to the laboratory, the only option for culturing *Coelomomyces* spp. is *in vivo* cycling between its two hosts (if these are known). Because of the high degree of host specificity exhibited by many species, and the difficulty of copepod taxonomy, it is best to rely on material collected directly from the habitat where larval infection has been documented. Details of maintaining these fungi in the laboratory are summarized in the next section.

Lagenidium giganteum can either be maintained *in vivo* as described in the following section, or it can be isolated from infected larvae on agar media. Protocols for isolation have been described by Brey & Remaudiere (1985), and the following description is a variation on methods described in that reference. The parasite apparently relies on its fast growth rate to colonize its host, since it does not produce appreciable quantities of antimicrobial compounds (Domnas & Warner, 1991; Kerwin, unpublished observations). It is common to find a cadaver infected with this parasite to have nothing remaining of the original host tissue except its cuticle. These arc selected for surface sterilization, using 5% ethanol or antibiotic solutions (up to 500 mg/l each of streptomycin and ampicillin or a suitable substitute). Infected larvae should have no or minimal numbers of protozoa associated with them since it is very difficult to obtain axenic cultures in the presence of protozoans. Short-term exposure (5 min) to dilute ethanol or longer incubation (1–2 h) in antibiotic solutions is followed by washing in sterile distilled water and streaking the larva on a suitable nutrient medium (see Appendix) supplemented with 200–500 mg/l ampicillin or neomycin plus 200–500 mg/l streptomycin. Chlortetracycline or chloramphenicol can also be used, but are toxic at concentrations over *c.* 25 mg/l. At least two successive transfers onto antibiotic-containing media are recommended to insure an axenic culture. The amounts and types of antibiotics used can be varied, but care must be taken because many antibacterial compounds will also inhibit *L. giganteum* growth at high concentrations.

3 PROPAGATION

A *Lagenidium giganteum*

In vivo maintenance of the fungus is possible, and much early work was done recycling *L. giganteum in vivo* in the laboratory and in the field for months or years (Umphlett & Huang, 1972; McCray *et al.*, 1973a,b; Fetter-Lasko & Washino, 1983). This involves placing 5–10 infected larvae in a small volume (50–100 ml) of clean water with 10–20 larvae. In 2–3 days all or most of the larvae should become moribund. Microscopic confirmation of infection is the next step, and infected larvae with a minimum of associated bacteria, other fungal species and protozoa are picked out and the infection cycle initiated once more. This is labour intensive, however, and subject to the vagaries of mosquito colonies. The parasite apparently does not synthesize compounds that inhibit the growth of other micro-organisms, so attempts to recycle *L. giganteum* using larvae co-infected with a high titre of bacteria are often unsuccessful. Protozoans often associated with mosquito colonies also interfere with growth of the parasite (Woodring *et al.*, 1995).

Since *L. giganteum* has fairly non-specific nutrition requirements for vegetative growth (Willoughby, 1969; McInnis, 1971; Domnas, 1981), and can be cultured on a wide variety of defined and complex media (Domnas *et al.*, 1982; Jaronski *et al.*, 1983; Guzman & Axtell, 1986; Kerwin *et al.*, 1986; Su & Guzman, 1990), *in vitro* culture is preferred. Isolation attempts from field material can use a wide

Table 1 Selected media for culture of *Lagenidium giganteum.*

Medium/culture protocol	*General attributes*	*References*[a]
Defined media		
Agar or liquid shake	Physiological/biochemical studies	Gleason (1968), Willoughby (1969), McInnis (1971), Kerwin *et al.* (1995)
Undefined media		
Agar	Stock maintenance; laboratory bioassay; small scale field tests	Domnas *et al.* (1977), Jaronski *et al.* (1983), Kerwin & Washino (1983), Guzman & Axtell (1986)
Liquid shake	Laboratory bioassay; small scale field tests	Domnas *et al.* (1982), (Kerwin *et al.* (1986a), Su & Guzman (1990)
Fermentation	Large scale field tests	Kerwin & Washino (1986b, 1987, 1988)

[a] Media recipes are included in the Appendix.

range of standard microbiological media (Table 1, Appendix). Complications arise primarily from the sterol auxotrophic nature of *L. giganteum*. Simply adding sterols to growth media will not ensure that the parasite will enter its reproductive cycle. Although these developmental processes are morphologically simple, the underlying physiological processes are complex, and zoosporogenesis and oosporogenesis are affected by a variety of nutritional components (Hohl,1983). Since zoospores are required for mosquito infection, it is best to use one of the media listed in Table 1 and described in detail in the Appendix.

The standard agar medium used for routine maintenance of *L. giganteum* cultures is PYG (Emerson, 1958). Although cultures can be maintained on this medium for months or years, it has been noted by several authors that prolonged maintenance on sterol-free media will result in gradual loss by the cultures of their ability to enter the sexual and asexual cycles (Kerwin & Washino, 1983; Lord & Roberts, 1986). For this reason sterols should be included in all culture media for *L. giganteum*. This can be done by adding a vegetable oil such as corn or wheat germ, or by adding purified sterols solubilized in lecithin or detergents such as the Tween series (see Appendix). Although vegetable oils consist primarily of triglycerides, there are usually trace quantities of sterol in these oils that are sufficient to induce reproduction. The pH of most agar media does not have to be adjusted, but growth is optimum from *c.* pH 6–8. There is also an absolute requirement for calcium. Although most undefined media will contain trace quantities of calcium, addition of 5 mM $CaCl_2.2H_2O$ will ensure there is sufficient quantity of this mineral to support growth and reproduction.

Larger-scale liquid culture of *L. giganteum*, especially in 10–1000 l fermentation culture, has used yeast extract-based media. Yeast extract will acidify most media to below pH 5, so the pH has to be adjusted for these cultures. As production is scaled up, especially in stirred tank fermentors, the configuration and design of individual machines will have significant effects on growth and morphogenesis, and published media recipes may have to be altered. Another complication in larger scale (100 l or greater) fermentation of *L. giganteum* is its incredibly rapid growth during the log growth phase. This causes extreme foaming in the fermentor tank, and some anti-foaming agents adversely affect sporulation (Kerwin *et al.*, 1994). Alternatives to the use of anti-foaming agents include increasing the head pressure and/or reducing the aeration rate.

Zoosporogenesis by *L. giganteum* is induced by nutrient deprivation; therefore, whether using *in vivo* or *in vitro* material, it is necessary to use relatively clean water to dilute the cells, and the cell density must be low. Using distilled or deionized water reduces variablity that can arise from using water from natural habitats. Tap water can be used in some instances, but ions commonly found in domestic water supplies such as chlorine or boron can reduce or completely inhibit sporulation. Studies have been done on the effects of salinity, organic load, pesticides and other water quality parameters on the induction of zoosporogenesis and zoospore survival (Jaronski & Axtell, 1982; Merriam & Axtell, 1982; Lord & Roberts, 1985); however, as pointed out by Woodring & Kaya (1992), causal relationships between specific water quality parameters and infectivity are difficult to establish. There is at this time no alternative to preliminary laboratory evaluation of *L. giganteum* sporulation in a given source of water when using sources other than distilled water. Care must be taken even with distilled water since, depending on how the purification is done, extremes of pH, usually on the acid end of the scale, can be encountered.

Water from ponds, streams, tree holes and other natural habitats can also be used as long as the salt and organic loads are relatively low. If water from natural habitats is used, filtration with or without subsequent autoclaving will remove protists and algal, fungal and oomycete spores that might interfere with interpretation of subsequent bioassays. If the purpose of a study is to simulate natural conditions, these steps will obviously be omitted.

There is a tendency to add too many cells to a small volume of water to achieve high levels of infection, but, unlike deuteromycetous entomopathogenic fungi where a higher spore density will often increase infection, there is a distinct threshold for *L. giganteum* above which no sporulation, and, therefore, no infection will occur. The optimum dilution will vary greatly depending on the medium used to culture the parasite, but generally a minimum of 1 : 100 dilution of a mature liquid culture will result in good zoospore release. As an example, the presporangia produced in 3–8 l of fermentation broth is sufficient to treat 1 hectare of mosquito breeding habitat (Kerwin & Washino, 1988; Kerwin *et al.*, 1994). Initial attempts at inducing zoosporogenesis should evaluate several different water sources, and serial dilution of mature cells over at least 3 orders of magnitude.

Mycelia (presporangia) must be mature before zoospores can be induced. For agar cultures, 1–4-week-old cultures can be used, depending on what medium is used, the incubation temperature, and how the plates are inoculated. If plates are inoculated with liquid cultures or sterile zoospore suspensions, growth will be accelerated and can be used within a week. Most *L. giganteum* growth on agar media occurs on the surface, so if agar cultures are used, scraping the surface with a spatula to remove mycelia, rather than cutting out blocks of agar, will result in improved zoospore release. When using liquid cultures of the parasite, culture media should be removed from the cells by gentle filtration using paper or nylon filters before dilution in water. Depending on culture protocols, liquid cultures from 3 days to several weeks old will produce zoospores, with 7–10-day-old cultures usually the most reliable.

Oospores, which can be stored in some instances for over 7 years (Kerwin *et al.*, 1986), can also be used as a source of inoculum. These dormant spores can be stored in the original culture media or as a dry powder. Their longevity is highly variable, and is largely determined by the medium used to culture the parasite. The main problem with this spore is breaking dormancy without causing premature abortion. Although some oospores will germinate, especially in the field, within several days after they mature, many will remain dormant for months or years. Cycles of hydration and rehydration will activate some oospores. An alternate strategy that we have employed is to use a two-stage activation process pioneered by researchers working with plant pathogenic Oomycetes. The first step is a prolonged (1 week to 2 month) incubation in soil, soil extract, or 0.5% dimethylsulphoxide (DMSO + 5 mM calcium, that can activate the metabolically dormant spores. During this period the thick inner oospore wall is dissolved, leaving the spore in a state similar to that of the 'go stage' described for *Coelomomyces* in the next section. The oospores, which are referred to as converted spores, are filtered from this incubation mixture, and resuspended in water. If they do not abort at some stage in this process – and this happens with regularity – *c.* 40–80% of the converted oospores will germinate within 48 h.

It should be noted that most Oomycetes, including *L. giganteum*, do not survive well using liquid nitrogen, and that technique is not recommended for long-term preservation of this group of organisms.

B *Coelomomyces*

It is obviously difficult to maintain these fungi in the laboratory due to their obligate parasitism of two different organisms in disparate taxa. Techniques for mosquito rearing are well established (American Mosquito Control Agency Bulletin, 1994), but some species cannot be colonized; therefore, species of *Coelomomyces* that will only infect mosquito species that cannot be maintained in the laboratory would be even more difficult to maintain. Some copepods and ostracods can be reared using a number of simple feeding regimens, e.g. powdered alfalfa meal (Padua *et al.*, 1986), or a solution of brewer's yeast and egg yolk solution (Federici & Chapman, 1977). Procedures have been developed for laboratory maintenance for a number of species, including *C. dodgei* (Federici, 1980), *C. psorophorae* (Zebold *et al.*, 1979), *C. stegomyiae* (Padua *et al.*, 1986, Lucarotti, 1987) and *C. punctatus* (Federici & Roberts, 1976).

Perhaps the easiest species to culture *in vivo* is *C. stegomyiae. Phyllognathopus viguieri*, its copepod host, can be reared in the laboratory with minimal effort. The fungus infects the yellow fever mosquito, *Aedes aegypti*, some laboratory colonies of which are several decades old. Although separate rearing and infection trays can be set up for the mosquito and copepod infections (Lucarotti, 1987), it is also possible to combine the two hosts and obtain a fairly constant low level of infection. Distilled water, autoclaved pond water, or many dilute salt solutions, e.g. variations of DS described in Fuller & Jaworski (1987, see Appendix), can be used to rear mosquito and copepod hosts. It is best to become proficient in rearing the mosquito and the crustacean hosts before attempting to recycle *Coelomomyces* spp. in the laboratory.

Resistant sporangia (RS) are usually chosen as a convenient starting point because they can be stored indefinitely in a refrigerator, and are the easiest stage of this parasite to recognize. The RS of many species will gradually dehisce over several weeks or months when incubated at room temperature in one or more of the water sources listed in the previous paragraph. If the RS have been stored in larval cadavers, repeated drawing of the mosquitoes through a pasteur pipette will separate most spores from the larvae and enhance germination. RS from 6–12 larvae can be placed directly into copepod rearing trays (Lucarotti, 1987) and infected animals will be evident for *C. stegomyiae* and a number of other species within 6–10 days. An alternative is to remove 50–100 copepods to Petri plates or other small containers, and add the infected mosquito larvae. By placing the copepods in a smaller volume of clean water, it is easier to monitor death and infection of the animals, which is helpful especially when using smaller species such as *Phyllognathopus viguieri*, the alternate host for *C. stegomyiae*.

To obtain infected mosquitoes, infected copepods can be added directly to mosquito rearing pans, or removed to smaller containers as described above. The number of infected copepods used for infection will vary with the size of the container, the number of mosquitoes to be infected, species susceptibility to infection, and for those species in which the fungus can survive in adults, whether infected larvae or adults are desired. Generally 5–10 copepods are sufficient to guarantee infection of a minimum of 40–50 larvae, and can often infect several hundred mosquitoes. Infected larvae will usually die in 6–10 days. Larvae with RS can then be stored in a refrigerator (0–8 °C) on damp filter paper for several months. Alternately, RS can be removed from infected larvae by homogenizing the cadavers (mortar and pestle, mechanical shearing using a microblender or forcing through a syringe) and filtering the spores from cellular debris using no. 5 Whatman filter paper. If necesssary the RS can be centrifuged at low speed through water or dilute salt solutions to further clean up the homogenate. These spores can then be stored as described for the intact cadavers. Longer-term RS storage using liquid nitrogen is described elsewhere in this book.

For a more reliable supply of the fungus for laboratory experimentation or field trials, it is possible to synchronize several stages of the *Coelomomyces* life cycle (Federici *et al.*, 1985). RS of many species, which can be stored for at least several months on damp filter paper at 0–5 °C, can be triggered to germinate by preincubation at 4 °C in a dilute salt solution for 7–14 days in the dark, followed by reduction of oxygen tension by bubbling nitrogen into the spore suspension for 20–30 min (Whisler *et al.*, 1983). During the preincubation period a discharge crack opens in the spore wall, and it can remain at this stage for fairly extended periods of time. This allows a fairly high percentage of spores to be converted into this 'go stage' (Whisler *et al.*, 1972), which will then rapidly differentiate into and release meiospores upon exposure to anaerobic conditions.

Gametogenesis and gamete release from copepod/ostracod hosts can also be highly synchronized, in this instance by photoperiod (Federici, 1983). Gating of gametogenesis is species-specific, with the development of many *Coelomomyces* spp. responding to the onset of the dark period (Federici, 1983; Lucarotti & Federici, 1984; Whisler, 1985). For instance, *C. dodgei*-infected copepods, following a 6–10-day parasite maturation period, can be induced to release gametes by a dark period as short as 2 h. There is synchronous release 16–19 h after onset of the dark period (Federici, 1983). This phenomenon can be used to provide a reliable source of large numbers of spores for physiological investigations and small-scale inundative field trials.

There are two reports of limited *in vitro* culture of *C. psorophorae* and *C. punctatus* (Shapiro & Roberts, 1976; Castillo & Roberts, 1980). Both studies used complex mixtures of vertebrate and/or invertebrate tissue culture media. Mycelial growth was slow and differentiation into reproductive structures did not occur or was very limited. Further progress in this area will require detailed physiological studies on both the host and parasite. This would involve profiling of selected classes of compounds likely to be involved in parasite development, and preparative scale isolation of these products, followed by monitoring their subsequent metabolism *in vivo* and *in vitro* as the fungus proceeds through its developmental cycle. Such a project should not be undertaken unless resources, and the technical skills required to utilize those resources, are available for years of concerted effort.

Unless material is obtained from a laboratory where a given species or strain of the fungus has been maintained, establishing a reliable protocol for culture of *Coelomomyces* is likely to require extensive time and labour. Familiarity with the variety of approaches taken by investigators when working with these fungi will increase the chance of success.

The major problems encountered usually revolve around maintaining healthy colonies of both hosts.

4 BIOASSAY

A *Lagenidium*

Since sporulation is required for mosquito infection by *Coelomomyces* spp. and *L. giganteum*, attention must be focused on culture protocols promoting zoosporogenesis. For both species zoospore density (and RS density for Coelomomyces spp.) can be assessed using a haemocytometer after immobilizing the spores with 0.5–1% formalin (Nestrud & Anderson, 1994). This is difficult to do when using water from natural habitats that has not been sterilized, due to the large numbers of motile spores of similar size and morphology present in many of these samples. An experienced observor can often differentiate fungal and Oomycete spores from those of other taxa based on swimming patterns. The ability to differentiate fungal zoospores from other motile spores when examining field samples is necessary to minimize confusion. Unfortunately, among the most ubiquitous inhabitants of freshwater habitats frequented by mosquitoes are species of *Achlya* and *Saprolegnia*, both Oomycetes that have large biflagellate spores (8–10 μm) similar in size and morphology to those of *L. giganteum*. Another common mistake is to confuse epiphytic protozoans on larval cuticle with sporangial vesicles of this parasite. The parasite is most definitively identified by its large yeast-like morphology. Cells are best seen either in the larval head capsule, anal gills or siphon. A final complication is that other zoosporic fungi and Oomycetes with morphology similar to *L. giganteum* can infect mosquitoes, either as primary parasites (*Crypticola clavulifera*; Frances *et al.*, 1989) or saprophytes.

Many of the basic techniques required for bioassays of this parasite are described in previous sections. Quantification of zoosporogenesis is relatively easy in the laboratory as long as field water is not being used. Similar data are all but impossible for field studies. The usual approach is to quantify the number of mycelial cells that are going to be applied using a haemocytometer, and then estimate the percentage of cells that are going to germinate and the average number of spores released from each sporangium. This can be done by evaluating germination of the parasite in water collected from a field site and brought back to the laboratory; however, this only provides an estimate, and we have often found that the parasite was unexpectedly much more virulent under field conditions.

Zoospores, which lack a cell wall, are too fragile to be applied in natural habitats; therefore, concentrated preparations of either mycelia (presporangia) or oospores are applied, with sporulation subsequently induced by dilution in the larval breeding habitat. Agar media (or, in those rare instances where *in vivo*-cultured preparations are used, infected larvae) are homogenized in distilled, tap or other clean water prior to application. This increases sporulation by separating the prosporangia from the nutrient-rich culture media. Mycelia cultured in liquid shake or fermentation culture can either be applied directly, or the media filtered out and the mycelia resuspended in tap or distilled water. The parasite can be manually applied in smaller-scale plots using dissemination from flasks or buckets. Larger scale applications are made using low pressure backpack sprayers, and multi-hectare aerial applications can be made using Beecomist or similar air-driven systems that are commonly used to apply preparations of the bacterial insecticide *Bacillus thuringiensis* (Kerwin & Washino, 1986b, 1988).

Care must be taken when using any spray apparatus due to the common use of filters in one or more places between the holding tank and the spray nozzle. Cells of the parasite can be up to 200 μm long, and many filters used for chemical applications are much smaller than this. Some nozzle orifices are too small to allow free passage of mycelia. Even if the mycelia can pass through a given mesh, if the mechanical shear is excessive, the mycelia will not survive to sporulate. This is not as much a problem if oospores rather than presporangia are applied. A final precaution when doing aerial applications is to minimize holding tank agitation, not only because of mechanical shear, but also because water temperatures can quickly reach levels that are deleterious to the mycelia.

The main advantage of using *L. giganteum* for operational mosquito control is its ability to recycle at appreciable levels for weeks, months or even years following a single field application (Fetter-Lasko & Washino, 1983; Jaronski & Axtell, 1983). This complicates experimental design and data interpretation of bioassays. Using early instar larvae and high zoospore densities, the parasite can consume its host and begin producing zoospores in as little as 24 h after initiating infection. A 2–3-day recycling time is more common, but under suboptimal conditions of temperature or water quality, it can take as long as 7 days for sporulation to occur. If the goal of a study is to establish quantitative zoospore density–larval mortality relationships, larvae have to be removed at appropriate intervals so that sporulation from infected larvae does not amplify the observed mortality. This interval has to be determined empirically for each habitat and mosquito species. This is relatively easy for laboratory evaluations, but differentiation between the effects of the parasite originally applied to a field site and subsequent recycling due to infection of the indigenous larval population is not possible. The only exception involves floodwater (primarily *Aedes*) univoltine mosquitoes that develop very rapidly (egg hatch to adult emergence in less than 4 days) and are often the only species present in a given habitat.

B *Coelomomyces*

In addition to the larval infections previously discussed, adult mosquito infections have also occasionally been observed (Kellen *et al.*, 1963; Taylor *et al.*, 1980; Lucarotti, 1987). As is the case with most stages of *Coelomomyces*, these infections will not be obvious to the casual observer. Resistant sporangia are the most recognizable stage of these fungi. Even RS can be confused with several other organisms, including oospores of smaller species of Oomycetes, some encysted protozoa, or even microsporidia. Infections in copepods and ostracods will not be obvious until the very end of the ca. week-long infection period. It is recommended that groups interested in working with *Coelomomyces* first become proficient in the laboratory before attempting any field research.

The only option for applications of this fungus is to use either infected mosquitoes, infected crustacean hosts, or both. Maceration of infected mosquitoes prior to application is recommended to increase the rate and percentage of RS germination. Infected copepods have to be applied intact. As discussed for *L. giganteum*, zoospores are too fragile for field application.

Bioassays for these fungi involve variations of techniques described in previous sections. Quantification of zoospore and RS density using a haemocytometer was discusssed in the previous section, as were complications of bioassays using an organism which can recycle. Data interpretation is simpler for *Coelomomyces* due to its slow development in both hosts, and its obligate alternation of hosts. It is possible to maintain both hosts simultaneously in a single rearing pan in which RS are allowed to germinate gradually over a long period of time, but that type of protocol is not amenable to quantitative analyses. Mixed cultures can be useful for long-term maintenance, and to provide material for light microscopic or ultrastructural analyses of the interaction of host and parasite. However, even when determining something as straightforward as host range, it is better to maintain separate control and infected colonies of the two hosts (Toohey *et al.*, 1982). If at all possible, quantitative studies should attempt to use synchronized spore release from one or both hosts using techniques previously described.

If synchronized germination is not possible, or when evaluating material newly isolated from the field, there is no alternative to longer-term assays lasting for weeks or months. The most common initial source of *Coelomomyces* is mosquito larvae. Since the RS are the equivalent of a natural slow-release chemical formulation, germination will be unpredictable. The general approach to these assays by Toohey *et al.* (1982) involved establishment of colonies of different species of copepods. After holding 150–200 animals for up to several weeks, infected larvae with mature RS (the equivalent of the number found in about three heavily infected late instar larvae) were added to each cup. After 10–12 days 20 early instar mosquito larvae were placed in each cup. Dead or moribund larvae, pupae and adult mosquitoes were removed daily and examined for infection. If no infection was noted within one month, a second group of larvae was added to the cups.

One approach to simplify the experimental system used larval exuviae rather than intact larvae to monitor host recognition (Kerwin, 1983). This can also be used for *L. giganteum* host recognition assays because the zoospores of both species encyst on larvae only after recognizing specific chemical signals on the cuticular surface. Although the endocuticle is extensively modified during the moulting process, the outermost layers are not extensively degraded. The advantage of using exuviae is that chemical treatments and isolation of cuticular components can be completed without the interference of extraneous tissue not involved in the host recognition process. As with all laboratory mosquito colonies and field-collected material, there can be extensive contamination of the larval surface with bacteria and protozoa. Heating (35–50 °C) in 0.05% HCl for 5–30 min is usually sufficient to remove most epiphytic organisms without appreciably altering the surface properties of exuviae.

The taxonomy of one common alternate host, copepods, is notoriously difficult. It is advisable that an expert in copepod taxonomy examine representative field collections, especially for *Coelomomyces* spp. that have restricted host ranges. Incorrect identification of native species can lead to a negative assessment of field efficacy without realizing that an appropriate alternate host was not present.

5 CONCLUSIONS

Methods have been presented describing the identification, isolation and culture of *L. giganteum* and species of *Coelomomyces*. The former parasite has been registered as an operational mosquito control agent with the United States Environmental Protection Agency and by several states. The latter species, despite its complicated life cycle, has shown field efficacy in natural epizootics (Pillai & Smith, 1968; Chapman, 1973). Development of resistance by mosquitoes to available insecticides continues unabated, and there is little financial incentive to develop new classes of pesticides for mosquito control. This problem in conjunction with increasing regulatory restrictions on chemical applications, especially in aquatic habitats, should encourage further work with these two microbial pest control agents.

ACKNOWLEDGEMENTS

This work was supported in part by a grant from the National Institutes of Health (AI 34339). We thank B. Federici for providing micrographs of *Coelomomyces* , and V. Kerwin and M. Semon for technical drawing.

REFERENCES

American Mosquito Control Association (1994) Bulletin no. 5. *Manual for Mosquito Rearing and Experimental Techniques.*

Barr, D. J. S. (1992) Evolution and kingdoms of organisms from the perspective of a mycologist. *Mycologia* **84**, 1–11.

Brey, P. T. (1985) Observations of *in vitro* gametangial copulation and oosporogenesis in *Lagenidium giganteum. J. Invertebr. Pathol.* **45**, 276–281.

Brey, P. T. & Remaudiere, G. (1985) Recognition and isolation of *Lagenidium giganteum* Couch. *Bull. Soc. Vector Ecol.* **10**, 90–97.

Case, T. J. & Washino, R. K. (1979) Flatworm control of mosquito larvae in rice fields. *Science* **206**, 1412–1414.

Castillo, J. M. & Roberts, D. W. (1980) In vitro studies of *Coelomomyces punctatus* from *Anopheles quadrimaculatus* and *Cyclops vernalis. J. Invertebr. Pathol.* **35**, 144–157.

Chapman, H. C. (1973) Assessment of the potential of some pathogens and parasites of biting flies. *Proc. Symp. Biting Fly Control Environ. Qual.*, 1972, pp. 71–77.

Copeland, H. F. (1956) *The Classification of Lower Organisms.* Frontis, Pacific Books, Palo Alto, 302 pp.

Couch, J. N. (1935) A new saprophytic species of *Lagenidium*, with notes on other forms. *Mycologia* **27**, 376–387.

Couch, J. N. & Bland, C. E. (eds) (1985) *The Genus Coelomomyces.* Academic Press, New York, 399 pp.

Dick, M. W. (1996) Stramenopilous fungi – A new classification for the biflagellate fungi and their uniflagellate relatives, with particular reference to lagenidiaceous fungi. *Mycol. Papers* (in press).

Domnas, A. J. (1981) Biochemistry of *Lagenidium giganteum* infection in mosquito larvae. In *Pathogenesis of Invertebrate Microbial Diseases* (ed. E. W. Davidson) pp. 425–449. Allanheld, Osmun, Totowa, New Jersey.

Domnas, A. J. & Warner, S. A. (1991) Biochemical activities of entomophagous fungi. *Crit. Rev. Microbiol.* **18**, 1–13.

Domnas, A., Giebel, P. E. & McInnis, Jr, T. M. (1974) Biochemistry of mosquito infection: preliminary studies of biochemical change in *Culex pipiens quinquefasciatus* following infection with *Lagenidium giganteum. J. Invertebr. Pathol.* **24**, 293–304.

Domnas, A. J., Srebro, J. P. & Hicks, B. F. (1977) Sterol requirement in the mosquito-parasitizing fungus *Lagenidium giganteum. Mycologia* **69**, 875–886.

Domnas, A. J., Fagan, S. M. & Jaronski, S. (1982) Factors influencing zoospore production in liquid cultures of *Lagenidium giganteum*. *Mycologia* **74**, 820–825.

Elliot, C. G., Hendrie, M. R. & Knights, B. A. (1966) A steroid growth factor requirement in a fungus. *Nature* **203**, 427–428.

Emerson, R. (1958) Mycological organization. *Mycologia* **50**, 589–621.

Federici, B. A. (1980) Production of the mosquito-parasitic fungus, *Coelomomyces dodgei*, through synchronized infection and growth of the intermediate copepod host, *Cyclops vernalis*. *Entomophaga* **25**, 209–217.

Federici, B. A. (1982) Inviability of interspecific hybrids in the *Coelomomyces dodgei* complex. *Mycologia* **74**, 555–562.

Federici, B. A. (1983) Species-specific gating of gametangial dehiscence as a temporal reproductive isolating mechanism in *Coelomomyces*. *Proc. Natl. Acad. Sci. USA* **80**, 604–607.

Federici, B. A. & Chapman, H. C. (1977) *Coelomomyces dodgei*: establishment of an in vivo laboratory culture. *J. Invertebr. Pathol.* **30**, 288–297.

Federici, B. A. & Roberts, D. W. (1976) Experimental laboratory infection of mosquito larvae with fungi of the genus *Coelomomyces*. II. Experiments with *Coelomomyces punctatus* in *Anopheles quadrimaculatus*. *J. Invertebr. Pathol.* **27**, 333–341.

Federici, B. A., Tsao, P. W. & Lucarotti, C. J. (1985) *Coelomomyces* (Fungi) *Bull. Am. Mosquito Control Assoc.* **6**, 75–86.

Fetter-Lasko, J. L. & Washino, R. K. (1983) *In situ* studies on seasonality and recycling pattern in California of *Lagenidium giganteum* Couch, an aquatic fungal pathogen of mosquitoes. *Environ. Entomol.* **12**, 635–640.

Frances, S. P., Sweeney, A. W. & Humber, R. A. (1989) *Crypticola clavulifera* gen. et sp. nov. and *Lagenidium giganteum:* Oomycetes pathogenic for dipterans infesting leaf axils in an Australian rain forest. *J. Invertebr. Pathol.* **54**, 103–111.

Fuller, M. S. (1978) *Lower Fungi in the Laboratory*. University of Georgia, Athens, 213 pp.

Fuller, M. S. & Jaworski, A. (1987) *Zoosporic fungi in teaching and research*. Southeastern Publishing Company, Athens, Georgia, 303 pp.

Gleason, F. H. (1968) Nutritional comparisons in Leptomitales. *Am. J. Bot.* **55**, 1003–1010.

Guzman, D. R. & Axtell, R. C. (1986) Effect of nutrient concentration in culturing three isolates of the mosquito fungal pathogen, *Lagenidium giganteum* (Oomycetes: Lagenidiales), on sunflower seed extract. *J. Am. Mosq. Control Assoc.* **2**, 196–200.

Hendrix, J. W. (1964) Sterol induction of reproduction and stimulation of growth of *Pythium* and *Phytophthora*. *Science* **144**, 1028–1029.

Hohl, H. R. (1983) Nutrition of *Phytophthora*. In: *Phytophthora: its biology, taxonomy, ecology and pathology* (eds D. C. Erwin, S. Bartnicki-Garcia & P. H. Tsao) pp. 41–54. American Phytopathological Society, St Paul, Minnesota.

Jaronski, S. & Axtell R. C. (1982) Effects of organic water pollution on the infectivity of the fungus *Lagenidium giganteum* (Oomycetes: Lagenidiales) for larvae of *Culex quinquefasciatus*. *J. Med. Entomol.* **19**, 255–262.

Jaronski, S. & Axtell R. C. (1983) Persistence of the mosquito fungal pathogen *Lagenidium giganteum* (Oomycetes: Lagenidiales) after introduction into natural habitats. *Mosq. News* **43**, 332–337.

Jaronksi, S. T. & Axtell, R. C. (1984) Simplified production system for the fungus *Lagenidium giganteum* for operational mosquito control. *Mosq. News* **44**, 377–381.

Jaronski, S., Axtell R. C., Fagan, S. M. & Domnas, A. J. (1983) *In vitro* production of zoospores by the mosquito pathogen *Lagenidium giganteum* (Oomycetes: Lagenidiales) on solid media. *J. Invertebr. Pathol.* **41**, 305–309.

Karling, J. S. (1977) *Chytridiomycetarum iconographia: an Illustrated and Brief Descriptive Guide to the Chytridiomycetous Genera with a Supplement of the Hyphochytriomycetes*. J. Cramer, FL-9490 Vaduz, Germany, 414 pp.

Karling, J. S. (1981) *Predominantly Holocarpic and Eucarpic Simple Biflagellate Phycomycetes*. J. Cramer, FL-9490 Vaduz, Germany, pp. 89–154.

Kellen, W. R., Clark, T. B. & Lindegren, J. E. (1963) A new host record for *Coelomomyces psorophorae* Couch in California (Blastocladiales: Coelomomycetaceae). *J. Insect Pathol.* **5**, 167–173.

Kerwin, J. L. (1983) Biological aspects of the interaction between *Coelomomyces psorophorae* zygotes and the larvae of *Culiseta inornata*: host-mediated factors. *J. Invertebr. Pathol.* **41**, 233–237.

Kerwin, J. L. & Washino, R. K. (1983) Sterol induction of sexual reproduction in *Lagenidium giganteum*. *Exp. Mycol.* **7**, 109–115.

Kerwin, J. L. & Washino, R. K. (1986a) Regulation of oosporogenesis by *Lagenidium giganteum*: promotion of sexual reproduction by unsaturated fatty acids and sterol availability. *Can. J. Microbiol.* **32**, 294–300.

Kerwin, J. L. & Washino, R. K. (1986b) Ground and aerial application of the sexual and asexual stages of *Lagenidium giganteum* (Oomycetes: Lagenidiales) for mosquito control. *J. Amer. Mosq. Control Assoc.* **2**, 182–189.

Kerwin, J. L. & Washino, R. K. (1987) Ground and aerial application of the asexual stage of *Lagenidium giganteum* for the control of mosquitoes associated with rice culture in the Central Valley of California. *J. Am. Mosq. Control Assoc.* **3**, 59–64.

Kerwin, J. L., & Washino, R. K. (1988) Field evaluation of *Lagenidium giganteum* (Oomycetes:Lagenidiales) and description of a natural epizootic involving an apparently new isolate of the fungus. *J. Med. Entomol.* **25**, 452–460.

Kerwin, J. L., Simmons, C. A. & Washino, R. K. (1986) Oosporogenesis by *Lagenidium giganteum* in liquid culture. *J. Invertebr. Pathol.* **47**, 258–270.

Kerwin, J. L., Duddles, N. D. & Washino, R. K. (1991) Effects of exogenous phospholipids on lipid composition and sporulation by three strains of *Lagenidium giganteum*. *J. Invertebr. Pathol.* **58**, 408–414.

Kerwin, J. L., Dritz, D. D. & Washino, R. K. (1994) Pilot scale production and application in wildlife ponds of *Lagenidium giganteum* (Oomycetes: Lagenidiales). *J. Am. Mosq. Control Assoc.* **10**, 451–455.

Kerwin, J. L., Tuininga, A. R., Wiens, A. M., Wang, J. C., Torvik, J. J., Conrath, M. L. & MacKichan, J. K. (1995) Isoprenoid-mediated changes in the glycerophospholipid molecular species of the sterol auxotrophic fungus *Lagenidium giganteum*. *Microbiology* **141**, 399–410.

Lord, J. C. & Roberts, D. W. (1985) Effects of salinity, pH, organic solutes, anaerobic conditions, and the presence of other microbes on production and survival of *Lagenidium giganteum* (Oomycetes: Lagenidiales) zoospores. *J. Invertebr. Pathol.* **45**, 331–338.

Lord, J. C. & Roberts, D. W. (1986) The effects of culture medium quality and host passage on zoosporogenesis, oosporogenesis, and infectivity of *Lagenidium giganteum* (Oomycetes: Lagenidiales). *J. Invertebr. Pathol.* **48**, 355–361.

Lucarotti, C. J. (1987) *Coelomomyces stegomyiae* infection in adult *Aedes aegypti*. *Mycologia* **79**, 362–369.

Lucarotti, C. J. & Federici, B. A. (1984) Gametogenesis in *Coelomomyces psorophorae* Couch (Blastocladiales, Chytridiomycetes). *Protoplasma* **121**, 65–76.

MacKichan, J. K., Tuininga, A. R. & Kerwin, J. L. (1994) Preliminary characterization of phospholipase A_2 in *Lagenidium giganteum*. *Exp. Mycol.* **18**, 180–192.

McCray, E. M., Umphlett, C. J. & Fay, R. W. (1973a) Laboratory studies on a new fungal pathogen of mosquitoes. *Mosq. News* **33**, 54–60.

McCray, E. M., Womeldorf, D. J., Husbands, R. C. & Eliason, D. A. (1973) Laboratory observations and field tests with *Lagenidium* against California mosquitoes. *Proc. Calif. Mosq. Control Assoc.* **41**, 123–128.

McInnis, Jr, T. M. (1971) A physiological and biochemical investigation of the aquatic phycomycete *Lagenidium giganteum*, a facultative parasite of mosquito larvae. PhD Dissertation, University of North Carolina, Chapel Hill.

McInnis, Jr, T. M., Schimmel, L. & Noblet, R. (1985) Host range studies with the fungus *Leptolegnia*, a parasite of mosquito larvae (Diptera: Culicidae). *J. Med. Entomol.* **22**, 226–227.

Machlis, L. (1953) Growth and nutrition of water molds in the subgenus *Euallomyces*. I. Growth factor requirements. *Am. J. Bot.* **40**, 189–195.

Merriam, T. L. & Axtell R. C. (1982) Salinity tolerance of two isolates of *Lagenidium giganteum* (Oomycetes: Lagenidiales), a fungal pathogen of mosquito larvae. *J. Med. Entomol.* **19**, 388–393.

Nestrud, L. B. & Anderson, R. L. (1994) Aquatic safety of *Lagenidium giganteum*: effects on freshwater fish and invertebrates. *J. Invertebr. Pathol.* **64**, 228–233.

Orduz, S., Zuluaga, J. S., Diaz, T. & Rojas, W. (1992) Five isolates of the mosquito pathogenic fungus *Lagenidium giganteum* (Oomycetes : Lagenidiales) from Colombia. *Mem. Inst. Oswaldo Cruz* **87**, 597–599.

Padua, L. E., Whisler, H. C., Gabriel, B. P. & S. L. Zebold. (1986) In vivo culture and life cycle of *Coelomomyces stegomyiae*. *J. Invertebr. Pathol.* **48**, 284–288.

Pillai, J. S. & Smith, J. M. B. (1968) Fungal pathogens of mosquitoes in New Zealand. I. *Coelomomyces opifexi* sp. n. on the mosquito *Opifex fuscus* Hutton. *J. Invertebr. Pathol.* **11**, 316–320.

Seymour, R. L. (1984) *Leptolegnia chapmanii*, an oomycete pathogen of mosquito larvae. *Mycologia* **76**, 670–674.

Shapiro, M. & Roberts, D. W. (1976) Growth of *Coelomomyces psorophorae* mycelium in vitro. *J. Invertebr. Pathol.* **27**, 399–402.

Sparrow, Jr, F. K. (1960) *Aquatic Phycomycetes*, 2nd edn. University of Michigan Press, Ann Arbor.

Su, X. & Guzman, D. R. (1990) Studies on oospores of *Lagenidium giganteum* (Oomycetes: Lagenidiales) II. The use of SFE in artificial production of oospores. *J. Guiyang Med. Coll.* **15**, 88–91.

Taylor, B. W., Harlos, J. A. & Brust, R. A. (1980) *Coelomomyces* infection of the adult female mosquito *Aedes trivittatus* (Coquillet) in Manitoba. *Can. J. Zool.* **58**, 1215–1219.

Toohey, M. K., Prakash, G., Goettel, M. S. & Pillai, J. S. (1982) *Elaphoidella taroi:* the intermediate host in Fiji for the mosquito pathogenic fungus *Coelomomyces*. *J. Invertebr. Pathol.* **40**, 378–382.

Umphlett, C. J. & Huang, C. S. (1972) Experimental infection of mosquito larvae by a species of the aquatic fungus *Lagenidium*. *J. Invertebr. Pathol.* **20**, 326–331.

Weiser, J. & McCauley, V. J. E. (1971) Two *Coelomomyces* infections of chironomidae (Diptera) larvae in Marion Lake, British Columbia. *Can. J. Zool.* **49**, 65–68.

Whisler, H. C. (1985) Life history of species of *Coelomomyces*. In *The Genus Coelomomyces* (eds J. N. Couch & C. E. Bland). pp. 9–22. Academic Press, New York.

Whisler, H. C., Shemanchuk, J. A. & Travland, L. B. (1972) Germination of the resistant sporangia of *Coelomomyces psorophorae*. *J. Invertebr. Pathol.* **19**, 139–147.

Whisler, H. C., Zebold, S. L., & Shemanchuk, J. A. (1974) Alternate host for the mosquito parasite *Coelomomyces*. *Nature* **251**, 715–716.

Whisler, H. C., Zebold, S. L. & Shemanchuk, J. A. (1975) Life history of *Coelomomyces psorophorae*. *Proc. Natl. Acad. Sci. USA* **72**, 693–696.

Whisler, H. C., Wilson, C. M., Travland, L. B., Olson, L. W., Borkhardt, B., Aldrich, J., Therrien, C. D. & Zebold, S. L. (1983) Meiosis in *Coelomomyces*. *Exp. Mycol.* **7**, 319–327.

Willoughby, L. G. (1969) Pure culture studies on the aquatic phycomycete, *Lagenidium giganteum*. *Trans. Br. Mycol. Soc.* **52**, 393–410.

Woodring, J. L. & Kaya, H. K. (1992) Infectivity of *Lagenidium giganteum* to *Culex tarsalis* (Diptera: Culicidae) in rice field waters: laboratory evaluation. *Environ. Entomol.* **21**, 183–190.

Woodring, J. L., Kaya, H. K. & Kerwin, J. L. (1995) *Lagenidium giganteum* in *Culex tarsalis* larvae: Production of infective propagules. *J. Invertebr. Pathol.* **66**, 25–32.

Zebold, S. L., Whisler, H. C., Shemanchuk, J. A. & Travland, L. B. (1979) Host specificity and penetration in the mosquito pathogen *Coelomomyces psorophorae*. *Can. J. Bot.* **24**, 2766–2770.

APPENDIX

Media for culture of *Lagenidium giganteum*

Solid media

1. *Peptone–yeast–glucose* (*PYG*) (Fuller & Jaworski, 1987)

Peptone	1.25 g
Yeast extract	1.25 g
Distilled water	1000 ml
Glucose	3.0 g
Agar	15–20 g

This medium is a variation of a medium developed by Emerson (1958). Some researchers will add 1.36 g/l KH_2PO_4 and 0.71 g/l Na_2HPO_4, especially when using this medium for liquid culture (Gleason, 1968). In order to promote sporulation, $CaCl_2{\cdot}2H_2O$ and $MgCl_2{\cdot}6H_2O$ are often added at concentrations of 0.5–5.0 mM. Sterol deprivation will result in the gradual loss of the ability of cultures to sporulate; therefore, PYG is commonly supplemented with vegetable oils or purified sterols with a suitable solubilizing agent. Commonly used vegetable oils include soybean, safflower and corn oil. Wheat germ, linseed and cod liver oil have also been used, usually in conjunction with corn oil. Oils are usually added at a concentration of 0.5–2 ml/l (Kerwin *et al.*, 1986). Purified sterols such as cholesterol, ergosterol and sitosterol (10–100 mg/l) can be added with oil, or solubilized using Tween 20 or crude preparations of lecithin (phosphatidylcholine). Lecithin (50–100 mg/l is solubilized in 20–25 ml distilled water using a stir bar and gentle heating, and added to culture media after complete dissolution. If this is not done, the lecithin will form large clumps during autoclaving, and the sterol, which forms complexes with this lipid, will not be solubilized in a form that can be utilized by *L. giganteum*. A variation of this medium (PYG4S, Frances *et al.*, 1989) used a mixture of sterols – 0.0625% each of cholesterol, lanosterol, ergosterol and cholestan-ß-ol – and 5% lecithin added to basal PYG medium. Difco Laboratories (Detroit, Michigan) is a common source for peptone and yeast extract. We have used a variety of commercial sources for yeast extract. Peptones, however, vary greatly in source, preparation and composition, and unless preliminary evaluations prove otherwise, Difco peptone should be used.

2. *Hemp seed agar* (*HSA*) (Domnas *et al.*, 1974)

Hemp seed (5 g) is ground in 100 ml of 0.05 M phosphate buffer, pH 7, and the suspension stirred for several hours to solubilize fibrous material. The suspension is filtered through cheesecloth and incorporated into agar at a concentration of 1 mg protein/ml. One mM calcium chloride enhances sporulation. Use 15–20 g agar/l.

3. *Sunflower seed extract* (*SFE*) (Jaronski & Axtell, 1984)

Shelled sunflower seeds are ground to a fine powder and mixed with distilled water (10 g/100 ml). After blending for 60 s, the suspension is filtered through cheesecloth, the residue resuspended in 100 ml water, and the process repeated. The two filtrates are combined and can be frozen in small aliquots until use. This provides a stock solution with approximately 10–12 mg protein/ml. This stock is diluted to provide 1 mg protein/ml in the culture media, using 15–20 g agar/l.

4. $D_{6.5}$ *medium* (Kerwin *et al.*, 1991)

Yeast extract	1.5 g
Corn oil	1 ml
$CaCl_2{\cdot}2H_2O$	2 mM
pH adjusted to	6.5
Glucose	1.0 g
Cholesterol	25 mg
$MgCl_2{\cdot}6H_2O$	0.5 mM
Agar	15–20 g

The corn oil can be replaced by 50 mg/l lecithin solubilized as described above. There are a number of sources of crude lecithin. We have found that preparations from either soybean or egg yolk provide consistent sporulation.

Defined media

Defined media are useful for physiological and biochemical studies. These media can be used for zoospore production following supplementation with appropriate sterols, but yields are more variable and unpredictable than when more complex media are used.

1. *Gleason's defined medium* (Machlis, 1953, as modified by Gleason, 1968)

	g/l
KH_2PO_4	1.36
Na_2HPO_4	0.71
$MgSO_4{\cdot}7H_2O$	0.12
$CaCl_2{\cdot}2H_2O$	0.07
$FeCl_3{\cdot}6H_2O$	4.84×10^{-3}
$MnCl_2{\cdot}4H_2O$	1.80×10^{-3}
H_3BO_4	2.86×10^{-3}
$CuSO_4{\cdot}5H_2O$	0.39×10^{-3}
$(NH_4)_6Mo_7)O_{24}{\cdot}4H_2O$	0.37×10^{-3}
$CoCl_2{\cdot}6H_2O$	0.81×10^{-3}
thiamine	0.10×10^{-3}
$ZnSO_4{\cdot}7H_2O$	0.44×10^{-3}

The pH is adjusted to 6.6, and appropriate carbon and nitrogen sources added. Glutamic acid (1–2 g/l) and glucose (2–3 g/l) work well. It often takes the parasite several passages to become adapted to growth on defined media. Suitable sterol-containing components have to be added if sporulation is desired.

2. *DM2* (Kerwin *et al*, 1995)

	g/l
Glucose	4.0
Asparagine	1.0
KH_2PO_4	0.05
K_2HPO_4	0.05
$MgSO_4{\cdot}7H_2O$	0.1
$CaCl_2{\cdot}2H_2O$	0.15
$Fe(NO_3)_2{\cdot}9H_2O$	4.84×10^{-3}
$MnCl_2{\cdot}4H_2O$	0.036
$ZnSO_4{\cdot}7H_2O$	0.01
Na glutamate	2.0
Na_2EDTA	0.05
H_3BO_4	0.063
$CuSO_4{\cdot}5H_2O$	0.008
$(NH_4)_6Mo_7)O_{24}{\cdot}4H_2O$	0.01
$CoCl_2{\cdot}6H_2O$	0.017
thiamine·HCl	0.10×10^{-3}
methionine	0.05
KNO_3	0.1

250 μl tergitol NP-40/ethanol (1 : 1, v:v)
Adjust to pH 7.0 with 1 N NaOH.

Liquid media

1. Any of the media listed under *Solid media* can be used for liquid culture by omitting the agar. The pH of liquid media should be adjusted to ca. 6.5–7.5. Calcium is an obligate requirement for growth and morphogenesis. If there is not sufficient trace calcium in crude media components (1–5 mM is usually optimum), additional quantities of this mineral have to be added. This requirement is usually met in solid media by trace quantities present in agar, but it may occasionally be necessary to add calcium to those media also.

2. *SEX/A medium* (Kerwin *et al*., 1991)

	g/l
Yeast extract	1.5
Hydrolysed lactalbumin	0.5
Cholesterol	0.025
Calcium	2 mM
Glucose	1.0
Lecithin	0.15
Corn oil	1.0
Magnesium	0.5 mM
pH 6.5	

SEX/C medium (Kerwin & Washino, 1987) consists of the same components plus 0.5 g/l dehydrated egg yolk. A third medium consisted of SEX/C, but instead of corn oil, 0.25 ml cod liver oil and 0.6 ml wheat germ oil were added (Kerwin & Washino, 1986a). These two media have been successfully scaled-up for fermentation production of *L. giganteum* in 10–130 l fermentors.

3. *Z medium* (Domnas *et al*., 1982)

	g/l
Yeast extract	1.25
Powdered wheat germ	3.2
Glucose	1.2

hemp seed extract as described above at 250 mg/l soluble protein.

4. *Alternate fermentation medium* (Kerwin & Washino, 1988).

	g/l
Yeast extract	2.5
Proflo cottonseed extract	3.0
Cholesterol	0.1
Calcium	5 mM
Glucose	2.0
Fish meal	8
Proflo oil	2.0
Magnesium	1 mM
pH 6.5–7.0	

This medium can be supplemented with homogenized fresh chicken eggs (2–20 ml egg/l).

5. *PB2 medium* (MacKichan *et al.*, 1994)

	g/l
Dehydrated egg yolk	1.0
Proflo cottonseed extract	2.0
Fish peptone	0.1
$CaCl_2 \cdot 2H_2O$	0.225
Glucose	1.5
Lactalbumin hydrolysate	0.5
$Fe(NO_3)_2 . 9H_2O$	0.008
$MgCl_2 \cdot 6H_2O$	0
pH 7.5	

Dilute salts solution

Stock solution volume	*Weight (g)*
1. 500 ml	
$(NH_4)_2 \cdot HPO_4$	66.04
KH_2PO_4	68.05
K_2HPO_4	87.09
$MNCl_2 \cdot 4H_2O$	1.8
$ZnSO_4 \cdot 7H_2O$	0.44
H_3BO_3	2.86
2. 250 ml	
$CaCl_2 \cdot 2H_2O$	18.38
$MgCl_2 \cdot 6H_2O$	25.42

Use 0.5 ml of stock solution (1) and 0.1 ml of stock solution (2) per 5 l distilled water. A simpler variation developed for *C. psorophorae* (Whisler *et al.*, 1975) consists of 0.5 g $NaHCO_3$, 0.25 g $MgSO_4 \cdot 7H_2O$, 0.1 g KCl and 0.5 g $Ca(NO_3)_2 \cdot 4H_2O$ per litre distilled water.

CHAPTER V-5

Fungi: Preservation of cultures

RICHARD A. HUMBER

USDA-ARS Plant Protection Research Unit, US Plant, Soil & Nutrition Laboratory, Tower Road, Ithaca, New York 14853-2901, USA

1 INTRODUCTION

All research or applied studies using live organisms requires a constant supply of them in a suitable condition. Work with fungi usually requires keeping cultures, a task that is both easier and facilitates more possible research approaches than dealing, for example, with migratory birds, marine mammals, mountain gorillas, mature redwood trees, or even many insects.

The isolation and growth of microbial cultures are dealt with elsewhere in this book. Although this chapter focuses on fungi, the techniques described here apply equally for nearly all other types of entomopathogens.

No matter why or how one may store cultures, all preservation techniques increase the time between transfers to periods ranging from several weeks or months to many years with a minimal loss of viability or other key properties of the organism during storage. Each of these preservation techniques has strengths and weaknesses (Table 1). Once one's needs to store cultures go beyond the most casual level, it is very important to choose the preservation technology that best fits one's needs with a convenient, affordable level of technological sophistication. Much time, anguish and money can be saved by carefully weighing the real purposes and needs to preserve cultures before committing one's effort and financial and physical resources to any specific storage technique. The demands for space, materials, record keeping and labour are much lower for researchers maintaining a few cultures being used in current research or teaching than for laboratories keeping small archival collections with dozens to several hundred cultures, or for general service culture collections that are actively acquiring, storing and distributing large numbers of cultures.

Although it is not usually recognized as such, formulations of microbial biocontrol agents usually serve to preserve a living infective virus, bacterium (except, possibly, for *B. thuringiensis*), microsporidium or fungus. The 'active' ingredient of a formulation remains in a quiescent but viable state during shelf storage to be activated upon application.

MANUAL OF TECHNIQUES IN INSECT PATHOLOGY
ISBN 0-12-432555-6

Table 1 Advantages and disadvantages of preservation techniques.

Preservation method	*Advantages*	*Disadvantages*
Storage temperature ≥0 °C		
Serial transfer (stored at 4 °C)	• Technologically simple • Allows continuous monitoring of phenotype	• Basic (phenotypic) characters may change • Continuing need for materials and labour • Some fungi do not tolerate cold
Mineral oil	• Inexpensive and technologically simple	• Space intensive; tubes must be stored upright • Must inspect periodically for contaminants
Distilled water stasis	• Inexpensive and technologically simple • Needs little space if using small vials	• Must check for water levels and contaminants
Lyophil (Freeze-dried)	• Standard methods can be used for many fungi • Dried cultures can be mailed	• Equipment is relatively expensive • Ampoules should be refrigerated • Not suitable for some fungi
Storage temperature <0 °C		
Silica gel	• Inexpensive and technologically simple • Uses standard appliance-type freezers • Can store only one tube per culture • These may be kept at room temp. or –20 °C	• Long-term success depends on security of screw-top seal
Electric (mechanical) freezers		
Standard (–20 °C)	• Inexpensive and readily available	• Not recommended (except for silica gel storage) • Subject to losses in event of power failures
Ultracold (–80 or –120 °C)	• Convenient and available in many labs • Long-term cryostorage without liquid N_2	• Freezer temperatures favour ice crystal formation • Subject to losses during power failures • Nitrogen back-up is recommended
Liquid nitrogen (LN_2) freezers	• Unexcelled for duration of storage • Long-term viability is maximized • All types of culturable fungi can be stored	• High initial and continuing costs • Continuous access to LN_2 is vital
Vapour phase (–120 to –196 °C)	• LN_2 supply cannot contaminate vials • No risk of vials exploding when thawed	• Temperature is relatively unstable • LN_2 levels must be monitored continuously • Automatic LN_2 level control depends on uninterrupted electrical service
Liquid phase (–196 °C)	• Temperature is stable during immersion • Ice crystal formation is unlikely • Monitoring of nitrogen levels is minimal • Unaffected by loss of electrical service	• Exclusion of nitrogen from entering plastic cryovials cannot be guaranteed • Vial contents may be contaminated by nitrogen leakage during storage • Vials containing LN_2 may explode when thawed

Examples of this approach include techniques to prepare dried mycelium that resumes sporulation when rehydrated (McCabe & Soper; 1985; Rombach *et al.*, 1989; Verkleij *et al.*, 1992) or the encapsulation of spores and/or hyphae in alginate and starch (Pereira & Roberts, 1991; Salgado & Campos, 1993). The production and formulation of entomopathogens is not covered further in this book.

There is a rich literature on the preservation of fungal cultures. This chapter is not a comprehensive review of all possible techniques but it does treat a range of techniques used for fungal pathogens of invertebrates. More detailed discussions of these techniques can be found in (uncited) journal articles and in compendia on fungal preservation (Onions, 1983; Simione & Brown, 1991; Smith & Onions, 1994).

The protocols discussed here are not fixed in stone; they can be adapted to meet individual needs and limitations. The exact reproduction of published preservation protocols should not the ultimate goal; that goal must be to preserve viable cultures by the most successful, practical and reasonable means suited to one's operating conditions.

2 PRESERVATION AT TEMPERATURES ABOVE FREEZING

Storing fungi at ambient temperatures rather than in refrigeration may require fewer resources than any other preservation strategy, but it also seems to be the least dependable and shortest-term technique. Several differing techniques – including serial transfer, storage under mineral oil or distilled water, on silica gel crystals, and even after lyophilization (freeze-drying) – may be used to prepare fungi for storage at ambient temperatures. The phenotypic and genotypic stability of the culture will be affected by the choice of preparative method used and by both the overall range and stability of temperature during the storage period.

A Serial transfer

This most obvious way to store cultures is best suited for relatively short-term studies (over a few weeks or months), although it can serve adequately to maintain small numbers of cultures for many years.

There are several potentially serious drawbacks to relying on serial transfer of fungi. Repeated subculturing may lead to such deleterious changes as losses of pathogenicity, virulence, or sporulation, but there is no way to predict if or when repeated subculturing might result in losses of vital characters. Cultures of invertebrate pathogens must be inspected carefully for such morphological changes and bioassayed periodically for changes of essential pathological characteristics. Fennell (1960) and Onions (1971) discuss what fungal structures (hyphae, spores, etc.) constitute appropriate inoculum for serial transfers but the staff of the USDA-ARS Collection of Entomopathogenic Fungal Cultures (ARSEF; Ithaca, NY) uses both hyphae and, if available, spores as the preferred inoculum.

B Mineral oil

Storing culture slants under a layer of sterile mineral oil is one of the oldest, simplest and least expensive methods for long-term culture preservation. This approach is still widely used, especially for fungi that do not tolerate freeze drying or where cryogenic storage is too costly.

The oil both prevents desiccation and diminishes gas exchange, thus reducing fungal metabolism to a very low level. If space is available to store racks of tubes, this is a common alternative for the storage of a very diverse range of fungi. Onions (1971) warns that McCartney bottles or other screwcap bottles with rubber gaskets should be avoided unless the gaskets are removed since oil-soluble components in the rubber may be toxic to cultures.

Cultures kept under mineral oil may remain viable for decades (Cavalcanti, 1991; Silva *et al.*, 1994). Pathogenicity of entomopathogenic fungi may be undiminished after several months of storage (Balardin & Loch, 1988), but whether pathogenicity or virulence decline after many years of storage under mineral oil remains uncertain.

Set up for storage

- Use vigorously growing culture slants in glass culture tubes.
- Autoclave a supply of heavy mineral oil and re-autoclave 24–48 h later to kill any bacterial spores activated by the first autoclaving.
- Aseptically cover the culture slant with sterile oil to the depth of 1 cm.
- Cover tubes with tight caps or plugs and apply a couple of layers of paraffin film as a further vapour barrier.
- Store tubes upright in racks. Viability may be retained longer if refrigerated than if kept at room temperature.

Culture recovery

- With a sterile scalpel, loop, or needle, recover an explant from the submerged culture.
- Drain excess oil from the explant and place on fresh medium.
- Reseal and return the tube to long-term storage.
- Monitor the culture for viability and/or contamination.

C Distilled water stasis

Storage of metabolically inactive fungi under sterile distilled water may be least technologically demanding of any preservation techniques discussed here. This method is useful for many diverse fungi, but may be especially welcomed by those needing to preserve fungi that cannot be freeze-dried (e.g. Oomycetes; Clark & Dick, 1974). This method has been used successfully with a wide range of fungi including human or plant pathogens (Castellani, 1967; Figueiredo & Pimentel, 1975). Viability of some fungi for up to 20 years has been documented with this method (Hartung de Capriles *et al.*, 1989) although many fungi lose viability much sooner. No literature specifically describes using this technique for entomopathogens but it can be used with the Entomophthorales (Humber, unpublished). Water stasis is probably a suitable storage technique for nearly all fungal pathogens of invertebrates.

This technique requires having only sterile screw-cap tubes or vials and sterilized water (tap water can be used if distilled or deionized water is unavailable). Material in water stasis can be kept at room temperature or refrigerated.

The most serious problems of this technique are easily avoided. Too much inoculum for the volume of water may jeopardize the ability of the fungus to withstand long-term storage; this can be avoided by using a volume of water at least 40 times greater than that of the inoculum blocks. Evaporative loss of water from poorly sealed storage tubes can be overcome by periodically adding sterile water to those storage units that might need it.

Set up for storage

- Use vigorously growing, relatively young cultures for inoculum.
- Autoclave a supply of distilled water and screw-cap vials or tubes (or use presterilized screwcap tubes, tissue culture flasks, or centrifuge tubes).
- Dispense sterile water into sterile storage tubes.
- Inoculate tubes with small (ca. 1 mm^3) blocks of a culture and/or an aqueous suspension of spores. The volume of water must be ≥40 times the total volume of the culture inoculum preserved in it.
- Cover tubes with tight fitting caps. For more security, also seal with paraffin film or by dipping tube tops in melted paraffin.
- Store tubes upright at room temperature or in refrigerator.

Culture recovery

- Recover a block from the tube with a sterile scalpel, loop, or needle and place (fungus side down) onto fresh medium.
- Monitor the culture for viability and/or contamination.
- Reseal the tube and return to long-term storage.

3 PRESERVATION BY FREEZE-DRYING (LYOPHILIZATION)

Lyophilization may be the most widely used of the 'technologically sophisticated' approaches to preserving fungal germplasm and is the primary technique used at most general service culture collections. Cultures can be processed for storage in a relatively short time after which the lyophilized ampoules can be filed and then sent out for revival by the recipient.

Freeze-drying is effective for nearly all conidial fungi (Hyphomycetes and Coelomycetes), ascomycetes, and basidiomycetes. It is not usually useful for fungi with very 'watery' cells (with large vacuolar volumes) such as many Oomycetes and the Entomophthorales.

Many different sorts of ampoules or other containers may be used to hold cultures during lyophilization (Simione & Brown, 1991; Smith & Onions, 1994). Most small- to medium-scale users rely on manifold or centrifugal freeze-dryers and use commercial lyophil ampoules or glass tubing sealed at one end to make small tubes. Large-scale operations may use serum bottles dried on shelves in a large vacuum chamber and sealed under vacuum by lowering a pressure plate onto the unseated tops of the bottles.

The following discussion on methods for

lyophilizing fungi is a condensed summary of the process; detailed protocols are needed only in laboratories with access to a lyophilizer and are readily available in general reference works on culture preservation (e.g. Simione & Brown, 1991; Smith & Onions, 1994). However, because nearly any laboratory studying fungi may receive and need to revive lyophilized cultures, detailed instructions for doing so are included below.

Set up for storage

- Use sporulating cultures for inoculum. (**Note:** Spores may retain viability better than hyphae but non-sporulating cultures can also be lyophilized.)
- Sterilize and dry cotton-plugged ampoules.
- Cover sporulating cultures with sterilized skim milk solution or another nutritionally complex carrier, and suspend spores and hyphae. Transfer small quantities of this suspension to sterile ampoules and 'cure' the contents for several hours in a refrigerator. (**Note:** If left overnight in the milk solution, some fungi such as *Metarhizium anisopliae* may clear this carrier but do so without affecting the viability of the lyophilized culture.)
- *Rapidly* freeze the preparations in a mixture of dry ice and either ethanol or propylene glycol or by placing ampoules in an ultracold freezer or in the vapour phase in a liquid nitrogen dewar).
- Attach frozen ampoules to a strong vacuum on the lyophilizer. Preparations must remain frozen during the initial stages of vacuum desiccation. After desiccation, ampoules are flame-sealed while still under vacuum.
- Store ampoules at ca. 4 °C. They may also be kept at ambient temperature if necessary but viability will probably decline sooner than if refrigerated.

Culture recovery

- Most culture collections that send out lyophilized cultures provide detailed directions on how to open and reconstitute such preparations.
- If an ampoule is not pre-scored, score the neck with a file or diamond pencil.
- Surface sterilize in 70% ethanol or sodium hypochlorite solution (e.g. a 1 : 1 dilution of commercial bleach), wrap the scored ampoule in a sterile paper wipe moistened (but not soaking!) with ethanol and break at the scoring.
- Add sterile water or liquid medium (the type and quantity of liquid will usually be indicated by the culture's sender) to reconstitute the culture.
- Rest the material in a sterile hood for 1–30 min to soften and to rehydrate the dried pellet. Resuspend and pipette the reconstituted mixture onto fresh culture medium.
- Monitor the culture for viability and/or contamination.

4 PRESERVATION IN A FROZEN STATE

Standard domestic freezers (operating at –20 °C) might seem to be an ideal and economical tool for keeping frozen cultures. However, such freezers are not recommended for preserving cultures although Carmichael (1962) successfully kept peptone–yeast extract cultures of a wide range of fungi at –20 °C. The most reliable long-term cryopreservation techniques require much colder temperatures; it is critical to understand that the colder the storage temperature, the greater the likelihood will be for long-term viability.

A Silica gel

Storage of spores on sterile, anhydrous silica gel crystals is the only storage technique for microorganisms that recommends use of –20 °C freezers. It is possible to store fungus-inoculated silica gel at room temperature if freezer space is unavailable but fungal viability will usually decrease sooner. Fungal spores stored on silica gel may remain viable for more than ten years (Windels *et al.*, 1993), and this method is inexpensive, simple and reliable for many fungi (Smith, 1993) including entomopathogens (Bell & Hamalle, 1974).

The use of anhydrous silica gel crystals as a carrier for culture propagules is limited to aerobic bacteria and fungi that grow on solid culture media. Fungi with high ratios of vacuolar to cytoplasmic volumes (e.g. Oomycetes and Entomophthorales), viruses, microsporidia and anaerobes are unsuited for storage on silica gel.

Although several grades of silica gel are available commercially, only some of them are acceptable for storing fungi. The crystals should be relatively large and uncoloured (with no indicator dye); the blue indicator dye that turns pink when hydrated can be toxic to fungi (Perkins, 1962).

Because the uptake of water by anhydrous silica gel is a strongly exothermic reaction, silica gel tubes must be put in an ice bath or to chilled in a freezer to –20 °C to dissipate the heat of hydration during their inoculation.

Set up for storage

- Use sporulating cultures for inoculum.
- Fill 25 × 200 mm screw-cap glass tubes one-third full with 6–12 mesh, grade 40, uncoloured anhydrous silica gel crystals.
- Sterilize the tubes and silica gel in an oven at 160–180 °C for 1–6 h to assure that the silica gel is both sterile and fully anhydrous.
- Dispense 1–5 ml sterile water – or an autoclaved solution of 5–7% (v/v) skim milk – into a sporulating culture. Cap and agitate the tube or rub the surface of the plate with a sterile glass rod.
- Dispense 1 ml of the suspension by drops onto cold silica gel crystals. Tilt tubes during inoculation to expose the greatest possible surface area and rotate or agitate tubes while adding the inoculum suspension.
- Hold inoculated tubes at room temperature for a few days, rotating or agitating them periodically until all water has been absorbed and the crystals are separated.
- Store at –20 °C although tubes may also be kept at room temperature.
- Check viability and sterility of the stored preparation after ca. 1–2 weeks.

Culture recovery

- Sprinkle a few granules of inoculated silica gel from a tube onto fresh culture medium.
- Tightly reseal tube and return to long-term storage.

B Cryogenic preservation in mechanical or liquid nitrogen freezers

1. General considerations and protocols

The widespread proliferation of ultracold freezers in biological laboratories has enabled growing numbers of biologists to store cultures of their research organisms cryogenically. Storing cultures in liquid nitrogen dewars, while theoretically still more secure and longer-term than storage in ultracold freezers, is both more expensive and wholly dependent on a continuous supply of liquid nitrogen, a commodity that is either unavailable or prohibitively expensive for most laboratories.

Virtually every fungus that can be preserved by any of the techniques noted above or that can be cultured *in vitro* can be preserved cryogenically. Further, many fungi that cannot be preserved by such techniques as lyophilization can be stored cryogenically.

Despite the versatility of cryogenic techniques, they are no panacea; all techniques to preserve germplasm have flaws and weaknesses (see Table 1). The viability of frozen cultures may be strongly affected by such factors as the choice of cryoprotectant, the rate of freezing, and temperature stability during storage; these same factors may also affect the viability of freeze-dried cultures (Tan *et al.*, 1995). The physical factors affecting the freezing, storage and recovery of living cells demand an awareness of some basic principles of chemistry and physics; these points and other general considerations about cryopreservation are synthesized and discussed by Calcott (1978).

What happens to water in and around cells during freezing, storage and thawing may be the most critical such factor: Water freezes in one of two states, either the amorphous (glassy or vitreous) state or the crystalline (icy) state. Cryoprotectants serve to favour the freezing of intracellular water as a glass rather than as ice crystals whose growth may disrupt or destroy a cell's membrane systems. Frozen water can convert from the glassy into the crystalline state (leading to the initiation and growth of ice crystals while remaining frozen!). This devitrification, the shift out of the glassy state by ice nucleation, is increased by fluctuation of the storage temperature (discussed below) and is favoured at a few temperatures that are troublingly close to those in ultracold mechanical freezers. One of these glass transition temperatures for frozen protein solutions is –80 °C (Chang & Randall, 1992). In water–glycerol mixtures, ice nucleation (without much crystal growth) is favoured throughout the temperature range of –93 to –123 °C (Vigier & Vassoille, 1987).

a. Cryoprotectant

Many different cryoprotectants can be used to help prevent the formation of ice crystals during freezing, storage and thawing of cultures. Some of the cryoprotectants that are widely used in various laboratories include glycerol, dimethylsulphoxide (DMSO), polyethylene glycol and propylene glycol. In practice, however, most laboratories preserving fungi use only 10% glycerol (v/v); this dependence on a single cryoprotectant is based on the convenience and suitability of this cryoprotectant rather than from any experimental demonstration of its optimal effectiveness (see Sanskär & Magalhães, 1994). Glycerol is used for virtually every fungus in the ARSEF collection (more than 300 species from several fungal classes); the few fungi that have been frozen in 6% (v/v) DMSO are entomophthoraleans. DMSO is not usually used as the cryoprotectant for fungi, however, until after repeated failures to freeze an isolate with glycerol as a cryoprotectant but while varying the inoculum by using very young or somewhat senescent cultures or cultures grown on a different medium from that routinely used for the fungus. In many instances, isolates that resist repeated attempts to freeze them are those that grow poorly in culture, and which tend to die after several serial transfers.

The choice of cryoprotectant determines the temperature, the heat of fusion, at which the cryoprotectant freezes with a strongly exothermic reaction. Electronically controlled cell freezers can dissipate this heat while allowing material to continue freezing at a highly uniform overall rate, usually ca. –1 °C per minute.

Regardless of the cryoprotectant used, cryoprotectant solutions should be made and filter sterilized (not autoclaved!) in relatively small batches to restrict contamination hazards since a contaminated cryoprotectant stock solution will contaminate every culture treated with it.

b. Cryogenic storage units

Plastic cryogenic vials, plastic straws, or glass ampoules are used to contain material to be frozen. The choice among these storage units is driven by the racking system chosen for the physical containment of individual storage units, by convenience in handling, by cost and by safety considerations. Although they are not the least expensive alternative, commercially available plastic cryogenic vials are probably the most commonly used storage units; polypropylene drinking straws (which may be the least expensive storage containers) are less commonly used. Glass ampoules or tubes are rarely chosen since they are more difficult to use during the freezing process and may be especially hazardous if they shatter when recovering material from storage.

The exclusion of liquid nitrogen from the interior of storage units is the most critical issue for choosing a unit to be stored by immersion in liquid nitrogen. The leakage of nitrogen into storage units may result in contamination of the preserved material as well as explosions of these units during thawing. Sections of polypropylene drinking straws whose ends are heat-sealed are an inexpensive alternative to screwcap cryovials (Challen & Elliott, 1986; Stalpers *et al.*, 1987) and, if their end seals are secure, can be immersed safely in liquid nitrogen. Cryogenic vials can be protected from nitrogen infiltration during immersed storage by sheathing entire sets of vials or individual vials in plastic tubing that can be heat-shrunk onto the vials and sealed by heat-crimping either end of the tubing section before freezing the material. However, cultures delicate enough to require such sheathing may be unable to tolerate the heating received during the sealing process. It has been the experience of the ARSEF collection staff that those cultures grown in tissue culture media – the isolates that are most vulnerable to contamination during storage immersed in liquid nitrogen – are the cultures least able to tolerate this heat-sealing.

c. Auxiliary equipment for culture freezing

An inexpensive, commercially produced polycarbonate freezing container (Nalgene™ Cryo 1 °C Freezing Container) can be a very suitable low-technology tool for producing a semi-controlled rate of freezing approximating –1 °C/min. This device has been used successfully to freeze entomophthoralean protoplasts (A.E. Hajek, unpublished), a growth form that is among the most difficult to freeze among all of the entomopathogenic fungi. This commercial freezing device accommodates 18 cryovials. After conditioning the loaded cryovials at refrigerator temperature for several hours, the rack containing them is placed on a pool of isopropanol in this freezing container and the whole assembly put in an ultracold freezer for 24 h. The temperature in the vials in this container drops at about 1 °C/min.

Styrofoam boxes with sides 2.5 cm thick and tops and bottoms 1.25 cm thick can be used as freezing containers, but these may have much faster freezing rates than an isopropanol-mediated freeze or those in electronically controlled (and vastly more expensive) cell freezing units.

Electronically controlled, programmable freezing units are most useful in large-scale operations because of their high capacity and precise controls of freezing curves. These expensive freezing units are driven with liquid nitrogen vapours and are best suited for use in laboratories that depend heavily on cryogenic storage of germplasm or on highly controlled freezes to ensure viability of valuable but relatively delicate germplasm.

McLaughlin *et al.* (1990) suggested that there was little difference in viability of human spermatozoa frozen in an uncontrolled manner in nitrogen vapour or in an electronically controlled freezing unit. It has been the experience of the ARSEF collection staff that most entomopathogenic fungi, especially hyphomycetes, have excellent recoverability after having been frozen by direct plunging into liquid nitrogen.

d. Freezing and thawing protocols

Set up for storage

- *For cultures grown on solid media:* Dispense sterile cryoprotectant into storage units (vials, straws, etc.) and then add 2–4 cubes (1–3 mm on a side); make sure that there is ≥40 : 1 ratio of cryoprotectant to inoculum (as in Section 2C for water stasis).
- *For cultures grown in liquid media:* Add an appropriate volume of undiluted, sterile cryoprotectant directly to cultures and disperse quickly by gentle agitation to minimize osmotic damage. Aliquot directly into sterile cryovials.
- 'Cure' fungus in cryoprotectant at 4 °C for 2–48 h to allow uptake of the cryoprotectant. Most ARSEF cultures are cured overnight (ca. 12 h) before freezing.
- Freeze at an uncontrolled rate by direct placement into the storage freezer or in a controlled freezing device (see Section 4A.1c). With controlled and semi-controlled freezes, cultures are usually frozen to between –40 and –50 °C and held there for 30–120 min before being moved to the storage facility.
- 1–7 days after freezing, thaw a unit of each culture to check for viability, growth rate, morphology, etc. If this viability check does not meet all expectations, the entire lot should be discarded after a new lot is frozen with acceptable post-recovery characteristics.

Culture recovery

- Thaw all cryogenically stored material directly in 37 °C water. Leave units in the warm water only until the ice is completely melted. (**IMPORTANT SAFETY WARNING:** *For material stored immersed in liquid nitrogen, visually check transparent or translucent storage units upon removal from storage for moving liquid (nitrogen) inside. If motion is seen, allow nitrogen to vaporize by warming at room temperature for a minute or two under a metal container. Placing a glass or plastic unit enclosing liquid nitrogen directly into warm water can result in a small but violent explosion, flying shards of the storage unit (and risks of serious bodily injury), and the widespread dispersion around a laboratory of a microbial culture.* **When thawing frozen cultures, *always* use proper protective gear – lab coat, cryogenic gloves and a full face mask – and keep unprotected workers 2–3 m away from thawing vials.**)

e. Checking viability

No matter what preservation method is chosen, always check the viability of a preserved sample relatively few days after its preparation. If the test sample is viable and uncontaminated upon recovery, the remaining samples should be reliable for long-term storage. If the sample is not viable or is contaminated, it is likely that rest of the lot is similarly unacceptable. Once a new, successful freeze is completed and confirmed, the older, inviable material should be discarded. If a viability check indicates contamination or any loss of a culture's essential properties, it is best to check the original culture and to pull a second confirmatory unit from storage. If the second unit also fails to meet the required standards, try a new freeze procedure and, when the isolate has been successfully frozen, discard all unsuccessfully preserved material.

f. Stability of temperature during storage

The stability of temperature during frozen storage may be more important than the actual temperature at which storage occurs. The fewer and smaller the temperature fluctuations during frozen storage, the less likely it will be that ice crystals can form and grow to damaging sizes in the preserved cells.

MacFarlane *et al.* (1992) suggest that optimal storage conditions require both the coldest possible temperature and greatest possible temperature stability. The optimal conditions for long-term storage of living material should be, therefore, immersion in liquid nitrogen at –196 °C. However, the exclusion from storage units of nitrogen and nitrogen-borne contaminants is a challenging problem for material immersed in liquid nitrogen. Storage in nitrogen vapour over a pool of liquid nitrogen avoids this contamination hazard but is neither so cold nor so stable as immersion in the liquid phase. In a nitrogen dewar, the temperature in the vapour phase ranges from –196 °C at the vapour/liquid interface up to ca. –120 °C at the top of the dewar.

In addition to the absolute temperature during storage, the physical racking system in which storage units are placed within a freezer is also critical. This racking dramatically affects the overall stability of temperature for individual units during storage. Vials or straws clipped to vertical aluminium canes stored immersed in liquid nitrogen need never be removed from liquid nitrogen except to remove the topmost unit; such storage conditions are the theoretical optimum. To facilitate the location of specific vials, canes are usually grouped in tall boxes that are, in turn, arranged in a single palisade layer. Stacked racking systems like those usually illustrated in sales literature for large cryogenic storage systems – sets of vertical stainless steel racks holding vertical stacks of covered boxes that are internally divided to hold individual cryogenic units – should be avoided because the manner in which they must be handled assures enormous temperature fluctuations for stored material every time any material is added or removed from the stack. In such a system, the most temperature-stable positions are those at the top where a stack may be lifted a minimal distance in order to withdraw a box, place it on top of the other stacks, remove its top and the desired storage units from it, and then to close and return the box to its permanent storage position. To recover material from the lowest boxes in a stack (in the coldest storage positions), all the boxes in the stack must be lifted out of nitrogen (or its vapours) until the desired box is accessible and removed; after material is recovered, the stack must be lifted out again to return the box to its position. In this worst-case scenario, material in the stack will be exposed to temperatures from 70 °C to potentially more than 200 °C hotter than the long-term storage temperature each time an individual stack of boxes is retrieved from storage, and may be submitted to lesser warming each time the stack is partially raised to gain access to higher-placed boxes in it. Even though each of these warming periods may only last a few seconds, it may still be enough time and temperature change to damage the least robust fungi. For an instructive comparison about the magnitude of such temperature fluctuations, consider that plunging your hand into boiling water for only a few seconds might entail a relatively smaller temperature change for a comparable time span to that experienced by frozen material every time it is pulled from a nitrogen dewar or ultracold freezer. Despite the theoretical possibility for damage due to large and rapid temperature fluctuations, practical experience with cryogenic storage in many laboratories suggests that this sort of handling-induced temperature fluctuation does not routinely cause obvious damage to most cultures.

2. *Comments about specific cryopreservation facilities*

a. *Mechanical freezers*

Standard freezers (–20 °C). Except for cultures preserved on silica gel (see Section 4A), stand-alone standard (upright or chest) freezers or the freezing units of two-door refrigerator–freezer combination units are not recommended for preserving fungal cultures. The freezing units inside single-door refrigerator–freezer combination appliances may only be able to achieve temperatures of ca. –5 °C. The temperature of all of these freezers is too high (too close to freezing) and too unstable to be truly reliable for germplasm storage. Frost-free freezer units undergo regular, periodic heating cycles to maintain a frost-free interior; this fluctuating temperature, as noted above for cryogenic storage, is potentially harmful to cultures and should be avoided.

Ultracold freezers (–80 or –120 °C). Many laboratories store cultures in ultracold mechanical freezers. Chest-type freezers are inherently more temperature stable and preferable to upright designs. Although ultracold freezers may be very adequate for storing many fungi, liquid nitrogen back-up facilities are recommended to be attached to them to maintain the temperature during power failures. Further, as was noted above, the temperatures in these freezers are close to those which pose some risks for long-term viability of frozen cells.

b. Liquid nitrogen freezers

Storage in or over liquid nitrogen is the most expensive if most nearly optimal of all available preservation methodologies. Dependence on such an approach to germplasm preservation requires a serious commitment in time and resources and presupposes both ready and uninterruptable access to the required nitrogen. The laboratory that relies on nitrogen storage facilities has made a long-term and expensive commitment to maintain (usually large numbers of) cultures perceived as having a very high intrinsic value.

REFERENCES

Balardin, R. S. & Loch, L. C. (1988) Methods for inoculum production and preservation of *Nomuraea rileyi* (Farlow) Samson. *Summa Phytopathol.* **14**, 144–151.

Bell, J. V. & Hamalle, R. J. (1974) Viability and pathogenicity of entomogenous fungi after prolonged storage on silica gel at –20°C. *Can. J. Microbiol.* **20**, 639–642.

Calcott, P. H. (1978) *Freezing and thawing microbes*. Meadowfield Press, Ltd., Durham, UK. 68 pp.

Carmichael, J. W. (1962) Viability of mold cultures stored at –20°C. *Mycologia* **54**, 432–436.

Castellani, A. (1967) Maintenance and cultivation of common pathogenic fungi of man in sterile distilled water. Further researched. *J. Trop. Med. Hyg.* **70**, 181–184.

Cavalcanti, M. A. D. Q. (1991) Viability of Basidiomycotina cultures preserved in mineral oil. *Rev. Latinoam. Microbiol.* **32**, 265–268.

Challen, M. P and Elliott, T. J. (1986) Polypropylene straw ampoules for the storage of microorganisms in liquid nitrogen. *J. Microbiol. Methods* **5**, 11–22.

Chang, B. S. & Randall, C. S. (1992) Use of subambient thermal analysis to optimize protein lyophilization. *Cryobiology* **29**, 632–656.

Clark, G. & Dick, M. W. (1974) Long-term storage and viability of aquatic Oomycetes. *Trans. Br. Mycol. Soc.* **63**, 611–612.

Fennell, D. I. (1960) Conservation of fungus cultures. *Bot. Rev.* **26**, 80–141.

Figueiredo, M. B. & Pimentel, C. P. V. (1975) Métodos utilizados para conservação de fungos na micoteca da seção de micologia fitopatológica do Instituto Biológico. *Summa Phytopathol.* **1**, 299–302.

Hartung de Capriles, C., Mata, S. & Middelveen, M. (1989) Preservation of fungi in water (Castellani): 20 years. *Mycopathologia* **106**, 73–80.

MacFarlane, D. R., Forsyth, M. & Barton, C. A. (1992) Vitrification and devitrification in cryopreservation. In *Advances in low-temperature biology* (ed. P. L. Steponkus), vol. 1, pp. 221–278. JAI Press, London.

McCabe, D. E. & Soper, R. S. (1985) Preparation of an entomopathogenic fungal insect control agent. US Patent No. 4,530,834 (23 July 1985).

McLaughlin, E. A., Ford, W. C. L. & Hull, M. G. R. (1990) A comparison of the freezing of human semen in uncirculated vapor above liquid nitrogen and in a commercial semi-programmable freezer. *Hum. Reprod.* **5**, 724–728.

Onions, A. H. S. (1971) Preservation of fungi. In *Methods in microbiology* (ed. C. Booth), Vol. 4, pp. 113–151. Academic Press, London.

Onions, A. H. S. (1983) Preservation of fungi. In *The filamentous fungi* (eds J. E. Smith, D. R. Berry & B. Kristiansen), vol. 4, 373–390. Edward Arnold, London.

Pereira, R. M. & Roberts, D. W. (1991) Alginate and cornstarch mycelial formulations of entomopathogenic fungi *Beauveria bassiana* and *Metarhizium anisopliae*. *J. Econ. Entomol.* **84**, 1657–1661.

Perkins, D. D. (1962) Preservation of *Neurospora* stock cultures with anhydrous silica gel. *Can. J. Microbiol.* **8**, 591–594.

Rombach, M. C., Rombach, G. M. & Roberts, D. W. (1989), Growth *in vitro* of the entomogenous fungi *Hirsutella citriformis* and *Hirsutella versicolor*. *J. Pl. Prot. Trop.* **6**, 193–204.

Salgado, S. M. D. L. & Campos, V. P. (1993) Formulation of the fungus *Arthrobotrys conoides* in sodium alginate to control nematodes. *Nemat. Bras.* **17**, 140–151.

Sanskär, B. & Magalhães, B. (1994) Cryopreservation of *Zoophthora radicans* (Zygomycetes, Entomophthorales) in liquid nitrogen. *Cryobiology* **31**, 206–213.

Silva, A. M. M. D., Borba, C. M. & Oliveira, P. C. D. (1994) Viability and morphological alterations of *Paracoccidioides brasiliensis* strains preserved under mineral oil for long periods of time. *Mycoses* **37**, 165–169.

Simione, F. P. & Brown, E. M. (eds) (1991) *ATCC preservation methods: freezing and freeze-drying*. American Type Culture Collection, Rockville, MD, 42 pp.

Smith, C. (1993) Long-term preservation of test strains (fungus). *Int. Biodeterior. Biodegrad.* **31**, 227–230.

Smith, D. & Onions, A. H. S. (1994) *The preservation and maintenance of living fungi*, 2nd edn. IMI Technical Handbooks, No. 2. International Mycological Institute. CAB International, Wallingford, UK, 122 pp.

Stalpers, J. A., Hoog, G.de & Vlug, Ij. (1987) Improvement of the straw technique for the preservation of fungi in liquid nitrogen. *Mycologia* **79**, 82–89.

Tan, C. S., van Ingen, C. W., Talsma, H., van Miltenburg, J. C., Steffensen, C. L., Vlug, Ij. A. & Stalpers, J. A. (1995) Freeze-drying of fungi: influence of composition and glass transition temperature of the cryoprotectant. *Cryobiology* **32**, 60–67.

Verkleij, F. M., van Amelsvoort, P. A. M. & Smits, P. H. (1992) Control of the pea weevil *Sitona lineatus* L. (Col., Curculionidae) by the entomopathogenic fungus *Metarhizium anisopliae* in field beans. *J. Appl. Entomol.* **113**, 183–193.

Vigier, G. & Vassoille, R. (1987) Ice nucleation and crystallization in water-glycerol mixtures. *Cryobiology* **24** 345–354.

Windels, C. E., Burnes, P. M. & Kommendahl, T. (1993) *Fusarium* species stored on silica gel and soil for ten years. *Mycologia* **85**, 21–23.

CHAPTER VI

Techniques in insect nematology

HARRY K. KAYA* & S. PATRICIA STOCK†

* Department of Nematology, University of California, Davis, CA 95616 USA and † CEPAVE, College of Natural Sciences and Museum, National University of La Plata, La Plata, 1900, Argentina

1 INTRODUCTION

Nematodes are non-segmented animals with excretory, nervous, digestive, reproductive and muscular systems but lacking circulatory and respiratory systems. Although nematodes are non-segmented, they may have superficial annulations on their non-cellular, elastic cuticle. Most are more or less cylindrical and elongate and are often referred to as roundworms, eelworms or threadworms. The alimentary tract consists of a mouth followed by the buccal cavity or stoma, oesophagus, intestine, rectum and anus. The sexes are usually separate. The male's reproductive system opens ventrally into the rectum forming a cloaca, has one or two testes, and has spicules that are used as a copulatory structure. The one or two cuticularized (sclerotized) spicules may or may not be guided by another cuticularized structure called the gubernaculum. The adult female has one or two ovaries with the vulva located ventrally and near mid-body or more posteriorly. These and other structures are shown in Figure 1.

Many nematode species are associated with insects, and these insect–nematode relationships may range from fortuitous to parasitic. Commensalism is one of the most common associations between nematodes and insects. The nematodes may be found externally on various areas of the insects' exoskeleton or internally in the reproductive, respiratory, digestive or excretory system. Commensal nematodes may also be found in the haemocoel where they do very little, if any, damage to their host. Most of these nematodes, as well as free-living ones, may utilize a dead insect as a nutrient resource, feeding on the microflora or microfauna associated with the cadaver. Thus, examination or dissection of living hosts or cadavers may disclose nematodes that may be erroneously classified as parasitic. Proper identification of such nematodes can frequently establish their lack of parasitic relationships with the insect hosts, but the fulfilment of Koch's postulates is a prerequisite for proving or disproving parasitism.

Parasitism by nematodes has variable deleterious effects on their insect hosts including sterility, reduced fecundity, reduced longevity, reduced flight

MANUAL OF TECHNIQUES IN INSECT PATHOLOGY
ISBN 0–12–432555–6

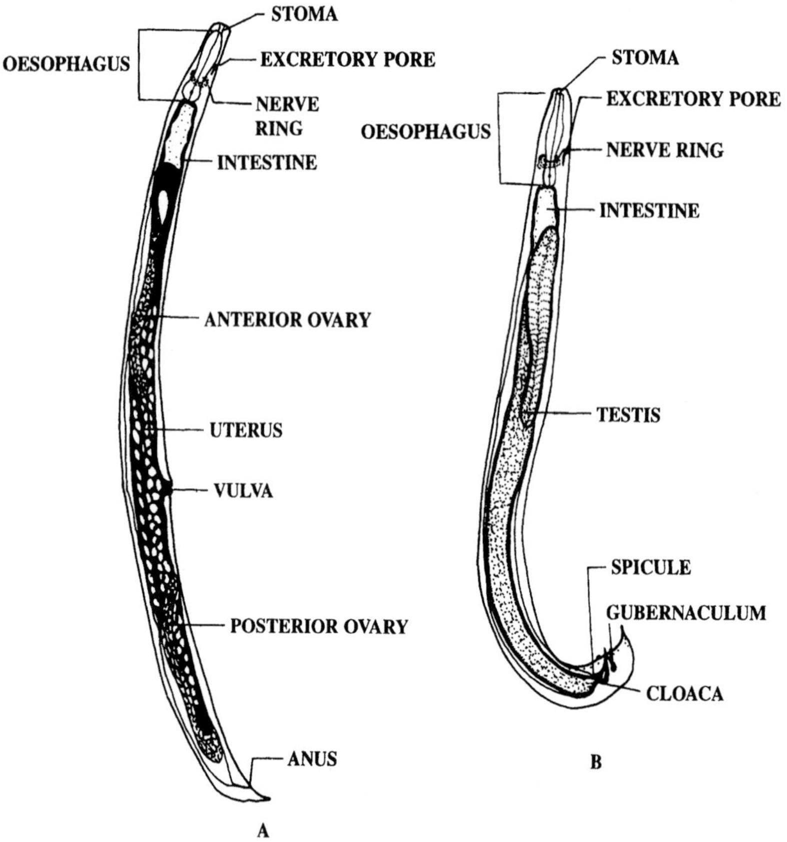

Figure 1 Adult female (A) and male (B) of a rhabditoid nematode showing general structures.

activity, delayed development, or other behavioural, physiological and morphological changes (Poinar, 1979). Some nematodes (i.e. steinernematids and heterorhabditids) have symbiotic relationships with bacteria that kill their hosts quickly (within 48 h). Because of this rapid kill, the term 'entomopathogenic' is used to describe these nematodes.

This chapter's focus is on the techniques used for identifying, isolating, propagating, assaying and preserving nematodes that are parasitic in or pathogenic to insects. In many cases, the techniques for the parasitic forms (e.g. mermithids, tetradonematids, sphaerulariids, and allantonematids) involve rearing the insect hosts, but insect rearing techniques are not discussed in this chapter. Rather, we will concentrate on the nematodes with particular emphasis placed on the steinernematids and heterorhabditids because they are the most widely studied group at this time. In addition, Ayoub (1980), Zukerman *et al.* (1985) and Barker *et al.* (1985) present detailed methods on handling plant-parasitic nematodes from soil, and the techniques described therein are applicable for insect-parasitic nematodes that spend part of their life cycle in soil. Poinar (1975, 1979), Poinar & Thomas (1984), Woodring & Kaya (1988) and Grewal (1992) provide useful information on techniques, bionomics and literature of entomophilic, entomogenous or entomopathogenic nematodes. Finally, many techniques in this chapter use chemicals and/or equipment that require safety measures. Be sure that personal protection devices (eyeglasses, gloves, laboratory coat and apron, fume hoods, etc.) are used and instructions from the manufacturers for handling chemicals and equipment are followed.

2 GENERALIZED LIFE CYCLE OF NEMATODES

Most nematodes have a simple life cycle (Figure 2). The mated female lays eggs and usually the nema-

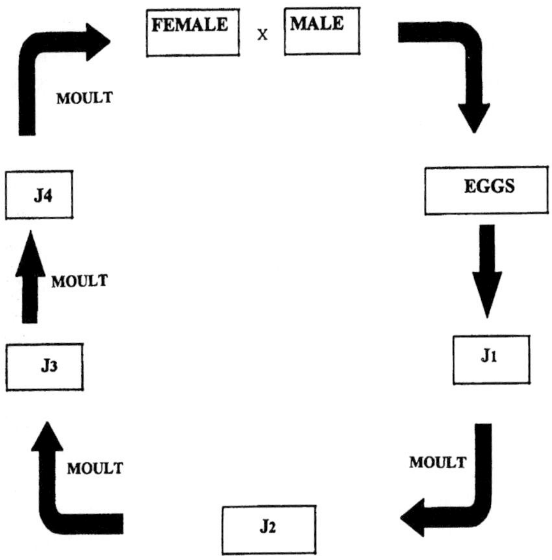

Figure 2 General life cycle of nematodes.

todes develop to the second juvenile (J2) stage before hatching. (We refer to the immature stages as juveniles rather than larvae to avoid confusion with the immature stages of insects.) Each subsequent juvenile stage feeds and moults to the next stage. The J2 moults to J3 and then to J4 which moults to the adult stage. Most nematodes are amphimictic (requiring males and females), but some groups are hermaphroditic or parthenogenetic (requiring females only) or have alternate gamogenetic and parthenogenetic life cycles.

The stage of entomogenous and entomopathogenic nematodes that is infective varies depending on the group. In some cases, the infective stage is the egg (e.g. thelastomatids), the J2 (e.g. mermithids) or the J3 which is often referred to as the dauer or infective juvenile (e.g. steinernematids and heterorhabditids) or the mated female (e.g. allantonematids, sphaerulariids, and phaenopsitylenchids). In the latter group, the adult males are non-infective and die after mating.

3 IDENTIFICATION

A Initial preparation

A good stereomicroscope is essential for nematode identification and should have a range of magnification between 10 and 100×, a fairly flat field, and good resolution (see Chapter VIII-I). Illumination by transmitted light should be as even as possible. Handling most nematodes in distilled water presents no problem; however, the use of a saline solution (i.e. Ringer's solution) (see Appendix) is highly recommended to avoid osmotic shock. For living material, the nematodes can be mounted in Ringer's solution or water on a glass slide with supports and a cover-slide sealed with paraffin or nail polish (see Section 3C,2).

1. Perform dissections in a Petri dish with Ringer's solution and under a stereomicroscope.
2. Lift individual specimens out of the solution with the help of a handling needle (Figure 3A-D) of which there are many types: nylon toothbrush bristle, toothpick handling 'L-shaped' needle, dentist's probe, steel needle, bamboo splinter, etc. For smaller nematodes, an eyebrow hair glued on to the end of a mounted needle is very useful and does not damage the nematodes. Beginners will have some difficulty in picking up nematodes. Use the lowest microscope magnification (to give good depth of focus and working distance) and start with nematodes that are near the centre of the dish.

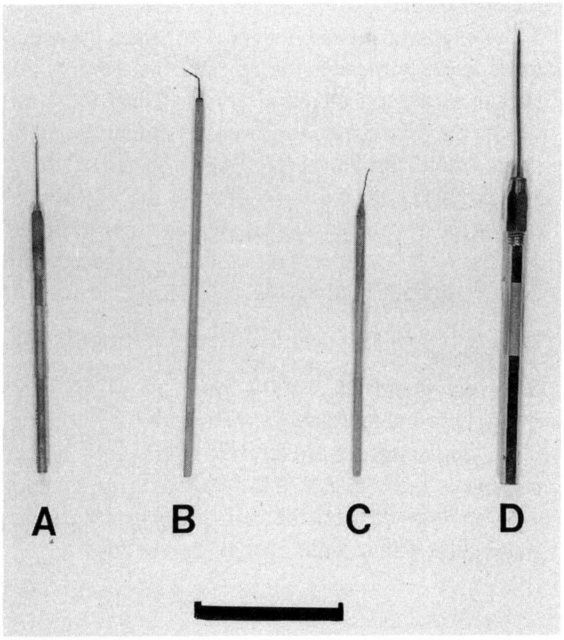

Figure 3 Various types of nematode picks for handling an individual nematode. A, Dentist's probe; B, Toothpick handling L-shaped needle; C, nylon toothbrush bristle; D, steel needle. Bar = 2 cm.

3. Place the needle underneath the nematode and lift up quickly so that the nematode is pulled through the meniscus.

The time frame for doing dissections may vary according to the different species/isolates. For steinernematids, dissections should be done 2–3 days after infection to recover first generation adults (males and females), and 5–6 days after infection for second generation adults. For heterorhabditids, dissections should be done 3–4 days after infection to recover first generation hermaphrodites, and 6–9 days after infection for second generation adults (males and amphimictic females). Infective juveniles of both steinernematids and heterorhabditids should be recovered during the first 3 days after initial emergence.

B Killing and fixing

1. Heating

Nematodes may be killed by heating in a water bath at 60 °C for 2 min. Once the specimens are killed, add the fixative (TAF) (see Appendix) to an equal amount of Ringer's solution. The temperature of the fixative and the Ringer's solution should be approximately 60 °C.

An improved method that kills and fixes the nematodes in one process is that of Seinhorst (1966). The specimens are collected in a very small drop (ca. 1 ml) of water in a Syracuse watchglass or similar deep concave vessel. The fixative (i.e. TAF) is heated to 100 °C and an excess (3–4 ml) is added to the vessel containing the nematodes. Nematodes remain in this solution for 12 h, and then the solution is replaced with double-strength fixative (see Appendix).

2. Cooling

Store the sample in a refrigerator (ca. 4 °C) or in water with ice to relax the nematodes by cooling. Once the nematodes are relaxed, add fixative heated previously at 65 °C. Allow the fixative to infiltrate for at least 24 h. Transfer the nematodes to a Syracuse watchglass with as little fixative as possible.

C Permanent mounts

The gonads and other structures of fixed nematodes may be obscured by the granular appearance of the intestine. Specimens can be cleared by processing to lactophenol or glycerin. Although both are good mounting media, nematodes will keep almost indefinitely if they are processed to glycerin.

1. Processing to glycerin

1. Transfer fixed nematodes to a Syracuse watchglass containing 0.5 ml of Solution I (20 parts 95% ethanol, 1 part glycerin, 79 parts distilled water) (Seinhorst, 1959) (Figure 4A).
2. Place the Syracuse watchglass in a desiccator and add 95% ethanol to the desiccator so that the space below the holding shelf is half full (Figure 4B).
3. Place the desiccator in an oven for at least 12 h at 35 °C to allow slow evaporation of the ethanol from solution I in the watchglass (Figure 4C).
4. Remove the watchglass from the desiccator (Figure 4D).
5. Fill the watchglass with solution II (5 parts glycerin, 95 parts of 95% ethanol) (Figure 4E), place in a Petri dish which should be partially opened to allow slow evaporation of the ethanol, and put it back in an oven for 3 h at 40 °C (Figure 4F). After processing the nematodes in pure glycerin, they are ready for mounting.

2. Mounting and labelling

Aluminium double-coverslip slides (Cobb, 1917) are suitable for permanent mounts (Figure 5A-G) because they allow viewing of specimens from either side, are durable, and are less likely to crack during shipping or rough handling. If they are not available, good quality glass slides (76 × 25 × 1 mm) can be used. When using an aluminium slide, a square coverglass must be slipped in the aluminium carrier between two pieces of cardboard. The edges of the carrier must then be pressed to hold the cardboard and the coverglass in place.

The following procedure is used to make a permanent slide.

1. Place a drop of glycerin in the centre of the slide (i.e. on the coverglass on the aluminium slide or a regular glass slide).
2. Add short pieces of glass fibre to the glycerin to give support.

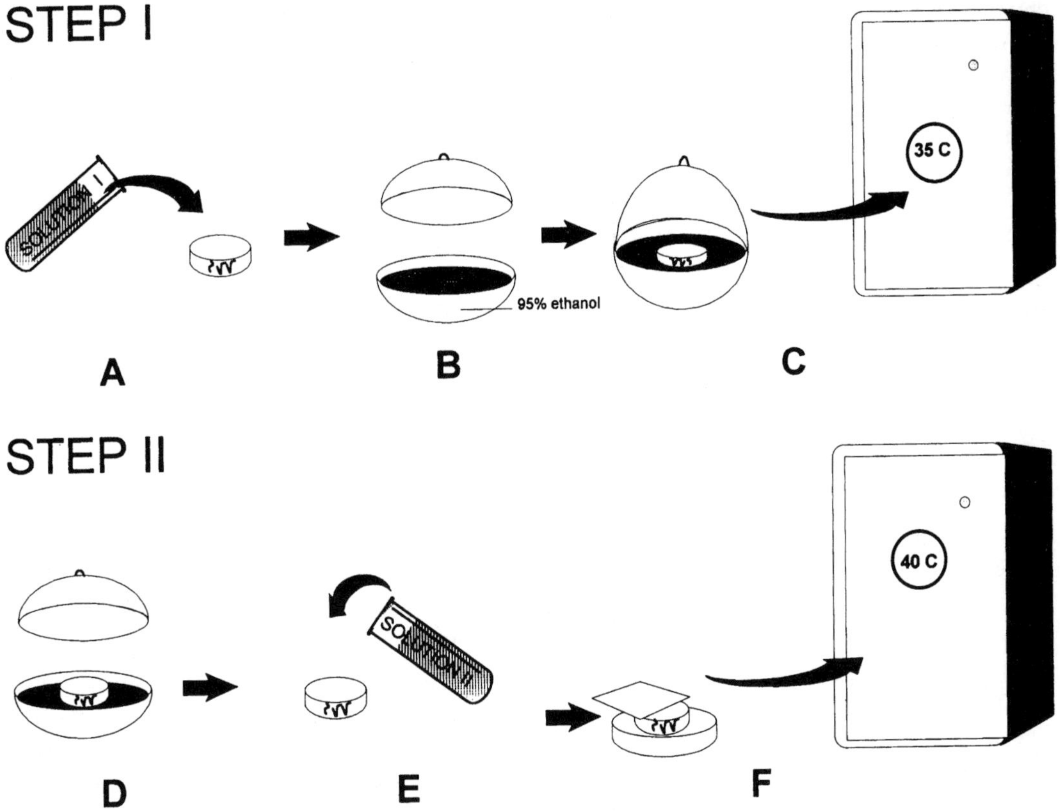

Figure 4 Schematic diagram for processing nematodes into glycerin (see text for details).

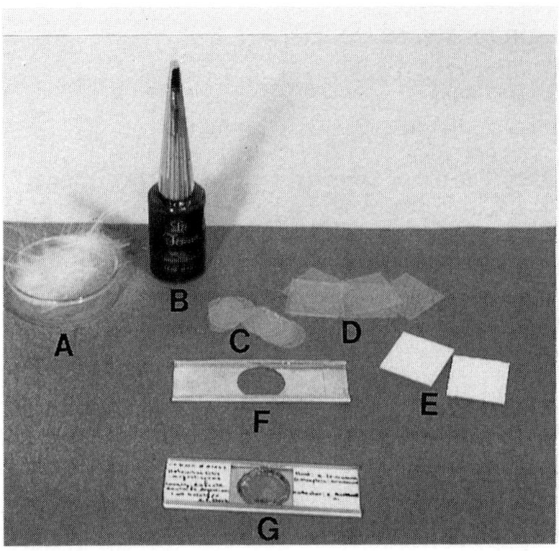

Figure 5 Components of an aluminium double-coverslip slide (Cobb, 1917) for permanent mounting of nematodes after being processed through glycerin. A, Glass fibre; B, nail polish; C, round coverglass; D, square coverglass; E, cardboards; F, aluminium double-coverslip slide; G, specimen in permanent mount and properly labelled (see text for details).

3. Place processed nematode specimens (usually 1–5) into the glycerin. Be sure that the nematodes are aligned and do not overlap.
4. Place a round coverglass on top of the glycerin.
5. Seal the round coverglass with a double ring of clear nail polish. (A substance called Zut was previously recommended, but it is no longer commercially available.)

An alternative method is to seal the round coverglass with paraffin. Use the following procedure to seal the round coverglass with paraffin.

1. Heat a 1.7 mm diameter metal tube with polished rim using a Bunsen burner or alcohol lamp.
2. Press the hot tube into the paraffin, remove and hold it against the coverglass on the aluminium slide to leave a ring of paraffin (the thickness of the ring should be in accordance with the size of the nematodes to be mounted).
3. Place a drop of glycerin in the centre of the ring and place a few nematodes in the bottom of the

drop making sure that the nematodes are aligned and do not overlap.
4. Place the slide on a hot plate and drop a round coverglass on top as soon as the paraffin ring melts and immediately take the slide off the hot plate. Once the paraffin sets, the slide is sealed and contains a small central area of glycerin and nematodes with a surrounding rim of paraffin.

3. Mounting cross sections

The cephalic structures and the number of longitudinal chords are diagnostic characters for generic or specific determination of certain groups of nematodes (i.e. rhabditids and mermithids). These structures are best seen by making cross sections.

1. Place nematodes that have already been processed to glycerin in a drop of glycerin and cut into small sections (anterior end or midbody) with a razor blade or a small oculist's scalpel.
2. Transfer the section, with the help of a needle, to a drop of melted glycerin jelly (see Appendix) in the centre of a coverglass and orient in the proper position.
3. Invert the coverglass and gently place on a coverglass on the aluminium slide or directly on a glass slide with pieces of glass fibre to prevent the jelly from touching the slide.
4. Seal the mount with paraffin or nail polish.

D Scanning electron microscopy

Scanning electron microscopy (SEM) (see Chapter VIII-1) is a very useful tool for the visualization and interpretation of certain nematode features that cannot be appreciated with a light microscope but are important for taxonomic identification.

1. Prepare nematodes by placing them in a water bath at 60 °C for 2 min to kill them.
2. Rinse nematodes three times in Ringer's solution (pH 7.3) or phosphate buffer (pH 7.4) (5 min each change).
3. Prefix in 8% glutaraldehyde (glutaraldehyde 25% EM grade, diluted in Ringer's solution).
4. Leave nematodes overnight in this solution.
5. Rinse nematodes three times in Ringer's solution (5 min each change) and once in distilled water (5 min).
6. Post-fix in 1% osmium tetroxide (OsO_4) for 2 h.
7. Rinse post-fixed nematodes three times in water (5 min each) and dehydrate them using a series of ethanol washes (30, 50, 70, 90, 95 and 100%).
8. Dry the nematodes to critical point with liquid CO_2.
9. Mount on SEM stubs, and coat with gold for 1 h.

4 TYPE SPECIMENS

A Preparation of type and voucher specimens

Specimens used for the diagnosis of a new species/genus must be deposited in a public collection. Accordingly, holotype, allotype and paratypes should be prepared as permanent slides (see Section 3), with the following information accompanying the specimens.

- Slide number or code (usually given by the collection's curator once description of the new specimen has been accepted for publication)
- Nematode species name
- Type designation (holotype, paratype or allotype) and the number of specimens per slide
- Type host
- Type locality
- Author's name
- Collector's name

B Deposition of type and voucher specimens

Holotype, allotype, paratype and voucher specimens should be deposited immediately in a national depository where they can be loaned, used and properly maintained. In the United States, the US Department of Agriculture Nematode Collection at Beltsville, Maryland and the Nematode Collection (UCDNC) at the Department of Nematology, University of California, Davis, are the sites where the largest nematode collections are housed, whereas in Europe the collection at the Museum National d'Histoire Naturelle, Paris, France, is a major site for type specimens. If sufficient material is available, specimens should be placed in established collections on different continents.

Usually primary types (i.e. holotypes) are unavailable for loan, but guidelines may vary with each institution. Other type specimens (i.e. paratypes) are loaned to recognized specialists for up to 1 year. Researchers are not permitted to change labels or remount specimens. When returning slides, the following recommendations should be followed:

- Alert the curator that you are mailing the slides before the package is mailed.
- Place slides into the mailer and put a small piece of cotton on the top of each label to prevent the slides from moving.
- Place the mailer inside a larger box, with at least 5–8 cm of packing material between the mailer and the box.
- Send the package by registered mail.

5 MAJOR GROUPS OF NEMATODES ASSOCIATED WITH INSECTS

The classification of most nematode groups, including those associated with insects is not yet stable at both the family and generic level. For instance, Maggenti (1991) places four families (Allantonematidae, Iotonchiidae, Sphaerulariidae and Fergusobiidae) in the superfamily, Sphaerularioidea, whereas Remillet and Laumond (1991) have six families (the four families of Maggenti plus Parasitylenchidae and Phaenopsitylenchidae). The recognition of new families has resulted in concomitant changes at the generic level. In this chapter, we have modified the older and newer classifications and placed the families Iotonchiidae and Parasitylenchidae in the Allantonematidae. We have kept the Phaenopsitylenchidae separate because it contains some important species that have been used in biological control of woodwasps (Siricidae). This scheme will allow the researcher who uses the key to get to the Allantonematidae and further resolution can be obtained by reading Remillet & Laumond's (1991) paper.

There are more than 30 nematode families associated with insects including animal and plant parasites that use insects as vectors (Table 1). Animal parasitic nematode families having species that use insects as vectors include Filariidae (e.g. mosquitoes for elephantiasis of humans and dog heartworm), Onchocercidae (e.g. black flies for onchocerciasis of humans) and Thelaziidae (e.g. *Musca* spp. for several species of eyeworms) (Geden & Stoffolano, 1984). The nematode family, Aphelenchidae, contains two important species of plant parasites that use

Table 1 Classes, orders and families in the Phylum Nemata (syn. Nematoda) associated with insects (modified after Maggenti, 1991). The description in parentheses after each family indicates the insect–nematode association. Families with biological control potential are capitalized.

Class Adenophorea (syn. Aphasmidia)		
	Order Mononchida	
	Family	Plectidae (Phoretic)
	Order Stichosomida	
	Family	MERMITHIDAE (Obligate Parasite)
		TETRADONEMATIDAE (Obligate Parasite)
Class Secernentea (syn. Phasmidia)		
	Order Rhabditida	
	Family	Carabonematidae (Obligate Parasite)
		Cephalobidae (Phoretic)
		Chambersiellidae (Phoretic)
		HETERORHABDITIDAE (Obligate Pathogen)
		Oxyuridae (Obligate Parasite)
		Panagrolaimidae (Phoretic)
		Rhabditidae (Phoretic, Facultative Parasite)
		STEINERNEMATIDAE (Obligate Pathogen)
		Syrphonematidae (Obligate Parasite)
		Thelastomatidae (Obligate Parasite)
	Order Spirurida (Animal Parasites with Insects as Vectors or Intermediate Hosts)	
	Family	Filariidae (Obligate Parasite)
		Onchocercidae (Obligate Parasite)
		Physalopteridae (Obligate Parasite)
		Syngamidae (Obligate Parasite)
		Spiruridae (Obligate Parasite)
		Subuluridae (Obligate Parasite)
		Thelaziidae (Obligate Parasite)
	Order Diplogasterida	
	Family	Diplogasteridae (Phoretic, Facultative Parasite)
		Cylindrocorporidae (Phoretic)
	Order Tylenchida	
	Family	ALLANTONEMATIDAE[a] (Obligate Parasite)
		Aphelenchidae (Phoretic)
		Aphelenchoididae (Phoretic, Facultative Parasite)
		Entaphelenchidae (Obligate Parasite)
		Fergusobiidae (Obligate Parasite)
		PHAENOPSITYLENCHIDAE (Facultative Parasite)
		SPHAERULARIIDAE (Obligate Parasite)
		Tylenchidae (Phoretic)

[a] Includes species from the families Iotonchiidae and Parasitylenchidae (Remillet & Laumond, 1991).

insects as vectors (i.e. cerambycids for pine wilt and palm weevils for red ring disease of coconut) (Giblin-Davis, 1993).

Table 1 shows nematode families with species that have a phoretic relationship with insects and have low pathogenicity (e.g. Oxyuridae, Thelastomatidae, Rhabditidae) or are rarely encountered (Carbonematidae, Entaphelenchidae, Syrphonematidae). It also shows nematode families that have potential for biological control of insects (Allantonematidae, Diplogasteridae, Heterorhabditidae, Mermithidae, Phaenopsitylenchidae, Sphaerulariidae, Steinernematidae, and Tetradonematidae).

A Key to major orders and families of nematode parasites of insects

1. Stylet usually present, at least in infective stages. 2

1a. Stylet absent. Juveniles and rarely adults in the body cavity of their hosts 3

2. Stage exiting the host elongate or thread-like. Oesophagus a long narrow tube and its posterior part modified into a stichosome or a tetrad of cells (stichocytes) (Figure 6B). Adults with degenerate intestine, forming a storage organ (trophosome). Infective stage with a stylet . . . . .Order Stichosomida 4

2b. Stage exiting the host never elongated or thread-like. Oesophagus not a narrow tube, but divisible into three parts: corpus, isthmus and posterior glandular bulb. Infective stage females with a stylet or spear (Figure 6C) . . . . .Order Tylenchida 5

3. Stoma often armed with movable teeth or fossoria. Oesophagus divided into corpus, isthmus and basal bulb. Corpus often muscular and metacorpus, with few exceptions, valvate. Basal bulb glandular, never valvate (Figure 6D). Obligate or facultative parasites; juveniles and adults in the intestinal tract of living hosts . . . . .Order Diplogasterida Family Diplogasteridae

3a. Stoma generally tubular with small denticle. Oesophagus divided into corpus, isthmus and basal bulb that is muscular and valvate (Figure 6A). Facultative parasites or obligate pathogens associated with mutualistic bacteria . . . . .Order Rhabditida 7

4. Thread-like nematodes, ranging in length from 1 to 30 cm (usually between 1 and 10 cm). Stichosome well developed. Only juveniles in the host. Post-parasitic stage juveniles reach adulthood in the host's environment. Aquatic and terrestrial invertebrate species parasitized, but mostly insect hosts . . . . . . Family Mermithidae

4a. Nematodes not thread-like, generally less than 1 cm in length. Stichosome reduced to 3–4 cells. Adult stage and mating in the host's body cavity. Parasites of insects (Diptera, Coleoptera and Hymenoptera) . . . . Family Tetradonematidae

5. Only one heterosexual generation known 6

5a. Alternation of generations. One or several free-living heterosexual or parthenogenetic generations . . . . Family Phaenopsitylenchidae

6. Free-living stages with two oesophageal glands with their outlets in the middle of the oesophageal tube. Large swollen parasitic females with everted hypertrophied uteri . . . . Family Sphaerulariidae

6a. Free-living stages with three oesophageal glands, the dorsal outlet opening near base of stylet. Large swollen parasitic females without everted uteri . . . . .Family Allantonematidae

7. Obligate pathogens, killing host; 2–3 amphimictic generations in the cadaver. Cadaver ochre, yellow–brown or black with no luminescence in the dark. Infective juveniles with excretory pore anterior to the nerve ring. Males with no bursa; 21–23 genital papillae . . . . . .Family Steinernematidae

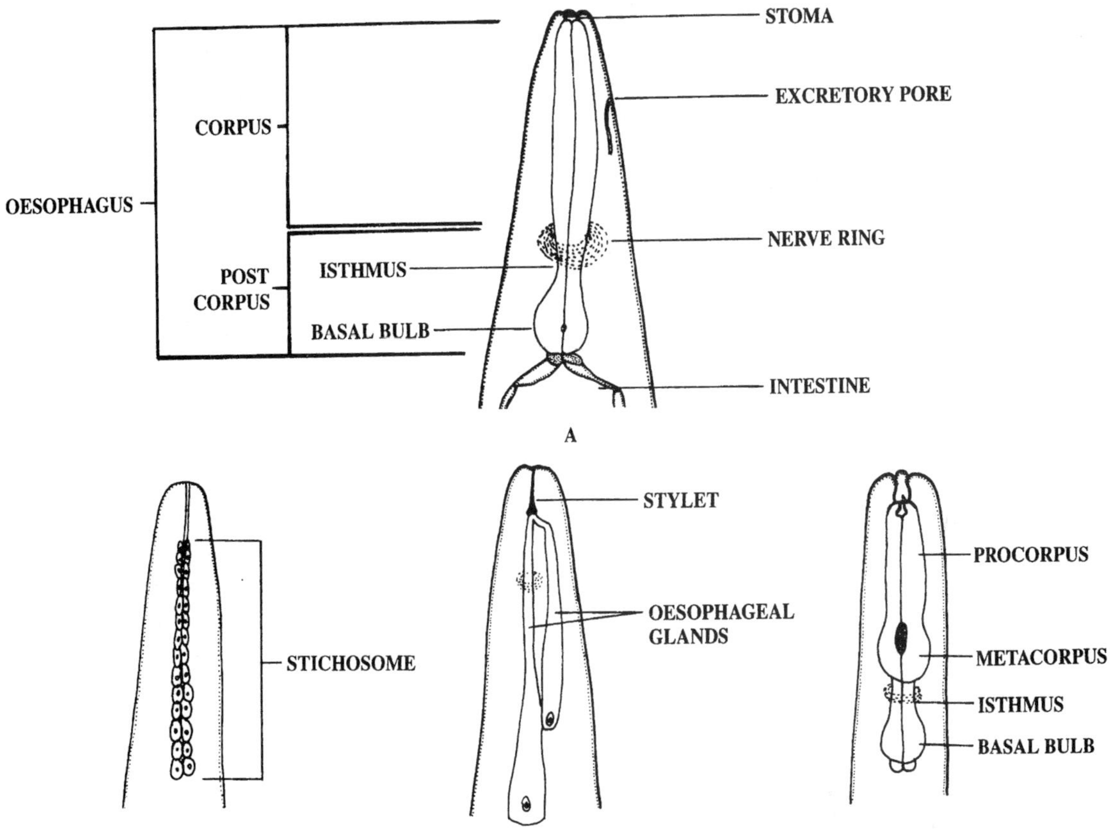

Figure 6 Different types of oesophagi. A, Oesophagus of a steinernematid (rhabditoid) nematode indicating various structures; B, stichosome of a mermithid; C, tylenchoid oesophagus; D, diplogasteroid oesophagus.

7a. Obligate pathogens; killing host; two generations in the cadaver. Hermaphroditic in first generation, and amphimictic in the second generation. Cadavers red, brick–red, purple orange or sometimes green with luminescence in the dark. Infective juveniles with excretory pore posterior to the nerve ring. Males with a bursa; nine pairs of genital papillae . . . . Family Heterorhabditidae

B Diagnosis and bionomics of major families of insect-parasitic nematodes

We provide the following information about the diagnosis and bionomics of the more common nematode parasites of insects. More detailed information can be obtained from Poinar (1975, 1979) and Nickle (1984, 1991).

1. Mermithidae

a. Diagnostic characters

Long slender nematodes sometimes reaching a length of 50 cm, but usually 1–10 cm; cuticle smooth or containing criss-cross fibres; anterior end containing 2, 4 or 6 cephalic papillae and rarely a pair of lateral mouth papillae; amphids tube-like or modified pouch like; oesophagus modified into a slender tube that is surrounded posteriorly by stichosomal tissue with each cell called a stichocyte; intestine modified into a trophosome or food storage organ that becomes a blind sac soon after the nematode enters a host; preparasitic juvenile (J2) possesses a functional stylet and a pair of penetration glands that become rudimentary after host invasion; ovaries paired; muscular vagina straight or curved; males with a single fused or paired spicules; gubernaculum and bursa absent; several rows of genital papillae are generally present. Most of the diagnostic characteristics for determining the genera and species of mermithids occur on the adult forms. Accordingly, the post-parasitic stage that emerges from the host should be allowed to mature for proper identification (see below).

b. Bionomics

Mermithids are obligate parasites of many invertebrate species. In most species, the life cycle is as follows: eggs are deposited in the host's environment and embryonic development occurs after oviposition. The J2 (= preparasitic juvenile) emerges and seeks its hosts. The juvenile penetrates through the host's cuticle using its stylet and enters the haemocoel. Usually one nematode enters each host, but when nematode population densities are high, several nematodes may penetrate an insect. (When superparasitism occurs in amphimictic species, the majority of the resulting adults will be males.) During the parasitic phase, the nematode absorbs nutrients from the host directly through its cuticle and stores them in the trophosome. The oral cavity becomes highly cuticularized so that the stylet is no longer recognizable, and the stichosome grows into an organ of considerable size. When the parasitic phase is completed, the J4 emerges from the host and starts its free-living stage. Host death occurs because of the trauma caused by the exiting nematode through the insect's cuticle. Upon emergence the mermithid is called a post-parasitic juvenile (= J4) that moults to the adult stage which mates, and oviposits eggs to complete its life cycle.

For identification, it is important to have the adult stage, especially males. Some species may remain in the J4 stage over the winter months, whereas others reach the adult stage in 2–4 weeks. Therefore, the post-parasitic stage should be placed in the proper environment for maturation to occur (e.g. water or moist sand for aquatic mermithids and moist soil for terrestrial nematodes).

Mermithids parasitize many different insect groups as well as other arthropods (i.e. ticks, spiders and crustaceans) and can often be seen within the haemocoel of living arthropods that have transparent cuticles. Insect orders that contain hosts include Orthoptera, Dermaptera, Hemiptera, Lepidoptera, Diptera, Coleoptera and Hymenoptera. Examples of mermithids that have received significant attention include *Romanomermis culicivorax*, a parasite of mosquito larvae (Petersen, 1984, 1985), *Filipjevimermis leipsandra*, a parasite of larval banded cucumber beetles, *Diabrotica balteata* (Creighton & Fassuliotis, 1981), and *Mermis nigrescens*, a parasite of grasshoppers (Webster & Thong, 1984). Unfortunately, the biocontrol potential of mermithids is limited because they have to be mass produced *in vivo*.

Mermithids are often confused with gordian worms (Phylum Nematomorpha) that commonly parasitize crickets and other orthopterans. Gordian worms (also known as horsehair worms) differ from nematodes by their asymmetrical oesophagus and the gonads opening at the posterior end of the body through a cloaca in both sexes.

2. Tetradonematidae

a. Diagnostic characters

Lips rudimentary or absent; juvenile with a minute stylet that disappears in later stages; cephalic papillae reduced; oesophagus consists of a simple hollow tube and associated tetrad or 4-celled structure; anus absent; ovaries and testes paired; vulva near body, spicule single; bursa absent. The nematode matures to the adult stage and mates in the body cavity of the insect host.

b. Bionomics

Unlike mermithids, these nematodes develop to the adult stage and mate in the insect's haemocoel. Sometimes the eggs are released in the body cavity of the host, but generally the gravid female leaves the host and deposits eggs in the host's environment. After hatching the infective stage juvenile enters a new host. As with the mermithids, these nematodes kill their hosts at the time of emergence. The hosts probably die as a result of nutrient depletion and the entry of microbial agents through the exit hole made by the nematode. Tetradonematids have been primarily described from Diptera such as the fungus gnats (Sciaridae), scavenger flies (Sepsidae), midges (Chironomidae), chaoborid flies (Chaoboridae) and black flies (Simuliidae). They have also been found in Coleoptera and Hymenoptera. The most extensively studied species is *Tetradonema plicans*, an endoparasite of greenhouse sciarid flies (Kaya, 1993).

3. *Sphaerulariidae*

a. Diagnostic characters

Adults lacking median bulb and substylet dorsal gland opening. Oesophagus composed of a non-muscular oesophageal tube and two or three basal oesophageal glands. Orifice of dorsal gland at the base of stylet, not always distinct. Free-living adult with a stylet; male with paired separate spicules; gubernaculum present, bursa absent or present. Free-living female with stylet well developed, base slit into three flanges or knobs; uterus very long in fertilized female. Parasitic female with hypertrophied uterus, more or less everted.

b. Bionomics

Sphaerulariids have a single heterosexual generation. They rarely kill their hosts but cause sterility, reduced fecundity, delayed development, and/or altered behaviour. Members of this family are parasites of bumble bees (Bombycidae), flies (Sciaridae and Cecidomyiidae), and bark beetles (Scolytidae). Unlike most sphaerulariids, *Tripius sciarae* often causes fatal infections of larval sciarid flies. The fertilized female, ensheathed in the fourth stage cuticle, penetrates directly through the host's integument by using its stylet and glandular secretions. The female nematode partially everts its uterus through the vulva allowing absorption of nutrients from the host's haemolymph. Eggs are deposited in the haemocoel where they develop to J4s which exit the host through the digestive tract. Maturation of J4s occurs in 2 days. After mating, males die but the females can live for 2 weeks in moist soil while searching for a host.

4. *Allantonematidae*

a. Diagnostic characters

Stylet present but often reduced in males and juvenile stages; valvate median bulb absent; with three long oesophageal glands that generally overlap the intestine. Ovary single. Infective female enters host after fertilization and assumes a swollen shape characteristic for the genus/species. Male with paired arcuate spicules; gubernaculum rarely absent; bursa present or absent.

b. Bionomics

Only one heterosexual generation. Some groups (Iotonchiidae) have a parthenogenetic and gamogenetic generation within the host. Because of the complexity of their life cycles and the difficulty in maintaining cultures of these obligate parasites, most nematode species in this group have not been exploited for biological control. In general, the life cycle of these nematodes is as follows: the fertilized female enters the host's body cavity by directly penetrating through the cuticle. The female develops into a swollen mature parasitic female that deposits eggs and/or juveniles into the host's haemocoel. The juveniles develop to the third or fourth stage (rarely to adult stage) and leave the host through the insect's intestine or reproductive tract. Once in the environment, the nematodes moult, mature and mate, and the fertilized females seek new hosts.

5. *Phaenopsitylenchidae*

a. Diagnostic characters

Parasitic generation: free-living female with a strong stylet, three well-developed oesophageal glands, ovary reduced, excretory pore anterior to the hemizonid; male with bursa and minute sperm. Free-living generation: female with fine stylet, three oesophageal glands but two are reduced, ovary normally developed; male similar to that of parasitic generation but with large sperm.

b. Bionomics

One parasitic heterosexual generation and one or several free-living heterosexual or parthenogenetic generations occur in this family. The family is composed of two genera: *Beddingia* (= *Deladenus*) and *Phaenopsitylenchus* which are parasites of hymenopterans and coleopterans, respectively.

Beddingia spp. are characterized by the presence of one parasitic generation that can be obligatory (in curculionids) or facultative (in siricids) with several heterosexual free-living generations (mycetophagous or phytoparasitic). *Phaenopsitylenchus* spp., parasites of coleopterans, are characterized by one parasitic heterosexual generation (parasitic female, male and infective female) and one free-living parthenogenetic generation.

6. Steinernematidae

a. Diagnostic characters

Stylet absent; lips fused; stoma short and wide, esophagus composed of a cylindrical corpus; a slightly swollen metacorpus and a valvate basal bulb; ovaries paired, opposite; testis single; spicules paired and separate; gubernaculum present, nipple genital papillae present; bursa absent. Infective juvenile (IJ) which is the J3, with excretory pore located anterior to nerve ring.

The family is composed of two genera, *Steinernema* with currently 17 species (Table 2) and *Neosteinernema* with only one species, *N. longicurvicauda* (see footnote in *Steinernema* key, below, for proposed name change). In addition, a number of isolates of these steinernematid nematodes exist, some representing new species that are being or need to be described.

Table 2 List of recognized species of *Steinernema* and synonyms. The authority is provided after the scientific names for these nematodes.

Species	*Synonyms*
S. kraussei (Steiner)	*Aplectana kraussei* Steiner
S. glaseri (Steiner)	*Neoaplectana glaseri* Steiner
S. feltiae (Filipjev)	*N. feltiae* Filipjev *Steinernema bibionis* (Bovien)
S. affinis (Bovien)[a]	*N. affinis* Bovien
S. carpocapsae (Weiser)	*N. carpocapsae* Weiser *N. carpocapsae pieridarum* Stanuszek *N. feltiae* sensu Filipjev nec Stanuszek
S. anomali (Kozodoi)[a]	*N. anomali* Kozodoi
S. intermedia (Poinar)[a]	*N. intermedia* Poinar
S. *rara* (Doucet)[a]	*N. rara* Doucet
S. kushidai Mamiya	——
S. ritteri Doucet and Doucet	——
S. scapterisci Nguyen and Smart	*N. carpocapsae* Uruguay strain Nguyen and Smart
S. longicaudum Shen and Wang	——
S. neocurtillis Nguyen and Smart	——
S. riobravis Cabanillas, Poinar and Raulston[a]	——
S. cubana Mracek, Hernandez and Boemare[a]	——
S. puertoricensis Roman and Figueroa[a]	——
S. bicornutum Tallosi, Peters and Ehlers	——

[a] Specific names of these species have been amended (Stock *et al.*, 1996).

b. Bionomics

Steinernematids are obligate pathogens in nature and are characterized by their mutualistic association with bacteria in the genus *Xenorhabdus*. This nematode/bacterium complex is pathogenic to many insects in the laboratory, but in the field, the host range is limited (Kaya & Gaugler, 1993). The non-feeding IJ carries the bacteria in the ventricular portion of its intestine. Once a suitable host is found, the IJ enters the host through natural openings (mouth, anus, spiracles) and penetrates into the haemocoel. The bacteria are released into the haemolymph, propagate, and kill the host by septicaemia within 48 h. The nematodes feed on the bacteria and host tissues, produce 2–3 amphimictic generations, and emerge from the cadaver as IJs to search for new hosts.

Of all nematodes studied for biological control of insects, the steinernematids together with the heterorhabditids have received the most attention because they possess many of the attributes of effective biological control agents.

c. Key to Steinernema species

Figure 7 shows the morphology of males tail, spicule and gubernaculum. Ratio d = distance from head to excretory pore divided by distance from head to oesophagus base; ratio e = distance from head to excretory pore divided by tail length.

1. Infective juveniles' average total body length more than 1000 μm 2
1a. Infective juveniles' average body length less than 1000 μm 6

2. Infective juveniles' average distance from anterior end to excretory pore 85 μm (range 72–92.5 μm) 3
2a. Infective juveniles' average distance from anterior end to excretory pore more than 85 μm (range 90–110 μm) 4

3. Males' spicules with swollen tip; ratio d about 0.93 . . . . *S. anomali**
3a. Males' spicules with notched tip under light microscope; ratio d about 0.62 . . . . *S. longicaudum*

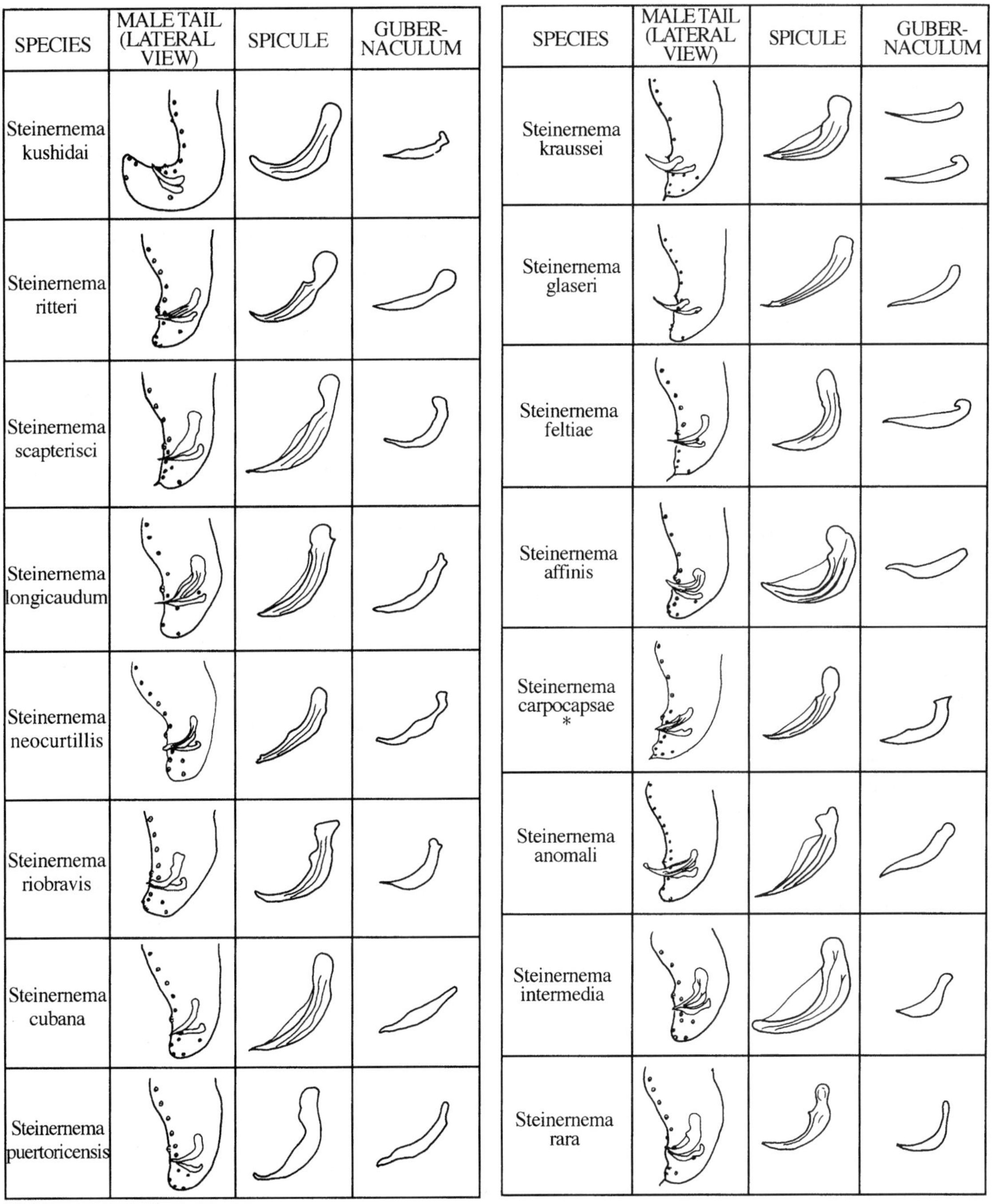

Figure 7 Schematic drawings showing male tail, spicule and gubernaculum of 16 *Steinernema* species. (**S. bicornutum* and *S. carpocapsae* share the same male morphological characters.) Not drawn to scale.

4. Infective juveniles' average tail length 90 μm or more (range 88–107 μm); females with double flapped epiptygma .... *S. puertoricensis**

4a. Infective juveniles' average tail length less than 90 μm (range 61–87 μm); females without double flapped epiptygma 5

5. Infective juveniles' ratio e about 1.60; males spicule tip not notched; females with asymmetric vulva . . . *S. cubana**
5a. Infective juveniles' ratio e about 1.31; males' spicule notched; females without asymmetric vulva . . . . *S. glaseri*

6. Infective juveniles' average total body length more than 800 μm 7
6a. Infective juveniles' average total body length less than 800 μm 9

7. Infective juveniles' distance from anterior end to excretory pore less than 50 μm (range; 14–22 μm); average ratio e about 23 . . . . .*S. neocurtillis**
7a. Infective juveniles' distance from anterior end to excretory pore more than 50 μm (range 53–67 μm), average ratio e more than 23 8

8. Male's average length of mucro 4–13 μm; spicules manubrium elongate, about twice as long as wide, rostrum absent . . . .*S. feltiae*
8a. Male's average length of mucro 1–4 μm; spicules' manubrium rounded, about as long as wide, rostrum present . . . . *S. kraussei*

9. Infective juveniles with horn-like papillae in the labial region . . . .*S. bicornutum*
9a. Infective juveniles without horn-like papillae in the labial region 10

10. Infective juveniles' average distance from anterior end to excretory pore less than 50 μm 11
10a. Infective juveniles' average distance from anterior end to excretory pore more than 50 μm 13

11. Males' spicules with velum; testis reflection more than 300 μm (range 283–1000 μm) 12
11a. Males' spicules without velum; testis reflection less than 300 μm (range 185–257 μm) . . . . *S. riobravis**

12. Infective juveniles with refractile spine in tail tip; males spicules about 70 μm long . . . .*S. affinis**
12a. Infective juveniles without refractile spine in tail tip; males spicules about 49 μm long . . . . *S. intermedia**

13. Infective juveniles' average distance from anterior end to excretory pore less than 40 μm 14
13a. Infective juveniles' average distance from anterior end to excretory pore more than 40 μm 15

14. Males' spicules average length more than 75 μm (range 72–92 μm); females with double flapped epiptygma . . . . *S. scapterisci*
14a. Males' spicules average length less than 75 μm (range 58–72μm); females without double flapped epiptygma . . . . *S. carpocapsae*

15. Infective juveniles' average ratio e less than 0.85 (range 0.63–0.80) . . . . *S. rara**
15a. Infective juveniles' average ratio e more than 0.85 (range 79–97 μm) . . 16

16. Males' average distance from anterior end to excretory pore less than 75 μm (range 53–78 μm); average tail length less than 30 μm (21–32 μm); spicules' tip with hooked appearance . . . . *S. ritteri*
16a. Males' average distance from anterior end to excretory pore more than 75 μm (range 71–105 μm); average tail length more than 30 μm; spicules' tip blunt . . . . *S. kushidai*

* According to the Zoological Code of Nomenclature, the endings of some specific names in the genus *Steinernema* are incorrect. Therefore, the denoted *Steinernema* species have been amended (see Stock *et al.*, 1996). The proposed corrected names are as follows with the synonyms given within the parentheses: *S. affine* (*S. affinis*); *S. anomali* (*S. anomalae*); *S. cubanum* (*S. cubana*); *S. intermedium* (*S. intermedia*); *S. neocurtillae* (*S. neocurtillis*); *S. puertoricense* (*S. puertoricensis*); *S. rarum* (*S. rara*); *S. riobrave* (*S. riobravis*). In addition, the proposed corrected name for the species in the genus *Neosteinernema* is *N. longicurvicaudatum* (*N. longicurvicauda*).

7. Heterorhabditidae

a. Diagnostic characters

Stylet absent. Head truncate or slightly rounded. Six distinct lips present which may be partially fused at base; each lip with a single labial papilla; two additional papillae at the base of each submedial lip; lateral lip with a single cephalic papilla and a circular amphidial opening. Cheilorhabdions present as a refractile ring in anterior portion of the stoma (Figure 8). Posterior portion of the stoma collapsed, with reduced pro-, meso- and metarhabdions. Procorpus of oesophagus wide and cylindrical. Basal bulb with reduced valve. Nerve ring distinct usually located near middle of isthmus in female or basal bulb in male. Hermaphrodites with sperm in proximal portion of ovotestis and functional vulva. Amphimictic females with amphidelphic ovaries with reflexed portions often extending past the vulva opening. Sperm in proximal portion of oviduct and vulva non-functional for egg passage. Tail pointed with postanal or anal swelling. Rectal glands present. Males with single reflexed testis. Bursa present, open peloderan or pseudoleptoderan (Figure 9A,B). Spicules paired and separate (Figure 9C). Gubernaculum present (Figure 9D). Nine pairs of genital papillae. IJs with excretory pore located posterior to nerve ring. The family is composed of one genus, *Heterorhabditis* with eight currently described species (Table 3).

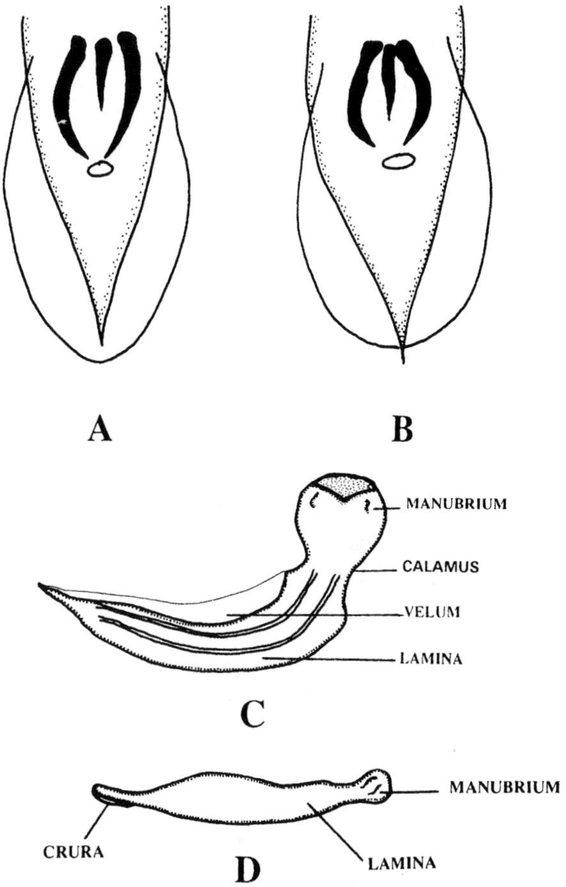

Figure 9 Ventral views of male tails (A and B) and lateral views of spicule (C) and gubernaculum (D). A, Tail with peloderan bursa; note that bursa goes beyond tail and is complete; B, tail with pseudopeloderan bursa; note that tail extends beyond bursa and that the bursa is incomplete; C, spicule showing various structures; D, gubernaculum showing various structures.

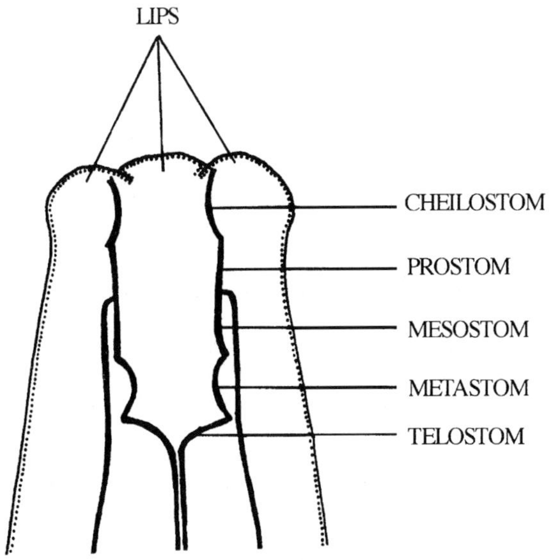

Figure 8 Anterior end of a rhabditoid nematode showing the general structure of the stoma.

b. Bionomics

Heterorhabditis has a similar life cycle to the steinernematids, but major differences also exist. IJs invade the haemocoel, release the bacterium *Photorhabdus luminescens*, and kill their host within 48 h. Insects killed by heterorhabditids turn red, brick-red, purple, orange, yellow and sometimes green. In the dark, heterorhabditid-infected insects luminesce due to the presence of the mutualistic bacterium. Adults resulting from the IJs are hermaphrodites rather than being male and female as in the steinernematids. Therefore, only one IJ is needed to enter the host for progeny production. Eggs laid by the hermaphrodites produce juveniles that develop

Table 3 List of recognized species of *Heterorhabditis* and synonyms.

Species	*Synonym*
H. bacteriophora Poinar	*Chromonema heliothidis* Khan, Brooks & Hirschmann *H. heliothidis* (Khan, Brooks and Hirschman)
H. megidis Poinar, Jackson and Klein	——
H. zealandica Poinar	
H. indicus Poinar, Karunakar and David	——
H. argentinensis Stock	——
H. hawaiiensis Gardner, Stock & Kaya	——
H. brevicaudis Liu	——
H. marelatus Liu and Berry	——

into males and females or IJs. The males and females mate and produce eggs that hatch and become IJs. If adequate moisture is present, the IJs will leave the host and infect another insect host.

c. Key to Heterorhabditis *species*

Ratio c = length divided by tail length; ratio e = distance from head to excretory pore divided by tail length; ratio f = width divided by tail length. When appropriate, the range is provided in parentheses for the mean values.

1. Infective juveniles' average total body length more than 700 μm; males with pseudopeloderan bursa . . . . *H. megidis*

1a. Infective juveniles' average total body length less than 700 μm; males with peloderan bursa 2

2. Infective juveniles' average tail length less than 80 μm (range 68– 80 μm) . . . . *H. brevicaudis*

2a. Infective juveniles' average tail length more than 80 μm (82–119 μm) 3

3. Males' spicules with a rostrum 4

3a. Males' spicules without a rostrum 5

4. Males' average distance from anterior end to excretory pore less than 150 μm (range 71–146 μm); infective juveniles' mean ratio c less than 6.5 (range 5.8–6.4) . . . . *H. hawaiiensis*

4a. Males' average distance from anterior end to excretory pore more than 150 μm (range 145–172 μm); infective juveniles' mean ratio c more than 6.5 (range 6.7–8.1 μm) . . *H. argentinensis*

5. Infective juveniles' mean ratio c less than 6.0 (range 4.5–5.6) . . . . *H. indicus*

5a. Infective juveniles' mean ratio c more than 6.0 (range 5.5–7.0) 6

6. Infective juveniles' mean ratio e less than 1.0 (range 0.89–1.10) . . . . . *H. marelatus*

6a. Infective juveniles' mean ratio e more than 1.0 (range 1.03–1.30) 7

7. Males' average length of testis reflection less than 100 μm (range 61–89 μm); spicules' length less than 45 μm (31–41 μm) . . . . *H. bacteriophora*

7a. Males' average length of testis reflection more than 100 μm (range 88–173 μm); spicules' length more than 45 μm (range 46–52 μm) . . *H. zealandica*

6 SAMPLING TECHNIQUES FOR NEMATODES

The purpose of sampling will determine where and how many samples are taken. Samples can be taken from the host's habitat (soil, manure, tree, insect gallery, etc.), or the insects may serve as the sampling unit. Familiarity with the life cycle of the nematode and/or the insect will make selection of the proper time for sampling easier. If the prevalence of nematode parasitism of the larval, pupal or adult insects is to be determined, sampling should commence when these stages are present in the field and should be conducted over a period of time.

A Soil samples

Stratified random soil sampling should be taken from an area for intensive study (Campbell *et al.*, 1995) or over a geographical area for extensive study (Akhurst & Bedding, 1986; Hominick & Briscoe, 1990a,b; Griffin *et al.*, 1991; Hara *et al.*, 1991). In the former case, soil samples from a transect should be taken from the study area over time. In the latter case, samples should include a full range of elevations, soil textures and habitats (e.g. cultivated fields, forests, pastures, parks, seashores, riparian areas). At a given site, soil samples should be taken to a depth of at least 15 cm using a soil corer, trowel, post-hole digger, auger, sampling tube, hand shovel, etc. The sample volume should contain a minimum of 1000 cm^3. At each area, collect at least three random samples over an area of 2–4 m^2. Within a sample, three subsamples (e.g. 10 × 10 × 15 cm) should be taken whenever possible. Depending on the objective, the samples may be combined or kept separate. Place the subsamples in a plastic bag, mix, and keep in a cooler (8–15 °C) during transport to the laboratory. Between samples, the collecting tool should be wiped with 70% alcohol or 0.5% bleach solution or thoroughly washed with water to prevent contamination of the next sampling unit.

As a minimum, collect the following information from each sampling site:

- Date
- Habitat
- Site location
- Temperature
- Associated vegetation
- Elevation

B Insect samples

Sampling for insects can be done according to procedures detailed by Southwood (1992). Insects in soil, litter, manure or plant material can be collected by dry or wet sieving, flotation, centrifugation sedimentation or Berlese funnel or through combinations of one or more of the listed methods. Insects in aquatic habitats can be sampled using a mosquito dipper, sampling vegetation, or sampling with various types of nets and traps. For aerial insects, the following traps can be employed: sticky, water, visual or light, host plant, sound and pheromone traps. Soil insects can be captured in pitfall; food, carbon dioxide; carrion or dung; and animal traps, or an insect net or an aspirator can be also be used.

Many of the above sampling methods have been used to collect nematode-infected insects or study nematode infections in insect populations. For example, pheromone traps or white sheets illuminated with ultraviolet light have been employed for noctuid moths parasitized by *Noctuidema* spp. (Rogers *et al.*, 1993), sticky traps, sweep nets, bloody boards or dung traps for face flies infected with *Paraiotonchium autumnale* (= *Heterotylenchus autumnalis*) (Jones & Perdue, 1967; Kaya & Moon, 1978; Kaya *et al.*, 1979), infested logs in emergence cages for bark beetles (Choo *et al.*, 1987), and aspirators for leafhoppers parasitized by *Agamermis unka* (Choo *et al.*, 1995). In addition, pitfall traps have been used to assess the impact of applied nematodes on non-target insects (Georgis *et al.*, 1991). However, the researcher must be careful in interpreting the data from a single sampling technique because nematode-infected insects may have aberrant behaviours that may bias the estimate of nematode prevalence (Wheeler, 1928; Poinar & van der Laan, 1972; Poinar *et al.*, 1976; Kaya *et al.*, 1979).

C Aquatic samples

According to Poinar (1991), aquatic nematodes can be sampled quantitatively or qualitatively. The former method involves the use of auger or probe to remove core samples of measurable size from the bottom of standing water (lakes, ponds, pools, etc.). Techniques used for marine nematodes can be applied for freshwater forms (Holme & McIntyre, 1971; Downing & Rigler, 1984). Qualitative methods vary and depend on the water flow and the zone to be sampled. In sampling nematodes from beds of fast flowing streams, a net, a set of sieves and instruments to turn over rocks are needed. Filipjev & Schuurmans-Stekhoven (1959) described a number of devices to collect fresh water nematodes. These include dredges, bottom catchers, plankton nets, sledge trawls, mudsuckers and worm nets.

7 NEMATODE EXTRACTION TECHNIQUES

Extraction methods for insect nematodes are derived from techniques developed with plant-parasitic nematodes. The most common methods are the Baermann funnel, sieving (gravity-screening), elutriation (Byrd *et al.*, 1976) and centrifugal flotation (Jenkins, 1964). Southey (1986) and Hooper & Evans (1993) provide examples of modifications of the above methods as well as other methods that are not covered in this chapter. One method that is unique for entomopathogenic nematodes is the "trap" insect method developed by Bedding & Akhurst (1975).

A Baermann funnel technique

This technique is useful for motile nematodes. It has the advantage of being simple and inexpensive, and nematode recovery from a small sample is very good. The disadvantages are the poor recovery of inactive nematodes, poor movement or mortality of nematodes due to lack of aeration, and the small amount of sample that can be processed.

Construction of a Baermann funnel is easy (Figure 10).

1. Insert a piece of rubber tubing to the funnel stem and close with a pinch clamp.
2. Place the funnel in a suitable rack, place a stainless steel wire screen that serves as a basket and support inside the funnel.
3. Wrap the material that contains the nematodes with a piece of muslin or tissue paper.
4. Place the wrapped material on to the wire basket.
5. Add water to the funnel until the wrapped material is touching the water. The motile nematodes will emerge from the material and settle into the rubber tubing.
6. Loosen the clamps after a few hours and collect the water containing the nematodes into a beaker or test tube.

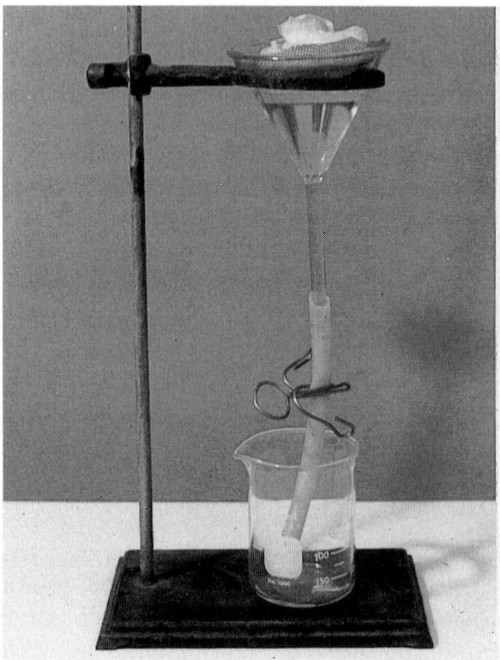

Figure 10 A Baermann funnel set up for extracting nematodes. (See text for details.)

A modification of the Baermann funnel technique is to use a mister.

1. Place the funnel stem (without the tubing) directly into a receptacle (test tube, beaker, etc.).
2. Spray a fine tepid mist intermittently (e.g. 1 min in every 10 min) over the funnel containing the material with the nematodes.
3. Collect nematodes from the beaker or tube.

The advantages are that aeration is not limiting, there is no build up of toxins in the water, the downward movement of water allows for all nematodes to be collected in the receptacle, and it does not require continual monitoring by an individual to collect nematodes. The disadvantage is that a more elaborate system is required and therefore, it is more costly to set up and maintain.

B Sieving (gravity-screening) technique

This technique and its variations uses the difference in size and specific gravity between nematodes and soil components. It is not dependent on nematode movement and large samples can be processed relatively quickly. The nematodes can be obtained within 30 min, but the initial investment in equipment is greater and an experienced worker is needed.

The method is described in detail by Hooper & Evans (1993) and Ayoub (1980). Various sieves (2 mm down to 38 μm), two beakers (250 and 600 ml), Syracuse watchglass or small Petri dishes, two stainless steel pans or plastic buckets, 4 litres or greater in capacity, and a rubber hose attached to a cold water tap are needed (Figure 11). The idea is to capture nematodes on the sieves so that large nematodes (>250 μm), average size nematodes (>90 μm), small adults and larger juvenile nematodes (>45 μm) and very small juveniles (>38 μm) are trapped on the 250, 90, 45 and 38 μm aperture sieves, respectively. Therefore, the size of the nematodes being sampled

Figure 11 Components for extracting nematodes using the sieving technique (see text for details). Arrow indicates that rubber hose needs to be attached to a cold water tap to wash sieves and pans.

determines which series of sieves are used. In addition, Peloquin & Platzer (1993) obtained *Tetradonema plicans* eggs by rearing infected sciarid hosts in the laboratory. The nematode eggs were extracted from the compost by passing them through 149, 53 and 43 μm sieves and collecting the eggs on a 25 μm sieve.

1. Place a soil sample (ca. 200–400 cm^3) that has been thoroughly mixed into a 600 ml beaker.
2. Place the soil sample into one of the stainless steel pans or plastic buckets (Pan A).
3. Add 2–3 l of cold water and mix the soil and water by hand.
4. Thoroughly agitate, and pour the soil–water mixture through the largest sieve (2 mm) into the second pan or bucket (Pan B).
5. Wash the sieve with water from the rubber hose over Pan B.
6. Discard any sediments in Pan A and wash the pan for the next step.
7. Agitate the material in Pan B, let it set for 10–60 s.
8. Pour the material into the next smaller sieve (710 μm) into clean Pan A.
9. Wash the sieve over Pan A. Heavy sediments in Pan B are discarded.
10. Repeat the process with smaller aperture sieves (250, 90, 63 and 38 μm).
11. Depending upon the size of the nematode, place the final sieved material into the 250 ml beaker and allow to settle for 1–2 h.
12. Remove the supernatant carefully by decanting or siphoning.
13. Leave about 40 ml of the material for examination. However, sample can be further processed using the centrifugal flotation technique (see Section 7D).

C Elutriation technique

This technique uses an upward current of water to separate nematodes from soil particles and hold them in suspension. It provides cleaner extraction and larger soil samples can be processed. The equipment and procedure are more elaborate and combine sugar flotation and sieving techniques. Those interested in elutriators should refer to Byrd *et al.* (1976) or Southey (1986).

D Centrifugal flotation technique

This technique, developed by Jenkins (1964), is very useful after large samples have been processed with the sieving (Kung *et al.*, 1990) and/or elutriation

(Hooper & Evans, 1993) techniques. The concept is to separate the nematodes from the soil particles and organic debris by floating them out in a solution of specific gravity greater than their own. The technique can provide nematodes for examination in a few minutes, is generally more efficient than other extraction techniques, and can extract living as well as dead nematodes. It is not very useful for large nematodes and requires expensive equipment, and the extraction solution can distort or kill the nematodes because of osmotic stress.

1. Place the extract containing the nematodes into a 50 ml centrifuge tube.
2. Add sufficient water until the centrifuge tube is filled up to 10 cm from the top.
3. Spin in a centrifuge at 2900 ***g*** for 5 min.
4. Pour off and discard supernatant.
5. Add a solution of sucrose at a specific gravity of about 1.18 (484 g of sucrose dissolved in water and made up to 1 litre) to the centrifuge tube containing the pellet until it is filled up to 10 cm from the top.
6. Mix the pellet and sugar solution thoroughly with a stirring rod or vortex mixer.
7. Centrifuge at 2900 ***g*** for 1 min.
8. Pour the supernatant through a fine sieve (5–38 μm, depending on the size of the nematodes).
9. Rinse the nematodes from the sieve into a beaker. (The nematodes should be removed from the sucrose solution as quickly as possible to avoid killing them.)

Other than sucrose, solutions of NaCl, $MgSO_4$, or $ZnSO_4$ can be used (Curran & Heng, 1992; Hooper & Evans, 1993).

E Visual technique

The visual technique is only useful with large nematodes, especially post-parasitic and adult mermithids in soil or water. They can be easily removed by searching through moist soil (Poinar & Gyrisco, 1964; Choo *et al.*, 1989) or water.

F Trap insect technique

The standard method for collecting entomopathogenic nematodes from soil involves the use of trap insects such as the greater wax moth (*Galleria mellonella*) (Bedding & Akhurst, 1975). Other insects such as crickets or mealworms may also be used (Rueda *et al.*, 1993). These insects are easily reared or are readily obtained from commercial bait companies in many countries. The advantages of this method are that it is simple, is selective for entomopathogenic nematodes (but if the insect dies, free-living nematodes may colonize the cadaver) and inexpensive. The disadvantages are that it is only useful for entomopathogenic nematodes and can be time consuming, and if the entomopathogenic nematode tends to be host specific, the nematode may not infect the trap insect.

1. Add water to the soil to bring the moisture level to –7 to –10 kPa (kiloPascal) water potential (see Section 12). In addition, after nematode extraction, the soil (usually for positive samples) should be sent to a soil analysis laboratory to characterize it for sand, silt and clay content and organic matter, pH and electrical conductivity.
2. Place 250–500 cm^3 of moist soil in a clean container.
3. Place 5–10 larvae of the greater wax moth (or other insect species) on the soil surface of each sample and cover with a lid.
4. Invert the containers and hold at 22–24 °C (if possible, another temperatures should be used; a cooler temperature in temperate (e.g. 15 °C) and a warmer one in tropical (e.g. 30 °C areas)).
5. Remove dead larvae from the soil every 3–4 days. (Depending upon the purpose of the study, healthy larvae can be added to the soil sample to extract more nematodes from the soil.)
6. Rinse cadavers in sterile water and place cadavers from each sample separately on filter paper in a Petri dish (60 × 15 mm).
7. After 2–3 days in the Petri dish (5–7 days after initial soil exposure), place dead larvae with signs of infection by entomopathogenic nematodes (Poinar, 1979) on a modified White trap (White, 1927) or Whitehead and Hemming tray (Whitehead & Hemming, 1965) for collecting IJs.

The modified White trap consists of an inverted Petri dish cover (60 mm diameter) or similar size watchglass placed within a Petri dish (100 × 15 mm).

1. Place the *G. mellonella* cadavers on a 55 mm moist piece of filter paper in the Petri dish cover.
2. Fill the outer Petri dish (100 × 15 mm) with ca.

20 ml of sterile distilled water. Do not place water into the dish holding the cadavers.

If a number of cadavers are being processed, a larger Petri dish (150 × 20 mm) can be used (Figure 12). The Whitehead and Hemming tray consists of a sieve to support the insects. The sieve is made by removing the bottom of a container and replacing it with a plastic mesh (90 μm) that has large enough openings for nematodes to pass through.

1. Place a double layer of paper tissue or fine mesh nylon cloth on the sieve.
2. Place the cadavers on the tissue or cloth.
3. Place the sieve on a collecting tray.
4. Add water to the tray so that the cloth or paper tissue is moist. The insects should not be covered with water as the nematodes within will die from anoxia.
5. Hold cadavers on the sieve for 1 week for IJs to migrate into the water. (For further information on harvesting nematodes, see Section 9B). Emergence of IJs varies according to the nematode group: for steinernematids, 8–10 days after infection and for heterorhabditids, 14–15 days after infection.
6. Expose the collected IJs to new *G. mellonella* larvae to confirm pathogenicity and complete Koch's postulates.

G Extraction of nematodes from aquatic samples

Although the majority of freshwater nematodes are too small to view without magnification, the free-living stages of many aquatic mermithids are large enough to see and can be picked from the samples with an 'L-shaped needle' (Figure 3B). Other small mermithids or insect-parasitic nematodes that escape visual detection can be collected with various extraction procedures.

The extraction methods used for soil and plant-parasitic nematodes (i.e. Baermann funnel, sieving, centrifugal flotation methods, etc.) are suitable for freshwater forms. Quantitative methods of extraction for monitoring nematode populations over a period of time often require the use of elutriators which have been discussed above.

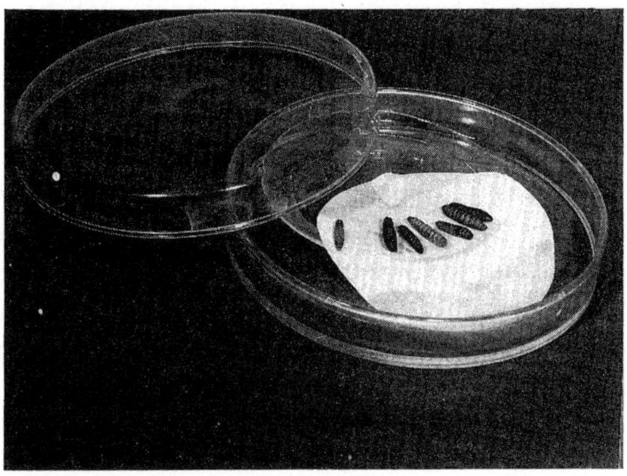

Figure 12 A White trap to collect infective juveniles from insect cadavers.

8 EXTRACTION EFFICIENCY

Extraction efficiency depends on a number of factors including the method, soil type, nematode species and size, and the laboratory personnel (Barker, 1985). Relative extraction efficiency for various plant-parasitic nematode species have been determined (Barker, 1985), but little information is available for nematodes associated with insects.

Saunders & All (1982) compared the efficiency of recovering *Steinernema carpocapsae* IJs from soil using the Baermann funnel, centrifugation flotation and flotation-sieving methods. They found that the Baermann funnel technique was superior to centrifugation flotation and flotation sieving methods. Kung *et al.* (1990) used the centrifugal flotation technique and obtained an extraction efficiency of 52% for *S. carpocapsae* with a 45% sucrose solution and 65% for *S. glaseri* with a 75% sucrose solution. Extraction efficiencies were used for studying persistence (survival) of the nematodes over time in spiked soil (addition of a known number of nematodes) as follows:

$$\text{extraction efficiency} = \frac{\text{live nematodes extracted}}{\text{nematodes inoculated} \times \text{nematode extraction efficiency}}$$

Before 1992, most researchers used the percentage of trap insects infected in a given soil sample as an indication of presence or absence of entomopathogenic nematodes. Generally, a high percentage indicated that many nematodes were present and a low percentage indicated that very few nematodes were present. This information was qualitative and not quantitative.

Fan & Hominick (1991) developed an efficiency assay for determining the numbers of steinernematid and heterorhabditid IJs capable of infecting a host insect. This assay is accomplished as follows:

1. Place known numbers of IJs of a given species into soil or use field soil known to have IJs.
2. Add a trap insect to the soil to determine the actual numbers of nematodes capable of infecting the host insect.
3. After a specified time (24–72 h depending upon temperature), remove the trap insect and rinse in water.
4. Dissect the insect immediately in Ringer's solution or hold on moistened filter paper.

The intention is to allow the nematodes sufficient time to kill the host and develop to adults (which facilitate counting) but before progeny production (the presence of progeny makes counting difficult and timing the dissection before this happens is critical). Another trap insect can be added to the soil and the process repeated until no more nematodes are trapped. It was found that it took nearly 36 days of consecutive sampling before infection ceased. Two or three consecutive assays over 12 days did recover 75% or more of the nematodes. The Fan & Hominick method is widely used to provide quantitative data of the number of nematodes that could penetrate and establish in the hosts and is a useful parameter for comparing the efficacy of different nematodes under specific conditions.

To facilitate counting the number of nematodes, Mauleon *et al.* (1993) developed a pepsin enzyme technique that has made dissection of the trap insect an easier task (see Appendix).

1. Place the cadaver into a Petri dish (100 × 15 mm).
2. Pour the pepsin solution (ca. 15–20 ml) to cover the insect.
3. Dissect the insect.
4. Place the Petri dish in an incubator at 37 °C on a shaker (120 rpm min^{-1}) for 2 h. Do not let the digestion go beyond 2 h without close monitoring because the nematodes may also become digested. The pepsin will digest away most of the insect tissue but will keep the nematodes intact. After digestion, the dishes can be held at 4–15 °C for several days until counting can be done. In this case, dilute with an equal volume of water before storage.
5. Count and sex the nematodes using a dissecting stereomicroscope.

The advantages of this method are that a large number of insects can be processed quickly and counted immediately or at a later time, and it is not necessary to spend time searching through the insect's tissues to find the nematodes.

9 NEMATODE CULTURE

Many species of obligate nematode parasites, such as *Romanomermis culicivorax*, *Heleidomermis magnapapula*, *Tetradonema plicans*, *Tripius sciarae* and *Paraiotonchium autumnale*, can be cultured *in vivo* in their respective insect host. Poinar (1979) has summarized culturing techniques for many species of nematodes that show biological control potential. In addition, Petersen (1984) and Petersen & Willis (1972) discuss mass production of *R. culicivorax* in mosquitoes, Mullens & Velten (1994) cover the laboratory culture of *H. magnapapula* in midges, and Peloquin & Platzer (1993) provide detailed methods for rearing *T. plicans* in sciarid flies.

A *In vivo* rearing of *Romanomermis culicivorax*

1. Flood eggs of *R. culicivorax* that are stored in pans with dechlorinated water (add one drop of 5% sodium thiosulphate/l of tap water to inactivate the chlorine).
2. Add J2 mermithids at the desired rate to newly hatched mosquito larvae in the rearing pans (*Culex pipiens* can be reared in the laboratory and is susceptible to the mermithid). The ratio of J2s to mosquito larvae can vary from 2.5 : 1 to 12 : 1. A ratio of 5 : 1 can give a 95% parasitization of mosquitoes in small containers, but for mass production a ratio of 12 : 1 is needed to achieve the same level of parasitization. Too high a ratio of mermithids to host will result in a sex ratio that is skewed towards males because male nematodes predominate when several individuals occur in the same host.
3. Incubate the rearing pans at 27 °C.
4. Feed the mosquito larvae finely ground rabbit chow as needed. Do not overfeed because the

rearing containers will become cloudy and foamy. This situation can be minimized by aerating the rearing pan.

5. Transfer the larval mosquitoes to a modified Whitehead & Hemming apparatus 6 days after nematode exposure. This apparatus is two containers, one nested into the other. The upper container is adapted with a 500 μm screen that allows the post-parasitic mermithids (J4s) to settle to the bottom container. The uninfected mosquitoes and cadavers are retained in the upper container. The J4s, which will begin emerging on the 7th day, can be collected from the bottom container.
6. Concentrate the J4s in a small container.
7. Allow the J4s to settle and discard the supernatant.
8. Repeat the procedure several times to eliminate most of the debris.
9. Transfer the J4s to a pan containing clean coarse sand (1.5 cm deep) covered to a depth of 1 cm with dechlorinated water. Fine sand is not recommended because it becomes tightly compacted and inhibits nematode movement.
10. Cover the pan with a loose fitting lid and store at room temperature.
11. Remove any visible dead nematodes and decant the water 3 weeks after setting the mermithids in the coarse sand.
12. Absorb the excess water in the pan with paper towels.
13. Store the pan for an additional 4–15 weeks before use. Mermithid eggs will be laid in the sand, and J2s can be obtained by flooding the pan. The best hatching of J2s occurs after 6 weeks of storage and declines after 20 weeks (Petersen & Willis, 1972).

B *In vivo* culture of entomopathogenic nematodes

Most steinernematids and heterorhabditids can be produced in *G. mellonella* larvae or other suitable insect hosts. *Steinernema kushidai* and *S. scapterisci* are not adapted to lepidopteran larvae, and scarab larvae and mole crickets or house crickets, respectively, should be used. The following *in vivo* culture method can be used for most steinernematids and heterorhabditids (Woodring & Kaya, 1988), and can be modified to fit varying requirements. The *in vivo* culturing technique is useful for laboratory and small field trials but is not practical for large-scale nematode production.

1. Adjust the number of IJs in suspension to 200 ml^{-1}.
2. Evenly distribute 1 ml of the nematode suspension on a 9-cm piece of filter paper in the lid of a 100 × 15 mm plastic Petri dish.
3. Add 10 *Galleria* larvae to the dish. The goal is to have about 20 IJs/larva; too many IJs per larva produces few progeny because of competition and/or contamination with foreign bacteria.
4. Cover the lid with the inverted Petri bottom.
5. Label the Petri dish and store it in a plastic bag (to conserve moisture) at room temperature. Be sure that the Petri dish is not tightly covered because the nematodes will die without sufficient air. (Try to do only one nematode species per day; if it is necessary to handle more than one species, be sure to wash your hands thoroughly and use sterilized tools because you can cross contaminate the cultures very easily.)
6. After 5–7 days, place the cadavers on to the White trap (Figure 12) or Whitehead & Hemming tray as described in Section 7F. If the cadavers smell putrid, the culturing was not successful.
7. Harvest nematodes from the White trap (use the 150 × 20 mm Petri dish) by removing the inner dish that contains the cadavers and pouring the suspension of IJs into a beaker.
8. Replace the dish containing the cadavers into the larger Petri dish and add more water to the larger dish. The IJs in the beaker will settle to the bottom in 15–20 min.
9. Decant the water and add new water to the beaker. This should be done three times.

Two other faster methods can be used. The first is as follows:

1. Centrifuge the nematodes at a low speed (250 rpm) for 2–3 min.
2. Decant the supernatant.
3. Add water to the nematodes and repeat the process three times.

The second method is as follows:

1. Vacuum filtrate the nematode suspension using a 5 μm membrane filter (retains nematodes, but allows bacteria through) in a vacuum filter unit (Figure 13).

Figure 13 A vacuum filter unit to harvest nematodes and process for storage or other uses. Arrow indicates that hose is attached to a vacuum pump.

2. Resuspend in sterile water.
3. Repeat the process three times.
4. Resuspend the nematodes in water for storage.

Some species, especially *S. glaseri*, emerge from the cadavers as 'pre-IJs.' If the pre-IJs emerge directly into the water, they will not develop successfully into IJs. Good results have been obtained by forcing the pre-IJs to travel some distance. One method is as follows; to prepare the Petri dish by filling with a thin layer (2 mm) of plaster of Paris, allowing the plaster to dry for at least a day, and then rewetting the plaster just prior to use. The plaster should be moist, but there should be no standing water. Place cadavers on the specially prepared Petri dish. The pre-IJs emerge, complete their development on the plaster substrate and then migrate within 1–2 days into the water.

Another method is to have the pre-IJs emerge from the cadavers on to a moist filter paper inside the bottom of a Petri dish within a White trap. They complete their development and migrate over the Petri dish lip into the water. They can be harvested as described above.

In preparation for storage, examine the IJs for activity. Most species are active at room temperature. However, IJs of *S. carpocapsae* may remain still (Figure 14) and can be probed for activity. Another good indication that these IJs are not dead is to determine if they have a characteristic 'J' shape. IJs that

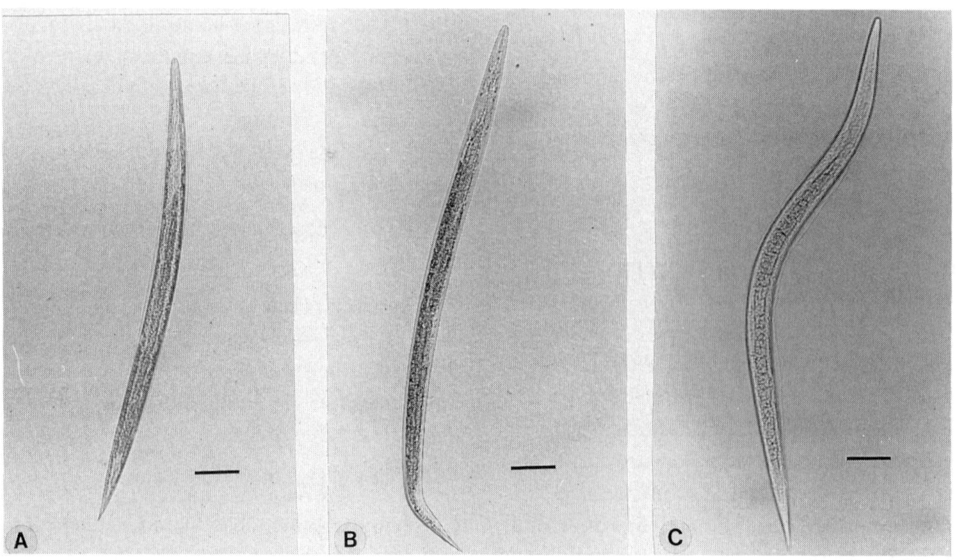

Figure 14 Infective juveniles of *Steinernema carpocapsae* showing dead (A), and living nematodes (B and C). B, living J shaped nematode, C, actively moving nematode.

are straight without the 'J" shape are probably dead. If host tissues or large numbers of dead IJs are present, a separatory step should be performed before storage.

1. Separate active IJs by using the Baermann funnel technique. Non-infective nematode stages can be killed by rinsing the nematodes in 0.4% methylbenzethonium chloride solution for 15 min.
2. Rinse IJs using the technique described under harvesting.
3. Store IJs in tissue culture flasks for several months. Usually, about 1000–2000 IJs ml^{-1} is a good number to store in a flask. Lower concentrations should be used for larger nematodes such as *S. glaseri* (300–500 IJs ml^{-1}).
4. Add a drop of Triton X-100 (wetting agent) to prevent the IJs from sticking to the surface of the containers. To prevent the formation of rosettes (clumps) of heterorhabditids, a few drops of sodium bicarbonate solution (1g NaH_2CO_3/50 ml H_2O) may be added (Woodring & Kaya, 1988)
5. Store the tissue culture flask flat to allow air exchange. The amount of water in a tissue culture flask should be no more than 1 or 2 mm deep when the tissue culture flask is stored flat. If IJs are stored in an Erlenmeyer flask, they should be aerated (an aquarium pump works well), in which case a higher concentration can be stored.
6. Store IJs between 4 and 15 °C. However, some species store better at warmer temperatures. It is important to store IJs in more than one incubator because the incubators can malfunction, killing the stored IJs.

C *In vitro* culture of entomopathogenic nematodes

In vitro production has been successfully accomplished with steinernematids and heterorhabditids. In the past, steinernematids have been cultured on a variety of substrates (see Woodring & Kaya, 1988; Friedman, 1990). The ingredients for nematode culture include a source of nutrients for the symbiotic bacterium and a sterol source for the nematodes. Our focus will be on monoxenic culture in the laboratory rather than on commercial production and on solid rather than liquid culture. Source of monoxenic cultures is presented in Section 9C,2. Reference should be made to Friedman (1990) for axenic cultural methods.

1. Culture in test tubes or Petri dish

1. Place 1 g of sterile rabbit kidney on 10 ml of 1% agar to rear *S. glaseri* and other steinernematids. A small piece of autoclaved pig kidney placed aseptically on to peptone–glucose–agar has also been used for steinernematids. Another artificial medium for steinernematid and heterorhabditid production uses dry dog food as the main ingredient (Hara *et al.*, 1981; Ogura & Haraguchi, 1993) (see Appendix).
2. After sterilization and cooling of the medium, seed with Phase I bacterium from the species to be cultured (see Section 16).
3. Let the bacterium grow for 1–2 days and inoculate with surface-sterilized IJs or eggs or from another monoxenic culture. Alternatively, surface-sterilized IJs can be placed on the medium without the Phase I bacterium (see Section 9C,3); the IJs will release the bacterial cells from their intestines but development will be delayed by 2–3 days.
4. After 2–3 weeks, harvest the IJs or, if in test tubes, store at cooler temperatures until needed.

To harvest from Petri dishes, proceed as follows:

1. Use a modified version of the White trap.
2. Remove the cover from the Petri dish, place the bottom of the Petri dish into a larger dish.
3. Add water to the larger dish.
4. Place four pieces of filter paper on the medium and allow a part of each filter paper to touch the water. The IJs will migrate into the water using the filter paper as a bridge.
5. Collect IJs from the water as previously mentioned.

2. Culture in flasks

Bedding (1981, 1984) developed a technique whereby large numbers of nematodes may be produced using chicken offal medium on a porous foam substrate such as polyether polyurethane. This substrate provides the largest surface-to-volume ratio while providing adequate interstitial space. Glass flasks serve as rearing containers (see Appendix), but autoclavable bags have also been used.

1. After autoclaving, allow the medium to cool and inoculate with Phase I bacterium of the nematode species to be cultured.
2. Inoculate with 5 ml of nutrient broth with 2- to 3-day-old bacterium.
3. One day after bacterial inoculation, inoculate with ca. 1000 surface-sterilized nematodes or nematodes from another monoxenic culture. Care should be taken to maintain monoxenicity during the transfer.
4. Harvest IJs from the flasks in 2–4 weeks.

Harvesting IJs from the foam can be accomplished as follows:

1. Pile foam 5 cm deep on a 20-mesh (833 μm) sieve.
2. Place the sieve in a pan of water with the water level adjusted so that the foam is just submerged. If a mist chamber is available, the pan and sieve may be placed in it for 2–24 h. Do not pour water over the foam as this washes particles of homogenate into the nematode suspension. Within 2 h, 95% of the IJs will migrate into the water.
3. Sediment IJs and rinse to remove the particulate matter by passing through a sieve (90–38 μm depending on nematode size).
4. Rinse with water several times until it is clear.

A drop in production can be due to a number of problems. These include, but are not limited to, contamination, reversion of Phase I bacterium to Phase II (see Section 16B), unsuitable incubation temperatures, improper moisture content, and poor genetic stock. Contamination will sometimes be visually evident in the form of fungal or bacterial colonies or unusual coloration or exudates. Often this will be associated with putrid or unusual odours. If contamination is suspected, purity can be routinely verified using NBTA or MacConkey agar (see Appendix).

Some nematode species do better at lower temperatures and placing the flask at temperatures below 25 °C will increase production. If the foam is too dry or wet, adjust the amount of medium placed on the foam. If the medium starts to dry out during culture, holding the flask at higher humidity may help.

Production levels can remain adequate while quality, measured in terms of infectivity or activity drops. *In vivo* passage is recommended after several generations of *in vitro* culture. Another method is to store IJs in liquid nitrogen and use this source to initiate new cultures (see Section 15B).

It is difficult to foresee every potential problem. With some knowledge of nematode and bacterium biology, solutions for most troubles should not be too difficult to pinpoint. However, *in vitro* production is not easy, especially for the novice, and trial and error will improve the system.

3. *Surface-sterilized IJs and nematode eggs*

IJs can be surface-sterilized using four methods including: (a) Hyamine 10X (methylbenzethonium chloride) or Hyamine 1622 (benzethonium chloride); (b) formaldehyde; (c) merthiolate (with or without antibiotics) or (d) sodium hypochlorite. Be sure that the nematode suspension is completely free of foreign particulate matter. If particulate matter is present, the surface sterilization procedures will fail.

1. The first method uses Hyamine as follows. Place IJs into 0.1% Hyamine 10X or 1622 solution for 15–30 min. Rinse three times with sterile distilled water.
2. The second method uses formaldehyde as follows. Place IJs into 0.1% formaldehyde solution for 30 min. Make two subsequent transfers into 0.1% formaldehyde solution for 30 additional min each. Rinse three times in sterile distilled water.
3. The third method involves merthiolate. Place IJs overnight in Ringer's solution. Transfer to a mixture of 0.1% merthiolate solution (w/v) plus streptomycin (5000 units/ml) for 2–3 h. Triple rinse with sterile Ringer's solution. A modification of this method is as follows. Place IJs in 0.1% merthiolate solution for 1 h. Transfer to fresh 0.1% merthiolate solution for 3 h. Rinse three times with sterile water.
4. The fourth method uses 0.1% sodium hypochlorite solution following the same duration as described for the modified merthiolate method described in the preceding paragraph.

A simple way to sterilize IJs is to set up a test tube with 10–15 ml of Hyamine solution.

1. Place ca. 500 IJs into 0.1% Hyamine 10X or 1622 solution.
2. Concentrate IJs with a centrifuge at low speed (250 rpm). IJs can also be concentrated with a

Buchner funnel by placing a piece of filter paper in the Buchner funnel, pouring the IJ suspension into the funnel with the vacuum on, washing with several rinses of sterile distilled water, and picking up the IJs with a camel hair brush.
3. Place IJs into the Hyamine solution.
4. Have three other test tubes with sterile distilled water and using a sterile pipette, transfer the IJs from the Hyamine solution to the first sterile water test tube.
5. Transfer IJs after they have settled to the bottom of the tube and complete the procedure by transferring the IJs sequentially into the next two tubes containing sterile distilled water. Do not try to recover all IJs. Start with enough IJs so that there are sufficient numbers at the end.

An effective way to obtain axenic or monoxenic nematodes is to sterilize the eggs. Adult females or hermaphrodites can be obtained from *in vitro* monoxenic cultures or from 3-to 6-day-old insect cadavers. For steinernematids, sufficient time should elapse before collection to ensure that the female is mated.

1. Suspend adults in 30 ml of 0.4 N NaOH.
2. Gently agitate for 15 min. This procedure kills the adults and separates eggs from adults.
3. Neutralize the NaOH with 1 ml of 50% phosphoric acid.
4. Separate the eggs by filtration through a 20 μm membrane filter and let stand for 30 min.
5. Decant the supernatant, add 25 ml of 2% sodium hypochlorite, and let stand for 5 min.
6. Rinse the eggs three to five times in sterile distilled water.
7. Check the sterilization process by incubating some of the eggs on nutrient broth for 5–7 days at 25 °C.

10 QUANTIFICATION METHODS

A Nematodes

Nematodes can be quantified in one of three ways: direct count, dilution or volumetric estimation (Woodring & Kaya, 1988).

1. Direct counts

This method is useful for numbers up to 50 nematodes.

1. Count nematodes individually into a positive displacement micropipette or microdispenser while observing them with the aid of a dissecting microscope.
2. Transfer the nematodes to the desired place.
3. Check the micropipette or microdispenser to verify that all nematodes were expulsed.

For 1 to 50 nematodes, direct counting is much more accurate and is preferred over dilution. However, for numbers above 50, the time saved using a dilution is justified over the small loss in accuracy.

2. Dilution

1. Prepare a dilution to facilitate counting. About 100 nematodes/ml is ideal.
2. Mix the initial suspension as homogeneously as possible (a stir bar in a beaker suspension works well).
3. Remove 1 ml and add to x ml water, where x is estimated to yield a new suspension of about 100 nematodes/ml. This is a $1{:}x + 1$ dilution.
4. Count the nematodes in the dilution under a dissecting microscope.
5. Take samples from a thoroughly mixed, diluted suspension.
6. Count on a special counting slide that holds a known volume and provides grids to aid counting. An 'eelworm counter' in which 1 ml of nematode suspension occurs on a grid is commercially available. Conventionally, nematodes touching or crossing the top and right-hand sides of a grid are counted whereas those touching or crossing the bottom or left-handed side are not counted. In this manner, the same nematode will not be counted twice. Alternatively, 1 ml of suspension may be spread on a microscope slide or Petri dish and counted. If the dilution has too few (<50/ml) or too many nematodes to count easily, make a new dilution.

Statistically, an acceptable sample size is 10 counts; however, most workers probably count 3–5 samples from two to three dilutions. To determine the concentration of nematodes in the original suspension, the following formula is used:

$$N \times \frac{1}{M} \times (x + 1) = S$$

where N = average number of nematodes per counted sample; M = number of ml per counted sample; S = concentration (nematodes/ml) in initial suspension; $x + 1$ = dilution,

To prepare dilutions with a given number of nematodes per ml, dilute the initial suspension or the counted suspension. Statistically, it would be preferable to use the counted suspension, but practically it may contain far too few nematodes/ml. The following formula is used:

$$A = \frac{D \times C}{B}$$

where A = ml of suspension of known concentrations, the suspension to be diluted; B = number of nematodes/ml in suspension; C = final volume, in ml, of new dilution; D = desired concentration in new dilution.

Then: $C - A$ = ml of water to be added to make a new dilution.

3. Volumetric counts

Large numbers of nematodes can be estimated on the settled volume. This technique is useful when the nematodes are similar in size, such as IJs of entomopathogenic nematodes. Calibrate the system in the following way:

1. Add 10 ml of a known (high) concentration of nematodes to a graduated centrifuge tube,
2. Centrifuge at a given rpm for a set period of time (e.g. 300 rpm for 1 min),
3. Divide the number of nematodes in the tube by the volume (in ml) that the nematodes now fill to obtain the number of nematodes/ml (V).

V should be estimated several times from different samples to verify that the system is relatively consistent for the given species and procedure. Once the system is calibrated, the number of nematodes may be estimated by multiplying the volume taken up by the appropriate V. However, this method is the least precise and repeatability may be a problem, particularly involving different workers.

B Bacteria

Two quantification methods for the symbiotic bacteria of steinernematids and heterorhabditids are provided. More detailed descriptions of methods for quantifying bacteria are given in Chapter III.

1. Microscopic counts

A haemocytometer can be used to determine the concentration of bacterial cells in a liquid medium. It is ideal for counting cells grown in shake cultures. If bacteria grown on agar are suspended in a liquid medium, vortex the suspension to break up clumps of bacteria. Because cells of *Xenorhabdus* and *Photorhabdus* are motile, they should be killed before counting by adding an equal volume of 20% formalin to the suspension. Haemocytometers come with their own instructions on dilutions and counting. This method gives an absolute count and does not differentiate between living and dead cells.

2. Macroscopic counts using the spread plate method (Akhurst, personal communication)

1. Prepare YS agar or Medium X agar Petri dishes (see Appendix) 24 h prior to inoculation of the bacterium, allow to set, and store in the dark
2. Transfer aseptically a known quantity of liquid containing the bacterial suspension to a known quantity of sterile Ringer's solution which should be at the same temperature as the bacterial suspension (*Xenorhabdus* and *Photorhabdus* are very sensitive to sudden temperature changes).
3. Vortex the suspension to break up clumps of bacterial cells.
4. Make serial dilutions with sterile Ringer's solution.
5. Pipette 100 μl aliquot from a serial dilution on to the agar (made in step 1) and spread with a sterile glass hockey stick or sterile glass beads that are tipped off the agar when spreading is complete. Label the Petri dish with the dilution.
6. Set out and label two or more Petri dishes for each dilution.
7. Count the number of bacterial colonies within 1–2 days.
8. Select dishes with 30–300 colonies for counting.
9. Mark each colony with a dot as it is counted to avoid repeats. Each colony should correspond to one live cell or colony forming unit (cfu) in the suspension or dilution used to make the dish. The concentration of cfu in the original inoculum can

be determined by multiplying the number of cfus times the inverse of the dilution.

11 BIOASSAY METHODS

A universal standard bioassay for entomogenous and entomopathogenic nematodes has not yet been developed. We will focus on bioassay techniques developed for the mermithid parasite of mosquitoes, *Romanomermis culicivorax*, and for steinernematids and heterorhabditids.

A *Romanomermis culicivorax* assay

Infectivity of J2 mermithid can be tested against first instar *Culex pipiens* or other susceptible mosquito species as follows:

1. Capture mosquito larvae with an eye dropper.
2. Count as they are released from the dropper on to a 4 cm^2 of silk bolting cloth.
3. Transfer 20 mosquito larvae by immersing the bolting cloth into 250 ml plastic containers containing 125 ml of dechlorinated tap water.
4. Feed larvae ca. 75 mg of finely ground rabbit chow and Brewer's yeast (3 : 1 ratio).
5. Add J2s to the container with mosquito larvae.
6. Cover container with a perforated lid for aeration and place in the dark at the desired temperature for 24 h.
7. Terminate the infection process by pouring the contents of each container on a screen with 180 μm openings which retains the mosquito larvae but allows the J2 mermithids in the water to pass through.
8. Return the mosquito larvae to the container with 125 ml dechlorinated water.
9. Feed and maintain for 10 days at 27 °C.

The number of pupae or adult mosquitoes and number of post-parasitic mermithids in each container can be counted.

B Entomopathogenic nematode assays

For steinernematids and heterorhabditids, the one-on-one (Miller, 1989), the Petri dish (Woodring & Kaya, 1988) and sand barrier (Woodring & Kaya, 1988) assays are the most commonly used with *G. mellonella* larvae as hosts. Although *Galleria* is too susceptible to be an ideal bioassay host, it is widely available and continues to be used. Other insects that have been used for bioassays include other available lepidopteran hosts, crickets, *Tenebrio molitor*, and maggots (house fly and sheep blow fly). Be sure to include adequate controls when doing these bioassays.

Except for the one-on-one assay that uses one IJ and one insect in a tissue culture well, the range of nematode concentrations and other parameters are not given because they must be chosen in accordance with the given situation. Be sure to have sufficient replication to account for variation in the data. In all cases, begin with IJs that have been equilibrated with the test conditions.

1. One-on-one assay

1. Obtain a 24-well multiwell tissue culture plate with lid (8 × 12.2 cm) with each well having an area of ca. 2 cm^2 (Figure 15).
2. Place a piece of filter paper inside each well. A small amount of sand can also be used instead of the filter paper.
3. Place an IJ in each well. This can be done using a 25 μl microdispenser (see Section 10A,1). An alternative is to mouth-pipette individual

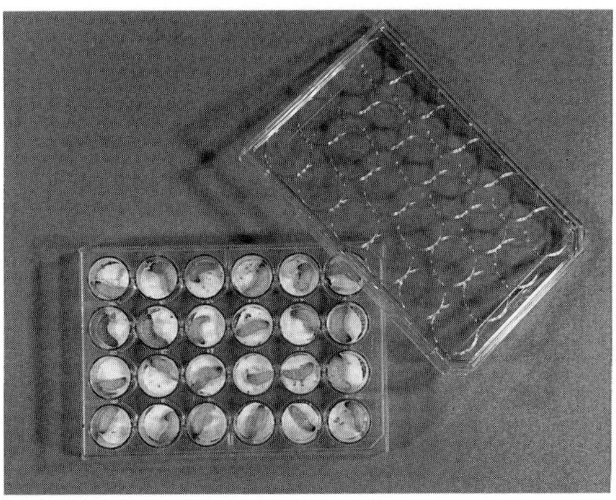

Figure 15 One-on-one assay method for entomopathogenic nematodes using a 24 multi-well tissue culture plate with a piece of filter paper and a *Galleria mellonella* larva in each well.

nematodes using a 25 μl micropipette fitted with a rubber tubing.

4. Place a last instar *Galleria* larva into each well.
5. Replace lid and incubate at the desired temperature.
6. Record mortality at 24-h intervals.

2. Petri dish assay

1. Place two 9 cm pieces of filter paper in the lid of a 100 mm Petri dish.
2. Apply the required number of IJs evenly over the filter paper in 1 ml of water.
3. Add one (or more) insect host(s) and cover with inverted Petri dish lid.
4. Place the Petri dish in a plastic bag to conserve moisture and incubate at the desired temperature.
5. Record mortality at 24-h intervals. Remove and rinse dead insect. Dissect to verify the number and presence of nematodes before progeny is produced.

3. Sand barrier assay

1. Use sieved sand (or soil). Particle size is dependent upon purpose.
2. Use 5 cm × 5 cm (4.4 cm inner diam.) polyvinylchloride (PVC) cylinders (tubes) (Figure 16). Covers of 6-cm plastic Petri dishes make excellent bottoms and lids for the cylinders.

Figure 16 Sand barrier assay method for entomopathogenic nematodes using a 5 × 5 cm cylinder with: A, *Galleria mellonella* larva (arrow) on top of the cylinder, and B, cylinder turned upside down with *Galleria mellonella* in the bottom. (See text for details.)

3. Put a dish bottom on a tube and fill it with moist sand (–15 kPa or 8% moisture).
4. Place one insect host at the bottom of the tube. To avoid crushing the insect, first make a small depression in the sand. Replace bottom.
5. Apply the required number of IJs in 0.5 ml of water at the top of the tube (end opposite to insect).
6. Follow procedures as in the Petri dish assay from 4. It may be difficult to determine whether the insect host is dead and its condition may only be recorded during removal of the insect host from the tube.

12 HANDLING SOIL

As most nematodes associated with insects spend part of their life cycle in the soil, it is important to have some understanding of soil science. A good general textbook on soils such as Brady (1990) should be consulted. As stated earlier, the soil texture should be analysed.

Many laboratories use sand as a test medium. Sterilized sand used for children's sandboxes is commercially available in the United States. Potting soil with varying amounts of organic matter is available from nurseries. Field soil can also be used. The soil should be screened before use to eliminate unwanted materials (rocks, debris, etc.). The soil can be used raw or can be pasteurized or sterilized. If raw soil is used, there may be unwanted antagonists that can affect the nematodes (Kaya & Koppenhöfer, 1996). Soil may be pasteurized at 62 °C for 2 h. This will eliminate the microfauna and many fungi and bacteria but may not kill all micro-organisms. If soil is sterilized, its properties will change (Skipper & Westerman, 1973; Lotrario *et al.*, 1995). Autoclaved soil should not be used immediately because toxic components from organic matter and micro-organisms may occur. Although pasteurization is less drastic than autoclaving, both soils should be air dried and kept at room temperature for a few weeks before using.

Soil moisture has important effects on nematode activity. Much of the earlier nematode studies evaluating soil moisture effects on entomopathogenic nematodes cannot be interpreted because soil moisture content was usually expressed as percentage of waterholding capacity or percentage of soil dry weight. These soil moisture measurements are not related to the availability of water to the nematodes. Water potential is a more meaningful measure biologically because two soils with the same water potential make water equally available to nematodes, even if their water content, measured as a percentage of soil dry weight differ. Water potential is defined as the chemical potential of water per unit volume and has the dimensions of pressure. The units of measurements are usually expressed in bars, megaPascals (MPa), or kiloPascals (kPa) and are negative values. Saturated soil has a water potential at nearly 0 kPa. As the soil becomes drier, the water potential becomes increasingly negative. Soil water potential has two components, the matric potential which is the capillary and absorption forces associated with the soil matrix, and the solute potential which is the osmotic forces caused by the salt in the soil solution. The matric potential is essentially equal to the water potential as long as the soil is low in salts. For most soils, the matric potential is the important component.

Water potential of soils can be determined using a pressure plate apparatus (Studdert *et al.*, 1990), but this method takes several days to conduct and requires specialized equipment. Fawcett & Collis-George (1967) and Hamblin (1981) developed a rapid method for determining water potential by using Whatman no. 42 5.5 cm diameter filter paper. The filter paper only absorbs the water that is available and not bound to the soil.

1. Obtain the dry weight of the filter paper before placing into the soil because the weight of the filter paper varies by 10%.
2. Sandwich the weighed filter paper between two other pieces of filter paper and place it in the soil.
3. Allow filter paper to equilibrate with the water in soil. Equilibration time of the paper in the soil varies with wet soils equilibrating within a few minutes and dry soils taking up to 36 h.
4. Retrieve the weighed filter paper after equilibrium has been reached.
5. Reweigh the paper and calculate the percentage moisture as follows:

$$\frac{\text{wet weight} - \text{dry weight}}{\text{dry weight}} = \%\ \text{moisture of filter paper}$$

6. Determine the water potential (kPa) of any soil type by reading off the graph (Figure 17).

13 STAINING NEMATODES

A Live nematodes

Stained IJs can be used for ecological studies and may be a means to separate them from naturally occurring IJs. The offal artificial medium can be supplemented with stains (0.5% Sudan II, Sudan III, or oil soluble blue) that results in IJs of steinernematids with colour present in the intestine and fat body (Yang & Jian, 1988). The stain in the IJs will last for several months and is lost when they infect a host. No adverse effect on nematode fecundity or

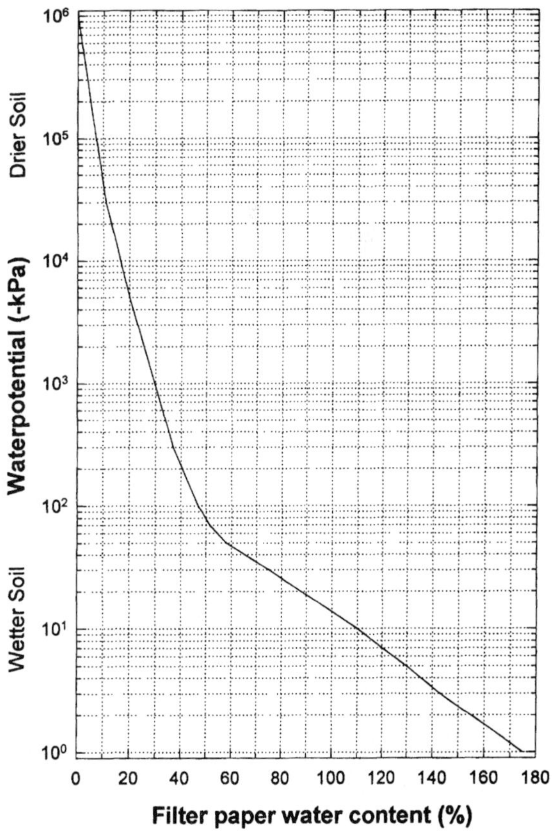

Figure 17 Calibration curve for Whatman no. 42 filter papers showing water potential in kiloPascals (kPa) of soil against filter paper water content (modified after Hamblin, 1981).

infectivity was observed. Ogura (1993), using a different artificial diet supplemented with 0.5% Sudan III, could not obtain stained *S. kushidai*, but the incorporation of 0.4% neutral red in the medium produced stained IJs. The red pigment was discernible in the intestinal tract of the IJs. Survival of the stained *S. kushidai* IJs appeared to be less than that of the unstained ones.

B Nematodes in insects

There are occasions when the location of nematodes within a host is needed. Dissection may reveal the tentative location of the nematodes, but a more exact method is to stain the nematode *in situ*. This method is useful for those insects with soft or transparent cuticle.

1. *Staining*

1. Fix the parasitized larva (or adult) in 70% ethanol (EtOH) for at least 24 h. If a beetle larva is being processed, puncture it to allow stains to penetrate efficiently.
2. Transfer the larval specimen to the Grenacher's borax carmine (see Appendix II).
3. Leave for at least 12 h.
4. After this time, remove specimens from the stain and rinse excess stain with several changes of 70% EtOH.

2. *Destaining*

1. Destain in acid alcohol (70% EtOH + 2% HCl) until the nematodes can be seen inside the insect's body. In insects with very thin integument, destaining takes a few seconds, whereas those with thicker cuticle may take 15–30 min.
2. Stop destaining by placing the insect in plain 70% EtOH; if more staining is needed, reinitiate by placing into the acid alcohol.

3. *Dehydration*

1. Place the insect into plain 70% EtOH for 2 h to remove acid from the tissues.
2. Transfer the insect to 80% EtOH for 2 h and then to 95% EtOH for 2 h.

4. *Counterstaining*

1. Counterstain by adding 1 or 2 drops of the stock solution of 1% Fast green in 95% EtOH (see Appendix) in a Syracuse watchglass.
2. Dip specimen into the diluted stain for 1–5 s while observing under a dissecting microscope.
3. Remove immediately to 95% EtOH and examine. Repeat if necessary. The cuticle of the insect should have a green blush.
4. Place specimen into absolute EtOH for 1–2 h. Do not leave longer than 2 h as the green will discolour and fade.

5. *Clearing and mounting*

1. Transfer specimen directly into methyl salicylate for ca. 5 min.
2. Remove for mounting as soon as the specimen has cleared.
3. Mount specimen directly from methyl salicylate into permount, damar balsam, or neutral Canada balsam.

14 CROSS-BREEDING STUDIES WITH ENTOMOPATHOGENIC NEMATODES

Within the steinernematids and heterorhabditids, cross-breeding tests may be used as a tool for species identification. According to the group, methods vary and are described below.

A Lipid agar plates

1. For each cross, place 10 virgin females and 10 males of the appropriate strain or species on each of five lipid agar plates.
2. Inoculate these plates and preincubate with Phase I bacteria isolated from the nematode strain or species from which the female partner in the cross is derived.
3. Incubate the plates at 25 °C.
4. Three days later, place 10 additional males on each plate to ensure that viable males are always available to fertilize the females.
5. Monitor the plates under a microscope every few days for the appearance of outcross progeny. In

successful crosses, progeny will be visible after 2–3 days.

6. Collect IJs from these crosses after 2 weeks.
7. Transfer to fresh lipid agar plates so that a hybrid line can be established.

Include the following controls for each cross:

- Virginity/self fertility test: 10 virgin females are placed without males on to each of five lipid agar plates.
- Mating test/self-cross: 10 virgin females and 10 males of the same strain or species are placed on lipid agar plates, and 10 additional males are added 3 days later.

The result of a cross between different isolates is taken as valid if (i) there are no progeny in the virginity test, and (ii) there are progeny in the self-cross.

B *Galleria mellonella* or other insect cadavers

Insect cadavers provide a better nutritional source than the lipid agar plates, and the richer environment can help to establish a hybrid line when the progeny have low viability (Griffin *et al.*, 1994).

1. Maintain *G. mellonella* larvae that are to be used for injection at 15 °C without food for 1–2 weeks prior to injection. (Commercially obtained *G. mellonella* can be used immediately because they have been heat-treated and starved.)
2. Prepare the cadavers by injecting the appropriate strain of the symbiotic bacterium into the haemocoel of *G. mellonella* larvae.
3. Suspend one loop of Phase I symbiotic bacterium isolated from the appropriate nematode strain in 2 ml of sterile Ringer's solution.
4. Inject 10 ml of this bacterial suspension into the haemocoel of a *G. mellonella* larva at the base of the last proleg using a 1 ml syringe and a small bore hypodermic needle (27 or 30 gauge).
5. Incubate the injected larvae at 20 °C, and check for mortality after 3 days.
6. Store cadavers at 15 °C for up to 14 days before using for nematode crosses.
7. Inject five virgin females and five males of the appropriate strains (by aspiration using micro-capillary pipette) into each of 10 *G. mellonella* cadavers killed by the proper isolate of symbiotic bacterium for each cross. That is, the cadavers should have been prepared using bacteria isolated from the nematode strain of the female partner in the cross.
8. Incubate cadavers at 25 °C on a damp filter paper in a Petri dish.
9. Inject additional males 3–5 days later.
10. Always prepare appropriate controls for virginity/self fertility and for self-crossing/mating test. One week later, dissect 3–5 cadavers in Ringer's solution to evaluate the success of the cross. In crosses where progeny are established, the IJs from the remaining cadavers can be collected on white traps.

C Hanging blood drop

This technique is as follows:

1. Surface sterilized IJs by immersing them in 0.1% Hyamine 10X or 1622 solution for 15–20 min (see Section 9C3.).
2. Triple rinse IJs through sterile distilled water.
3. Place a drop of haemolymph (obtained from surface-sterilized *G. mellonella* larvae in 95% ethanol) on a coverglass. To prevent the drop from drying, add 10 μl of serum-free medium for insect tissue culture to the drop and mix (Stock, unpublished data).
4. Place 30–50 IJs in the drop.
5. Turn the coverglass upside down, and gently place on a well slide (deep concave slide).
6. Place the slide in a Petri dish (100 × 15 mm) on a filter paper saturated with water.
7. Wrap the dish in a plastic bag and incubate at 25–27 °C.
8. Place separate pre-adult males and females of the isolates to be tested in new hanging drops with pre-adults of the opposite sex of the other isolates (ratio of 5 males : 5 females is recommended).

Evaluation of the mating should be done over a 10-day period. Controls consist of crosses of the same isolates. The presence of progeny is considered positive whereas its absence means negative results. Crosses should have sufficient replications to validate the results.

D Cross-mating within heterorhabditids

Because IJs of *Heterorhabditis* species always develop into hermaphroditic females, the second generation amphimictic adults are used for the cross-mating studies (Dix, 1994). Cross-mating with heterorhabditids must be done with adequate controls because the amphimictic females may have already mated, the second generation females may produce progeny without mating, or sterile offspring may be produced.

1. Obtain second generation adults from *G. mellonella* that have been incubated at 25 °C for 6–7 days post-infection with IJs of the appropriate strain or species.
2. Dissect the *G. mellonella* cadavers in Ringer's solution or M9 buffer (Brenner, 1974) (see Appendix).
3. Select female nematodes with immature gonads using a stereomicroscope at 50× magnification.
4. Collect immature females using a nematode pick (Figure 3) or by aspiration, using a microcapillary pipette that is drawn out to give an external diameter of 150–250 μm.
5. As an additional precaution, collect virgin females only from those cadavers in which no more than 5–10% of the second generation females have started oogenesis. Virgin females can be used immediately for cross-breeding or can be placed on a lipid agar plate (see Section 14A and Appendix) for 1–2 days to confirm that they are producing only unfertilized eggs. Males are collected in the same manner as females; however, their age or state of development are not critical.
6. It is advisable to infect a series of *G. mellonella* with IJs at 2-day intervals to ensure that female nematodes at a suitable stage for cross-breeding are available.

15 PRESERVATION

A Short-term nematode preservation

Entomopathogenic nematodes can be stored (with no aeration) in distilled water in tissue culture flasks (see Section 9B). Generally, steinernematids can be stored at 4–15 °C for 6–9 months and heterorhabditids for 3–4 months before subculturing is needed. For tests, use IJs that are no more than 1 month old after harvesting.

B Long-term nematode preservation

Temperatures below −130 °C (the recrystallization point of ice) can assure long-term, and possibly indefinite preservation of certain biological specimens (White & Wharton, 1984). Successful cryopreservation, however, almost always requires: (i) an appropriate pretreatment of the specimens with substances (cryoprotectants) that minimize intracellular and/or intercellular crystal formation; (ii) a precisely controlled rate of cooling of the specimens, at least during the early stages of freezing; and (iii) a controlled rate of thawing (Triantaphyllou & McCabe, 1989).

Popiel & Vasquez (1991) developed a reliable method for the cryopreservation of IJs of *Steinernema carpocapsae* and *Heterorhabditis bacteriophora*. Curran *et al.* (1992) introduced some modifications.

1. Place specimens for cryopreservation in prepared glycerol solution at 2× final concentration, and mix equal weights (final weights from 10 to 100 g) of nematode suspension and glycerol solution by stirring on a magnetic stirrer.
2. Pour the suspension into a Petri dish and incubate for 24, 48 or 72 h at 23 °C.
3. After incubation, remove excess glycerol solution by suction filtration of the IJs on to a Whatman no. 42 filter paper supported in a Buchner funnel.
4. Apply suction until IJs no longer appear glossy.
5. Rinse IJs under suction with ca. 15 ml 70% methanol (23 °C).
6. Wash IJs off the filter paper with cool 70% methanol (5 °C) into tapered 15 ml centrifuge tube on ice.
7. Incubate IJs for 10 min on ice.
8. Resuspend in cool 70% methanol, agitate after 5 min, and allow to sediment.
9. Transfer the IJ pellet to cool 2 ml round-bottomed polypropylene cryotubes with silicone rubber seals.
10. Plunge immediately into liquid nitrogen.
11. Thaw IJs by immersing the opened cryotube in 15–30 ml of Ringer's solution at 23 °C for 24 h.

C Nematode preservation in substrates

Nematodes can also be stored on various substrates such as charcoal, alginate, or clay. Georgis & Manweiler (1994) cite the primary references, and Woodring & Kaya (1988) detail the alginate process. The approach is primarily commercial, and the reader is referred to the above publications for further details.

D Short-term and long-term bacterial preservation

Bacterial colonies can be stored at 12–25 °C and routinely subcultured at monthly (12 °C) or twice weekly (25 °C) intervals on NBTA or MacConkey agar (see Appendix) so that colonies reverting to Phase II can be recognized and not used. For long-term storage proceed as follows.

1. Disperse a loopful of Phase I bacterial cells from a 24-h YS agar culture (see Appendix) in 5 ml of nutrient broth containing 17% (v/v) glycerol in Martney bottles or other suitable containers.
2. Store immediately at –70 °C or in liquid nitrogen (Akhurst, personal communication).

For use, the suspension can be rapidly thawed in a water bath at 60 °C prior to use (Bedding, 1984) or rather than thawing, scrape a sample from the Martney bottle with a sterile swab stick (Akhurst, personal communication).

16 SYMBIOTIC BACTERIA

The symbiotic bacteria associated with entomopathogenic nematodes are responsible for the death of the insect host (Akhurst, 1993; Forst & Nealson, 1996). Some of the more important techniques for working with the bacteria are provided here. For more details see Chapter III.

A Isolation techniques for symbiotic bacteria

1. Isolation from surface-sterilized IJs (Akhurst, 1980)

1. Suspend surface-sterilized IJs (see Section 9C3) in yeast salts or nutrient broth.
2. Homogenate suspension in a sterile tissue homogenizer and aseptically transfer to a test tube.
3. Allow bacteria to multiply in the yeast salts or nutrient broth, preferably on a shaker for 24–48 h.

2. Isolation from the haemolymph of infected hosts (Akhurst, 1980)

1. Place 10 last instar *G. mellonella* larvae (or other suitable host) in a Petri dish with moist filter paper and inoculate with ca. 100 IJs/insect.
2. At 12–48 h, surface sterilize the *G. mellonella* larvae either by dipping them into 1.0% sodium hypochlorite and rinse three times in sterile distilled water or by dipping into 95% ethanol, igniting and plunging into sterile distilled water.
3. Open cadaver with a sterile scissors or needle, being careful not to rupture the midgut.
4. Streak a drop of haemolymph on to MacConkey, NBTA or nutrient agar (see Appendix).
5. If no contaminants can be identified after subculture, transfer colony into YS broth and incubate for 1–3 days at 25 °C in the dark.
6. Fill a 2.2 ml cap with 0.3 ml sterile glycerol.
7. Add ca. 1.7 ml bacterial suspension.
8. Store caps at –20 °C.

3. Isolation from insect haemolymph (Poinar & Thomas, 1966)

1. Surface sterilize IJs and *G. mellonella* larvae.
2. Aseptically transfer a drop of haemolymph to a sterile microscope slide cover.
3. Add a few IJs (ca. 5–10) to the drop of haemolymph.
4. Place slide cover on a sterile depression slide so that the drop is suspended in the depression. As the IJs begin to develop and ingest haemolymph, they will release bacterial cells that will rapidly multiply.

B Characterization of Phase I and II variants

1. Adsorption of neutral red from MacConkey agar

1. Streak bacteria on MacConkey agar (see Appendix) so that individual colonies will be formed (Akhurst, 1986a).
2. Incubate at 25–28 °C for 24 h.

The colony morphology of Phase I is granulated, convex, opaque and circular with irregular margins. Colonies also have a "sticky" consistency and red, bright pink or red-brown are Phase I colonies. Cells are small to middle sized and the majority have ovoid and/or rhombic or rectangular inclusion bodies. The antibiotic activity of Phase I is positive. For *P. luminescens*, the bioluminescence in the dark is very strong. They support nematode propagation very well in comparison to Phase II bacteria.

Phase II colonies will be flat, translucent with irregular margins and usually have a greater diameter. Their coloration is usually light yellow, brown or grey, depending on the strain. Phase II bacteria have low or no antibiotic activity. *P. luminescens* does not bioluminesce as strongly as Phase I and may not always be detectable by the human eye. Cells are usually long and inclusion bodies are rarely found.

2. Adsorption of Bromothymol Blue from NBTA

According to Gerritsen & Krasomil-Osterfeld (1994), NBTA is less suitable to compare Phase I and Phase II *Photorhabdus* strains than *Xenorhabdus* strains since several Phase II *P. luminescens* forms show differences in their luminescence intensity.

1. Streak bacteria on NBTA (see Appendix) and obtain formation of individual colonies (Akhurst, 1986a,b; Nishimura *et al.*, 1994).
2. Incubate at 25–28 °C for 3–5 days.

All Phase I colonies are surrounded by cleared zones in the agar because the bromothymol blue (BTB) has been adsorbed. As BTB is adsorbed, Phase I colonies take on a characteristic colour. The exception is *Xenorhabdus poinarii* whose colonies are red for both Phase I and II because they do not adsorb BTB (Akhurst, 1986a).

Xenorhabdus beddingii	blue
X. bovienii	greenish to blue green
X. japonicus	blue
X. nematophilus	blue to deep purple
X. poinarii	red
Photorhabdus luminescens	greenish with reddish-brown centres.

Phase I absorbs and reduces triphenyltetrazolium chloride (TTC), so some reddish colour will sometimes be evident even when reading the test at 3–5 days. Phase II will be shaded from red to rust from adsorption and reduction of TTC. There will be no cleared zones in the agar plates around Phase II bacteria because BTB is not adsorbed. Morphology of Phase I and II in NBTA is the same as that described for MacConkey agar.

REFERENCES

Akhurst, R. J. (1980) Morphological and functional dimorphism in *Xenorhabdus* spp., bacteria symbiotically associated with the insect pathogenic nematodes *Neoaplectana* and *Heterorhabditis*. *J. Gen. Microbiol.* **121**, 303–309.

Akhurst, R. J. (1986a) *Xenorhabdus nematophilus* subsp. *poinarii*: its interaction with insect pathogenic nematodes. *Syst. Appl. Microbiol.* **8**, 142–147.

Akhurst, R. J. (1986b) *Xenorhabdus nematophilus* subsp. *beddingi* (Enterobacteriaceae): a new subspecies of bacteria mutualistically associated with entomopathogenic nematodes. *Int. J. Syst. Bacteriol.* **36**, 454–457.

Akhurst, R. J. (1993) Bacterial symbionts of entomopathogenic nematodes – the power behind the throne. In *Nematodes and the biological control of insect pests.* (eds R. Bedding, R. Akhurst, & H. Kaya). pp. 127–135. CSIRO Publications, East Melbourne, Victoria.

Akhurst, R. J. & Bedding, R. A. (1986) Natural occurrence of insect pathogenic nematodes (Heterorhabditidae and Steinernematidae) in soil in Australia. *J. Aust. Entomol. Soc.* **25**, 241–244.

Ayoub, S. M. (1980) *Plant Nematology an Agricultural Training Aid.* NemaAid Publications, Sacramento, CA.

Barker, K. R. (1985) Nematode extraction and bioassays. In *An advanced treatise on* Meloidogyne, *Volume II Methodology* (eds K. R. Barker, C. C. Carter & J. N. Sasser), pp. 20–35. North Carolina State University Graphics, Raleigh, NC.

Barker, K. R., Carter, C. C., & J. N. Sasser, J. N. (1985) *An advanced treatise on* Meloidogyne, Volume II *Methodology.* North Carolina State University Graphics, Raleigh, NC.

Bedding, R. A. (1981) Low cost *in vitro* mass production of *Neoaplectana* and *Heterorhabditis* species (Nematoda) for field control of insect pests. *Nematologica* **27**, 109–114.

Bedding, R. A. (1984) Large-scale production, storage and transport of the insect-parasitic nematodes *Neoaplectana* spp. and *Heterorhabditis*. *Ann. Appl. Biol.* **101**, 117–120.

Bedding, R. A. & Akhurst, R. J. (1975) A simple technique for the detection of insect parasitic rhabditid nematodes in soil. *Nematologica* **21**, 109–110.

Brady, N. C. (1990) *The nature and properties of soils*, 8th edn. Macmillian Publishing Co., New York.

Brenner, S. (1974) The genetics of *Caenorhabditis elegans*. *Genetics* **77**, 71–94.

Byrd, D. W., Jr, Barker, K. R., Ferris, H., Nusbaum, C. J., Griffin, W. E., Small, R. H. & Stone, C. A. (1976) Two semi-automatic elutriators for extracting nematodes and certain fungi from soil. *J. Nematol.* **8**, 206–212.

Campbell, J. F., Lewis, E., Yoder, F. & Gaugler, R. (1995) Entomopathogenic nematode (Heterorhabditidae and Steinernematidae) seasonal population dynamics and impact on insect populations in turfgrass. *Biol. Control* **5**, 598–606.

Choo, H. Y., Kaya, H. K., Shea, P. & Noffsinger, E. M. (1987) Ecological study of nematode parasitism in *Ips* beetles from California and Idaho. *J. Nematol.* **19**, 495–502.

Choo, H. Y., Kaya, H. K. & Kim, J. B. (1989) *Agamermis unka* (Mermithidae) parasitism of *Nilaparvata lugens* in rice fields in Korea. *J. Nematol.* **21**, 254–259.

Choo, H. Y., Kaya, H. K. & Kim, H. H. (1995) Biological studies on *Agamermis unka* (Nematoda: Mermithidae), a parasite of the brown planthopper *Nilaparvata lugens*. *Biocontrol Sci. Technol.* **5**, 209–223.

Cobb, N. A. (1917) Notes on nemas. *Contribution Science Nematology* no. **5**, 117–128.

Creighton, C. S. & Fassuliotis, G. (1981) A laboratory technique for culturing *Filipjevimermis leipsandra*, a nematode parasite of *Diabrotica balteata* (Insecta: Coleoptera) *J. Nematol.* **13**, 226–227.

Curran, J., Gilbert, C. & Buttler, K. (1992) Routine cryopreservation of isolates of *Steinernema* and *Heterorhabditis* spp. *J. Nematol.* **24**, 269–270.

Curran, J. & Heng, J. (1992) Comparison of three methods for estimating the number of entomopathogenic nematodes present in soil samples. *J. Nematol.* **24**, 170–176.

Dix, I. (1994) Use of second generation *Heterorhabditis* females for genetic crosses. In *Genetics of entomopathogenic nematodes-bacterium complexes* (eds A. M. Burnell, R. U. Ehlers & J. P. Masson). Biotechnology Proceedings of a Symposium and Workshop, COST 812, pp. 190–193. St. Patrick's College, Maynooth, Co. Kildare.

Dunphy, G. B. & Webster, J. M. (1989) The monoxenic culture of *Neoaplectana carpocapsae* DD-136 and *Heterorhabditis heliothidis*. *Rev. Nematol.* **12**, 113–123.

Downing, J. A. & Rigler, F. H. (1984) *A manual on the methods for the assessment of secondary production in freshwaters*. Blackwell, Oxford.

Fan, X. & Hominick, W. M. (1991) Efficiency of the *Galleria* (wax moth) baiting technique for recovering infective stages of entomopathogenic rhabditids (Steinernematidae and Heterorhabditidae) from sand and soil. *Rev. Nematol.* **14**, 381–387.

Fawcett, R. G. & Collis-George, N. (1967) A filter-paper method for determining the moisture characteristics of soil. *Aust. J. Exp. Agric. Anim. Husb.* **7**, 162–167.

Filipjev, I. N. & Schuurmans-Stekhoven, J. H., Jr. (1959) *Agricultural Helminthology*. Brill, Leiden.

Forst, S. & Nealson, K. (1996) Molecular biology of the symbiotic-pathogenic bacteria *Xenorhabdus* spp. and *Photorhabdus* spp. *Microbiol. Rev.* **60**, 21–43.

Friedman, M. J. (1990) Commercial production and development, In *Entomopathogenic nematodes in biological control* (eds R. Gaugler & H. K. Kaya), pp. 153–172. CRC Press, Boca Raton, FL.

Geden, C. J. & Stoffolano, J. G., Jr (1984) Nematode parasites of other Dipterans. In *Plant and insect nematodes* (ed. W. R. Nickle), pp. 849–898. Marcel Dekker, New York.

Georgis, R., Kaya, H. K. & R. Gaugler, R. (1991) Effect of steinernematid and heterorhabditid nematodes (Rhabditida: Steinernematidae and Heterorhabditidae) on nontarget arthropods. *Environ. Entomol.* **20**, 815–822.

Georgis, R. & Manweiler, S. A. (1994) Entomopathogenic nematodes: a developing biological control technology. In *Agricultural zoology reviews*, Vol. 6 (ed. K. Evans), pp. 63–94. Intercept, Andover.

Gerritsen, L. J. M. & Krasomil-Osterfeld, K. C. (1994) Characterization of *Xenorhabdus* and *Photorhabdus* strains and form variants. In *Genetics of entomopathogenic nematodes–bacterium complexes* (eds A. M. Burnell, R. U. Ehlers & J. P. Masson), Biotechnology Proceedings of a Symposium and Workshop, COST 812, pp. 194–203. St. Patrick's College, Maynooth, Co. Kildare.

Giblin-Davis, R. (1993) Interactions of nematodes with insects. In *Nematode interactions* (ed. M. W. Khan), pp. 302–344. Chapman & Hall, London.

Götz, P., Boman, A. & Boman, H. G. (1981) Interaction between insect immunity and an insect-pathogenic nematode with symbiotic bacteria. *Proc. R. Soc. Lond. B* **212**, 333–350.

Grewal, P. S. (1992) Laboratory techniques for studying insect-parasitic rhabditid nematodes. In *Recent advances in nematology* (ed. B. K. Dwivedi), pp. 51–66. Bioved Research Society, Allahabad.

Griffin, C. T., Moore, J. F. & Downes, M. J. (1991) Occurrence of insect-parasitic nematodes (Steinernematidae, Heterorhabditidae) in the Republic of Ireland. *Nematologica* **37**, 92–100.

Griffin, C. T., Joyce, S. A., Dix, I., Burnell A. M. & Downes, M. J. (1994) Characterization of the entomopathogenic nematode *Heterorhabditis* (Nematoda: Heterorhabditidae) from Ireland and Britain by molecular and cross-breeding techniques, and the occurrence of the genus in these islands. *Fund. Appl. Nematol.* **17**, 245–253.

Hamblin, A. P. (1981) Filter-paper method for routine measurement of field water potential. *J. Hydrol.* **53**, 355–360.

Hara, A. H., Gaugler, R., Kaya, H. K. & LeBeck, L. M. (1991) Natural populations of entomopathogenic nematodes (Rhabditida: Steinernematidae and

Heterorhabditidae) from the Hawaiian Islands. *Environ. Entomol.* **20**, 211–216.
Hara, A. H., Lindegren, J. E. & Kaya, H. K. (1981) Monoxenic mass production of the entomogenous nematode, *Neoaplectana carpocapsae* Weiser on dog food/agar medium. Science and Education Administration, Advances in Agricultural Technology, Western Series No. 16.
Holme, N. A. & McIntyre, A. O. (1971) *Methods for the study of marine benthos*. IBP Handbook no. 16. Blackwell, Oxford.
Hominick, W. & Briscoe, B. R. (1990a) Survey of 15 sites over 28 months for entomopathogenic nematodes (Rhabditida: Steinernematidae). *Parasitology* **100**, 289–294.
Hominick, W. & Briscoe, B. R. (1990b) Occurrence of entomopathogenic nematodes (Rhabditida: Steinernematidae and Heterorhabditidae) in British soils. *Parasitology* **100**, 295–302.
Hooper, D. J. & Evans, K. (1993) Extraction, identification and control of plant parasitic nematodes. In *Plant parasitic nematodes in temperate agriculture* (eds K. Evans, D. L. Trudgill & J. M. Webster), pp. 1–59. CAB International, Wallingford.
Jenkins, W. R. (964) A rapid centrifugal-flotation technique for separating nematodes from soil. *Plant Dis. Rep.* **48**, 692.
Jones, C. M. & Perdue, J. M. (1967) *Heterotylenchus autumnalis*, a parasite of the face fly. *J. Econ. Entomol.* **60**, 1393–1395.
Kaya, H. K. (1993) Entomogenous and entomopathogenic nematodes in biological control. In *Plant parasitic nematodes in temperate agriculture* (eds K. Evans, D. L. Trudgill & J. M. Webster). pp. 565–591. CAB International, Wallingford.
Kaya, H. K. & Gaugler. R. (1993) Entomopathognic nematodes. *Annu. Rev. Entomol.* **38**, 181–206.
Kaya, H. K. & Koppenhöfer, A. M. (1996) Effects of microbial and other antagonistic organisms and competition on entomopathogenic nematodes. *Biocontrol Sci. Technol.* **6**, 357–371.
Kaya, H. K. & Moon, R. D. (1978) The nematode *Heterotylenchus autumnalis* and face fly *Musca autumnalis*: a field study in northern California. *J. Nematol.* **10**, 333–341.
Kaya, H. K., Moon, R. D. & Witt, P. L. (1979) Influence of the nematode, *Heterotylenchus autumnalis*, on the behavior of the face fly, *Musca autumnalis*. *Environ. Entomol.* **8**, 537–540.
Kung, S-P., Gaugler, R. & Kaya, H. K. (1990) Soil type and entomopathogenic nematode persistence. *J. Invertebr. Pathol.* **55**, 401–406.
Lotrario, J. B., Stuart, B. J., Lam, T., Arands, R. R., O'Connor, O. A. & Kosson, D S. (1995) Effects of soil: implications for sorption isotherm analyses. *Bull. Environ. Contam. Toxicol.* **54**, 668–675.
Maggenti, A. R. (1991) Nemata: higher classification. In *Manual of agricultural nematology* (ed. W. R. Nickle). pp. 147–187. Marcel Dekker, New York.
Mauleon, H., Briand, S., Laumond, C. & Bonifassi, E. (1993) Utilisation d'enzyme digestives pour l'etude du parasitisme des *Steinernema* et *Heterorhabditis* envers les larves d'insectes. *Fundam. Appl. Nematol.* **16**, 185–191.
Miller, R. W. (1989) Novel pathogenicity assessment technique for *Steinernema* and *Heterorhabditis* entomopathogenic nematodes. *J. Nematol.* **21**, 574.
Mullens, B. A. & Velten, R. K. (1994) Laboratory culture and life history of *Heleidomermis magnapapula* in its host, *Culicoides variipennis* (Diptera: Ceratopogonidae). *J. Nematol.* **26**, 1–10.
Nickle, W. R. (1984) *Plant and insect nematodes*. Marcel Dekker, New York.
Nickle, W. R. (1991) *Manual of agricultural nematology*. Marcel Dekker, New York.
Nickle, W. R. & MacGowan, J. B. (1992) Grenacher's borax carmine for staining nematodes inside insects. *J. Helminthol. Soc. Wash.* **59**, 231–233.
Nishimura, Y., Hagiwara, A., Suzuki, T. & Yamanaka, S. (1994) *Xenorhabdus japonicus* sp. nov. associated with the nematode *Steinernema kushidai*. *World J. Microbiol. Biotechnol.* **10**, 207– 210.
Ogura, N. (1993) A method to produce neutral-red-labelled infective juveniles of *Steinernema kushidai*. *Jpn. J. Nematol.* **23**, 37–38.
Ogura, N. & Haraguchi, N. (1993) Xenic culture of *Steinernema kushidai* (Nematoda: Steinernematidae) on artificial media. *Nematologica* **39**, 266–273.
Peloquin, J. J. & Platzer, E. G. (1993) Control of root gnats (Sciaridae: Diptera) by *Tetradonema plicans* Hungerford (Tetradonematidae: Nematoda) produced by a novel culture method. *J. Invertebr. Pathol.* **62**, 79–86.
Petersen, J. J. (1984) Nematode parasitism of mosquitoes. In *Plant and Insect Nematodes* (ed. W. R. Nickle), pp. 797–820. Marcel Dekker, New York.
Petersen, J. J. (1985) Nematodes as biological control agents: Part I. Mermithidae. *Adv. Parasitol.* **24**, 307–344.
Petersen, J. J. & Willis, O. R. (1972) Procedures for the mass rearing of a mermithid parasite of mosquitoes. *Mosq. News.* **32**, 226–230.
Poinar, G. O., Jr (1975) *Entomogenous nematodes*. E. J. Brill, Leiden.
Poinar, G. O., Jr (1979) *Nematodes for biological control of insects*. CRC Press, Boca Raton, FL.
Poinar, G. O., Jr (1991) Nematoda and Nematomorpha. In *Ecology and classification of North American freshwater invertebrates* (eds J. N. Thorp & A. P. Covich), pp. 249–283. Academic Press, San Diego, CA.
Poinar, G. O., Jr & Gyrisco, G. G. (1964) Studies on the bionomics of *Hexamermis arvalis* Poinar and Gyrisco, a mermithid parasite of the alfalfa weevil, *Hypera postica* (Gyllenhal). *J. Insect Pathol.* **4**, 469–483.
Poinar, G. O., Jr & Thomas, G. M. (1966) Significance of *Achromobacter nematophilus* sp. nov. (Achromobacteriaceae: Eubacteriales) associated with a nematode. *Int. Bull. Bacteriol. Nomencl. Taxon.* **15**, 249–252.

Poinar, G. O., Jr & Thomas, G. M. (1984) *Laboratory guide to insect pathogens and parasites*. Plenum Press, New York.

Poinar, G. O., Jr & van der Laan, P. A. (1972) Morphology and life history of *Sphaerularia bombi*. *Nematologica* **18**, 239–252.

Poinar, G. O., Jr, Lane, R. S. & Thomas, G. M. (1976) Biology and redescription of *Pheromermis pachysoma* (V. Linstow) n. gen. comb. (Nematoda: Mermithidae), a parasite of yellowjackets (Hymenoptera: Vespidae). *Nematologica* **22**, 360–370.

Popiel, I. & Vasquez, E. M. (1991) Cryopreservation of *Steinernema carpocapsae* and *Heterorhabditis bacteriophora*. *J. Nematol.* **23**, 432–437.

Remillet, M. & Laumond, C. (1991) Sphaerularioid nematodes of importance in agriculture. In *Manual of agricultural nematology* (ed. W. R. Nickle). pp. 967–1024. Marcel Dekker, New York.

Rogers, C. E., Marti, O. G., Jr & Simmons, A. M. (1993) *Noctuidonema guyanense* (Nematoda: Aphelenchoididae): host range and pathogenicity to the fall armyworm, *Spodoptera frugiperda* (Lepidoptera; Noctuidae). In *Nematodes and the biological control of insect pests* (eds R. Bedding, R. Akhurst & H. Kaya), pp. 27–32. CSIRO Publications, East Melbourne, Victoria.

Rueda, L. M., Osawaru, S. O., Georgi, L. L. & Harrison, R. E. (1993). Natural occurrence of entomogenous nematodes in Tennessee nursery soils. *J. Nematol.* **25**, 181–188.

Saunders, M. C. & All, J. N. (1982) Laboratory extraction methods and field detection of entomophilic rhabditoid nematodes from soil. *Environ. Entomol.* **11**, 1164–1165.

Seinhorst, J. W. (1959) A rapid method for the transfer of nematodes from fixative to anhydrous glycerin. *Nematologica* **4**, 117–128.

Seinhorst, J. W. (1966) Killing nematodes for taxonomic study with hot F. A. 4 : 1. *Nematologica* **12**, 178.

Skipper, H. D. & Westerman, D. L. (1973) Comparative effects of propylene oxide, sodium azide, and autoclaving on selected soil properties. *Soil Biol. Biochem.* **5**, 409–414.

Southey, J. F. (1986) *Laboratory methods for work with plant and soil nematodes*, 6th edn. Ministry of Agriculture, Fisheries and Food. Her Majesty's Stationary Office, London.

Southwood, T. R. E. (1992) *Ecological methods*, 2nd edn. Chapman & Hall, London.

Stock, S. P., Choo, H. Y. & Kaya, H. K. (1996) An entomopathogenic nematode, *Steinernema monticolum* sp. n. (Rhabditida: Steinernematidae) from Korea with a key to other species. *Nematologica*. in press.

Studdert, J. P., Kaya, H. K. & Duniway, J. M. (1990) Effect of water potential, temperature, and clay-coating on survival of *Beauveria bassiana* conidia in a loam and peat soil. *J. Invertebr. Pathol.* **55**, 417–427.

Triantaphyllou, A. C. & McCabe, E. (1989) Efficient preservation of root-knot and cyst nematode in liquid nitrogen. *J. Nematol.* **21**, 423–426.

Webster, J. M. & Thong, C. H. S. (1984) Nematode parasites of Orthopterans. In *Plant and insect nematodes* (ed. W. R. Nickle), pp. 697–726. Marcel Dekker, New York.

Wheeler, W. M. (1928) *Mermis* parasitism and intercastes among ants. *J. Exp. Zool.* **50**, 165–237.

White, G. F. (1927) A method for obtaining infective nematode larvae from cultures. *Science* **66**, 302–303.

White, W. & Wharton, K. L. (1984) Development of a cryogenic preservation system. *Am. Lab.* June, 65–76.

Whitehead, A. G. & Hemming, J. R. (1965) A comparison of some quantitative methods of extracting small veriform nematodes from soil. *Ann. Appl. Biol.* **55**, 25–38.

Woodring, J. L. & Kaya, H. K. (1988) *Steinernematid and heterorhabditid nematodes: a handbook of techniques*. Southern Cooperative Series Bulletin. 331. Arkansas Agricultural Experiment Station, Fayetteville, AK. 30 pp.

Yang, H. & Jian, H. (1988) Labelling living entomopathogenic nematodes with stains. *Chin. J. Biol. Control* **4**, 59–61. (In Chinese, English Abstract.)

Zukerman, B. M., Mai, W. F. & Harrison, M. B. (1985) *Plant nematology laboratory manual*. University of Massachusetts Agricultural Experiment Station. Amherst, MA.

GLOSSARY

Adanal. Situated in the proximity of the anus.

Adanal genital papillae. The adanal supplements.

Adult. A sexually mature individual. That stage following the 4th (juvenile) and final moult.

Allotype. A paratype of the opposite sex of the specimen designated as the holotype. (See holotype.)

Amphid. Paired lateral sense organs which generally open to the exterior on or near the lip region; variable in size and shape according to taxa.

Amphid opening. The opening or aperture leading to the pouch of the amphid through which stimuli are received.

Amphidelphic. Pertaining to uteri opposed; position and direction of the uteri, not the ovary.

Amphidial pouch. The anterior cavity or chamber of the amphid that contains the sensilla.

Amphimixis. The union of two gametes in sexual reproduction, as opposed to automixis.

Anal opening. The orifice or aperture to the exterior at the terminus of the rectum and delimited by the anus.

Anus. The extremity of the rectum. The posterior opening of the alimentary canal.
Arcuate. Curved like a bow. Arched.
Automixis. Obligatory self-fertilization; egg and sperm being derived from the same individual, as opposed to amphimixis.
Basal bulb. An enlargement of the oesophageal wall, muscular or glandular, at the posterior of the oesophagus.
Body cavity. The hollow within the body that contains the internal organs. The principal cavity between the body wall and internal organs of an organism. See pseudocoel.
Buccal aperture. The oral opening.
Bursa. Wing-like extensions of the lateral cuticle at the caudal end of the male.
Calamus. The shaft of the spicule.
Capitulum. Head or manubrium of the spicules. In the gubernaculum, a sclerotized guiding piece on the ventral cloacal surface.
Caudal. Belonging to or like a tail. Situated on or near the tail.
Caudal alae. Lateral cuticular extensions on the posterior end of male nematodes; the bursa.
Caudal papillae. In mermithids the terminal portion of the larval tail; papillae located on the tail.
Cephalic. Belonging to or situated in, on, or near the anterior end.
Cephalic papillae. Papillae of the outer circlet about the stoma.
Cheilorhabdions. The cuticularized walls of the cheilostom.
Cheilostom. The lip cavity of the stoma, delimited anteriorly by the oral aperture, posteriorly by the prostom.
Cloaca. In the male a common chamber lined with cuticle that receives the products of the intestinal and reproductive tracts and empties to the exterior via the anus.
Corpus. The most anterior segment of the oesophagus, of cylindrical form, and of moderate width. In some cases, the corpus can be divide into procorpus, mesocorpus and metacorpus.
Cross-breeding. To hybridize.
Cross-section. A transverse cut.
Crura. Strongly cuticularized longitudinal pieces that strengthen the lamina of the gubernaculum.
Cuticle. The non-cellular external covering of the body wall of nematodes and other invertebrates.
Dauer juvenile. A quiescent stage entered by juveniles while enclosed in the cast cuticle of the previous stage. In steinernematids, the cast cuticle (sheath) is easily lost.
Denticle. Small, tooth-like projections.
Epiptyma. Anterior and posterior cuticular flaps associated with the vulval opening of some nematodes.
Excretory duct. A tube or canal by which the excretory products of the excretory system are conveyed to the exterior via the excretory pore.
Excretory pore. A ventral opening in the cuticle by which the waste products of the excretory system are emitted to the exterior.
Filiform. Having the shape of a thread or filament.
Flaccid. Limp, limber.
Fossoria. The cheilostomal, outwardly movable teeth.
Gamogenetic. Sexual reproduction
Generation. The length of time from any given stage in the life cycle of an organism to the same stage in the offspring. The period of time in which one set of progeny follows another.
Genital papillae. Tactile papillae or setae in the anal region of the male. They may be pre-anal, adanal or postanal in position.
Gonad. A primary sex gland. An ovary, ovotestis, or testis.
Gravid. Bearing eggs.
Gubernaculum. A grooved cuticularized structure in male nematodes that guides the spicules.
Hemizonid. A nerve commissure from the nerve ring that is highly refractive at the point it joins the ventral nerve cord near the excretory pore.
Hermaphrodite. An individual with both functional male and female reproductive organs.
Holotype. The individual specimen selected by the author as type of a species.
Host. The organism that is invaded or parasitized by a disease-producing agent and from which the parasite obtains its nourishment.
Infect. To invade and establish a parasitic relationship within the proper host.
Infective. Having the qualities necessary to enter a host and produce a disease condition.
Infective juvenile. The period of development in the life cycle of a parasitic nematode in which it possesses the qualities enablingit to infect a host.
Inner labial papilla. Sensory papilla situated toward the apex of the lips forming a circlet of six in primitive forms.

Intestine. A simple tube composed of a single layer of epithelial cells in which digestion of food takes place. Gut.
Isthmus. The segment of the oesophagus between the medium bulb, a narrow section of the oesophagus.
Juvenile. The immature or larval stages. To avoid confusion with immature insect stages, juvenile is the preferred term.
Labial papilla. Papilla located on the lips.
Lamina. The main body of the male spicule or gubernaculum; the blade.
Lip region. The cuticular area from the basal ring forward.
Longitudinal cords. The longitudinal thickenings of the hypodermis.
Longitudinal striation. In the cuticle a groove parallel to the longitudinal axes.
Manubrium. The enlarged cephalated proximal portion of the spicule.
Median bulb. See metacorpus.
Mesorhabdions. The walls of the mesostom.
Mesostom. A subdivision of the prostom as distinguished by clefts in the rhabdions, located posteriorly to the prostom and anteriorly by the metastom.
Metacorpus. The posterior subdivision of the corpus taking an ovate form and being preceded by a cylindrical anterior segment. The median bulb.
Metarhabdions. The walls of the metastom.
Metastom. The posterior subdivision of the prostom as distinguished by clefts in the rhabdions.
Monorchic. Having one testis.
Monoxenic. Pertaining to rearing of an organism with only one known species of associated organism.
Morphology. The study of form and structure of organisms.
Morphometry. Measurement of external form.
Mucron. A small knob-like or spine-like ending on a terminus.
Nerve ring. Any ring of nerve fibres, may be around the mouth, oesophagus, etc. The nerve ring represents the dorsal and ventral connections between the lateral ganglia.
Oesophageal glands. Elongated glands of simple or branched tubules located in the oesophageal sectors, the secretions of which are of enzymatic nature.
Oesophagus. The tube that leads the stoma or stylet base to the intestine
Outer labial papilla. Sensory papilla situated distally on each lip segment forming a circlet of six in primitive forms. This circlet is considered cephalic by some authors.
Ovary. The reproductive gland of the female that produces eggs.
Oviduct. A tube which serves for the passage of eggs from the ovary to the uterus.
Oviparous. Producing eggs that hatch after expulsion from the body.
Ovotestis. Hermaphroditic reproductive gland; an organ that produces both spermatozoa and ova at the same or at different periods of the life cycle.
Ovoviviparous. Producing eggs that are incubated and hatched within the uterus.
Papilla. Nipple-like or pimple-like projections from the cuticle; simple sensory organ.
Paratype. All specimens remaining after the selection of the holotype and allotype and so designated by the original author.
Parthenogenetic. The development of an individual from an egg without fertilization.
Peloderan. Type of bursa where the caudal alae (caudal expansions of the cuticle) do meet completely at the tip of the male's tail. Pertaining to caudal alae that meet posterior to the male's tail.
Phasmid. Paired postanal lateral sensory organs.
Phoretic. A form of symbiotic relationship when the symbiont, the phoront, is mechanically carried about by its host; neither being physiologically dependent on the other.
Postcorpus. Posterior region of the oesophagus where the oesophageal glands are located.
Procorpus. The anterior subdivision of the corpus taking a cylindrical form and generally being terminated by an oval posterior segment.
Prodelphic. Uteri parallel and anteriorly directed.
Prorhabdion. The wall of the prostom.
Prostom. The anterior subdivision of the protostom.
Protostom. In rhabditoid-like nematodes, the cylindrical midportion of the stoma, delimited anteriorly by the cheilostom and posteriorly by the telestom.
Pseudocoel. A body cavity not lined with a mesodermal epithelium.
Pseudopeloderan. Type of bursa where the caudal alae (caudal expansions of the cuticle) do not meet completely at the tip of the male's tail.
Rectum. The hindgut. A narrow tube flattened in dorsoventral direction and separated from the intestine by a sphincter muscle.
Rhabdion. The wall of the stoma.
Rostrum. Distal projection of the spicule.

Sclerotized. Hardened, dense, refractive parts of the nematode body. Cuticularized.
Septicemia. A morbid condition caused by pathogenic bacteria and their toxin within the host's body.
Setaceous. Bristlelike, slender.
Shaft. The main body of the spicule, usually curved and bearing a thin flange of membranous extensions termed velum.
Spermatheca. The enlarged portion of the female reproductive system which functions as a reservoir in receiving and holding sperm from the male.
Sperm duct. The vas deferens.
Spicule. Male intromittent structures, which are extrusible through the cloacal opening and functioning during copulation for the transfer of sperm.
Stichocyte. An individual cell of a stichosome.
Stichosome. A longitudinal series of cells (stichocytes) that form the posterior oesophageal glands of some nematodes.
Stoma. The mouth cavity; it is the segment of the digestive tract between the oral opening and the oesophagus including oral aperture and the stoma walls.
Stylet. A relatively long, rather slender, hollow feeding structure. A spear.
Testis. The reproductive organ of the male that produces the spermatozoa.
Telostom. A short valve that connects the stoma with the oesophagus.
Trophosome. A modification of the intestine in which walls and lumen disappear resulting in a syncytic structure.
Type. A zoological specimen that serves as the base for the name of a taxon.
Type host. The designated organism from which a type specimen has been collected.
Type locality. The place from which a type specimen has been collected.
Type specimen. A single example of a nematode designated as a type individual.
Uterus. A region of the oviduct modified to function as a place of development of the eggs.
Vagina. The terminal portion of the female reproductive tract that opens to the outside.
Vector. An organism which transmits or disseminates pathogens or other organisms.
Velum. Delicate membranous extensions on the inner side of the spicule of some male nematodes.
Vulva. The female genital opening and its delimiting margins of the cuticle.
Water potential. The chemical potential of water per unit volume with the dimensions of pressure.

APPENDIX

Media

1. A Dog food/agar medium for entomopathogenic nematodes (Hara *et al.*, 1981)

100 g of pulverized dry dog food
500 ml of dissolved 1% agar

1. Mix dog food and dissolved agar by hand or in a blender.
2. Autoclave the dog food/agar mixture at 121 °C for 20 min.
3. Pour aseptically into sterile Petri dishes. (The dog food/agar mixture can also be poured into test tubes and after autoclaving made slanted to provide sufficient surface area for the nematodes to grow.)

A variation of the medium by Ogura & Haraguchi (1993) is powdered dog food (8.8%), peptone (1.2%), agar (0.2%), and Sörensen phosphate buffer (33 mM KH_2PO_4; 33 mM Na_2HPO_4; 7 : 3) (pH 6.5).

2. Offal medium for entomopathogenic nematodes (Bedding, 1981, 1984)

The method outlined below is for a 500-ml wide-mouth Erlenmeyer flask.

1. Obtain chicken, duck or turkey offal from local slaughterhouses; discard the gall bladder as the bile salts are detrimental to nematode production.
2. Cook offal to soften (autoclaving 10 l of the offal at 121 °C for 20–25 min is effective) and place into a blender for several minutes.
3. For *Steinernema* species, mix 8 parts of offal homogenate to 2 parts of water. For *Heterorhabditis* species, mix 7 parts of homogenate to 2 parts of water and 1 part of beef fat.
4. Impregnate small pieces (<1 cm diameter) of foam with the offal homogenate. The best ratio is 12.5 parts medium to 1 part foam by weight. The pores of the foam should still be clearly visible, but medium should ooze out when the foam is

squeezed. (Wearing rubber gloves during this procedure is highly recommended.)

5. Fill the flask with foam homogenate mixture to the 250–300 ml mark (about 100 g).
6. Wipe the mouth of the flasks well.
7. Plug with non-absorbent cotton wrapped in cheesecloth.
8. Autoclave for 20 min at 121 °C.

3. Lipid agar plates (modified from Dunphy & Webster, 1989):

10 g corn syrup
5 g yeast extract
25 g nutrient agar
2.5 ml cod liver oil
2 g $MgCl_2 \cdot 6H_2O$
1 litre distilled H_2O

Autoclave at 121 °C for 15 min. Pour aseptically into Petri dishes.

4. M9 Buffer (Brenner, 1974)

3 g KH_2PO_4
6 g Na_2HPO_4
5 g NaCl
1 ml $MgSO_4$ (1 M)
1 litre distilled H_2O

5. Media for symbiotic bacteria associated with entomopathogenic nematodes (Akhurst, 1980, 1986a,b; Götz *et al.*, 1981; Nishimura *et al.*, 1994; Woodring & Kaya, 1988)

Most media are commercially available and separate ingredients do not have to be purchased (e.g. nutrient agar, nutrient broth and MacConkey agar).

Nutrient agar
3 g beef extract
5 g peptone
8 g NaCl, 15 g agar
Nutrient broth is without the agar.

1. Mix 31 g of the prepared ingredients into 1 l distilled water.
2. Heat in a boiling water bath until agar is dissolved. (With nutrient broth, heat in boiling water and dispense into test tubes; omit step 4 below)
3. Autoclave at 121 °C for 15 min.
4. Aseptically dispense into sterile Petri dishes when cool but not yet solid.

MacConkey agar
17 g peptone
3 g proteose peptone
10 g lactose
1.5 g bile salts no. 3
5 g NaCl
13.5 g agar
0.03 g neutral red
0.001 g crystal violet
Follow protocol for nutrient agar by using 50 g of the prepared material in 1 l of distilled water.

Yeast Salts or YS Broth (Akhurst,1980)
0.5 g K_2HPO_4
0.5 g $NH_4H_2PO_4$
0.2 g $MgSO_4 \cdot 7H_2O$
5.0 g Yeast extract
1 l distilled H_2O

Dispense into tubes or flasks and autoclave at 121 °C for 15 min. To make Y/S agar, add 15 g agar, melt agar in water bath and then autoclave. Dispense into sterile Petri dishes.

NBTA (Akhurst, 1980, 1986a)
1 litre nutrient agar
0.04 g triphenyltetrazolium chloride (TTC)
0.025 g Bromothymol Blue (BTB)

1. Mix nutrient agar and BTB.
2. Melt agar in water bath.
3. Autoclave at 121 °C for 15 min.
4. Add the TTC, just before pouring into dishes. If TTC stock is aseptic, carefully weigh out TTC into sterile container and pour it into the autoclaved medium (<50 °C). If sterility of TTC is in question, add TTC to 2–3 ml sterile distilled water and run through a millipore filter (0.2 μm) before adding to the cooled medium (<42 °C). TTC will break down if added when agar is too hot.
5. Swirl to mix well.
6. Dispense into sterile Petri dishes.

Medium X (Götz *et al.*, 1981)
4 g bacteriological peptone
5 g NaCl
4 g glucose
15 g agar
1 M Tris buffer, at pH 7.4

1. Place dry ingredients into a 2 l flask
2. Add 1 M Tris buffer to make 1 l of suspension.

3. Melt agar in water bath.
4. Autoclave at 121 °C for 15 min.
5. Dispense into sterile Petri dishes.

Stains and fixatives

1. Grenacher's borax carmine stain (Nickle & MacGowan, 1992)

1 g carmine
2 g borax
50 ml H_2O

1. Boil the ingredients in a covered vessel for 30 min or until the carmine dissolves.
2. Add 50 ml of 70% ethyl alcohol (EtOH).
3. Allow the solution to stand for 1–2 days and filter through filter paper.
4. Use the filtrate for staining.

Counterstain

1 g Fast green in 99 ml 95% EtOH (1% Fast green)

2. Pepsin solution:

8 g Pepsin
23 g NaCl
20 ml HCl
940 ml distilled water

1. Stir the mixture until the ingredients go into solution.
2. Keep the solution at 4 °C until used, but prolonged storage can inactive the enzyme. If there is any doubt about the freshness of the solution, prepare a new batch or prepare only enough solution needed for the situation.

3. Ringer's solution (Woodring & Kaya, 1988)

9 g NaCl
0.4 g KCl
0.4 g $CaCl_2$
0.2 g NaH_2CO_3
1 litre distilled H_2O

4. Fixatives (Seinhorst, 1959, 1966; Southey, 1986; Woodring & Kaya, 1988)

Numerous fixatives have been recommended for studying different features of nematode anatomy. Double-strength fixatives are prepared using half the amount of water indicated below. The commonly used fixatives are as follows:

TAF

7 ml formalin (40% formaldehyde)
2 ml triethanolamine
91 ml distilled H_2O

F.A. 4 : 1 or F.A. 4 : 10

10 ml formalin
1 or 10 ml glacial acetic acid
up to 100 ml distilled H_2O

F. A. A.

20 ml 95% ethanol
6 ml formalin
1 ml glacial acetic acid
40 ml distilled H_2O

Buffered formalin

1000 ml 4·0% formaldehyde
4 g $NaH_2PO_4 \cdot H_2O$
6.5 g Na_2HPO_4

5. Recipe for hard glycerin jelly (Southey, 1986)

1. Soak 20 g of gelatin in 40 ml of distilled water for 2 h.
2. Add 50 ml of glycerin and 1 ml of phenol.
3. Place in water bath (70–80 °C) for 10–15 min and stir until the mixture is homogeneous.

CHAPTER VII

Testing the pathogenicity and infectivity of entomopathogens to mammals

JOEL P. SIEGEL

Center for Economic Entomology, 607 East Peabody Drive, 172 Natural Resource Building, Champaign, IL 61820, USA

1 INTRODUCTION

Initially, one might be surprised to find a chapter on conducting mammalian safety testing of entomopathogens in this manual. After all, there is a long chain of events between the isolation of an entomopathogen from an insect cadaver and its subsequent identification, to the production of that isolate on a commercial scale. Furthermore, it may seem a matter of common sense that known vertebrate pathogens will not be used in agriculture or in vector control, so that the need for these tests is questionable. However, if an entomopathogen is going to be mass produced, either in a government-sponsored laboratory or by industry, and/or disseminated over large areas, safety concerns will inevitably be expressed. Safety testing is one way to produce the data needed to address these concerns. Although in many countries safety testing is the province of specialized laboratories, in the long run an understanding of the philosophy behind these tests will facilitate communication between the investigator and the testing laboratory.

Mammalian safety screens are a subset of a larger grouping of tests that assess the effects of a microbial pest control agent (MPCA) on non-target organisms (NTO). Non-target organisms include plants, fish, beneficial arthropods, birds and mammals. All of these screens cannot be covered in this chapter due to space constraints. However, there is a shared philosophy behind many of these protocols and the same principles used in evaluating the potential infectivity of an MPCA to mammals can be applied to testing infectivity in birds and fish. Ultimately, NTO testing can be viewed as attempts to manipulate a candidate organism into doing something it would not do in nature, either by providing access to hosts outside its natural range or by varying both the dose and route of exposure in order to produce infection and/or mortality.

When conducting mammalian safety screens, one expects that all of the infectivity and the majority of the toxicity tests will be negative because of the specificity of entomopathogens to arthropods. In the most favourable testing scenario, all of the mammalian safety tests produce negative data and a relatively straightforward decision to proceed can be

MANUAL OF TECHNIQUES IN INSECT PATHOLOGY
ISBN 0-12-432555-6

made. However, if mortality occurs after an invasive route of administration, its significance must be judged in the context of the experimental protocol. An adverse outcome in any one screen does not automatically mean that a candidate is unsafe. Usually, findings of toxicity necessitate further tests to quantify the effect. It is my goal in this chapter to summarize briefly the history and philosophy of the safety testing of entomopathogens, discuss issues associated with animal procurement and test design, and then present a series of single exposure short-term tests that can be used to assess the mammalian safety of a candidate organism. The tests presented are based on the registration requirements of the United States Environmental Protection Agency, subdivision M (Anonymous, 1988) and guidelines published by the World Health Organization (WHO) (Anonymous, 1981). These screens may serve as a starting point for the registration of an entomopathogen in a country that does not have established regulations.

Currently, there is no single safety protocol accepted worldwide. In 1993, the WHO attempted to review the testing protocols of the United States, Canada and the European Economic Community, with the goal of drafting guidelines for the registration of MPCAs. There have been no documents issued at this current time and there does not appear to be any interest in pursuing this project in the near future. That does not mean that there is a shortage of published guidelines. Summaries of the registration requirements for North America (Betz *et al.*, 1990), the European Economic Community (Quinlan, 1990), the former USSR and Eastern Europe (Kandybin & Smirnov, 1990) and Japan (Aizawai, 1990) have been published. Detailed explanations of the United States Guidelines have also been published (Briggs & Sands, 1992; Campbell & Sands, 1992; Fisher & Briggs, 1992; Kerwin, 1992; Siegel & Shadduck, 1992; Spacie, 1992). These chapters are an excellent starting point, but they are no substitute for obtaining the actual regulations of the country or countries in which the MPCA will be registered. Furthermore, the regulatory process is dynamic and both the European Economic Community and Canadian testing requirements are currently being rewritten; the former in order to standardize the requirements of the individual member countries and the latter to harmonize the Canadian requirements with those of the United States. Additionally, with the break-up of the former USSR, it seems likely that some of the regulations currently in place in these new states will be revised. In most countries, the published requirements provide a framework for guiding a registrant, but the actual tests performed are decided in a series of meetings with the government regulators assigned to the applicant. Specific tests may be waived by regulatory agencies on a case-by-case basis, providing data already exist, or additional tests may be required.

2 DEFINITION OF INFECTION

Although both insecticidal chemicals and MPCAs share safety concerns with respect to toxicity, the major characteristic that differentiates MPCAs from chemical toxicants is their ability to multiply and cause infection. Infection may occur not only in the arthropod targets of the MPCA but in non-target species as well, hence the need for infectivity testing. However, determining infectivity is complicated by the fact that the term infection is not defined in most testing guidelines. Infection actually has several definitions and these definitions affect the interpretation of test data. Historically, there have been two schools of thought concerning the definition of infection, and each definition has a different implication for the relationship between micro-organism and host. One school defines infection as simple colonization by micro-organisms that may or may not be detrimental to the host. Colonization occurs shortly after birth and the micro-organisms are restricted by host defences to areas where they can be tolerated, such as the gastrointestinal tract or the upper respiratory tract. Infectious disease is then a special condition that results when host defences are breached and micro-organisms are introduced into areas where they cannot be tolerated (Siegel & Shadduck, 1992). The term pathogenicity denotes the intrinsic capability of a micro-organism to penetrate host defences, and the term virulence refers to the speed by which this is accomplished (Davis *et al.*, 1973).

The second school of thought links infection directly to disease. Living agents cause a disruption to the host either by multiplication in tissue, toxin production, or both. In my opinion, from the viewpoint of safety testing, this second definition, which links the presence of a micro-organism to tissue damage, is the more useful one.

Simple recovery of an MPCA is not indicative of infection. Safety tests by design introduce MPCAs into mammals by a variety of routes and transient disturbances in the normal flora should be expected. Assuming that host immune response is constant, recovery of a portion of an inoculum from host tissue, as well as redistribution of the inoculum among host tissues, may occur over a variable length of time, depending on both the route of administration and the magnitude of the dose. Under these circumstances infection is demonstrated when there is evidence of multiplication of an MPCA (recovery of an amount greater than injected, recovery of vegetative forms when spores were injected, failure to clear over time) coupled with tissue damage. This coupling is essential because the occurrence of tissue damage alone, or even death, can be caused by the physical nature of the inoculum and/or the host response to foreign protein. Proper controls, such as an autoclaved inoculum, are useful in infectivity studies in order to determine if an adverse outcome is due merely to the presence of foreign protein.

3 DEFINITION OF PERSISTENCE

In this chapter the term 'persistence' will be used to describe the ability of an MPCA to remain viable in mammalian tissue without multiplying. Persistence occurs because clearance of an MPCA from a test animal is not instantaneous. Adlersberg *et al.* (1969) reported that intravenously injected radioiodinated latex particles were recovered from mice as long as six months after administration. The particles were distributed between the lungs, spleen and liver during this period, and the proportion of the inoculum recovered from these three organs changed over time. One would expect that when environmentally resistant life-stages of micro-organisms such as some spores are injected, they will be redistributed in tissue and be recovered for a variable amount of time, depending on the dose and the route of injection. Viable spores of *Bacillus thuringiensis* subsp. *israelensis* have been recovered as long as 70 days after injection and oospores of *Lagenidium giganteum* have been observed in spleen tissue four weeks after intraperitoneal injection (Siegel & Shadduck, 1990). Consequently, assessing the infectivity of an MPCA is complicated by the need to distinguish between infection and persistence. To reiterate, simple recovery of an MPCA from a mammal should not automatically be construed as infection, although the significance of prolonged persistence must be addressed on a case-by-case basis.

4 BACKGROUND OF MAMMALIAN SAFETY TESTS AND PHILOSOPHY

Initially, MPCAs underwent the entire battery of tests required of chemical toxicants, including long-term dietary studies (e.g. two-year rat dietary study for *Heliothis* nuclear polyhedrosis virus and *B. thuringiensis* as well as a 23-month hen feeding study for *B. thuringiensis*) (Heimpel, 1971; Ignoffo, 1973; Burges, 1981). MPCAs also underwent infectivity studies that lasted as long as 10 weeks. This approach not only placed MPCAs at a disadvantage compared to their chemical counterparts, because of the expense of the additional testing, but also failed to address fundamental differences between MPCAs and chemical toxicants. Shadduck (1983) noted that four assumptions underlay standard chemical pesticide safety tests and that these assumptions were not applicable to MPCAs. First, most chemical safety test designs assume that a measurable biological effect can be achieved if the dose is high enough. Typically, this is expressed as the median lethal dose, LD_{50}. In contrast, it may be impossible to achieve an adverse effect with an MPCA, especially if it is delivered by conventional means such as aerosol, dermal or oral application. A dose high enough to obtain an effect by these routes often suffocates an animal or blocks its gastrointestinal tract. Second, chemical safety tests assume that the material administered is metabolized, excreted, or both, and that toxic effects can be predicted if one knows the routes by which the primary metabolite(s) are eliminated. Currently, there is no evidence to indicate that MPCAs are metabolized or genetically altered by passage through mammals. Toxic effects from the metabolic breakdown products of MPCAs are unlikely. Third, chemical safety tests assume that a chemical may accumulate and exert its effects over a long period of time; the amount of material stored in body tissues increases with prolonged exposure. Acute and long-term studies are therefore necessary. In contrast, there is no evidence to date that MPCAs colonize mammals or have teratogenic or carcinogenic effects, rendering long-term feeding studies unnecessary. Finally, chemical safety testing assumes that knowledge of the chemical structure of a toxicant enables one to make educated guesses about the toxicity and/or carcinogenicity of other molecules belonging to the same family. This assumption is not applicable to MPCAs. For example, despite the close genetic relationship between *B. thuringiensis* and *B. anthracis* (anthrax), *B. thuringiensis* is not a mammalian pathogen.

A different philosophy of testing known as maximum challenge arose that recognized the unique characteristics of MPCAs. This approach advocated the use of invasive exposure routes (intracerebral, intraocular, intraperitoneal injection) targeted against vulnerable organ systems in short-term tests (2–3 week observation period) in order to achieve a measurable biological effect without the use of massive doses of the MPCA. This approach is parallel to the LD_{50} concept, but the route of exposure is varied rather than the dose. In a maximum challenge test, the highest dose possible (dependent on the physical nature of the material) is given by the route that most severely compromises an animal's natural defences (Shadduck, 1983). Selection of target sites is based in part on a literature review as well as knowledge of vulnerable mammalian sites such as the central nervous system. If there are records of the MPCA or related species infecting vertebrates, attempts are made to reproduce the infection by targeting the organs reported infected.

The use of immunocompromised animals may also be considered part of the maximum challenge philosophy. The rationale for these studies is that the ability of certain MPCAs (bacterial spores) to persist at mammalian body temperatures raises concerns about their fate in immunocompromised mammals. Additionally, with the increase in the number of humans infected with the Human Immunodeficiency Virus (HIV), as well as the increasing proportion of the population undergoing immunosuppressive cancer therapies, there is concern about a possible threat posed to this segment of the population by MPCAs. This question can be addressed by the use of immunocompromised animals. If a researcher demonstrates that multiplication of an MPCA is blocked in immunocompromised animals, it is further evidence that the MPCA is not a mammalian pathogen.

Maximum challenge tests are controversial because of the difficulty in interpreting mortality data and extending the results to human safety. The greatest value of an intracerebral injection study is that it clearly is a 'worst case' scenario and a negative result provides strong evidence of an MPCA's safety. However, non-pathogenic organisms can cause death when injected intracerebrally, and this in turn can lead to premature rejection of a candidate. Likewise, although the failure of an MPCA to infect an immunocompromised animal provides convincing evidence of safety, these tests are opposed by researchers who argue that immunocompromised humans will succumb to opportunistic infections by a variety of micro-organisms before MPCAs could cause infection (Burges, 1981). In addition, the various methods of chemical immune suppression used in these studies as well as the variety of laboratory animal strains with genetic immunological defects makes comparison between studies difficult. Some chemical immunosuppressants such as cyclophosphamide have a greater effect on B-lymphocytes, whereas glucocorticosteroids primarily effect T-lymphocytes (Dumont, 1974; Parrillo & Fauci, 1979). Consequently, there is a host of secondary effects associated with these agents that must be evaluated together with the data on potential infectivity of the MPCA. Currently, the use of immunocompromised animals is not required in the United States, Canadian or European Economic Community guidelines and these tests are not likely to be required in the foreseeable future. However, some of these maximum challenge studies may serve as an inexpensive preliminary screen in order to determine if an MPCA is worth pursuing. An organism that is innocuous when injected intracerebrally or when injected intraperitoneally into immunocompromised animals, will most likely be harmless when test animals are exposed to the MPCA by less invasive routes.

Current testing guidelines in the United States combine elements of conventional toxicological testing such as oral administration and dermal exposure with more invasive routes of exposure such as intravenous or intraperitoneal injection, depending on the MPCA. Intravenous injection is the most invasive test and intraperitoneal injection is reserved for MPCAs of a large particle size that would collapse or plug the veins when injected intravenously. Although maximum challenge tests are not required, elements of this philosophy, specifically an emphasis on short-term testing (2–4 weeks), have been adopted.

The WHO was the first organization to propose a tiered testing strategy (Anonymous, 1981) to evaluate the hazard posed by MPCAs to mammals, Figure 1. The first level, Tier 1, consisted of acute oral, inhalation and intraperitoneal administration of an unformulated MPCA, combined with *in vitro* mutagenicity screens of unformulated material. The tests for formulated MPCAs consisted of dermal, ocular and allergenicity tests with a maximum observation period of 28 days. The endpoints of interest in this series of assays were persistence, infection and irritation caused by the MPCA. If infectivity was clearly indicated in this tier, the candidate organism would be rejected. If there was uncertainty concerning the significance of persistence, the next phase of testing, Tier 2, was initiated. The tests proposed included multiple exposure tests, expanded mutagenicity tests, and studies to determine the time elapsed before the MPCA cleared from tissue. The observation period for these studies could last as long as 90 days. If there were further questions, the candidate organism would enter Tier 3. These assays included conventional two-year feeding studies and teratogenicity testing. The safety tests proposed for entomopathogenic viruses were similar to those described above, except that in Tier 1, *in vitro* assays were included that assessed the ability of viral DNA to incorporate into the host genome. Tier 2 included procedures to assess the ability of entomopathogenic

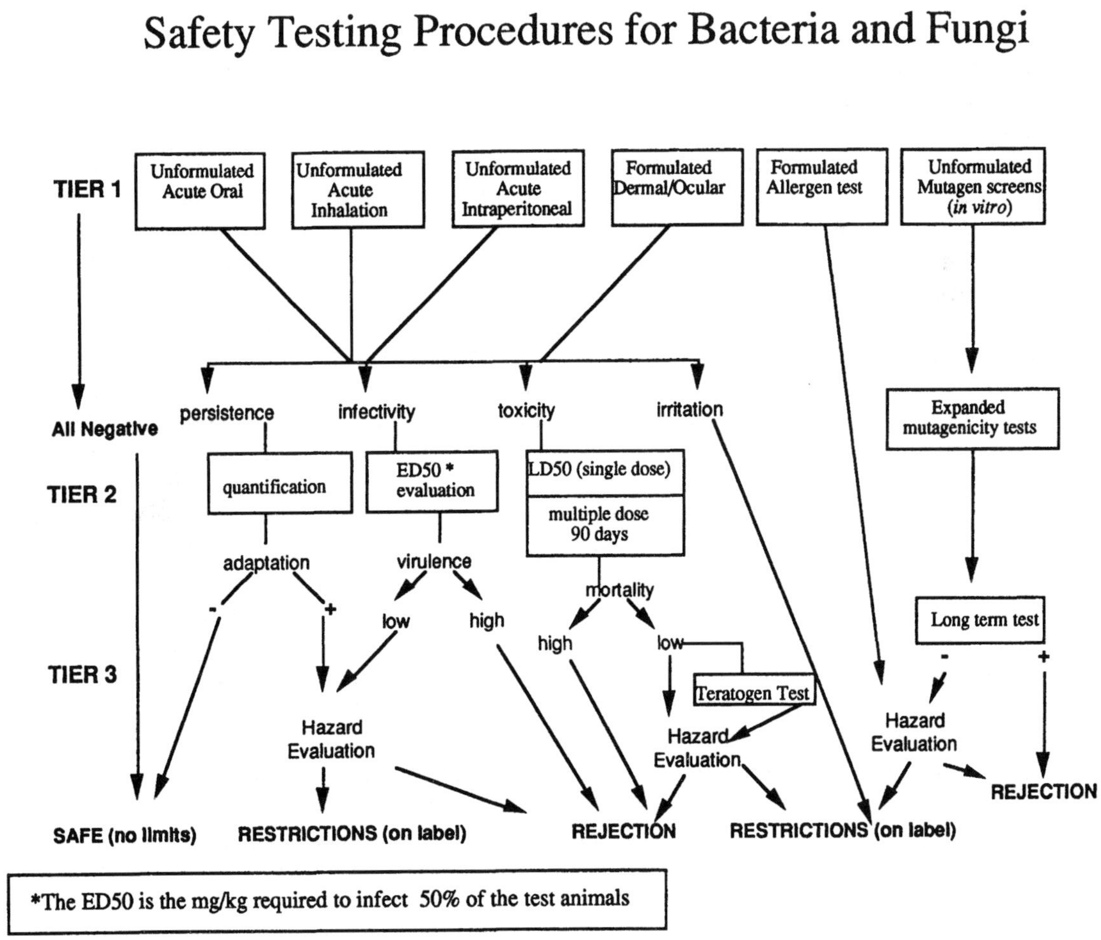

Figure 1 Three-tier testing scheme for bacterial and fungal agents for use in vector control. (Reproduced with permission from *Bulletin of WHO* **59**, 857-863.)

viruses to transform host cells. There were no Tier 3 studies proposed for entomopathogenic viruses.

This three-tier format was adopted by the United States, and elements of these protocols can also be found in the testing requirements of Canada and the member nations of the European Economic Community. In practice, it is unlikely that an MPCA that must undergo Tier 3 testing will ever be registered in the United States or any Western European country. It is worth noting that this protocol tested both formulated and unformulated product, which the United States guidelines also require. Other nations have questioned the necessity of testing formulated product. Recently, the Canadian regulatory authorities have advocated the elimination of formulated MPCA from testing requirements for the following two reasons: First, formulated product is more dilute than the manufacturing grade product (due to the addition of inert ingredients), so in essence a lower dose is actually being tested. The Canadian regulators felt that it was more prudent to raise the test dose for the manufacturing grade product than to include formulated MPCAs in the testing scheme. Second, there is no evidence to suggest that formulated MPCAs clear differently than unformulated MPCAs from mammals, so it is unnecessary to include them in the testing protocols. Finally, formulated product is not included in the safety protocols for chemical toxicants, and there is no empirical evidence to support this additional requirement for microbial insecticides.

Currently, there is no international standard for test dose, therefore, doses vary between countries. The WHO guidelines recommended that the oral dose approach that of a 70-kg man exposed to a hypothetical one hectare dose, and that the total dose not exceed 5 g/kg body weight of the test animal. In contrast, the current Subdivision M guidelines require a dose of 10^8 units of MPCA per mouse or rat. The intraperitoneal dose suggested in the WHO guidelines is set at 10^6 organisms per mouse and 10^7 organisms for the rat. The United States guidelines require a minimum dose of 10^7 units MPCA regardless of species. No doses are specified for the inhalation, ocular and dermal exposure experiments in the WHO guidelines whereas the United States guidelines require minimum doses of 10^8 units of MPCA, 10^7 units of MPCA, and 2 g (dry weight) MPCA respectively. It is interesting to note that the 1988 United States guidelines did not take into account the tenfold weight difference between mice and rats. A dose of 10^7 units of MPCA in mice is equivalent to a dose of 10^8 units in a rat. If one wanted to lessen the likelihood of an adverse outcome when following the United States guidelines, rats should be the test animal of choice because they will receive a proportionately smaller dose on a milligram per kilogram basis than mice. Despite these differences regarding dose, the influence of the WHO guidelines is quite evident in the United States regulations.

5 ANIMAL PROCUREMENT AND HOUSING

Animals should be procured from reputable suppliers in order to ensure their overall quality and freedom from disease. When a qualified vendor is not used, a consultant must be engaged to certify the animals' health status. I strongly discourage the use of wild animals. It is a common practice to overstate the animal order by 10% in order to have a surplus in case of accident or disease. I recommend the use of outbred rodent lines, but consideration may be given to the use of inbred animals under certain circumstances (to optimize uniformity of response). The animals used in any experiment should be as homogeneous as possible with regard to sex, weight and age. Infection by other micro-organisms complicates the interpretation of test data, as does the use of stressed animals. If one chooses to use immunocompromised animals in a testing protocol, special care must be taken to reduce their risk of infection. Animals must be housed in such a way as to minimize stress and should be treated ethically in full compliance with all pertinent government regulations (Anonymous, 1985; McGregor, 1986). In summary, the use of reputable animal suppliers, proper housing and a standardized diet will help ensure the uniformity of the test animals and minimize aberrant results

Housing rodents is a special concern. As a general rule, female mice are less likely to fight than males, and littermates housed together are less likely to fight. Male mice must be carefully monitored, especially mice on the lowest level of the social hierarchy, which may need to be removed if they become injured. These mice are more stressed than their cagemates and may respond abnormally when exposed to the MPCA. When mice are used, it is important to remove dead mice promptly because they may be eaten by their cagemates. This practice may destroy tissue of interest or otherwise disrupt an experiment.

Housing control animals can also pose a challenge. With some MPCAs, it might be important to prevent the control animals from being exposed to the MPCA by copraphagous activity or communal grooming. Separate cages are a simple solution to minimize control animal exposure to the MPCA. If the MPCA is shed in the faeces, ingestion of faeces by the treated animals may complicate clearance studies, because of additional exposure to the MPCA.

6 MODIFIED FIRST TIER SCHEME

The rest of this chapter presents a series of acute (single dose) tests that can be used as an initial screen for testing an unformulated MPCA. Toxicity in these tests may be caused by the constituents of the MPCA or by its metabolic by-products. If toxicity is observed without infectivity, conventional chemical safety tests can then be used to quantify the effect. Infectivity in these tests is indicated by the failure of an MPCA to clear after administration and/or tissue damage. These tests will include suggestions on the number of animals to test and tissues to collect for histopathology. These tests do not last longer than four weeks. Although it may take an inoculum more than twice this amount of time to clear completely, clearance curves can be calculated for the timeframe of the test period. It is also important to note that prior to testing an organism, sufficient quantity of the MPCA should be available so that a single lot is used for all tests. This same lot will form the 'mother culture' for subsequent production of the agent. If an MPCA is produced *in vivo*, consideration should be given to the method by which the batches will be standardized. Additional characterization of insect fragments and other micro-organisms present in the test material may be necessary in order to register MPCAs that are produced *in vivo* instead of by fermentation.

A Acute oral administration test

The purpose of this test is to determine if a single dose of 10^8 units of the MPCA, suspended in an appropriate vehicle and administered with a curved feeding needle (in a volume not to exceed 2 ml/100 g body weight) produces mortality in rats. The observation period is 21 days. Two groups of animals must be used in this test, consisting of an experimental group administered the MPCA and a control group administered the autoclaved MPCA. An additional control group consisting of rats administered the sterile carrier may also be included. Exposure to the autoclaved inoculum is necessary to identify lesions and/or mortality arising from the presence of heat-stable toxins and/or protein in the MPCA. Each group shall consist of at least 10 young male and 10 young female rats, for a total of 40 animals. The feeding needle should be placed into the rat's mouth either over the tongue or to one side of the mouth. The feeding needle should then be slowly inserted until approximately 10 cm is in the rat's esophagus, then the contents should be slowly expelled. Should any resistance be felt during insertion, the feeding needle should be immediately withdrawn and reinserted. The rat should be carefully observed for any sign of respiratory distress during this procedure (Paget & Thomson, 1979).

The rats should be weighed one day before administration of the inoculum, and then denied food and water 16 h before treatment. After the MPCA has been administered, food should be withheld for an additional 3–4 h. If it is not practical to administer a single dose, the dose may be given in increments over a 24-h period. If the dose is administered over this prolonged period, it will be necessary to allow the animals access to food and water. All animals should be observed daily following administration of the MPCA and the observations should include notes on the condition of the skin, fur, eyes and mucous membranes, as well as observations on respiration, behaviour and somatomotor activity. Findings of interest include tremors, convulsions, diarrhoea, lethargy, salivation and coma. Transient effects such as diarrhoea, emesis or loss of appetite may occur and should be noted, but these effects are not grounds for rejecting a candidate MPCA. All of the animals should be weighed weekly and weight change calculated at the end of the experiment for each rat. If practical, faeces should be collected weekly and cultured in order to determine if viable MPCA is shed.

Upon termination of the experiment, all rats are killed and a gross necropsy conducted on each animal (Paget & Thomson, 1979). At a minimum, the spleen, liver, mesenteric lymph node, stomach and portions of the intestine should be examined for lesions. If lesions are observed, histopathology should be conducted on the affected tissue which is fixed in buffered 10% formalin. Tissues should be trimmed and then embedded in paraffin blocks and stained with haematoxylin and eosin. Where applicable, the MPCA should be enumerated from the kidney, liver, lung and spleen by plating on to appropriate media. All animals that die during the experiment should be promptly necropsied. If mortality occurs in the treated rats during the first three days of the experiment, half the remaining animals in each group should be killed and necropsied one week after administration of the MPCA.

An MPCA that cannot clear this test is unlikely to be a suitable candidate for further development. If toxin production by an MPCA is suspected as the cause of mortality rather than infection, the toxic components should be isolated and characterized if possible. These components can then be tested according to conventional chemical safety protocols. If the MPCA is recovered from tissue three weeks after administration, additional testing is necessary to determine the rate of clearance. The researcher should be aware that regulatory agencies may require long-term testing to quantify this effect when the MPCA is submitted for registration. The null hypothesis in this experiment is a one-tailed hypothesis, that there is no difference in response between the experimental group and control group. Response in the context of this test includes weight change, illness and mortality. Given the sample size and the likelihood that if the MPCA truly has a deleterious effect the symptoms will be dramatic, this test has sufficient power to detect a true difference between the experimental and control group 96% of the time (Cohen, 1988).

B Acute abraded dermal toxicity test

The purpose of this test is to determine if a single dermal exposure to 2 g of the MPCA over a 24-h period causes infection or mortality in albino rabbits. The observation period is 14 days. Dermal toxicity testing is a standard toxicology protocol, but I suggest modifying the protocol by applying the MPCA to abraded skin. Since dermal toxicity is unlikely, this test then assesses the role played by intact skin in preventing infection by the MPCA.

Ten young rabbits, consisting of five males and five females are used in this test. No controls are necessary unless the characteristics of the MPCA are completely unknown. Fur should be clipped (as close to the skin as possible without irritating it) from the dorsal and ventral area of the rabbits 24 h before the test. About 10% of the animal's body surface area should be cleared for application of the MPCA. If the abraded skin protocol is followed, the abrasion should penetrate the stratum corneum but not the dermis. The inoculum should be evenly applied in a liquid or cream carrier, over the shaved area, and held in contact with the animal by porous gauze and non-irritating tape. Care must be taken to prevent the animals from ingesting the inoculum, but complete immobilization of the animal is not recommended. At the end of the exposure period, any residual MPCA should be removed with water.

The rabbits should be weighed one day before administration of the inoculum, and at weekly intervals thereafter. Weight change should be calculated for each rabbit. All animals should be observed daily following administration of the MPCA and the observations should include notes on the condition of the skin, fur, eyes and mucous membranes, as well as observations on respiration, behaviour and somatomotor activity. Findings of interest include tremors, convulsions, diarrhoea, lethargy, salivation and coma. Particular attention should be paid to skin lesions and irritancy (irritancy can be evaluated according to McCreesh & Steinberg, 1983), weight loss, behavioural abnormalities and mortality. Gross necropsies should be performed on all animals that die during the course of the experiment and samples of skin (6 × 6 cm) should be collected for histopathology. The skin is fixed in buffered 10% formalin. The null hypothesis in this experiment is that no mortality or irritancy will arise after administration of the MPCA. If irritancy or mortality occurs, the test should be repeated using autoclaved inoculum. This test is not considered invasive and an MPCA that produces deleterious effects when administered by this route is not likely to be a suitable candidate.

C Acute pulmonary infectivity test

The purpose of this test is to determine if a single exposure to 10^8 units of the MPCA, instilled intranasally or intratracheally (Nicholson &

Kinkead, 1982) and suspended in an appropriate vehicle such as sterile distilled water (not to exceed 0.3 ml/100 g body weight) clears from the lungs of rats over a 21-day period. The evidence for infection in this study is failure of the inoculum to clear and/or mortality. Mortality may also occur in this study due to mechanical obstruction of the airways or the occurrence of foreign body pneumonia. A trial study should first be conducted, in order to determine if the dose and/or method of administration will cause mortality (four animals is suggested). If mortality is excessive, the dose should be reduced by at least a factor of 10 and the reason for reduction documented for justification to the regulatory agency.

Two groups of animals are used in this test, consisting of an experimental group administered the MPCA and a control group administered the autoclaved MPCA. The control group will enable the investigator to determine if any lesions observed in lung tissue are due to the presence of foreign protein, or the introduction into the lungs during instillation of the normal microbial flora present in the pharynx. The treated group should consist of at least 10 young male and 10 young female rats, and the control group should consist of five young male and five young female rats, for a total of 30 animals (200–250 g in weight). Extra rats should be administered the MPCA in case of mortality. All rats should be weighed one day before administration of the MPCA and at weekly intervals thereafter. Weight change should be calculated for all rats. It is important to note that in many strains of rats female weights do not exceed 250 g, and that this sex-related difference in weight gain should be considered when evaluating the results (male weights can exceed 600 g).

In my experience, the majority of clearance occurs during the first two weeks after administration of the inoculum. Therefore, in order to accurately determine the rate of clearance, most of the sampling should occur during this period. One male and one female rat should be killed one hour, one day, two days, three days, four days, five days, six days, one week, two weeks and three weeks after administration. Gross necropsies should be performed on all of the rats. The lungs, liver and spleen should be removed, weighed, and a portion retained for enumerating the MPCA. At a minimum, the MPCA should be enumerated from the lungs by homogenizing a known weight of lung tissue in sterile distilled water (1 : 9 weight/volume), serially diluting the homogenate, and then plating the MPCA on to suitable growth media. The number of colony forming units (cfu) per gram of lung tissue can then be calculated after incubation. In addition, samples of tissue from the lungs, spleen and liver should be saved for histology and fixed in buffered 10% formalin. The control rats should be killed 3 weeks after administration of the MPCA. Gross necropsies should be performed on the rats and tissue from the lungs of the control group should be collected for histology. The purpose of the control group in this study is to determine the background level of lesions in the lungs resulting from administration of the inoculum or due to environmental irritants.

If mortality does not occur in the treated group, the endpoint of interest in this study is the clearance of the MPCA, with a primary focus on clearance from the lungs. The clearance rate of the MPCA from the lungs on a CFU/g basis should be calculated by simple linear regression. Logarithmic, exponential, or power transformations may be appropriate in order to calculate the clearance curve with the best fit (determined by the coefficient of determination, r^2). The same methodology can be applied to calculate the rate of clearance from the liver and spleen. Failure of the MPCA to clear from the lungs, indicated by a regression that is not significant ($P > 0.05$), should be regarded as evidence of infection. A MPCA that produces infection when administered by this route is not a suitable candidate for development.

D Acute intraperitoneal infectivity test

This study evaluates the ability of an inoculum consisting of 10^7 units of the MPCA, suspended in 0.1 ml of carrier, to clear from the spleens of outbred female mice in a 28-day period. The inoculum is injected using the smallest practicable hollow needle (26 gauge is preferable, but a wider bore may be necessary if the inoculum clogs the needle). The mouse should be held so that the ventral surface is facing the person performing the injection. The needle should be introduced rapidly into a point slightly left or right of the midline, and halfway between the pubic symphysis and the xiphisternum, of the ventral surface of the mouse. The syringe contents should be expelled by firmly depressing the plunger (Paget & Thomson, 1979). When injecting a female mouse, it is important to avoid the mammary glands. The site of injection should be noted in a logbook.

The MPCA is enumerated from the spleen because this organ is more efficient than the liver at filtering particles and micro-organisms from the blood on a milligram per milligram basis. The spleen is collected aseptically, weighed, and then homogenized (1 : 9 weight/volume) in sterile distilled water to liberate the MPCA. A known amount of homogenate is then collected, serially diluted, and the cfu/g spleen calculated. Based on my experience with entomopathogenic bacteria, most of the inoculum is recovered during the first two weeks. In this protocol, mice are followed for a total of four weeks in order to evaluate fluctuations in the recovery of the MPCA that may arise from liberation of the MPCA from extrasplenic sites and subsequent filtration by the spleen. If mortality occurs in this test, the toxic factors should be identified when possible. It may be necessary to inject autoclaved MPCA by this route in order to determine whether the toxicity is due to heat-labile compounds.

In this clearance experiment a total of 33 mice of the same age and sex are injected and three mice are killed on days 1, 2, 4, 6, 8, 10, 14, 16, 20, 24 and 28 after exposure. When applicable, heart blood should be collected from the mice killed, and cultured for the MPCA. Additional mice may be injected in order to ensure that there are three spleens available for each specified time period, in case some of the injected mice die. The spleens should be collected and homogenized as previously described. The clearance rate of the MPCA from the spleen on a unit per gram basis should be calculated by simple linear regression. Logarithmic, exponential or power transformations may be appropriate in order to calculate the clearance curve with the best fit determined by comparing the coefficient of determination, r^2 (Montgomery & Peck, 1992).

It is quite likely that some of the inoculum may be recovered from the spleen 28 days after injection, but this should not necessarily be interpreted as evidence of infection, given a significant regression with a negative slope. However, if the regression is not significant, this should be regarded as evidence of possible infection and a follow-up study assessing clearance over a 90-day period is warranted. In this study, the same amount of mice are used but the intervals of sacrifice are stretched over the longer time period. If the MPCA is present in heartblood at the end of this second study, this is evidence that units of the MPCA are still circulating in the bloodstream and these data may indicate that multiplication of the MPCA occurred. Follow-up studies may then be specified by the regulatory authorities.

Intravenous and intraperitoneal injection are the most invasive routes of administration in Tier 1 of Subdivision M. These tests evaluate the likelihood of infection by an MPCA when the skin is bypassed as a barrier. In the United States guidelines, intravenous injection is the preferred route of exposure. Intraperitoneal injection is reserved for an MPCA such as an entomopathogenic fungus that has a large particle size (in order to prevent embolisms). In contrast, the WHO protocol specified intraperitoneal injection as the route of administration in the original three tier scheme. I believe that intraperitoneal injection is a more challenging route than intravenous injection and should be used for the following reasons. First, the relatively anoxic environment of the abdominal cavity may allow the MPCA to produce toxins that would not ordinarily be expressed in the more oxygenated bloodstream. Second, the bulk of an intravenously injected inoculum is rapidly filtered by the spleen and liver within 4 h (Adlersberg *et al.*, 1969). An inoculum injected intraperitoneally, must first pass through the lymphatic system before it can be filtered from the bloodstream by the spleen and liver. This delay in filtration in turn provides an additional opportunity for an MPCA to multiply. Since the intraperitoneal route of administration maximizes the opportunity for an MPCA to cause deleterious effects, I believe that it is a more conservative route and is preferable.

7 CONCLUSION

The basic challenge of safety testing is determining when an MPCA is reasonably safe. The acute tests suggested in this protocol should enable a researcher to determine the hazard posed to mammals by a candidate MPCA. One could always require an ever increasing series of invasive tests that do not address the biology of the MPCA, but in the long run, it will be almost impossible to evaluate the significance of the data. Ultimately, unrealistic standards will drive MPCAs out of the marketplace. Burges (1981) eloquently summarized the difficulty of evaluating any MPCA when he stated that

> a no-risk situation does not exist, certainly not with chemical pesticides, and even with biological agents, one cannot absolutely prove a negative. Registration of a chemical is essentially a statement of usage in which risks are acceptable, and the same must be applied to biological agents.

The question then shifts from hazard, to how much risk is acceptable? From a researcher's point of view, the answer to this question lies with the regulatory authorities, although at a different level the answer to this question lies both in the scientific and political arena. Ironically, although there is a vast quantity of safety data in government archives, few of these studies are publicly available, which in turn may fuel public concern. In order to allay public concern, additional tests may be required to register an MPCA.

In the United States there have been 149 products registered over a 47-year period that have microbial insecticides listed as the active ingredient. In 1995 alone, 14 products containing *B. thuringiensis* subsp. *kurstaki*, *Beauveria bassiana*, *Candida oleophila* and *Pseudomonas syringae* were registered. Most of the information contained in the studies used to register these organisms is proprietary. The paucity of published mammalian and NTO studies is due in part, to the fact that there is no incentive for a company to publish its safety data concerning a particular MPCA. Published data may be cited by competitors, who in turn can avoid conducting costly tests. Additionally, if safety data are taken out of context, it is possible that a product may be unfairly labelled as unsafe. I believe it unlikely that corporate policies concerning publication will change in the near future, consequently the majority of the studies conducted on any MPCA will remain confidential.

It is quite possible that the same safety question will be answered repeatedly, at a cost of many animal lives, as well as time and money. That is the nature of the system. However, the knowledge gained over these past decades is available to regulators, and in fact has aided in formulating guidelines that acknowledge the difference between microbial insecticides and chemical toxicants. What I hope to have communicated in this chapter in addition to acute testing protocols, is the fact that safety requirements are not static, and will inevitably change in the future due to inputs from both science, government and the general public.

ACKNOWLEDGEMENT

I am indebted to Dr John A. Shadduck and thank him for the many hours of conversation that we have had regarding safety testing. I also thank the World Health Organization Special Program for Research and Training in Tropical Diseases for helping to lay the foundation for our present mammalian safety testing requirements.

REFERENCES

Adlersberg, L., Singer, J. M. & Ende, E. (1969) Redistribution and elimination of intravenously injected latex particles in mice. *J. of the Reticulo-endothel. Soc.* **6**, 536–560.

Aizawai, K. (1990) Registration requirements and safety considerations for Microbial Pest Control Agents in Japan. In *Safety of microbial insecticides* (eds M. Laird, L. Lacey & E. Davidson), pp. 31–42. CRC Press, Boca Raton.

Anonymous (1981) Mammalian safety of microbial control agents for vector control: a WHO Memorandum. *Bull. Wld. Hlth. Org.* **59**, 857–863.

Anonymous (1985) *Guide for the care and use of laboratory animals.* US Department of Health and Human Services, NIH publication no. 86–23, revised 1985. United States National Institute of Health. 83 pp.

Anonymous (1988) *Toxicology guidelines for microbial pest control agents, subdivision M.* United States Environmental Protection Agency. Office of Pesticide and Toxic Substances, 303 pp.

Betz, F. S., Forsyth, S. F. & Stuart, W. E. (1990) Registration requirements and safety considerations for Microbial Pest Control Agents in North America. In *Safety of microbial insecticides* (eds M. Laird, L. Lacey & E. Davidson), pp. 3–10. CRC Press, Boca Raton.

Briggs, J. D. & Sands, D. C. (1992) Overview: The effects of Microbial Pest Control Agents on Nontarget Organisms. In *Microbial ecology: principles, methods, and applications*, (eds M. A. Levin, R. J. Seidler & M. Rogul), pp. 685–688. McGraw-Hill, New York.

Burges, H. D. (1981) Safety, safety testing, and quality control of microbial pesticides. In *Microbial control of pests and plant diseases, 1970–1980*, (ed. H. D. Burges), pp. 738–769. Academic Press, New York.

Campbell, C. L. & Sands, D. C. (1992) Testing the effects of microbial agents on plants. In *Microbial ecology: principles, methods, and applications* (eds M. A. Levin, R. J. Seidler & M. Rogul), pp. 689–705. McGraw-Hill, New York.

Cohen, J. (1988) *Statistical power analysis for the behavioral sciences*, 2d edn. Lawrence Erlbaum Associates, New Jersey.

Davis, B. D., Dulbecco, R., Eisen, H. N., Ginsberg, H. S. & Wood, W. B. (1973) *Microbiology*, 2d edn. Harper & Row, New York.

Dumont, F. (1974) Destruction and regeneration of lymphocyte populations in the mouse spleen after cyclophosphamide treatment. *Int. Arch. Allergy* **47**, 110–123.

Fisher, S. W. & Briggs, J. D. (1992) Testing of microbial pest control agents in nontarget insects and acari. In *Microbial ecology: principles, methods, and applications*, (eds M. A. Levin, R. J. Seidler & M. Rogul), pp. 761–777. McGraw-Hill, New York.

Heimpel, A. M. (1971) Safety of Insect Pathogens for Man and Vertebrates. In *Control of insects and mites* (eds H. D. Burges & N. W. Hussey), pp. 469–489. Academic Press, New York.

Ignoffo, C. M. (1973) Effects of entomopathogens on vertebrates. *Ann. New York Acad. Sci.* **217**, 141–164.

Kandybin, N. V. & Smirnov, O. V. (1990) Registration requirements and safety considerations for Microbial Pest Control Agents in the USSR and adjacent Eastern European countries. In *Safety of microbial insecticides* (eds M. Laird, L. Lacey & E. Davidson), pp. 19–30. CRC Press, Boca Raton.

Kerwin, J. L. (1992) Testing the effects of micro-organisms on birds. In *Microbial ecology: principles, methods, and applications* (eds M. A. Levin, R. J. Seidler, & M. Rogul), pp. 729–744. McGraw-Hill, New York.

McCreesh, A. H. & Steinberg, M. (1983) Skin irritation testing in animals. In *Dermatotoxicity*, 2d edn. (eds F. N. Marzuli & H. I. Maibach), pp. 147–166. Hemisphere Publishing, New York.

McGregor, D. (1986) Ethics of Animal Experimentation. *Drug Metab. Rev.* **17**, 349–361.

Montgomery, D. C. & Peck, E. A. (1992) *Introduction to linear regression analysis*. 2nd edn. Wiley, New York.

Nicholson, J. W. & Kinkead, E. R. (1982) A simple device for intratracheal injections in rats. *Lab. Anim. Sci.* **32**, 509–510.

Paget, G. E. & Thomson, R. (1979) *Standard operating procedures in pathology*. University Park Press, Baltimore.

Parrillo, J. E. & Fauci, A. S. (1979) Mechanisms of unformulated action on immune processes. *Annu. Rev. Pharmacol. Toxicol.* **19**, 179–201.

Quinlan, R. J. (1990) Registration requirements and safety considerations for Microbial Pest Control Agents in the European Economic Community. In *Safety of microbial insecticides* (eds M. Laird, L. Lacey & E. Davidson), pp. 11–18. CRC Press, Boca Raton.

Shadduck, J. A. (1983) Some considerations on the safety evaluation of nonviral microbial pesticides. *Bull. WHO.* **61**, 117–128.

Siegel, J. P. & Shadduck, J. A. (1990) Safety of microbial insecticides to vertebrates-humans. In *Safety of microbial insecticides* (eds M. Laird, L. A. Lacey & E. W. Davidson), pp. 102–112. CRC Press, Boca Raton.

Siegel, J. P. & Shadduck, J. A. (1992) Testing the effects of microbial pest control agents on mammals. In *Microbial ecology: principles, methods, and applications* (eds M. A. Levin, R. J. Seidler & M. Rogul), pp. 745–759. McGraw-Hill, New York.

Spacie, A. 1992. Testing the effects of microbial agents on fish and crustaceans. In *Microbial ecology: principles, methods, and applications* (eds M. A. Levin, R. J. Seidler & M. Rogul), pp. 707–728. McGraw-Hill, New York.

CHAPTER VIII-1

Complementary techniques: Preparations of entomopathogens and diseased specimens for more detailed study using microscopy[1]

JAMES J. BECNEL

* CMAVE, USDA, ARS, PO Box 14565, Gainesville, FL 32604, USA

1 INTRODUCTION

The science of Insect Pathology encompasses a diverse assemblage of pathogens from a large and varied group of hosts. Microscopy techniques and protocols for these organisms are complex and varied and often require modifications and adaptations of standard procedures. The objective of this chapter is to provide the researcher with some of the basic techniques and protocols used to study insect pathogens realizing that the guidelines must be tailored for specific needs. Many specialized protocols have been developed and for an extensive, current review of the literature on techniques for light and electron microscopy refer to Adams & Bonami (1991). Recommended texts on general histological techniques for light microscopy are by Barbosa (1974) and Luna (1960) with specific protocols for the diagnosis of insect diseases found in Poinar & Thomas (1984) and Thomas (1974). Procedures for electron microscopy can be found in Aldrich & Todd (1986), Glauert (1974) and Hayat (1986).

2 LIGHT MICROSCOPY

The first evidence for the presence of a pathogen is often observed with either a stereoscopic or compound microscope. These observations are often crucial for making the initial diagnosis which leads to the specific approach required depending on the type of pathogen found. Specific protocols for identification and preparation of specimens for the various types of pathogens are detailed in the previous

[1] Mention of a commercial or proprietary product in this paper does not constitute an endorsement of this product by the United States Department of Agriculture.

MANUAL OF TECHNIQUES IN INSECT PATHOLOGY
ISBN 0-12-432555-6

chapters. This chapter deals with some general and specialized techniques for light microscopy and detailed procedures for the preparation and analysis of diseased specimens for electron microscopy.

There are a number of general protocols for the handling of entomopathogens and diseased individuals regardless of the host or pathogen. Because many micro-organisms are found associated with healthy insects, good laboratory practices are essential to prevent contamination. This can be accomplished by maintaining good sanitation practices at all times and good sterilization protocols when required. Common sense dictates that all working surfaces and instruments be kept clean.

A General remarks

The two most common light microscopes used for the study of entomopathogens are bright-field and phase-contrast. The type of microscope used is determined by the preparation of the specimen examined. In general, bright field microscopy is suited for specimens of high contrast such as Giemsa-stained preparations or stained histological sections. Phase-contrast microscopy is useful for examination of living cells in what is normally called 'fresh preparations'. Phase-contrast is also extremely useful for the examination of 1 μm thick unstained plastic sections of material prepared for electron microscopy. Proper alignment of the phase-contrast microscope is essential for optimum performance.

Another more specalized, but very useful type of optics, is differential interference contrast microscopy (commonly referred to as Nomarski-interference). Nomarski-interference provides an apparent three-dimensional quality to the image. This technique is especially useful for examining surface structure important for taxonomic purposes.

B Histological methods

1. Dissecting fluids

Preliminary dissection and preparation of insect tissues for further examination requires that tissues be kept moist without damage or disruption to the tissues. Typically, saline solutions have been employed for this purpose with the most simple being a 0.85% NaCl solution. A commonly used dissecting fluid is Ringer's solution recommended as a normal salt solution for insect tissues. Several specialized physiological solutions have been developed, for example, Eide and Reinecke's Saline (Eide & Reinecke, 1970) for muscoid fly sperm. Other specialized saline solutions can be found elsewhere (Barbosa, 1974).

2. Chemical fixation

This is the process of stabilizing the cellular integrity of tissues for detailed histological examinations. The fixative must penetrate quickly to preserve the tissues in a natural state with a minimum of artefacts due to swelling, shrinkage, leaching or other detrimental effects. This process often requires the preservation of whole insects or dissected tissues which then are embedded, sectioned and stained. The selection of a fixing agent depends on the purpose for which the tissue is intended. Generally, the live specimen is dissected in one of the saline solutions given above and the tissues of interest are removed and placed into the fixative. Alternatively, the specimen may be dissected directly in the fixative if none of the tissue is intended for other purposes that may be adversely affected by the fixative. If whole specimens are to be fixed, it is often necessary to immerse the live specimen into the fixative and carefully make additional openings in the cuticle to allow for better infiltration of the fixative. Vacuum can also be used for difficult to fix tissues of whole specimens with hard cuticle. Tissues should be placed in at least ten times its volume of the fixative.

There are many fixatives developed for specific purposes but a few general purpose fixatives are commonly used in insect pathology. Buffered neutral formalin is a good overall fixative that acts quickly and allows for long-term storage of tissues. The tissues must be thoroughly washed in distilled or deionized water prior to further processing. Formaldehyde is dangerous and must be used under a fume hood. Perhaps two of the most commonly used fixatives are Carnoy's and Bouin's. Carnoy's is an excellent general insect fixative because it penetrates rapidly and acts quickly. Fixation is complete for normal sized tissues (< 1cm) in 3 h and whole specimens in 12–24 h. Rinsing is in 70% alcohol (commonly ethanol) and can be stored in this solution for extended periods prior to additional processing. Bouin's is also a good general fixative with fixation completed in 4–12 h depending on the size of the tissue. It is critical that the tissues are washed thoroughly in 50% alcohol for 4–6 h (preferably agitated) to remove the picric acid. Failure to do this can adversely affect the staining of the tissue. Properly washed specimens can be stored in 70% ethanol for extended periods. The specialized fixative TAF is suggested for nematodes (Southey, 1970).

3. Dehydration and paraffin embedding

Following fixation, the tissue must be dehydrated, infiltrated with paraffin and embedded in paraffin prior to sectioning and staining. The general procedure given is an example of the process but an experienced technician should be consulted prior to the undertaking of a specific project.

a. Embedding in Paraplast™ (a paraffin-plastic mixture)

1. Fix living insect host or freshly dissected tissue sample in Carnoy's or Bouin's fixative for 2–4 h.
2. Rinse in 70% ethanol for 1 h, soak in ethanol overnight. At this point, tissue may be stored in 70% ethanol.
3. Dehydrate tissue to tertiary-butyl alcohol and infiltrate with paraffin:
   - (a) 80% ethanol, 2 h
   - (b) 95% ethanol, 2 h
   - (c) 100% ethanol, 1 h
   - (d) 100% ethanol, 1 h
   - (e) Absolute ethanol : butanol (1 : 1), 2 h

   Steps (f)–(h) must be done at a temperature > 25.5 °C, the melting point of t-butyl alcohol.
   - (f) 100% butanol, 2h
   - (g) 100% butanol, 2h
   - (h) 100% butanol, 2h

   Steps (i)–(k) are done in a 60°C oven.
   - (i) Butanol : paraffin (1 : 1), 2 h
   - (j) 100% paraffin, 2 h
   - (k) 100% paraffin (under vacuum), 2 h
4. Embed in fresh paraffin, with the tissue sample near but not on the bottom of the container ('boat'). This is done by pouring a bit of the paraffin into the container, and allowing it to harden slightly before adding the tissue sample. The paraffin containing the sample should be cooled rapidly.
5. After the paraffin has hardened, remove the container, trim the block to expose the tissue which is now ready to section.

Sectioning and transfer of the sections to slides is probably the most tedious and difficult part of the process. Training by an experienced technician is strongly suggested but detailed procedures can be found in manuals such as Luna (1960).

6. Section the faced block with a microtome to obtain the thinnest sections possible (approximately 5 μm).
7. Place ribbons of sections on clean slides warmed on a slide warming tray set at 5 °C below the melting point of the paraffin to flatten and fix them to the slide. A water bath set below the melting temperature of paraffin can be used to transfer sections to slides. Float sections in the water bath and then pick them up with the warm slides, remove excess water and dry.

Before staining, sections must be deparaffinized. The paraffin is removed from the sections with a solvent (e.g. xylene or Hemo-De™, a natural citrus by-product, can be used in place of xylene in many cases) and the tissue rehydrated. This is done by hydration through a decreasing ethanol concentration series to distilled water; 3 min in each solution should be sufficient. The slides should not be allowed to dry.

1 xylene : 1 absolute ethanol
Absolute ethanol
95% ethanol
70% ethanol
50% ethanol
Distilled water

4. Staining

Haematoxylin is one of the most common and valuable histological stains used. There are many different formulas for this stain depending on the specific objectives of the study. Many variations have been developed (Luna, 1960; Barbosa, 1974). The procedure below is a common one used for study of diseases in insects.

a. Heidenhain's haematoxylin
This is a classic procedure that stains nuclei a blue–

black to brown–black colour. It is time consuming but the stain is durable, actually improving with age. This procedure can be used on either wet smears or sectioned material.

Wet smears

1. Rinse insect in distilled water; blot dry.
2. Smear the insect or selected tissue on a coverslip and drop it immediately into aqueous Bouin's fixative. The cover slip should float by surface tension, smear side down on the fixative.
3. Fix for at least 2 h.
4. Rinse coverslip three times in 70% ethanol and then soak overnight in 70% ethanol.
5. After all of the yellow colour (from the picric acid in the fixative) is gone, rinse the coverslip in distilled water for 2–3 min and proceed with staining without allowing the smear to dry.

Stain

1. Pretreat (mordant) in iron alum for at least 5 h.
2. Rinse in distilled water for 3–4 min.
3. Stain in Heidenhain haematoxylin solution overnight.
4. Rinse in slowly running tap water for 5 min.
5. Destain in iron alum until nuclei stand out sharp against a grey–tan background. Monitor the destaining process by occasional examination under a microscope, rinsing the slides well in distilled water before examination.
6. After destaining, rinse the coverslip briefly (10 s) in tap water containing a few drops (approx. 5 drops/100 ml) of concentrated ammonium hydroxide. Then rinse in slowly running tap water for 30–45 min.
7. Dehydrate in graded ethanol and xylene.
   (a) 70% ethanol, 0.5 min
   (b) 95% ethanol, 0.5 min
   (c) 100% ethanol, 1 min
   (d) 100% ethanol, 1 min
   (e) Ethanol : xylene (1 : 1), 3 min
   (f) Xylene, 3 changes, 3 min each
8. Mount in Permount™ or other suitable mounting medium.

Tissue sections

1. Deparaffinize sections (as described above).
2. Stain as described above for wet smears.
3. Rinse in slowly running tap water for 30 min.
4. Counterstain in Eosin Y for 1.5 min (this step can be omitted).
5. Dehydrate slides to xylene and mount as above.

C Specialized stains and protocols

Many different stains are utilized depending upon the pathogen group under study. Some of the most common stains for each group will be provided, however, other staining procedures are often required and can be found in the chapters of this manual dealing with the specific pathogens and in more specialized references (Adams & Bonami, 1991). Recipes for the stains mentioned in this section are found in the appendix.

1. Protozoa. (see also Chapter IV and Lee et al., 1985)

a. Microsporidia

(i) *Giemsa-stain.* This stain was originally designed to examine blood for the presence of malarial parasites. It has become the most widely used stain for the identification of vegetative stages of microsporidia. The staining methods described below are methods successfully used in different laboratories and are offered here without explanation of differences in rinses and pH. Some experimentation with this stain is usually necessary to adapt it to different microsporidia, hosts and even laboratory water quality. A procedure for utilizing Giemsa-stain for viruses can be found in Chapter II.

Air-dried tissue smears for Giemsa's stain can be made on either slides or cover slips. In either case they should be clean.

Smearing procedure

1. If the host to be sampled is an aquatic one, excess water must be blotted from the surface of the organism before dissection.
2. Dissect a sample of tissue from a large host. If small, the entire organism can be crushed.
3. (a) Using a pair of forceps, press the sample against the slide with sufficient force to disrupt the host cells and release the microsporidian cells. Draw the tissue over the slide in a circular, spiral manner without

passing over the same area twice. Make several small smears on a slide.

(b) Or, dissect a small piece of tissue from the insect and macerate it on a slide in a small drop of haemolymph or physiological saline then, with the forceps, transfer a drop of the macerate to a coverslip and spread it thinly.

4. The ideal smear is a monolayer of dispersed, disrupted cells. In reality, the cells will be sufficiently dispersed to be usable in some areas and in other regions the sample will be too thick.
5. If a frosted slide is used, label it with no. 2 pencil or India ink, otherwise use a diamond marking pen. Ink used for the label must be insoluble in both absolute methanol and water.
6. Set slide with smear side up, on slide holder and allow to air dry.
7. Float absolute methanol on the slide and fix for 5 min. After 5 min, pour off the excess methanol and stain the slide. Alternatively, allow the slides to dry and then stain, preferably within 24 h.

Staining procedure

1. On a staining rack, place slides horizontally and flood with 10% Giemsa stain in pH 7.4 buffer for 10–20 min.
2. Rinse the slides in running tap water and blot dry with bilbulous paper.
3. Examine after drying (usually a coverslip will be required on the dry slide) using a 16–40 × dry objective and bright field optics. For more detailed observation immersion oil can be placed directly on stained smear for use of a higher power, oil emersion objective.
4. Nuclei stain red and cytoplasm stains blue. Acidophilic organelles in the cytoplasm will also stain red. The pH of the stain and rinse is important for proper colour development. Lower pH shifts the colours toward red.
5. Apply a mounting medium (Histoclad™, Permount™ or Pro-Texx™ are appropriate; Canada Balsam cannot be used) and a coverslip to the dried slides to improve longevity of the stains.

Alternative staining procedure

1. Make a smear on a coverslip, dry
2. Place in absolute methanol for 7–10 min in a Coplin jar.
3. Remove and air dry.
4. Place with Giemsa stain solution plus 1 drop 0.01 M phosphate buffer, pH 7.0, in Coplin jars for ca.1.5 h.
5. Remove and rinse briefly in tap water buffered with a few drops of phosphate buffer, 6.8 pH.
6. Air dry.
7. Place coverslip, stain side down, on a drop of Pro-texx™ on a clean slide.
8. Harden for 2 days.

The above procedure can also be used with smears on slides. Buffers often need adjustment because of local differences in pH of the tap water.

(ii) *HCl–Giemsa* (Weiser, 1976). Acid hydrolysis prior to Giemsa's stain will reveal the number of nuclei in the spore.

1. Heat 1 N HCl to 60 °C.
2. Lower the smear into the hot HCl.
3. The time in HCl will vary with the species; try 30, 60, 90 s. This can be done on one slide with a long smear by lowering the slide into the acid a bit at a time. (Alternatively, place a drop of the 1N HCl on the smear and heat it gently over a flame, moving the slide frequently – maximum of 30 s – until the first tiny bubbles appear.)
4. Rinse for several minutes in distilled water.
5. Fix with methanol and stain with Giemsa's stain as usual.

Note: Hot HCl destains Giemsa, therefore, HCl–Giemsa can be used on slides that have been previously stained with Giemsa.

(iii) *Giemsa – colophonium* (Short & Cooper, 1948). This is an adaptation of Giemsa for staining paraffin sections.

1. Start with deparaffinized sections (see Embedding in paraffin, p. 339).
2. Fix for 5 min in absolute methanol.
3. Stain for 20–30 min in 10% Giemsa stain (staining time can vary with tissues).
4. Wash briefly in tap water.
5. Destain (differentiate) in colophonium resin (gum rosin, 15 g in 100 ml acetone) for at least 15 s, checking occasionally under the microscope. Renew the colophonium solution if a film forms on its surface.
6. Transfer the slides to 70% acetone – 30% xylene solution to remove the colophonium and stop differentiation.

7. Pass the slides through several changes of xylene until the sections clear.
8. Apply mounting medium and coverslip.

(iv) *Calcofluor white* (Vavra & Chalupsky, 1982). The optical brightener, Calcofluor M2R binds to the chitinous layer of the microsporidian spore wall and makes them fluoresce in UV light. This can be a useful diagnostic technique.

One drop of water with spores is mixed with Calcofluor white (10^{-4} dilution) dissolved in distilled water. The slide is immediately observed under a fluorescence microscope. The spores exhibit a bright green fluorescence. This stain also works on methanol-fixed spores and in deparaffinized histological sections. If spores stored for a long time have to be visualized, the use of an alkaline solution (Calcofluor in 0.1 N NaOH) is recommended.

(v) *Burri ink* (Vavra & Maddox, 1976). The simplest way to preserve the size and shape of spores on smears is to mix the spores with a solution of water-soluble nigrosine stain (Burri Ink) and allow to dry. Nigrosine stain can also be applied to a smear that has already dried. The spores appear colourless on a grey background. The shape of the spores and any external appendages are revealed.

(vi) *Wheatley's modified Gomori trichrome* (Alger, 1966). This is a simple alternative to Heidenhain for staining microsporidian infected specimens.

1. Slides with deparaffinized sections (see 'Embedding in paraffin') are dipped ten times in 50% ethanol–HCl solution (0.1 ml conc. hydrochloric acid per 10 ml, 50% ethanol).
2. Stain for 10 min in undiluted Wheatley's stain.
3. Dip 10 times in 90% ethanol containing 0.1 ml glacial acetic acid per 10 ml 90% ethanol.
4. Dip 10 times in 90% ethanol.
5. Dip 10 times in 100% ethanol.
6. 3 min in 100% ethanol.
7. 3 min in 1 : 1, ethanol : xylene.
8. 3 min or longer in xylene.
9. Apply a mounting medium and a coverslip.

(vii) *India ink test for presence of a mucocalyx* (Lom & Vavra, 1963). Some microsporidian spores of aquatic hosts are surrounded by a mucocalyx that is thought to reduce their density, extending their time in the feeding zone of the host. This mucous layer is detected by mixing a drop of spores with a small amount of India ink on a microscope slide under a coverslip. The layer of fluid between the slide and coverslip must be thin, not much thicker than the spores. The small carbon particles will be held away from the spores by the mucocalyx, revealing it as a clear area around the spore.

(viii) *Lacto-aceto-orcein chromosome squashes.* This stain has been adapted to examine chromosomes of microsporidia. A modification of this stain for Fungi can be found in Chapter V-1.

1. Clean slides (not siliconized) with 45% acetic acid.
2. Dissect infected tissue out into 45% acetic acid. Mince up and remove excess hard tissues that may prevent good spread.
3. Place siliconized coverslip and flatten by applying direct pressure to the squash to prevent smearing of the cells (try both hard and soft pressure).
4. Put on dry ice, freeze, pop off cover slip and let air dry.
5. Place a drop of 2% lacto-aceto-orcien in 45% acetic acid onto the squash, cover slip, heat over alcohol lamp briefly (until fog leaves).
6. Cool a bit, place slide into ethanol, remove cover slip and add Euperol™ or Protexx™ mounting media and new coverslip.

Alternative lacto-aceto-orcein procedure

1. Dissect infected tissue in small drop of water.
2. Add 1 drop of Carnoy's fixative. Fix for 1 min.
3. Remove fixative by carefully absorbing excess.
4. Add 4 drops of stock lacto-aceto-orcein stain; gently place cover glass.
5. After 5 min, apply direct pressure for 5 s with slide placed into folded filter paper to adsorb the excess stain. (Experiment with time and amount of pressure for proper staining and spread of chromosomes.)

b. Ciliates

(i) *Fixing protozoa on a slide for permanent mounts* (Farmer, 1980). Ciliary structures and nuclei will be clearly differentiated against a grey background.

1. Place a drop of protozoa culture on a clean slide.
2. Pipette a drop of slide affixative from a height of 2–3 cm onto the sample.

3. Carefully remove excess with a pipette. Repeat steps 2 and 3, three times.
4. After approximately 15 s, move the slide through a dehydrating series of ethyl alcohols, 35% to 100%.
5. Clear in xylene and cover with mounting medium and cover slip.

(ii) *Klein's silver stain* (Farmer, 1980). This stain reveals the tubules and other supporting structure for the cilia.

1. Place a drop of ciliates on a slide and let dry, or use the fixing method described previously.
2. Immerse for 20 min in 3% silver nitrate solution at 5–10 °C.
3. Wash the slides in cold distilled water.
4. Submerge in water, expose to sunlight for 30 min (or an equivalent time under a UV lamp).
5. Dehydrate in a graded ethanol series into xylene and mount.

2. *Bacteria* (see Chapter III)

a. Gram stain (Poinar & Thomas, 1984)
An important bacteriological stain for diagnostic identification. Gram-positive organisms retain the violet stain and appear blue–violet; Gram negative organisms are coloured with the counterstain and appear red. A variation of this procedure can be found in Chapter III.

Procedure

1. Air dry smears, lightly heat fix in flame (smear side up).
2. Flood slide with ammonium oxalate crystal violet for 1 min.
3. Rinse in tap water for 5 s.
4. Rinse with Gram's iodine then flood with this solution for 1 min.
5. Rinse in tap water for 5 s.
6. Rinse slide in three changes of *n*-propyl alcohol in coplin jars, 1 min each.
7. Rinse in tap water for 5 s.
8. Rinse with safranin counterstain then flood with counterstain for 1 min.
9. Rinse in tap water for 5 s, then air dry.
10. Examine under oil immersion.

b. Flagella stain (Poinar & Thomas, 1984)
This is used to visualize bacterial flagella.

Culture. Grow test organisms in 3 ml of a phosphate-enriched broth medium for 16 h or less at 20 °C.

Fixation. Add 6.0 ml of 10% formalin to the 3 ml of culture.

Wash

1. Dilute the fixed culture with distilled water and centrifuge at 3000 rpm for 30 min.
2. Decant and discard the supernatant, resuspend the pellet in distilled water and centrifuge again. Repeat.
3. Suspend the pellet in distilled water until barely turbid.

Slide preparation

1. Clean slides overnight in hot (70–80 °C) sulphuric acid saturated with potassium dichromate.
2. Rinse slides thoroughly in tap water, then distilled water and then air dry. Slides must be kept grease free so handle only with clean forceps. Store in a clean, dry, airtight container.
3. Just prior to use, heat a slide in the flame of a Bunsen burner (the side to be used against the flame) and draw a line with a wax pencil across the slide about one third of the distance from one end. Handle slide only on the short end.
4. Place a drop of the final bacterial suspension on the distal end of the cooled slide, tilt the slide to cause the suspension to run down to the wax line. After the slide has air dried, it is ready to be stained.

Staining procedure

1. Place the prepared slide on a staining rack and flood with the flagellar staining solution for 5–15 min (shorter time for new and/or warm stain, longer for old and/or cold stain).
2. Wash all stain off the slide with running tap water.
3. Air dry and examine under oil for flagella.

3. *Fungi*

For additional information on staining Fungi, see Chapter V.

a. Aceto-orcein nuclear stain
See Chapter V.

b. Lactophenol Cotton Blue (Lipa, 1975)
Hyphae and spores stain blue; fat substances stain orange–red. This is used as both a mounting media and stain for fungi. A variation of this stain can be found in Chapter V.

1. Prepare lactophenol and add 0.5% methyl blue.
2. Place the fungal preparation into a drop of the stain on a glass slide.
3. Cover with a cover glass and heat slightly to enhance staining.
4. Cool and examine.

4. Viruses

For additional information on staining viruses see Chapter II.

a. Buffalo Black 12 B
See Chapter II.

b. Giemsa stain
See Chapter II.

c. Sudan III stain for virus polyhedrosis inclusions (Thomas, 1974)
This is used to differentiate virus polyhedra from fat droplets. Fat droplets stain red while polyhedra remain unstained.

1. Air dry smear.
2. Stain for 10–15 min in saturated aqueous Sudan III.
3. Rinse for 5–10 s in running tap water.
4. Air dry and examine under oil.

d. Modified azan staining technique (Hamm, 1966)
This is used for detection of occlusion body viruses (NPV, CPV, GV and Entomopox virus).

Staining procedure for paraffin sections

1. Toluene via alcohols to water
2. 50% acetic acid, 5 min.
3. Distilled water rinse, 2 min.
4. Azocarmine (solution 1), 15 min.
5. Distilled water rinse, 5 s.
6. Aniline, 1% in 95% alcohol, 30 s. (aniline should be distilled and kept in the freezer).
7. Distilled water rinse, 5 s (change often).
8. Counterstain (solution 2) 15 min.
9. 50% Alcohol, 10 s.
10. Absolute alcohol, two changes, 30 s each
11. Toluene, two changes
12. Mount in neutral, synthetic mounting medium

Results

- Virus inclusion bodies – red
- Epicutile – red
- Endocuticle – blue
- Muscle – light blue to blue–green
- Epidermal cells – yellowish-green
- Fat body – yellowish-green with darker green nuclei
- Nerve tissue – light blue
- Silk gland – green, contents red or blue
- Midgut epithelium – green and blue

e. Negative staining
This procedure can be used to detect small non-occluded viruses with transmission electron microscopy by creating a darker background around the virus particle. This is done by using a 2% (w/v) aqueous phosphotungstic acid (PTA) adjusted to pH 7.5 with 1 N NaOH or KOH. This should be made fresh for each use. A drop of the viral suspension is placed onto a formvar coated grid for approximately 1 min depending on the size of the particles. The excess is removed from the slide with a sliver of filter paper. A drop of the PTA is then placed on the slide for 1 min and removed and the grid allowed to air dry prior to viewing. Alternatively, the viral suspension and PTA can be mixed together and then placed onto the grid. After 1 min (time will vary) remove excess and allow to dry. Modification of this procedure is usually required and references for additional information can be found in Adams & Bonami (1991).

5. Nematodes

Detailed procedures for staining living and fixed nematodes can be found in Chapter VI.

a. Permanent mounts (Woodring & Kaya, 1988)
Fix the nematodes in TAF for 4–5 days. Process to glycerin via the evaporative method of Poinar (1975). Make certain specimens are free of dust and

dirt. Filter solutions if necessary. Put fixed specimens in an ethanol–glycerine–water solution in a small dish. Cover all but ⅛ of the surface area for 2 days and then all but ¼ for 7 days. The alcohol and water will evaporate to leave the nematodes in pure glycerin. Mount as described by Southey (1970).

3 ELECTRON MICROSCOPY

A Transmission electron microscopy (TEM)

Electron microscopy of biological materials places rather strict requirements on specimen preparation in order to obtain high quality micrographs for detailed study. Protocols for preservation, dehydration and embedding of tissues in a suitable medium must be carefully followed but modifications are often required depending upon the host and pathogen under investigation. Once this has been accomplished, thin sections (approximately 90–150 nm) are mounted on grids, stained and then viewed and photographed with the electron microscope. This is a tedious process involving many steps making problem resolution a difficult task. Although experience and practice are key to successful electron microscopy, the protocols and procedures given below are intended to provide a basic foundation for initiating studies utilizing the electron microscope.

1. Tissue preparation

This process involves the fixation of the tissue (hardening and preservation), dehydration and infiltration with a medium that can be hardened to give a material suitable for thin sectioning. The main goal of this process is to stabilize and preserve the fine structural details of the cells to a state near to that in the living tissues.

a. Fixation

This is the first step in the preparation of biological specimens for examination by electron microscopy. This process must be accomplished as soon as possible after sacrifice so that post-mortem changes are kept to a minimum. Aldehydes and osmium tetroxide (OsO_4) are the most effective fixatives for TEM. Fixatives cross-link macromolecules causing them to become immobilized and insoluble. One standard procedure involves double fixation using glutaraldehyde as the primary fixative followed by osmium tetroxide (OsO_4). Glutaraldehyde stabilizes tissues by cross-linking proteins. Osmium tetroxide reacts with lipids and certain proteins but also provides electron density to the tissue. Therefore, OsO_4 acts as both a postfixative and an electron stain (Figure 1A). Without OsO_4 or if the OsO_4 is bad, nuclear membranes and cytoplasmic membranes of the endoplasmic reticulum, Golgi and other organelles will not be preserved (Figure 1B). Some procedures also involve a third fixative, uranyl acetate, before or during dehydration often to enhance the electron

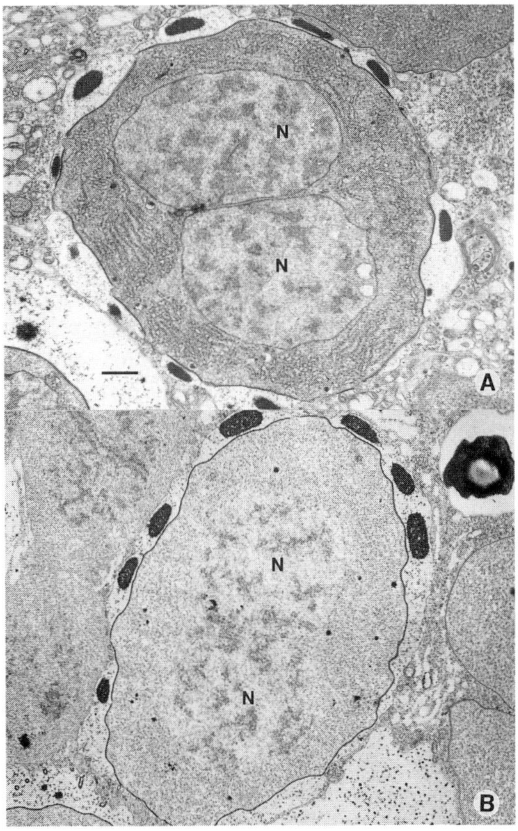

Figure 1 Transmission electron micrographs of diplokaryotic sporonts of the microsporidium *Amblyospora californica* from the mosquito *Culex tarsalis*. **A.** Double fixation with glutaraldehyde–osmium. Membranes of nuclei (N) and cytoplasmic membranes (endoplasmic reticulum, Golgi) are well preserved. **B.** Fixation with glutaraldehyde only. The two nuclei (N) are evident by the presence of chromatin but nuclear membranes and cytoplasmic membranes are not preserved. Bar = 1 μm.

density of the material which is therefore referred to as 'en block stain'. It also acts as a fixative particularly for lipid components. Good fixation is usually measured by the continuity of membrane structures and the lack of obvious distortions and discontinuity in cytoplasmic details (Figure 1A). One group of organelles to examine carefully are the mitochondria which should have clearly defined cisternae without swelling or lysis. The following is a general procedure for the fixation of insect tissues infected with a pathogen.

General procedure

1. Dissect specimen in 2.5% glutaraldehyde (it is critical that the specimens be living when sacrificed) After 5–15 min, the specimen can be cut into smaller pieces (1–2 mm^3).
2. Transfer pieces to fresh glutaraldehyde and fix for a total of 2.5 h at room temperature or overnight in the refrigerator.
3. Wash in 0.1 M cacodylate buffer (pH 7.2–7.3) three times at 15 min each (for a total of 45 min). Rinses are important to prevent any reaction between the primary and postfixative.
4. Postfix in 1.0% OsO_4 (pH 7.5) for 1 h 45 min. to 2 h. This should be done at room temperature with the vials wrapped in foil. Osmium should be handled with great care and only under a fume hood. Gloves should be used.
5. Double distilled water washes – three times at 15 min each (for a total of 45 min).
6. Begin dehydration or, for extended storage, use sucrose buffer.

An alternative to chemical fixation is freeze-substitution. This protocol was developed to avoid the many artefacts associated with conventional chemical fixation. The term freeze-substitution refers to the dissolution of ice in a frozen specimen by an organic solvent at low temperatures. The sample is quickly frozen by one of several ultrarapid freezing techniques and then the water in the sample is substituted by an organic fluid, such as methanol, ethanol or acetone, at very low temperatures. Usually, the solvent contains a chemical fixative, such as OsO_4, with substitution requiring 48 h at –75 to –85 °C. The sample is then brought to room temperature and infiltrated and embedded conventionally. Excellent results have been obtained with this method but the sample size is critical and is usually limited to specimens made up of individual cells and cell layers. An excellent discussion of this technique for use with fungal cells is provided by Hoch (1986).

b. Problem tissues

Processing certain tissues is often difficult due to either the small size of the specimens or problems with tissues that do not readily sink in the fixatives. In most cases, small specimens (cells, spores, eggs, etc.) can be embedded in agar and handled like pieces of tissue. The specimens are first fixed (at least through glutaraldehyde), and the fixative removed by centrifugation. The specimens are then washed at least twice in buffer and the specimens resuspended in warm 2% agar. After hardening, the agar with the specimens can be cut into small pieces and handled like pieces of tissue to complete processing.

For tissues that will not sink in the fixative, small carriers can either be purchased or constructed from Beem capsules and small wire mesh (Adams & Bonami, 1991). The tissues are placed into the holders that will sink in the fixative and can usually be removed prior to osmium fixation. Make sure that no air bubbles are trapped around the tissues in the holder so the fixative is in contact with the tissue. A simple alternative is to overfill the vial containing the tissues with glutaraldehyde until you have a positive meniscus (the tissue will be floating on the surface). Carefully stretch a piece of parafilm over the top of the vial trapping the tissue and removing all the air. Tighten the cap and the tissue should sink to the bottom of the vial. Process as normal.

c. Dehydration

After fixation, tissues must be dehydrated and embedded. Dehydration is achieved by transferring the material through an ascending alcohol or acetone series into absolute alcohol or acetone. A sample dehydrating protocol is given below but can be modified to reduce the steps by using increments of 25% (for example 25, 50, 75, 95% for 10 min each).

1. 10% ethanol (ETOH), 10 min
2. 30% ETOH, 10 min
3. 50% ETOH, 10 min
4. 70% ETOH, 10 min (good point for en block staining, wrap in foil and hold overnight)
5. 80% ETOH, 10 min
6. 90% ETOH, 10 min

7. 95% ETOH, 10 min
8. 100% ETOH, 15 min
9. 100% ETOH, 15 min
10. 100% Acetone, 15 min
11. 100% Acetone, 15 min

Immediately put specimen into plastic dilutions.

Note: Absolute alcohols and acetone must be stored over molecular sieve to ensure the absence of water.

Quick dehydration protocols using 2,2-dimethoxypropane (DMP)

1. Add 1–2 ml DMP + 1–2 ml distilled H_2O + 3–4 drops of 0.2 N HCl. Shake; should turn cold. Hold 5–15 min.
2. Remove solution, add 2–3 ml DMP + 3–4 drops HCl (no distilled H_2O). Shake; hold 5–15 min (two changes).
3. Absolute acetone three times, 15 min each.

d. Infiltration and embedding

The final process in tissue preparation is to infiltrate the specimens with a liquid embedding medium which is then polymerized to produce a solid block. Epoxy resins are perhaps the most commonly used media and a general protocol is given below that is easy to use and provides uniform blocks that are easy to section and stain. Other resins are available for specific purposes such as Spurr's resin which is less viscous but is more difficult to section and stain and is not as stable under the beam of the electron microscope. Embedding media should be handled with caution, and paying careful attention to the safety data sheets is essential.

A combination of two epoxy resins, Araldite and Epon, is easy to prepare and has been shown to be highly reliable. After dehydration, specimens are infiltrated with the embedding medium by passing them through a series of solutions until the dehydrating agent has been completely replaced by the final embedding medium. This is done in small vials on a shaker at room temperature. Activator must be included in all dilutions. After the pure resins, tissues are transferred to capsules, filled with pure resins and polymerized in an oven.

1. 25% resin: 75% absolute acetone, 4 h to overnight.
2. 50% resin: 50% absolute acetone, 4 h.
3. 75% resin: 25% absolute acetone, 4 h.
4. Pure resin overnight.
5. Pure resin (change vials), all day (≈6 hours) (see note below).
6. Embed in Beem™ capsules which have dried for at least 24 h in a 60 °C oven. A small drop of fresh plastic is put into the tip of the Beem capsule and the tissue is placed into the drop and the capsule filled with resin. Make sure to include label with block number when embedding. Leave in oven (uncovered) overnight. Be sure no air bubbles are below the tissue.
7. Remove the embedded blocks next morning and allow to cool (best for 24 h) prior to sectioning.

Note: For better infiltration of difficult tissues, extend the specimen in pure resin for another day (overnight) or for several days changing daily. Embed as usual.

2. Sectioning and staining

a. Remarks on thick and thin sectioning

Prior to facing and thin sectioning, thick sections (0.5–1 μm thick) can be removed from the block using a glass knife. These sections can be transferred directly to a slide and mounted with Pro-Texx™ and a cover slip. Sections can be examined directly (without staining) with phase contrast to locate areas of interest for the final trimming (facing). Alternatively, the sections can be stained prior to covering. Once the area of interest has been determined, the block is trimmed until a 'face' of the appropriate size is obtained. The trimmed block is mounted in a holder on the ultramicrotome and automatically advanced to be sectioned by either a glass or diamond knife. Sections are floated onto water and transferred to a grid for thin sections. Thin sections are generally in the range 90–150 nm which can be judged from the interference colours shown by the sections as they float on the water surface. Sections in this thickness range will generally appear gold with light gold sections thinner and dark gold sections thicker. Thin sections are transferred to grids or grids coated with formvar for added stability under the beam. After drying, the sections are ready for post-staining prior to viewing in the electron microscope.

b. Post-staining

This process serves to increase the contrast in thin sections and is usually performed immediately prior to viewing. A two-step staining protocol is commonly employed with excellent results. Grids are floated onto a drop of uranyl acetate (section side down) for 5 min. The time will vary depending on whether aqueous or methanolic uranyl acetate is used and the thickness of the sections (thicker sections take less time). The grids are passed through three rinses in deionized water (hold grid and quickly dip in water), blot and immediately submerge into a drop of lead citrate, section side up, for 5 min. Grids are rinsed three times in deionized water, blotted on filter paper and allowed to dry before viewing. For difficult to stain material, time in the uranyl acetate can be extended or various concentrations of methanolic uranyl acetate used. For extremely difficult tissues, 1% dimethylsulphoxide (DMSO) in 100% methanolic uranyl acetate has proven useful. A possible problem when post-staining with 100% methanolic uranyl acetate is the loss of sections from the grid. This can be prevented by passing the grid under the electron beam at low intensity to adhere the sections to the grid before post-staining.

One of the most common problems encountered in post-staining is the presence of lead precipitate on the sections. A simple solution is to restain the sections in uranyl acetate which will remove the precipitate. It is then necessary to restain in freshly made lead citrate. Another common problem is the presence of uranyl acetate precipitate, which can be removed with oxalic acid (Avery & Ellis, 1978).

B Scanning electron microscopy (SEM)

Scanning electron microscopy has been used in the study of insect pathology primarily for examining surface morphology of microsporidian spores and fungal conidia and developmental stages. Some applications have also been useful for bacteria and viruses. Usually the process involves fixation of the material, dehydration and drying followed by mounting onto a grid or stub and applying a conductive coating. An extensive reference section on SEM is found in Adams & Bonami (1991).

1. Specimen preparation

Although some specimens can be viewed without fixation, results are generally improved by fixing in both glutaraldehyde and osmium similar to the procedures for TEM. Specimens are then dehydrated and can be mounted and air dried. Often this results in artefacts caused by shrinkage and collapse of the specimens. Best results are usually obtained when specimens are critically point dried. Specimens must be placed into a carrier, dehydrated and critically point dried to reduce damage to the tissue. A chemical method of drying soft tissues has also been developed (Nation, 1983).

Low-temperature scanning electron microscopy examines samples that are rapidly frozen (frozen-hydrated) and maintained under vacuum. This is an alternative to chemical fixation and provides excellent results but does require a specialized SEM. For an excellent discussion of the procedures and protocols see Beckett & Read (1986).

2. Mounting and coating

Depending on the size, the specimen can be mounted on grids or stubs. Larger specimens can be adhered to a stub with conductive silver paint after critical point drying. There are many methods for handling small specimens such as collecting them on a filter disk after fixation. The disk can then be used to carry the specimen through dehydration and critical point drying. The disk is then mounted onto a stub with conductive silver paint and coated. For SEM, a coating of a conductive metal layer (usually gold or palladium) is required. This is usually applied to the mounted specimens with a sputter coater. Experimentation is necessary to obtain a coating of suitable thickness. Specimens are then ready for examination with the SEM.

REFERENCES

Adams, J. R. & Bonami, J. R. (1991) *Atlas of invertebrate Viruses*. CRC Press, Boca Raton, 684 pp.

Aldrich, H. C. & Todd, W. J. (1986) *Ultrastructure techniques for Microorganisms*. Plenum Press, New York, 533 pp.

Alger, N. E. (1966) A simple, rapid, precise stain for intestinal Protozoa. *Amer. J. Clin. Pathol.* **45**, 361–362.

Avery, S. W. & Ellis, E. A. (1978) Methods for removing uranyl acetate from ultra-thin sections. *Stain Technol.* **53**, 137.

Barbosa, P. (1974) *Manual of basic techniques in insect histology*. Autumn Publishers, Amherst, 245 pp.

Beckett, A. & Read, N. D. (1986) Flow-temperature scanning electron microscopy. In *Ultrastructure techniques for microorganisms* (eds H. C. Aldrich & W. J. Todd), pp. 45–86. Plenum Press, New York.

Eide, P. E. & Reinecke, J. P. (1970) A physiological saline solution for sperm of the house fly and the black blow fly. *J. Econ. Entomol.* **63**, 1006.

Farmer, J. N. (1980) *The Protozoa: introduction to protozoology*. C. V. Mosby, St Louis, 732 pp.

Glauert, A. M. (ed.) (1974) *Practical methods in electron microscopy* vol. 2. North-Holland, Amsterdam, 353 pp.

Hamm, J. J. (1966) A modified azan staining technique for inclusion body viruses. *J. Invertebr. Pathol.* **8**, 125–126.

Hayat, M. A. (1986) *Basic techniques for transmission electron microscopy.* Academic Press, New York, 411 pp.

Hoch, H. C. (1986) Freeze-substitution of fungi. In *Ultrastructure techniques for microorganisms* (eds H. C. Aldrich & W. J. Todd), pp. 183–212. Plenum Press, New York.

Lipa, J. J. (1975) *An outline of insect pathology*. PWRiL, Warszawa, 342 pp.

Lee, J. J., Small, E. B., Lynn, D. H. & Bovee, E. C. (1985) Some techniques for collecting, cultivating and observing protozoa. In *Illustrated guide to the protozoa* (J. J. Lee, S. H. Hutner & E. C. Bovee, eds), pp. 1–7. Society of Protozoologists, Lawrence.

Lom, J. & Vavra, J. (1963) Mucous envelopes of spores of the subphylum Cnidospora (Dolfein, 1901). *Vestn. Cesk. Spol. Zool.* pp. 274–276.

Luna, L. G. (ed.) (1960) *Manual of histologic staining methods of the Armed Forces Institute of Pathology*, 3rd edn. McGraw–Hill, New York, 258 pp.

Nation, J. L. (1983) A new method using hexamethyldisilazane for preparation of soft insect tissues for scanning electron microscopy. *Stain Technol.* **58**, 347.

Poinar, G. O. Jr (1975) *Entomogenous nematodes*. E. J. Brill, Leiden, Netherlands, 317 pp.

Poinar, G. O. Jr. & Thomas, G. M. (1984) *Laboratory guide to insect pathogens and parasites*. Plenum Press, New York, 392 pp.

Short, H. E. & Cooper, W. (1948) Staining of microscopical sections containing protozoal parasites by modification of McNamara's method. *Trans. R. Soc. Trop. Med. Hyg.* **41**, 427–428.

Southey, J. F. (ed.) (1970) *Laboratory methods for work with plant and soil nematodes*. Ministry of Agriculture, Fisheries and Food, Technical Bulletin 2. HMSO, London.

Thomas, G. M. (1974) Diagnostic techniques. In *Insect Diseases*, vol. 1 (G. E. Cantwell, ed.), pp. 1–48. Marcel Dekker, New York.

Vavra, J. & Chalupsky, J. (1982) Fluorescence staining of microsporidian spores with the brightener "Calcofluor White M2R". *J. Protozool.* **29**, 503 (Abstract no. 121).

Vavra, J. & Maddox, J. V. (1976) *Methods in micro sporidiology. Comparative Pathobiology*, vol. 1. *The Biology of the Microsporidia*, pp. 271–319. Plenum Press, New York.

Weiser, J. (1976) Staining of the nuclei of microsporidian spores. *J. Invertebr. Pathol.* **28**, 147–149.

Woodring, J. L. & Kaya, H. K. (1988) *Steinernematid and heterorhabditid nematodes: A handbook of biology and techniques*. Southern Cooperative Series Bulletin 331, Fayetteville, AR, 30 pp.

APPENDIX

Dissecting fluids

Ringer's solution

Sodium chloride (NaCl)	8.0 g
Calcium chloride ($CaCl_2$)	0.25 g
Potassium chloride (KCl)	0.25 g
Sodium bicarbonate ($NaHCO_3$)	0.25 g
Distilled water to make 1000 ml	

Note: The amount of sodium chloride can vary from 6.5 g to 9.0 g depending on the organisms under study.

Simple physiological saline

Sodium chloride (NaCl)	0.85 g
Distilled water to make 100 ml	

Eide & Reinecke's Physiological Saline (Eide & Reinecke, 1970)

Sodium chloride (NaCl)	0.453 g
Magnesium chloride hexahydrate ($MgCl_2 \cdot 6H_20$)	0.3 g
Sodium bicarbonate ($NaHCO_3$)	0.035 g
Dextrose ($C_6H_{12}0_6$)	1.155 g
Potassium chloride (KCl)	0.107 g
Monosodium phosphate ($NaH_2PO_4H_2O$)	0.04 g
Sodium acetate ($C_2H_3O_2Na$)	0.025 g
Distilled water to make 100 ml	

Fixatives for light microscopy

Buffered neutral formalin

Formalin (CH_2O, 37–40%)	100.0 ml
Distilled water	900.0 ml
Sodium phosphate monobasic ($NaH_2PO_4 \cdot H_2O$)	4.0 g
Sodium phosphate dibasic (Na_2HPO_4)	6.5 g

Carnoy's fixative

Absolute ethanol (C_2H_5OH)	60 ml
Chloroform ($CHCl_3$)	30 ml
Glacial acetic acid ($C_2H_4O_2$)	10 ml

Bouin's fixative

Saturated aqueous picric acid	75 ml
Formalin (CH_2O, 37–40%)	25 ml
Glacial acetic acid ($C_2H_4O_2$, add just before use)	5 ml

TAF (Southey, 1970)

Formalin (CH_2O, 37–40%)	7 ml
Triethanolamine ($C_6H_{15}NO_3$)	2 ml
Distilled water	91 ml

See Chapter VI for additional fixatives for nematodes.

Stains for light microscopy

Heidenhain's haematoxylin stain

Distilled water	90 ml
Absolute ethanol	10 ml
Haematoxylin (Harleco no. 234 – Haematoxylin stain CI no. 75290)	0.5 g

Dissolve haematoxylin in alcohol, add water and age in the dark for at least 6 weeks. Store in the dark. To use, dilute 1 : 1 with distilled water and add 3 drops saturated lithium carbonate (Li_2CO_3)/100 ml.

Mordant

Iron alum (ferric ammonium sulphate ($FeNH_4(SO_4)_2 \cdot 2H_2O$) (Iron alum crystals should have a violet-pinkish colour)	2.5 g
Distilled water	100 ml

Wheatley's modified Gomori trichrome stain

Distilled water	100 ml
Glacial acetic acid ($C_2H_4O_2$)	1.0 ml
Phosphotungstic acid ($12WO_3 \cdot H_3PO_4 \cdot H_2O$)	0.7 g
Chromotrope 2R ($C_{16}H_{10}N_2Na_2O_8S_2$)	0.4 g
Bright green SF, certified	0.3 g
Bismarck brown, certified	0.1 g

Combine ingredients and let stand 24–48 h before use. A sediment develops but causes no problem. The stain is stable for 1 year.

Eosin Y stain

Eosin Y	5 g
Distilled water	100 ml

Filter the eosin solution and add a few drops of glacial acetic acid just before use.

Slide affixative

Saturated mercuric chloride ($HgCl_2$)	10 ml
Glacial acetic acid ($C_2H_4O_2$)	2 ml
Formalin (CH_2O, 37–40%)	2 ml
Tertiary butyl alcohol ($(CH_3)_3COH$)	10 ml

Giemsa-stain

Giemsa stain	1 part
0.01M phosphate buffer, pH 7.4	9 parts

Good results have been obtained with the Fisher Scientific, Baker Chemicals products and a new product from Sigma that needs only to be diluted with distilled water because it is already buffered. Phosphate buffer at pH 7.4 (premixed packets can be obtained from Fisher Scientific). The stain solution must be prepared fresh for each use.

Gram stain

1. Ammonium oxalate crystal violet

Solution A

Crystal violet (90% dye content)	4 g
Dissolve in 40 ml 95% ethanol	

Solution B

Ammonium oxalate ($C_2H_8N_2O_4 \cdot H_2O$)	1.6 g
Dissolve in 160 ml distilled water	

Mix solutions A and B 48 h before use.

2. Gram's iodine

Potassium iodine (KCl)	2 g
Iodine	1 g

Grind in a mortar for 5–10 s. Add 1 ml distilled water and grind until all ingredients are in solution. Add 10 ml water and mix. Rinse into a reagent bottle and bring the volume to 200 ml.

3. Counterstain

Safranin (86% dye content)	0.5 g
Ethanol (95%)	20 ml

Mix. Add to 180 ml distilled water.

Flagella stain

A. Basic fuchsin 1.2 g
Dissolve in 100 ml of 95% ethanol.

B. Tannic acid 3.0 g
Dissolve in 100 ml of distilled water

C. Sodium chloride (NaCl) 1.5 g
Dissolve in 100 ml of distilled water

Prepare the stain by mixing equal parts of the three stock solutions. The stain solution may be stored for 1 week at room temperature, 1–2 months under refrigeration and indefinitely if frozen.

Lactophenol and cotton blue

Phenol crystals ($C_6H_6O_2$)	100 g
Lactic acid (USP 85%)	80 ml
Glycerin	159 ml
Distilled water	100 ml

Mix ingredients and heat until hot; add 0.5% cotton blue.

Aqueous eosin

Eosin Y (C.I. 45380)	1 g
Distilled water	100 ml

Mix ingredients and filter. Add several drops of glacial acetic acid to staining solution before use.

Lacto-aceto-orcein stock (2%)

Orcein	2 g
Glacial acetic acid (45%)	50 ml
Lactic acid (85%)	50 ml

Place orcein and acids into flask and plug with cotton. Heat to near boil (do not boil!) and hold for 30 min. Filter while hot. Cool and dilute stock 1 : 3 with 45% acetic acid for the final stain.

Modified Azan staining technique (Hamm, 1966)

Solution 1

Azocarmine G	0.1 g
Glacial acetic acid	2 ml
Distilled water	100 ml

Dissolve azocarmine G in water and boil for 5 min. Cool and add acid. Filter before use.

Solution 2

Phosphotungstic acid	1.0 g
Aniline blue (water soluble)	0.1 g
Orange G	0.5 g
Fast green FCF	0.2 g
Distilled water	100 ml

Dissolve all ingredients in water.

Ethanol–glycerine–water solution (Poinar, 1975)

95% Ethanol	15 parts
Glycerin	1 part
Distilled water	5 parts

Fixatives, buffers and stains for electron microscopy

Working solutions

0.2 M Cacodylate buffer

Cacodylate buffer stock	50 ml
0.2 M HCl	6 ml
Double distilled water to make	100 ml

2.5% Glutaraldehyde

8% Glutaraldehyde	10 ml
0.2 M Cacodylate buffer	16 ml
Double distilled water	6 ml
Calcium chloride ($CaCl_2$)	32 mg

1% OsO_4

4% OsO_4	1 ml
0.3 M Sucrose	1 ml
0.2 M Cacodylate buffer	2 ml

Wrap vial in foil during fixation.

0.1 M Cacodylate buffer/sucrose

0.2 M Cacodylate buffer	5 ml
Double distilled water	5 ml
Sucrose	0.1 g

En bloc stain

Uranyl acetate (UrAc)	0.1 g
70% Ethanol	20 ml

Wrap in foil.

Stocks

0.4 M Cacodylate buffer

Cacodylate Acid ($Na(CH_3)_2AsO_2{\cdot}3H_2O$)	42.8 g
Double distilled water to make	500 ml

0.3 M Sucrose

Sucrose	5.1 g
Double distilled water to make 50 ml	

Store in refrigerator.

4% Osmium tetroxide stock

Osmium tetroxide (OsO_4)	1 g
Double distilled water	25 ml

Wrap in foil; dissolve at room temp., usually 24 h; store in refrigerator.

0.2 M HCl

Hydrochloric acid (HCl)	1.6 ml
Double distilled water to make 100 ml	

Store in refrigerator.

Epon–Araldite

Epon 812	2 g
Araldite 502	1 g
DDSA, Hardner	4.5 g
DMP-30, Activator	4 drops

Plastic tripour beakers (50 or 100 ml) are used to mix the plastics. The Epon 812, Araldite 502 and the DDSA are weighed out on a top loading balance into the tripour beaker. This mixture is usually placed into a 55–60 °C oven for 1–2 min to facilitate the mixing of the resins. Four drops of DMP-30 are added with a medicine dropper and mixed immediately by swirling the components. Attempt to avoid too many bubbles, but this is not crucial. The plastic will darken but should not turn orange. If the plastic turns orange then the DDSA used is probably not good. The plastics should last a long time. Only enough plastic is mixed for each use. Beem capsules are used for embedding the tissues. Do not put the lids on these during the curing process. Cure overnight in a 62–65 °C oven. Larger batches can be made by multiples of the ingredients, but not more than four times the basic formula.

Standard Reynolds lead citrate

Lead nitrate ($Pb(NO_3)_2$)	1.33 g
Sodium citrate ($Na_3(C_6H_5O_4) \cdot 2H_2O$	1.76 g
Freshly boiled and cooled distilled water	

1. Dissolve lead nitrate completely in 30 ml distilled H_2O.
2. Add sodium citrate. A heavy white precipitate will form.
3. Add 8 ml of 1 N NaOH (1g/25ml) and dilute to 50 ml with boiled then cooled water.
4. Mix until precipitate is dissolved.
5. pH should be 12.

This stain can be stored for several months in the refrigerator if sealed properly. Discard when precipitate forms or when contamination is found on stained grids.

2.5% Uranyl Acetate in 50% Methanol

Uranyl acetate (UrAc)	0.5 g
Absolute methanol	10 ml
Distilled water	10 ml

Wrap in foil, shake until dissolved, store in refrigerator.

This is a standard uranyl acetate stain but can be easily modified to suit individual needs for different plastics or section thickness. For easily stained sections, aqueous or 25% methanol can be used. More difficult sections can be stained in 75 or 100% methanol. In some cases when additional staining is needed for particularly difficult material, 1% DMSO can be added to the 100% methanol.

Formvar coated grids

Wash a glass microscope slide in 95% ethanol. Air dry for 1–2 min. Soak the business end of the slide in dilute dishwashing detergent for 2–3 min. Wipe the slide partially dry with a Kimwipe™ but leave some of the detergent on the slide so that when it dries a detergent residue remains. When dry, wipe the slide vigorously with a dry Kimwipe™. It will feel slightly slick and waxy but will look clean. Dip the slide in 0.25% formvar dissolved in ethylene dichloride or chloroform and dry. Scrape edges of slide with a razor blade to free film from slide. Release onto water by inserting slide slowly under a water surface at a 45° angle. Place grids face down onto the floating film. Pick the film up on an index card and dry overnight in a Petri dish cracked open placed on the top of a 60° oven. (Contributed by Henry C. Aldrich, University of Florida.)

Safety, hazards and precautions

Many of the reagents utilized to prepare insect pathogens for study are potentially hazardous.

Material safety data sheets are provided with all reagents and should be made available to all individuals who handle the material. Preventative protocols should always be followed when appropriate, such as working under a fume hood or wearing gloves and lab coats. Remediation procedures should be in place in the event of an accidental spill or exposure to toxic substances. Safety training should be a part of every laboratory's general operating procedures.

CHAPTER VIII-2

Complementary techniques: Fluorescence microscopy

TARIQ M. BUTT

IACR-Rothamsted, Harpenden, Hertfordshire AL5 2JQ, UK

1 INTRODUCTION TO FLUORESCENCE MICROSCOPY

Considerable advances have been made in recent years in the development of new fluorescent dyes which, in turn, have been used to investigate different aspects of fungal development. It is recognized that fluorescence microscopy (FM) offers a powerful means for observing the interplay of ions, organelles and other cell components during growth and differentiation. This chapter briefly describes the basic principles of FM and provides protocols for its use in the study of invertebrate mycopathogens.

A Fluorescence and fluorochromes

Fluorescence is the luminescence of a substance excited by radiation. When radiation of relatively high energy falls on a substance the latter absorbs and/or converts a certain small part of the energy into heat. Most of the energy which is not absorbed by the substance is re-emitted at a longer wavelength, a process which is referred to as 'fluorescing'. Some substances (e.g. some oils, waxes and chlorophyll) autofluoresce. By using the correct staining and filter combinations it is possible to exclude autofluorescence and encourage useful fluorescence.

Fluorochromes are dyes that fluoresce when excited by light of a specific wavelength. They are either absorbed by cell organelles, or bind to specific residues inside or on the cell. They can also be conjugated to probes like certain drugs, antibodies and lectins which exhibit high affinity for specific cell components (Butt *et al.*, 1989). A list of some useful fluorochromes and their properties are given in Table 1.

B The fluorescence microscope

In fluorescence microscopy there are three possible ways of illuminating the specimen:

- dia-illumination by a substage bright-field condenser;

MANUAL OF TECHNIQUES IN INSECT PATHOLOGY
ISBN 0–12–43255–6

Table 1 Solubility, spectral and other properties of selected fluorochromes.

Fluorescent Dye	*Source*	*Molecular weight*	*Excitation filter*	*Barrier filter*	*[conc] Solvent*	*Applications*
Calcofluor white M2R = Tinopal LPW	Sigma	960.9	340–370	420–530	0.1% H_2O	Cell wall
Auromine O	Sigma	304.0	355–440	530–580	0.1% H_2O	Cell wall
Acridine orange	Sigma	301.82	410–490	510–530	H_2O, Ethanol	Nucleic acids, mucus
4′6-diamidino-2-phenylindole (DAPI)	Sigma	350.0	340–370	420–530	0.1–5 µg/ml H_2O	AT specific DNA probe
Hoechst 33342 Bis-benzimide	Hoechst	454.56	355	468	H_2O	AT specific DNA probe
Hydroethidine	Polysciences	315.42	370/535	420–585	2 µg–7 mg/ml H_2O, ethanol, DMSO	Cytoplasm is blue, chromatin stains red/orange
Nile red	Molecular probes	380.83	450–500 515–560	528–608	organic solvents	Lipid specific
3,3′-Dihexyloxacarbocyanine Iodide ($DiOC_6$)	Molecular probes	572.53	450–497	526	0.5–5 µg/ml DMSO, H_2O	Cell membrane potentials, endoplasmic reticulum, mitochondria
Rhodamine 123	Molecular probes	380.83	470 515–560	590	3–10 µg/ml ethanol, H_2O	Mitochondria
Chlorotetracycline (CTC)	Sigma	515.3	365 450–490	420 520	10^{-4}M H_2O	Extracellular calcium
Fluorescein diacetate	Molecular probes	416.4	450–490	520	0.1–0.4 µg/ml acetone	Indicator of cell viability, esterases
Propidium iodide	Sigma	668.4	515–560	590	20–60 µg/ml H_2O	Indicator of disrupted plasma membrane/dead cells

- oblique illumination by a substage dark-field condenser;
- epi-illumination by a dichroic beam splitter placed above the objective.

The best, and probably most widely used, system is the last of these which is also referred to as reflected light fluorescence, incident-light excitation and epi-fluorescence. In this system, the excitation light, selected using appropriate excitation filters, is directed by a dichroic prism (= beam splitter) through the objective lens to the specimen. Most of the light not absorbed by the specimen passes through. Fluorescent light and some exciting light reflected by the glass surface enter the objective but the exciting light is cut off by a barrier filter (a special coating on the dichroic prism). The prism simultaneously transmits the reflected, fluorescent light to the eyepiece and/or camera. The barrier filter ensures that only the longer wavelength, fluorescent light passes through and that shorter wavelength light, including harmful UV, is excluded.

In epifluorescence, the objective acts as a condenser which need not be centred but yet concentrates light precisely onto the field of view. Because illumination and observation of the specimen are made from the same direction this supplies more brightness and better image quality especially of thick or opaque objects. Excitation and emitted fluorescence radiation are well separated and therefore do not interfere with the fluorescence image. Epifluorescence can be used simultaneously with phase, differential interference contrast or bright-field illumination not only reducing exposure times when taking photographs but also yielding more information on the spatial relationship of fluorescent and non-fluorescent cell components.

1. Light source

The function of the light source is to provide light at a wavelength corresponding to an excitation maximum of the fluorochrome. High pressure mercury lamps are widely used because they have strong emission at specific wavelengths (e.g. 365, 406, 436, 546, 577 nm) which can be readily isolated by a narrow band filter to give relatively monochromatic light of high intensity. For simultaneous visualization of metachromatic (= fluorochromatic) fluorochromes, which are dyes with two or more excitation and emission spectra, a broad-band excitation filter is required while the barrier filter is chosen on the basis of the orthochromatic form which has the shortest wavelength.

2. Choice of optics

Only objectives made exclusively of non-fluorescent materials should be used, as should lenses with wide numerical apertures because these increase the amount of fluorescent light reaching the eye or camera. It is also important to use non-fluorescing immersion oils in conjunction with the immersion lens.

Most standard objectives can be used for epi-fluorescence with violet, blue or green excitation. For genuine UV excitation (336 nm Hg line) Plan Neofluar or normal Neofluar objectives are recommended. Other lens types may absorb UV light.

3. General information on filters

The role of filters is extremely important in FM. An incorrect selection can cause the background to be too bright to distinguish specific from non-specific fluorescence and can prevent the use of double-staining. The choice of filters depends primarily upon the dye. Filters are employed to remove unwanted transmission and protect filters and the specimen from heat. The most important are the excitation and barrier filters which have already been discussed.

The best approach when selecting a filter combination for a specific application is to try several combinations and see which one gives the best results. First establish the excitation peak of the dye and then choose an excitation filter which selects the nearest line to the desired wavelength. Next, determine the emission spectrum; the barrier filter must block excitation but transmit most of the fluorescence. For example, 4′6-diamidino-2–phenylindole (DAPI) has excitation and emission peaks of 365 and 450 nm, respectively. Since the excitation maximum corresponds to the mercury line of 365, a filter which transmits only this light should be selected together with a barrier filter with a cut-off below 450 nm.

2 PHOTOMICROGRAPHY IN FLUORESCENT LIGHT

Some tips for improving fluorescence photomicrography skills are given below:

- Apply incident instead of transmitted-light excitation wherever possible. Epifluorescence is the simplest, effective mode of FM.
- Align light sources according to the operating instructions.
- Ensure there is no groundglass in the illuminating beam path.
- Condenser aperture diaphragm should be completely open.
- Use objectives of high numerical aperture. Note that fluorescence intensity increases exponentially with the increase in numerical aperture.
- Use microscopes with short light paths. Long distances between light source and specimen involve light losses at all free lens surfaces and reflecting mirrors.
- Use high speed film material for photomicrography such as 400 ASA. Film of slower speed will give less grainy pictures but would only be suitable in instances where fluorescence is intense and relatively stable. For rapidly fading images or weakly fluorescing specimens 800 ASA film may be needed.
- Retard fading by mounting the specimen in an 'antiquenching' agent.
- Use bottom illumination in conjunction with acetate filters complementing fluorescence light.
- A darkened room greatly enhances the image because of reduced glare.
- Locate the microscope in a vibration-free area (i.e. where there is no traffic of people, or heavy machinery in operation), this will ensure photographs are 'in focus'.
- Specimen drift can be remedied by attaching cells to poly-L-lysine-coated coverslips or drawing-off excess liquid from beneath the coverslip with a piece of filter paper.

3 ANTIQUENCHING ('ANTIFADE') AGENTS

Quenching can be caused by a variety of chemicals such as oxygen or excess fluorochrome. Emission is also strongly dependent on the pH of the mountant. For example, fluorescein isothiocyanate (FITC) has an optimum at about pH 9.0. The mounting medium should be non-fluorescent (e.g. water, inorganic buffer solutions, glycerol). Many dyes break down in aqueous solutions so these should be freshly made.

Recipes for selected antifading agents and mounting media are given in the Appendix.

The rapid fading of most fluorochromes can be a serious problem, but the use of antiquenching agents such as *n*-propyl gallate, *p*-phenylenediamine or 1,4-diazobicyclo-(2,2,2)-octane in glycerol markedly retard fading. Commercial preparations of antiquenching agents are available such as Citifluor (Marivac, Halifax, NS, Canada) and Slowfade (Molecular Probes) are available. The latter appears to act as a free-radical scavenger that extends the time of useful fluorescence emission (Haughland, 1992).

4 USES OF FLUOROCHROMES

A Cell wall stains

The fluorochromes Calcofluor White M2R, Uvitex BOPT, and Tinopal LPW, are widely used to stain fungal cell walls. These dyes bind to sugars in the cell wall and can be invaluable in certain mycological studies. For example, identification of fungal propagules, spore counts, and study of fungal propagules in the soil or on plant and insect surfaces. Their major contribution to invertebrate mycopathology has been in the study of fungal infection processes (e.g. Schreiter *et al.*, 1994). Combined with bioassays and biochemical studies, it is possible to investigate the role of specific pathogenicity determinants and to identify vulnerable sites on the host cuticle and barriers to infection using these stains (e.g. Butt *et al.*, 1988, 1995).

Cell wall stains have assisted in monitoring the *in vivo* development of invertebrate pathogens, demonstrating that some fungi multiply as protoplasts within their respective hosts.

In contrast to conventional light microscopy, fluorescing fungal elements can be readily located on dark as well as light transmissible surfaces (Butt, 1987). There is little evidence that it interferes with infection processes. Recently formed septa, appressoria, germ tubes, hyphal tips and buds fluoresce more intensely, possibly due to the loose lattice structure of the newly synthesized wall. Occasionally, it is possible to observe the extracellular matrix surrounding hyphal tips and appressoria. Melanized or pigmented propagules do not fluoresce, presumably because the binding sites are masked by pigments. However, on hydration they may expand and expose the underlying cell wall layer containing β-glucans. This is particularly true for *Metarhizium anisopliae* (Plate 18).

Cell wall stains, when incorporated in formulations for entomopathogens, offer protection against harmful B-UV radiation presumably by absorbing and translating the high energy radiation to a form less damaging to fungal cells.

These dyes are used at low concentrations (0.01–1% w/v), are water soluble, and are stable for several months provided they are kept in the dark. Yellow crystals may form during long-term storage, but these can be filtered through Whatmans filter paper and the solution re-used. It is not necessary to prepare dyes in alkaline buffers. Most of these dyes are compatible with fluorochromes specific for other cellular components such as lipids and nuclei (Plate 19).

1. Methods for staining of cell walls and/or nuclei

a. Method 1

- Prepare stock solutions:
  (a) 0.01% w/v aqueous solution of Primulin or Calcofluor or Uvitex.
  (b) Hydroethidine (7 mg/ml) in dimethyl-sulphoxide (DMSO).
- Add a drop of cell wall stain (e.g. calcofluor) to cells mounted on microscope slide and rinse off excess. Then add 2–5 μl hydroethidine but do not rinse.
- Examine using appropriate filter sets.

b. Method 2

- Add 20–50 μl hydroethidine (7 mg/ml) to 10 ml buffer or culture medium. At high concentrations (>30 μg/ml) the dye is toxic to several fungi. Incubate for 5–30 min at room temperature.
- Harvest cells and rinse once with buffer.
- Spread cells in solution of cell wall stain and examine using appropriate filter sets.

B Nuclear stains

Fungal nuclei and chromosomes are tiny and difficult to detect at the light microscope level compared with other organisms. Although stains such as aceto-orcein and Giemsa are moderately effective, fluorescent stains such as DAPI and mithramycin are highly specific, even to the point of differentiating between A-T and G-C rich regions of DNA (Butt *et al.*, 1989).

Mithramycin is a yellow crystalline antibiotic which binds specifically to guanine bases in double-stranded DNA. Its fluorescence is directly proportional to DNA content, and like DAPI, it does not stain nucleoli. DAPI is more readily taken up by living cells because of its smaller size (mol.wt. 350) and it will not fluoresce unless bound to DNA. It also stains mitochondrial DNA. It is usually used at concentrations of 0.25–5 μg/ml and can be stored at –20 °C until needed. At a concentration of 1 μg/ml in the final wash, it makes a very useful 'counterstain' to rhodamine immunofluorescence preparations. DAPI absorbs UV light (360 nm) and emits blue light (Plate 20).

Other dyes such as acridine orange, Hoechst 33342, propidium iodide, and ethidium bromide have been used occasionally as nuclear stains but are inferior in several respects to DAPI.

Hydroethidine is an uncharged racemic fluorescent compound produced by the reduction of ethidium bromide. This small molecule (mol.wt. 315) is readily taken up by living cells where it gives a blue fluorescence in the cytoplasm until it is enzymatically dehydrogenated to form ethidium ions which intercalate into the DNA forming red fluorescent DNA–ethidium complexes especially in the nucleus (Plate 19). Although not widely used it is an effective fungal nuclear stain (St Leger *et al.*, 1989).

Hydroethidine has a broad excitation–emission spectrum and does not bleach quickly. It will, in some organisms, stain other cell components (e.g. vacuoles and lipids) various shades of blue and so it is not component specific. The stock solution has a comparatively long shelf life (>6 months) at room temperature. Cells can be stained with 2–5 μl hydroethidine (7 mg/ml in DMSO or dimethylacetamide) or 20–50 μl can be added to 10 ml buffer or culture medium. At high concentrations (>30 μg/ml) the dye is toxic to several fungi including *Entomophaga maimaiga*, *Erynia pieris* and *Massospora cicadina*.

C Lipid stains

Nile Red (9-diethylamino-5H-benzo[a] phenoxazine-5-one), a hydrophobic probe, preferentially dissolves and strongly fluoresces in lipid. It is readily soluble in organic solvents such as acetone and xylene but less so in water (<1 μg/ml). Lipid globules can be localized with this dye in a range of fungal cell types.

Not only does Nile Red have a wide spectral range, but it is compatible with other dyes including cell wall stains. Staining of lipids in most fungal cells is both rapid and briefly intense following excitation. Nile red added to liquid cultures (1–100 μg/ml) is non-toxic and stable since cells will fluoresce several days post-staining. At higher concentrations (>100 μg/ml) the dye is toxic and fluorescence so intense that cytological details are obscured.

Benzapyrene-caffeine stains total lipids and sterols. It is stable when exposed to relatively long periods of UV illumination and a solution prepared according to Jensen (1962) will keep for several months at 4 °C. Fungal cells can either be suspended in benzapyrene-caffeine or a drop of this stain can be added to the specimen just before examination. Lipids stained with benzapyrene-caffeine fluoresce blue to yellow–white when excited with UV light.

1. Method for lipid staining

- Prepare stock solution of Nile Red, dissolve in organic solvent, i.e. acetone, methanol or xylene (<1 mg/ml).
- Add Nile Red to liquid cultures (1–100 μg/ml) or directly to specimen on a microscope slide and examine. If fluorescence is intense then rinse specimen in buffer before mounting.

D Stains for localizing endoplasmic reticulum (ER) and mitochondria

The lipophilic, cationic, cyanine dyes 3,3′-dihexylocarbocyanine iodide ($DiOC_6(3)$) and rhodamine 123 selectively stain mitochondria in living (and chemically fixed) cells in response to their high membrane potential (Plate 21). At high concentrations (i.e. >0.5 μg/ml) $DiOC_6(3)$ will also stain ER (Butt *et al.*, 1989).

Stock solutions of $DiOC_6(3)$ and rhodamine 123 (0.5–5 mg/ml w/v in ethanol or DMSO) are stable for several months at room temperature. The final working concentrations are usually 0.5–5 μg/ml and 3–10 μg/ml respectively. Both dyes are generally non-toxic to fungal growth, and have similar excitation-emission spectra.

Rhodamine 123 can be toxic if cells are exposed to high concentrations over long periods of time. At 3 µg/ml it has little impact on the growth of entomophthoralean fungi in liquid culture; at 30 µg/ml, however, growth is markedly slow. The standard staining procedure of up to 10 µM for 10 min produces no obvious effect on cellular function or mitochondrial morphology.

1. Method for staining ER and mitochondria

- Stock solution: 0.5 mg/ml $DiOC_6$ in ethanol. Store in dark at room temperature.
- Use live or fixed cells (3–5 min using 0.25% glutaraldehyde in 0.1 M sucrose in 0.1 M sodium cacodylate, pH 7.4).
- Expose cells for 10 s to $DiOC_6$ 2.5 µg/ml in sodium cacodylate buffer.
- Wash cells with sodium cacodylate buffer.
- Examine using appropriate filter system (Table 1).

E Vacuole and liposomal stains

Scopoletin (= 7-Hydroxy-6-methoxy-2H-1-benzopyran-2-one; 6-methoxyumbelliferone) and 5(6)-carboxyfluorescein are liposomal stains which are also taken up by vacuoles.

1. Method for staining vacuoles

- Dissolve scopoletin in warm water (30–40 °C) and add 0.2 ml (1 mg/ml) to 25 ml culture. Check uptake of dyes into vacuoles/liposomes using 360 nm excitation 470 nm emission).
- Dissolve 5(6)-carboxyfluorescein in water and add 0.2 ml (0.1 mg/ml) to 25 ml culture. Check fluorescence using 495 nm excitation filter and 515 nm barrier filter.
- Both dyes can be excited at 400 nm with emission at 470 nm.

F Localization of calcium (Ca^{2+}) and calmodulin

Secondary messengers like Ca^{2+} play a vital role in relaying complex intracellular signals (St Leger *et al.*, 1990). This is usually mediated *via* the Ca^{2+} binding protein, calmodulin. Widely used fluorescent Ca^{2+} probes include the antibiotic chlortetracycline (CTC) and the Ca^{2+} chelators Fura 2 and Quin-2. Whereas CTC, also known as aureomycin, stains membrane-bound calcium, the chelator-dyes (e.g. Fura-2) locate free intracellular calcium. CTC is comparatively easy to use and inexpensive; it is water soluble, used at the final concentration of 10^{-4}M, and can be added directly to the medium. The same is also true for the calmodulin antagonists W-7 (= *N*-(6-aminohexyl)-5-chloro-1-naphthalene-sulphonamide), fluphenazine-2 HCl and chlorpromazine-HCl though their hydrophobicity may preclude passage through membranes. The calmodulin inhibitors bind reversibly and rather specifically to the calcium–calmodulin complex. They can be photo-oxidized to fluorescent derivatives which facilitates visualization of calmodulin. These antagonists, used at the final concentration of 2×10^{-7}–2×10^{-5} M, can locate the Ca^{2+}-calmodulin complex in most eukaryote cells.

Specimens stained with CTC or calmodulin antagonists need to be incubated in the stain for 1–10 min before visualization. Interestingly, the intensity of fluorescence using CTC, W-7 or fluphenazine-2/chlorpromazine increases in brightness during observation before fading dramatically. CTC, W-7 and fluphenazine fluoresce yellow, whereas chlorpromazine fluoresces blue–yellow when excited by UV. These compounds were used to localize Ca^{2+} and calmodulin in the apices (growing regions) of germlings of the hyphomycete fungus *Metarhizium anisopliae* (St Leger *et al.*, 1990).

The chelator-dyes are most generally used with video image analysis equipment to detect changes in fluorescence intensity or spectral shifts produced by binding Ca^{2+}. Fura-2 is used in microscopy by ratioing two excitation wavelengths (generally 340 or 352 nm without Ca^{2+} and 380 nm with Ca^{2+}).

1. Method for locating calmodulin

- Use live or fixed cells (0.1–2% v/v glutaraldehyde in 0.1 M sodium cacodylate, pH 7.1 for 10 min at room temperature and permeabilize with 0.01% Triton X-100 for 5 min).
- Stained cells on a slide in a drop of 2×10^{-7} to 2×10^{-5} M fluphenazine-2 HCl or chlorpromazine-HCl dissolved in the subject specific medium.
- Observe slide immediately using the following filter combination: G365, FT395, LP420.

2. Method for locating Ca^{2+}

- Dissolve chlortetracycline in medium or water to final concentration of 10^{-4} M. Add to cell suspension.
- Observe 1–10 min after exposure using blue or green excitation filters.

G Fluorochromes as indicators of cell viability

The vital stain fluorescein diacetate (FDA) is an excellent indicator of cell viability. The principle of staining with FDA relies on the non-polar FDA molecule crossing the plasma membrane and its ester bonds being hydrolysed in the cytoplasm to release fluorescein. The polar fluorescein molecule accumulates in the cytoplasm because it cannot pass through either the plasma membrane or the tonoplast of living cells. Living cells are therefore distinguished by their bright, stable, yellowish-green fluorescence when illuminated with blue light. In healthy cells, vacuoles do not take up the stain and therefore appear dark. Staining of vacuoles indicates that the tonoplast is ruptured. In contrast, propidium iodide (PI) an analogue of ethidium bromide, binds to both DNA and RNA, is water soluble, and supposedly only penetrates dead cells.

1. Method for determining cell viability

- Prepare stock solution of FDA (5 mg/ml in acetone), store at −20 °C until needed.
- Use FDA at the final concentration of 0.01% (w/v). Water, buffer or culture medium can be used to dilute the stock solution. The working solution can be kept for several days if refrigerated, however, fresh acetone may have to be added just before use.
- When FDA is used in conjunction with PI (100 μg/ml) it is possible simultaneously to observe and count dead and living cells (Firstencel *et al.*, 1990). FDA and PI fluoresce green and red, respectively when excited by blue light. However, only PI fluoresces (red) when excited by green light. By employing a double-exposure method to record the fluorescence from cells stained with both FDA and PI, dead and living cells can be distinguished on the basis of fluorochromasia (Plate 22).

5 INDIRECT VISUALIZATION OF CELL COMPONENTS

Fluorochromes can be attached to compounds which exhibit specificity for a particular cell component.

A Fluorochrome-labelled phallotoxins

Actin is a major cytoskeletal protein which takes part in cytokinesis, septation, penetration of the host cuticle, hyphal tip growth and many other cellular processes (Butt & Humber, 1989). It may assist in signal transduction and may even generate the force in the penetration peg which enables the peg to pierce the host cuticle. Filamentous actin can be localized in cells using FITC or rhodamine-conjugated phalloidin (Plate 23).

1. Method for labelling with rhodamine–phalloidin and DAPI

- Fix cells in 3% PBS-buffered formaldehyde (see Appendix) at room temperature for 30–60 min. Wash several times with PBS, and permeablize cells using 1% PBS-buffered Triton X-100 for 30 s.
- Apply cell slurry to poly-L-lysine-coated coverslips/microscope slides and allow to dry.

  Coating with poly-L-lysine: apply drop of 1% aq (w/v) poly-L-lysine (mol.wt. 400 000) to surface of clean glass microscope slides or coverslips, remove excess and leave to air dry. Use as needed.
- Dissolve rhodamine–phalloidin (Sigma) in methanol (25 μg/ml), dilute in PBS to 2.5 μg/ml and add to cells. Suspend cells in this dye before applying as slurry OR apply drop to cells attached to a coverslip or microscope slide.
- Wash with PBS several times to remove free rhodamine–phalloidin and allow to dry or partially dry.
- Add a drop of 100 μg/ml DAPI in mounting medium containing antifade then examine using the appropriate filters (Table 1).

B Fluorochrome-labelled lectins

Carbohydrates and glycosylated proteins form a major constituent of fungal mucilage and may play a key role in spore adhesion. They can also be used to discriminate between species and monitor physiological changes in hyphae (Pendland & Boucias, 1984). Sugars in fungal cell walls are known to play a major role in activating the invertebrate immune system and allow host discrimination of non-pathogenic and pathogenic fungi. Lectins are proteins or glycoproteins which may agglutinate animal blood cells and/or precipitate complex carbohydrates. The agglutination activity of these highly specific carbohydrate binding proteins is usually inhibited by simple monosaccharides, but for some lectins di-, tri-, and even polysaccharides are required (Table 2). Fluorochrome- labelled lectins have been useful for:

- determining which carbohydrates are present in cell walls and extracellular matrix;
- establishing which sugar groups elicit host defence responses;
- investigating the mechanism of spore adhesion to biological and non-biological surfaces;
- determining the sugars present on the surfaces of artificial and naturally occurring fungal protoplasts;
- estimating the surface carbohydrates on fungal infection structures by microscope photometry;
- use as taxonomic markers.

1. Method for localizing sugar groups using lectins

- Harvest and wash cells with PBS (or any other suitable buffer), pH 6.8 and resuspend pellet in PBS (5× volume of pellet).
- Mix equal volumes of cell suspension and lectin conjugate (0.2–0.5 mg/ml in PBS, pH 6.8) and incubate at room temperature for 60–100 min with gentle agitation.
- Wash cells twice with PBS, resuspend in PBS (equal volume as pellet) and apply slurry to poly-L-lysine-coated coverslips and examine using appropriate filter system.
- Control – preincubate lectin conjugate with the correct inhibitor sugar at least 20–30 min before testing with fungal cells; 0.2 M sugar solution is used if using 0.5–1.0 mg/ml of lectin conjugate. Add equal volumes of 0.4 M inhibitor sugar to lectin conjugate at 0.25–0.5 mg/ml so that final concentration of sugar and lectin are 0.2 M and 0.125–0.25 mg/ml, respectively.

C Immunofluorescence

Antigens can be visualized by direct or indirect immunofluorescence. The latter involves first labelling the antigen with a primary antibody followed by labelling with a fluorochrome-conjugated secondary antibody. For direct immunofluorescence the primary antibody is conjugated with a dye and used directly for labelling the antigen. Various recipes are available for immunofluorescence of fungal cells (Butt & Humber, 1989).

Often, immunofluorescence techniques are used together with conventional staining methods and it is possible to visualize the spatial-temporal relationship between two or more different cell components, (e.g. tubulin and DNA during nuclear division) (Plate 24).

1. Methods for immunolocalization of antigens

a. Method 1

This method is useful if you have cells growing in liquid shake culture.

- Fix sample for 30 min in 3% formaldehyde at room temperature (shaking). Add 1 ml 30% formaldehyde to 9 ml culture/buffer so final concentration is 3%. Normally, <30 min fixation time is insufficient whereas >60 min results in high background fluorescence.

Table 2 Some examples of lectins, sugar specificity and recommended buffers.

Lectin	*Inhibitory sugar = Sugar specificity*	*Mol.wt. of sugar*	*Recommended buffer*
Con A	*a*-D-glucose	180.2	Tris buffer.
	a-D-mannose	180.2	
PNA	D-galactose	180.2	PBS pH 6.8
SBA	*N*-acetyl-D-galactosamine	221.2	PBS + Mn^{2+} and Ca^{2+}
WGA	*N*-acetyl-D-glucosamine	221.2	PBS

Con A = concanavalin agglutinin; PNA = peanut agglutinin; SBA = soybean agglutinin; WGA = wheat germ agglutinin.

- Pellet cells by centrifugation in an Eppendorf bench centrifuge. Only a short (1–3 min) gentle spin is required.
- Wash 5–6 times with buffer (PBS, pH 7.4)
- Incubate cells for ca. 10 min in 1 mg/ml buffered Novozym 234 (Novo Biolabs) to digest the cell wall.
- Extract for 10 min in 0.1% Triton X-100 in buffer. The detergent makes holes in the cell membrane.
- Wash with buffer five times (5 min washes).
- Add primary antibody. Dilute antibody in PBS and incubate for 2–4 h at 37 °C. However, pretreat with 2% (w/v) buffered non-fat dried milk or bovine serum albumin if incubation exceeds 6 h and add sodium azide to 0.01% if left overnight. If possible ensure cells are shaken or kept in suspension.
- Wash with buffer five times (5 min washes).
- Incubate with secondary antibody. Dilute in PBS. Incubate for 2–4 h at 37 °C. If possible ensure cells are shaken or kept in suspension.
- Wash with buffer five times (5 min washes).
- Resuspend cells in PBS and apply to coverslips pretreated with poly-L-lysine (1 mg/ml in water). Remove excess and air dry.
- Add antifade in mounting medium and examine using appropriate filter system in fluorescence microscope.

b. *Method 2*

- Fix cells at room temperature (0.6 ml 37% formaldehyde + 5 ml phosphate buffer) for 2 h.
- Wash cells 3 × 5 min with buffer.
- Remove cell wall using 55 µl gluculase + 1 ml sorbitol/phosphate for 2 h at 37 °C.
- Wash with sorbitol/phosphate 2 × 1 min.
- Resuspend in buffer and apply slurry to poly-L-lysine (mol.wt. 400 000)-coated glass microscope slides.
- Air dry slides. Cells must be firmly attached to glass.
- Immerse slide in 100% methanol at –20° C for 6 min.
- Transfer the slide to acetone at –20 °C for 30 s. Remove slide and air dry at room temperature.
- Place 5 µl primary antiserum (at appropriate dilution) over cells and place slide in humid chamber at room temperature or 37 °C for 1.5–3 h.
- Wash slide with buffer 8 × and aspirate excess fluid.
- Add secondary conjugated antiserum over cells and incubate in humid chamber at room temperature or 37° C for 1.5–3 h.
- Wash slide with buffer 8 times and dry.
- Place a drop of mounting medium and antifade on specimen and examine. Or store at –20 °C in the dark if they are to be observed later.

6 CHECKLIST FOR FLUORESCENCE MICROSCOPY

1. Ensure that the correct set of filters for the dyes to be used are available.
2. Ensure mercury vapour bulb is emitting light. Most bulbs have a 200 h life span. Once switched off, do not switch back on again for at least 20–30 min otherwise the life span of the bulb is greatly reduced.
3. Light source must be centred and appropriate apertures opened.
4. Fluorochromes must be in correct solvent and diluted to correct concentration.
5. Check for autofluorescence of specimen (± antifade + mounting medium).
6. Check that film and camera ASI settings are correct.
7. Black-out environment to reduce glare.
8. Record images on film and in notebook. Make labelled sketch of image. Record filters, magnification, etc.
9. Reduce exposure time by using bottom, bright-field illumination in combination with contrasting filters.
10. Make sure there is no stage vibration or specimen drift before taking photograph.
11. Ensure lens is cleaned of immersion oil after use.

REFERENCES

Butt, T. M. (1987). A fluorescent microscopy method for the rapid localization of fungal spores and penetration sites on insect cuticle. *J. Invertebr. Pathol.* **50**, 72–74.

Butt, T. M. & Humber, R. A. (1989). An immunofluorescence study of mitosis in a mite-pathogen, *Neozygites* sp. (Zygomycotina: Entomophthorales). *Protoplasma* **151**, 115–123.

Butt, T. M., Wraight, S., Galaini-Wraight, S., Humber, R. A., Roberts, D. W. & Soper, R. S. (1988). Humoral encapsulation of the fungus *Erynia radicans* (Entomophthorales) by the potato leafhopper *Empoasca fabae* (Hemiptera). *J. Invertebr. Pathol.* **52**, 49–56.

Butt, T. M., Hoch, H. C., Staples, R. C. & St Leger, R. J. (1989). Use of fluorochromes in the study of fungal cytology and differentiation. *Exp. Mycol.* **13**, 303–320.

Butt, T. M., Ibrahim, L., Clark, S. J. & Beckett, A. (1995). The germination behaviour of the entomogenous fungus *Metarhizium anisopliae* on the surface of the aphid and flea beetle cuticles. *Mycol. Res.* **99**, 945–950.

Firstencel, H., Butt, T. M. & Carruthers, R. I. (1990). A fluorescence microscopy method for determining viability of entomophthoralean fungal spores. *J. Invertebr. Pathol.* **55**, 258–264.

Haughland, R. P. (1992) *Molecular Probes. Handbook of fluorescent probes and research chemicals 1992–1994. 5th edn.* Molecular Probes, Inc., Eugene, OR, 421 pp.

Jensen, W. A. (1962) *Botanical histochemistry.* W. H. Freeman, San Francisco.

Pendland, J. C. & Boucias, D. G. (1984). Use of labeled lectins to investigate cell wall surfaces of the entomogenous hyphomycete *Nomuraea rileyi. Mycopathologia* **87**, 141–148.

Schreiter, G., Butt, T. M., Beckett, A., Moritz, G. & Vestergaard, S. (1994). Invasion and development of *Verticillium lecanii* in the Western Flower Thrips, *Frankliniella occidentalis. Mycol. Res.* **98**, 1025–1034.

St Leger, R., Butt, T. M., Goettel, M. S., Staples, R. & Roberts, D. W. (1989). Production *in vitro* of appressoria by the entomopathogenic fungus *Metarhizium anisopliae*. *Exp. Mycol.* **13**, 274–288.

St Leger, R. J., Butt, T. M., Staples, R. C. & Roberts, D. W. (1990). Second messanger involvment in differentiation of the entomopathogenic fungus, *Metarhizium anisopliae*. *J. Gen. Microbiol.* **136**, 1779–1789.

GLOSSARY

Absorption curve, absorption spectrum. The relative tendency of a material to absorb a specific colour or energy of light, typically plotted as intensity, transmission, or optical density versus wavelength.

Autofluorescence. The natural fluorescence of some tissue components.

Bandpass filter. Optical coating assembly that transmits a specific range of colour or energy of light while attenuating, by reflection and/or absorption, other colours or energies within the free spectral range.

Barrier filters. Long pass filters that transmit light with a wavelength longer than a certain value.

Dichroics. Optical coatings that are controlled for both transmitted light over a defined region and reflected light over a defined region.

Epi-illumination. A system in which the objective is also the condenser. In addition, a dichroic mirror or beam splitter is employed to reflect the exciting light to the specimen and simultaneously transmit the fluorescent light to the eyepiece.

Excitation filters. Bandpass filters which have a single large transmission peak.

Fluorochromasia. Occurs when stained tissue fluoresces at two different wavelengths.

Optical density. The inverse natural log of transmission.

Stokes' shift. The low frequency emission spectrum of electromagnetic energy emitted as a result of absorbed energy.

Stokes' Law. States that the wavelength of the emitted fluorescence light is longer than that of the exciting radiation.

Wavelength distortion. The degree of disruption of an optical wavefront, measured by viewing the interference fringes of a two-arm interferometer with the component at test in one arm and a known reference in the other.

APPENDIX

Preparation of antiquenching agents

Preparation A

Dissolve 50 mg *p*-phenylenediamine (4-benzenediamine) free base crystalline (Sigma) in 5 ml PBS and adjust to pH 9. Add 45 ml glycerol and stir until homogeneous. Store at –20 °C in the dark.

Preparation B

Dissolve 50 mg *p*-phenylenediamine in 5 ml PBS pH 8 (= 10× conc). Thoroughly mix 20 μl phenylenediamine with 180 μl polyvinyl alcohol (PVA) or glycerol mounting medium (see below) before use.

Preparation C

0.1% (w/v) *n*-propyl gallate crystalline FW 212.2 (Sigma, = 3,4,5-trihydroxybenzoic acid *n*-propyl ester). Dissolve in a mixture of 70% glycerol + 10% KPO_4 + 20% H_2O, adjust to pH 8.8 before bringing to final volume. Store at 4 °C in the dark.

Preparation D
2.5% (w/v) 14-diazobicyclo-[2.2.2] = octane (DABCO) dissolved in gelvatol or other suitable mounting medium (see below).

Mounting media (temporary/permanent mounts)

A *PVA/Glycerol*

- Dissolve 1 g PVA in 4 ml PBS pH 8.0. Stir for 16 h. Keep in dark.
- Add 2 ml high purity, water-free glycerol and stir for another 16 h. Keep in dark.
- Spin the white suspension in Eppendorf bench centrifuge at 12 000 rpm or 15 000 ***g*** for 15 min.
- Aliquot the clear solution to 1.5 ml Eppendorf vials. Store at –20 °C.

B *Gelvatol (= polyvinyl alcohol) or Mowiol*

- Add 2.4 g Gelvatol 20–30 (Monsanto) or Mowiol 4–88 (Hoechst) to 6 g glycerol. Stir to mix.
- Add 6 ml water and leave for several hours at room temperature (RT).
- Add 12 ml 0.2 M Tris (pH 8.5), and heat to 50 °C for 10 min with occasional mixing.
- After the Gelvatol or Mowiol dissolves, clarify by centrifugation at 5000 ***g*** for 15 min. Add 14-diazobicyclo-[2.2.2] = octane (DABCO) to 2.5% to reduce fading.
- Aliquot in air-tight containers and store at –20 °C. Stocks of these mounts are stable at RT for several weeks after thawing.
- Mounting medium can be used directly on the washed specimen.
- Add a small drop to specimen but ensure no air bubbles are trapped. Mounting media will set overnight.

Preparation of 3% formaladehyde fixative

First prepare 30% formaldehyde solution. Dissolve 3 g paraformaldehyde in 10 ml distilled water and store at 4 °C. If required, heat to 60 °C or until sediment has dissolved. Concentrated NaOH may also be used to settle the solution. Cool before use. Heat formaldehyde in fume hood and wear gloves when handling formaldehyde-containing solutions.

Preparation of phosphate buffered saline (PBS) buffer

Dissolve 8 g NaCl + 0.2 g KCl + 1.14g Na_2HPO_4; make up to 1 l following adjustment of pH to 7.4.

Preparation of Tris buffer

0.01 M Tris + 0.015 M NaCl + 0.01 M $MnCl_2$ + 0.01 M $CaCl_2$, pH 7.0.

CHAPTER VIII-3

The application of molecular techniques to insect pathology with emphasis on entomopathogenic fungi

RAYMOND J. ST. LEGER & LOKESH JOSHI

Boyce Thompson Institute, Tower Road, Cornell University, Ithaca, NY 14853, USA

1 INTRODUCTION

Any consideration of the suitability of a microorganism for commercial purposes inevitably leads to the possibility of improving its performance. This is particularly true for traditional biopesticides (viruses, bacteria and fungi) as their performance is commonly perceived to be poor compared to chemical pesticides. However, registration requests to date have been for naturally occurring fungi obtained by standard selection procedures and improved as pathogenic agents by developing the technologies required for optimizing production and stability of the inoculum. Improvements have also been attempted through parasexual crossing and protoplast fusion (Heale *et al.*, 1989). With these techniques available, it may be possible to enhance pathogenicity without genetic engineering given that tremendous variability exists in field isolates of these fungi (Bidochka *et al.*, 1994; St Leger *et al.*, 1992a) and that the relative pathogenicity varies between the strains. A recombinant strain produced by somatic hybridization techniques with enhanced native genes would bypass some of the difficulties in gaining regulatory approval for release of engineered strains. Nevertheless, genetic engineering using transformation and gene cloning for pathogenicity traits that are regulated by single genes or gene clusters provides the most targeted and flexible approach to dissect, and eventually alter, the physiology of entomopathogenic fungi without the co-transfer of possibly undesirable linked characteristics.

Entomopathogenic fungi represent an unconsidered, and therefore untapped, reservoir of pesticidal genes for the production of advanced engineered pesticides; an important consideration, given that the lack of 'useful' pesticidal genes for transfer has been a major constraint in the implementation of biotechnology in crop protection (Gatehouse *et al.*, 1992).

This chapter has two aims. The first is to provide a brief didactic overview of the current usage of molecular techniques in understanding entomopathogenicity. The second is to provide an introduction to the state of knowledge of the molecular biology which could be relevant to entomopathogenic fungi,

MANUAL OF TECHNIQUES IN INSECT PATHOLOGY
ISBN 0-12-432555-6

so that workers may make an initial judgement as to the applicability of these techniques to their research problems, and potential applied scientists can see which avenues of research might fruitfully be followed. Sambrook *et al.* (1989) and Ausubel *et al.* (1991) have written extensive compilations of current techniques. They also reference recent advances in technology in more detail than we are able to in this overview.

A The potential of biotechnology

The traditional approach to discovering a biopesticide has included searching natural ecosystems to discover an organism that attacks the target pest. The advanced-engineered biopesticide approach would begin by designing the ideal biocontrol organism using genetic engineering or other techniques (Stowell, 1994). Recently, molecular biology methods have been applied to elucidate pathogenic processes in *Metarhizium;* several genes have been cloned which are expressed when the fungus is induced by starvation stress to alter its saprobic growth habit, develop a specialized infection structure – the appressorium – and attack its insect host. These studies have illustrated how the isolation and *in situ* manipulation of such genes allows the biochemical role of their encoded products to be defined. In the first genetically improved entomopathogenic fungus, additional copies of the gene encoding the regulated cuticle-degrading protease (*Pr1*) were inserted into the genome of *M. anisopliae* such that the gene was constitutively overexpressed (St Leger *et al.*, 1996a). The toxicity of *Pr1*, expressed in the haemolymph, caused a significant reduction in time of death of infected lepidopterous larvae and reduced food consumption compared to the wild-type fungus. Accordingly, we have demonstrated the potential application of this enzyme gene in biocontrol methods.

The development of a native strain of *M. anisopliae* that constitutively expresses a homologous gene should not change host range and is unlikely to raise public concern. There is an inherent uncertainty because of the paucity of our knowledge concerning the fate of fungal genotypes at the population and ecosystem level. In fact, there is no information available on survival of individual genotypes (clones) of entomopathogenic fungi in nature, nor is there experimentally derived information on gene transfer from populations of genetically engineered entomopathogenic fungi to wild-type or other fungal species. Such information will be crucial to the US EPA in assessing risk in evaluating registration requests for entomopathogenic fungi, particularly recombinant strains (Faust & Jayaraman, 1990).

B Elements of fungal gene structure

The availability of over 100 fungal gene sequences, the majority of them from Ascomycete or Deuteromycete species allows some generalizations to be made which are relevant to studies on entomopathogens and hold for those genes cloned to date from *M. anisopliae* (St Leger *et al.*, 1992 b,c, and unpublished) :

1. Gene expression in filamentous fungi is regulated primarily at the level of initiation of transcription. Sequences recognizable as TATA boxes occur in the 5′ untranslated regions of many, but by no means all fungal genes, and CAAT boxes are also found (see Davies, 1992 and Unkles, 1992 for review).
2. Introns commonly occur in filamentous fungal genes, with consensus splice-junction and lariat sequences similar to those of yeast and higher eukaryotes (Ballance, 1991). Introns in highly expressed *M. anisopliae* genes are usually less than 100 base pairs (bp) in length (unpublished data).
3. Ballance (1991) has compiled data on filamentous fungal sequences in the region immediately around the AUG which is important for efficient initiation of translation.
4. Bias in codon usage occurs in filamentous fungi, with a preference for codons ending in pyrimidines or G. As a rule in *M. anisopliae* genes, whenever possible, a T or a C is preferred at the third position. For example, of all codons used in the *Pr1* gene, 43% end with a C and 36% end in a T, 16% end with a G and only 4% used end with an A (St Leger *et al.*, 1992b). A similar codon usage occurs for the *Pr1* gene of *Beaveria bassiana* (Joshi *et al.*, 1995).

C An introduction to genetic engineering

Genetic engineering may be defined as the genetic alteration of cells or organisms by methods that require the *in vitro* modification of DNA. The cloning, analysis and modification of DNA fragments is accomplished by a small but powerful set of techniques. The field of recombinant DNA began its rapid growth after the purification and characterization of a set of DNA cutting and modifying enzymes. The restriction enzymes are used to cut DNA at specific recognition sites, whereas the enzyme ligase is used to rejoin these DNA fragments. Other enzymes are used to label DNA (e.g. T4 polynucleotide kinase for 5′ labelling with radioactive nucleotides), modifying the ends of DNA (e.g. Klenow fragment for removal of 3′ protruding ends) or DNA sequencing (e.g. Taq polymerase). The application of these enzymes to practical problems requires the use of a variety of gel electrophoresis methods for the resolution and separation of RNA and DNA molecules. Electrophoresis methods are used for the purification of RNA or DNA molecules, for restriction site mapping, for DNA sequencing, and for many other procedures. Finally, Southern and Northern hybridization procedures allow the detection of specific DNA and RNA species, respectively after gel electrophoresis.

2 PROCEDURES FOR ISOLATING PATHOGEN DNA FOR DIFFERENT PURPOSES

The application of molecular techniques to pathogen genomes often depends on the ability to purify high molecular weight DNA. There are various methods of doing this and a correspondingly wide range of commercial kits is available. Researchers should also read the *Fungal Genetics Newsletter*. Every edition has one or more papers describing improved methods for purifying small or large amounts of DNA. This chapter describes methods which we know work well for purifying high-molecular-weight DNA from entomopathogenic fungi. This section will also describe purification of plasmid DNA, a procedure crucial for manipulating DNA fragments into plasmids for purposes such as sequencing, labelling and cloning.

All the protocols describing the purification of high-molecular-weight DNA involve lysis of cells and solubilizaton of the DNA followed by one or more chemical or enzymatic processes to remove contaminating proteins, RNA and other macromolecules. The standard protocol involves the use of phenol to remove contaminants from solutions. Chloroform is added for further denaturation of proteins, extraction of lipids and stabilization of the boundary between the aqueous and the phenolic phases. Isoamyl alcohol may be added as it also aids phase separation and prevents foaming upon vortexing the solution. DNA is precipitated from the aqueous phase with either isopropanol or ethanol. The resultant DNA pellet is then dissolved in nuclease-free sterile water or TE (Tris-EDTA, pH 8.0) buffer if it is to be stored indefinitely. Large aqueous volumes of DNA can be concentrated by extraction with 2-butanol. This process requires mixing and centrifuging the aqueous solution of DNA sample with 2-butanol until the desired volume of aqueous solution is achieved, followed by extraction of the DNA from the lower aqueous phase with 25 : 24 : 1 phenol:chloroform:isoamyl alcohol and ethanol precipitation.

Two major factors influencing satisfactory purification of high-molecular-weight DNA are shearing and the presence of nucleases. To avoid shear forces, the lysate (containing DNA) should be treated gently and pipetting with narrow pipette tips should be avoided. To prevent nuclease activity, the cells can be frozen (preferably under liquid nitrogen) and should only be thawed in the extraction buffer.

Some specific methods are described below.

A Preparation of DNA from protoplasts (after Bidochka et al., 1994)

1. Prepare protoplasts by shaking mycelia with 0.8% Novozyme 234 in 1.2 M sorbitol, 10 mM Tris–HCl (pH 7.5) for up to 3 h. The protoplasts are filtered through two layers of sterile cheese cloth and centrifuged.
2. Burst protoplasts by resuspending them in 0.01 M Tris, 2% sodium dodecyl sulphate (SDS), 1 mM EDTA and treat sequentially with RNase A and proteinase K.
3. Extract proteins with phenol:chloroform, precipitate DNA with 1/10 vol 3M sodium acetate and 1 vol isopropanol. Purified DNA is washed

sequentially with ice-cold 100% and 80% ethanol and air dried or dried under a speed vac.

B Preparation from mycelia

We have also purified genomic DNA by lyophilizing mycelia and homogenizing under liquid nitrogen (St Leger *et al.*, 1992b). The homogenate is further processed as described above. Homogenization of the mycelia eliminates the protoplast formation step but is laborious, particularly when DNA is being isolated from many fungal samples.

C Extraction of DNA from an infected insect

Identifying the presence of a particular pathogen genotype using specific DNA probes (see Section 12 B) is greatly facilitated by the ability to extract fungal DNA from infected insects or soil. A method to extract DNA from dead grasshoppers (Bidochka *et al.*, 1995) appears to be broadly applicable.

1. Freeze dry each soil, insect or plant sample, grind under liquid nitrogen, extract with phenol–chloroform and precipitate DNA with 0.3 M sodium acetate/isopropanol.
2. Remove polysaccharides and polyphenols which co-precipitate with DNA by selective precipitation of DNA with polyethylene glycol (Rowland & Nguyen, 1993).
3. Blot samples onto nitrocellulose following standard protocols (Sambrook *et al.*, 1989).
4. Incubate the blots in laundry detergent (LaFrance; Dial Corp.) to remove contaminating proteins.

If necessary, it is also possible to use PCR (polymerase chain reaction) techniques to improve the sensitivity of DNA blotting. The DNA is separated from polyphenols which could interfere with PCR and amplified using universal primers (Bidochka *et al.*, 1994).

D Purification of Plasmid DNA

This always involves three steps; growth of the bacterial culture containing the plasmid, harvesting and lysis of the bacteria at appropriate cell concentration, and purification of the plasmid DNA. Plasmids are most commonly prepared from bacterial cultures (usually XL1–Blue and DH5 alpha *E. coli* strains) grown in the presence of selective agents such as an antibiotic. Detailed accounts of growing bacterial cultures under selective pressure for plasmid DNA purification are given by Sambrook *et al.* (1989) and Ausubel *et al.* (1991).

After harvesting, the cells are lysed in the presence of NaOH/SDS and RNase A and the cellular lysate is neutralized by adding acidic potassium acetate. The plasmid DNA is then purified either by conventional phenol:chloroform extraction followed by ethanol precipitation or by using affinity resin columns where plasmid DNA is bound and released under suitable salt and pH conditions. The Qiagen plasmid kit functions on this principle. We use this system to provide DNA pure enough to be sequenced using an automatic sequencer.

3 PROCEDURES FOR ISOLATING PATHOGEN RNA FOR DIFFERENT PURPOSES

Clean RNA is a fundamental requirement in molecular biology, particularly in the gene cloning processes and in analysing gene expression. It is absolutely crucial that precautions are taken to minimize RNase contamination (e.g. the use of DEPC (diethyl pyrocarbonate)-treated deionized water, baking of glassware at 300°F (149°C) for 4–6 h). We recommend sterile, disposable plastic ware as this is free of RNases and can be used for the preparation and storage of RNA without any pretreatments. All the work should be done in a flow hood the surface of which has been treated with RNase inactivating agents such as RNase Away (Molecular Bio-products, CA).

There are several reliable methods and kits available to purify total RNA and poly(A) RNA from cells. Most procedures for isolating RNA involve the lysis and denaturing of the cells to release the nucleic acids. Additional steps are taken to remove DNA and protein contamination from RNA. We have found that Tri Reagent® (Molecular Research Center, OH) gives good results. This method is based on the property of RNA to remain water soluble in a solution containing phenol and guanidine thiocyanate at acidic pH. Under these conditions DNA and proteins stay in organic phase. Guanidine thiocyanate is one of the most potent protein denaturants known and is a necessary reagent when purifying RNA from fungal cultures growing in nutrient-deprived conditions or on insect cuticles as these conditions enhance synthesis of high levels of a potent extracellular RNase.

One pre-requisite for RNA purification is the use of fresh cells or cells frozen at –80 °C in liquid nitrogen immediately after harvesting. The cells with thick cell walls are homogenized under liquid nitrogen whereas thin walled cells or protoplasts can be used as a suspension to purify RNA. The homogenized cells are added to the Tri Reagent® and stirred on a magnetic stirrer (15 min). The suspension is centrifuged at 12 000 × ***g*** for 10 min at 4 °C to remove insoluble material from the homogenate. Chloroform or bromochloropropane (BCP) is added to the supernatant to separate the suspension into the aqueous and organic phases. After a brief incubation at room temperature and centrifugation at high speed the mixture separates itself into a lower organic phase, an interface and topmost aqueous phase. The DNA and proteins stay in the interface and organic phase, respectively. RNA remains exclusively in the aqueous phase and can be precipitated by mixing with isopropanol. The RNA pellet should not be allowed to dry completely, as this decreases the solubility. RNA can be dissolved by heating at 60 °C for few minutes with occasional gentle vortexing. For long storage, RNA should be kept in ethanol, but for short storage RNA can be dissolved in DEPC-treated water or SDS solution.

This procedure yields total RNA, which is primarily composed of ribosomal RNA (rRNA) and transfer RNA (tRNA). Only 1–5% of the total cellular RNA fraction is messenger RNA (mRNA). Many techniques in molecular biology require messenger RNA (poly (A^+) RNA) that is free of contamination from rRNA and tRNA. For example, isolation of pure and undegraded mRNA is essential for good quality *in vitro* translation and for the construction of cDNA libraries.

Two new methods have made the task of purifying mRNA much easier. Both utilize the distinguishing poly(A) tail at the 3′ end of mRNA. In the first approach, separation of mRNA from structural RNA is achieved by hybridizing the poly(A) tails of mRNA molecules to oligo-dT primers which are coupled to a solid matrix. RNA species lacking poly(A) tails fail to bind to the oligo-dT and are removed. mRNA is subsequently eluted from the oligo-dT cellulose columns by applying low ionic strength buffer that destabilizes dT-rA hybrids. The Oligotex-dT™ (Qiagen) kit functions on this principle.

The second method uses a biotinylated oligo-dT primer to hybridize to the poly(A) tail. The oligo-dT–poly(A) hybrids are captured and washed using streptavidin coupled to paramagnetic particles and then the reaction mixture is placed on a magnetic separation stand. The mRNA is eluted from the paramagnetic particles by adding ribonuclease free deionized water. The PolyATtract® mRNA isolation system (Promega) utilizes this approach to purify good quality mRNA.

The concentration of total RNA and mRNA can be determined by spectrophotometry. The ratio between A_{260}/A_{280} provides an estimate of the purity of mRNA. Pure RNA preparations have an absorbance ratio of 2.0, a ratio of more than 1.7 is considered as highly pure, but it is never absolute so we suggest that the researcher also runs a series of RNA dilutions on an agarose gel to confirm the concentration and integrity of the samples. On such a gel, ethidium bromide stained mRNA appears as a smear ranging from 200 bp to 7–8 kb. Most of the mRNA should lie between 1.5 kb (kilobases) and 2.5 kb. In total RNA gels the fluorescence intensity of 25S rRNA (3360 nucleotides in yeast) should be approximately twice as strong as 17S rRNA (1650 nucleotides). If this is not the case, the samples will not provide northern blot data suitable for quantitative mRNA measurement (Piper, 1994). All purified total RNA and mRNA should be stored at –80 °C in aliquots.

4 RESOLUTION AND ANALYSIS OF DNA AND RNA

A Running gels

Most analytical and preparative protocols in molecular biology require, at some point, a fractionation and purification method involving gel-electrophoresis. The use of gel electrophoresis to resolve nucleic acids is usually simpler than the separation of proteins because nucleic acids are uniformly negatively charged. The factors which affect the migration and separation of nucleic acids on gels, e.g. the conformation of the nucleic acid, the pore size of the gel, voltage applied, salt concentration of the buffer used and the type of agarose used, are covered by Ausubel *et al.* (1991) and Sambrook *et al.* (1989).

B Southern blotting

This is the process of capillary transfer of DNA fragments from gels to various types of support membranes. Until recently, Southern blotting was performed by upward movement, i.e. the nylon membrane was placed on top of the agarose gel which was in contact with the transfer buffer (Ausubel *et al.*, 1991). This method takes 6–12 h and sometimes results in poor transfer. We have been using the downward alkaline transfer of DNA for Southern blotting (Chomczynski, 1992). The composition of a blot is shown in Figure 1. Briefly, the gel is incubated in 0.25 N HCl for 15–20 min or until the bromophenol dye turns yellow. The incubation is continued in 0.4 M NaOH for a similar length of time, while the transferring set-up is prepared. On a stack of paper towels (6–8 inches high), four filter papers (Gel blot paper, GB 002, Schleicher & Schuell) are placed. The filter papers and paper towels are cut to the same size as the gel. A filter paper wetted with 0.4 M NaOH is placed on top of the other filter papers followed by careful placing of the membrane which has been soaked in 0.4 M NaOH for 10 min. The agarose gel with the area above the loading slots removed is put on the membrane. The area around the gel is covered with a clear plastic food wrap or parafilm to stop any transfer of the transfer buffer from around the gel. On top of the gel, four more layers of filter papers soaked in 0.4 M NaOH are positioned, followed by a long, wet filter paper strip to form a wick that carries the transfer buffer from the reservoir to the blotting set-up. It should be ensured that there are no air bubbles trapped between the membrane and gel or in the filter papers because the air pockets created between layers cause failure of transfer at that particular spot. Then, the top of the transfer set-up is covered with a light weight (e.g. a plastic cover) to prevent evaporation. Transfer is usually completed within 2–4 h. Thinner gels can be transferred more rapidly than thicker gels and similarly, smaller fragments are transferred more rapidly than larger fragments of DNA. After the transfer is complete, the position of different lanes of DNA is marked with a black ball point pen on the membrane. The membrane is incubated briefly in 2 × standard saline citrate (SSC) to neutralize and to remove any traces of agarose attached to it. After the transfer, the DNA on the membrane is usually immobilized by either baking at 80 °C for 2 h under vacuum or with UV treatment but we found that this is not a crucial step because alkaline transfer allows the DNA to bind irreversibly to the membrane. The membrane is then prehybridized and hybridized with specific probes in conditions designed to minimize non-specific binding (Table 1), followed by detection of signals. The advantages of using nylon membranes over nitrocellulose membranes is that the former are sturdier and can be used for multiple hybridizations.

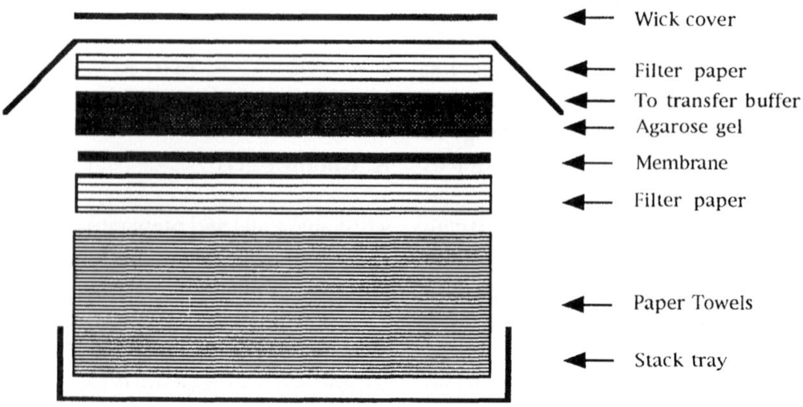

Figure 1 Schematic representation of DNA and RNA transfer set-up.

Table 1 Conditions for probing DNA libraries and blots with radiolabelled probes.

	High-stringency (homologous probe)	*Low-stringency (heterologous probe)*
Hybridization solution composition	50% Formamide 30% 6 × SSPE[a] 1% SDS 10 mg salmon sperm DNA/100 ml	30–40% Formamide 30% 6 × SSPE 1% SDS 10 mg salmon sperm DNA/100 ml
Prehybridization	42 °C, minimum 1 h	42 °C, minimum 1 h
Hybridization	42 °C, overnight[b]	37 °C, up to 72 h
Wash Buffers	2 × SSC[c] at RT, 15 min 2 × SSC + 1% SDS, 20 min 42–65 °C 2 × SSC + 0.1% SDS, 20 min 42–65 °C	2 × SSC at RT, 15 min 2 × SSC + 1% SDS, 20 min, RT

[a] 20 × SSPE: Dissolve 175.3 g of NaCl, 27.6 g of $NaH_2PO_4 \cdot H_2O$ and 7.4 g of EDTA in 800 ml of H_2O. Adjust the pH to 7.4 with NaOH. Make up to 1 l with H_2O. Sterilize by autoclaving (Sambrook *et al.*, 1989).
[b] The filters are hybridized with a ^{32}P-labelled nucleic acid probe with a specific activity of at least 10^7 cpm/µg.
[c] 20 × SSC: Dissolve 175.3 g of NaCl and 88.2 g of sodium citrate in 800 ml of H_2O. Adjust the pH to 7.0 with NaOH. Make up to 1 l with H_2O. Sterilize by autoclaving (Sambrook *et al.*, 1989).

C Northern blotting

This is used to detect the presence and size of a specific RNA species. RNA is resolved by a denaturing agarose gel. Either formaldehyde, glyoxal/DMSO or methyl mercuric hydroxide are used as denaturants. The electrophoresis apparatus should always be cleaned with detergents, sterile water and then filled with 3% H_2O_2 solution to denature RNase activity. The electrophoresis tank can also be rinsed with DEPC-treated water prior to casting the agarose gel. The separated RNA is transferred to a nitrocellulose or nylon membrane. The membrane is prehybridized and hybridized using similar protocols as for DNA hybridization. Water and any buffer used in RNA analysis should be treated with DEPC to inhibit RNase activity. For casting the gel and electrophoresis, MOPS (3-[*N*-morpholino]propanesulphonic acid) or sodium phosphate buffers are used and 1–2% filter sterilized formaldehyde is added. RNA is applied in a sample buffer containing formaldehyde which is heated at 60 °C for 15 min. RNA gels are run relatively fast and should always be kept cool (4 °C) by placing ice jackets around the gel apparatus. RNA can also be denatured by using glyoxal and dimethyl sulphoxide (DMSO). Glyoxal/ DMSO gels take longer than formaldehyde gels and it is necessary to recirculate the buffer to maintain the pH of the buffer at 7.0. At pH 8.0 or higher, glyoxal dissociates from RNA. Following electrophoresis, the transfer set-up is similar to Southern transfer except that RNA is not transferred under alkaline conditions, instead 10 or 20 × SSC is used as the transfer buffer.

Northern blot experiments should be designed carefully. High levels of probe are required to achieve optimum hybridization signals that accurately quantify levels of target mRNA. Reprobing the blot for a constitutively expressed transcript can be used to show that adjacent wells have been loaded with approximately equal amounts of RNA. Probing for actin mRNA sequence is often used as an indication of uniform RNA loading (Adams & Gross, 1991).

D Pulse field gel electrophoresis

The utility of conventional agarose gel electrophoresis drops sharply for DNA molecule > 25 kb, as electrophoretic mobilities become increasingly independent of molecular size. Pulse field gel electrophoresis (PFGE) has been developed to overcome the constraint of the mobility of larger DNA molecules. In PFGE, molecules are subjected to electric fields applied alternatively in two different directions. It has been suggested that separation of molecules is achieved because in order to migrate, the molecules must first reorient themselves in response to each changing orientation of the electric field, the time taken by the molecules to reorient themselves depends on the molecular size (Bustamante *et al.*, 1993).

Most of the PFGE protocols for electrophoretic karyotyping have been developed empirically. The counter-clamped homogeneous field (CHEF) PFGE system has been used to determine the size and number of chromosomes of certain entomopathogenic fungi (Shimizu *et al.*, 1993). This method has also been used in our laboratories to compare the electrophoretic karyotypes of various entomopathogenic fungi. As such it offers a major advance over the traditional time-consuming strategy of linkage analysis. Following electrophoresis, the gels can be blotted and hybridized successively to a series of probes to establish their chromosomal location. These techniques have great potential for genome mapping and elucidating the nature of the variation between the races of a pathogen. The protocol we followed is given here.

1. Preparation of protoplasts

Protoplasts are prepared from young mycelia growing in liquid shake cultures. The mycelia are centrifuged and washed with osmotic stabilizer containing varying amounts of sorbitol/ $NaCl/MgSO_4$. Novozyme 234 is added to the osmoticum and incubated at 30°C with gentle shaking for 60–90 min. This usually yields 10^7–10^8 protoplasts/ml. The protoplasts are filtered through two layers of sterile cheese cloth, centrifuged and then resuspended and washed three times with STC solution (sorbitol, Tris-HCl and calcium chloride).

2. Preparation of agar plugs

Each protoplast suspension is equilibrated at 42°C and added to an equal amount of molten 1.4% low melting agarose (PFGE grade agarose, Bio-Rad) in a protoplast mold. The protoplast plugs are incubated overnight at 50°C with proteinase K solution to dissolve the protoplast membrane and release nuclear contents. The plugs are then washed with EDTA to remove excess proteinase K and can be stored for up to 4 months at 4°C before use.

3. Run conditions of CHEF gel

CHEF analysis is carried out using Bio-Rad's CHEF Drive III. The gel is cast and loaded as described in the manufacturer's manual. Different voltage, switching intervals and total run time conditions were tried. Consecutive 1800 and 2500 s switching time intervals, 1.8–2.5 V voltage/cm, 105°–120° angle and total run time of 72–96 h gave reasonable separation of chromosomes from different strains of *Metarhizium anisopliae* and *Metarhizium flavoviride*.

5 TRANSFORMATION SYSTEMS

Techniques developed to transform filamentous fungi have followed the same three conceptual steps applied to all organisms. The first step is the preparation of 'competent' cells (i.e. cells capable of taking up foreign DNA). The second step is inducing the cells to take up the transforming DNA, and the third step is applying a selective pressure to the cells so that only those cells which take up and express the DNA (i.e. transformants) are capable of growing. Transformation systems for *M. anisopliae* (Bernier *et al.*, 1988, Goettel *et al.*, 1990), are based on the original work of Tilburn *et al.* (1983) with *Aspergillus nidulans*. Competence is achieved in part by digesting away the cell wall (which is a barrier to DNA entry) from log phase mycelium using a complex enzyme mixture such as Novozyme 234. Transformation is achieved by mixing osmotically stabilized protoplasts (in 0.6 M KCl or 1.2 M sorbitol) with DNA in the presence of buffered 10–50 mM $CaCl_2$ and polyethylene-glycol which induces membrane fusions allowing $CaCl_2$ precipitated DNA to be internalized. The following generalizations can be made about the results. Transforming DNA is integrated into the genome, frequently in multiple copies, sometimes at a homologous locus, and most systems rely on the expression of a foreign gene for the selection of transformants.

Filamentous fungi are often very permissive with respect to the expression of foreign genes (Covert & Cullen, 1992) and this is also demonstrated by *M. anisopliae*. For example, the utilization of the benomyl resistance gene of *Neurospora crassa* to transform *M. anisopliae* to benomyl resistance. Also, promoters which function in one filamentous fungus can often be used to confer expression of a molecular gene in other genera or species, e.g. the glyceraldehyde phosphate dehydrogenase promoter of *A. nidulans* can drive expression of the bacterial GUS gene or the *M. anisopliae Pr1* gene in *M. anisopliae*. This has important consequences, not only for the development of transformation systems, but for our ability to produce improved pathogens secreting foreign proteins and probably reflects an underlying genetic relatedness between Ascomycete and Deuteromycete fungi and their molecular genetics. This may not apply to the distantly related Zygomycete fungi. In contrast to the results obtained for ascomycetous and deuteromycetous fungi in which vector DNA becomes integrated into the genome of the host after transformation, in zygomycetous fungi, like *Mucor circinelloides* (Benito *et al.*, 1995a), autonomous replication of vectors was observed in most cases, though some elements of chromosomal origin may be capable of promoting stable mitotic segregation of plasmids (Burmester *et al.*, 1992). The vectors used in these studies could presumably form the basis for a transforming vector for entomopathogenic Zygomycetes. Plasmids harbouring the *P. blakesleeanus pyrG* gene have been used to transform *M. circinelloides* by autonomous replication (Benito *et al.*, 1995b), indicating conservation among Zygomycetes of at least some elements involved in autonomous replication.

Autonomously replicating plasmids containing telomeric sequences (ARSs) are reported to improve transformation frequencies in plant pathogenic Ascomycetes (Covert & Cullen, 1992). ARSs were identified by screening libraries (constructed in a vector that normally integrates) for clones that gave an increased transformation frequency. A special feature is that the ARS vector containing the selectable gene should be easily lost under non-selective conditions, a feature that is useful when testing the correlation between a function and the presence of a particular gene. However, stable integration is essential for strain construction of improved pathogens for commercial uses in which case the non-selective pathogenicity related gene must be in a plasmid that integrates stably into the chromosomal DNA.

A Problems with transformation systems for entomopathogenic fungi

Transformation frequencies utilizing traditional Ca/PEG technology are usually 10–100-fold less than the 10^3 per μg DNA routinely achieved in *A. nidulans*. Furthermore, the efficiencies of protoplast-mediated procedures varies between strains of *M. anisopliae* and *Verticillium lecanii*, indicating that optimization of conditions may be required for each strain which is very time consuming. Also, the stability of transforming DNA in the genome can vary markedly between strains. For example, the vector, pNOM-102, containing the GUS expression gene stably transforms some, but not all, strains of *M. anisopliae* and many strains exhibit transient expression (unpublished data). The following unpublished hints may be useful to workers attempting to optimize transformation conditions.

Batch-dependent variations will be noted in the concentrations of Novozyme required to generate 95% protoplasts within 3 h. In addition, and more worrying, transformation frequencies vary greatly with different lots of enzyme so that the type and batch of the enzyme used to make protoplasts can be a critical factor in generating transformable cells. The ability to generate protoplasts that can regenerate gives no assurance whatsoever that they are competent to take up DNA, so when in doubt try another batch of enzyme. In general, workers should incubate cells with enzyme for the minimum time necessary in order to obtain protoplasts. Some strains are resistant to protoplasting using Novozyme alone. An incubation with mercaptoethanol or dithiothretol (DTT) (10 mM) for 20 min preceding enzyme treatment will facilitate the action of Novozyme, particularly if supplemented with additional β-glucuronidase. The developmental stage of the fungus can also be crucial. For example, the blastospores of *V. lecanii* are extremely resistant to protoplasting.

Aurin tricarboxylic acid added to protoplast washes or during PEG treatment facilitates transformation presumably by inhibiting endogenous nucleases. This seems particularly important when only small amounts of transforming DNA are being used. Avoid using carrier DNA mixed with transforming DNA to 'mop up' nucleases. Although transformation frequencies are improved, many transformants are non-pathogenic, presumably due to multiple disruptions of important fungal genes by integration of the carrier DNA.

B Standardized transformation procedures

A simple standardized method for transformation of essentially any entomopathogenic species would be novel and useful. We have tried several alternative methods to Ca/PEG such as electroporation and biolistic transformation to deliver DNA into cells (St Leger *et al.*, 1995). Compared to the Ca/PEG method, no significant improvements of transformation frequency were observed but they are less laborious, although treatments to weaken the cell wall are also required for electroporation. The biolistic process was comparatively straightforward and enabled co-transformations of *M. anisopliae* and *V. lecanii*, albeit at comparatively lower rates compared to the protoplast method. This technique has some promise for routine transformations of entomopathogens, particularly as conditions optimized for *M. anisopliae* worked almost equally well for *V. lecanii*.

C Protocol modifications for entomopathogenic fungi

For both electroporation and biolistic transformation, regeneration media (RM) consist of 1% potato dextrose broth (Difco), 0.2% yeast extract, 0.5 M $MgSO_4$ (or 1.2 M sorbitol) and 0.3% $NaNO_3$. RM is solidified with 1.2% noble agar (RA) or 1.2% sea plaque low-gelling-temperature agarose (RLGA). X-glucuronide (5-bromo-4-chloro-3-indoyl glucuronide, X-gluc, Clontech, Palo Alto) is included in RA or Czapek Dox agar (CDA, Oxoid) at 20 μg/ml. For selection of benomyl-resistant (BenR) transformants, RLGA plates are overlaid with molten (42 °C) RLGA containing 5 μg/ml benomyl (DuPont Co.) from a stock solution of 5 mg/ml in dimethyl sulphoxide (DMSO).

1. Transformation by electroporation

Published protocols were optimized for *M. anisopliae* (St Leger *et al.*, 1995). Using this method with some combinations of fungal strains and vectors gives high levels of unstable (transient) transformants.

1. Germinate 10-day-old conidia for up to 9 h in 0.2% yeast extract at 27 °C with shaking.
2. When germ-tubes begin to emerge add 1 mg/ml crude cellulase (Sigma, type C-1184) and 1 mg/ml β-glucuronidase (Sigma, type H1) and resume incubation for up to 2 h.
3. Centrifuge germlings before they produce protoplasts, wash twice in electroporation buffer (1 mM HEPES, 50 mM mannitol, pH 7.5, 1% (wt/vol) PEG 6000) containing 10 μg/ml aurin tricarboxylic acid and resuspend in electroporation buffer at approximately 5×10^7 spores/ml.
4. Gently mix aliquots of 100 μl with 10 μg plasmid DNA and keep on ice for 15 min.
5. Deliver high-voltage 5 ms pulses (25 μFD capacitor and a field strength of 12.5 kV/cm) to samples in disposable cuvettes (Bio-Rad Laboratories; inter-electrode distance of 0.2 cm) by using a gene pulse apparatus (Bio-Rad Laboratories).
6. Immediately following electroporation mix the spore suspension with 1 ml of potato dextrose broth (PDB) plus 0.5 M $MgSO_4$ and incubate at 25 °C for 3 h.
7. Plate aliquots containing 2×10^5 germlings onto 10 ml PDA and incubate for 10 h before selection using an RLGA overlay containing 5 μg/ml benomyl. Transfer colonies arising on transformation plates to PDA containing benomyl and X-gluc (20 μg/ml).

2. Transformation by particle bombardment

High-velocity ballistic transformation of *M. anisopliae* germlings can be performed with the DuPont particle delivery system PDS 1000/HE by modifying methods used with other filamentous fungi as described (St Leger *et al.*, 1995).

1. Mix 25 μl of a tungsten (M10 particles, mean diameter 1 μm) suspension (60 mg/ml) with 2.5 μg (2 μg/μl) of plasmid DNA, 25 μl of 2.5 M $CaCl_2$ and 10 μl of 0.1 M spermidine.
2. Vortex the particle–DNA mixture at 4 °C for 15 min and incubate for 10 min on ice. Collect the particles by centrifugation and wash sequentially with water, 70% ethanol and absolute ethanol.
3. Resuspend the particles in 20 μl of ethanol by sonication in a sonicator water bath (Branson 100) for 5 s.

4. Spread 6 μl of coated particles on a kapton flying-disc and use for bombardment.
5. Propel the particles toward the germinating conidia by release of helium at 1200 psi.

Before DNA is introduced into conidia, they should be germinated for 9 h at 27 °C with shaking (100 rpm) in 0.2% yeast extract. For transformation, spread germlings onto RA containing 0.1% Igepal (Alltech Associates, Inc.). Once seeded, place the plates into the chamber of the particle delivery system at a distance of 14 cm from the launch site. Bombard the germlings once or twice with tungsten particles coated with mixes of the plasmid DNAs. Selection for benomyl resistance can be achieved with an overlay applied 10 h after bombardment.

D Selective markers

1. Nutritional selective markers

The first group of selective markers, and the most widely used in the transformation of *A. nidulans*, are the nutritional selective markers (see May, 1992, for review). Transformation with nutritional markers is based on having auxotrophic mutant strains that can be transformed to prototrophy for the selective marker being used. Mutants in general and auxotrophs in particular will be difficult to obtain in many fastidious entomopathogens of economic importance. Although stable auxotrophs are readily obtained by chemical mutagenesis of *M. anisopliae* (Al-Aidroos, 1980) this raises the possibility of additional alterations in the genome which could influence virulence. It is possible to select for spontaneous mutants in the *niaD* and *pyrG* genes thus eliminating this potential problem and providing an attractive system in genetically poorly characterized species. In the case of *pyrG* they can be selected by resistance to 5-fluoro-orotic acid (Goosen *et al.*, 1987) and in the case of *niaD* by resistance against chlorate (Unkles *et al.*, 1989). Such an approach was applied to *B. bassiana* using the *niaD* gene for nitrate non-utilization, but the high level of natural resistance to chlorate shown by the wild-type fungus will probably limit the applicability of this technique. This is a pity, because since it is possible to select both for and against the mutant and wild-type phenotype. These markers are also particularly useful for genetic manipulation strategies, such as gene disruption studies. In short, if a mutation in the fungus has been identified and a heterologous gene is available from another fungus, then the likelihood is that a transformation system can be developed. If an appropriate mutation is not evident then it is probably not worth the time it takes to isolate one. The broad host range of fungal promoters will probably allow employment of one of the already existing dominant markers for experiments.

2. Dominant selective strategy

The most straightforward and industrially available approach for the transformation is to develop a dominant selective strategy. This involves transforming an existing wild-type cell, such as one sensitive to a drug, with a cloned gene that is selectable in that cell, such as a gene for drug resistance. The only requirement is that the gene be dominant or semi-dominant. A gene can be used to transform cells even if the gene has no eukaryotic origin of DNA replication. This is possible because the transforming DNA can integrate into the chromosomal DNA of the recipient cell, either by homologous recombination at the locus of the resident gene or by non-homologous recombination elsewhere in the genome (May, 1992). The dominant selective markers used to transform many filamentous fungi are resistance genes for the antibiotics (oligomycin, bleomycin, G418, hygromycin and phleomycin) and mutant β-tubulin genes that give resistance to the antimicrotubule compound benomyl. Unfortunately, except for *Verticillium lecanii* which is susceptible to hygromycin, most entomopathogens are naturally resistant to antibiotics and can only be transformed to benomyl resistance. This raises the pragmatic issue that there may be more transgenes to introduce into the fungus than suitable selection markers available. In which case it will be necessary to develop a strategy for removing a selectable marker gene so that the same optimized selection procedure can be used in subsequent transformations.

E Modifying plasmid vectors

The plasmid vectors used to date have not had homologous counterparts in *M. anisopliae*, and therefore constitute a target for improvement to optimize DNA delivery systems. Having cloned genes from *M.*

anisopliae we developed a homologous system of transformation by incorporating a disrupted *Pr1* gene into a suitable vector (see Figure 2). Although transformation frequencies of 50 per μg DNA were routinely obtained and integrants were highly stable, 50–65% of transformants had transforming DNA solely integrated at the marker locus. This may not be desirable in development of an improved pathogen, since chromosomal location is a determinant of expression level, and one wants to access as much of the genome as possible when isolating high-expression strains. In addition, the vector, as intended, was

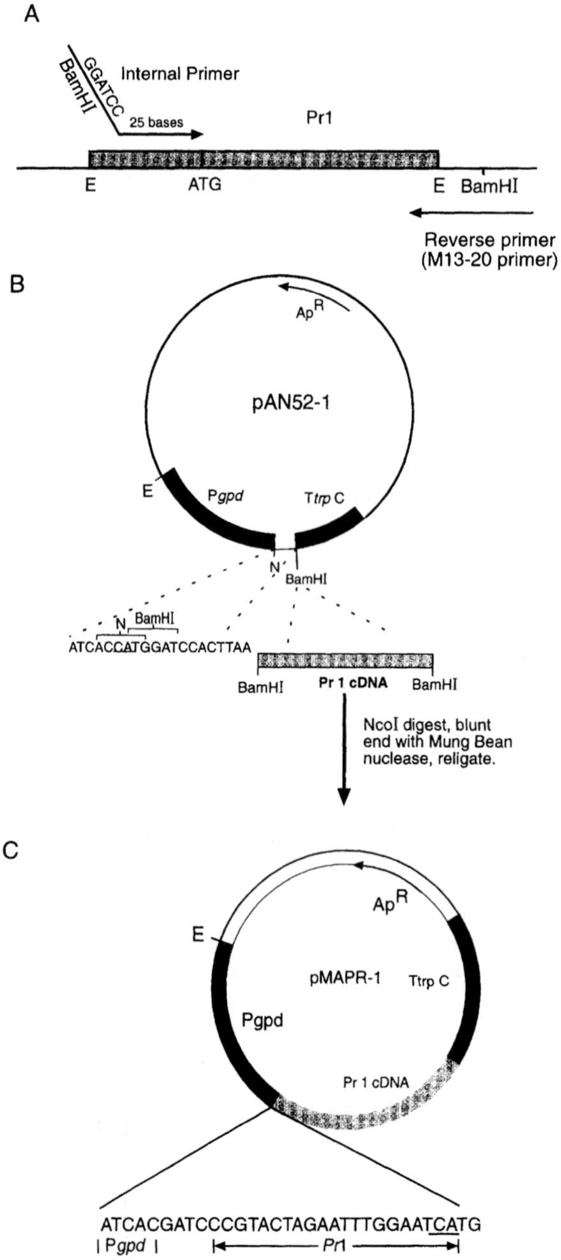

positioned within the *Pr1* gene disrupting its function. Further development of plasmids will involve positioning the marker and cloned genes outside any genes within the fungus chromosomal fragment to ensure that the transplacing fragment does not impair fungal function. Integrative fragment vectors could also be targeted for multicopy integration, thereby combining efficient transformation with high copy number. In yeast systems this has been achieved by targeting integrations to DNA that is multiply reiterated in the genome, such as the ribosomal DNA repeats or transposable elements (May, 1992). We have already mentioned that transformation frequencies can be increased by including autonomously replicating sequences (ARSs). It is not yet clear from the published literature whether autonomous vectors will in fact transform all fungi with a higher frequency than their integrating counterparts.

F Co-transformation

This is very effective in *M. anisopliae*, with around 60–80% of colonies transformed with DNA carrying the selectable benomyl resistance marker gene also being transformed for any unselected DNA present in the transformant mix. This has several practical advantages for strain construction. It also suggests a simple system for the elimination of selectable marker genes if the co-transformed DNAs (one incorporating a selectable marker gene and the other the gene of interest) integrate at genomic locations sufficiently unlinked to allow effective recovery of recombinant events, i.e. use the parasexual cycle to cross with the parent strain to obtain segregants containing only the gene of interest.

6 CONSTRUCTION OF CLONING AND EXPRESSION VECTOR COMPONENTS AND MARKERS

The availability of different gene transfer systems with different characteristics permits a molecular-genetic study of all kinds of biologically interesting processes by isolation, characterization and functional analysis of the genes and gene products involved. To perform these studies, specific vectors are constructed which facilitate genetic manipulation such as cloning of a gene, gene disruption or gene replacement. Although there already exists a wide selection of filamentous fungi vectors encoding marker proteins for vector studies, it may be that the exact vector for a particular purpose has not been made and so a tailor-made vector has to be constructed. Perhaps for example, this will involve introducing a gene that encodes a toxin or enzyme, or a new marker or promoter. Thus, besides knowing how to employ existing vectors in transformation systems, it is necessary to know how to make new ones, when circumstances dictate. Here we will consider the options available for making such constructions, along with the circumstances under which they may be employed. To illustrate the possibilities of genetic manipulation for molecular genetic studies examples will be given of research on *Metarhizium* that is progressing in our laboratory.

Figure 2 Precision DNA fragment trimming via PCR to create a new promoter expression construction. (A) A useful general strategy using a fragment cloned into a polylinker of an M13 type plasmid. If the trimmed end is to be blunt, then the 5′ terminal base of the internal primer will represent that of the fragment generated. Alternatively, if a restriction site is to be introduced, the recognition site bases are introduced at the 5′ end of the primer. To create a new promoter expression construct for the expression of *Pr1* we used the pAN52-1 vector. pAN52-1 comprises the constitutive *gpd* promoter and the terminator region of the *trp C* gene (both from *A. nidulans*) separated by BamHI and NcoI sites (B). A full length cDNA clone of *Pr1* in a Bluescript vector was amplified by RT-PCR with the forward primer 5′-AAGGATCCCGTACTAGAATTTGGAAT CATG-3′ containing the BamHI recognition site bases at the 5′ end GGATCC. We added two additional bases at the 5′ end, as some restriction enzymes fail to cut efficiently if the site is right at the end. However, addition of bases reduces the specificity of the primer so not too many should be added. Having the fragment cloned into the Bluescript polylinker enabled us to use a universal sequencing primer (5′-GTAAAACGACGGCCAGT-3′; universal M13-20 primer) at the non-deletion end. The amplified cDNA was cloned into vector pCRII (Invitrogen) and subsequently excised by digestion with BamHI, which cuts at the sites incorporated into the forward primer and 19 nucleotides downstream of the *Pr1* insert in the amplified portion of the Bluescript vector. The 1.2 kb BamHI fragment was cloned into BamHI-cut pAN52-1. The resulting vector was digested with NcoI, blunt ended with Mung bean nuclease to remove the *gpd* ATG translation start codon, and religated. Sequence analysis confirmed the orientation of the insert in the recombinant vector and that the *Pr1* initiator ATG was in frame with the *gpd* promoter (C).

A Vector construction

Basic vector construction starts with a precursor plasmid, the new DNA component to be ligated in and a procedure to identify the new construction. The major procedures involved in this process are the isolation of specific DNA fragments from gels and the cloning of them into the burgeoning vector construction.

1. The isolation of DNA fragments from gels

For agarose gels, Tris-acetate electrophoresis buffer is used, the gel is stained with ethidium bromide, and the DNA bands are viewed under UV illumination in the standard way (Sambrook *et al.*, 1989). Using a scalpel, the appropriate fragment is then excised in a minimum-sized gel slice, and this is then processed to extract the DNA. There are a number of ways to do this. The methods that work well are using either the QIAEX kit from QIAGEN or low-melt agarose gel electrophoresis. We recommend the QIAEX kit because of its ease of use and low cost. Other methods, such as Geneclean (which only works well for DNA larger than 300 bp) or filtration, result in low yield.

2. Cloning manipulations

It is important to set up the cloning ligation with the correct concentration, and relative molar ratio of vector and insert fragment. How to calculate these conditions is outside the scope of this manual, but has been described elsewhere (Sambrook *et al.*, 1989; Ochman *et al.*, 1990). After selecting the appropriate concentration conditions for the ligation, it can be set up and undertaken as described in Ausubel *et al.* (1991) using T4 DNA ligase (5–7 Weiss units), ligation buffer and vector and fragment DNA (diluted in 5 mM Tris–HCl, pH 7.6).

3 Fusion construct

We have prepared constructs from a variety of expression plasmids. These contain an isolated promoter fragment that has a restriction site positioned immediately downstream of the transcription initiation site, with no associated translation initiation codon, this is then followed by a terminator-containing fragment, which provides discretely sized mRNA. Such constructs are therefore suitable for the cloning and expression of cDNAs.

There is a wide choice of possible promoters to drive heterologous gene expression (reviewed by Davies, 1992). Of those commonly used, the phosphoglycerate kinase (PGK), alcohol dehydrogenase 1 (ADH1) and glyceraldehyde 3–phosphate dehydrogenase (GAPDH) promoters generally give a good level of constitutive expression. These genes encode glycolytic enzymes that are some of the most abundant mRNA and protein species within the cell and therefore their promoters are obvious choices for directing foreign gene expression. Other promoters, however, have also been found to function well in this regard, with some, such as the cross-pathway control (CPC) showing higher levels of heterologous gene expression (Orbach, 1994). Of the regulated promoters, the GAL series are very tightly repressed by glucose and induced by galactose. Expression signals of the glucamylase (GLA) gene are induced by starch (Davies, 1992). The inducible *Neurospora crassa* quinic acid (QA) promoter has been used to drive expression of genes of the plant pathogen *Colletotrichum trifolli* and may have broad applicability (Martin Dickman, personnel communication).

A typical example of an expression vector is pAN52–A (Figure 2) which utilizes the *gpd* promoter and the terminator region of the *trp C* gene (both from *A. nidulans*). We recently used this vector to constitutively express the *Pr1* gene. This required tailoring of the end of the *Pr1* gene to obtain a precise in-frame junction. There are two major methods for doing this: PCR (polymerase chain reaction) and Bal31 exonuclease digestion. We adopted PCR as this allows precise tailoring with ease and efficiency. PCR requires the availability of a DNA thermocycler and primer oligonucleotides, based on a known DNA sequence. When PCR works correctly, it gives rise to a single DNA band visible on a gel. However, if the primers are not specific enough, several bands may be visible. With luck, these may be eliminated by optimizing the Mg^{2+} concentration (a kit sold by Invitrogen is very useful for this purpose) or increasing the annealing temperature. For the 17-mer universal primers of the pUC or pIC plasmids, typical PCR conditions are: denature at 94 °C for 20 s, anneal at 55 °C for 20 s, and extend at 72 °C for 30 s for a total of 25–30 cycles. It is usually best to start with a modest concentration of $MgCl_2$, such as 2.5 mM, and then alter if non-specific primings occur.

Optimization of PCR has been reviewed by Sambrook *et al.* (1989). If using a 'Universal' primer, these are available from Pharmacia, Gibco-BRL, New England Biolabs and others.

The basic PCR strategy we have adopted is shown in Figure 2. Since two primers are needed to generate a PCR fragment, it was convenient to have the *Pr1* gene in the Bluescript vector so that one of the primers could be a universal M13 sequencing primer that primes from the vector itself. Thus only the specific internal *Pr1* sequence needed to be synthesized. This primer was designed to generate a precise 5′ end for fusion to the BamH1 site in pAN52. This was done by introducing the recognition site bases at the 5′ end of the primer. The 3′ end of the fragment was generated by BamH1 cleavage within the polylinker sequence of the vector present in the fragment due to the external priming of the universal primer.

If it is not possible to use PCR because the sequence is not known, Bal31 provides a double-stranded exonuclease activity that will progressively remove terminal nucleotides resulting in a spread of termini. Treatment with Klenow large-fragment DNA polymerase in the presence of nucleotides will fill in any 5′ recessed ends. Alternatively, we have used mung bean nuclease to obtain blunt ends (Figure 2). Linkers (short-chain double-stranded oligonucleotides that encode a restriction endonuclease site) can be added to blunt ends to introduce terminal restriction sites (Sambrook *et al.*, 1989). Subsequent cleavage with the appropriate restriction enzyme then generates an artificial restriction fragment with the appropriate ends. The laborious feature of using Bal31 is that, at some stage, the termini of the clones obtained will have to be sequenced to find one with a suitable end-point.

7 GENE CLONING STRATEGIES

Isolation of single copy genes from a complex genome and isolation of rare cDNA clones from a complex mRNA population requires the generation of recombinant DNA libraries which contain complete representation of genomic or cDNA sequences. Consequently, the main problem in generating a useful library is the creation of the huge population of clones necessary to ensure that the library contains at least one version of every sequence of interest. The solutions to this problem are essentially similar for genomic and cDNA libraries. The genomic DNA or cDNA are first prepared for insertion into the chosen vector. The vector and target DNA are then ligated together and introduced into *E. coli* by packaging into phage lambda heads *in vitro*. These procedures are described in detail in Sambrook *et al.* (1989) and Ausubel *et al.* (1991) but much the easiest way to get a library is to get one from someone else; several laboratories have genomic or cDNA libraries for important entomopathogens. Alternatively, use one of the comprehensive kits available from Stratagene or Clontech. It takes about two weeks to prepare a library using a kit. These companies will also make a custom library from DNA and total or poly(A) RNA provided to them. Stratagene has the greater choice of vectors including the very useful unidirectional ZAP vectors, but Clontech is generally less expensive.

A Identifying pathogenicity related genes

The characteristics of a disease may be defined differently at the molecular level, the cellular level, the whole insect/pathogen level and the population level. Thus the processes and gene products which are involved in the ability of the fungus to cause disease will be many and varied and the insect pathologist could be interested in several broad classes of pathogenicity genes. Some genes will encode receptors that detect either directly or indirectly the presence of the host (e.g. a GTP-regulated adenylate cyclase, tyrosine protein kinases, serine and threonine protein kinases and phosphoprotein phosphatase (St Leger, 1993)). These act to change second messenger levels or are themselves activated by second messengers to trigger differentiation (St Leger, 1993). Activation of such receptors and signal transduction pathways may result in the induction of the generic pathogenicity genes. Another class of pathogenicity genes may inactivate host defences. Other pathogenicity genes may encode toxins that are required for disease symptoms, e.g. destruxins and hydrophobins. A fourth category of pathogenicity genes encode enzymes that allow the fungus to overcome host barriers. Much of our research has concentrated on these genes. A fifth category of pathogenicity genes may be a heterogeneous group, some of which have little or no generality in insect–fungal interactions. To determine whether such pathogenicity genes exist and what are the characteristics of each class, it will probably be necessary to characterize multiple genes conferring pathogenicity to diverse hosts, preferably from several pathogen species.

Several approaches have been used to isolate genes from plant pathogens (Garber, 1992; Turner, 1991) which are also broadly applicable to entomopathogens.

One can broadly distinguish between cloning strategies involved when the protein product of the gene to be cloned is known and identified biochemically as playing a role in disease, and isolation of genes with unknown products. The later approach can identify whole new, previously unsuspected strategems of pathogen attack. In cases where there is evidence that an mRNA of interest is transcribed at a particular stage of development, or in response to a specific induction signal, it is possible to take advantage of differential gene expression to isolate a cDNA clone. Two approaches have been successfully applied to *M. anisopliae*, differential hybridization and differential display. We will describe the latter in more detail as the results have not yet been published.

B Differential hybridization

The analysis and characterization of changes in pathogen or host genes expressed at particular stages of differentiation or under certain circumstances is of prime interest in host–pathogen study. The first pathogenicity-related genes cloned from *M. anisopliae* were obtained by probing a cDNA library in the lambda gt10 vector constructed from cells transferred from nutrient-rich to nutrient-deprived conditions to enrich for messages subject to catabolite repression (St Leger *et al.*, 1992b,c). Replica filters were prepared of plaques from the library. Each set of filters was hybridized with one of two probes made of total cDNA from the nutrient-deprived or nutrient-rich conditions. Clones of starvation stress genes (ssg) hybridized more strongly to the probe made of cDNA from the nutrient-deprived state and thus represent differentially expressed genes. By this means we obtained the *Pr1* gene, a previously suspected pathogenicity determinant, and four other clone families including a hydrophobin gene (identified by its resemblance to published sequences of hydrophobins) not previously suspected of being involved in pathogenicity (St Leger *et al.*, 1992c) but shown to be involved in building appressorial walls in a plant pathogen (Talbot *et al.*, 1993). We still utilize the technique of differential hybridization but are now using libraries prepared in the lambda ZAP II vectors available from Stratagene Cloning systems. Inserts are cloned in to lambda ZAP within the plasmid Bluescript SK (-) which it contains. The library can be screened with a cDNA probe or with antibodies because clones are expressed as fusions between an *E. coli* β-galactosidase protein and the protein encoded by the insert. The Bluescript plasmid can be excised from the lambda ZAP vector and the insert directly sequenced. Using lambda gt10, the insert has to be excised with EcoR1 and subcloned into Bluescript for sequencing. Details of probing these libraries with radiolabelled DNA or oligonucleotide probes are provided in Sambrook *et al.* (1989), Ausubel *et al.* (1991) and Joshi *et al.* (1995). However the following may be useful:

1. Plate out a library at high density using standard procedures (Sambrook *et al.*, 1989). Allow plaques to become visible but not too large (5–6 h at 37 °C). Chill plates for 1 h before carrying out plaque lifts.
2. Carry out duplicate plaques lifts. First lift has 30 s contact with plate, second lift 2 min. First lift is always probed with the – probe (in our example mycelial cDNA from nutrient-rich cultures). Second lift is probed with the + probe (from mycelial cDNA from nutrient-deprived cultures).
3. Colony or plaque lifts for screening with radioactive probes are best made using NEN nylon hybridization membranes. These allow researchers working with multiple lifts to combine the lysing (of *E. coli* cells) and fixing (of DNA) steps by autoclaving lifts for 1 min at 100 °C with quick exhaust. In our experience this simple time-saving procedure does not work well with membranes from other suppliers. For hybridization conditions refer to Table 1.
4. Radiolabelled cDNA probes are made by carrying out a first strand cDNA reaction with 1 μg poly(A) RNA and ^{32}P-CTP using one of the available kits from Clontech, Invitrogen, etc. The probe can be purified through a Biogel p60 column or other sizing medium (this will reduce background, but is not essential) and is denatured with boiling before use.

C The use of reverse transcription-differential display-PCR (RT-DD-PCR) technique to identify differentially regulated genes

Differential display (DD) has been developed as a tool to detect and characterize differentially expressed transcripts in eukaryotic cells that have been subjected to different environmental and developmental conditions. This method also allows an estimate of the number of genes that are differentially expressed under different circumstances.

Differential display does not have the limitations of other methods such as subtractive and differential hybridization, which require large amounts of RNA, are rather difficult to establish and are usually less reproducible. Subtractive and differential hybridization methods are mainly qualitative and do not detect quantitative changes, differential display detects all mRNA species expressed by the cell. Comparing the expressed mRNA patterns from different cells makes it possible to detect both qualitative and quantitative changes. The experimental design involves using anchored oligo-dT primers which anneal to the beginning of the poly(A) tails of mRNA. These are used in conjunction with arbitrarily defined decamer oligonucleotides (AP) for subsequent PCR amplifications. The products are radioactively labelled during the PCR amplifications. The amplified fragments of cDNA are then separated by size on large denaturing polyacrylamide gels. The differentially expressed bands are cut out, amplified and used in northern blots to confirm their mode of expression. Suitable cDNA fragments are ligated into PCR cloning vectors and are either sequenced directly or used to pull out the genomic or cDNA clone from a suitable library (Figure 3). Most laboratories use RNAmap™ or RNAimage™ (Genhunter, Brookline, MA, USA) kits, when commencing their studies. However, if partial sequences are available specific primers can be ordered and used instead of arbitrary primers.

The general protocol for differential display is described below:

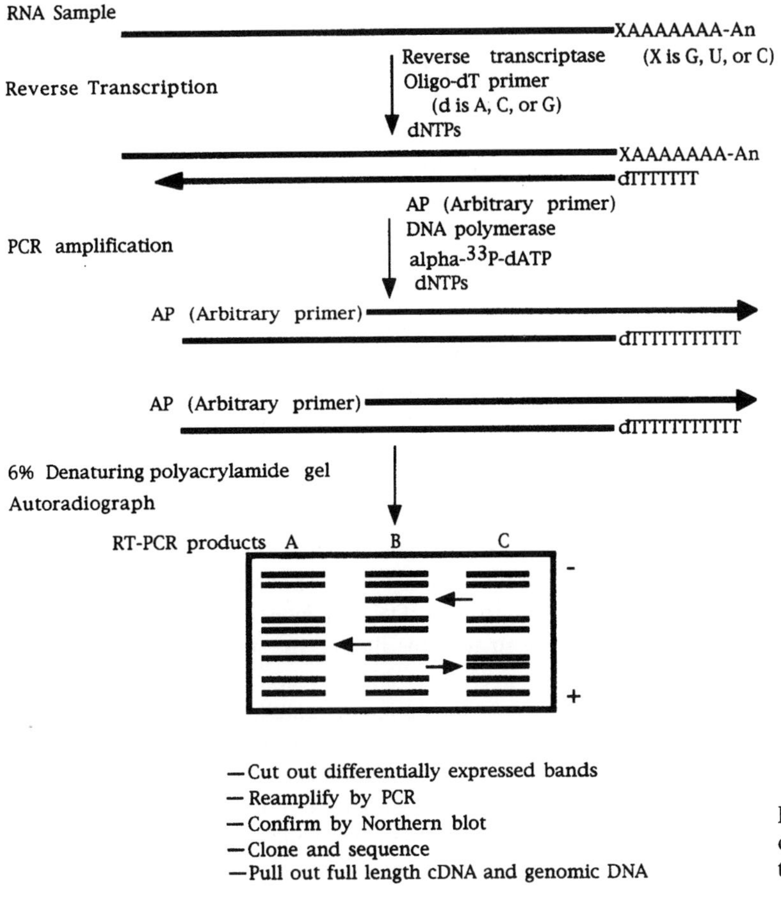

Figure 3 Schematic representation of reverse transcription DD-PCR technique.

1. Purification of total RNA and DNase treatment

The integrity of the RNA preparation is the most important factor in RNA-DD. In Section 3 we described how to purify good quality total RNA. Although mRNA can be used in differential display, total RNA is preferred as the substrate because of its cleaner background signal, the relative ease of purification and the ease with which it allows for verification of total RNA.

DNase treatment of the total RNA is often a prerequisite to differential display because any traces of DNA increase the chances of having false positives by this method.

DNase I and ribonuclease inhibitor are added to total RNA and incubated at 37 °C for 15 min in a water bath. RNA is purified with phenol: chloroform: isoamyl alcohol and precipitated by adding 3 M sodium acetate and 95% ice cold ethanol. The RNA pellet is resuspended in 40 μl DEPC-treated water.

2. Differential display

Typically, 0.02–2 μg RNA will achieve a clear band pattern with RT (Song *et al.*, 1995). It is best to run duplicates of each RNA sample to minimize errors in the PCR amplification which could lead to spurious results. RNA is incubated with one of the oligo-dT primers, dNTP and reverse transcriptase. The reverse transcription reaction is carried out for 1 h at 37 °C. The anchored oligo-dT primers consist of 11 or 12 Ts plus one or two additional 3′ bases which provide specificity. The use of one-base anchored oligo-dT primers instead of two-base anchored oligo-dT primers reduces the number of reverse transcription reactions for each RNA species due to the degeneracy of the primers (Liang *et al.*, 1994).

The commonly used cycling parameters are: 94 °C for 30 s, 40 °C for 2 min, 72 °C for 30 s for 40 cycles followed by 72 °C for 5 min (RNAimage™, GenHunter). In the PCR amplification α-[^{35}S]dATP can be used instead of ^{33}P but ^{35}S can diffuse out from the PCR tubes and contaminate the thermocycler block. Amplification with either isotope generates the same patterns but ^{33}P requires less exposure time, gives higher sensitivity and better resolution than ^{35}S.

The amplified PCR products are separated on a 6% denaturing sequencing polyacrylamide gel. The gel is blotted onto a filter paper and dried under vacuum at 80 °C for 2 h. The dried gel is marked with photoluminescent markers or radioactive ink and exposed to an X-ray film overnight.

False positives can sometimes count for a significant percentage of the total number of bands observed. One solution to this is to run duplicates, or triplicates if possible and to repeat the experiment for the lanes in which potential candidate cDNA bands are observed. Inosine will pair with similar binding strength to all four bases (Ausubel *et al.*, 1991). Consequently, the use of inosine at ambiguous positions during the synthesis of primers will increase the rate of reproducibility since many of the false positives are caused by imperfect annealing of primers to sequences within mRNA (Rohrwild *et al.*, 1995).

3. Recovery, reamplification and cloning of cDNA Fragments

After developing the X-ray film, cDNA bands of interest are marked and cut out. Each gel slice is soaked in water and boiled to facilitate diffusion of the cDNA from the polyacrylamide gel. The cDNA is purified by precipitating with 3M sodium acetate and 100% ethanol, using glycogen as a carrier. Reamplification is done with the same primer set and PCR conditions as the first PCR reaction except that no radioisotope is added. Part of the reamplified cDNA is run on a 1.5% agarose gel and the rest is stored at −20 °C for further use. The gel bands are cut out and used as probes on northern blots to confirm the differential expression of the transcripts of interest.

Smearing or multiple bands often necessitate gel purification which may result in decreased ligation efficiency. For gel purification, all nuclease contamination must be removed. Spin columns for gel purification should be avoided since they can result in loss of ligation efficiency. Electroelution and agarase are satisfactory. The reamplification of cDNA fragments sometimes fails or a smear of bands is produced due to the arbitrary primer not being totally complementary to the cDNA of interest. This situation can be dealt with by altering the annealing and elongation temperatures during reamplification, thus reducing the annealing temperature has helped in some reamplification reactions (T.E. Battle, personnel communication). Reamplified cDNA probes are cloned into various vectors but the most commonly used one is the pCR™II (TA® cloning kit, Invitrogen) vector. Taq polymerase has a non-template-dependent activity which adds a single deoxyadenosine (A) to the 3′ ends of PCR products. The linearized TA vectors (such as pCR™ II) have a single 3′ deoxythymidine (T) residue. This allows PCR inserts to ligate efficiently with the vector. It is best to use fresh PCR inserts since the inserts gradually loose the 3′ A-overhangs resulting in reduced ligation efficiency. Competent *E. coli* cells are transformed and recombinant colonies are selected. To determine the sequence and orientation of the insert, selected colonies are grown overnight and the plasmid is isolated and sequenced either using a manual sequencing kit or an automated sequencer.

4. Modifications

Since the method of differential display was first described, there have been several reports of modifications in the basic protocol with each having certain advantages. Some of these are mentioned here:

1. Replacement of the oligo-dT primer with a second AP primer in the PCR step to allow amplification along the entire length of the cDNA, with reverse transcription as per usual differential display (Haag & Raman, 1994).
2. Use of specific primers to replace arbitrary primers in the PCR step. We have used the primers from the most conserved regions of subtilisin-like enzymes and certain protein kinases to clone and analyse all the possible genes with sequence homologies (Joshi *et al.*, in press).
3. Differential display has been used to detect transcripts from very few cells using the whole cell lysate, without RNA being purified. The cells are suspended in phosphate buffer saline in the presence of tRNA which acts as a substrate for RNase activity limiting the degradation of transcripts of interest. The cells are lysed, incubated with proteinase K and heated at 95 °C. The supernatant is used for RT reactions (O'Brien *et al.*, 1994).
4. RNA finger printing is rather similar to differential display except that instead of using different oligo-dT primers, a single reverse transcription reaction is performed for each RNA sample. In RNA fingerprinting, the subdivision of the RNA population takes place at the PCR amplification step. Three initial cyclings are performed at low stringency in the presence of arbitrary primers allowing annealing and initial DNA synthesis. Due to the low stringency conditions, each arbitrary primer binds to many cDNA sites with imperfect and/or incomplete matches. The products are then amplified using the same arbitrary primers and downstream T primers with high stringent conditions in PCR cycling. These PCR reactions produce characteristic 'fingerprints' of the starting RNA. There are some kits available to perform RNA fingerprinting, e.g. Clontech's Delta RNA Fingerprinting Kit.
5. mRNA is reverse transcribed with the use of a fully degenerate 6-mer oligonucleotide into cDNA. Internal regions of the cDNA are amplified in a PCR reaction using two or three longer primers with arbitrary but defined sequences and (usually) several different distinct cDNA bands are displayed by agarose gel electrophoresis. By repeated reamplification of the cDNA obtained in the first step with different primers, large numbers of different cDNA fragments are obtained from the same mRNA sample. These different bands are cut out and used for subcloning and northern blot analysis as in differential display (Sokolov & Prockop, 1994).

D Heterologous probing

The simplest method to clone a gene is to 'clone by phone', that is to use a gene obtained from one organism as a heterologous probe to pull out the related gene from a DNA library of a second organism. This requires that the nucleotide sequences of the equivalent genes be similar which is more likely if the two organisms are related. Thus the *Pr1* protease gene from *M. anisopliae* was used to clone the corresponding protease gene from *B. bassiana* using relaxed hybridization and washing conditions (Joshi *et al.*, 1995), but it is less likely that the equivalent bacterial gene could have been employed in the same way. Some genes are much more strongly conserved than others over broad taxonomic distances, heat shock genes, ribosomal RNA, histone and tubulin genes, and genes involved in signal transduction (e.g. calmodulin and some kinases) being good examples. For instance, *Drosophila* and chicken genes have been used as heterologous probes to identify fungal heat shock and tubulin genes (May 1992).

Probing a library with a heterologous probe is essentially standard (Sambrook *et al.*, 1989). We recommend using a random primer labelling method for the probe. The BRL random primers labelling system provides a very convenient way of labelling nucleic acids to high specific radio activity with ^{32}P. It can be easily adapted to label DNA in agarose gel as follows. The reaction mixture consists of 25 μl of DNA in low-melting-point agarose (this should have been diluted with two volumes of water, boiled for 3 min and then incubated for 15 min at 37 °C), 15 μl of random primers buffer, 6 μl mix of ATP, GTP and TTP (provided with the kit), 3 μl of ^{32}P CTP and 1 μl of Klenow fragment. The reaction mixture should be mixed, incubated overnight and boiled (5 min) before adding to the hybridization buffer.

E Synthetic probes

If the protein product of the gene is known and is available in purified form, it can be sequenced, and the genetic code can be deduced from the sequence. The sequence generally contains many ambiguities, because of the degeneracy of the genetic code. The least ambiguous region of 11–20 bases (up to seven amino acids) is chosen and a synthetic oligonucleotide (or pool of related oligonucleotides to increase the possibility that at least a portion of the probe will match the DNA sequence) of complementary sequence is then chemically synthesized (Garber 1992). This synthetic DNA can be labelled with polynucleotide kinase (many companies sell kits containing all the necessary ingredients) for use as a hybridization probe. The big disadvantage of this technique is that many incorrect clones may hybridize. To some extent this problem may be overcome by the use of carefully controlled hybridization conditions, or by the use of two probes derived from different parts of the protein sequence (see Ausubel *et al.* (1991) for a thorough discussion). However, in part because of these problems the use of oligonucleotides as probes has been largely superseded by PCR techniques.

F Antibody detection

Several screening methods that rely on the expression of the cloned gene have been devised. Most of these use antibodies against the gene product of expression. Expression is obtained by cloning cDNA (to avoid the problem of introns) in expression vectors so that they are expressed as β-galactosidase fusion proteins. To screen clones for expression, protein from plaques is fixed to a nitrocellulose filter, and the filter is treated first with antibody, and then with a labelled second antibody to detect the binding of the first antibody. Again, several companies (e.g. Amersham, Promega) provide kits containing secondary antibodies (usually labelled with alkaline phosphatase or horseradish peroxidase) and full instructions on their use. The most efficient and easily screened vectors are those that allow directional cDNA cloning (e.g. the Uni-ZAP XR or lambda ZAP express unidirectional vectors from Stratagene) as these double the number of clones detected by antibody screening.

The primary prerequisite for this technique is a protein of sufficient purity to generate antibodies. The most convenient way we know to separate a mix of polypeptides is by SDS-PAGE. The Coomassie-blue stained band of interest can then be cut out and injected directly into a rabbit. It is also possible to electroblot the protein onto immobilon-P™ (Millipore, USA) membranes, quick-stain with Coomassie blue and use the band to obtain sequencing data directly with an automated sequencer (St Leger *et al.*, 1996b).

G Mutant complementation

Many genes of pathological importance cannot be isolated by the foregoing approaches because their protein products are currently unknown, and prospects for finding an abundant mRNA are low. The classic way to isolate such genes is by complementation of a mutant lacking the gene function with a library of DNA fragments prepared from an isolate that has the gene function. The presence of the gene in question is recognized by its ability to cause the mutant to behave like the wild type. This is sometimes called 'cloning by transformation'. Typically, cosmid vectors are used for this purpose. They contain one or two bacteriophage -cos sites to allow *in vitro* assembly of phage particles containing large (30–40 kb) cloned inserts. In addition, they should be 'shuttle vectors' including genes for selection in both *E. coli* and the eukaryotic host and one or more restriction sites for insertion of foreign DNA fragments. The vectors should also allow transformation frequencies high enough to generate thousands of independent transformants, because even with the small haploid genomes typical of fungi, a minimum of 2000 transformants is required to adequately represent an entire genome when cosmid clones are used for transformation (Yoder *et al.*, 1986). Of the vectors currently in use, pSV50 would appear to be the most suitable as it already incorporates a benomyl resistance gene which is selectable in *M. anisopliae* and *B. bassiana*, and has been used to clone genes from *N. crassa* (Vollmer & Yanofsky, 1986). A sophisticated new vector, pMOcosX, has a hygromycin marker for fungal library construction and screening (Orbach, 1994) and unfortunately many entomopathogens are naturally resistant to hygromycin. Recovery of a gene from a transformed strain for further analysis is straightforward in those few systems where autonomous replication of transforming DNA can be achieved: it is a simple matter to transform *E. coli* directly with total DNA isolated from the fungal cell. Alternatively, when the transforming DNA integrates into the genome, recovery can be achieved by constructing a DNA library from the transformed strain and using adjacent vector sequences as a hybridization probe, since vector DNA will usually not hybridize to DNA of the untransformed fungal cell.

H Restriction enzyme-mediated integration

A new development which is likely to supersede this laborious approach is to tag genes by restriction enzyme-mediated integration (REMI) of transforming plasmid. This technique was developed following Schiestl & Petes (1991) findings that introducing the restriction enzyme BamHI together with a BamHI-cut DNA fragment increased the number of transformed yeast cells and the DNA fragment was often integrated into BamHI sites of the host genome. It was later demonstrated that REMI generates insertions into genomic restriction sites in an apparently random manner, some of which cause mutations (Lu *et al.*, 1994). This provided a simple method of insertional mutagenesis to tag genes based on their mutant phenotypes. The integrated plasmid (which contains an *E. coli* selectable gene) along with flanking genomic DNA can be excised from some of these mutants (by cutting genomic DNA with a restriction enzyme that does not cut within the plasmid), cloned in *E. coli* cells and the flanking DNA sequenced to identify the disrupted gene. Obviously, the ability to clone and identify the tagged mutations makes the technique preferable to untagged, chemically induced mutant.

The starting point for development of our protocol for *M. anisopliae* was the reported method of Lu *et al.* (1994) and was developed with advice from Dr Alice Churchill.

1. REMI transformation vector

The plasmid to be used as the vector should include genes for selection in both *E. coli* and the eukaryotic host, and possess no detectable homology to the recipients genome. We have used the pBENA3 plasmid.

2. Preparation of vector-restriction enzyme mixture

1. Digest the plasmid with an excess of a restriction enzyme which has a unique site in the plasmid, e.g. pBENA3 with XbaI.
2. Check for complete digestion by electrophoresis.

3. Prepare fungal protoplasts using an enzyme–osmoticum mixture

Protoplasts of *M. anisopliae* are prepared by shaking mycelia with 0.8% Novozyme 234 in 1.2 M sorbitol, 10 mM Tris–HCl (pH 7.5) for up to 3 h. The protoplasts are filtered through two layers of sterile cheese cloth, centrifuged and then resuspended and washed three times with STC solution (sorbitol, Tris–HCl and calcium chloride).

4. Transformation

The following steps should be performed on ice.

1. Mix 100 μl each of the protoplast preparation and plasmid-restriction enzyme mixture and incubate for 15 min.
2. Add PEG (10 mM Tris–HCl (pH 7.5), 50 mM $CaCl_2$, 70% polyethylene glycol (MW 3350)) in three aliqots of 200 μl, 200 μl and 800 μl each and incubate for 10 min after each addition. Add 1 ml STC to each tube.
3. Plate protoplasts by adding a 200 μl aliquot to

20 ml molten regeneration media (detailed in section 7b) and incubate overnight.

4. Overlay each plate with 10 ml of 1% agar containing 10 μg benomyl.

5. Recovery of insertion point flanks from the genome

1. Digest 2 μg genomic DNA from a colony with a mutant phenotype with excess restriction enzyme that has no site in pBENA3.
2. Remove enzyme by phenol/chloroform extraction and ethanol precipitation.
3. Ligation: 15 μl (1 μg) digested genomic DNA, 25 μl 10× ligase buffer (Promega), 3 μl T4 DNA ligase (Promega), 50 μl 25% PEG, 157 μl water. Incubate at 22 °C overnight, stop reaction with 12.5 μl 0.5 M EDTA.
4. Precipitate DNA with 0.1 vol. 3 M Sodium-acetate and 2 vol. 100% ethanol.
5. Transform entire sample into XL1-Blue cells (Stratagene) and distribute to LB + ampicillin plates. Incubate overnight. The efficiency of transformation should be about 5 × 10^8 colonies/μg DNA.

8 GENE DISRUPTION

Once a fungal gene has been cloned, molecular techniques allow a rigorous determination of whether the gene product is required for pathogenicity. If a gene is altered or eliminated, then pathogenicity will be reduced to an extent dependent on the importance of the gene in pathogenicity. This ability changes forever our ideas about what constitutes an acceptable standard of evidence about the role of a molecule in disease.

Two methods regularly used in gene disruption are indicated in Figure 4. In both cases the disruption vector contains a non-functional copy of the chromosomal gene to be disrupted. The method indicated in Figure 4A requires knowledge about the exact position of the gene in the cloned frag-

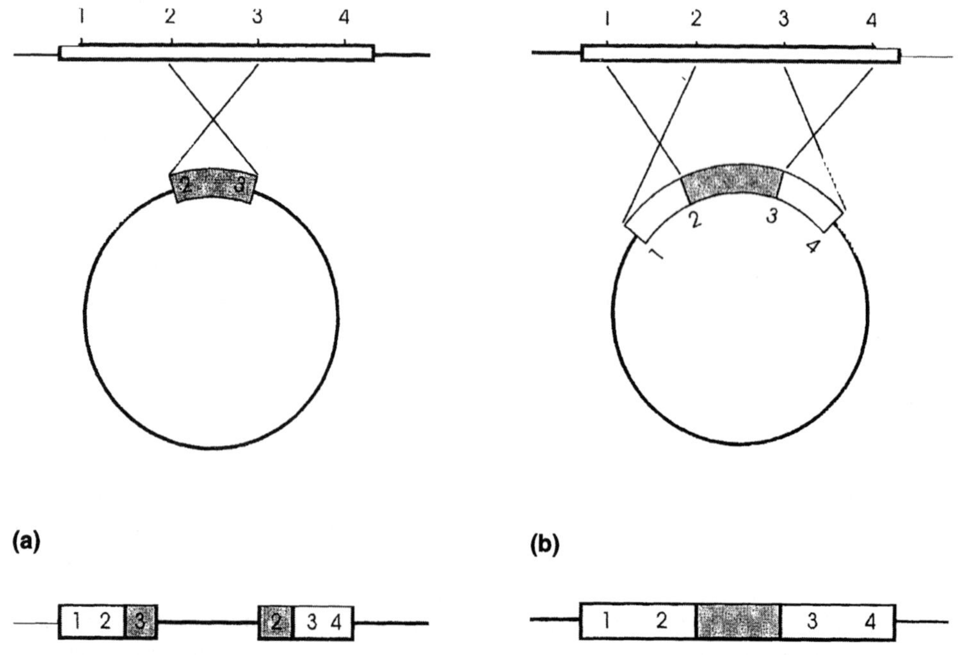

Figure 4 Models for gene replacement and gene disruption. (A) a single crossover event can generate two mutated copies (123--and 234) of the gene if the vector bears a gene fragment (23) and the integration occurs by homologous recombination. (B) Transformation vector containing a copy of a sequence (1234) which differs from the chromosomal copy (1234) can be replaced by the altered copy if a double crossover event which spans the difference between the chromosomal copy and the vector-born copy occurs by homologous recombination.

ment, whereas for the second method this is not necessary. To obtain an *M. anisopliae* strain in which the *Pr1* gene was disrupted, the second method was used. A plasmid was constructed in which a 400 bp fragment of the coding region of the gene was replaced with a larger fragment containing the benomyl resistance unit from pBENA3. Transformation of *M. anisopliae* with this *Pr1* disruption plasmid resulted in a number of benomyl resistant colonies. Southern blot analysis revealed that about 15% of these had undergone a gene replacement. Further analysis showed that these transformants demonstrated only a reduction in pathogenicity, indicating that *Pr1* is a determinant of virulence but is not essential for pathogenicity. This is likely to be because other subtilisin-like enzymes as well as other proteases produced by the pathogen can, in part substitute for *Pr1* activity. It appears that *M. anisopliae* rarely produces just a single enzyme capable of hydrolysing a particular chemical bond present in the cuticle (St Leger, 1995), thus single-gene disruption experiments are unlikely to reveal the role of these proteins in pathogenicity. The various hydrolases produced during growth on cuticle will need to be identified, characterized and localized during the infection process to provide an accurate representation of cuticle degradation.

9 POPULATION STUDIES

It is becoming increasingly apparent that identification to species level is not adequate for ecological studies. Within a species, strains are quite host specific and selective isolation does not provide any information on the pathogenicity of the isolates. Up to now, definitive identification of strains has not been possible; rather, isolates have been coded and characterized by methods such as isozymes, serology, cultural characteristics, and host specificity (Glare & Milner, 1991).

Allozymes provided the first unambiguous markers available in sufficient numbers to enable reliable genetic studies of entomopathogenic fungi (St Leger *et al.*, 1992a). More recently, random amplification of polymorphic DNA (RAPD) has provided additional markers for population analysis (Bidochka *et al.*, 1994). Strain-specific DNA probes have been developed to track wild-type strains, e.g. the Zygomycete pathogen, *Entomophaga grylli* (Bidochka *et al.*, 1995).

A Differentiation of species and strains of entomopathogenic fungi by random amplification of polymorphic DNA (RAPD)

Genomic variability can be easily assessed based on length and sequence differences in PCR-amplified segments generated with primers of arbitrary nucleotide sequence. The following protocol is taken from Bidochka *et al.* (1994).

Amplification reactions: 10 mM Tris–HCl (pH 8.3), 50 mM KCl, 7 mM $MgCl^2$, 0.0001% gelatin containing 100 μM each of dCTP, dATP, dGTP and dTTP (Perkin–Elmer–Cetus), 10 ng of primer, 0.5 μl of Taq DNA polymerase and approximately 10 ng of genomic DNA in a final volume of 50 μl. The samples should be overlaid with 90 μl of mineral oil.

Amplification conditions: Amplification is performed in a Perkin–Elmer–Cetus DNA Thermal cycler programmed for 97 °C for 5 min, cooled to 25 °C at which time the Taq polymerase is added, then brought to 94 °C for 3 min, and followed by 45 cycles of 94 °C for 1 min, 37 °C for 1.5 min and 72 °C for 2 min. Amplified fragments are electrophoresed in 1.2 % TAE agarose gels and detected by staining with ethidium bromide (Sambrook *et al.*, 1989).

It is necessary to screen primers for RAPD-PCR to identify those which produce consistent and distinguishable fragment patterns. Oligonucleotide primers useful for *Metarhizium spp.* include 5′-TTATGTAAAACGACGGCCAGT (universal M13 primer), 5′-(GACA)4 and 5′-CGACTGTCGG. Also, as assessment of phylogenetic relationships by RAPD analysis involves the assumption that bands of similar size are homologous, it is necessary to minimize the possibility of faulty conclusions (derived if different DNA fragments have similar size) by using several primers which will collectively allow comparisons between a large number of bands. Also, Southern analysis can be performed to confirm that some of the same-sized fragments are homologous in different fungal strains.

B Identification of strain-specific DNA probes

RFLP and PCR-based methods require the isolation and growth of the pathogen away from the infected host insect and also require the isolation of intact DNA in relatively large amounts of at least 100 ng for PCR methods and significantly more for Southern hybridizations. For field trials it is much preferred to produce a sensitive and strain-specific DNA probe which could allow the detection of the fungus within a single infected insect in the presence of host DNA (see Section 2C for extraction of DNA). Strain-specific DNA probes for isolates of fungi can be generated as follows (Bidochka *et al.*, 1995).

1. Digest total DNA from the fungus with restriction enzymes such as EcoRI or HindIII. This generates random DNA fragments, generally less than 5000 bp.
2. Ligate the DNA into a pBluescript plasmid (Stratagene) and use it to transform *E. coli* (strain BB4, also obtainable from Stratagene). Identify recombinant clones as white colonies on LB agar containing ampicillin, isopropyl-b-D-galactoside (IPTG) and 5-bromo-4-chloro-3-idoyl-b-D-thiogalactoside (X-gal). Non-recombinants are blue.
3. In the initial screening, the recombinant clones should be picked up, grown in LB broth containing ampicillin and the plasmids isolated and used to prepare duplicate dot blots on nylon membranes (procedures are described in Sambrook *et al.*, 1989).
4. Individual blots should be hybridized with total *M. anisopliae* DNA from several strains that is digested with EcoRI or HindIII and radioactively labelled by random primer labelling with [^{33}P]dCTP. Controls should contain pBluescript DNA. This initial screening will identify clones whose inserts hybridize to DNA from the isolate in question, but not to DNA from other *M. anisopliae*.
5. DNA can then be dot blotted in duplicates and the putative strain-specific inserts labelled and hybridized to the dot blots. Probes which are highly repetitive as well as strain specific are desired in order to increase sensitivity. To test the iteration frequency of the DNA probes, DNA from *M. anisopliae* should be restriction-enzyme digested, electrophoretically separated, transferred to nylon membranes and hybridized with radioisotope labelled probe.

C Tagging pathogens with marker genes

An advantage of nucleic acid techniques is that they avoid sampling biases that can result from laboratory cultivation of field-collected microbes. However, since such probes give no indication of the viability or activity of a specific population, they are poor tools for predicting the environmental impact of an inoculum if used directly on soil or foliage samples without culturing the fungi present. The application of molecular markers (i.e. introduced genes conferring distinctive phenotype properties) has greater potential for increasing our understanding of environmental microbiology.

The β-glucuronidase (GUS) gene fusion system has been found to be a powerful tool for identifying and localizing marked strains of saprophytic and plant pathogenic fungi under laboratory conditions and was recently used to analyse competition for root colonization between pathogenic and non-pathogenic strains of *Fusarium oxysporum* using sterilized soils in growth chambers (Eparvier & Alabouvette, 1994). To date, only naturally occurring marker genes (e.g. spore colour) have been used to track plant pathogens under field conditions and these often interfere with pathogenicity or fitness (Fry *et al.*, 1984).

We found that the use of GUS-expression in entomopathogenic fungi and X-gluc as a histochemical substrate is a practical means of assessing gene activity in transgenic fungi and allows identification and localization of the active biomass of marked strains under different experimental conditions (St Leger *et al.*, 1995). By co-transformation with pBENA3 (containing a benomyl resistance gene), we have transformed several strains of *M. anisopliae* as well as *V. lecanii* to the GUS phenotype with pNOM102 (containing a bacterial GUS gene with the *A. nidulans* GPD promoter). GUS is not a common enzyme; there is little endogenous activity in entomopathogenic fungi or insect haemolymph facilitating the detection of even low levels of GUS in transformed fungi or infected tissues (St Leger *et al.*, 1995). Discrimination and selection from background populations are easily provided by distinctive blue coloration of hyphae during growth on minimal medium containing X-gluc, or squashes of fungus-infected insects in X-gluc and by the histochemical detection of transformants directly in soil or in insect tissues (Plate 25). In combination with techniques such as DNA probes and pulsed-field gel electrophoresis analysis, marker genes will allow analysis of potential gene transfer to indigenous strains of *M. anisopliae*, i.e. by determining whether fungi retain the marker elements in their original form. To this end, it would probably be a wise precaution to construct transformants containing two or three different marker genes in the genome. Integrative transformants are very stable when grown for long periods in the absence of selection in pure culture under lab conditions (Goettel *et al.*, 1990; St Leger *et al.*, 1995). However, stability may be different in a complex environment; in which case, it is unlikely that multiple markers would be all lost at once. The frequency of loss of each gene relative to the others could be determined, and there should usually be at least one marker remaining to distinguish positively a transformant from an extraneous organism. It might still be prudent to test such a transformant with a less ambiguous marker (the strain-specific DNA) to confirm the identification.

REFERENCES

Adams, C. C. & Gross, D. S. (1991) The yeast heat shock response is induced by conversion of cells to spheroblasts and by potent transcriptional inhibitors. *J. Bacteriol.* **173**, 7429–7435.

Al-Aidroos K. (1980) Demonstration of a parasexual cycle in the entomopathogenic fungus *Metarhizium anisopliae. Can. J. Genet. Cytol.* **22**, 309–314.

Ausubel, F. M., Brent, R., Kingston, R. E., Moore, D. D., Seidman, J. G., Smith, J. A. & Struhl, K. (eds) (1991) *Current protocols in molecular biology*, Wiley Interscience, New York.

Ballance, D. J. (1991) Transformation systems for filamentous fungi and an overview of fungal gene structure. In *Molecular industrial mycology: systems and applications for filamentous fungi* (eds S. A. Leong & R. M. Berka), pp. 1–29. Marcel Dekker, New York.

Benito, E. P., Diaz-minguez, J. M., Iturriaga, E. A., Campuzano, V. & Eslava, A. P. (1995a) Cloning and sequence analysis of the *Mucor circinelloides* pyrG gene encoding ortodine-5'-monophosphate decarboxylase: use of pyrG for homologous transformation. *Gene* **116**, 59–67.

Benito, E. P., Campuzano, V., Angeles Lopez-matas, M., De Vicente, J. I. & Eslava, A. P. (1995b) Isolation, characterization and transformation, by autonomous replication, of *Mucor circinelloides* OMP decase-deficient mutants. *Mol. Gen. Genet.* **248**, 126–135.

Bernier, L., Cooper, R. M., Charnley, A. K. & Clarkson, J. M. (1988) Transformation of the entomopathogenic fungus *Metarhizium anisopliae* to benomyl resistance. *FEMS Microbiol. Lett.* **60**, 261–266.

Bidochka, M. J., St Leger, R. J. & Roberts, D. W. (1994) Differentiation of species and strains of entomopathogenic fungi by random amplification of polymorphic DNA (RAPD). *Curr. Genet.* **25**, 107–113.

Bidochka, M. J., Walsh, S. R. A., Ramos, M. E., St Leger, R. J., Silver, J. C. & Roberts, D. W. (1995) Pathotypes in the Entomophaga grylli species complex of grasshoppers pathogens differentiated by molecular methods. *Appl. Environ. Microbiol.* **61**, 556–560.

Burmester, A., Wostemeyer, A., Arnau, J. & Wostemeyer, J. (1992) The SEG1 element: a new DNA region promoting stable mitotic segregation of plasmids in the zygomycete Absida glauca *Mol. Gen. Genet.* **235**, 166–172.

Bustamante, C., Gurrieri, S. & Smith, S. B. (1993) Towards molecular description of pulse-field gel electrophoresis. *Trends Biotechnol.* **II**, 23–30.

Chomczynski, P. (1992) One hour downward alkaline capillary transfer for blotting of DNA and RNA. *Anal. Biochem.* **201**, 134–139.

Covert, S. F. & Cullen, D. J. (1992) Heterologous protein expression in filamentous fungi. In *Frontiers in Industrial Mycology* (G. F. Leatham, ed.), pp. 66–77. Chapman & Hall, New York.

Davies, R. W. (1992) Genetic engineering of filamentous fungi for heterologous gene expression and protein secretion. In *Transgenesis, applications of gene transfer* (ed. J. A. H. Murray). pp 81–104. John Wiley, Chichester.

Eparvier, A. & Alabonvette, C. (1994) Use of Elisa and GUS-transformed strains to study competition between pathogenic and non-pathogenic *Fusarium oxysporum* for root colonization. *Biocontrol Sci. Technol.* **4**, 33–47.

Faust, R. M. & Jayaraman, K. (1990) Current trends in the evaluation of the impact of deliberate release of microorganisms in the environment: A case study with a bioinsecticidal bacterium. In *Introduction of genetically modified organisms into the environment. scope 44* (eds H. A. Mooney, & G. Bernardi). pp. 167–189. John Wiley, New York.

Fry, W. E., Yoder, O. C. & Apple, A. E. (1984) Influence of naturally occuring marker genes on the ability of *Cochliobolus heterostrophus* to induce field epidemics of southern corn leaf blight. *Phytopathology* **74**, 175–178.

Garber, R. C. (1992) Molecular approaches to the analysis of pathogenicity genes from fungi causing plant disease. In *The fungal spore and disease initiation in plants and animals* (eds G. T. Cole & H. H. Hoch), pp. 483–502. Plenum Press, New York.

Gatehouse, A. M. R., Boulter, D., & Hilder, V. A. (1992) Potential of plant-derived genes in the genetic manipulation of crops for insect resistance. In *Plant genetic manipulation for crop protection* (eds A. M. R. Gatehouse, V. A. Hilder, & D. Boulter). pp. 155–182. CAB International, Wallingford.

Glare, T. R. & Milner, R. J. (1991) Ecology of entomopathogenic fungi. In *Handbook of applied mycology* (eds D. K. Arora, L. Ajello & K. G. Mukerji), pp. 547–612. Marcel Dekker, New York.

Goettel, M. S., St Leger, R. J., Bhairi, S., Jung, M. K., Oakley, B. R., Roberts, D.W. & Staples, R. C. (1990) Pathogenicity and growth of *Metarhizium anisopliae* stably transformed to benomyl resistance. *Curr. Genet.* **17**, 129–132.

Goosen, T., Bloemheuvel, G., Gysler, C., De Bie, D. A., Van den Broek, H. W. J. & Swart, K. (1987) Transformation of *Aspergillus niger* using the homologous orotidine-5'-phosphate-decarboxylase gene. *Curr. Genet.* **11**, 499–503.

Haag, E. & Raman, V. (1994) Effects of primer choice and source of Taq DNA polymerase on the bending patterns of differential display RT-PRC. *Biotechniques* **17**, 226–228.

Heale, J. B., Isaac, J. E. & Chandler, D. (1989) Prospects for strain improvement in entomopathogenic fungi. *Pesticide Sci.* **26**, 79–92.

Joshi, L., St Leger, R. J. & Bidochka, M. J. (1995) Cloning of a cuticle-degrading protease from the entomopathogenic fungus, *Beauveria bassiana. FEMS Microbiol. Lett.* **125**, 211–218.

Joshi, L., St. Leger, R. J. & Roberts, D. W. Isolation of a cDNA encoding a novel subtilisin-like protease ($Pr_1\beta$) from the entomopathogenic fungus, *Metarhizium anisophae* using differential display RT-PCR. *Gene*. In press.

Liang, P., Zhu, W., Zhang, X., Gou, Z., O'Conell, R. P., Averboukh, L., Wang & Pardee, A. B. (1994) Differential display using one-base anchored oligo-dT primers. *Nucleic Acid Res.* **22**, 5763–5764.

Lu, S., Lyngholm, L., Yang, G., Bronson, C., Yoder, O. C. & Turgeon, B. G. (1994) Tagged mutations at the Tox1 locus of *Cochliobolus heterostrophus* by restriction enzyme-mediated integration. *Proc. Natl. Acad. Sci. USA* **91**, 12649–12653.

May, G. (1992) Fungal technology. In *Applied molecular genetics of filamentous fungi* (eds J. R. Kinghorn & G. Turner), pp. 1–27. Blackie, Glasgow.

O'Brien, D. P., Billadeau, D. & Van Ness, B. (1994) RT-PCR Assay for detection of transcripts from very few cells using whole cell lysates. *Biotechniques*, **16**, 586–589.

Ochman, H., Medhora, M. M., Garza, D. & Hartl, D. C. (1990) Amplification of flanking sequences by inverse PCR. In *PCR protocols: a guide to methods and applications* (eds M. A. Innis, D. H. Gelfrand, J. J. Sironsky & T. K. White), pp. 219–227. Academic Press, New York.

Orbach, M. J. (1994) A cosmid with a HyR marker for fungal library construction and screening. *Gene* **150**, 159–162.

Piper, P. W. (1994) Measurement of transcription. In *Molecular genetics of yeast, A practical approach* (ed. J. R. Johnston) pp. 135–146. IRL Press, Oxford.

Rothrwild. M., Alpan, R. S., Liang, P. & Pardee, A. B. (1995). Inosine-containing primers for mRNA differential display. *Trends Genet.* **11**, 300.

Rowland, L. J. & Nguyen, B. (1993) Use of polyethylene glycol for purification of DNA from leaf tissue of woody plants. *Biotechniques* **14**, 734–736.

Sambrook, J., Fritsch, E. F. & Maniatis, T. (1989) *Molecular cloning. A laboratory manual*, 2nd edn. Cold Spring Harbor Laboratory Press, Cold Spring Harbour, New York.

Schiestl, R. H. & Petes, T. D. (1991) Integration of DNA fragments by illegitimate recombination in *Saccharomyces cerevisiae. Proc. Natl. Acad. Sci. USA* **88**, 7585–7589.

Shimizu, S., Yoshioka, H. & Matsumoto, T. (1993) Electrophoretic karyotyping of the entomogenous fungus *Paecilomyces fumoroseus*. Lett. in *Appl. Microbiol.* **16**, 183–186.

Sokolov, B. P. and Prockop, J. (1994) A rapid and simple PCR-based method for the isolation of cDNAs from differentially expressed genes. *Nucleic Acids Res.* **22**, 4009–4010.

Song, P., Yamamoto. E. & Allen, R. D. (1995) Improved procedure for differential display of transcripts from cotton tissues. *Plant Mol. Biol. Reporter.* **13**, 174–181.

St Leger, R. J. (1993) Biology and mechanisms of invasion of deuteromycete fungal pathogens. In *Parasites and pathogens of Insects* vol. 2. (eds N. C. Beckage, S. N. Thompson & B. A. Federici), pp. 211–229. Academic Press. San Diego.

St Leger, R. J. (1995) The role of cuticle-degrading proteases in fungal pathogens of insects. *Can. J. Bot.* **73**, Suppl. 1, s1119–s1125.

St Leger, R. J., Allee, L. L., May, B. & Roberts, D. W. (1992a) World-wide distribution of genetic variation among isolates of Beauveria spp. *Mycol. Res.* **96**, 1007–1015.

St Leger, R. J., Roberts, D. W., & Staples, R. C. (1992b) Molecular cloning and regulatory analysis of the cuticle-degrading protease structural gene from the entomopathogenic fungus Metarhizium anisopliae. *Eur. J. Biochem.* **204**, 991–1001.

St Leger, R. J., Staples, R. C. & Roberts, D. W. (1992c) Cloning and regulatory analysis of ssgA: A gene encoding a hydrophobin-like protein from the entomopathogenic fungus, *Metarhizium anisopliae. Gene* **120**, 119–124.

St Leger, R. J., Shimizu, S., Joshi, L., Bidochka, M. J. & Roberts, D. W. (1995) Co-transformation of *Metarhizium anisopliae* by electroporation or using the gene gun to produce stable GUS transformants. *FEMS Microbiol. Lett.* **131**, 289–294.

St Leger, R. J., Joshi, L., Bidochka, M. J. & Roberts, D. W. (1996a) Constuction of an improved mycoinsecticide over-expressing a toxic protease. *Proc. Natl. Acad. Sci. USA* **93**, 6349–6354.

St Leger, R. J., Joshi, L., Bidochka, M. J., Rizzo, N. W. & Roberts, D. W. (1996b) Characterization and ultrastructural localization of chitinases from *Metarhizium anisopliae*, *M. flavoviride*, and *Beauveria bassiana* during fungal invasion of host (*Manduca sexta*) cuticle. *Appl. Environ. Microbiol.* **62**, 907–912.

Stowell, L. J. (1994) Factors influencing acceptance and development of biopesticides. In *Advances in engineered pesticides* (L. Kim, ed.), pp. 249–260. Marcel Dekker, New York.

Talbot, N. J., Ebbole, D. J. & Hamer, J. E. (1993) Identification and characterization of MPG1, a gene involved in pathogenicity from the rice blast fungus *Magnaporthe grisea*. The *Plant Cell.* **5**, 1575–1590.

Tilburn, J., Scazzacchio, C., Taylor, C. G., Zabicky-Zissman, J. H., Lockington, R. A. & Davis, R. W. (1983). Transformation by integration in *Aspergillus niger*. *Gene.* **26**, 205–221.

Turner, G. (1991) Strategies for cloning genes from filamentous fungi. In *Applied molecular genetics of fungi* (eds J. F. Peberdy, C. E. Caten, J. E. Ogden & J. W. Bennett) pp. 29–43. Cambridge University Press, Cambridge.

Unkles, S. E. (1992) Gene organization in industrial filamentous fungi. In *Applied molecular genetics of filamentous fungi* (eds J. R. Kinghorn & G. Turner). pp. 28–53. Blackie, Glasgow.

Unkles, S. E., Campbell, E. I., De Ruiter-Jacobs, Y. M. J. T., Broekhuijsen, M., Macro, J. A., Carrez, D., Contreras, R., Van den Hondel, C. A. M. J. J. & Kinghorn, J. R. (1989) The development of a homologous transformation system for *Aspergillus oryzae* based on the nitrate assimilation pathway: A convenient and general selection system for filamentous fungal transformation. *Mol. Gen. Genet.* **218**, 99–104.

Vollmer, S. V. & Yanofsky, C. (1986) Efficient cloning of genes of *Neurospora crassa Proc. Natl. Acad. Sci. USA.* **83**, 4869–4873.

Yoder, O. C., Weltring, K., Turgeon, B. G., Garber, R. C. & VanEtten, H. D. (1986) Technology for molecular cloning of fungal pathogenicity genes. In *Biology and molecular biology of plant–pathogen interactions* (J. Bailey, ed.), pp. 371–384. Springer-Verlag, Berlin.

GLOSSARY

Amplification. Selective replication of a gene to produce more than the normal single copy in a haploid genome.

Autoradiography. Detection of radioactive label by exposure to photographic film.

Auxotroph. An organism that requires one or more substances in addition to minimal medium.

Base. A cyclic, nitrogen-containing compound linked to deoxyribose in DNA and to ribose in RNA.

CCAAT box. An upstream motif, having the sequence CCAAT, found in many eukaryotic promoters recognized by RNA polymerase II.

cDNA. A DNA copy of an RNA, made by reverse transcription.

cDNA library. A collection of cDNA molecules that were generated *in vitro* from mRNA of a single type of cell population.

Codon. A sequence of three adjacent nucleotides encoding an amino acid or termination of translation.

Colony hybridization. A procedure for selecting a bacterial clone containing a gene of interest. DNAs from a large number of clones are simultaneously tested with a labelled probe that hybridizes to the gene of interest.

Constitutive. A constitutively expressed gene is always turned on.

Electrophoresis. Separation of charged molecules such as DNA, RNA and proteins, in an electric field.

Exon. A region of a gene that is ultimately represented in that gene's mature transcript.

Gene. The fundamental unit of heredity. It contains the information for making one RNA.

Gene expression. The process by which gene products are made.

Genome. A haploid set of chromosomes of an organism.

Heterokaryon. A cell containing two or more nuclei of different origin.

Homologous recombination. Recombination that requires extensive sequence similarity between the recombining DNAs.

Hybridization. A process in which double stranded structures are formed from two nucleotides (either DNA or RNA) from different sources.

Hybridization stringency. The combination of factors (temperature, salt, organic solvent, detergent, etc.) that influences the ability of two polynucleotide strands to hybridize.

Intron. A region that interrupts the transcribed part of a gene. An intron is transcribed, but is removed by splicing during maturation of the transcript.

Karyotype. A pictorial representation of all the chromosomes in a given organism.

Kilobase pair. One thousand base pairs.

Klenow fragment. A fragment of DNA polymerase I, created by cleaving with protease, that lacks the 5′-3′ exonuclease activity.

Lambda phage. A phage of *E. coli*. It can replicate lytically or lysogenically.

Lysis. Rupturing the membrane of a cell, as by a virulent phage.
Lysogen. A bacterium containing a prophage.
Marker. A gene or mutation that serves as a signpost at a known location in the genome.
Open reading frame (ORF). A reading frame that is uninterrupted by translation stop codons.
Phage. A virus capable of infecting and multiplying in bacteria.
Plaque. A clearing zone that a phage makes on a layer of growing bacterial cells by infecting and either killing or slowing their growth.
Plasmid. A circular DNA that replicates independently of the cell's DNA.
Primer. A polynucleotide sequence that serves as a growing point for polymerization.
Promoter. A region of DNA to which RNA polymerase binds prior to initiation of translation.
Prophage. Inactive state of a bacteriophage maintained as a part of chromosomal DNA in a host cell.
Recombinant DNA. The product of recombination between two or more fragments of DNA. It can either occur naturally or can be performed *in vitro*.
Restriction endonucleases. A class of enzymes that recognize specific base sequences in DNA and cuts at those specific sites.
Restriction fragment. A piece of DNA cut from a larger DNA by the action of restriction endonucleases.
Restriction map. A map that shows the locations of restriction sites in a region of DNA.
Stop codon. One of the three codons (UAG, UAA and UGA) that code for the termination of translation.
TATA box. A consensus sequence that is found about 25 base pairs upstream from the start of transcription in most eukayotic promoters, recognized by RNA polymerase II.
Transformation. The genetic modification induced by the incorporation or a foreign DNA into a cell.

Index

Note: Page references in *italics* refer to Figures; those in **bold** refer to Tables